Microbial Secondary Metabolites and Biotechnology

Microbial Secondary Metabolites and Biotechnology

Editors

Mireille Fouillaud
Laurent Dufossé

MDPI • Basel • Beijing • Wuhan • Barcelona • Belgrade • Manchester • Tokyo • Cluj • Tianjin

Editors
Mireille Fouillaud
University of La Réunion
France

Laurent Dufossé
Université de La Réunion
France

Editorial Office
MDPI
St. Alban-Anlage 66
4052 Basel, Switzerland

This is a reprint of articles from the Special Issue published online in the open access journal *Microorganisms* (ISSN 2076-2607) (available at: https://www.mdpi.com/journal/microorganisms/special_issues/microbial_secondary_metabolites_biotechnology).

For citation purposes, cite each article independently as indicated on the article page online and as indicated below:

LastName, A.A.; LastName, B.B.; LastName, C.C. Article Title. *Journal Name* **Year**, *Volume Number*, Page Range.

ISBN 978-3-0365-3246-2 (Hbk)
ISBN 978-3-0365-3247-9 (PDF)

Contents

About the Editors ix

Preface to "Microbial Secondary Metabolites and Biotechnology" xi

Mireille Fouillaud and Laurent Dufossé
Microbial Secondary Metabolism and Biotechnology
Reprinted from: *Microorganisms* **2022**, *10*, 123, doi:10.3390/microorganisms10010123 1

Yujin Jeong, Sang-Hyeok Cho, Hookeun Lee, Hyung-Kyoon Choi, Dong-Myung Kim, Choul-Gyun Lee, Suhyung Cho and Byung-Kwan Cho
Current Status and Future Strategies to Increase Secondary Metabolite Production from Cyanobacteria
Reprinted from: *Microorganisms* **2020**, *8*, 1849, doi:10.3390/microorganisms8121849 11

Charifat Said Hassane, Mireille Fouillaud, Géraldine Le Goff, Aimilia D. Sklirou, Jean Bernard Boyer, Ioannis P. Trougakos, Moran Jerabek, Jérôme Bignon, Nicole J. de Voogd, Jamal Ouazzani, Anne Gauvin-Bialecki and Laurent Dufossé
Microorganisms Associated with the Marine Sponge *Scopalina hapalia*: A Reservoir of Bioactive Molecules to Slow Down the Aging Process
Reprinted from: *Microorganisms* **2020**, *8*, 1262, doi:10.3390/microorganisms8091262 35

Nor Hawani Salikin, Jadranka Nappi, Marwan E. Majzoub and Suhelen Egan
Combating Parasitic Nematode Infections, Newly Discovered Antinematode Compounds from Marine Epiphytic Bacteria
Reprinted from: *Microorganisms* **2020**, *8*, 1963, doi:10.3390/microorganisms8121963 59

Ali Nawaz, Rida Chaudhary, Zinnia Shah, Laurent Dufossé, Mireille Fouillaud, Hamid Mukhtar and Ikram ul Haq
An Overview on Industrial and Medical Applications of Bio-Pigments Synthesized by Marine Bacteria
Reprinted from: *Microorganisms* **2021**, *9*, 11, doi:10.3390/microorganisms9010011 79

Sunghoon Hwang, Ly Thi Huong Luu Le, Shin-Il Jo, Jongheon Shin, Min Jae Lee and Dong-Chan Oh
Pentaminomycins C–E: Cyclic Pentapeptides as Autophagy Inducers from a Mealworm Beetle Gut Bacterium
Reprinted from: *Microorganisms* **2020**, *8*, 1390, doi:10.3390/microorganisms8091390 105

Yara I. Shamikh, Aliaa A. El Shamy, Yasser Gaber, Usama Ramadan Abdelmohsen, Hashem A. Madkour, Hannes Horn, Hossam M. Hassan, Abeer H. Elmaidomy, Dalal Hussien M. Alkhalifah and Wael N. Hozzein
Actinomycetes from the Red Sea Sponge *Coscinoderma mathewsi*: Isolation, Diversity, and Potential for Bioactive Compounds Discovery
Reprinted from: *Microorganisms* **2020**, *8*, 783, doi:10.3390/microorganisms8050783 121

Yannik Schneider, Marte Jenssen, Johan Isaksson, Kine Østnes Hansen, Jeanette Hammer Andersen and Espen H. Hansen
Bioactivity of Serratiochelin A, a Siderophore Isolated from a Co-Culture of *Serratia* sp. and *Shewanella* sp.
Reprinted from: *Microorganisms* **2020**, *8*, 1042, doi:10.3390/microorganisms8071042 139

About the Editors

Mireille Fouillaud is a senior lecturer at the University of Reunion and a researcher at the Laboratory of Chemistry and Biotechnology of Natural Products (ChemBioPro). She graduated in 1990 as an industrial microbiology engineer. Following a Ph.D. in Cell Biology and Microbiology from the University of Aix-Marseille I, defended in 1994, she was recruited at the Faculty of Sciences and Technology of the University of Reunion. Between 2009 and 2020, she joined the Ecole Supérieure d'Ingénieurs Réunion Océan Indien (ESIROI), in the food engineering department. She taught biology and microbiology applied to agribusiness. In 2020, she came back to the Faculty of Sciences of University of La Réunion to teach microbiology and project management. Her main research interests range from the microbial diversity of ecosystems and organisms to the production of metabolites of interest for industries, through biotechnology. Some years ago, she decided to focus on pigmented metabolites obtained from marine-derived filamentous fungi, with potential applications in the food or dyeing industries. She also studies actinomycetes and their production of specialized bioactive metabolites.

Laurent Dufossé has held the position of Professor of Food Science and Biotechnology at Reunion Island University, which is located on a volcanic island in the Indian Ocean, near Madagascar and Mauritius, since 2006. The island is one of France's overseas territories, with almost one million inhabitants, and the university has15,000 students. Previously, Professor Dufossé was a researcher and senior lecturer at the Université de Bretagne Occidentale, Quimper, Brittany, France. He attended the University of Burgundy, Dijon, where he received his PhD in Food Science in 1993 and has been involved in the field of biotechnology of food ingredients for more than 30 years. His main research in the last 20 years has focused on the microbial production of pigments and studies are mainly devoted to aryl carotenoids, such as isorenieratene, C50 carotenoids, azaphilones and anthraquinones. This research has a close relationship with applications in food science and technology within the cheese industry, the sea salt industry, etc.

Preface to "Microbial Secondary Metabolites and Biotechnology"

Technological advances have led to in-depth comprehension of the microbial world, still unimaginable a century ago.

In the worthy tradition of Louis Pasteur, many research teams are working today to demonstrate that microorganisms can be our daily partners, through the great diversity of biochemical transformations and molecules that they can produce.

This Special Issue highlights several facets of the production of microbial metabolites of interest. From the discovery of new strains or new bioactive molecules issued from novel environments, to the increase of their synthesis by traditional or innovative methods, different levels of biotechnological processes are addressed. Combining the new dimensions of "Omics" sciences, such as genomics, transcriptomics, or metabolomics, microbial biotechnologies are opening up incredible opportunities for discovering and improving microorganisms and their productions. There is no doubt that industries will be able to seize this opportunity to take new steps in the knowledge and control of biological processes for the benefit of mankind.

The dissemination of science being a necessary step for the appropriation of knowledge, we would like to thank the MDPI book staff and the Editorial Team of the journal "Microorganisms". We also thank all the contributing authors and reviewers without whom this book could not exist.

Mireille Fouillaud, Laurent Dufossé
Editors

microorganisms

Commentary

Microbial Secondary Metabolism and Biotechnology

Mireille Fouillaud [1,*] and Laurent Dufossé [1,2]

1 Laboratoire de Chimie et de Biotechnologie des Produits Naturels—CHEMBIOPRO, Université de la Réunion, 15 Avenue René Cassin, CS 92003, F-97744 Saint-Denis, Ile de la Réunion, France; laurent.dufosse@univ-reunion.fr
2 Ecole Supérieure d'Ingénieurs Réunion Océan Indien—ESIROI Agroalimentaire, 2 Rue Joseph Wetzell, F-97490 Sainte-Clotilde, Ile de la Réunion, France
* Correspondence: mireille.fouillaud@univ-reunion.fr; Tel.: +33-262-262-483-363

Abstract: In recent decades scientific research has demonstrated that the microbial world is infinitely richer and more surprising than we could have imagined. Every day, new molecules produced by microorganisms are discovered, and their incredible diversity has not yet delivered all of its messages. The current challenge of research is to select from the wide variety of characterized microorganisms and compounds, those which could provide rapid answers to crucial questions about human or animal health or more generally relating to society's demands for medicine, pharmacology, nutrition or everyday well-being.

Keywords: secondary metabolites; microorganisms; biotechnology; screening; production; extraction; bioactive properties; perspectives

Citation: Fouillaud, M.; Dufossé, L. Microbial Secondary Metabolism and Biotechnology. *Microorganisms* **2022**, *10*, 123. https://doi.org/10.3390/microorganisms10010123

Academic Editor: Alexander I. Netrusov

Received: 21 December 2021
Accepted: 7 January 2022
Published: 7 January 2022

1. Microorganisms and Metabolites: An Incredible World of Novelty for Biotechnologists, New Opportunities for Industries

Microbial secondary metabolites, now named as specialized metabolites, often have unusual structures and many have demonstrated major effects on the health, nutrition and economics of our society [1]. These compounds are usually low molecular mass products of secondary metabolism, which take place out of step with the main microbial growth phase. They include antibiotics, pigments, toxins, enzyme inhibitors, immunomodulators, effectors of ecological competition or symbiosis, or compounds with hormonal activity or particular effects on lipids or carbohydrates metabolism. Some have already established themselves as antimicrobials, antivirals, antioxidants, antitumorals, vaso-relaxants or contractants, diuretics or laxatives. Others are used as colorants, pesticides, or growth promoters for animals or plants [2]. Approximately 53% of the FDA-approved natural products-based drugs are originating from microorganisms [3].

The synthesis of specialized metabolites is finely adjusted by nutritive sources, growth conditions, feedback control, enzyme induction or inactivation. Their regulation is often influenced by specific low molecular mass compounds, also transfer RNA, σ factors and gene products formed during post-exponential development. Recent research demonstrated that the production of specialized metabolites is mostly coded by clustered genes on chromosomal DNA rather than by plasmidic DNA. However, the related pathways are still not fully clarified and thus provide a new theoretical frontier for academic researchers in enzymology, control and differentiation [4].

Omic sciences such as genomic, transcriptomic or metabolomic applied to industrial microorganisms now offer new opportunities for strain discovery characterization and improvement. Thus, great potential exists for the development of novel compounds for pharmaceutical, nutraceutical, dyeing, or agricultural industries.

In this special issue secondary metabolites produced by bacteria, actinomycetes, fungi or microalgae are investigated under different facets linked with bioactivity or other interesting properties. Biotechnology combines a range of scientific fields that connect basic and applied research (Figure 1). This is the initial but fundamental step towards the resolution of technical problems, anticipating industrial exploitation.

The era of microbial natural products manufactured through biotechnology has just begun.

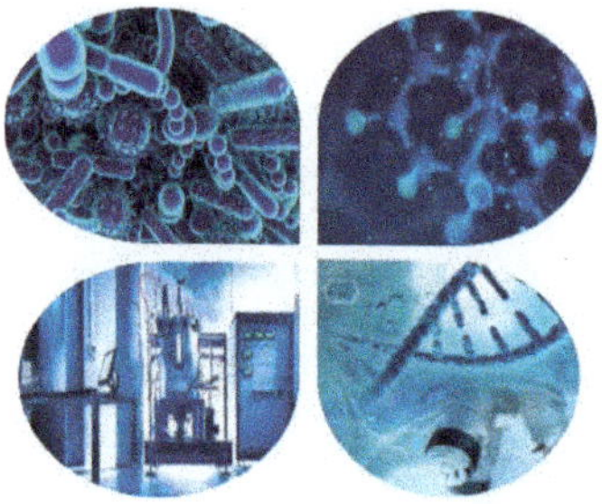

Figure 1. Illustration of the diversity of scientific fields underlying microbial biotechnology (clockwise): microbiology, chemistry, genomic and industrial technologies.

2. From the Beginning: Screening and Characterization of Valuable Strains

Not so long ago, scientists discovered that the marine realm is an incredibly rich biotope for the discovery of new microorganisms and subsequently the characterization of valuable new molecules.

Sponges and their associated microbiota have been found to produce a wide variety of bioactive secondary metabolites. In the pharmaceutical field, several natural products extracted from marine organisms have already demonstrated their capacity to delay aging and/or extend lifespan. However, the biodiversity from the Southwest of the Indian Ocean is much less studied, especially regarding anti-aging activities. In the study presented by Saïd-Hassane et al. (2020) [5], the microbial diversity of the marine sponge *Scopalina hapalia* was investigated by metagenomic analysis. Twenty-six bacterial and two archaeal phyla were recovered from the sponge, of which the Proteobacteria phylum was the most abundant. Thirty isolates affiliated to the genera *Bacillus*, *Micromonospora*, *Rhodoccocus*, *Salinispora*, *Aspergillus*, *Chaetomium*, *Nigrospora* and or related to the family Thermoactinomycetaceae were cultivated for secondary metabolites production. Crude extracts from selected microbial cultures were found to be active against elastase, tyrosinase, catalase, sirtuin 1, CDK7 (Cyclin-dependent kinase 7), Fyn kinase and proteasome. These results highlight the potential of microorganisms associated with a marine sponge from the Indian Ocean to produce anti-aging compounds.

Marine epiphytic bacteria are also highly diverse, and ubiquitous in marine biotopes where they have to survive constant pressures coming from physicochemical parameters (hydrostatic pressure, low oxygenation, salinity …), biotic competition or predation (e.g., protozoans and nematodes). These marine organisms have developed defense strategies to face adversity, for example by producing toxic bioactive compounds. Several active metabolites have been identified from these microorganisms coming from specific genes expression. The review of Salikin et al. (2020) [6] focuses on the potential of marine epiphytic bacteria to be a new platform for novel anti-nematode drug development. Emerging strategies, including culture-independent high-throughput screening to discover new strains are highlighted.

The review by Nawaz et al. (2020) [7] focuses on pigments and provides an overview of marine bacteria synthetizing bio-pigments, along with their applications. It highlights a range of molecules already valued at the industrial level such as prodigiosin, astaxanthin, violacein, zeaxanthin, lutein or lycopene.

Insect-associated bacteria, are supposed to be involved in the life cycle of their host due to the panel of secondary metabolites they are able to produce. These strains are certainly one of the less explored sources of new active compounds. With the aim of efficiently discovering new bioactive molecules, the diversification of the culture conditions of one strain may induce the activation of diverse biosynthetic gene clusters. This OSMAC approach (one strain many compounds) is based on the fact that some microbial metabolites are not produced under certain set of physicochemical parameters and may appear when the conditions are modified. Inspired by these two approaches, the production of the cyclic pentapeptides pentaminomycins C, D and E was significantly improved in the culture of the *Streptomyces* sp. GG23 strain, an actinobacteria isolated from the guts of *Tenebrio molitor* (the mealworm beetle) [8]. The analysis of the non-ribosomal peptide biosynthetic gene cluster suggested that the unicity of the compounds, based on the structural variations, originates from the low specificity of the adenylation domain in the non-ribosomal peptide synthetase (NRPS) module 1, indicating that macrocyclization can be catalyzed non-canonically by penicillin binding protein (PBP)-type TE. Additionally, pentaminomycins C and D exhibited significant autophagy-inducing activities and were cytoprotective against menadione-induced oxidative stress in vitro.

One original domain of exploration is the analysis of secondary metabolites obtained from microbial co-cultures, via metabolome tools. Among the rich microbiote isolated from the Red Sea sponge *Coscinoderma mathewsi* (23 isolates), three actinomycetes strains were defined as novel species of the genera *Micromonospora*, *Nocardia*, and *Gordonia* [9]. This study of Shamikh et al. (2020) [9] demonstrated that *Micromonospora* sp. UA17 co-cultured with two mycolic acid-containing actinomycetes namely *Gordonia* sp. UA19 and *Nocardia* sp. UA 23, or supplemented with pure mycolic acid could produce metabolites such as a chlorocardicin, neocopiamycin A, and chicamycin B that were not found in the respective monocultures. This implies a mycolic acid effect on the induction of cryptic natural product biosynthetic pathways and reveals that silent biosynthetic gene clusters can show their unusual capacities for the production of secondary metabolites, under unclassical cultivation conditions.

Siderophores, as serratiochelins, are specialized compounds with high affinity for ferric iron, that are produced by the opportunistic pathogen *Serratia marcescens*. The siderophores are of pharmaceutical interest as they can be used in their native form to treat iron overload diseases or facilitate uptake of antibiotics by bacteria, through siderophore-antibiotic drug conjugate. In the study of Schneider (2020) [10], the rare siderophore serratiochelin A was extracted with high yields from an iron-depleted co-culture of *Serratia* sp. and *Shewanella* sp. (a strain from marine environment also involved in iron cycle). As this molecule was not observed in axenic cultures of *Shewanella* or *Serratia*, the co-culturing may induce the production of the compound, possibly because of the competition for iron between the two strains in the culture medium. The serratiochelin A antibacterial effect was tested and was specific towards *S. aureus*. The molecule also exhibited toxic effects on both eukaryotic cell lines A2058 and MRC5.

Thereafter, the construction of microbe consortia opens up an endless avenue of exploration towards the production of new bioactive molecules.

3. New Molecules

Streptomyces is a very studied genus, historically known for the production of bioactive compounds. However, strains of unusual origins are able to furnish new natural products. As an example, the *Streptomyces* sp. strain IB201691-2A from the Lake Baikal endemic mollusk *Benedictia baicalensis* synthesizes three new angucyclines (baikalomycins A–C), as well as large quantities of rabelomycin and 5-hydroxy-rabelomycin [11]. Baikalomycins A–C exhibited varying levels of anticancer properties. Rabelomycin and 5-hydroxy-rabelomycin also showed antiproliferative activities. The gene cluster for baikalomycins biosynthesis was identified, cloned and expressed in *S. albus* J1074. Heterologous expression and deletion experiments allowed the glycosyltransferase functions implicated in the synthesis of these original compounds to be specified.

The crude extract from a culture of *Curvularia papendorfii*, an endophytic fungus isolated from *Vernonia amygdalina*, a medicinal plant from Sudan, revealed an important antiviral effect against the human coronavirus HCoV 229E and the feline coronavirus

FCV F9 [12]. Additionally, a selective antibacterial activity against *Staphylococcus* sp. was observed, as well as an interesting antiproliferative competence against the human breast carcinoma MCF7 cell line. Twenty-two metabolites were identified from this extract and two major pure compounds were characterized including a new polyhydroxyacid: kheiric acid. Kheiric acid showed effective inhibition capacities against methicillin-resistant *Staphylococcus aureus* (MRSA). Hence, endophytes merit more attention, as a treasure trove of new bioactive compounds.

4. Increasing Knowledge on Bioactive Properties

Given the societal implications, the search for bioactive properties of bio-based molecules mobilizes considerable energy among research teams.

The marine microalgae *Aurantiochytrium* sp. is considered a promising source for docosahexaenoic acid (DHA) production. DHA is a n-3 long-chain polyunsaturated fatty acid and is critical for cellular processes involved in maintaining health. In the study of Liu et al. (2020) [13], UV mutagenesis was utilized to obtain competitive *Aurantiochytrium* sp. strain with enhanced DHA production (1.90-fold higher than wild strain). The key genes related to the increasing DHA accumulation were explored by comparing the transcriptome between the mutant and the parent strain. The mRNA expression levels of *CoAT, AT, ER, DH,* and *MT* genes, are linked to the increased intercellular production of DHA, and can be manipulated to control DHA yields in *Aurantiochytrium* sp. The genetic improvement of microbial strains is thus only at the beginning of its immense possibilities.

Biofilms, composed of microbial secreted exopolysaccharides, protects bacteria against adverse environmental conditions but favor the contamination of surfaces as diverse as foodstuffs or medical material. A classical approach to eliminate biofilms is to use natural anti-microbial compounds. However, new chemical synergies are still emerging, as developed by Argüelles et al. (2020) [14], such as the joint use of carnosic acid (obtained from rosemary) and propolis (from honeybees' panels), to control the pathogenic yeast *Candida albicans*. Recent advances in biofilm eradication also involve the intervention of lytic phages. In the study of Papaianni et al. (2020) [15] the lytic activity of Xccφ1 (*Xanthomonas campestris* pv. *campestris*-specific phage) was evaluated in combination with 6-pentyl-α-pyrone (a secondary metabolite produced by *Trichoderma atroviride* P1) associated with hydroxyapatite. The results demonstrated that Xccφ1, alone or in combination with 6PP and HA (φHA6PP complex), interferes with the gene pathways involved in the formation of biofilm, by modulating the genes involved in the biofilm formation and stability (*rpf, gumB, clp* and *manA*). This approach can be used as a model to fight against other biofilm-producing bacteria.

With the hypothesis that the use of antimicrobial textiles may significantly reduce the risk of nosocomial infections, Hamed et al. (2020) [16] focused their attention on endophytic fungi isolated from marine organisms, collected from saline environments. The antimicrobial and antioxidant activities of 32 fungal isolates were examined against a panel of pathogenic bacteria and fungi. The ethyl acetate crude extracts of 21 strains initially possessed antimicrobial or antioxidant activities. As an innovation, the surface of cellulosic fabrics was functionalized by grafting of MCT-βCD (monochlorotriazinyl β-cyclodextrin), creating core-shaped hydrophobic cavities. Furthermore, inclusion of the three most active fungal extracts (*Aspergillus calidoustus* M113, *Aspergillus terreus 7S4, Alternaria alternata 13A*) into the hydrophobic cavities was achieved. The experience demonstrated that *Aspergillus calidoustus* strain M113 exhibited the most promising improvement in the antimicrobial functionality, and the second best in the UV protection of this novel generation of fabrics. A low/weak toxicity against normal human skin fibroblasts was determined. Large-scale production of this bioactive extracts as well as the industrial application of the process to develop eco-friendly multifunctional textiles, is the first step to determine the economic feasibility.

Based on in silico techniques that have recently gained attention in drug discovery programs (e.g., structure and ligand-based virtual screening, docking and molecular dynamics), the team of Sayed (2020) worked on the protein Mpro [17]. This important protease of the severe acute respiratory syndrome coronavirus 2 (SARS-CoV-2) is a key of the viral infection, ensuring the cleavage of the two replicase proteins pp1a and pp1ab. Mpro was subjected to hyphenated pharmacophoric-based and structural-based virtual screenings. The Natural Products Atlas (N. P. Atlas), a data base of more than 24,000 microbial natural compounds, was screened to find out analogues exhibiting antiviral activity, by adapting to the catalytic site of Mpro. Using Lipinski's rules, 9933 drug-like candidates were detected. Top-scoring hits were further filtered out depending on their ability to show appropriate binding affinities towards the molecular dynamic simulation (MDS)-derived enzyme's conformers. Six compounds exhibiting high potential as anti-SARS-CoV-2 drug candidates were consequently detected. Further in vitro testing of the selected molecules is a promising starting point for the rapid development of medicament candidates against SARS-CoV-2.

5. Challenges in Production Steps

Omic sciences now make it possible to refine our knowledge on the biological transformations underlying biotechnologies.

The wine industry is making significant progress thanks to research carried out on the yeast *Saccharomyces cerevisiae*. The work of Gonzáles-Jiménez et al. (2020) [18] makes it possible to draw a proteomic map in order to distinguish the protein content of sparkling wines produced by different strains of *Saccharomyces cerevisiae*. The study shows that the proteins in the yeasts, that are responsible for the production of the volatile compounds released during sparkling wine elaboration, are quite similar in a flor yeast and a conventional one, except for the proteins Adh1p, Fba1p, Tdh1p, Tdh2p, Tdh3p, and Pgk1p. These proteins are present in higher concentrations in the flor yeast versus the conventional strain. The higher concentration of these proteins may explain the different organoleptic properties obtained when ageing using flor yeasts. The mannoproteins released as well as compounds derived from autolysis and enzymes are involved in reactions that affect some aroma precursors and thus the specific volatilome. Proteomic analysis can thus be the prerequisite for a more precise characterization of the specific quality of a wine.

Genetic modifications today provide science with incredible tools to promote the production of molecules of interest in host organisms other than the original producers. The use of bacteria as productive hosts makes it possible to no longer depend on the geoclimatic and spatial constraints, inherent in plant cultures. Moreover, the molecules can be produced in low-cost media at relatively high growth rates, allowing the utilization of sunlight as an energy source for sustainable cultivation and production processes. In the work of Hilgers et al. (2020) [19], the heterologous host *Rhodobacter capsulatus* was used to improve the plant sesquiterpenoid pathway leading to β-caryophyllene. A maximum production of 139 ± 31 mg·L^{-1} was obtained. As the sesquiterpene usually demonstrates beneficial anti-phytopathogenic capacities, the bioactivity of β-caryophyllene and its oxygenated derivative was determined against diverse phytopathogenic fungi. The molecules significantly inhibited the growth of the 2 phytopathogenic *Sclerotinia sclerotiorum* and *Fusarium oxysporum*, while other were left unaffected, including some plant growth-promoting bacteria. Thus, the production of β-caryophyllene and β-caryophyllene oxide in *Rhodobacter capsulatus* can be considered as promising process for the production of natural compounds for the management of some plant pathogenic fungi in agricultural crop production.

The safe production of amino acids for the food and feed industry has already been widely established with *Corynebacterium glutamicum* as the major production host. The challenge of the work of Walter et al. (2020) [20] was to favor a specific pathway for the production of a N-functionalized amino acid N-methylanthranilate (NMA). This amino-acid is a major precursor of bioactive compounds such as anticancer acridone alkaloids and other peptide-based drugs. However, the current routes of chemical synthesis or biosynthesis of N-alkylated amino acids cannot be easily exploited, due to insufficient yields or low profitability. The research work develops a fermentative NMA production by metabolic engineering of genome-reduced chassis strain *C. glutamicum* C1 combined with the introduction of a SAM-dependent ANMT (N-methyl transferase enzyme) gene, originating from *Ruta graveloens*. With this engineered strain the production of NMA reached a final titer of 0.5 $g{\cdot}L^{-1}$ with a yield of 4.8 $mg{\cdot}g^{-1}$ glucose, which has never been obtained before. Very specific transformations adapted to market needs are thus possible.

After strain characterization or improvement and metabolite production, the extraction step is a key step toward the industrial production. A microbial product's extraction is often solvent consuming and inconsistent with the principle of sustainable environment. In the study of Lebeau et al. (2020) [21], the azaphilone red pigments and ergosterol derivatives produced by a wild type of marine derived *Talaromyces* species are described. Alternative extraction process of the fungal pigments is developed, using a high pressure with eco-friendly solvents. These fungal pigments could be of interest due to their applications in the design of new pharmaceutical products.

However, Herculean work must be undertaken for the optimization of the molecules' extraction. It would be pointless to promote the production and use of microbial natural molecules, while simultaneously releasing a greater quantity of pollutants into ecosystems.

6. Improving the Biotechnological Process and Opening to Unusual Fields

Improving the biotechnological processes is one of the cornerstones of the transition from pilot-scale fermenters to industrial applications.

Jawan et al. (2020) [22] aim to optimize the growth parameters of *Lactococcus lactis* Gh1 in regard with the production of BLIS (bacteriocin-like inhibitory substances). Indeed, lactic acid bacteria (LAB) bacteriocins are considered good bio-preservative agents due to their non-toxic, non-immunogenic and thermo-resistance characteristics as well as broad bactericidal activity. The large-scale production of BLIS can offer promising opportunities for the development of efficient food bio preservation strategies. The parameters such as the age and size of the inoculum, initial pH of culture media and the nature of nitrogen and carbon sources, were studied in the study. Thanks to this work, the best set of parameters was defined, based on the production of BLIS. It can be concluded that bacteriocin production can be notably improved by altering the cultivation conditions of the bacterium. Results could be used for subsequent application in process design and optimization at larger scale.

Oslan et al. (2020) [23] worked on improving the growth and viability of the mutant *gdhA Pasteurella multocida* B:2. As they have demonstrated that *gdhA P. multocida* B:2 growth was inhibited by the accumulation of by-product ammonium in the culture medium, they developed a 2 L stirred-tank bioreactor integrated with an internal column using cation-exchange adsorption resin. This way, they demonstrated a significant improvement in growth performance and viability of the bacteria, by continuously reducing the in situ inhibitory effect of ammonium.

Efficiently increasing the yields of specific metabolites undeniably requires the use of computer and statistical tools.

The work of Venkatachalam et al. (2020) [24] fully involves computational tools to optimize the physico-chemical parameters (initial pH, temperature, agitation speed and fermentation time), for the production of pigments and biomass in submerged fermentation by a marine-derived strain of *Talaromyces albobiverticillius*. To achieve the optimization, a Box–Behnken experimental design (BBED) based on a three-interlocking, 22 factorial matrix was applied, reducing notably the number of experiments required. The results were statistically analyzed, and a prediction of the optimal conditions was proposed using a response surface modeling (RSM) process. This methodology allowed easier consideration of multifactorial interactions between a set of diverse parameters and enabled a rapid selection of the best culture conditions, in regard with the various objectives set. Therefore, the predictive model was validated and the optimal conditions for the maximum production of pigments (red and orange) as well as biomass, were determined as follows: initial pH 6.4, 24 °C, agitation speed of 164 rpm for a fermentation time of 149h. This methodology greatly facilitates and accelerates the selection of the best experimental conditions for the production of metabolites of interest.

In the field of environment depuration, sugarcane distillery spent wash (DSW) is one of the most polluting industrial effluents. Its acidic pH, high mineral matters and chemical oxygen demand (COD) are involved in deep environmental disruptions. High added-value products generated from waste are particularly encouraged through the European directive 2009/28/CE [3] and are an economically favorable way to perform bioremediation. Thus, Chuppa-Tostain et al. (2020) [25] selected 37 strains of yeasts and filamentous fungi to test their abilities in reducing the organic load of vinasses, through measuring their impact on COD, pH, minerals and OD_{475nm} parameters. Among the strains tested, the species from *Aspergillus* and *Trametes* genus offered the best results in the depollution of DSW. The increase in soil and water pollution will certainly require, in the coming years, a concentration in research efforts in the field of biological purification.

The agro-food-processing industrial effluents generated in the maize (*Zea mays*) cooking process, are also considered as environmental pollutants. Some scientific programs aim at studying the utilization of these wastewater as raw substrate for microorganism's growth and bioactive metabolite production. The research of Bacame-Valenzuela et al. (2020) [26] focused on the culture conditions for the production of pyocyanin (a redox metabolite of the phenazine group) by *Pseudomonas aeruginosa*. The parameters were first optimized using statistical design and RSM (response surface methodology) in a defined medium. The optimized parameters were then applied in a culture using an alkalinized maize-based effluent. The production of pyocyanin was up to 3.25 $\mu g \cdot mL^{-1}$, higher than in initial defined medium. In this way, valorization of lime-cooked maize wastewater used as a substrate for microbial growth was demonstrated in the production of a value-added compound.

In a complementary way, Zhang et al. [27] focused on the process for lactic acid production from corn stover, a common agricultural residue. Lignocellulose residues are one of the most abundant renewable feedstocks. They selected a *Pediococcus acidilactici* safe strain (PA204) with a high-efficiency for the utilization of xylose, a good ability to ferment sugar derived from lignocellulose, and a high temperature tolerance. To improve the lactic acid production, they developed a fed-batch simultaneous saccharification and fermentation (SSF) process (at 37 °C, pH 6.0) using corn stover and corn steep powder as carbon sources, in a 5 L bioreactor. The culture produced up to 104.11 $g \cdot L^{-1}$ lactic acid and a yield up to of 0.77 $g \cdot g^{-1}$ in regard with the NaOH-pretreatment applied on corn stover and the addition of cellulases. This study developed a feasible and efficient fed-batch process for lactic acid production from corn stover, and provides a promising candidate strain for high-titer and -yield lignocellulose-derived lactic acid production.

7. Future Strategies and Perspectives

Due to technical and scientific progress, the future is rich with new opportunities to go further in the use of microorganisms as controlled cell factories, producing high value-added metabolites for our daily needs.

As an example, even if filamentous actinobacteria are historically known and used as antibiotic producers, novel approaches are on the way. Protoplast fusion is an ancient technique for genome recombination that drastically increases microbial genetic diversity and favors the expression of silent or poorly expressed genes. However, the method requires multiple fusion and regeneration phases. Moreover, cells face a short time frame for recombination, a consequence of the necessity for protoplasts to regenerate their cell wall, and additionally, the recombinants are commonly fickle. These disadvantages are clear limitations for the production and the exploitation of original valuable metabolites. The article of Shitut et al. (2020) [28] highlights the potential of new engineered wall-deficient bacteria having the capacity to proliferate without their cell wall. Unlike protoplasts, L-forms are able to stabilize multiple chromosomes over many cycles of division. After the fusion, the period of recombination would thus be lengthened. Gene expression could also be based on the two parental genomes. These advances would significantly contribute to increase the chemical diversity of molecules produced by the engineered cells. These new L-forms open a perspective for the discovery of novel compounds, especially in the field of antibiotic research.

Cyanobacteria, as well as all photosynthetic organisms, have the capacity to produce secondary metabolites based on solar energy and carbon dioxide utilization, fitting a current sustainable development philosophy. Metabolic engineered-based approaches are now widely applied to cyanobacteria for the industrial production of value-added compounds, including specific metabolites and non-natural biochemicals. However, the review of Jeong et al. (2020) [4] covers a large synthetic and systems biology approach for advanced metabolic engineering of cyanobacteria. This involves an overview of known biosynthetic clusters coding for essential compounds, and knowledge of heterologous expression for cyanobacterial secondary metabolite production. As a perspective, the review highlights the development of next-generation sequencing (NGS) techniques, combined with the collection of omics data such as transcriptome, translatome, proteome, metabolome or interactome, which enrich the worldwide databases, opening huge opportunities to better manipulate and control cyanobacteria productions. The development of an in silico model at the genome scale (GEM) could quickly make it possible to remedy the current limitations of cyanobacterial engineering.

As a conclusion, we are very happy that we received 25 reviews/perspectives/original papers for publication in this Special Issue. We wish to thank all the authors and reviewers for their significant contributions and for making it a highly successful and timely collection of studies.

Author Contributions: Conceptualization, M.F. and L.D.; writing—original draft preparation, M.F.; writing—review and editing, L.D. and M.F. All authors have read and agreed to the published version of the manuscript.

Funding: This research received no external funding.

Institutional Review Board Statement: Not applicable.

Informed Consent Statement: Not applicable.

Conflicts of Interest: The authors declare no conflict of interest.

References

1. Monciardini, P.; Iorio, M.; Maffioli, S.; Sosio, M.; Donadio, S. Discovering new bioactive molecules from microbial sources. *Microb. Biotechnol.* **2014**, *7*, 209–220. [CrossRef]
2. Fouillaud, M.; Venkatachalam, M.; Girard-Valenciennes, E.; Caro, Y.; Dufossé, L. Anthraquinones and derivatives from marine-derived fungi: Structural diversity and selected biological activities. *Mar. Drugs* **2016**, *14*, 64. [CrossRef]
3. Patridge, E.; Gareiss, P.; Kinch, M.S.; Hoyer, D. An analysis of FDA-approved drugs: Natural products and their derivatives. *Drug Discov. Today* **2016**, *21*, 204–207. [CrossRef]
4. Jeong, Y.; Cho, S.-H.; Lee, H.; Choi, H.-K.; Kim, D.-M.; Lee, C.-G.; Cho, S.; Cho, B.-K. Current Status and Future Strategies to Increase Secondary Metabolite Production from Cyanobacteria. *Microorganisms* **2020**, *8*, 1849. [CrossRef]
5. Said Hassane, C.; Fouillaud, M.; Le Goff, G.; Sklirou, A.D.; Boyer, J.B.; Trougakos, I.P.; Jerabek, M.; Bignon, J.; de Voogd, N.J.; Ouazzani, J.; et al. Microorganisms Associated with the Marine Sponge *Scopalina hapalia*: A Reservoir of Bioactive Molecules to Slow Down the Aging Process. *Microorganisms* **2020**, *8*, 1262. [CrossRef] [PubMed]
6. Salikin, N.H.; Nappi, J.; Majzoub, M.E.; Egan, S. Combating Parasitic Nematode Infections, Newly Discovered Antinematode Compounds from Marine Epiphytic Bacteria. *Microorganisms* **2020**, *8*, 1963. [CrossRef] [PubMed]
7. Nawaz, A.; Chaudhary, R.; Shah, Z.; Dufossé, L.; Fouillaud, M.; Mukhtar, H.; ul Haq, I. An Overview on Industrial and Medical Applications of Bio-Pigments Synthesized by Marine Bacteria. *Microorganisms* **2021**, *9*, 11. [CrossRef]
8. Hwang, S.; Le, L.T.H.L.; Jo, S.-I.; Shin, J.; Lee, M.J.; Oh, D.-C. Pentaminomycins C–E: Cyclic Pentapeptides as Autophagy Inducers from a Mealworm Beetle Gut Bacterium. *Microorganisms* **2020**, *8*, 1390. [CrossRef]
9. Shamikh, Y.I.; El Shamy, A.A.; Gaber, Y.; Abdelmohsen, U.R.; Madkour, H.A.; Horn, H.; Hassan, H.M.; Elmaidomy, A.H.; Alkhalifah, D.H.M.; Hozzein, W.N. Actinomycetes from the Red Sea Sponge *Coscinoderma mathewsi*: Isolation, Diversity, and Potential for Bioactive Compounds Discovery. *Microorganisms* **2020**, *8*, 783. [CrossRef] [PubMed]
10. Schneider, Y.; Jenssen, M.; Isaksson, J.; Hansen, K.Ø.; Andersen, J.H.; Hansen, E.H. Bioactivity of Serratiochelin A, a Siderophore Isolated from a Co-Culture of *Serratia* sp. and *Shewanella* sp. *Microorganisms* **2020**, *8*, 1042. [CrossRef]

11. Voitsekhovskaia, I.; Paulus, C.; Dahlem, C.; Rebets, Y.; Nadmid, S.; Zapp, J.; Axenov-Gribanov, D.; Rückert, C.; Timofeyev, M.; Kalinowski, J.; et al. Baikalomycins A-C, New Aquayamycin-Type Angucyclines Isolated from Lake Baikal Derived *Streptomyces* sp. IB201691-2A. *Microorganisms* **2020**, *8*, 680. [CrossRef]
12. Khiralla, A.; Spina, R.; Varbanov, M.; Philippot, S.; Lemiere, P.; Slezack-Deschaumes, S.; André, P.; Mohamed, I.; Yagi, S.M.; Laurain-Mattar, D. Evaluation of Antiviral, Antibacterial and Antiproliferative Activities of the Endophytic Fungus *Curvularia papendorfii*, and Isolation of a New Polyhydroxyacid. *Microorganisms* **2020**, *8*, 1353. [CrossRef]
13. Liu, L.; Hu, Z.; Li, S.; Yang, H.; Li, S.; Lv, C.; Zaynab, M.; Cheng, C.H.K.; Chen, H.; Yang, X. Comparative Transcriptomic Analysis Uncovers Genes Responsible for the DHA Enhancement in the Mutant *Aurantiochytrium* sp. *Microorganisms* **2020**, *8*, 529. [CrossRef] [PubMed]
14. Argüelles, A.; Sánchez-Fresneda, R.; Guirao-Abad, J.P.; Belda, C.; Lozano, J.A.; Solano, F.; Argüelles, J.-C. Novel Bi-Factorial Strategy against *Candida albicans* Viability Using Carnosic Acid and Propolis: Synergistic Antifungal Action. *Microorganisms* **2020**, *8*, 749. [CrossRef]
15. Papaianni, M.; Ricciardelli, A.; Fulgione, A.; d'Errico, G.; Zoina, A.; Lorito, M.; Woo, S.L.; Vinale, F.; Capparelli, R. Antibiofilm Activity of a *Trichoderma* Metabolite against *Xanthomonas campestris* pv. *campestris*, Alone and in Association with a Phage. *Microorganisms* **2020**, *8*, 620. [CrossRef] [PubMed]
16. Hamed, A.A.; Soldatou, S.; Qader, M.M.; Arjunan, S.; Miranda, K.J.; Casolari, F.; Pavesi, C.; Diyaolu, O.A.; Thissera, B.; Eshelli, M.; et al. Screening Fungal Endophytes Derived from Under-Explored Egyptian Marine Habitats for Antimicrobial and Antioxidant Properties in Factionalised Textiles. *Microorganisms* **2020**, *8*, 1617. [CrossRef] [PubMed]
17. Sayed, A.M.; Alhadrami, H.A.; El-Gendy, A.O.; Shamikh, Y.I.; Belbahri, L.; Hassan, H.M.; Abdelmohsen, U.R.; Rateb, M.E. Microbial Natural Products as Potential Inhibitors of SARS-CoV-2 Main Protease (Mpro). *Microorganisms* **2020**, *8*, 970. [CrossRef]
18. González-Jiménez, M.d.C.; García-Martínez, T.; Mauricio, J.C.; Sánchez-León, I.; Puig-Pujol, A.; Moreno, J.; Moreno-García, J. Comparative Study of the Proteins Involved in the Fermentation-Derived Compounds in Two Strains of *Saccharomyces cerevisiae* during Sparkling Wine Second Fermentation. *Microorganisms* **2020**, *8*, 1209. [CrossRef] [PubMed]
19. Hilgers, F.; Habash, S.S.; Loeschcke, A.; Ackermann, Y.S.; Neumann, S.; Heck, A.; Klaus, O.; Hage-Hülsmann, J.; Grundler, F.M.W.; Jaeger, K.-E.; et al. Heterologous Production of β-Caryophyllene and Evaluation of Its Activity against Plant Pathogenic Fungi. *Microorganisms* **2021**, *9*, 168. [CrossRef]
20. Walter, T.; Al Medani, N.; Burgardt, A.; Cankar, K.; Ferrer, L.; Kerbs, A.; Lee, J.-H.; Mindt, M.; Risse, J.M.; Wendisch, V.F. Fermentative N-Methylanthranilate Production by Engineered *Corynebacterium glutamicum*. *Microorganisms* **2020**, *8*, 866. [CrossRef] [PubMed]
21. Lebeau, J.; Petit, T.; Fouillaud, M.; Dufossé, L.; Caro, Y. Alternative Extraction and Characterization of Nitrogen-Containing Azaphilone Red Pigments and Ergosterol Derivatives from the Marine-Derived Fungal *Talaromyces* sp. 30570 Strain with Industrial Relevance. *Microorganisms* **2020**, *8*, 1920. [CrossRef]
22. Jawan, R.; Abbasiliasi, S.; Tan, J.S.; Mustafa, S.; Halim, M.; Ariff, A.B. Influence of Culture Conditions and Medium Compositions on the Production of Bacteriocin-Like Inhibitory Substances by *Lactococcus lactis* Gh1. *Microorganisms* **2020**, *8*, 1454. [CrossRef]
23. Oslan, S.N.H.; Tan, J.S.; Abbasiliasi, S.; Ziad Sulaiman, A.; Saad, M.Z.; Halim, M.; Ariff, A.B. Integrated Stirred-Tank Bioreactor with Internal Adsorption for the Removal of Ammonium to Enhance the Cultivation Performance of *gdhA* Derivative *Pasteurella multocida* B:2. *Microorganisms* **2020**, *8*, 1654. [CrossRef] [PubMed]
24. Venkatachalam, M.; Shum-Chéong-Sing, A.; Dufossé, L.; Fouillaud, M. Statistical Optimization of the Physico-Chemical Parameters for Pigment Production in Submerged Fermentation of *Talaromyces albobiverticillius* 30548. *Microorganisms* **2020**, *8*, 711. [CrossRef] [PubMed]
25. Chuppa-Tostain, G.; Tan, M.; Adelard, L.; Shum-Cheong-Sing, A.; François, J.-M.; Caro, Y.; Petit, T. Evaluation of Filamentous Fungi and Yeasts for the Biodegradation of Sugarcane Distillery Wastewater. *Microorganisms* **2020**, *8*, 1588. [CrossRef]
26. Bacame-Valenzuela, F.J.; Pérez-Garcia, J.A.; Figueroa-Magallón, M.L.; Espejel-Ayala, F.; Ortiz-Frade, L.A.; Reyes-Vidal, Y. Optimized Production of a Redox Metabolite (pyocyanin) by *Pseudomonas aeruginosa* NEJ01R Using a Maize By-Product. *Microorganisms* **2020**, *8*, 1559. [CrossRef] [PubMed]
27. Zhang, Z.; Li, Y.; Zhang, J.; Peng, N.; Liang, Y.; Zhao, S. High-Titer Lactic Acid Production by *Pediococcus acidilactici* PA204 from Corn Stover through Fed-Batch Simultaneous Saccharification and Fermentation. *Microorganisms* **2020**, *8*, 1491. [CrossRef] [PubMed]
28. Shitut, S.; Bergman, G.Ö.; Kros, A.; Rozen, D.E.; Claessen, D. Use of Permanent Wall-Deficient Cells as a System for the Discovery of New-to-Nature Metabolites. *Microorganisms* **2020**, *8*, 1897. [CrossRef]

microorganisms

MDPI

Review

Current Status and Future Strategies to Increase Secondary Metabolite Production from Cyanobacteria

Yujin Jeong [1], Sang-Hyeok Cho [1], Hookeun Lee [2], Hyung-Kyoon Choi [3], Dong-Myung Kim [4], Choul-Gyun Lee [5], Suhyung Cho [1,*] and Byung-Kwan Cho [1,*]

1 Department of Biological Sciences and KAIST Institutes for the BioCentury, Korea Advanced Institute of Science and Technology, Daejeon 34141, Korea; mist@kaist.ac.kr (Y.J.); graysky@kaist.ac.kr (S.-H.C.)
2 Institute of Pharmaceutical Research, College of Pharmacy, Gachon University, Incheon 21999, Korea; hklee@gachon.ac.kr
3 College of Pharmacy, Chung-Ang University, Seoul 06911, Korea; hykychoi@cau.ac.kr
4 Department of Chemical Engineering and Applied Chemistry, Chungnam National University, Daejeon 34134, Korea; dmkim@cnu.ac.kr
5 Department of Biological Engineering, Inha University, Incheon 22212, Korea; leecg@inha.ac.kr
* Correspondence: shcho95@kaist.ac.kr (S.C.); bcho@kaist.ac.kr (B.-K.C.)

Received: 29 October 2020; Accepted: 23 November 2020; Published: 24 November 2020

Abstract: Cyanobacteria, given their ability to produce various secondary metabolites utilizing solar energy and carbon dioxide, are a potential platform for sustainable production of biochemicals. Until now, conventional metabolic engineering approaches have been applied to various cyanobacterial species for enhanced production of industrially valued compounds, including secondary metabolites and non-natural biochemicals. However, the shortage of understanding of cyanobacterial metabolic and regulatory networks for atmospheric carbon fixation to biochemical production and the lack of available engineering tools limit the potential of cyanobacteria for industrial applications. Recently, to overcome the limitations, synthetic biology tools and systems biology approaches such as genome-scale modeling based on diverse omics data have been applied to cyanobacteria. This review covers the synthetic and systems biology approaches for advanced metabolic engineering of cyanobacteria.

Keywords: cyanobacteria; photosynthesis; secondary metabolites; metabolic engineering; synthetic biology; systems biology; genome-scale model

1. Introduction

Cyanobacteria are oxygenic photosynthetic bacteria that can produce various secondary metabolites. Given the ability to utilize sunlight and atmospheric carbon dioxide (CO_2) as a part of the renewable photosynthetic process, cyanobacteria are considered sustainable bioproduction hosts [1]. A number of secondary metabolites naturally synthesized by cyanobacteria, such as carotenoids, phycocyanins, and squalene, are used in the pharmaceutical, cosmetic, and healthcare industries [2–4]. In addition, owing to their rapid growth and increased scope for engineering, multiple efforts have been made to utilize cyanobacteria as production hosts for valuable biochemicals by introducing heterologous pathways [5,6].

While continuous development has been reported in metabolic engineering strategies for producing biochemicals in bacterial hosts, the synthetic biology approach accelerated the development by providing diverse genetic parts and engineering tools. For other model platforms such as *Escherichia coli*, there is an abundant catalog of genetic parts including synthetic promoters and ribosome binding sites (RBSs), which have been successfully introduced to improve gene expression in heterologous pathways [7]. However, owing to the lack of genetic parts for pathway engineering in cyanobacteria, application of metabolic engineering tools is limited [8]. Thus, development of various tools for pathway engineering and subsequent engineering strategies are required for industrial-scale production of target compounds in cyanobacteria.

With the recent progress in systems biology, genome-wide information of diverse layers such as the genome, transcriptome, translatome, proteome, metabolome, and interactome are being constantly accumulated [9]. Massive amounts of data formed the basis for establishment and development of an in silico genome-scale model (GEM) [10]. It is expected that the application of system-level approaches with the integration of omics data and GEM would address the existing limitations of cyanobacterial engineering. The current review not only describes the value-added secondary metabolites produced by cyanobacteria and current metabolic engineering approaches for their production but also introduces the synthetic and systems biology approach for further development.

2. Secondary Metabolite Production by Cyanobacteria

Bacteria produce two kinds of metabolites: primary metabolites essential for survival and secondary metabolites required for auxiliary purposes, such as stress responses, defense mechanisms, metal carrying, and signaling [11]. Secondary metabolites include terpenes, alkaloids, polyketides (PKs), non-ribosomal peptides (NRPs), and ribosomally synthesized and post-translationally modified peptides (RiPPs), which are produced via biosynthetic gene clusters (BGCs). BGCs are clusters of genes positioned in approximate proximity to each other for the production and processing of a compound. Cyanobacteria, being rich in BGCs, are capable of producing diverse secondary metabolites for various purposes, including toxins for defenses or protectants for relieving photodamage and oxidative stress (Table 1).

Table 1. Bioactive secondary metabolites produced in cyanobacteria.

Class	Metabolite	Bioactivity	Producing Species	Ref.
Terpene	Phycocyanin	Antioxidant, anti-inflammatory, neuroprotective, hepatoprotective	All cyanobacteria	[12–16]
Terpene	Carotenoids	Antioxidant, sunscreen	All cyanobacteria	[17,18]
Terpene	Squalene	Antioxidant	*Phormidium*	[19]
Alkaloid	Saxitoxin	Neurotoxin	*Anabaena, Aphanizomenon, Cylindrospermopsis, Lyngbya, Planktothrix,*	[20–22]
Indole	Nostodione	Antifungal	*Nostoc*	[23]
Indole alkaloid	Scytonemin	Anti-inflammatory, sunscreen	*Scytonema, Nostoc*	[24–27]
Indole alkaloid	Hapalindole	Antibacterial, anti-tuberculosis, anticancer	*Hapalosiphon*	[28,29]
Alkaloid/Polyketide synthase (PKS)	Anatoxin-a	Neurotoxin, anti-inflammatory	*Anabaena, Aphanizomenon, Cylindrospermum, Oscillatoria, Planktothrix*	[30,31]
Alkaloid/PKS	Aplysiatoxin	Cytotoxin, antiviral	*Moorea*	[32,33]
Alkaloid/Non-ribosomal peptide synthetase (NRPS)	Lyngbyatoxin	Cytotoxin, dermatotoxin	*Moorea*	[34]
Alkaloid/PKS-NRPS	Cylindrospermopsin	Cytotoxin	*Aphanizomenon, Cylindrospermopsis, Oscillatoria, Raphidiopsis*	[35–37]
PKS	Fischerellin	Antifungal, antialgal, anti-cyanobacterial	*Fischerella*	[38]
NRPS	β-N-methylamino-L-alanine	Neurotoxin	*Anabaena, Nostoc*	[39]
NRPS	Cyanopeptolin	Protease inhibitor	*Planktothrix, Microcystis*	[40,41]
PKS-NRPS	Microcystin	Hepatotoxin	*Microcystis, Nostoc, Planktothrix, Anabaena*	[40,42–45]
PKS-NRPS	Nodularin	Hepatotoxin	*Nodularia*	[46]
PKS-NRPS	Apratoxin	Anticancer	*Lyngbya*	[47]
PKS-NRPS	Aeruginoside	Protease inhibitor	*Planktothrix*	[48]
PKS-NRPS	Aeruginosin	Protease inhibitor	*Microcystis, Planktothrix*	[40,49]
PKS-NRPS	Cryptophycins	Cytotoxin	*Nostoc*	[50]
PKS-NRPS	Nostophycins	Cytotoxin	*Nostoc*	[51]
PKS-NRPS	Curacins	Cytotoxin	*Moorea*	[52]
PKS-NRPS	Hectochlorin	Cytotoxin	*Moorea*	[53]
PKS-NRPS	Jamaicamides	Neurotoxin	*Moorea*	[54]

Table 1. *Cont.*

Class	Metabolite	Bioactivity	Producing Species	Ref.
PKS-NRPS	Dolastatin	Cytotoxin, anticancer, antiprotozoal	*Moorea, Lyngbya, Symploca*	[55,56]
Lipopeptide	Antillatoxin	Neurotoxin	*Moorea*	[57]
Lipopeptide	Carmabin	Antimalarial, anticancer, antiproliferative	*Moorea*	[58,59]
Lipopeptide	Lyngbyabellin	Cytotoxin, antifungal	*Moorea, Lyngbya*	[60,61]
Lipopeptide	Kalkitoxin	Neurotoxin	*Moorea*	[57]
Ribosomally synthesized and post-translationally modified peptide (RiPP)	Patellamide	Moderate cytotoxicity	*Prochloron*	[62]
RiPP	Microviridin	Protease inhibitor	*Microcystis, Planktothrix*	[63,64]
RiPP	Shinorin	Sunscreen	*Anabaena, Nostoc*	[65]
Fatty acid amide	Besarhanamide A	Moderate toxicity to brine shrimp	*Moorea*	[66]
Fatty acid amide	Semiplenamide	Toxicity to brine shrimp	*Lyngbya*	[67]
Lipopolysaccharide	Lipopolysaccharides	Endotoxin	All cyanobacteria	[68]
Polysaccharide	Polysaccharide	Antitumor, antiviral, antibacterial, anti-inflammatory, immunostimulant	All cyanobacteria	[69–71]
Nucleoside	Toyocamycin	Antifungal	*Tolypothrix*	[72]
Nucleoside	Tubercidin	Antifungal	*Tolypothrix*	[73]

2.1. Prediction of Biosynthetic Gene Clusters (BGCs) in Cyanobacterial Genomes

To investigate the secondary metabolites produced by cyanobacteria, 196 complete genome sequences of cyanobacteria available at the National Center for Biotechnology Information (NCBI) genome portal were inspected for BGCs using antiSMASH [74]. Thirty-three different types of BGCs were identified. The 196 complete genome sequences of cyanobacteria used in the BGC search were arranged according to the phylogenetic tree. The heatmap representing the numbers of each type of BGC found in each cyanobacterium showed that the cyanobacteria from the same genera had similar classes and numbers of BGCs (Figure 1A). It was evident that a single genome contained several BGCs with multiple occurrences. In particular, there were cyanobacteria with large number of bacteriocin, terpene, and non-ribosomal peptide synthetase (NRPS) BGCs, which accounted for 74.4% of the total predicted BGCs (n = 2119). For example, it was predicted that the genome of *Moorea producens* PAL-8-15-08-1 carries 18 NRPS BGCs. The most widely distributed BGC was the terpene BGC, which was found in all cyanobacteria except for two species (*Limnospira fusiformis* SAG 85.79 and *Nodularia spumigena* UHCC 0039). Terpene is essential for photosynthetic organisms. Undetected terpene BGCs in the two species could have resulted from the deviations in the BGC search criteria of antiSMASH. The 33 BGCs were classified according to their structural and functional similarities to the following categories: terpene, indole, PK synthase (PKS)/NRPS (type 1, 2, 3 PKSs, NRPS, cyclodipeptide synthase-based tRNA-dependent peptide, resorcinol, and siderophore), RiPP (bacteriocin, lanthidin, linear azole-containing peptide, microviridin, lasso peptide, cyanobactin, thiopeptide, trifolitoxin, proteusin, and lanthipeptide), lipid/saccharide/nucleoside (heterocyst glycolipid synthase, ladderane, arylpolyene, aminoglycoside/aminocyclitol, oligosaccharide, and nucleoside), and others (phosphonate, phenazine, ectoine, β-lactone, and homoserine lactone).

2.2. Terpenes

Terpene is a family of compounds with varying structures that occupies a large proportion of the natural products [75]. Terpenes are mainly produced by plants or fungi, as well as the bacterial species via mevalonate (MVA) pathway or methylerythritol-phosphate (MEP) pathway using acetyl-CoA or glyceraldehyde 3-phosphate and pyruvate as substrates [76]. While MVA and MEP pathways are mutually exclusive in most organisms, cyanobacteria mainly utilize the MEP pathway, using substrates generated during photosynthesis. The MEP pathway produces isomeric 5-carbon compounds, isopentyl pyrophosphate (IPP), and dimethylallyl pyrophosphate (DMAPP), which are further condensed into geranyl pyrophosphate (GPP), the building block in terpene biosynthesis. From the GPP, terpenes of varying structures can be generated. Terpenes conduct various cellular processes necessary for survival, such as the ubiquinone in the electron transport chain associated with cellular respiration, chlorophyll, carotenoids, and plastoquinones in photosynthetic processes, and hopanoids in cell membrane biosynthesis and stability (Figure 1B) [77]. In particular, photosynthetic cyanobacteria contain a wide variety of carotenoids. Most of the genome-sequenced cyanobacteria have β-carotene BGC. Production of other carotenoids, such as zeaxanthin and nostoxanthin are dependent on the presence of carotenogenesis pathway connected to β-carotene [3,78]. The terpene compounds, including the carotenoids obtained from cyanobacteria are of industrial value owing to their various applications. For example, β-carotene, astaxanthin, and canthaxanthin are used as color additives or animal feeds. Phycocyanin exhibits anti-oxidant, anti-inflammatory, neuroprotective, and hepatoprotective effects [2,13,79].

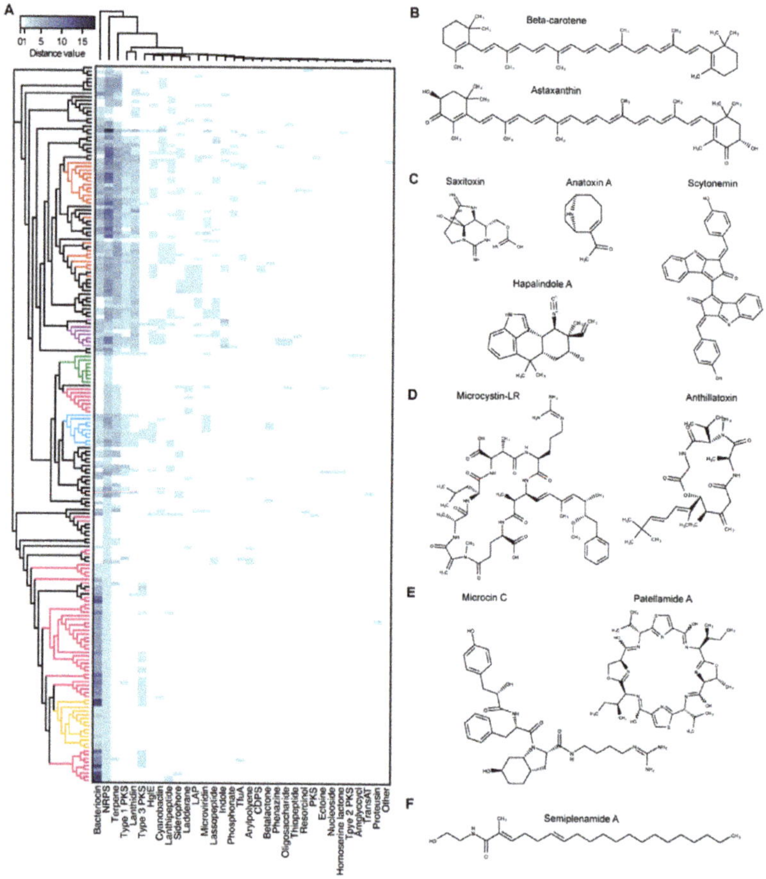

Figure 1. Cyanobacterial secondary metabolites. (**A**) Heatmap of the predicted cyanobacterial secondary metabolite biosynthstic gene clusters (BGCs). The left-most phylogenetic tree is constructed by up-to-date bacterial core gene (UBCG) phylogenetic analysis of the 196 cyanobacterial complete genome sequences. The evolutionary distances were provided by UBCG and plotted by RAxML [80,81]. The tree is not to scale. Red: *Nostoc*, purple: *Calothrix*, green: *Synechocystis*, pink: *Synechoccus*, blue: *Microcystis*, and yellow: *Prochlorococcus*. (**B–F**) Molecular structures of cyanobacterial secondary metabolites. (**B**) Terpenes, (**C**) alkaloids, (**D**) polyketides (PKs), non-ribosomal peptides (NRPs), (**E**) RiPPs, and (**F**) fatty acid amide. Abbreviations; NRPS, non-ribosomal peptide synthetase; HglE, heterocyst glycolipid synthase; LAP, linear azol(in)e-containing peptide; TfuA, ribosomally synthesized peptide antibiotic trifolitoxin; CDPS, cyclodipeptide synthase-based tRNA dependent peptide; PKS, polyketide synthase; Amglyccycl, aminoglycosides/aminocyclitols; TransAT, trans-acyltransferase type I PKS.

2.3. Alkaloids

Alkaloids comprise various nitrogen containing compounds that are produced from diverse organisms, including fungi, plants, bacteria, and animals. Alkaloids produced by cyanobacteria often show toxic characteristics. For example, the anatoxin-a produced by species of the *Anabaena* genera is a neurotoxin that binds irreversibly to nicotinic acetylcholine receptors causing paralysis or even death in fish and mammals (Figure 1C) [31]. Anatoxin-a is also categorized as a PK, which is synthesized by PKS [82]. Another well-known example, saxitoxin, blocks the sodium (Na^+) channels in shellfish and induces paralytic shellfish poisoning in humans on consumption of saxitoxin-accumulated seafood. The chemical derivatives carrying the indole rings are classified as indole alkaloids. They are biosynthesized using tryptophan as a precursor. Cyanobacterial indole alkaloids have diverse functions. For example, the hapalindole synthesized from cyanobacteria *Hapalosiphon fontinalis* exhibits antibacterial, anti-tuberculosis, and anticancer activities [83]. In addition, the scytonemin produced by *Scytonema* sp. renders photoprotective effects to the cyanobacterial cells by absorbing the harmful ultraviolet (UV)-A radiation [84].

2.4. Polyketides/Non-Ribosomal Peptide/Lipopeptides/Siderophores

PKS and NRPS are representatives of enzymes responsible for the biosynthesis of secondary metabolites in various organisms. Enzymes of these classes consists of at least three essential modular domains that facilitate chain elongation and modification [85]. First, the catalytic domain binds to and activates the building block, which then is transferred to the carrier protein domain. Second, the carrier protein domain loads the activated building block to the growing PK/NRP chain it holds. Third, the other catalytic domain catalyzes the bond formation between the growing chain and the newly loaded building block. PKS and NRPS differ in their use of precursors for the building block. While PKS utilizes malonyl-CoA or methylmalonyl-CoA, the NRPS uses proteinogenic and non-proteinogenic amino acid monomers. In addition, there are cases wherein compounds are synthesized via the PKS–NRPS hybrid system. A well-known example could be microcystin, the BGC of which contains two PKS, single PKS–NRPS, and three NRPS [42,86]. Microcystin produced from various cyanobacterial species belonging to the genus Microcystis, Nostoc, Planktothrix, and Anabaena, shows hepatotoxic activity in humans (Figure 1D). Various other toxins synthesized by the PKS, NRPS, or PKS–NRPS hybrid system includes lyngbyatoxin, apratoxin, and aplysiatoxin.

The NRPS includes lipopeptides owing to their lipid linked peptide structures synthesized by a combination of lipid tails and amino acids. Examples of lipopeptides include antillatoxin and carmabin from *M. producens*, and lyngbyabellin from *M. bouillonii* (Figure 1D). Antillatoxin and lyngbyabellin show neurotoxic activity and cytotoxicity, and carmabin exhibit anti-malarial activity. Siderophores are included in the NRPS-produced compounds. Iron is essential for bacterial survival. However, since it exists in an insoluble form in the environment, some bacteria have evolved to facilitate iron uptake by producing small molecules with high affinity to ferric iron, called siderophores.

2.5. Ribosomally Synthesized and Post-Translationally Modified Peptides

RiPP is a class of secondary metabolites that includes, as its name depicts, ribosomally synthesized and post-translationally modified peptides. Post-translational modifications include leader peptide hydrolysis, cyclization, and disulfide bond formation. RiPP BGC generally consists of a short precursor peptide with an *N*-terminal leader and a *C*-terminal core sequence, and post-translational modification (PTM) enzymes [87,88]. The PTM enzymes shape the linear peptide by several modifications that provide structural and functional diversity to the mature scaffold. Compounds that were previously classified as lanthipeptide, lasso peptide, microviridin, cyanobactin, and microcin are now re-classified under RiPP, which have a broad range of bioactivities such as protease inhibition, cytotoxicity, signaling, anti-cancer, and anti-human immunodeficiency virus (anti-HIV) (Figure 1E) [87]. For example, microviridin, which was first isolated from *M. viridis*, is a serine protease inhibitor, and patellamide A produced by *Prochloron didemni* has moderate cytotoxicity [62].

2.6. *Lipids/Saccharides/Nucleosides/Others*

Lipids, saccharides, and nucleosides are generally categorized as primary metabolites. However, there are exceptions, when they are considered as secondary metabolites instead of primary metabolites. For example, besarhanamide A and semiplenamide exhibiting toxicity against brine shrimp are fatty acid amides isolated from *M. producens* and *Lyngbya semiplena*, respectively (Figure 1F) [66,89]. It is known that cyanobacterium *Cyanothece* sp. 113 can produce up to 22 g/L of polysaccharide, which exceeds the producing ability of eukaryotic microalgae, such as *Dunaliella salina* [90,91]. Polysaccharides are generally used as stabilization or thickening agents for emulsions. In some cases, they are used as bioactive compounds owing to their antitumor, antiviral, antibacterial, anti-inflammatory, and immunostimulatory properties [92–95]. Toyocamycin and tubercidin are both anti-fungal nucleoside chemicals isolated from *Tolypothrix tenuis* [96]. In addition, a small number of phosphonate, phenazine, ectoine, and β-lactone BGC were also detected.

3. Engineering Cyanobacteria for Industrial Production of Secondary Metabolites

Engineering efforts have been made to increase the production of industrially important cyanobacterial natural compounds. The model cyanobacteria such as *Synechocystis* sp. PCC 6803 and *Synechococcus elongatus* PCC 7942 are often used as engineering hosts for the increased ease of genetic manipulation. The biosynthetic pathways of other cyanobacteria are adopted to these model species by heterologous expression for production of value-added compounds. In addition to the cyanobacterial natural products, cyanobacteria have also been identified as a suitable heterologous platform for the production of biofuels, such as ethanol, butanol, and 2,3-butanediol [5,97–99]. Episomal expression using a self-replicating vector is a popular method for introducing foreign genetic elements in other organisms such as *E. coli*. Compared to chromosomal integration through homologous recombination, the episomal expression is more advantageous owing to its higher expression level [100]. In addition, the genome polyploidy of cyanobacteria can cause problems in the natural recombination process by reversing the engineered genome copies back to the original sequence, resulting in poor engineering efficiency and a laborious selection process. However, in cyanobacteria, there are minimal options for vector systems; the only self-replicating vector origin currently available for application is RSF1010 (Figure 2A). Thus, chromosomal integration or deletion through homologous recombination is the most dominant method used in cyanobacteria to increase the production of natural compound or heterologous metabolites (Figure 2B). Recently developed clustered regularly interspaced short palindromic repeat (CRISPR)/Cas is an effective genome engineering tool that can target specific loci to generate a double-strand break, and thus it can solve the low engineering efficiency problem in polyploids (Figure 2C) [101]. Additionally, the repurposed CRISPR/Cas system, namely the CRISPR interference (CRISPRi), can repress the gene expression without nucleic acid strand excision, avoiding lethality caused by knock-out of essential genes.

3.1. *Heterologous Expression for Cyanobacterial Secondary Metabolite Production*

Genetic manipulation tools explained above have been applied in cyanobacteria to increase the production of secondary metabolites (Table 2). In recent years, multiple studies have targeted terpenes, such as squalene and limonene. Squalene has widespread applications in the healthcare, cosmetics, and pharmacological fields, and it is produced from several eukaryotes as well as the cyanobacteria, such as *Phormidium autumnale* [19,102]. However, squalene production from cyanobacteria is not sufficient for the industrial-scale production demands. Metabolic engineering efforts were made in cyanobacterium model, *S. elongatus* PCC 7942 [103–105]. *S. elongatus* PCC 7942 has the methylerythritol phosphate (MEP) pathway to biosynthesize diphosphate (IPP) and dimethylallyl diphosphate (DMAPP) from CO_2. DMAPP is converted to farnesyl diphosphate (FPP), a substrate for squalene biosynthesis, by FPP synthase (*ispA*). Heterogenous genes, including 1-deoxy-D-xylulose-5-phosphate synthase (*dxs*), isopentenyl diphosphate isomerase (*idi*), and *ispA* were introduced into the *S. elongatus* PCC 7942 genome by homologous recombination to increase the intracellular concentration of FPP. Next, a squalene synthase (*SQS*) was also introduced by homologous recombination, resulting in a maximum of 5.0 mg/L/OD_{730nm} squalene production [103]. The titer was further increased to 12.0 mg/L/OD_{730nm} by constructing a fusion protein of *SQS* with *cpcB1*, which encodes the β-subunit of phycocyanin and is highly expressed under the strong endogenous *cpcB1* promoter [104]. Recently, CRISPRi was applied to the squalene-producing *S. elongatus* PCC 7942 strain to repress two essential genes, *acnB* and *cpcB2* encoding aconitase and phycocyanin β-subunit, respectively, resulting in an improved squalene production [105]. The results of these previous studies suggest that there is sufficient potential to improve the production of target compounds in cyanobacteria through the discovery of new potent promoters or the selection of additional engineering targets.

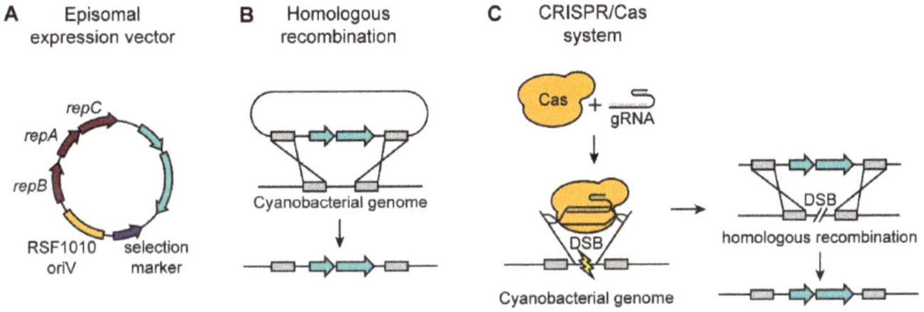

Figure 2. Genetic engineering tools. (**A**) Homologous recombination method using the recombination system in cyanobacteria. (**B**) RSF1010-derived vectors are self-replicating vectors used in episomal expression vector system. (**C**) CRISPR/Cas system utilizes Cas endonuclease to generate double-strand break to the gRNA-escorted loci inducing homologous recombination. Abbreviation; CRISPR, clustered regularly interspaced short palindromic repeat; gRNA, guide RNA; DSB, double-strand break.

Besides terpenes, a xanthophyll carotenoid called astaxanthin has been gaining significant attention in the healthcare field owing to the anti-inflammatory and antioxidant properties elucidated in human cells [106]. Astaxanthin production was enhanced through the engineering of *Synechocystis* sp. PCC 6803 [107]. First, the core biosynthetic genes, β-carotenoid ketolase (*crtW*) and β-carotene hydroxylase (*crtZ*) were integrated for astaxanthin production. The promoter combinations with diverse strength were tested for expressions of *crtW* and *crtZ*, because their relative expression level is known to be critical to produce astaxanthin in *E. coli* [107]. However, astaxanthin production was detected only when the super-strong promoter Pcpc560 was used for both genes, indicating that other tested endogenous promoters (PnirA, PpetE, and PrnpB) were not sufficient to express those genes. To test the relative expression level of the two genes, a promoter pool with more varied strength, including the stronger promoter than Pcpc560, was required. By using Pcpc560 for *crtZ* expression and a pea promoter PsbA for *crtW* expression, which showed two-fold higher activity than Pcpc560, resulting in improved production of astaxanthin. Further, based on liquid chromatography-mass spectrometry (LC-MS) metabolomics data, fructose-1,6-/sedoheptulose 1,7-bisphosphate (FBP/SBPase), which is involved in the Calvin–Benson–Bassham cycle, was found as an additional engineering target and overexpressed with episomal expression vector. Then, heterologous *dxs* and *ispA* gene was introduced into *Synechocystis* sp. PCC 6803 genome, and the engineered strain was finally able to produce astaxanthin of 29.6 mg/g dry cell weight, the highest level in the currently known engineered strain.

Table 2. Recent studies of engineering cyanobacteria for biochemical production.

Strategy [1]	Strain	Target [2]	Gene	Ref.
HR	*S. elongatus* PCC 7942	Isoprene	*ispGS, idi, dxr*	[6]
HR	*S. elongatus* PCC 7942	Succinate *	*ppc, gltA, kgd, gabD*	[108]
HR	*S. elongatus* PCC 7942	Amorpha-4,11-diene, Squalene *	*dxs, idi, ispA, dxr*	[103]
HR	*S. elongatus* UTEX 2973	Sucrose *	*cscB*	[109]
HR	*Synechocystis* sp. PCC 6803	Isoprene	*ispS*	[110]
HR	*S. elongatus* PCC 7942	Isopropanol *	*sadh, thl, atoAD', adc*	[111]
HR	*Synechocystis* sp. PCC 6803	Geranyllinalool	*NaGLS*	[112]
HR	*S. elongatus* PCC 7942	Squalene *	*dxs, idi, ispA, SQS*	[104]
HR	*S. elongatus* PCC 7942	Butyrate	*phaBJ, Ptb, buk, pte2, tesB, yciA*	[113]
HR	*Anabaena* sp. PCC 7120	Ethanol	*pdc, adhA, sigE, ald, invAB*	[114]
HR	*S. elongatus* PCC 7942	Sucrose *	*cscB, sps, glgC*	[115]
HR	*S. elongatus* PCC 7942	3-Hydroxybutyrate	*phaAB, tesB, nphT7, pptesB, yciA, pte1*	[116]
HR	*S. elongatus* PCC 7942	Heparosan	*galU, PmHS2*	[117]
HR	*Synechocystis* sp. PCC 6803	1-Butanol	*phaAB, nphT7, fadB, phaJ, ccr, ter, pduP, mhpF, yqhD, yjgB, pk, pta, adh, sigE*	[97]
HR	*Synechocystis* sp. PCC 6803	Sorbitol	*s6pdh, fbp, pnt, had1, had2*	[118]
HR	*Synechocystis* sp. PCC 6803	β-Phellandrene *	*GPPS, PHLS*	[119]
HR	*S. elongatus* PCC 7942	Acetone	*pdc, ald6, acs, pps, ppc, mmc*	[120]
HR	*S. elongatus* PCC 7942	Xylitol	*xylEFGH, XDH, DI, XR*	[121]
HR	*S. elongatus* PCC 7942	Trehalose *	*tpsp, Tret1, mts, glgCX, cscB, mth*	[122]
HR	*S. elongatus* PCC 7942	2,3-Butanediol	*alsD, alsS, adh, galP, zwf, edd, pgi, gnd, pfk, eda, cp12, rbcLXS, prk*	[5]
HR	*S. elongatus* PCC 7942	α-Farnesene	*AFS*	[123]
HR	*Synechocystis* sp. PCC 6803	Ethanol	*eno, pgk, pyk, prk*	[124]
HR	*S. elongatus* PCC 7942	Limonene *	*ls, GPPS, dxs*	[125]
HR	*Synechococcus* sp. PCC 7002	D-Lactate	*acsA*	[126]
epi	*Synechocystis* sp. PCC 6803	Isoprene	*ispS*	[127]
epi	*Synechocystis* sp. PCC 6803	p-Hydroxyphenylacetaldoxime, dhurrin	*CYP71E1, CYP79A1, UGT85B1*	[128]
epi	*Anabaena* sp. PCC 7120	Lyngbyatoxin A *	*ltxA-C, ltxA-D*	[129]
epi	*Synechocystis* sp. PCC 6803	Ethanol	*pdc, adh, rbcSC, 70glpX, tktA, fbaA*	[98]
epi	*Synechocystis* sp. PCC 6803	Shinorine *	*FsABCD, APPT*	[130]
epi	*S. elongatus* UTEX 2973	Hapalindole *	*famH1, famH2, famH3, aph3, famE2, famD2, famC1, famC2, famC3*	[131]
HR + epi	*Synechocystis* sp. PCC 6803	Astaxanthin*	*crtWZ, dxs, idi, ispA, F/SBPase, RuBisCO, rpe, tktA, psy*	[107]
HR + epi	*S. elongatus* PCC 7002	2,3-Butanediol	*alsDS, adh*	[132]
HR + epi	*Synechocystis* sp. PCC 6803	Isobutanol	*kivd, adh*	[133]
HR + epi	*Synechocystis* sp. PCC 6803	Limonene *	*lims, rpi, rpe, GPPS*	[134]
CRISPR	*S. elongatus* PCC 7942	Succinate *	*glgC, gltA, ppc*	[135]
CRISPR + epi	*Synechocystis* sp. PCC 6803	Fatty alcohol *	*Maqu2220, DPW, plsX, aar, ado, sll1848, sll1752, slr2060*	[136]
CRISPR	*Synechocystis* sp. PCC 6803	*N*-Butanol, ethanol	*adhA, pdc, pduP, phaJ, ter, phaBCE, nphT7, sth, yqhD, xfpk, PL22, SAS2203, gltA, odhB, ackA, pyrF, nrtA, ndhD*	[99]
CRISPR	*S. elongatus* PCC 7942	Squalene *	*acnB, cpcB2*	[105]

[1] HR, homologous recombination; epi, episomal expression [2], * cyanobacterial natural product.

3.2. Heterologous Expression for Biofuel Production

Along with the naturally produced cyanobacterial secondary metabolites, cyanobacteria also serve as an attractive platform for diverse heterologous biochemical production. For instance, cyanobacterial genome engineering was performed to use *S. elongatus* PCC 7942 as a host for producing 2,3-butanediol, a biochemical building block for plasticizers, liquid fuel additives, and industrial solvents [5]. In this study, homologous recombination was used for the integration of galactose-proton symporter (*galP*), glucose-6-phosphate dehydrogenase (*zwf*), and 6-phosphogluconate dehydrogenase (*gnd*) into neutral sites, and phosphoribulokinase (*prk*) and RuBisCO subunits (*rbcLXS*) into the *cp12* site, resulting in a maximum production of 12.6 g/L of 2,3-butanediol. In addition, in *Synechocystis* sp. PCC 6803, metabolic engineering was performed to enhance the biofuel production [99]. To improve ethanol production in the pyruvate decarboxylase (*pdc*)-inserted ethanol-producing *Synechocystis* sp. PCC 6803 strain, pyruvate dehydrogenase complex subunit (*odhB*) was repressed by CRISPRi. Repression of citrate synthase (*gltA*) in the phosphoketolase (*xfpk*) and acetoacetyl-CoA synthase (*nphT7*)-inserted n-butanol-producing strain increased *N*-butanol production.

Isoprene is a plant-derived building block, mainly used in the manufacturing of synthetic rubber. While most of the current synthetic rubber production depends on the petrochemical source, there have been efforts made to increase the cyanobacterial production of isoprene directly from CO_2. On introducing the isoprene synthase (*ispS*) obtained from various plants into the *Synechocystis* sp. PCC 6803 genome by homologous recombination, a maximum of 4.3 mg/L/h isoprene production rate was achieved with the aid of *dxs* and *idi* overexpression [6]. The heterologous inducible promoter Ptrc or endogenous promoters Pcpc and PpsbA2 were tested for expression of diverse *ispS*, and as a result, expression of *Eucalyptus globulus ispS* with Ptrc promoter showed the highest isoprene production. As demonstrated in various studies, cyanobacteria can produce industrially valuable biomaterials by utilizing light and CO_2. Therefore, cyanobacteria have the potential to be developed into an eco-friendly and economical photoautotrophic biofactory.

3.3. Improvement of Photosynthetic Efficiency

The value-added biochemicals that we have discussed are all products of photosynthesis. Thus, enhancement of the photosynthetic efficiency is crucial for supplying sufficient energy and reducing power for productivity increment. Some of the strategies for improving photosynthesis include expansion of the absorption spectra to capture more light energy, downsizing of the antenna to increase high illumination tolerance, and optimization of the electron transport chain. Furthermore, efficient use of photosystem-generated energy is another strategy that can be achieved by enhancing carbon fixation or reducing carbon loss [137].

3.4. Current Limitations in Engineering Cyanobacteria

Until now, we have presented studies showing the potential of cyanobacteria in producing various metabolites and that continuous engineering efforts can enhance the native and non-native metabolite production from cyanobacteria. However, despite the proposed and demonstrated potential as a production host, the production levels in cyanobacteria are not compatible with those in model organisms such as *E. coli* or *Saccharomyces cerevisiae*. In the case of *E. coli*, which is most widely used engineering host with a lot of information about genetic features and metabolic network, it is easy to apply knowledge-based engineering approaches such as enzyme structure modification, feedback inhibition removal, and precursor pool or cofactor level increasement [138]. In addition, a high-throughput screening technique through random mutagenesis is applicable using well-developed screening systems in *E. coli*. On the other hand, since cyanobacteria is a photoautotroph, more complex energy generation and distribution, and redox state should be considered when manipulating the metabolic network, and thus a more systematic insight is required. Additionally, when engineering multiple targets in the metabolic pathway, it is difficult to fine-tune the relative expression levels of the genes due to the lack of available bioparts such as neutral site, promoter, and RBS. In order to overcome these limitations, it is essential to systematically understand the complex metabolic network within the cell and to develop various genetic tools by genome-scale screening of native promoters and RBSs or constructing synthetic bioparts. In recent studies, a systematic approach through genome-scale modeling (GEM) has been successfully applied to engineering cyanobacteria [124,125]. For more effective and efficient engineering of cyanobacteria, the systematic approach should be further advanced.

4. Advanced Engineering Approaches through Synthetic and Systems Biology

Synthetic biology involves development of genetic parts, combination design to fulfill the desired function, and application of the combined tool into an organism. Quantification and standardization of the genetic parts represented by promoters, RBS, untranslated region (UTR) sequences, and terminator sequences are critical for proper employment of synthetic biology. Systems biology deals with the living system as an interactive network more than just a collection of reductive components. Therefore, understanding of the organism as a system is required for precise designing of the synthetic biology tools, and the introduction of synthetic biology tools into an organism affects the system, making the two biological approaches inseparable. The general synthetic and systems biology research flow is represented as the design–build–test–learn cycle (Figure 3). In the design step, the host for metabolite production is selected, and the biosynthesis pathway is designed using prior knowledge. Then, in the build step, a bioproduction host is engineered using either random, rational, or both methods. The constructed strain may undergo various tests for data generation. The data are then analyzed to produce and update the understanding of the bioproduction system. Systems and synthetic biology as an integrative approach, assisted the engineering of various organisms, including cyanobacteria.

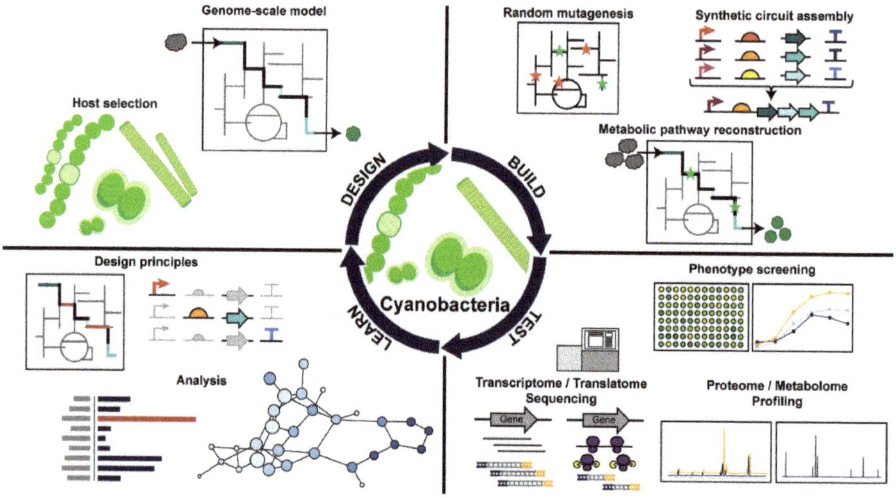

Figure 3. Schematic representation of design–build–test–learn cycle in cyanobacteria.

4.1. *Synthetic Biology*

Development of the genetic parts is critical for applying synthetic biology to metabolic engineering of cyanobacteria. While in other model species (e.g., *E. coli*), genetic parts such as promoters and RBS with varying strengths are available, there has been a significant lack of information and diversity concerning the cyanobacterial genetic parts. Currently used genetic parts are cyanobacterial endogenous promoters (P_{psbA} and P_{cpc}) and *E. coli* origin promoters (P_{trc}, P_{BAD}, P_{lac}, and P_{nrsB}) [139]. Promoters currently used in cyanobacteria are cataloged in previous literature [140]. However, to expand the promoter selection pool, several P_{T7} derivatives, P_{psbA}* derivatives, P_{trc}-based hybrid, and synthetic promoters have been developed and tested in *Synechocystis* sp. PCC 6803 [141,142]. Recently, a mutant promoter library from two popular promoters, P_{psbA} and P_{cpc} of *S. elongatus* PCC 7942, were generated to achieve promoters of varying strength [143]. A collection of 48 unique promoters were validated in three additional *S. elongatus* strains, expanding the cyanobacterial synthetic biology toolbox. The RBSs for cyanobacterial gene expression are mostly wild-type RBS associated with native promoter, or RBS of highly expressed genes such as *psbA2* and *rbcL* [144]. In addition, synthetic RBSs from BioBrick Registry of standard biological parts, and a newly designed RBS based on *Synechocystis* sp. PCC 6803 genome sequence (RBS*) are also used in several studies [145]. Recently, research has been conducted to diversify the RBS types used in cyanobacterial engineering. Twenty types of native RBS from *Synechocystis* sp. PCC 6803 were additionally identified, and 13 RBSs were rationally designed based on the known strong RBS sequences [142,146].

Riboswitches are another class of bioparts operated based on their RNA structures. A riboswitch comprises of an aptamer and an expression part. While the aptamer directly binds to a corresponding small molecule, the expression part regulates gene expression post-transcriptionally by causing structural changes in accordance with the small molecule binding. In *S. elongatus* PCC 7942, the operation of the synthetic theophylline riboswitch confirmed that it could control translation initiation [147]. Recently, this riboswitch was applied for flexible regulation of intracellular glycogen storage by controlling the expression of ADP-glucose pyrophosphorylase (glgC) [148]. The theophylline riboswitch was found to be operational in several other cyanobacterial species, including *Leptolyngbya* sp. BL 0902, *Nostoc* sp. 7120, and *Synechocystis* sp. WHSyn [149]. The theophylline riboswitch has also been used for chimeric riboswitch generation by combining it with a *Bacillus subtilis phuE* (adenine riboswitch). The chimeric riboswitch was validated in *Nostoc* sp. 7120 [150]. Overall, various synthetic biology toolboxes applicable to cyanobacteria are being developed to expand the pool of choice, which would contribute to effective engineering by enabling precise gene regulation of cyanobacteria.

4.2. Next-Generation Sequencing/Omics/Genome-Scale Model

Approximately 1500 cyanobacterial genome sequences have been registered in the NCBI genome database, and 196 of them, including the genome sequences of *Synechocystis* sp. PCC 6803 and *S. elongatus* PCC 7942, are completely assembled. In particular, the complete genome sequence of *Synechocystis* sp. PCC 6803 was reported as early as *E. coli* genome sequence, thus settling as a model organism among cyanobacteria [151]. Since then, with the development of diverse next-generation sequencing (NGS) techniques, various omics data such as transcriptome and translatome were generated based on the genome sequence [9]. Previously, transcriptome changes in response to stress conditions, such as temperature, light, and nutrition depletion, and effects of gene deletions were analyzed using cyanobacterium models, including *Synechocystis* sp. PCC 6803, *S. elongatus* PCC 7942, and *Nostoc* sp. 7120 [9]. In addition to the model cyanobacteria, recent transcriptome studies are being conducted in various non-model cyanobacterium species as well (Table 3). For example, in *Euhalothece* living in a hypersaline habitat, various salt resistance-related genes, such as Na^+ transporting multiple resistance and pH adaptation systems, and glycine betaine biosynthesis enzymes were highly upregulated [152]. In addition, differential RNA-seq revealed genome-wide transcription start sites (TSSs) in *S. elongatus* UTEX 2973 and two types of *Fischerella* strains, which were used to elucidate the differences in transcriptional regulation resulting in different phenotypes [153,154]. Additionally, the application of Ribo-seq to observe the translatome responses of *Synechocystis* sp. PCC 6803 under carbon starvation condition was reported in 2018 [155]. In order to observe the post-transcriptional responses in cyanobacteria, omics studies such as proteomics through liquid chromatography with tandem mass spectrometry (LC-MS/MS) analysis and metabolomics through gas chromatography-mass spectrometry (GC-MS) and ^{13}C isotopically nonstationary metabolic flux analysis have been implemented [5,156]. The massive omics data generated from the sequencing techniques described above served as the basis for understanding the cyanobacterial system under various conditions and also aided in the development of cyanobacterial GEMs.

GEM is an in silico tool that is useful for explaining the entire metabolism of an organism based on its genomic information. Given the ability to predict cellular metabolic behavior under given conditions or constraints, GEMs are mainly used for designing metabolic engineering strategies. GEM is an iteratively evolving system mended by new input information starting from the prior draft. Thus, more precise GEMs can be constructed by gathering greater types and amounts of genome information. GEM has been witnessing significant advancements with continuous accumulation of massive omics data supplied by various NGS techniques. Among cyanobacteria, *Synechocystis* sp. PCC 6803 has been studied most extensively, starting with the central carbon metabolic reconstruction under heterotrophic, mixotrophic, or autotrophic conditions [157,158]. The *Synechocystis* sp. PCC 6803 GEM was recently updated with more detailed photosynthesis and electron transport chain data [159,160]. GEM of other cyanobacterial species, such as *Arthrospira platensis*, *Cyanothece* sp., *Nostoc* sp., *S. elongatus* UTEX 2973, *S. elongatus* PCC 7942, and *Synechococcus* sp., are being established and further enhanced [139]. Through in silico GEM simulation, the bottleneck step can be selected for optimal engineering to enhance the production of value-added biochemicals. In the near future, the continuously evolving GEM would grow more useful in the metabolic engineering of cyanobacteria.

Table 3. Recent advances in omics studies of cyanobacteria.

Year	Omics Study	Strain	Ref.
2016	Genome-scale model (GEM) + Metabolome	*Synechococcus* sp. PCC 7002	[161]
2016	Metabolome	*S. elongatus* PCC 7942	[6]
2016	Metabolome + Transcriptome	*Synechocystis* sp. PCC 6803	[127]
2016	Proteome	*S. elongatus* PCC 7942	[162]
2016	Proteome	*Synechocystis* sp. PCC 6803	[163]
2016	Transcriptome	*S. elongatus* PCC 7942	[164]
2016	Transcriptome	*Synechocystis* sp. PCC 6803	[165]
2016	Transcriptome	*Prochlorococcus* NATL2A	[166]
2016	Transcriptome	*Nostoc* sp. PCC 7120	[167]
2016	Transcriptome	*S. elongatus* PCC 7942	[168]
2016	GEM	*S. elongatus* PCC 7942	[10]
2016	Transcriptome	*M. aeruginosa*	[169]
2017	Metabolome	*Synechococcus* sp. PCC 7002	[170]
2017	Metabolome	*S. elongatus* PCC 7942	[171]
2017	Metabolome	*S. elongatus* PCC 7942	[172]
2017	Metabolome	*S. elongatus* PCC 7942	[5]
2017	Metabolome	*Synechocystis* sp. PCC 6803	[173]
2017	Proteome	*Synechocystis* sp. PCC 6803	[174]
2017	Proteome	*Synechocystis* sp. PCC 6803	[175]
2017	Proteome	*Synechocystis* sp. PCC 6803	[176]
2017	Proteome	*Synechococcus* strains	[177]
2017	Proteome	*Prochlorococcus* strains	[178]
2017	Proteome	*P. marinus* SS 120	[179]
2017	Proteome	*Synechocystis* sp. PCC 6803	[180]
2017	Transcriptome	*Synechocystis* sp. PCC 6803	[181]
2017	Transcriptome + Interactome	*Synechocystis* sp. PCC 6803	[182]
2017	Transcriptome + Metabolome	*Synechococcus* sp. IU 625	[183]
2017	Transcription start site (TSS)	*F. muscicola* PCC 7414 and *F. thermalis* PCC 7521	[154]
2017	GEM	*Synechocystis* sp. PCC 6803	[184]
2017	GEM	*Nostoc* sp. PCC 7120	[185]
2017	GEM	*S. elongatus* UTEX 2973	[186]
2017	GEM	*Synechococcus* sp. PCC 7002	[161]
2018	Transcriptome	*M. aeruginosa*	[187]
2018	Transcriptome + Translatome	*Synechocystis* sp. PCC 6803	[155]
2018	TSS	*S. elongatus* UTEX 2973	[153]
2018	GEM	*Synechocystis* sp. PCC 6803	[161]

Table 3. *Cont.*

Year	Omics Study	Strain	Ref.
2019	Metabolome	*Synechococcus* sp. PCC 7002	[188]
2019	Proteome	*Synechocystis* sp. PCC 6803	[189]
2019	Transcriptome	*Prochlorococcus* MIT9313	[190]
2019	Transcriptome	*N. punctiforme* PCC 73102	[191]
2019	Transcriptome	*Leptolyngbya* sp. PCC 6406	[192]
2020	GEM	*Synechocystis* sp. PCC 6803	[160]
2020	Metabolome	*S. elongatus* PCC 11802 and PCC 11801	[193]
2020	Metabolome	*Nostoc* sp. UIC 10630	[194]
2020	Metabolome	*Leibleinia gracilis*	[195]
2020	Metabolome	*Synechocystis* sp. PCC 6803	[196]
2020	Metabolome	*S. elongatus* UTEX 2973	[197]
2020	Metabolome	*S. elongatus* PCC 11801	[198]
2020	Metabolome	*M. aeruginosa* PCC 7820 and PCC 7806	[199]
2020	Metabolome	*Synechocystis* sp. PCC 6803	[200]
2020	Metabolome	*Nodularia spumigena*	[201]
2020	Proteome	*Nostoc* sp. PCC 7120	[202]
2020	Proteome	*Synechococcus strains*	[203]
2020	Proteome	*Nodosilinea* strains	[204]
2020	Transcriptome	*Nostoc* sp. PCC 7120	[205]
2020	Transcriptome	*Euhalothece* sp. Z-M001	[152]
2020	Transcriptome	*Synechocystis* sp. PCC 6803	[206]
2020	Transcriptome	*N. punctiforme* PCC 73102	[207]
2020	Transcriptome	*Synechococcus* sp. PCC 7002	[208]
2020	Transcriptome + Metabolome	*Synechocystis* sp. PCC 6803	[107]
2020	GEM	*Synechococcus* sp. BDU 130192	[209]
2020	GEM	*A. variabilis* ATCC 29413	[210]

5. Conclusions and Future Perspectives

Cyanobacteria have significant industrial value owing to their ability to generate energy from photosynthesis and to produce various secondary metabolites. However, several improvements are required for cyanobacteria to meet the industry-level expectations and to establish themselves as a potential bioproduction platform. First, when using cyanobacterial native promoter or RBS, unexpected interaction may occur within the cell, which may reduce engineering efficiency. Therefore, development and application of the variety of orthogonal tools for engineering cyanobacteria is crucial. In addition, it is essential to obtain the precise metabolic network information to design strategies for the concise use of the synthetic biology tools. For example, the optimal production conditions can be discovered through promoter and RBS combination randomization, and the kind of neutral sites that can be used for chromosome integration can be expanded based on essential gene information found with transposon mutagenesis [143,146,211,212]. In addition, by applying the rapidly developing CRISPR application, it is possible to repress or activate multiple target genes at once, which can shorten the laborious and tedious engineering process caused by the polyploidy genome characteristic of cyanobacteria [101]. Systems biology enabled the discovery of various genetic tools by generating and accumulating massive omics data in cyanobacteria. In addition, development of GEM based on the accumulating omics and experimental data would lead to the development of a more accurate metabolic model.

Author Contributions: Conceptualization, Y.J., S.C. and B.-K.C.; writing—original draft preparation, Y.J., S.-H.C., S.C. and B.-K.C.; writing—review and editing, H.L., H.-K.C., D.-M.K., C.-G.L. and B.-K.C.; visualization, Y.J., S.-H.C. and S.C.; supervision, S.C. and B.-K.C.; project administration, S.C. and B.-K.C.; funding acquisition, S.C. and B.-K.C. All authors have read and agreed to the published version of the manuscript.

Funding: This research was funded by the Basic Core Technology Development Program for the Oceans and the Polar Regions of the National Research Foundation (NRF), funded by the Ministry of Science and ICT of Korea, grant numbers NRF-2016M1A5A1027458 and NRF-2016M1A5A1027455. The APC was funded by NRF-2016M1A5A1027458.

Conflicts of Interest: The authors declare no conflict of interest.

References

1. Singh, J.S.; Kumar, A.; Rai, A.N.; Singh, D.P. Cyanobacteria: A precious bio-resource in agriculture, ecosystem, and environmental sustainability. *Front. Microbiol.* **2016**, *7*, 529. [CrossRef] [PubMed]
2. Mogany, T.; Swalaha, F.M.; Kumari, S.; Bux, F. Elucidating the role of nutrients in C-phycocyanin production by the halophilic cyanobacterium *Euhalothece* sp. *J. Appl. Phycol.* **2018**, *30*, 2259–2271. [CrossRef]
3. Liang, C.; Zhao, F.; Wei, W.; Wen, Z.; Qin, S. Carotenoid biosynthesis in cyanobacteria: Structural and evolutionary scenarios based on comparative genomics. *Int. J. Biol. Sci.* **2006**, *2*, 197. [CrossRef] [PubMed]
4. Lee, H.J.; Lee, J.; Lee, S.-M.; Um, Y.; Kim, Y.; Sim, S.J.; Choi, J.-I.; Woo, H.M. Direct conversion of CO2 to α-farnesene using metabolically engineered *Synechococcus elongatus* PCC 7942. *J. Agric. Food Chem.* **2017**, *65*, 10424–10428. [CrossRef]
5. Kanno, M.; Carroll, A.L.; Atsumi, S. Global metabolic rewiring for improved CO_2 fixation and chemical production in cyanobacteria. *Nat. Commun.* **2017**, *8*, 14724. [CrossRef]
6. Gao, X.; Gao, F.; Liu, D.; Zhang, H.; Nie, X.; Yang, C. Engineering the methylerythritol phosphate pathway in cyanobacteria for photosynthetic isoprene production from CO_2. *Energy Environ. Sci.* **2016**, *9*, 1400–1411. [CrossRef]
7. Stephanopoulos, G. Synthetic biology and metabolic engineering. *ACS Synth. Biol.* **2012**, *1*, 514–525. [CrossRef]
8. Ramey, C.J.; Barón-Sola, A.N.; Aucoin, H.R.; Boyle, N.R. Genome engineering in cyanobacteria: Where we are and where we need to go. *ACS Synth. Biol.* **2015**, *4*, 1186–1196. [CrossRef]
9. Lin, W.-R.; Tan, S.-I.; Hsiang, C.-C.; Sung, P.-K.; Ng, I.-S. Challenges and opportunity of recent genome editing and multi-omics in cyanobacteria and microalgae for biorefinery. *Bioresour. Technol.* **2019**, *291*, 121932. [CrossRef]
10. Broddrick, J.T.; Rubin, B.E.; Welkie, D.G.; Du, N.; Mih, N.; Diamond, S.; Lee, J.J.; Golden, S.S.; Palsson, B.O. Unique attributes of cyanobacterial metabolism revealed by improved genome-scale metabolic modeling and essential gene analysis. *Proc. Natl. Acad. Sci. USA* **2016**, *113*, E8344–E8353. [CrossRef]
11. Kultschar, B.; Llewellyn, C. Secondary metabolites in cyanobacteria. In *Secondary Metabolites—Sources and Applications*; IntechOpen: London, UK, 2018; pp. 23–36.
12. Romay, C.; Armesto, J.; Remirez, D.; Gonzalez, R.; Ledon, N.; Garcia, I. Antioxidant and anti-inflammatory properties of C-phycocyanin from blue-green algae. *Inflamm. Res.* **1998**, *47*, 36–41. [CrossRef] [PubMed]
13. Romay, C.; Gonzalez, R.; Ledon, N.; Remirez, D.; Rimbau, V. C-phycocyanin: A biliprotein with antioxidant, anti-inflammatory and neuroprotective effects. *Curr. Protein Pept. Sci.* **2003**, *4*, 207–216. [CrossRef] [PubMed]
14. Benedetti, S.; Benvenuti, F.; Pagliarani, S.; Francogli, S.; Scoglio, S.; Canestrari, F. Antioxidant properties of a novel phycocyanin extract from the blue-green alga *Aphanizomenon flos-aquae*. *Life Sci.* **2004**, *75*, 2353–2362. [CrossRef] [PubMed]
15. Kuddus, M.; Singh, P.; Thomas, G.; Al-Hazimi, A. Recent developments in production and biotechnological applications of C-phycocyanin. *BioMed Res. Int.* **2013**, *2013*, 742859. [CrossRef]
16. Patel, A.; Mishra, S.; Ghosh, P.K. Antioxidant potential of C-phycocyanin isolated from cyanobacterial species *Lyngbya*, *Phormidium* and *Spirulina* spp. *Indian J. Biochem. Biophys.* **2006**, *43*, 25–31.
17. Stahl, W.; Sies, H. Antioxidant activity of carotenoids. *Mol. Asp. Med.* **2003**, *24*, 345–351. [CrossRef]
18. Wada, N.; Sakamoto, T.; Matsugo, S. Multiple roles of photosynthetic and sunscreen pigments in cyanobacteria focusing on the oxidative stress. *Metabolites* **2013**, *3*, 463–483. [CrossRef]
19. Fagundes, M.B.; Falk, R.B.; Facchi, M.M.X.; Vendruscolo, R.G.; Maroneze, M.M.; Zepka, L.Q.; Jacob-Lopes, E.; Wagner, R. Insights in cyanobacteria lipidomics: A sterols characterization from *Phormidium autumnale* biomass in heterotrophic cultivation. *Food Res. Int.* **2019**, *119*, 777–784. [CrossRef]
20. Kellmann, R.; Mihali, T.K.; Neilan, B.A. Identification of a saxitoxin biosynthesis gene with a history of frequent horizontal gene transfers. *J. Mol. Evol.* **2008**, *67*, 526–538. [CrossRef]
21. Mihali, T.K.; Kellmann, R.; Neilan, B.A. Characterisation of the paralytic shellfish toxin biosynthesis gene clusters in *Anabaena circinalis* AWQC131C and *Aphanizomenon* sp. NH-5. *BMC Biochem.* **2009**, *10*, 8. [CrossRef]
22. Murray, S.A.; Wiese, M.; Stuken, A.; Brett, S.; Kellmann, R.; Hallegraeff, G.; Neilan, B.A. *sxtA*-based quantitative molecular assay to identify saxitoxin-producing harmful algal blooms in marine waters. *Appl. Environ. Microbiol.* **2011**, *77*, 7050–7057. [CrossRef] [PubMed]

23. Burja, A.M.; Banaigs, B.; Abou-Mansour, E.; Burgess, J.G.; Wright, P.C.J.T. Marine cyanobacteria—A prolific source of natural products. *Tetrahedron* **2001**, *57*, 9347–9377. [CrossRef]
24. Rastogi, R.P.; Sonani, R.R.; Madamwar, D. Cyanobacterial sunscreen scytonemin: Role in photoprotection and biomedical research. *Appl. Biochem. Biotechnol.* **2015**, *176*, 1551–1563. [CrossRef] [PubMed]
25. Garcia-Pichel, F.; Castenholz, R.W.J.J.O.P. Characterization and biological implications of scytonemin, a cyanobacterial sheath pigment. *J. Phycol.* **1991**, *27*, 395–409. [CrossRef]
26. Proteau, P.J.; Gerwick, W.H.; Garcia-Pichel, F.; Castenholz, R. The structure of scytonemin, an ultraviolet sunscreen pigment from the sheaths of cyanobacteria. *Experientia* **1993**, *49*, 825–829. [CrossRef] [PubMed]
27. Soule, T.; Stout, V.; Swingley, W.D.; Meeks, J.C.; Garcia-Pichel, F. Molecular genetics and genomic analysis of scytonemin biosynthesis in *Nostoc punctiforme* ATCC 29133. *J. Bacteriol.* **2007**, *189*, 4465–4472. [CrossRef] [PubMed]
28. Klein, D.; Daloze, D.; Braekman, J.C.; Hoffmann, L.; Demoulin, V. New hapalindoles from the cyanophyte *Hapalosiphon laingii*. *J. Nat. Prod.* **1995**, *58*, 1781–1785. [CrossRef]
29. Moore, R.E.; Cheuk, C.; Patterson, G.M.L. Hapalindoles: New alkaloids from the blue-green alga *Hapalosiphon fontinalis*. *J. Am. Chem. Soc.* **1984**, *106*, 6456–6457. [CrossRef]
30. Mejean, A.; Mann, S.; Maldiney, T.; Vassiliadis, G.; Lequin, O.; Ploux, O. Evidence that biosynthesis of the neurotoxic alkaloids anatoxin-a and homoanatoxin-a in the cyanobacterium *Oscillatoria* PCC 6506 occurs on a modular polyketide synthase initiated by L-proline. *J. Am. Chem. Soc.* **2009**, *131*, 7512–7513. [CrossRef] [PubMed]
31. Rantala-Ylinen, A.; Kana, S.; Wang, H.; Rouhiainen, L.; Wahlsten, M.; Rizzi, E.; Berg, K.; Gugger, M.; Sivonen, K. Anatoxin-a synthetase gene cluster of the cyanobacterium *Anabaena* sp. strain 37 and molecular methods to detect potential producers. *Appl. Environ. Microbiol.* **2011**, *77*, 7271–7278. [CrossRef]
32. Moore, R.E.; Blackman, A.J.; Cheuk, C.E.; Mynderse, J.S.; Matsumoto, G.K.; Clardy, J.; Woodard, R.W.; Craig, J.C. Absolute stereochemistries of the aplysiatoxins and oscillatoxin A. *J. Org. Chem.* **1984**, *49*, 2484–2489. [CrossRef]
33. Gupta, D.K.; Kaur, P.; Leong, S.T.; Tan, L.T.; Prinsep, M.R.; Chu, J.J. Anti-Chikungunya viral activities of aplysiatoxin-related compounds from the marine cyanobacterium *Trichodesmium erythraeum*. *Mar. Drugs* **2014**, *12*, 115–127. [CrossRef] [PubMed]
34. Edwards, D.J.; Gerwick, W.H. Lyngbyatoxin biosynthesis: Sequence of biosynthetic gene cluster and identification of a novel aromatic prenyltransferase. *J. Am. Chem. Soc.* **2004**, *126*, 11432–11433. [CrossRef] [PubMed]
35. Mihali, T.K.; Kellmann, R.; Muenchhoff, J.; Barrow, K.D.; Neilan, B.A. Characterization of the gene cluster responsible for cylindrospermopsin biosynthesis. *Appl. Environ. Microbiol.* **2008**, *74*, 716–722. [CrossRef]
36. Stuken, A.; Jakobsen, K.S. The cylindrospermopsin gene cluster of *Aphanizomenon* sp. strain 10E6: Organization and recombination. *Microbiology (Reading)* **2010**, *156*, 2438–2451. [CrossRef] [PubMed]
37. Mazmouz, R.; Chapuis-Hugon, F.; Mann, S.; Pichon, V.; Mejean, A.; Ploux, O. Biosynthesis of cylindrospermopsin and 7-epicylindrospermopsin in *Oscillatoria* sp. strain PCC 6506: Identification of the cyr gene cluster and toxin analysis. *Appl. Environ. Microbiol.* **2010**, *76*, 4943–4949. [CrossRef] [PubMed]
38. Gross, E.M.; Wolk, C.P.; Jüttner, F. Fischerellin, a new allelochemical from the freshwater cyanobacterium *Fischerella Muscicola*. *J. Phycol.* **1991**, *27*, 686–692. [CrossRef]
39. Cox, P.A.; Banack, S.A.; Murch, S.J.; Rasmussen, U.; Tien, G.; Bidigare, R.R.; Metcalf, J.S.; Morrison, L.F.; Codd, G.A.; Bergman, B. Diverse taxa of cyanobacteria produce β-N-methylamino-L-alanine, a neurotoxic amino acid. *Proc. Natl. Acad. Sci. USA* **2005**, *102*, 5074–5078. [CrossRef]
40. Rounge, T.B.; Rohrlack, T.; Nederbragt, A.J.; Kristensen, T.; Jakobsen, K.S. A genome-wide analysis of nonribosomal peptide synthetase gene clusters and their peptides in a *Planktothrix rubescens* strain. *BMC Genom.* **2009**, *10*, 396. [CrossRef]
41. Tooming-Klunderud, A.; Rohrlack, T.; Shalchian-Tabrizi, K.; Kristensen, T.; Jakobsen, K.S. Structural analysis of a non-ribosomal halogenated cyclic peptide and its putative operon from *Microcystis*: Implications for evolution of cyanopeptolins. *Microbiology (Reading)* **2007**, *153*, 1382–1393. [CrossRef] [PubMed]
42. Tillett, D.; Dittmann, E.; Erhard, M.; von Dohren, H.; Borner, T.; Neilan, B.A. Structural organization of microcystin biosynthesis in *Microcystis aeruginosa* PCC7806: An integrated peptide-polyketide synthetase system. *Chem. Biol.* **2000**, *7*, 753–764. [CrossRef]

43. Kaneko, T.; Nakajima, N.; Okamoto, S.; Suzuki, I.; Tanabe, Y.; Tamaoki, M.; Nakamura, Y.; Kasai, F.; Watanabe, A.; Kawashima, K.; et al. Complete genomic structure of the bloom-forming toxic cyanobacterium *Microcystis aeruginosa* NIES-843. *DNA Res.* **2007**, *14*, 247–256. [CrossRef] [PubMed]
44. Rouhiainen, L.; Vakkilainen, T.; Siemer, B.L.; Buikema, W.; Haselkorn, R.; Sivonen, K. Genes coding for hepatotoxic heptapeptides (microcystins) in the cyanobacterium *Anabaena* strain 90. *Appl. Environ. Microbiol.* **2004**, *70*, 686–692. [CrossRef] [PubMed]
45. Christiansen, G.; Fastner, J.; Erhard, M.; Borner, T.; Dittmann, E. Microcystin biosynthesis in *planktothrix*: Genes, evolution, and manipulation. *J. Bacteriol.* **2003**, *185*, 564–572. [CrossRef]
46. Moffitt, M.C.; Neilan, B.A. Characterization of the nodularin synthetase gene cluster and proposed theory of the evolution of cyanobacterial hepatotoxins. *Appl. Environ. Microbiol.* **2004**, *70*, 6353–6362. [CrossRef]
47. Grindberg, R.V.; Ishoey, T.; Brinza, D.; Esquenazi, E.; Coates, R.C.; Liu, W.T.; Gerwick, L.; Dorrestein, P.C.; Pevzner, P.; Lasken, R.; et al. Single cell genome amplification accelerates identification of the apratoxin biosynthetic pathway from a complex microbial assemblage. *PLoS ONE* **2011**, *6*, e18565. [CrossRef]
48. Ishida, K.; Christiansen, G.; Yoshida, W.Y.; Kurmayer, R.; Welker, M.; Valls, N.; Bonjoch, J.; Hertweck, C.; Borner, T.; Hemscheidt, T.; et al. Biosynthesis and structure of aeruginoside 126A and 126B, cyanobacterial peptide glycosides bearing a 2-carboxy-6-hydroxyoctahydroindole moiety. *Chem. Biol.* **2007**, *14*, 565–576. [CrossRef]
49. Ishida, K.; Welker, M.; Christiansen, G.; Cadel-Six, S.; Bouchier, C.; Dittmann, E.; Hertweck, C.; Tandeau de Marsac, N. Plasticity and evolution of aeruginosin biosynthesis in cyanobacteria. *Appl. Environ. Microbiol.* **2009**, *75*, 2017–2026. [CrossRef]
50. Magarvey, N.A.; Beck, Z.Q.; Golakoti, T.; Ding, Y.; Huber, U.; Hemscheidt, T.K.; Abelson, D.; Moore, R.E.; Sherman, D.H. Biosynthetic characterization and chemoenzymatic assembly of the cryptophycins. Potent anticancer agents from cyanobionts. *ACS Chem. Biol.* **2006**, *1*, 766–779. [CrossRef]
51. Fewer, D.P.; Osterholm, J.; Rouhiainen, L.; Jokela, J.; Wahlsten, M.; Sivonen, K. Nostophycin biosynthesis is directed by a hybrid polyketide synthase-nonribosomal peptide synthetase in the toxic cyanobacterium *Nostoc* sp. strain 152. *Appl. Environ. Microbiol.* **2011**, *77*, 8034–8040. [CrossRef]
52. Chang, Z.; Sitachitta, N.; Rossi, J.V.; Roberts, M.A.; Flatt, P.M.; Jia, J.; Sherman, D.H.; Gerwick, W.H. Biosynthetic pathway and gene cluster analysis of curacin A, an antitubulin natural product from the tropical marine cyanobacterium *Lyngbya majuscula*. *J. Nat. Prod.* **2004**, *67*, 1356–1367. [CrossRef] [PubMed]
53. Ramaswamy, A.V.; Sorrels, C.M.; Gerwick, W.H. Cloning and biochemical characterization of the hectochlorin biosynthetic gene cluster from the marine cyanobacterium *Lyngbya majuscula*. *J. Nat. Prod.* **2007**, *70*, 1977–1986. [CrossRef]
54. Edwards, D.J.; Marquez, B.L.; Nogle, L.M.; McPhail, K.; Goeger, D.E.; Roberts, M.A.; Gerwick, W.H. Structure and biosynthesis of the jamaicamides, new mixed polyketide-peptide neurotoxins from the marine cyanobacterium *Lyngbya majuscula*. *Chem. Biol.* **2004**, *11*, 817–833. [CrossRef] [PubMed]
55. Nogle, L.M.; Williamson, R.T.; Gerwick, W.H. Somamides A and B, two new depsipeptide analogues of dolastatin 13 from a Fijian cyanobacterial assemblage of *Lyngbya majuscula* and *Schizothrix* species. *J. Nat. Prod.* **2001**, *64*, 716–719. [CrossRef] [PubMed]
56. Nogle, L.M.; Gerwick, W.H. Isolation of four new cyclic depsipeptides, antanapeptins A-D, and dolastatin 16 from a Madagascan collection of *Lyngbya majuscula*. *J. Nat. Prod.* **2002**, *65*, 21–24. [CrossRef]
57. Berman, F.W.; Gerwick, W.H.; Murray, T.F. Antillatoxin and kalkitoxin, ichthyotoxins from the tropical cyanobacterium *Lyngbya majuscula*, induce distinct temporal patterns of NMDA receptor-mediated neurotoxicity. *Toxicon* **1999**, *37*, 1645–1648. [CrossRef]
58. McPhail, K.L.; Correa, J.; Linington, R.G.; Gonzalez, J.; Ortega-Barria, E.; Capson, T.L.; Gerwick, W.H. Antimalarial linear lipopeptides from a Panamanian strain of the marine cyanobacterium *Lyngbya majuscula*. *J. Nat. Prod.* **2007**, *70*, 984–988. [CrossRef]
59. Hooper, G.J.; Orjala, J.; Schatzman, R.C.; Gerwick, W.H. Carmabins A and B, new lipopeptides from the Caribbean cyanobacterium *Lyngbya majuscula*. *J. Nat. Prod.* **1998**, *61*, 529–533. [CrossRef]
60. Choi, H.; Mevers, E.; Byrum, T.; Valeriote, F.A.; Gerwick, W.H. Lyngbyabellins K-N from two palmyra atoll collections of the marine cyanobacterium *Moorea bouillonii*. *Eur. J. Org. Chem.* **2012**, *2012*, 5141–5150. [CrossRef]

61. Han, B.; McPhail, K.L.; Gross, H.; Goeger, D.E.; Mooberry, S.L.; Gerwick, W.H.J.T. Isolation and structure of five lyngbyabellin derivatives from a Papua New Guinea collection of the marine cyanobacterium *Lyngbya majuscula*. *Tetrahedron* **2005**, *61*, 11723–11729. [CrossRef]
62. Schmidt, E.W.; Nelson, J.T.; Rasko, D.A.; Sudek, S.; Eisen, J.A.; Haygood, M.G.; Ravel, J. Patellamide A and C biosynthesis by a microcin-like pathway in *Prochloron didemni*, the cyanobacterial symbiont of *Lissoclinum patella*. *Proc. Natl. Acad. Sci. USA* **2005**, *102*, 7315–7320. [CrossRef] [PubMed]
63. Ziemert, N.; Ishida, K.; Quillardet, P.; Bouchier, C.; Hertweck, C.; de Marsac, N.T.; Dittmann, E. Microcyclamide biosynthesis in two strains of *Microcystis aeruginosa*: From structure to genes and vice versa. *Appl. Environ. Microbiol.* **2008**, *74*, 1791–1797. [CrossRef] [PubMed]
64. Philmus, B.; Christiansen, G.; Yoshida, W.Y.; Hemscheidt, T.K. Post-translational modification in microviridin biosynthesis. *Chembiochem* **2008**, *9*, 3066–3073. [CrossRef] [PubMed]
65. Balskus, E.P.; Walsh, C.T. The genetic and molecular basis for sunscreen biosynthesis in cyanobacteria. *Science* **2010**, *329*, 1653–1656. [CrossRef]
66. Tan, L.T.; Chang, Y.Y.; Ashootosh, T. Besarhanamides A and B from the marine cyanobacterium *Lyngbya majuscula*. *Phytochemistry* **2008**, *69*, 2067–2069. [CrossRef] [PubMed]
67. Essack, M.; Alzubaidy, H.S.; Bajic, V.B.; Archer, J.A. Chemical compounds toxic to invertebrates isolated from marine cyanobacteria of potential relevance to the agricultural industry. *Toxins (Basel)* **2014**, *6*, 3058–3076. [CrossRef] [PubMed]
68. Stewart, I.; Schluter, P.J.; Shaw, G.R. Cyanobacterial lipopolysaccharides and human health—A review. *Environ. Health* **2006**, *5*, 7. [CrossRef]
69. Chirasuwan, N.; Chaiklahan, R.; Ruengjitchatchawalya, M.; Bunnag, B.; Tanticharoen, M.J.A.; Resources, N. Anti HSV-1 activity of *Spirulina platensis* polysaccharide. *Kasetsart J. (Nat. Sci.)* **2007**, *41*, 311–318.
70. de Jesus Raposo, M.F.; De Morais, A.M.B.; De Morais, R.M.S.C. Marine polysaccharides from algae with potential biomedical applications. *Mar. Drugs* **2015**, *13*, 2967–3028. [CrossRef]
71. Delattre, C.; Pierre, G.; Laroche, C.; Michaud, P. Production, extraction and characterization of microalgal and cyanobacterial exopolysaccharides. *Biotechnol. Adv.* **2016**, *34*, 1159–1179. [CrossRef]
72. Moore, R.E. Toxins, anticancer agents, and tumor promoters from marine prokaryotes. *Pure Appl. Chem.* **1982**, *54*, 1919–1934. [CrossRef]
73. Banker, R.; Carmeli, S. Tenuecyclamides A–D, cyclic hexapeptides from the cyanobacterium *Nostoc spongiaeforme* var. *tenue*. *J. Nat. Prod.* **1998**, *61*, 1248–1251. [CrossRef] [PubMed]
74. Blin, K.; Shaw, S.; Steinke, K.; Villebro, R.; Ziemert, N.; Lee, S.Y.; Medema, M.H.; Weber, T. antiSMASH 5.0: Updates to the secondary metabolite genome mining pipeline. *Nucleic Acids Res.* **2019**, *47*, W81–W87. [CrossRef] [PubMed]
75. Gershenzon, J.; Dudareva, N. The function of terpene natural products in the natural world. *Nat. Chem. Biol.* **2007**, *3*, 408–414. [CrossRef]
76. Pattanaik, B.; Lindberg, P. Terpenoids and their biosynthesis in cyanobacteria. *Life* **2015**, *5*, 269–293. [CrossRef]
77. Belin, B.J.; Busset, N.; Giraud, E.; Molinaro, A.; Silipo, A.; Newman, D.K. Hopanoid lipids: From membranes to plant–bacteria interactions. *Nat. Rev. Microbiol.* **2018**, *16*, 304. [CrossRef]
78. Takaichi, S.; Mochimaru, M. Carotenoids and carotenogenesis in cyanobacteria: Unique ketocarotenoids and carotenoid glycosides. *Cell. Mol. Life Sci.* **2007**, *64*, 2607. [CrossRef]
79. Prasanna, R.; Sood, A.; Jaiswal, P.; Nayak, S.; Gupta, V.; Chaudhary, V.; Joshi, M.; Natarajan, C. Rediscovering cyanobacteria as valuable sources of bioactive compounds. *Appl. Biochem. Microbiol.* **2010**, *46*, 119–134. [CrossRef]
80. Na, S.I.; Kim, Y.O.; Yoon, S.H.; Ha, S.M.; Baek, I.; Chun, J. UBCG: Up-to-date bacterial core gene set and pipeline for phylogenomic tree reconstruction. *J. Microbiol.* **2018**, *56*, 280–285. [CrossRef]
81. Stamatakis, A. RAxML version 8: A tool for phylogenetic analysis and post-analysis of large phylogenies. *Bioinformatics* **2014**, *30*, 1312–1313. [CrossRef]
82. Mejean, A.; Mann, S.; Vassiliadis, G.; Lombard, B.; Loew, D.; Ploux, O. In vitro reconstitution of the first steps of anatoxin-a biosynthesis in *Oscillatoria* PCC 6506: From free L-proline to acyl carrier protein bound dehydroproline. *Biochemistry* **2010**, *49*, 103–113. [CrossRef] [PubMed]
83. Dittmann, E.; Gugger, M.; Sivonen, K.; Fewer, D.P. Natural product biosynthetic diversity and comparative genomics of the cyanobacteria. *Trends Microbiol.* **2015**, *23*, 642–652. [CrossRef] [PubMed]

84. Balskus, E.P.; Walsh, C.T. Investigating the initial steps in the biosynthesis of cyanobacterial sunscreen scytonemin. *J. Am. Chem. Soc.* **2008**, *130*, 15260–15261. [CrossRef] [PubMed]
85. Ansari, M.Z.; Yadav, G.; Gokhale, R.S.; Mohanty, D. NRPS-PKS: A knowledge-based resource for analysis of NRPS/PKS megasynthases. *Nucleic Acids Res.* **2004**, *32*, W405–W413. [CrossRef] [PubMed]
86. Méjean, A.; Ploux, O. A genomic view of secondary metabolite production in cyanobacteria. In *Advances in Botanical Research*; Elsevier: Amsterdam, The Netherlands, 2013; Volume 65, pp. 189–234.
87. Arnison, P.G.; Bibb, M.J.; Bierbaum, G.; Bowers, A.A.; Bugni, T.S.; Bulaj, G.; Camarero, J.A.; Campopiano, D.J.; Challis, G.L.; Clardy, J. Ribosomally synthesized and post-translationally modified peptide natural products: Overview and recommendations for a universal nomenclature. *Nat. Prod. Rep.* **2013**, *30*, 108–160. [CrossRef]
88. Montalbán-López, M.; Scott, T.A.; Ramesh, S.; Rahman, I.R.; van Heel, A.J.; Viel, J.H.; Bandarian, V.; Dittmann, E.; Genilloud, O.; Goto, Y. New developments in RiPP discovery, enzymology and engineering. *Nat. Prod. Rep.* **2020**. [CrossRef]
89. Han, B.; McPhail, K.L.; Ligresti, A.; Di Marzo, V.; Gerwick, W.H. Semiplenamides A–G, Fatty acid amides from a Papua New Guinea collection of the marina cyanobacterium *Lyngbya semiplena*. *J. Nat. Prod.* **2003**, *66*, 1364–1368. [CrossRef]
90. Chi, Z.; Su, C.; Lu, W. A new exopolysaccharide produced by marine *Cyanothece* sp. 113. *Bioresour. Technol.* **2007**, *98*, 1329–1332. [CrossRef]
91. Markou, G.; Nerantzis, E. Microalgae for high-value compounds and biofuels production: A review with focus on cultivation under stress conditions. *Biotechnol. Adv.* **2013**, *31*, 1532–1542. [CrossRef]
92. Delattre, C.; Vijayalakshmi, M. Monolith enzymatic microreactor at the frontier of glycomic toward a new route for the production of bioactive oligosaccharides. *J. Mol. Catal. B Enzym.* **2009**, *60*, 97–105. [CrossRef]
93. Kraan, S. Algal polysaccharides, novel applications and outlook. In *Carbohydrates-Comprehensive Studies on Glycobiology and Glycotechnology*; IntechOpen: London, UK, 2012.
94. Mišurcová, L.; Orsavová, J.; Vávra Ambrožová, J. Algal polysaccharides and health. *Polysacch. Bioactivity Biotechnol.* **2015**, *1*, 109–144.
95. Skjånes, K.; Rebours, C.; Lindblad, P. Potential for green microalgae to produce hydrogen, pharmaceuticals and other high value products in a combined process. *Crit. Rev. Biotechnol.* **2013**, *33*, 172–215. [CrossRef] [PubMed]
96. Swain, S.S.; Paidesetty, S.K.; Padhy, R.N. Antibacterial, antifungal and antimycobacterial compounds from cyanobacteria. *Biomed. Pharmacother.* **2017**, *90*, 760–776. [CrossRef]
97. Liu, X.; Miao, R.; Lindberg, P.; Lindblad, P. Modular engineering for efficient photosynthetic biosynthesis of 1-butanol from CO_2 in cyanobacteria. *Energy Environ. Sci.* **2019**, *12*, 2765–2777. [CrossRef]
98. Liang, F.; Englund, E.; Lindberg, P.; Lindblad, P. Engineered cyanobacteria with enhanced growth show increased ethanol production and higher biofuel to biomass ratio. *Metab. Eng.* **2018**, *46*, 51–59. [CrossRef]
99. Shabestary, K.; Anfelt, J.; Ljungqvist, E.; Jahn, M.; Yao, L.; Hudson, E.P. Targeted repression of essential genes to arrest growth and increase carbon partitioning and biofuel titers in cyanobacteria. *ACS Synth. Biol.* **2018**, *7*, 1669–1675. [CrossRef]
100. Xia, P.F.; Ling, H.; Foo, J.L.; Chang, M.W. Synthetic biology toolkits for metabolic engineering of cyanobacteria. *Biotechnol. J.* **2019**, *14*, e1800496. [CrossRef]
101. Behler, J.; Vijay, D.; Hess, W.R.; Akhtar, M.K. CRISPR-based technologies for metabolic engineering in cyanobacteria. *Trends Biotechnol.* **2018**, *36*, 996–1010. [CrossRef]
102. Fagundes, M.B.; Vendruscolo, R.G.; Maroneze, M.M.; Barin, J.S.; de Menezes, C.R.; Zepka, L.Q.; Jacob-Lopes, E.; Wagner, R. Towards a sustainable route for the production of squalene using cyanobacteria. *Waste Biomass Valorization* **2019**, *10*, 1295–1302. [CrossRef]
103. Choi, S.Y.; Lee, H.J.; Choi, J.; Kim, J.; Sim, S.J.; Um, Y.; Kim, Y.; Lee, T.S.; Keasling, J.D.; Woo, H.M. Photosynthetic conversion of CO_2 to farnesyl diphosphate-derived phytochemicals (amorpha-4, 11-diene and squalene) by engineered cyanobacteria. *Biotechnol. Biofuels* **2016**, *9*, 1–12. [CrossRef]
104. Choi, S.Y.; Wang, J.-Y.; Kwak, H.S.; Lee, S.-M.; Um, Y.; Kim, Y.; Sim, S.J.; Choi, J.-I.; Woo, H.M. Improvement of squalene production from CO_2 in *Synechococcus elongatus* PCC 7942 by metabolic engineering and scalable production in a photobioreactor. *ACS Synth. Biol.* **2017**, *6*, 1289–1295. [CrossRef] [PubMed]
105. Choi, S.Y.; Woo, H.M. CRISPRi-dCas12a: A dCas12a-mediated CRISPR interference for repression of multiple genes and metabolic engineering in cyanobacteria. *ACS Synth. Biol.* **2020**, *9*, 2351–2361. [CrossRef] [PubMed]

106. Farruggia, C.; Kim, M.-B.; Bae, M.; Lee, Y.; Pham, T.X.; Yang, Y.; Han, M.J.; Park, Y.-K.; Lee, J.-Y. Astaxanthin exerts anti-inflammatory and antioxidant effects in macrophages in NRF2-dependent and independent manners. *J. Nutr. Biochem.* **2018**, *62*, 202–209. [CrossRef] [PubMed]
107. Diao, J.; Song, X.; Zhang, L.; Cui, J.; Chen, L.; Zhang, W. Tailoring cyanobacteria as a new platform for highly efficient synthesis of astaxanthin. *Metab. Eng.* **2020**, *61*, 275–287. [CrossRef]
108. Lan, E.I.; Wei, C.T. Metabolic engineering of cyanobacteria for the photosynthetic production of succinate. *Metab. Eng.* **2016**, *38*, 483–493. [CrossRef]
109. Song, K.; Tan, X.; Liang, Y.; Lu, X. The potential of *Synechococcus elongatus* UTEX 2973 for sugar feedstock production. *Appl. Microbiol. Biotechnol.* **2016**, *100*, 7865–7875. [CrossRef]
110. Chaves, J.E.; Rueda-Romero, P.; Kirst, H.; Melis, A. Engineering isoprene synthase expression and activity in cyanobacteria. *ACS Synth. Biol.* **2017**, *6*, 2281–2292. [CrossRef]
111. Hirokawa, Y.; Dempo, Y.; Fukusaki, E.; Hanai, T. Metabolic engineering for isopropanol production by an engineered cyanobacterium, *Synechococcus elongatus* PCC 7942, under photosynthetic conditions. *J. Biosci. Bioeng.* **2017**, *123*, 39–45. [CrossRef]
112. Formighieri, C.; Melis, A. Heterologous synthesis of geranyllinalool, a diterpenol plant product, in the cyanobacterium *Synechocystis*. *Appl. Microbiol. Biotechnol.* **2017**, *101*, 2791–2800. [CrossRef]
113. Lai, M.J.; Lan, E.I. Photoautotrophic synthesis of butyrate by metabolically engineered cyanobacteria. *Biotechnol. Bioeng.* **2019**, *116*, 893–903. [CrossRef]
114. Ehira, S.; Takeuchi, T.; Higo, A. Spatial separation of photosynthesis and ethanol production by cell type-specific metabolic engineering of filamentous cyanobacteria. *Appl. Microbiol. Biotechnol.* **2018**, *102*, 1523–1531. [CrossRef] [PubMed]
115. Qiao, C.; Duan, Y.; Zhang, M.; Hagemann, M.; Luo, Q.; Lu, X. Effects of reduced and enhanced glycogen pools on salt-induced sucrose production in a sucrose-secreting strain of *Synechococcus elongatus* PCC 7942. *Appl. Environ. Microbiol.* **2018**, *84*, e02023. [CrossRef] [PubMed]
116. Ku, J.T.; Lan, E.I. A balanced ATP driving force module for enhancing photosynthetic biosynthesis of 3-hydroxybutyrate from CO_2. *Metab. Eng.* **2018**, *46*, 35–42. [CrossRef]
117. Sarnaik, A.; Abernathy, M.H.; Han, X.; Ouyang, Y.; Xia, K.; Chen, Y.; Cress, B.; Zhang, F.; Lali, A.; Pandit, R. Metabolic engineering of cyanobacteria for photoautotrophic production of heparosan, a pharmaceutical precursor of heparin. *Algal Res.* **2019**, *37*, 57–63. [CrossRef]
118. Chin, T.; Okuda, Y.; Ikeuchi, M. Improved sorbitol production and growth in cyanobacteria using promiscuous haloacid dehalogenase-like hydrolase. *J. Biotechnol. X* **2019**, *1*, 100002. [CrossRef]
119. Betterle, N.; Melis, A. Photosynthetic generation of heterologous terpenoids in cyanobacteria. *Biotechnol. Bioeng.* **2019**, *116*, 2041–2051. [CrossRef]
120. Lee, H.J.; Son, J.; Sim, S.J.; Woo, H.M. Metabolic rewiring of synthetic pyruvate dehydrogenase bypasses for acetone production in cyanobacteria. *Plant. Biotechnol. J.* **2020**, *18*, 1860–1868. [CrossRef] [PubMed]
121. Fan, E.S.; Lu, K.W.; Wen, R.C.; Shen, C.R. Photosynthetic reduction of xylose to xylitol using cyanobacteria. *Biotechnol. J.* **2020**, *15*, 1900354. [CrossRef] [PubMed]
122. Qiao, Y.; Wang, W.; Lu, X. Engineering cyanobacteria as cell factories for direct trehalose production from CO_2. *Metab. Eng.* **2020**, *62*, 161–171. [CrossRef] [PubMed]
123. Pattharaprachayakul, N.; Lee, H.J.; Incharoensakdi, A.; Woo, H.M. Evolutionary engineering of cyanobacteria to enhance the production of α-farnesene from CO_2. *J. Agric. Food Chem.* **2019**, *67*, 13658–13664. [CrossRef]
124. Nishiguchi, H.; Hiasa, N.; Uebayashi, K.; Liao, J.; Shimizu, H.; Matsuda, F. Transomics data-driven, ensemble kinetic modeling for system-level understanding and engineering of the cyanobacteria central metabolism. *Metab. Eng.* **2019**, *52*, 273–283. [CrossRef] [PubMed]
125. Wang, X.; Liu, W.; Xin, C.; Zheng, Y.; Cheng, Y.; Sun, S.; Li, R.; Zhu, X.-G.; Dai, S.Y.; Rentzepis, P.M. Enhanced limonene production in cyanobacteria reveals photosynthesis limitations. *Proc. Natl. Acad. Sci. USA* **2016**, *113*, 14225–14230. [CrossRef] [PubMed]
126. Selão, T.T.; Jebarani, J.; Ismail, N.A.; Norling, B.; Nixon, P.J. Enhanced production of D-lactate in cyanobacteria by re-routing photosynthetic cyclic and pseudo-cyclic electron flow. *Front. Plant. Sci.* **2019**, *10*, 1700. [CrossRef] [PubMed]
127. Pade, N.; Erdmann, S.; Enke, H.; Dethloff, F.; Dühring, U.; Georg, J.; Wambutt, J.; Kopka, J.; Hess, W.R.; Zimmermann, R. Insights into isoprene production using the cyanobacterium *Synechocystis* sp. PCC 6803. *Biotechnol. Biofuels* **2016**, *9*, 89. [CrossRef]

128. Wlodarczyk, A.; Gnanasekaran, T.; Nielsen, A.Z.; Zulu, N.N.; Mellor, S.B.; Luckner, M.; Thøfner, J.F.B.; Olsen, C.E.; Mottawie, M.S.; Burow, M. Metabolic engineering of light-driven cytochrome P450 dependent pathways into *Synechocystis* sp. PCC 6803. *Metab. Eng.* **2016**, *33*, 1–11. [CrossRef] [PubMed]
129. Videau, P.; Wells, K.N.; Singh, A.J.; Gerwick, W.H.; Philmus, B. Assessment of *Anabaena* sp. strain PCC 7120 as a heterologous expression host for cyanobacterial natural products: Production of lyngbyatoxin A. *ACS Synth. Biol.* **2016**, *5*, 978–988. [CrossRef] [PubMed]
130. Yang, G.; Cozad, M.A.; Holland, D.A.; Zhang, Y.; Luesch, H.; Ding, Y. Photosynthetic production of sunscreen shinorine using an engineered cyanobacterium. *ACS Synth. Biol.* **2018**, *7*, 664–671. [CrossRef]
131. Knoot, C.J.; Khatri, Y.; Hohlman, R.M.; Sherman, D.H.; Pakrasi, H.B. Engineered production of hapalindole alkaloids in the cyanobacterium *Synechococcus* sp. UTEX 2973. *ACS Synth. Biol.* **2019**, *8*, 1941–1951. [CrossRef]
132. Nozzi, N.E.; Case, A.E.; Carroll, A.L.; Atsumi, S. Systematic approaches to efficiently produce 2, 3-butanediol in a marine cyanobacterium. *ACS Synth. Biol.* **2017**, *6*, 2136–2144. [CrossRef]
133. Miao, R.; Liu, X.; Englund, E.; Lindberg, P.; Lindblad, P. Isobutanol production in *Synechocystis* PCC 6803 using heterologous and endogenous alcohol dehydrogenases. *Metab. Eng. Commun.* **2017**, *5*, 45–53. [CrossRef]
134. Lin, P.-C.; Saha, R.; Zhang, F.; Pakrasi, H.B. Metabolic engineering of the pentose phosphate pathway for enhanced limonene production in the cyanobacterium *Synechocysti* s sp. PCC 6803. *Sci. Rep.* **2017**, *7*, 1–10. [CrossRef] [PubMed]
135. Li, H.; Shen, C.R.; Huang, C.-H.; Sung, L.-Y.; Wu, M.-Y.; Hu, Y.-C. CRISPR-Cas9 for the genome engineering of cyanobacteria and succinate production. *Metab. Eng.* **2016**, *38*, 293–302. [CrossRef] [PubMed]
136. Kaczmarzyk, D.; Cengic, I.; Yao, L.; Hudson, E.P. Diversion of the long-chain acyl-ACP pool in Synechocystis to fatty alcohols through CRISPRi repression of the essential phosphate acyltransferase PlsX. *Metab. Eng.* **2018**, *45*, 59–66. [CrossRef] [PubMed]
137. Luan, G.; Zhang, S.; Lu, X. Engineering cyanobacteria chassis cells toward more efficient photosynthesis. *Curr. Opin. Biotechnol.* **2020**, *62*, 1–6. [CrossRef] [PubMed]
138. Yang, D.; Park, S.Y.; Park, Y.S.; Eun, H.; Lee, S.Y. Metabolic engineering of *Escherichia coli* for natural product biosynthesis. *Trends Biotechnol.* **2020**, *38*, 745–765. [CrossRef] [PubMed]
139. Santos-Merino, M.; Singh, A.K.; Ducat, D.C. New applications of synthetic biology tools for cyanobacterial metabolic engineering. *Front. Bioeng. Biotechnol.* **2019**, *7*, 33. [CrossRef]
140. Singh, S. Cyanoomics: An advancement in the fields cyanobacterial omics biology with special reference to proteomics and transcriptomics. In *Advances in Cyanobacterial Biology*; Elsevier: Amsterdam, The Netherlands, 2020; pp. 163–171.
141. Ferreira, E.A.; Pacheco, C.C.; Pinto, F.; Pereira, J.; Lamosa, P.; Oliveira, P.; Kirov, B.; Jaramillo, A.; Tamagnini, P. Expanding the toolbox for *Synechocystis* sp. PCC 6803: Validation of replicative vectors and characterization of a novel set of promoters. *Synth. Biol.* **2018**, *3*, ysy014. [CrossRef]
142. Wang, B.; Eckert, C.; Maness, P.-C.; Yu, J. A genetic toolbox for modulating the expression of heterologous genes in the cyanobacterium *Synechocystis* sp. PCC 6803. *ACS Synth. Biol.* **2018**, *7*, 276–286. [CrossRef]
143. Sengupta, A.; Madhu, S.; Wangikar, P.P. A Library of tunable, portable, and inducer-free promoters derived from cyanobacteria. *ACS Synth. Biol.* **2020**, *9*, 1790–1801. [CrossRef]
144. Thiel, K.; Mulaku, E.; Dandapani, H.; Nagy, C.; Aro, E.-M.; Kallio, P. Translation efficiency of heterologous proteins is significantly affected by the genetic context of RBS sequences in engineered cyanobacterium *Synechocystis* sp. PCC 6803. *Microb. Cell Fact.* **2018**, *17*, 34. [CrossRef]
145. Heidorn, T.; Camsund, D.; Huang, H.-H.; Lindberg, P.; Oliveira, P.; Stensjö, K.; Lindblad, P. Synthetic biology in cyanobacteria: Engineering and analyzing novel functions. In *Methods in Enzymology*; Elsevier: Amsterdam, The Netherlands, 2011; Volume 497, pp. 539–579.
146. Liu, D.; Pakrasi, H.B. Exploring native genetic elements as plug-in tools for synthetic biology in the cyanobacterium *Synechocystis* sp. PCC 6803. *Microb. Cell Fact.* **2018**, *17*, 1–8. [CrossRef] [PubMed]
147. Nakahira, Y.; Ogawa, A.; Asano, H.; Oyama, T.; Tozawa, Y. Theophylline-dependent riboswitch as a novel genetic tool for strict regulation of protein expression in cyanobacterium *Synechococcus elongatus* PCC 7942. *Plant. Cell Physiol.* **2013**, *54*, 1724–1735. [CrossRef] [PubMed]
148. Chi, X.; Zhang, S.; Sun, H.; Duan, Y.; Qiao, C.; Luan, G.; Lu, X. Adopting a theophylline-responsive riboswitch for flexible regulation and understanding of glycogen metabolism in *Synechococcus elongatus* PCC7942. *Front. Microbiol.* **2019**, *10*, 551. [CrossRef] [PubMed]

149. Ma, A.T.; Schmidt, C.M.; Golden, J.W. Regulation of gene expression in diverse cyanobacterial species by using theophylline-responsive riboswitches. *Appl. Environ. Microbiol.* **2014**, *80*, 6704–6713. [CrossRef]
150. Higo, A.; Ehira, S. Anaerobic butanol production driven by oxygen-evolving photosynthesis using the heterocyst-forming multicellular cyanobacterium *Anabaena* sp. PCC 7120. *Appl. Microbiol. Biotechnol.* **2019**, *103*, 2441–2447. [CrossRef]
151. Kaneko, T.; Tabata, S. Complete genome structure of the unicellular cyanobacterium *Synechocystis* sp. PCC6803. *Plant Cell Physiol.* **1997**, *38*, 1171–1176. [CrossRef]
152. Yang, H.W.; Song, J.Y.; Cho, S.M.; Kwon, H.C.; Pan, C.-H.; Park, Y.-I. Genomic survey of salt acclimation-related genes in the halophilic cyanobacterium *Euhalothece* sp. Z-M001. *Sci. Rep.* **2020**, *10*, 676. [CrossRef]
153. Tan, X.; Hou, S.; Song, K.; Georg, J.; Klähn, S.; Lu, X.; Hess, W.R. The primary transcriptome of the fast-growing cyanobacterium *Synechococcus elongatus* UTEX 2973. *Biotechnol. Biofuels* **2018**, *11*, 218. [CrossRef]
154. Koch, R.; Kupczok, A.; Stucken, K.; Ilhan, J.; Hammerschmidt, K.; Dagan, T. Plasticity first: Molecular signatures of a complex morphological trait in filamentous cyanobacteria. *BMC Evol. Biol.* **2017**, *17*, 1–11. [CrossRef]
155. Karlsen, J.; Asplund-Samuelsson, J.; Thomas, Q.; Jahn, M.; Hudson, E.P. Ribosome profiling of *Synechocystis* reveals altered ribosome allocation at carbon starvation. *mSystems* **2018**, *3*, e00126. [CrossRef]
156. Jahn, M.; Vialas, V.; Karlsen, J.; Maddalo, G.; Edfors, F.; Forsström, B.; Uhlén, M.; Käll, L.; Hudson, E.P. Growth of cyanobacteria is constrained by the abundance of light and carbon assimilation proteins. *Cell Rep.* **2018**, *25*, 478–486.e478. [CrossRef] [PubMed]
157. Yang, C.; Hua, Q.; Shimizu, K. Metabolic flux analysis in *Synechocystis* using isotope distribution from 13C-labeled glucose. *Metab. Eng.* **2002**, *4*, 202–216. [CrossRef] [PubMed]
158. Shastri, A.A.; Morgan, J.A. Flux balance analysis of photoautotrophic metabolism. *Biotechnol. Prog.* **2005**, *21*, 1617–1626. [CrossRef] [PubMed]
159. Yoshikawa, K.; Kojima, Y.; Nakajima, T.; Furusawa, C.; Hirasawa, T.; Shimizu, H. Reconstruction and verification of a genome-scale metabolic model for *Synechocystis* sp. PCC6803. *Appl. Microbiol. Biotechnol.* **2011**, *92*, 347. [CrossRef]
160. Toyoshima, M.; Toya, Y.; Shimizu, H. Flux balance analysis of cyanobacteria reveals selective use of photosynthetic electron transport components under different spectral light conditions. *Photosynth. Res.* **2020**, *143*, 31–43. [CrossRef]
161. Qian, X.; Kim, M.K.; Kumaraswamy, G.K.; Agarwal, A.; Lun, D.S.; Dismukes, G.C. Flux balance analysis of photoautotrophic metabolism: Uncovering new biological details of subsystems involved in cyanobacterial photosynthesis. *Biochim. Biophys. Acta Bioenerg.* **2017**, *1858*, 276–287. [CrossRef]
162. Guerreiro, A.C.; Penning, R.; Raaijmakers, L.M.; Axman, I.M.; Heck, A.J.; Altelaar, A.M. Monitoring light/dark association dynamics of multi-protein complexes in cyanobacteria using size exclusion chromatography-based proteomics. *J. Proteom.* **2016**, *142*, 33–44. [CrossRef]
163. Liberton, M.; Saha, R.; Jacobs, J.M.; Nguyen, A.Y.; Gritsenko, M.A.; Smith, R.D.; Koppenaal, D.W.; Pakrasi, H.B. Global proteomic analysis reveals an exclusive role of thylakoid membranes in bioenergetics of a model cyanobacterium. *Mol. Cell Proteom.* **2016**, *15*, 2021–2032. [CrossRef]
164. Choi, S.Y.; Park, B.; Choi, I.-G.; Sim, S.J.; Lee, S.-M.; Um, Y.; Woo, H.M. Transcriptome landscape of *Synechococcus elongatus* PCC 7942 for nitrogen starvation responses using RNA-seq. *Sci. Rep.* **2016**, *6*, 30584. [CrossRef]
165. Kizawa, A.; Kawahara, A.; Takimura, Y.; Nishiyama, Y.; Hihara, Y. RNA-seq profiling reveals novel target genes of LexA in the cyanobacterium *Synechocystis* sp. PCC 6803. *Front. Microbiol.* **2016**, *7*, 193. [CrossRef]
166. Lin, X.; Ding, H.; Zeng, Q. Transcriptomic response during phage infection of a marine cyanobacterium under phosphorus-limited conditions. *Environ. Microbiol.* **2016**, *18*, 450–460. [CrossRef] [PubMed]
167. Gonzalez, A.; Bes, M.T.; Peleato, M.L.; Fillat, M.F. Expanding the role of FurA as essential global regulator in cyanobacteria. *PLoS ONE* **2016**, *11*, e0151384. [CrossRef] [PubMed]
168. Hood, R.D.; Higgins, S.A.; Flamholz, A.; Nichols, R.J.; Savage, D.F. The stringent response regulates adaptation to darkness in the cyanobacterium *Synechococcus elongatus*. *Proc. Natl. Acad. Sci. USA* **2016**, *113*, E4867–E4876. [CrossRef] [PubMed]
169. Harke, M.J.; Jankowiak, J.G.; Morrell, B.K.; Gobler, C.J. Transcriptomic responses in the bloom-forming cyanobacterium *Microcystis* induced during exposure to zooplankton. *Appl. Environ. Microbiol.* **2017**, *83*, e02832. [CrossRef] [PubMed]

170. Hendry, J.I.; Prasannan, C.; Ma, F.; Möllers, K.B.; Jaiswal, D.; Digmurti, M.; Allen, D.K.; Frigaard, N.U.; Dasgupta, S.; Wangikar, P.P. Rerouting of carbon flux in a glycogen mutant of cyanobacteria assessed via isotopically non-stationary 13C metabolic flux analysis. *Biotechnol. Bioeng.* **2017**, *114*, 2298–2308. [CrossRef] [PubMed]
171. Hirokawa, Y.; Matsuo, S.; Hamada, H.; Matsuda, F.; Hanai, T. Metabolic engineering of *Synechococcus elongatus* PCC 7942 for improvement of 1, 3-propanediol and glycerol production based on *in silico* simulation of metabolic flux distribution. *Microb. Cell Fact.* **2017**, *16*, 1–12. [CrossRef] [PubMed]
172. Jazmin, L.J.; Xu, Y.; Cheah, Y.E.; Adebiyi, A.O.; Johnson, C.H.; Young, J.D. Isotopically nonstationary 13C flux analysis of cyanobacterial isobutyraldehyde production. *Metab. Eng.* **2017**, *42*, 9–18. [CrossRef]
173. Nakajima, T.; Yoshikawa, K.; Toya, Y.; Matsuda, F.; Shimizu, H. Metabolic flux analysis of the *Synechocystis* sp. PCC 6803 ΔnrtABCD mutant reveals a mechanism for metabolic adaptation to nitrogen-limited conditions. *Plant. Cell Physiol.* **2017**, *58*, 537–545. [CrossRef]
174. Sun, T.; Chen, L.; Zhang, W. Quantitative proteomics reveals potential crosstalk between a small RNA CoaR and a two-component regulator Slr1037 in *Synechocystis* sp. PCC6803. *J. Proteome Res.* **2017**, *16*, 2954–2963. [CrossRef]
175. Liberton, M.; Chrisler, W.B.; Nicora, C.D.; Moore, R.J.; Smith, R.D.; Koppenaal, D.W.; Pakrasi, H.B.; Jacobs, J.M. Phycobilisome truncation causes widespread proteome changes in *Synechocystis* sp. PCC 6803. *PLoS ONE* **2017**, *12*, e0173251. [CrossRef]
176. Ge, H.; Fang, L.; Huang, X.; Wang, J.; Chen, W.; Liu, Y.; Zhang, Y.; Wang, X.; Xu, W.; He, Q. Translating divergent environmental stresses into a common proteome response through the histidine kinase 33 (Hik33) in a model cyanobacterium. *Mol. Cell Proteom.* **2017**, *16*, 1258–1274. [CrossRef] [PubMed]
177. Mackey, K.R.; Post, A.F.; McIlvin, M.R.; Saito, M.A. Physiological and proteomic characterization of light adaptations in marine *Synechococcus*. *Environ. Microbiol.* **2017**, *19*, 2348–2365. [CrossRef] [PubMed]
178. Muñoz-Marín, M.d.C.; Gómez-Baena, G.; Díez, J.; Beynon, R.J.; González-Ballester, D.; Zubkov, M.V.; García-Fernández, J.M. Glucose uptake in *Prochlorococcus*: Diversity of kinetics and effects on the metabolism. *Front. Microbiol.* **2017**, *8*, 327. [CrossRef] [PubMed]
179. Domínguez-Martín, M.A.; Gómez-Baena, G.; Díez, J.; López-Grueso, M.J.; Beynon, R.J.; García-Fernández, J.M. Quantitative proteomics shows extensive remodeling induced by nitrogen limitation in *Prochlorococcus marinus* SS120. *MSystems* **2017**, *2*, 3. [CrossRef]
180. Fang, L.; Ge, H.; Huang, X.; Liu, Y.; Lu, M.; Wang, J.; Chen, W.; Xu, W.; Wang, Y. Trophic mode-dependent proteomic analysis reveals functional significance of light-independent chlorophyll synthesis in *Synechocystis* sp. PCC 6803. *Mol. Plant.* **2017**, *10*, 73–85. [CrossRef]
181. Wang, J.; Chen, L.; Chen, Z.; Zhang, W. RNA-seq based transcriptomic analysis of single bacterial cells. *Integr. Biol.* **2015**, *7*, 1466–1476. [CrossRef]
182. Giner-Lamia, J.; Robles-Rengel, R.; Hernández-Prieto, M.A.; Muro-Pastor, M.I.; Florencio, F.J.; Futschik, M.E. Identification of the direct regulon of NtcA during early acclimation to nitrogen starvation in the cyanobacterium *Synechocystis* sp. PCC 6803. *Nucleic Acids Res.* **2017**, *45*, 11800–11820. [CrossRef]
183. Newby, R., Jr.; Lee, L.H.; Perez, J.L.; Tao, X.; Chu, T. Characterization of zinc stress response in Cyanobacterium *Synechococcus* sp. IU 625. *Aquat. Toxicol.* **2017**, *186*, 159–170. [CrossRef]
184. Joshi, C.J.; Peebles, C.A.; Prasad, A. Modeling and analysis of flux distribution and bioproduct formation in *Synechocystis* sp. PCC 6803 using a new genome-scale metabolic reconstruction. *Algal Res.* **2017**, *27*, 295–310. [CrossRef]
185. Malatinszky, D.; Steuer, R.; Jones, P.R. A comprehensively curated genome-scale two-cell model for the heterocystous cyanobacterium *Anabaena* sp. PCC 7120. *Plant. Physiol.* **2017**, *173*, 509–523. [CrossRef]
186. Mueller, T.J.; Ungerer, J.L.; Pakrasi, H.B.; Maranas, C.D. Identifying the metabolic differences of a fast-growth phenotype in *Synechococcus* UTEX 2973. *Sci. Rep.* **2017**, *7*, 41569. [CrossRef] [PubMed]
187. Morimoto, D.; Kimura, S.; Sako, Y.; Yoshida, T. Transcriptome analysis of a bloom-forming cyanobacterium *Microcystis aeruginosa* during Ma-LMM01 phage infection. *Front. Microbiol.* **2018**, *9*, 2. [CrossRef] [PubMed]
188. Abernathy, M.H.; Czajka, J.J.; Allen, D.K.; Hill, N.C.; Cameron, J.C.; Tang, Y.J. Cyanobacterial carboxysome mutant analysis reveals the influence of enzyme compartmentalization on cellular metabolism and metabolic network rigidity. *Metab. Eng.* **2019**, *54*, 222–231. [CrossRef]
189. Choi, J.-S.; Park, Y.H.; Oh, J.H.; Kim, S.; Kwon, J.; Choi, Y.-E. Efficient profiling of detergent-assisted membrane proteome in cyanobacteria. *J. Appl. Phycol.* **2019**, *32*, 1–8. [CrossRef]

190. Fang, X.; Liu, Y.; Zhao, Y.; Chen, Y.; Liu, R.; Qin, Q.L.; Li, G.; Zhang, Y.Z.; Chan, W.; Hess, W.R.; et al. Transcriptomic responses of the marine cyanobacterium *Prochlorococcus* to viral lysis products. *Environ. Microbiol.* **2019**, *21*, 2015–2028. [CrossRef] [PubMed]
191. Gonzalez, A.; Riley, K.W.; Harwood, T.V.; Zuniga, E.G.; Risser, D.D. A tripartite, hierarchical sigma factor cascade promotes hormogonium development in the filamentous cyanobacterium *Nostoc punctiforme*. *mSphere* **2019**, *4*, e00231-19. [CrossRef] [PubMed]
192. Hirose, Y.; Chihong, S.; Watanabe, M.; Yonekawa, C.; Murata, K.; Ikeuchi, M.; Eki, T. Diverse chromatic acclimation processes regulating phycoerythrocyanin and rod-shaped phycobilisome in cyanobacteria. *Mol. Plant* **2019**, *12*, 715–725. [CrossRef] [PubMed]
193. Jaiswal, D.; Sengupta, A.; Sengupta, S.; Madhu, S.; Pakrasi, H.B.; Wangikar, P.P. A novel cyanobacterium *Synechococcus elongatus* PCC 11802 has distinct genomic and metabolomic characteristics compared to its neighbor PCC 11801. *Sci. Rep.* **2020**, *10*, 191. [CrossRef]
194. May, D.S.; Crnkovic, C.M.; Krunic, A.; Wilson, T.A.; Fuchs, J.R.; Orjala, J.E. (15)N Stable isotope labeling and comparative metabolomics facilitates genome mining in cultured cyanobacteria. *ACS Chem. Biol.* **2020**, *15*, 758–765. [CrossRef]
195. Solanki, H.; Pierdet, M.; Thomas, O.P.; Zubia, M. Insights into the metabolome of the cyanobacterium *Leibleinia gracilis* from the lagoon of Tahiti and first inspection of its variability. *Metabolites* **2020**, *10*, 215. [CrossRef]
196. Shi, M.; Chen, L.; Zhang, W. Regulatory diversity and functional analysis of two-component systems in cyanobacterium *Synechocystis* sp. PCC 6803 by GC-MS based metabolomics. *Front. Microbiol.* **2020**, *11*, 403. [CrossRef] [PubMed]
197. Cui, J.; Sun, T.; Li, S.; Xie, Y.; Song, X.; Wang, F.; Chen, L.; Zhang, W. Improved salt tolerance and metabolomics analysis of *Synechococcus elongatus* UTEX 2973 by overexpressing Mrp Antiporters. *Front. Bioeng. Biotechnol.* **2020**, *8*, 500. [CrossRef] [PubMed]
198. Sengupta, A.; Pritam, P.; Jaiswal, D.; Bandyopadhyay, A.; Pakrasi, H.B.; Wangikar, P.P. Photosynthetic co-production of succinate and ethylene in a fast-growing cyanobacterium, *Synechococcus elongatus* PCC 11801. *Metabolites* **2020**, *10*, 250. [CrossRef] [PubMed]
199. Georges des Aulnois, M.; Réveillon, D.; Robert, E.; Caruana, A.; Briand, E.; Guljamow, A.; Dittmann, E.; Amzil, Z.; Bormans, M.J.T. Salt shock responses of *Microcystis* revealed through physiological, transcript, and metabolomic analyses. *Toxins* **2020**, *12*, 192. [CrossRef] [PubMed]
200. de Alvarenga, L.V.; Hess, W.R.; Hagemann, M. AcnSP—A novel small protein regulator of aconitase activity in the cyanobacterium *Synechocystis* sp. PCC 6803. *Front. Microbiol.* **2020**, *11*, 1445. [CrossRef]
201. Popin, R.V.; Delbaje, E.; de Abreu, V.A.C.; Rigonato, J.; Dorr, F.A.; Pinto, E.; Sivonen, K.; Fiore, M.F. Genomic and metabolomic analyses of natural products in *Nodularia spumigena* isolated from a shrimp culture pond. *Toxins (Basel)* **2020**, *12*, 141. [CrossRef]
202. Koksharova, O.A.; Butenko, I.O.; Pobeguts, O.V.; Safronova, N.A.; Govorun, V.M. The first proteomics study of *Nostoc* sp. PCC 7120 exposed to cyanotoxin BMAA under nitrogen starvation. *Toxins (Basel)* **2020**, *12*, 310. [CrossRef]
203. Teoh, F.; Shah, B.; Ostrowski, M.; Paulsen, I. Comparative membrane proteomics reveal contrasting adaptation strategies for coastal and oceanic marine *Synechococcus* cyanobacteria. *Environ. Microbiol.* **2020**, *22*, 1816–1828. [CrossRef]
204. Romeu, M.J.L.; Dominguez-Perez, D.; Almeida, D.; Morais, J.; Campos, A.; Vasconcelos, V.; Mergulhao, F.J.M. Characterization of planktonic and biofilm cells from two filamentous cyanobacteria using a shotgun proteomic approach. *Biofouling* **2020**, *36*, 631–645. [CrossRef]
205. He, P.; Cai, X.; Chen, K.; Fu, X. Identification of small RNAs involved in nitrogen fixation in *Anabaena* sp. PCC 7120 based on RNA-seq under steady state conditions. *Ann. Microbiol.* **2020**, *70*, 4. [CrossRef]
206. Mironov, K.S.; Kupriyanova, E.V.; Shumskaya, M.; Los, D.A. Alcohol stress on cyanobacterial membranes: New insights revealed by transcriptomics. *Gene* **2020**, *764*, 145055. [CrossRef] [PubMed]
207. Arias, D.B.; Gomez Pinto, K.A.; Cooper, K.K.; Summers, M.L. Transcriptomic analysis of cyanobacterial alkane overproduction reveals stress-related genes and inhibitors of lipid droplet formation. *Microb. Genom.* **2020**, *6*, e000432. [CrossRef] [PubMed]

208. Gordon, G.C.; Cameron, J.C.; Gupta, S.T.P.; Engstrom, M.D.; Reed, J.L.; Pfleger, B.F. Genome-wide analysis of RNA decay in the cyanobacterium *Synechococcus* sp. strain PCC 7002. *mSystems* **2020**, *5*, e00224-20. [CrossRef] [PubMed]
209. Ahmad, A.; Pathania, R.; Srivastava, S. Biochemical Characteristics and a Genome-scale metabolic model of an Indian euryhaline cyanobacterium with high polyglucan content. *Metabolites* **2020**, *10*, 177. [CrossRef] [PubMed]
210. Malek Shahkouhi, A.; Motamedian, E. Reconstruction of a regulated two-cell metabolic model to study biohydrogen production in a diazotrophic cyanobacterium *Anabaena variabilis* ATCC 29413. *PLoS ONE* **2020**, *15*, e0227977. [CrossRef] [PubMed]
211. Rubin, B.E.; Wetmore, K.M.; Price, M.N.; Diamond, S.; Shultzaberger, R.K.; Lowe, L.C.; Curtin, G.; Arkin, A.P.; Deutschbauer, A.; Golden, S.S. The essential gene set of a photosynthetic organism. *Proc. Natl. Acad. Sci. USA* **2015**, *112*, E6634–E6643. [CrossRef]
212. Englund, E.; Liang, F.; Lindberg, P. Evaluation of promoters and ribosome binding sites for biotechnological applications in the unicellular cyanobacterium *Synechocystis* sp. PCC 6803. *Sci. Rep.* **2016**, *6*, 36640. [CrossRef]

microorganisms

MDPI

Article

Microorganisms Associated with the Marine Sponge *Scopalina hapalia*: A Reservoir of Bioactive Molecules to Slow Down the Aging Process

Charifat Said Hassane [1], Mireille Fouillaud [1], Géraldine Le Goff [2], Aimilia D. Sklirou [3], Jean Bernard Boyer [1], Ioannis P. Trougakos [3], Moran Jerabek [4], Jérôme Bignon [2], Nicole J. de Voogd [5,6], Jamal Ouazzani [2], Anne Gauvin-Bialecki [1,*] and Laurent Dufossé [1,*]

1 Laboratoire de Chimie et Biotechnologie des Produits Naturels, Faculté des Sciences et Technologies, Université de La Réunion, 15 Avenue René Cassin, CS 92003, 97744 Saint-Denis CEDEX 9, La Réunion, France; charifat.said-hassane@univ-reunion.fr (C.S.H.); mireille.fouillaud@univ-reunion.fr (M.F.); jean-bernard.boyer@univ-reunion.fr (J.B.B.)

2 Institut de Chimie des Substances Naturelles, CNRS UPR 2301, Université Paris-Saclay, 1, av. de la Terrasse, 91198 Gif-sur-Yvette, France; geraldine.legoff@cnrs.fr (G.L.G.); jerome.bignon@cnrs.fr (J.B.); Jamal.Ouazzani@cnrs.fr (J.O.)

3 Department of Cell Biology and Biophysics, Faculty of Biology, National and Kapodistrian University of Athens, 15784 Athens, Greece; asklirou@biol.uoa.gr (A.D.S.); itrougakos@biol.uoa.gr (I.P.T.)

4 Crelux GmbH, Am Klopferspitz 19a, 82152 Martinsried, Germany; Moran_Jerabek@wuxiapptec.com

5 Naturalis Biodiversity Center, Darwinweg 2, 2333 CR Leiden, The Netherlands; nicole.devoogd@naturalis.nl

6 Institute of Environmental Sciences, Leiden University, Einsteinweg 2, 2333 CC Leiden, The Netherlands

* Correspondence: anne.bialecki@univ-reunion.fr (A.G.-B.); laurent.dufosse@univ-reunion.fr (L.D.)

Received: 8 June 2020; Accepted: 17 August 2020; Published: 20 August 2020

Abstract: Aging research aims at developing interventions that delay normal aging process and some related pathologies. Recently, many compounds and extracts from natural products have been shown to delay aging and/or extend lifespan. Marine sponges and their associated microorganisms have been found to produce a wide variety of bioactive secondary metabolites; however, those from the Southwest of the Indian Ocean are much less studied, especially regarding anti-aging activities. In this study, the microbial diversity of the marine sponge *Scopalina hapalia* was investigated by metagenomic analysis. Twenty-six bacterial and two archaeal phyla were recovered from the sponge, of which the *Proteobacteria* phylum was the most abundant. In addition, thirty isolates from *S. hapalia* were selected and cultivated for identification and secondary metabolites production. The selected isolates were affiliated to the genera *Bacillus*, *Micromonospora*, *Rhodoccocus*, *Salinispora*, *Aspergillus*, *Chaetomium*, *Nigrospora* and unidentified genera related to the family *Thermoactinomycetaceae*. Crude extracts from selected microbial cultures were found to be active against seven targets i.e., elastase, tyrosinase, catalase, sirtuin 1, Cyclin-dependent kinase 7 (CDK7), Fyn kinase and proteasome. These results highlight the potential of microorganisms associated with a marine sponge from Mayotte to produce anti-aging compounds. Future work will focus on the isolation and the characterization of bioactive molecules.

Keywords: *Scopalina hapalia*; Actinomycetes; *Bacillus*; Fungi; elastase inhibition; tyrosinase inhibition; CDK7 inhibition; Fyn kinase inhibition; catalase activation; sirtuin 1 activation

1. Introduction

As the world's global population ages, an increase in the prevalence of a variety of age-related diseases such as inflammatory disorders, neurodegenerative and cardiovascular diseases, as well as cancer, is observed. Even though the use of the term "cause" can be debated, aging was identified as the main risk factor for many age-related diseases [1,2]. Over recent years, research about aging effects has experienced unprecedented advances, particularly with the discovery that delaying of aging (with genetic, dietary and pharmacological approaches) can impede the onset or progress of age-related diseases [3–6]. These interventions aim to slow down the effect(s) of normal aging, prevent age-related diseases and increase quality of life. In this context, we selected a set of biological targets relevant for our investigation on naturally occurring anti-aging agents. Elastase is a member of the chymotrypsin family of proteases and is primarily responsible for the degradation of elastin, which is an important protein giving elasticity to arteries, lungs and skin. Elastin breakdown through the action of elastase results in visible skin changes like wrinkles [7–9]. Tyrosinase is a major enzyme in melanin biosynthesis which determines the color of hair and skin. Increased tyrosinase activity was observed in hyperpigmentation disorders; conversely downregulation of tyrosinase activity led to melanogenesis inhibition [10,11]. Therefore, inhibition of elastase and tyrosinase is one of the most prominent strategies used to fight skin aging. Among the considered mechanisms of aging, the oxidative stress plays a substantial role [12]. As increased levels of reactive oxygen species (ROS) contribute to the pathogenesis of numerous diseases including age-related disorders, identifying compounds which have the capacity to increase the intracellular antioxidant defense can be useful to prevent these disorders [13]. Catalase is one of these antioxidant enzymes and prevents cell oxidative damage (i.e., the formation of oxidized proteins, lipids and/or DNA) by converting hydrogen peroxide (H_2O_2), which is continuously produced by metabolic reactions, to water (H_2O) and oxygen (O_2) [14]. Sirtuin 1 (Sirt1) belongs to the conserved sirtuin family (Sirt 1–7) of nicotinamide adenine dinucleotide (NAD^+) dependent protein deacetylases [15]. Identified as key regulators of caloric restriction (CR), these proteins represent one of the most promising targets for anti-aging approaches [16,17]. Indeed, CR is so far the only effective way known to extend the lifespan and healthspan of a number of organisms without genetic or pharmacological intervention [18,19]. Studies have shown that upregulation of these proteins alleviates the symptoms of aging as well as of age-related diseases and induces physiological changes that are similar to CR [20,21]. CDK7 belongs to the Cyclin-dependent kinases (CDKs), best known for their critical roles in cell cycle regulation. However, this protein is also involved in other physiological process like DNA repair and transcription [22–24]. Recent reports demonstrated that CDK7 is crucial for the pathogenesis of certain cancer types driven by RNA polymerase-II-based transcription like triple-negative breast cancer, high-grade glioma or peripheral T-cell lymphomas [25–28]. The proteasome is a multi-subunit enzyme that is essential in cell function as it plays a central role in the regulation of protein homeostasis [29,30]. Cancer cells rely on the proteasome activity to maintain the protein homeostasis required for their enhanced metabolism and unrestricted proliferation. Therefore, inhibition of proteasome function emerged as a powerful strategy for anti-cancer therapy, especially in haematological malignancies [31]. Fyn is a non-receptor tyrosine kinase belonging to the Src family kinase, which is an important class of molecules in human biology. Recent studies highlight its involvement in signaling pathways leading to the pathogenesis of Alzheimer's disease [32,33]. It has been demonstrated that Fyn interacts with both protein Tau and amyloid β-peptide, two key players responsible for the major pathologic hallmarks of Alzheimer's disease [34,35]. These lines of evidence identified Fyn kinase inhibition as a novel approach therapy of particular interest.

Amongst the compounds with anti-aging activities discovered recently, many are natural products, some of which are drugs used in the clinic [36,37]. For example, numerous natural products or extracts from plants and microorganisms are widely used as cosmetic or cosmeceutical ingredients because of their antioxidant, anti-elastase and anti-tyrosinase activities [38–42]. However, some of natural and synthetic anti-tyrosinase compounds used as hypopigmenting agents may induce side effects following chronic exposure [43]. It is also worth noting that the first potent sirtuin activating compounds (STACs) included plant-derived metabolites such as resveratrol [21]. Since then, molecules, structurally related to resveratrol or not, have been developed. Almost all sirtuin activators described to date target Sirt1. Furthermore, marine natural products have been shown to be an interesting source of kinase as well as proteasome inhibitors for cancer and neurodegenerative diseases [44–47]. Salinosporamide A, isolated from marine actinomycete *Salinispora tropica*, is one of the most potent proteasome inhibitors and is currently undergoing clinical trials for cancers [48]. So in order to widen the chemical diversity of bioactive compounds, exploration of understudied biological resources or geographical areas is gaining importance.

In our search for natural products with bioactivities against our set of selected biological targets, we investigated the marine sponges from Mayotte, a French island located in the Comoros archipelago, recognized as a hotspot of biodiversity. This island hosts a variety of marine tropical ecosystems which are of major ecological value and are yet poorly studied. During this investigation, a marine sponge *Scopalina hapalia* exhibited interesting anti-aging activities showing moderate elastase inhibition and high Fyn kinase inhibition. The few reported studies investigating sponges from this geographical area did not consider, thus far, their associated microorganisms and their biotechnological potential [49–56]. The present study aims at investigating the potential of microorganisms isolated from *S. hapalia* to produce secondary metabolites with anti-aging activity. First, the prokaryotic diversity associated with *S. hapalia* was characterized by targeted 16S rRNA sequencing. Then, a cultivation approach was carried out to isolate microorganisms from *S. hapalia* with a special attention given to actinomycetes and filamentous fungi. Members of the *Bacillales* order were also identified and included in this study because of well-known representative producers of important secondary metabolites [57,58]. Finally, their potential to produce bioactive compounds with anti-aging activities was assessed by screenings for elastase, tyrosinase, CDK7, proteasome and Fyn kinase inhibitory activity, as well as for catalase or sirtuin 1 activation.

2. Results and Discussion

2.1. Composition of the Prokaryotic Community of *Scopalina hapalia*

To explore the diversity of the microbial community associated with *Scopalina hapalia*, a culture-independent approach was implemented using targeted 16S rRNA and ITS2 metagenomics.

After quality-filtering, 16 353, 47 630, 37 213 and 46 370 sequences, respective to the V1–V3, V3–V4, V4–V5 regions of the 16S rRNA gene and to the ITS2 region, were obtained from the genomic DNA extracted from *Scopalina hapalia*. These sequences respectively clustered into 337, 408, 355 and 32 Operational Taxonomic Units (OTUs) at 97% similarity from the sequencing of V1–V3, V3–V4, V4–V5 and ITS2 regions (Table 1). Rarefaction curves were constructed and indicated a good recovery of OTUs from the sequencing depth (Figures S1–S3). The OTU richness and diversity estimations were obtained from rarefied data sets (Table 1). The alpha diversity analysis showed that the observed richness (observed OTUs) is slightly equivalent to the estimated richness (Chao1), suggesting that rare OTUs represent a small portion of the sponge microbial community (Table 1). The Shannon index from the three regions sequenced seems to indicate that *Scopalina hapalia* has a diversified microbial community.

Table 1. Alpha diversity of the microbial community associated with *Scopalina hapalia* as measured by the mean index (observed OTUs, Chao1 and Shannon) ± SD.

Targeted Region	No of Reads (SI 97%)	No. of Quality-Filtered Reads	No. of OTUs	Richness		Diversity
				Observed OTUs	Chao1	Shannon
V1–V3	20806	16353	337	286.74 ± 98.97	308.97 ± 94.37	5.65 ± 0.87
V3–V4	49235	47630	408	315.79 ± 119.56	361.43 ± 118.27	4.04 ± 0.69
V4–V5	39094	37213	355	313.23 ± 103.65	324.04 ± 101.00	5.39 ± 0.82
ITS2	46442	46370	32	26.88 ± 9.18	28.92 ± 9.21	0.52 ± 0.02

SI 97%: sequence similarity cutoff of 97%. OTUs: operational taxonomic units.

Unfortunately, the selected primer pair used to amplify fungal ITS2 (Internal Transcribed Spacer 2) region from the total genomic DNA of *S. hapalia* did not yield satisfactory results. Among the 46,370 (100%) quality-filtered reads, 43,669 (94.18%) reads were unclassified and 573 (1.24%) belonged to Metazoa. In total, 2128 (4.59%) reads of fungi were obtained and classified into three different phyla: Ascomycota (4.29%—8 different OTUs), Basidiomycota (0.15%—7 different OTUs) and other

(0.15%—7 different OTUs). Two OTUs belonged to the classes Sordariomycetes and Saccharomycetes (Ascomycota), while five OTUs belonged to the classes Agaricomycetes and Malasseziomycetes (Basidiomycota). Fifteen OTUs could not be classified either at the phylum or class level. Ascomycota and Basidiomycota divisions are widely distributed in marine sponges with Ascomycota being the dominant one [59]. Members of Sordariomycetes, Malasseziomycetes and Agaricomycetes were previously reported from specimens of *Scopalina* collected in Australia [59]. The very low number of fungal reads in comparison to the number of non-targeted DNA is in line with previous studies examining fungal communities in marine sponges by cultivation-independent methods [60–62]. One possible explanation for this result could be the lower abundance of fungi than that of bacteria in marine sponges. As a typical milliliter of seawater contains 10^3 fungal cells and 10^6 bacteria, without specific filtration from the sponge, the fungal cells could be much lower [63].

From the entire data sets, 26 bacterial and two archaeal phyla were identified (Figure S4). This is a high level of phyla diversity for a Low Microbial Abundance (LMA) sponge. Indeed, sponges belonging to the genera *Scopalina* (order Scopalinida) have been characterized as LMA sponges, which are generally assumed to have a less diverse bacterial communities (low phylum-level diversity) than high microbial abundance (HMA) sponges [64,65]. Nonetheless, our result is in agreement with the study by de Voogd et al. (2018), which reported that some LMA sponges from Mayotte harbored a more diverse bacterial community at the phylum level than HMA sponges (e.g., *Stylissa carteri*, order Scopalinida: 29) [66]. Our result supports their conclusion that a true dichotomy between HMA and LMA sponges does not appear to exist [66]. In our study, the *Scopalina hapalia* community was dominated by *Proteobacteria* (62.58%, 94.03% and 57.21% relative abundance in V1–V3, V3–V4 and V4–V5 data sets), *Cyanobacteria* (16.77%, 0.05% and 11.08%), *Planctomycetes* (11.71%, 0.06% and 9.90%), *Bacteroidetes* (4.77%, 1.24% and 17.99%) and *Actinobacteria* (1.68%, 2.82% and 1.30%). From the other 23 phyla, each contributed <1.16% to the *S. hapalia* data set sequences. On the genus and family level, the five most abundant taxa were an unidentified genus of the *Endozoicimonaceae* family (*Gammaproteobacteria*) (6.46%, 48.03%, 6.36%), a member of the uncultured order EC94 (*Betaproteobacteria*) (9.88%, 11.59%, 10.37%), *Ruegeria* (*Alphaproteobacteria*) (5.82%, 5.87%, 9.98%), *Labrenzia* (*Alphaproteobacteria*) (7.37%, 5.60%, 8.41%) and *Aquimarina* (*Flavobacteriia–Bacteroidetes*) (3.72%, 0.05%, 14.29%). Most of the sequences from these taxa corresponded to highly abundant OTUs.

Previous studies based on 16S rRNA gene sequencing or metagenomic analysis revealed that the prokaryotic microbiome of *Scopalina* is dominated by Proteobacteria (*Gammaproteobacteria* and *Betaproteobacteria*) and *Cyanobacteria* [65,67–69]. Even though the diversity of *Scopalina* symbionts has been addressed in some studies, it is difficult to do comparative work due to methodological, sample processing and data analysis differences. However, our results are consistent with previous results showing that *Scopalina* sponges are dominated by the LMA-indicator phylum *Proteobacteria* [65], and further support the notion that the community of LMA sponges is dominated by the classes *Alpha-*, *Beta-* and *Gammaproteobacteria* and *Flavobacteriia* [70].

2.2. *Isolation and Identification of Microorganisms Associated with Scopalina hapalia*

A number of studies investigating microbial communities of *S. ruetzleri* and *Scopalina* sp. have shown that these sponges host phyla (Actinobacteria, Cyanobacteria, Proteobacteria) with potential for natural products synthesis [67,68,71–73]. As the crude extract of *S. hapalia* exhibited interesting biological activities, isolation studies focusing on actinomycetes and filamentous fungi were carried out. Homogenates from *S. hapalia* were plated on a range of eight isolation media including selective ones for actinomycetes and fungi. In total, 124 microbial strains were isolated, among which 10 putative actinomycetes and three filamentous fungi were recovered. R2A media exhibited the highest recovery of isolates (31) and SFM recovered only three isolates. As for the selected media for the isolation of actinobacteria, MBA exhibited the highest recovery (4) followed by SCAM (3), R2A (2) and A1BFe+c (1). In terms of diversity, MBA and SCAM gave isolates belonging to three different species, followed by R2A (2) and A1BFe+c (1). Thirty microbial strains, including the three filamentous fungi, were selected for identification.

2.2.1. Identification of Culturable Actinomycetes Associated with Scopalina hapalia

A comparison of the partial 16S RNA gene sequences of the 10 putative actinomycetes strains against EzBioCloud 16S database exhibited 99–100% sequence similarities with validly described species (Table S1). The analysis revealed the phylogenetic affiliations to three different genera (*Micromonospora* (six strains), *Salinispora* (three strains) and *Rhodococcus* (one strain)) representing two families: Micromonosporaceae and Nocardiaceae.

Micromonospora and *Rhodococcus* have been reported to be amongst the dominant actinobacterial genera in marine environment along with *Streptomyces* [74]. *Salinispora* is an obligate marine actinomycete originally discovered in sediments, then found in sponges [75,76]. It is noteworthy to indicate the presence of the species *Micromonospora endophytica*, formerly known as *Jishengella endophytica* and cultivated from *Acanthus illicifolius* root collected from a mangrove zone in China [77,78]. To the best of our knowledge, this is the first report of the genus *Salinispora* in this geographical area and also the first one about the isolation of *M. endophytica* from any marine invertebrate, especially from sponges. As the Mayotte lagoon includes a mangrove, this result seems to make sense. Representatives of the genera *Micromonospora*, *Rhodococcus* and *Salinispora* have previously been isolated from the specimen *Scopalina ruetzleri* from the Caribbean [71], suggesting that these genera might be part of the specific symbionts associated to the genus *Scopalina*.

2.2.2. Identification of Bacillales Strains Isolated from Scopalina hapalia

The remaining 17 strains shared >98% sequence identity with validly described species within the order Bacillales (Table S2). Fifteen strains were closely related to the genus *Bacillus* and one strain was closely related to previously cultured bacteria DQ448769, for which the nearest type strain is *Laceyella sacchari* belonging to the Thermoactinomycetaceae family [79]. One strain (SH-39) shared <98% sequence identity with DQ448769 suggesting that this might represent a new phylotype. According to Kim and co-workers, a threshold of 98.65% 16S rRNA gene sequence similarity can be considerate to differentiate two species [80]. Most of the *Bacillus* isolates were affiliated either to *B. licheniformis* (AE017333) (7/15) or to *B. paralicheniformis* (KY694465) (7/15). Members of the *Bacillus* group, including *B. licheniformis*, are common inhabitants of marine environments and have been often isolated from sponges and other organisms [79,81]. *B. paralicheniformis*, closely related to *B. licheniformis* and also retrieve in sponges, was described in 2015 from soybean-based fermented paste [82,83]. Representative members of *Bacillus* were previously reported from *Scopalina* sp. [68].

It is worth noting that even though 16S rRNA gene sequencing has proved to be a powerful tool for bacterial identification, it tends to present limitations in distinguishing between closely related species like *Bacillus*, *Micromonospora* and *Salinispora* [84]. This is especially true for the genus *Salinispora*, as six novel species sharing >99% 16S RNA sequence similarity with the three currently recognized species were recently described [85]. Still, this method was chosen because it is a rapid and easy approach for the reliable identification of unknown strains.

2.2.3. Identification of Filamentous Fungi

From the culture-dependent approach, only three fungal strains were recovered. This low number of fungal isolates is consistent with a previous study on the cultivable fungi from Panama, where three fungal strains were recovered from three different specimens of *Scopalina* [86]. On the other hand, more than 19 fungal isolates were recovered from three replicates of *Scopalina* sp. collected in Australia [59]. Our results on the fungal community housed in this specimen of *Scopalina hapalia* are in line with previous works demonstrating that: (1) the number of fungal strains isolated per sponge could vary between locations [87] and (2) discrepancies can be obtained between the molecular and culture approach [59].

The molecular identification revealed a phylogenetic affiliation to the three different genera: *Aspergillus* (class Eurotiomycetes, order Eurotiales), *Chaetomium* (class Sordariomycetes, order Sordariales) and *Nigrospora* (class Sordariomycetes, order Trichosphaeriales) (Table S3). When compared to the ITS2-amplicon sequencing, which affords 22 fungal OTUs, it seems that further efforts would be needed to reveal the entire culturable diversity of fungi present in this sponge microbiome.

Previous works showed that many of the fungal strains isolated from sponges could be classified within the order Eurotiales as it included ubiquitous genera: *Aspergillus* and *Penicillium*, which are prolific sporulating and non-fastidious fungi, and the order Hypocreales (*Acremonium* and *Trichoderma*) [73–77]. Representative of the genus *Penicillium* (Eurotiales), *Acremonium* and *Trichoderma* (Hypocreales) were cultured from *Scopalina* sp. during previous studies [59,86]. The genera *Chaetomium* and *Nigrospora* were previously reported from studies of sponge-derived fungi [87–89], though to the best of our knowledge this is the first report of this newly described species *N. aurantiaca* from sponge [90].

2.3. Biological Assays

Thirty of the 124 isolates were selected for the production of biologically active metabolites. These isolates belong to genera known to produce novel bioactive metabolites or to genera/species that have not been extensively studied (Tables S1–S3). These 30 isolates were cultivated in liquid (LM) and solid (SM) Marine Broth media. Liquid- and solid-state fermentations were coupled with in situ solid-phase extraction using XAD-16 resin. After five and fifteen days of cultivation, the resin and biomass were extracted successively with ethyl acetate (EA) and methanol (MeOH) solvents. Overall, 118 microbial crude extracts were obtained and evaluated for their bioactivity(-ies) against seven different molecular targets.

2.3.1. Anti-Elastase Activities

Preliminary screening on purified elastase identified three extracts with moderate activities (25–50% inhibition) as compared to a known elastase inhibitor, namely elastatinal (produced by actinomycetes). Only crude extracts from isolate SH-82 (*Micromonospora fluostatini*) exhibited sufficient inhibition of elastase activity (Figure 1). The ethyl acetate extract (SH-82-EA-SM) showed the highest inhibition by 32.5% at 100 μg/mL.

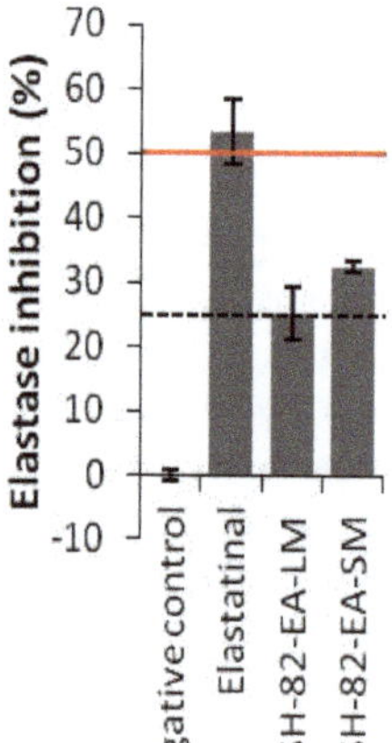

Figure 1. Percentage inhibition of elastase activity by selected microbial crude extracts. Only the most active crude extracts are shown. The code per isolates is outlined in Table S1. Bars, ±SD; $n \geq 2$.

2.3.2. Anti-Tyrosinase Activities

Crude extracts were tested on purified mushroom tyrosinase at the concentration of 150 µg/mL for tyrosinase inhibition activity. Their activities were compared to the commercially available tyrosinase inhibitor kojic acid. The screening revealed 29 extracts from 20 different isolates (nine *Bacillus*, five *Micromonospora*, two *Salinispora*, one *Aspergillus*, one *Chaetomium* and two Thermoactinomycetaceae) with moderate tyrosinase inhibitory activities (25–50% inhibition) (Figure 2). Only one extract from SH-89 isolate (*Micromonospora citrea*) exerted significant anti-melanogenic properties, as it showed 58.33% tyrosinase inhibition. In spite of the moderate inhibitory activity exhibited by these isolates, they represent a promising source for the discovery of new anti-melanogenic agents.

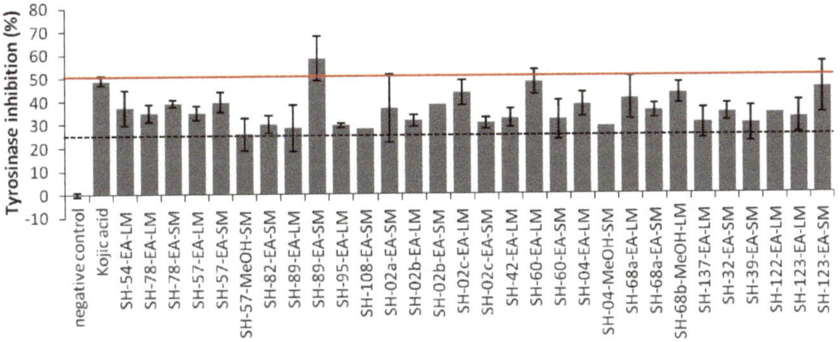

Figure 2. Percentage inhibition of tyrosinase activity by selected microbial crude extracts. Only the most active ones are shown. The code per isolates is outlined in Tables S1–S3. Bars, ±SD; $n \geq 2$.

2.3.3. Catalase Activities

All extracts were screened on purified catalase, at the concentration of 10 µg/mL, in order to assess their antioxidant properties through the catalase activation. Four extracts (SH-02a-MeOH-SM, SH-02b-MeOH-LM, SH-02c-EA-SM, SH-02a-MeOH-SM) obtained from *Bacillus* strains exhibited promising antioxidant capability (activation average > 150). These extracts were selected for a secondary screening, leading to the validation of SH-02b-MeOH-LM (*B. paralicheniformis*) and SH-02c-EA-SM (*B. paralicheniformis*) as catalase activators (Figure 3).

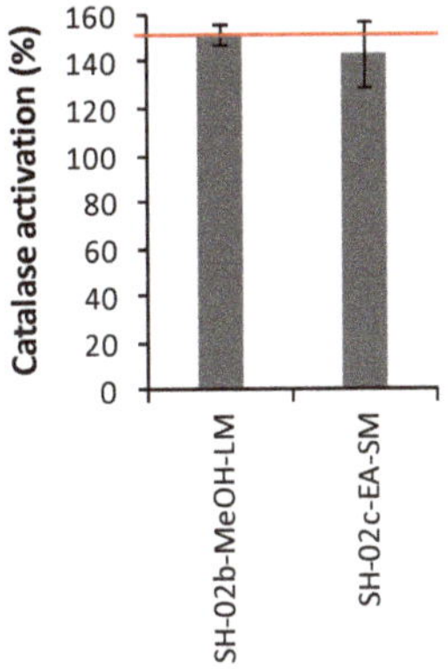

Figure 3. Activation average of catalase by selected microbial crude extracts. Only the most active extracts are shown. The codes of the isolates are outlined in Table S2. Bars, ±SD; $n = 4$.

2.3.4. Sirtuin 1 Activities

Activation of Sirtuin 1 by crude extracts was assessed on purified Sirtuin 1, at the concentration of 100 µg/mL, and expressed as an activation average in comparison to a reference compound (Figure 4). Of the 118 crude extracts tested, 13 extracts activated Sirt1; activation ranged between 170% and 130%. The most potent activators of Sirt1 activity were exhibited by SH-82 (*Micromonospora fluostatini*) and SH-100 (*Bacillus licheniformis*) isolates extracts. These three extracts (SH-82-EA-LM = 177%; SH-100-EA-LM = 190% and SH-100-MeOH-SM = 222%) increased Sirt1 activity with an activation average beyond 170%. During this primary screening, 10 strains were identified as producers of secondary metabolites capable of increasing the activity of Sirtuin 1. The most active ones are SH-82 (*Micromonospora fluostatini*) and SH-100 (*Bacillus licheniformis*). Three more strains of the *Bacillus* genus (*B. paralicheniformis*) gave moderate activities (SH-02a: 136%, SH-10: 137%, SH-22: 149%) along with two *Salinispora* strains (SH-54: 135%, SH-78: 132% and 144%), 2 *Micromonospora* strains (SH-36: 144%, SH-57: 148% and 168%) and one fungi of the *Nigrospora* genus (SH-53: 130% and 146%). Given the scarcity of naturally occurring sirtuin-activating compounds, these strains represent an interesting source for new STACs.

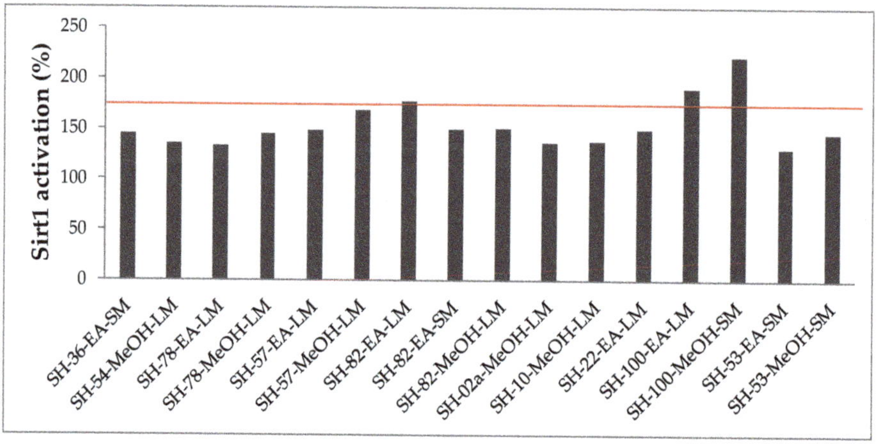

Figure 4. Activation rates of sirtuin 1 by selected microbial crude extracts. Only the most active extracts are indicated. The codes of the isolates are outlined in Tables S1–S3.

2.3.5. Anti-CDK7 Activities

The microbial crude extracts activity was assessed for inhibition of CDK7 activity in a dose dependent way against purified CDK7 and was compared to the reference compound staurosporine (produced by actinomycete such as *Salinispora*). Screening of all extracts for anti-CDK7 properties revealed three extracts from three different strains of *Salinispora arenicola* (SH-45, SH-54 and SH-78) with appreciable CDK7 inhibitory activities at the highest and medium concentrations tested (0.033 and 0.0033 µg/mL) (Table 2).

Table 2. Dose-response inhibition of CDK7 activity by microbial crude extracts. Only the most active extracts are shown. Code of isolates are outlined in Table S1.

Extracts	Dose-Response Inhibition		
	0.033 µg/mL	**0.0033 µg/mL**	**0.00033 µg/mL**
SH-45-EA-SM	Yes	Yes	No
SH-54-EA-SM	Yes	Yes	No
SH-78-EA-SM	Yes	Yes	No

2.3.6. Anti-Fyn Kinase Activities

Anti-Fyn activity was assessed on a purified protein Fyn in a dose dependent manner and was compared to the reference compound staurosporine. Nine strains were found to produce secondary metabolites with Fyn kinase inhibitory activities and ten anti-Fyn kinase extracts were identified (Table 3). The most actives were SH-78 *Salinispora arenicola* (SH-78-EA-SM) and SH-04 *Bacillus licheniformis* (SH-04-EA-SM) strains which inhibited Fyn activity at all three tested concentrations. Two more *S. arenicola* strains (SH-45 and SH-54) yielded crude extracts with good inhibitory activities at the highest and medium concentrations, as well as the *Rhodococcus* strain (SH-115), three more *B. licheniformis* strains (SH-68a, SH-68b and SH-116a) and the fungal strain *Chaetomium globosum* (SH-123).

Table 3. Dose-response inhibition of Fyn activity by microbial crude extracts. Only the most active extracts are shown. Code of isolates are outlined in Tables S1–S3.

Extracts	Dose-Response Inhibition		
	0.033 μg/mL	0.0033 μg/mL	0.00033 μg/mL
SH-45-EA-SM	Yes	Yes	No
SH-54-EA-SM	Yes	Yes	No
SH-78-EA-SM	Yes	Yes	Yes
SH-115-EA-SM	Yes	Yes	No
SH-04-EA-SM	Yes	Yes	Yes
SH-68a-EA-LM	Yes	Yes	No
SH-68b-EA-LM	Yes	Yes	No
SH-116a-EA-SM	Yes	Yes	No
SH-123-MeOH-LM	Yes	Yes	No
SH-123-MeOH-SM	Yes	Yes	No

2.3.7. Anti-Proteasome Activities

No dose-response was observed for this bioassay. The active crude extracts exhibited inhibitory activities only at the highest concentration tested (0.033 μg/mL). Of the 118 crude extracts, only five were active, corresponding to three *Salinispora arenicola* strains (SH-45, SH-54 and SH-78) and one *Bacillus licheniformis* strain (SH-99).

In short, 118 microbial crude extracts were screened against a set of molecular targets with potential pharmaceutical and cosmeceutical applications in the field of aging, for the identification of "hit" extracts. There were both quantitative and qualitative differences in the observed biological activities. Variations were observed not only with respect to the isolates but also according to the growth media (cultural conditions) and the extractive solvents used. Fifty-four of the 118 crude extracts (46%) were active in at least one bioassay (anti-elastase, anti-tyrosinase, anti-CDK7, anti-Fyn kinase, anti-proteasome, increased catalase or sirtuin 1 activities). Remarkably, anti-tyrosinase activity was the most common bioactivity. In total, 25% of the crude extracts (from 20 strains: bacteria and fungi) inhibited at least 25% of the tyrosinase activity, whereas only 1.7% of the extracts (from one *Micromonospora* strain: SH-82) inhibited at least 25% of elastase activity. Furthermore, 8.5% of the extracts (from nine strains) inhibited in a dose-depend manner the activity of the Fyn-kinase, while only 2.5% of the extracts (from three *Salinispora* strains) inhibited the activity of CDK7 in a dose-dependent manner. Of the crude extracts, 13.6% (10 strains) were found to increase Sirtuin 1 activity, with an activation average of at least 130%, whilst only 1.7% of the extracts (from four *Bacillus* strains) were found to activate catalase with an activation average of at least 150%. Extracts of only *Salinispora* and *Bacillus* strains (4%) were found to be active against the proteasome at the highest concentration tested. Although some of the isolates showed different biological activities, almost half of them exhibited unique biological activity. *Bacillus paralicheniformis* (SH-42, SH-60), *Micromonospora* (SH-89, SH-95, SH-108), *Aspergillus* (SH-122) as well as Thermoactinomycetaceae (SH-32, SH-39) strains were found to inhibit only tyrosinase activity. SH-36 (*Micromonospora chokoriensis*), SH-100 (*Bacillus licheniformis*), SH-53 (*Nigrospora aurantiaca*), SH-10 and SH-22 (*Bacillus paralicheniformis*) were active only on Sirtuin 1.

SH-115 (*Rhodococcus nanhaiensis*) and SH-116 (*Bacillus licheniformis*) inhibited solely the activity of Fyn kinase while SH-99 (*Bacillus licheniformis*) inhibited the proteasome activity. It is also interesting to note that only SH-82 (*M. fluostatini*) inhibited elastase, while solely *B. paralicheniformis* (SH-02a, SH-02b, SH-02c) and *B. berkeleyi* (SH-137) strains succeeded in stimulating catalase activity. All tested isolates, except one of *Bacillus* strain (SH-46), showed some bioactivity, suggesting the presence of bioactive compounds with potential anti-aging activity.

Most of the active strains isolated in this study belong to genera (*Salinispora, Micromonospora, Bacillus, Aspergillus* and *Chaetomium*), which are well known to be prolific metabolite producers. Biological activities in the natural products field are mainly focused on antimicrobial and anticancer properties. Nonetheless, some selective activities include anti-tyrosinase/anti-elastase activities, kinase and proteasome inhibition. Some compounds isolated from *Salinispora, Micromonospora, Bacillus* and *Chaetomium* with reported anti-aging activities are presented in Figure 5. For example, the genera *Micromonospora* and *Salinispora* are producers of staurosporine (**1**), a non-selective protein kinase inhibitor, and staurosporine derivatives. These compounds showed cytotoxic activity against a wide range of cancer cell lines [91]. Dereplication of the most active extract from *Salinispora* strain (SH-78) by low resolution mass spectrometry indicated the presence of staurosporine (**1**). The genus *Salinispora* (*S. tropica* and *S. pacifica*) also produce a potent proteasome inhibitor, salinosporamide A (**2**), one of the most promising anticancer agents, currently used in clinical trials for the treatment of multiple myeloma [48]. These compounds might explain the inhibitory activities against protein kinases (CDK7 and Fyn) and proteasome exhibited by *Salinispora* strains. However, to date, compounds in the salinosporamide class were not isolated from *S. arenicola* strains, as the production of this class of compounds appears to be very low in this species [92]. Interestingly, several *Micromonospora* and *Salinispora* strains exhibited anti-tyrosinase activity, while one *Micromonospora* strain showed anti-elastase activity. To our knowledge, no inhibitors of elastase or tyrosinase have been reported from *Micromonospora* or *Salinispora* genera. *Bacillus* species, of which several strains were isolated in this study, increased catalase activity and exhibited anti-tyrosinase, anti-Fyn kinase and anti-proteasome activities. Kinase and proteasome inhibitors with anticancer activity along with some anti-melanogenic compounds were reported from *Bacillus* strains (*Bacillus* spp.) [93,94]. Such an example is iturin A (**3**) (that is well-known for its strong anti-fungal activity) which inhibited protein kinases (MAPK and Akt) as well as baceridin (**4**), a new proteasome inhibitor [95,96]. The iturins are mainly produced by members of the *Bacillus subtilis* group; however, the production of iturins is strongly associated with *B. amyloliquefaciens* [57]. Poly-γ-glutamate (**5**), a natural polymer produced by different bacterial strains including *B. licheniformis*, exhibited strong anti-tyrosinase activity while increasing the catalase activity [93,97]. *Aspergillus* and *Chaetomium* strains (SH-122 and SH-123) were found to inhibit tyrosinase activity. Some fungal metabolites have been identified and reported for their inhibitory activity against tyrosinase. Kojic acid, for example, is produced by many species of *Aspergillus* and extensively used as a skin-whitening agent [39,98]. Curiously, only one isolate of *Chaetomium globosum* had been found to be a low-level producer of kojic acid [99]. *Chaetomium* strain also exhibited a Fyn kinase inhibitory activity that might be explained by the production of chaetominedione (**6**), a tyrosine kinase inhibitor isolated from an algicolous marine specimen of *Chaetomium* sp. [100]. The most remarkable result to emerge from our data was the identification of several crude extracts (16) from *Micromonospora, Salinispora, Bacillus* and *Nigrospora* strains that increased Sirtuin 1 activity. To date and to the best of our knowledge, few marine natural products or extracts that activate Sirtuin 1 have been reported. Conversely, some marine natural products with anti-Sirtuin 1 activity have been identified. As crude extracts were tested for biological activities, the chemical nature of the bioactive compounds was, up to now, unknown. To gain insight into the metabolic profile of these strains, the active crude extracts will be submitted to liquid chromatography-high resolution mass spectrometry (LC-HRMS/MS) analysis to assess strains with potential to produce novel/known bioactive metabolites. Furthermore, prior to secondary metabolites isolation, fermentation conditions will be optimized in order to improve the yield of produced bioactive metabolites.

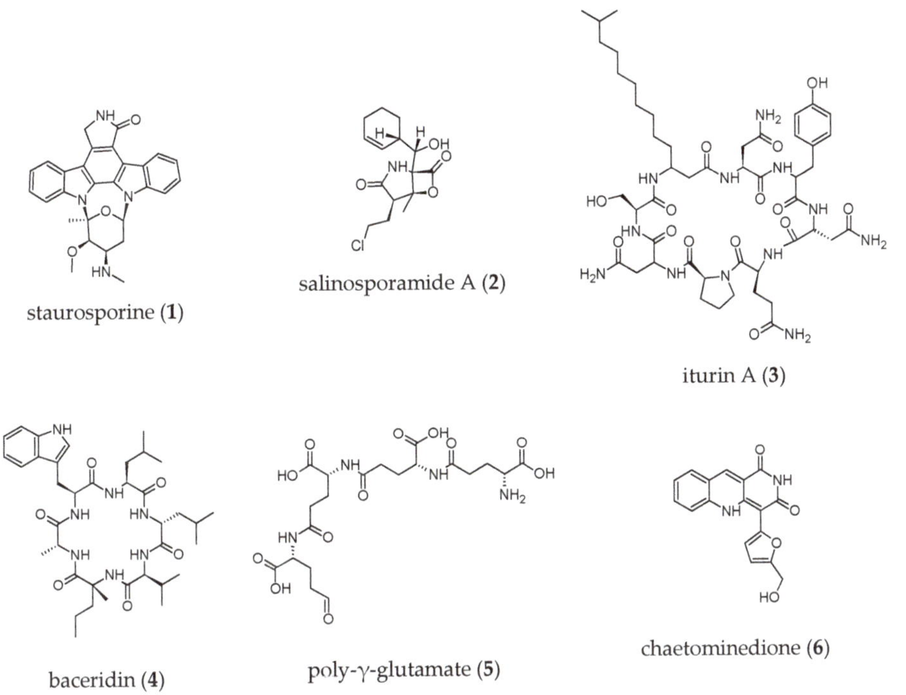

Figure 5. Structures of known microbial secondary metabolites with anti-aging activities isolated from *Bacillus, Chaetomium, Micromonospora* and *Salinispora* genera.

Another aspect of our future work will address strategies to activate cryptic genes with the aim of identifying new natural products. Indeed, investigation into the microbiome of Australian marine sponges by deep sequencing of NRPS (nonribosomal peptide synthetase) and PKS (polyketide synthase) genes found both NRPS and PKS biosynthetic pathways within *Scopalina* sp. microbiomes [101]. They further revealed within the microbiome of *Scopalina* sp. A great diversity of KS (PKS), the majority of which were trans-AT type and C (NRPS) domains. Trans-AT (*trans*-acyltransferase) type KS domains are an important group of PKS enzymes as they often lead to unique chemistry properties. Importantly, they identified the existence of novel biosynthetic pathways within these sponges, suggesting an untapped resource for natural products discovery. Products of PKS and NRPS pathways or hybrids of both represent a major class of the biologically active microbial natural products of interest to human health and industry. Surprisingly, few bioactive compounds were reported from *Scopalina* spp. In 2015, Vicente and co-workers reported the isolation of six new angucyclinone (aromatic polyketide) derivatives from a *Streptomyces* strain associated with *S. ruetzleri* [102]. Monacyclinone C (**7**) and E (**8**) exhibited moderate cytotoxic activities against rhabdomyosarcoma cancer cells, while monacyclinone F (**9**) showed the highest cytotoxic activity along with antibacterial property (Figure 6). This study represents the first report of the isolation of complex bioactive secondary metabolites from the genus *Scopalina*.

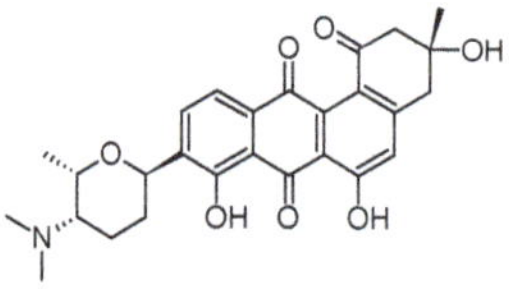

monacyclinone C (**7**)

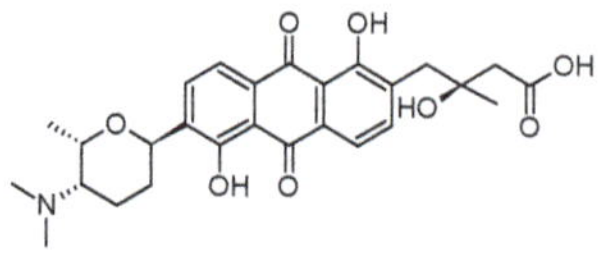

monacyclinone E (**8**)

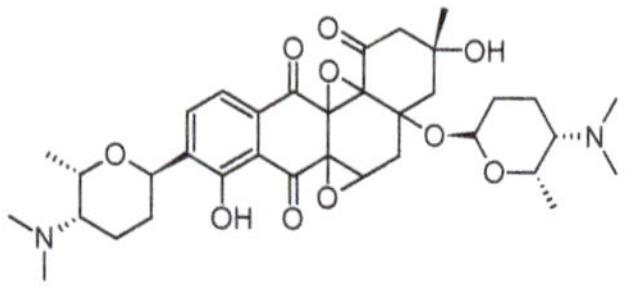

monacyclinone F (**9**)

Figure 6. Chemical structures of compounds isolated from *Streptomyces* sp. derived from *S. ruetzleri*.

Known to produce species-specific metabolites, studies have found site-specific secondary metabolite gene clusters in *Salinispora* strains collected at different locations. Exploring the secondary metabolism of the strains recovered in this study would be interesting for the discovery of new bioactive natural products. Furthermore, it is now accepted that even well-studied taxa like *Micromonospora* can harbor a wealth of biosynthetic pathways for which the products have yet to be discovered. Interestingly, some of the *Micromonospora* isolates recovered during our work belong to species barely investigated for secondary metabolites production (SH-36 *M. chokoriensis*, SH-89 *M. citrea*, SH-108 *M. endophytica*, SH-82 *M. fluostatini*, SH-95 *M. tulbaghiae*). Moreover, to the best of our knowledge there is no report of secondary metabolites isolated from *Nigrospora aurantiaca* (SH-53) and *Rhodococcus nanhaiensis* (SH-115). The occurrence of PKS and NRPS genes in actinomycetes, *Bacillus* and fungi further supports the high potential of microorganisms associated with the marine sponge *Scopalina hapalia* to produce interesting secondary metabolites. The idea that one microbial strain producing secondary metabolites often has the potential to produce various compounds is now well accepted. In order to reveal the chemical diversity of these strains, alteration of easily accessible cultivation parameters (media composition, adsorbent resins, pH, and temperature) will be carried out. The manipulation of the fermentation conditions, known as the OSMAC (One Strain Many Compounds) approach, represents an easy and effective way of activating silent or poorly expressed biosynthetic pathways [103].

3. Materials and Methods

3.1. Sponge Collection

Scopalina hapalia (ML-263) was collected in May 2013 by scuba diving at the depths of 2–10 m around the southeast coasts of Mayotte (Kani tip, Global Positioning System (12°57.624′ S; 45°04.697′ E)). The sponge sample was levered off with thin-bladed knife to prevent damage, transferred to a plastic bag and kept at -20 °C before being transported to the laboratory. For identification, a voucher specimen was preserved in 80% ethanol. The taxonomic identification was performed by Nicole de Voogd and voucher specimens were deposited at Naturalis Biodiversity Center, Leiden The Netherlands as RMNH POR.8332 and RMNH POR.8376.

3.2. Targeted 16S rRNA Gene Sequencing / Next Generation Sequencing

Sponge specimen was cut into pieces of ca. 1 g (1 cm^3), rinsed in ethanol 70% then in sterile artificial seawater (ASW) (Sea salts (Sel Instant Ocean, Aquarium système, Sarrebourg, France) 33 g/L) [104]. Genomic DNA was extracted from small pieces of sponge using a commercial genomic DNA extraction kit (Qiagen, Hilden, Germany) [105], as per manufacturer's instructions. DNA extract was visually checked for quality by agarose gel electrophoresis. The purity of the extraction was assessed, and the quantification was undertaken using a NanoDrop spectrophotometer (Thermo Fisher, Waltham, MA, USA). The 16S rRNA gene amplification, sequencing and taxonomic affiliation were performed by Genoscreen (Lille, France) according to their methodology Metabiote®. Briefly, 5 ng of genomic DNA sample were used for libraries preparation, and sequencing was performed using the Illumina MiSeq "paired-end" 2 × 300 bp technology, according to the Metabiote®protocol established by Genoscreen. The hypervariable regions V1–V3, V3–V4 and V4–V5 found in bacteria and archaea were targeted using specific primers. Data preprocessing and analysis of the sequence data were carried out using the Metabiote®v. 2.0 pipeline (Genoscreen, Lille, France), partially based on the QIIME v. 1.9.1 [106]. Raw forward and reverse sequence reads were assembled into contigs using 30 bp coverage (overlapping regions) and 97% sequence identity as parameters for merging with FLASH (Fast Length Adjustment of SHort reads) [107]. Chimeric sequences were identified and discarded by using a Genoscreen program based on Usearch 6.1. Filtered sequences were clustered into Operational Taxonomic Units (OTUs) at 97% sequence similarity using the algorithm Uclust v. 1.2.22q based on an open-reference OTU strategy [108]. The most abundant sequence for each OTU was used as reference and aligned to the Greengenes database v. 13.8 (greengenes.secondgenome.com). The phyla- and genus-level affiliation of the sequences was validated using the Ribosomal Database Project Classifier v. 2.2 [109].

3.3. *Alpha Diversity Analysis*

Bacterial richness and diversity estimators (observed OTUs, Chao1 and Shannon) were calculated with the script alpha_diversity.py QIIME v. 1.9.1 [110,111].

3.4. *Microbial Isolation*

Sponge specimen was cut into pieces of ca. 1 g (1 cm^3), rinsed in ethanol 70% and then in sterile ASW. After surface sterilization, the sample was thoroughly homogenized in a sterile mortar then transferred in a 50 mL Falcon vial containing 10 mL of sterile ASW. Two options were selected for the microbial isolation. Protocol 1 (P1): To allow the dissemination of the maximum revivable strains, the homogenate was diluted in ten-fold series (10^{-1}, 10^{-2} and 10^{-3}) and subsequently plated out on agar plates. Protocol 2 (P2): To eliminate fast growing/heat sensitive strains and favor the slow growing ones, the same protocol was repeated with a heat-shock pretreatment of the homogenate (30 min, 50 °C) before dilution (10^{-1}, 10^{-2} and 10^{-3}). Eight different media were used for the isolation of microorganisms A1BFe+C (starch (BD Difco™, Le Pont de Claix, France) 10 g, yeast extract (BD Bacto™, Le Pont de Claix, France) 4 g, peptone (BD Bacto™, Le Pont de Claix, France) 2 g, $CaCO_3$ (Carlo Erba, Val de Reuil, France) 1 g, $Fe_2(SO_4)_3$ (Carlo Erba, France) 40 mg, KBr (Carlo Erba, France) 100 mg, sea salts 30 g), LB (tryptone (Sigma Aldrich, St. Louis, MI, USA, Steinheim, Germany) 10 g, Yeast Extract 5 g, NaCl (Fisher Scientific Labosi, Elancourt, France) 10 g), Marine Broth (MB) (BD Difco™, Le Pont de Claix, France), MYA2 (malt extract (BD Difco™, Le Pont de Claix, France) 20 g, yeast extract 1 g, sea salts 30 g), Potato Dextrose Broth (PDB) (BD Difco™, Le Pont de Claix, France), R2A (Fisher Scientific, Waltham, MA, USA), SCAM (maize starch (Fisher Chemical, Loughborough, UK) 10 g, casein (VWR Chemicals, Leuven, Belgium) 1 g, sea salts 30 g), SFM (soybean flour (La Vie Claire, Montagny, France) 20 g, mannitol (Carlo Erba, Val de Reuil, France) 20 g). All media contained Difco™ Bacto agar (20 g/L) and were prepared in 1 L of purified water with pH adjusted to 7.2 ± 0.2, except for SFM. This last medium was prepared with tap water and no pH adjustment. The inoculated plates were incubated at 28 °C for 10 weeks. Distinct colony morphotypes were picked and re-streaked until visually free of contaminants. Isolates were inoculated on A1BFe+C for putative actinomycetes, on LB for bacteria and MYA2 for fungi. The isolates were maintained on plates at 4 °C for short-term conservation. For long-term strain collection, the microorganisms were stored at −80 °C in a cryoprotectant medium (skimmed milk (Régilait, Macon, France) 10% (*w*/*v*), glycerol (Carlo Erba, Val de Reuil, France) 10% (*v*/*v*) and sea salts 33 g/L). Bacterial strains were sorted into groups according to their morphological characteristics. Twelve putative actinomycetes and three filamentous fungi were selected for molecular identification and secondary metabolites production. Fifteen other bacteria types giving mucoid/smooth colonies adhering to the agar surface and/or showing cocci/bacilli morphology were included in this study.

3.5. *Molecular Identification*

Selected bacterial strains were cultured on A1BFe+c medium for two to 14 days and then were sent to Genoscreen (Lille, France). Targeted 16S rRNA gene amplification (V1–V3 and V3–V5) and sequencing were performed by Genoscreen. The obtained contigs were compared to EzBioCloud 16S database (www.ezbiocloud.net, accessed on 31 May 2019) for species-level identification, using sequence similarity searches [112].

The fungal strains were cultured on PDA medium and then sent to the Westerdijk Fungal Biodiversity Institute (Netherland) for identification. The PDA medium consisted of PDB supplemented with agar at the final concentration of 2%. On arrival, the strains were cultivated on malt extract agar (MEA) and dichloran 18% glycerol agar (DG18). DNA was extracted from one MEA plate after an incubation period of three days in the dark at 25 °C, using the Qiagen DNeasy Ultraclean™ Microbial DNA Isolation Kit (Qiagen, Germany). For strain SH-53, fragments containing the Internal Transcribed Spacer 1 and 2 regions including the 5.8S rDNA (ITS) and a partial β-tubulin gene (*BenA*) were amplified and sequenced. For strain SH-122, fragments containing a partial β-tubulin gene (*BenA*) and fragments containing a partial calmodulin gene (*CaM*) were amplified and sequenced. For strain SH-123, fragments containing the Internal Transcribed Spacer 1 and 2 regions including the 5.8S rDNA (ITS) and a partial ß-tubulin gene (*BenA*) were amplified and sequenced. The primers used were: ITS (SH-53 and SH-123): LS266 (GCATTCCCAAACAACTCGACTC) and V9G (TTACGTCCCTGCCCTTTGTA), *BenA* (SH-53 and SH-122): Bt2a (GGTAACCAAATCGGTGCTGCTTTC) and BT2b (GGTAACCAAATCGGTGCTGCTTTC), *CaM*: CMD5 (CCGAGTACAAGGARGCCTTC) and CMD6 (CCGATRGAGGTCATRACGTGG), *BenA* (SH-123): T1 (AACATGCGTGAGATTGTAAGT) and Tub4RD (CCRGAYTGRCCRAARACRAAGTTGTC). The PCR fragments were sequenced in both directions with the primers used for PCR amplification using the ABI Prism®Big DyeTM Terminator v. 3.0 Ready Reaction Cycle sequencing Kit (Thermo FisherFisher, USA). Samples were analyzed on an ABI PRISM 3700 Genetic Analyzer and contigs were assembled using the forward and reverse sequences with the program SeqMan from the LaserGene package. The sequences were compared on GenBank using BLAST (www.ncbi.nlm.nih.gov) and in the in-house sequence database of Westerdijk Fungal Biodiversity Institute.

3.6. Extracts Preparation

Thirty strains were selected based on their affiliation to taxa known to produce bioactive secondary metabolites with biotechnological interests. Prior to fermentation, each of the selected isolates were inoculated on two Petri plates (Ø 9 cm, Nunc™ Thermo FisherFisher, USA) containing MBA (Marine Broth supplemented with 2% agar) and incubated for 5–17 days depending on their growth rate at 28 °C. About 5 mL of ASW were spread on the agar layer, the culture was then scratched and transferred in a 50 mL Falcon vial containing 45 mL of ASW in order to prepare the inoculum. Then, 10 mL of the homogenate was used to inoculate a 2 L Erlenmeyer flask containing 1 L of MB mixed with 30 g/L of XAD-16 resin (Sigma, St. Louis, USA; Steinheim, Germany) (liquid-state fermentation coupled to in-situ solid-phase extraction) [113]. The remainder homogenate was mixed with 35 g of XAD-16 resin and uniformly spread on a large-surface Petri dish (25 × 25 cm, Nunc™ Thermo Fisher, USA) containing MBA (solid-state fermentation coupled to solid-phase extraction) [114]. The liquid cultures were grown for 5 days at 28 °C, then the biomass and the resins were recovered by Buchner filtration (filter paper Whatman®grade 4, Ø 110 mm). The solid cultures were grown for 15 days at 28 °C; then the mix of biomass/resins was peeled off. All the biomass/resins mixes were washed with distilled water, dried under vacuum on Buchner filter and successively extracted with ethyl acetate (EA) (Carlo Erba, France) and methanol (MeOH) (Carlo Erba, France). The solvents were removed by evaporation; the residues were weighed and used for biological assays.

3.7. Bioassays

Bioassays against selected targets involved in age-related diseases and disorders (elastase, tyrosinase, catalase, sirtuin 1, CDK7, Fyn and proteasome) were carried out as follows.

3.7.1. Elastase Activity Assay

Elastase enzyme activity was evaluated using elastase from porcine pancreas (PPE) type IV and *N*-succinyl-Ala-Ala-Ala-p-nitroanilide as substrate, as previously described [115]. The amount of released p-nitroaniline, which was hydrolyzed by elastase, was measured spectrophotometrically at 405 nm. The reaction mix was constituted of 70 μL Trizma-base buffer (50 mM, pH 7.5), 10 μL of the extract tested (the final concentration of the extracts was 100 μg/mL) and 5 μL of PPE (0.4725 U/mL), in a 96-well microplate. The samples were incubated for 15 min in room temperature, avoiding light exposure. Subsequently, 15 μL from 0.903 mg/mL *N*-succinyl-Ala-Ala-Ala-p-nitroanilide were added and the samples were incubated at 37 °C for 30 min . Then, the absorbance of p-nitroaniline production was measured in the reader Infinite 200 PRO series (Tecan). Elastatinal was used as positive control, whereas the negative control contained the Trizma-base buffer and the substrate. Experiments were performed in duplicates. The reagents of the assay were purchased from Sigma-Aldrich. The percentage of elastase inhibition was calculated as follows:

$$\text{Inhibition (\%)} = \{[(\text{Abs control-Abs control's blank}) - (\text{Abs sample-Abs sample's blank})] / (\text{Abs control-Abs control's blank})]\} * 100,$$

where Abs control is the absorbance of the elastase in Trizma base buffer, sample solvent and substrate, and Abs sample is the absorbance of the elastase in Trizma base buffer, extract or elastatinal and substrate. Blank experiments were performed for each sample with all the reagents except the enzyme.

3.7.2. Tyrosinase Activity Assay

In order to evaluate the inhibitory potency of the extracts against tyrosinase, the oxidation of L-DOPA to dopachrome was determined by an enzymatic method, as previously described [116]. Specifically, in a 96-well microplate, 80 μL of PBS (0.067 M, pH 6.8), 40 μL of the tested extract (the final concentration of the extracts was 150 μg/mL) and 40 μL of mushroom tyrosinase 92 U/mL were mixed and incubated for 10 min at room temperature, avoiding light exposure. Afterwards, 40 μL of 2.5 mM L-DOPA dissolved in PBS buffer were added and the mixture was incubated for 5 min before measurement of dopachrome formation at 475 nm using the reader Infinite 200 PRO series (Tecan, Salzburg, Austria). Kojic acid was used as positive control, whereas the negative control contained PBS and the substrate. Experiments were performed in duplicates. The reagents of the assay were purchased from Sigma-Aldrich. The tyrosinase inhibition percentage was calculated as follows:

$$\text{Inhibition (\%)} = \{[(\text{Abs control-Abs control's blank}) - (\text{Abs sample-Abs sample's blank})] / (\text{Abs control-Abs control's blank})]\} * 100,$$

where Abs control is the absorbance of tyrosinase enzyme in PBS, sample solvent, and substrate and Abs sample is the absorbance of tyrosinase enzyme in PBS, extract or Kojic acid, and substrate. Blanks contained all the aforementioned components except the enzyme.

3.7.3. Catalase Activation Assay

Catalase activity was measured using The Amplex®Red Catalase Assay Kit according to the manufacturer's instructions (Thermo Fisher Scientific, Waltham, MA, USA). Briefly, 2.5 μL of catalase (final concentration 62.5 mU/mL) first reacts with 5 μL of H_2O_2 (final concentration 10 μM) to produce water and oxygen (O_2). Next 10 μL of the Amplex Red reagent (final concentration 50 μM) reacts with any unreacted H_2O_2 in the presence of horseradish peroxidase (HRP) to produce the highly fluorescent oxidation product, resorufin. Primary and secondary screenings were performed respectively in monoplicate and quadruplicate. The fluorescence was measured on a Polar Star Omega (BMG Labtech, Ortenberg, Germany) plate reader. The results are typically plotted by subtracting the observed fluorescence from that of a no-catalase control.

3.7.4. Sirtuin 1 Activation Assay

Sirt1 activity was measured using SIRT 1 Fluorometric Drug Discovery Kit according to the manufacturer's instructions (Enzo Life Sciences, Farmingdale, NY, USA). Briefly, this assay uses a small lysine-acetylated peptide, corresponding to K382 of human p53, as a substrate. The lysine residue is deacetylated by SIRT1, and this process is dependent on the addition of exogenous NAD+. The assay was carried at 37 °C using Greiner white, small volume 384 well plates. First, 4 µL of substrate Fluor de Lys (final concentration 25 µM) were mixed with 4 µL of extract (stock concentration 10 mg/mL) previously diluted 1/100 in assay buffer and 2 µL of the enzyme were added. After an incubation of 15 min at 37 °C, 10 µL of Developer 1× solution (composed by buffer, developer 5× and Nicotinamide 50 mM) were added and incubated for 45 min at 37 °C. After 45 min, the fluorescence was measured on a Polar Star Omega (BMG Labtech, Ortenberg, Germany) plate reader. The fluorescence generated was proportional to the quantity of deacetylated Lysine (i.e., corresponding to Lysine 382). All measurements were performed in monoplicate and the final DMSO concentration was 0.1%. SIRT1 inhibitors nicotinamide (2 mM), suramin (100 µM), and sirtinol (100 µM) were used to confirm the specificity of the reaction. Calculation of net fluorescence included the substraction of a blank consisting of buffer containing no NAD+ and expressed as a percentage of control.

3.7.5. CDK7 Inhibition Assay

CDK7 activity was evaluated using CDK7 (Crelux construct CZY-3, PC09891). The inhibitory potency of the extracts against CDK7 was determined by using the ADP-Glo Kinase Assay (Promega, Madison, WI, USA) and CDKtide as substrate. The assay was carried out at 22 °C using Corning 4513 white low volume 384 well plates. All measurements were performed in singlicate and the final DMSO concentration was 3.3%. The assay buffer contained 20 mM Hepes pH 7.5, 150 mM NaCl, 10 mM MgCl2. First, 9.5 µL protein dilution (final concentration 300 nM) were mixed with 0.5 µL of either one of three extract dilutions (1, 0.1 or 0.01 mg/mL >> assay end conc. 0.033, 0.0033 or 0.00033 mg/mL) and preincubated for 120 min . Then 5 µL substrate/ATP mix were added (final concentration substrate 30 µM and ATP 125 µM) and the assay was incubated for another 2 h. Afterwards, 5 µL of the 15 µL assay reaction were transferred to new wells and 5 µL ADP Glo Reagent were added to terminate the kinase reaction and deplete the remaining ATP. After 40 min, 10 µL of Kinase Detection Reagent were added to convert ADP to ATP and allow the newly synthesized ATP to be measured using a luciferase/luciferin reaction. After another 40 min, the assay plates were measured in Luminescence mode on a Tecan M1000 plate reader. The light generated, i.e., luminescent signal, was proportional to the ADP concentration produced and was correlated with kinase activity. The IC50 was calculated using XLfit. Staurosporine, which inhibits CDK7, was used as positive control in order to assess the functionality of the assay.

3.7.6. FynB Inhibition Activity

FynB activity was evaluated using FynB wt (Crelux construct CTX4, PC09815-1). The inhibitory potency of the extracts against FynB was determined by using the ADP-Glo Kinase Assay (Promega) and Fyn kinase substrate (Enzo Life Sciences, P215). The assay was carried out at room temperature (22 °C) using Corning 4513 white low volume 384 well plates. All measurements were performed in singlicate and the final DMSO concentration was 3.3%. The assay buffer contained 20 mM Tris pH 8.0, 170 mM NaCl, 10 mM MgCl2. First, 9.5 µL protein dilution (final concentration 200 nM) were mixed with 0.5 µL of either one of three extract dilutions (1, 0.1 or 0.01 mg/mL >> assay end conc. 0.033, 0.0033 or 0.00033 mg/mL) and preincubated for 90 min . Then 5 µL substrate/ATP mix were added (final concentration substrate 10 µM and ATP 100 µM) and the assay was incubated for another 2 h. Afterwards, 5 µL of the 15 µL assay reaction were transferred to new wells and 5 µL ADP Glo Reagent were added to terminate the kinase reaction and deplete the remaining ATP. After 40 min, 10 µL of Kinase Detection Reagent were added to convert ADP to ATP and allow the newly synthesized ATP to be measured using a luciferase/luciferin reaction. After another 40 min, the assay plates were measured in Luminescence mode on a Tecan M1000 plate reader. The light generated, i.e., luminescent signal, was proportional to the ADP concentration produced and was correlated with kinase activity. The IC50 was calculated using XLfit. Staurosporine, which inhibits FynB kinase, was used as positive control in order to assess the functionality of the assay.

3.7.7. Proteasome Inhibition Assay

Proteasome activity was evaluated using yeast proteasome (TUM Groll group). The inhibitory potency of the extracts against yeast proteasome was assessed by using the Fluorescence Intensity Assay and Suc-Leu-Leu-Val-Tyr-AMC as substrate (Enzo Life Sciences, BML-P802-0005). The assay was carried out at room temperature (22 °C) using Corning 4514 black low volume 384 well plates. All measurements were performed in singlicate and the final DMSO concentration was 3.3%. The assay buffer contained 100 mM Tris pH 7.5 and 1 mM MgCl2. First, 9.5 µL protein dilution (final concentration 9 nM) were mixed with 0.5 µL of either one of three probe dilutions (1, 0.1 or 0.01 mg/mL >> assay end conc. 0.033, 0.0033 or 0.00033 mg/mL) and preincubated for 90 min . Then 5 µL substrate mix were added (final concentration 5 µM) and incubated for 60 min . The assay plates were measured in fluorescence mode on a Tecan M1000 plate reader (ex 380 nm, em 460 nm); the IC50 was calculated using XLfit. ONX-0914, which inhibits yeast proteasome, was used as positive control in order to assess the functionality of the assay.

4. Conclusions

The sponge-microbial associations from the southwest of Indian Ocean were not extensively explored for their potential to produce clinically relevant bioactive compounds. In this study, 30 isolates belonging to genera *Bacillus*, *Micromonospora*, *Rhodococcus*, *Salinispora*, *Aspergillus*, *Chaetomium*, *Nigrospora* and unidentified genera related to the family Thermoactinomycetaceae were recovered from the marine sponge *Scopalina hapalia* collected in Mayotte. In total, 54 of 118 microbial crude extracts from 29 different strains showed significant bioactivities against age-related molecular targets. These results demonstrate that the selected strains produce secondary metabolites with likely anti-aging activities. Taken together, these findings highlight the ability of marine sponge-associated microorganisms from Mayotte to produce secondary metabolites with cosmeceutical and therapeutic potential. Our marine microbial collection represents an important "eco-friendly" resource for the bioprospection of novel bioactive metabolites.

Supplementary Materials: The following are available online at http://www.mdpi.com/2076-2607/8/9/1262/s1, Figure S1. Rarefaction curves of observed OTUs for each targeted genomic DNA region, Figure S2. Rarefaction curves of Chao1 Index for each targeted genomic DNA region, Figure S3. Rarefaction curves of Shannon Index for each targeted genomic DNA region. Figure S4. Phylum distribution in *Scopalina hapalia* respective with V1–V3, V3–V4 and V4–V5 data sets. The bars represent the relative abundance of 16S rRNA sequences that were assigned to a given phylum in relation to the total number of sequences in each data set. Table S1. 16S rRNA taxonomic affiliation of 10 of the *Scopalina hapalia* associated actinomycetes. Table S2. 16S rRNA taxonomic affiliation of 17 Bacillales (order) strains isolated from *Scopalina hapalia*. Table S3. Taxonomic affiliation of the 3 fungal isolates from *Scopalina hapalia* after sequencing fragments containing ITS region as well as partial beta-tubulin and calmodulin genes.

Author Contributions: Conceptualization, A.G.-B., J.O. and G.L.G.; supervision, M.F. and L.D.; investigation (microbial isolation, extracts preparation and metagenomic data analysis), C.S.H. and J.B.B.; bioassays methodology and validation, A.D.S. and I.P.T. (elastase and tyrosinase assays); J.B. (catalase and sirt1 activities); M.J. (CDK7, FynB and proteasome assays); sponge identification, N.J.d.V.; writing—original draft preparation, C.S.H.; writing—review and editing, M.F., A.G.-B., C.S.H. and L.D. All authors have read and agreed to the published version of the manuscript.

Funding: This project was supported by the Regional Council of Reunion Island and by the TASCMAR project which is funded by the European Union under grant agreement number 634674.

Acknowledgments: The authors express their gratitude to Genoscreen (France) for their assistance in the study of *Scopalina hapalia* microbiome and the molecular identification of bacteria isolates. We are also thankful to the Westerdijk Fungal Biodiversity Institute (Netherland) for their contribution to fungal identification.

Conflicts of Interest: The authors declare no conflict of interest.

References

1. Niccoli, T.; Partridge, L. Ageing as a Risk Factor for Disease. *Curr. Biol.* **2012**, *22*, R741–R752. [CrossRef] [PubMed]
2. Kennedy, B.K.; Berger, S.L.; Brunet, A.; Campisi, J.; Cuervo, A.M.; Epel, E.S.; Franceschi, C.; Lithgow, G.J.; Morimoto, R.I.; Pessin, J.E.; et al. Geroscience: Linking aging to chronic disease. *Cell* **2014**, *159*, 709–713. [CrossRef] [PubMed]
3. Longo, V.D.; Antebi, A.; Bartke, A.; Barzilai, N.; Brown-Borg, H.M.; Caruso, C.; Curiel, T.J.; Cabo, R.; De Franceschi, C.; Gems, D.; et al. Interventions to slow aging in humans: Are we ready? *Aging Cell* **2015**, *14*, 497–510. [CrossRef] [PubMed]
4. Fontana, L.; Partridge, L.; Longo, V.D. Extending healthy life span-from yeast to humans. *Science* **2010**, *328*, 321–326. [CrossRef]
5. Davinelli, S.; Willcox, D.C.; Scapagnini, G. Extending healthy ageing: Nutrient sensitive pathway and centenarian population. *Immun. Ageing* **2012**, *9*, 9. [CrossRef]
6. Seals, D.R.; Justice, J.N.; LaRocca, T.J. Physiological geroscience: Targeting function to increase healthspan and achieve optimal longevity. *J. Physiol.* **2016**, *594*, 2001–2024. [CrossRef]
7. Robert, L.; Jacob, M.P.; Frances, C.; Godeau, G.; Hornebeck, W. Interaction between elastin and elastases and its role in the aging of the arterial wall, skin and other connective tissues. A review. *Mech. Ageing Dev.* **1984**, *28*, 155–166. [CrossRef]
8. Tsuji, N.; Moriwaki, S.; Suzuki, Y.; Takema, Y.; Imokawa, G. The role of elastases secreted by fibroblasts in wrinkle formation: Implication through selective inhibition of elastase activity. *Photochem. Photobiol.* **2001**, *74*, 283–290. [CrossRef]
9. Rijken, F.; Kiekens, R.C.M.; Bruijnzeel, P.L.B. Skin-infiltrating neutrophils following exposure to solar-simulated radiation could play an important role in photoageing of human skin. *Br. J. Dermatol.* **2005**, *152*, 321–328. [CrossRef]
10. Pillaiyar, T.; Manickam, M.; Namasivayam, V. Skin whitening agents: Medicinal chemistry perspective of tyrosinase inhibitors. *J. Enzym. Inhib. Med. Chem.* **2017**, *32*, 403–425. [CrossRef]
11. Bae-Harboe, Y.-S.C.; Park, H.-Y. Tyrosinase: A Central Regulatory Protein for Cutaneous Pigmentation. *J. Investig. Dermatol.* **2012**, *132*, 2678–2680. [CrossRef] [PubMed]
12. Hekimi, S.; Lapointe, J.; Wen, Y. Taking a "good" look at free radicals in the aging process. *Trends Cell Biol.* **2011**, *21*, 569–576. [CrossRef] [PubMed]
13. Liguori, I.; Russo, G.; Curcio, F.; Bulli, G.; Aran, L.; Della-Morte, D.; Gargiulo, G.; Testa, G.; Cacciatore, F.; Bonaduce, D.; et al. Oxidative stress, aging, and diseases. *Clin. Interv. Aging* **2018**, *13*, 757–772. [CrossRef] [PubMed]
14. Glorieux, C.; Calderon, P.B. Catalase, a remarkable enzyme: Targeting the oldest antioxidant enzyme to find a new cancer treatment approach. *Biol. Chem.* **2017**, *398*, 1095–1108. [CrossRef]
15. Michan, S.; Sinclair, D. Sirtuins in mammals: Insights into their biological function. *Biochem. J.* **2007**, *404*, 1–13. [CrossRef]
16. Longo, V.D.; Kennedy, B.K. Sirtuins in Aging and Age-Related Disease. *Cell* **2006**, *126*, 257–268. [CrossRef]
17. Bordone, L.; Guarente, L. Calorie restriction, SIRT1 and metabolism: Understanding longevity. *Nat. Rev. Mol. Cell Biol.* **2005**, *6*, 298–305. [CrossRef]
18. Masoro, E.J. Overview of caloric restriction and ageing. *Mech. Ageing Dev.* **2005**, *126*, 913–922. [CrossRef]
19. Most, J.; Tosti, V.; Redman, L.M.; Fontana, L. Calorie restriction in humans: An update. *Ageing Res. Rev.* **2017**, *39*, 36–45. [CrossRef]
20. Hubbard, B.P.; Sinclair, D.A. Small molecule SIRT1 activators for the treatment of aging and age-related diseases. *Trends Pharmacol. Sci.* **2014**, *35*, 146–154. [CrossRef]
21. Howitz, K.T.; Bitterman, K.J.; Cohen, H.Y.; Lamming, D.W.; Lavu, S.; Wood, J.G.; Zipkin, R.E.; Chung, P.; Kisielewski, A.; Zhang, L.-L.; et al. Small molecule activators of sirtuins extend Saccharomyces cerevisiae lifespan. *Nature* **2003**, *425*, 191–196. [CrossRef] [PubMed]

22. Roy, R.; Adamczewski, J.P.; Seroz, T.; Vermeulen, W.; Tassan, J.P.; Schaeffer, L.; Nigg, E.A.; Hoeijmakers, J.H.; Egly, J.M. The MO15 cell cycle kinase is associated with the TFIIH transcription-DNA repair factor. *Cell* **1994**, *79*, 1093–1101. [CrossRef]
23. Yee, A.; Nichols, M.A.; Wu, L.; Hall, F.L.; Kobayashi, R.; Xiong, Y. Molecular cloning of CDK7-associated human MAT1, a cyclin-dependent kinase-activating kinase (CAK) assembly factor. *Cancer Res.* **1995**, *55*, 6058–6062. [PubMed]
24. Kelso, T.W.R.; Baumgart, K.; Eickhoff, J.; Albert, T.; Antrecht, C.; Lemcke, S.; Klebl, B.; Meisterernst, M. Cyclin-dependent kinase 7 controls mRNA synthesis by affecting stability of preinitiation complexes, leading to altered gene expression, cell cycle progression, and survival of tumor cells. *Mol. Cell. Biol.* **2014**, *34*, 3675–3688. [CrossRef]
25. Wang, Y.; Zhang, T.; Kwiatkowski, N.; Abraham, B.J.; Lee, T.I.; Xie, S.; Yuzugullu, H.; Von, T.; Li, H.; Lin, Z.; et al. CDK7-dependent transcriptional addiction in triple-negative breast cancer. *Cell* **2015**, *163*, 174–186. [CrossRef] [PubMed]
26. Kwiatkowski, N.; Zhang, T.; Rahl, P.B.; Abraham, B.J.; Reddy, J.; Ficarro, S.B.; Dastur, A.; Amzallag, A.; Ramaswamy, S.; Tesar, B.; et al. Targeting transcription regulation in cancer with a covalent CDK7 inhibitor. *Nature* **2014**, *511*, 616–620. [CrossRef] [PubMed]
27. Cayrol, F.; Praditsuktavorn, P.; Fernando, T.M.; Kwiatkowski, N.; Marullo, R.; Calvo-Vidal, M.N.; Phillip, J.; Pera, B.; Yang, S.N.; Takpradit, K.; et al. THZ1 targeting CDK7 suppresses STAT transcriptional activity and sensitizes T-cell lymphomas to BCL2 inhibitors. *Nat. Commun.* **2017**, *8*, 14290. [CrossRef] [PubMed]
28. Greenall, S.A.; Lim, Y.C.; Mitchell, C.B.; Ensbey, K.S.; Stringer, B.W.; Wilding, A.L.; O'Neill, G.M.; McDonald, K.L.; Gough, D.J.; Day, B.W.; et al. Cyclin-dependent kinase 7 is a therapeutic target in high-grade glioma. *Oncogenesis* **2017**, *6*, e336. [CrossRef] [PubMed]
29. Adams, J. The proteasome: Structure, function, and role in the cell. *Cancer Treat. Rev.* **2003**, *29*, 3–9. [CrossRef]
30. Schmidt, M.; Finley, D. Regulation of proteasome activity in health and disease. *Biochim. et Biophys. Acta (BBA) Mol. Cell Res.* **2014**, *1843*, 13–25. [CrossRef]
31. Manasanch, E.E.; Orlowski, R.Z. Proteasome inhibitors in cancer therapy. *Nat. Rev. Clin. Oncol.* **2017**, *14*, 417–433. [CrossRef] [PubMed]
32. Nygaard, H.B.; Van Dyck, C.H.; Strittmatter, S.M. Fyn kinase inhibition as a novel therapy for Alzheimer's disease. *Alzheimers Res.* **2014**, *6*, 8. [CrossRef] [PubMed]
33. Kaufman, A.C.; Salazar, S.V.; Haas, L.T.; Yang, J.; Kostylev, M.A.; Jeng, A.T.; Robinson, S.A.; Gunther, E.C.; Van, C.D.; Nygaard, H.B.; et al. Fyn inhibition rescues established memory and synapse loss in Alzheimer mice. *Ann. Neurol.* **2015**, *77*, 953–971. [CrossRef] [PubMed]
34. Ittner, L.M.; Ke, Y.D.; Delerue, F.; Bi, M.; Gladbach, A.; Van Eersel, J.; Wölfing, H.; Chieng, B.C.; Christie, M.J.; Napier, I.A.; et al. Dendritic function of tau mediates amyloid-beta toxicity in Alzheimer's disease mouse models. *Cell* **2010**, *142*, 387–397. [CrossRef] [PubMed]
35. Um, J.W.; Nygaard, H.B.; Heiss, J.K.; Kostylev, M.A.; Stagi, M.; Vortmeyer, A.; Wisniewski, T.; Gunther, E.C.; Strittmatter, S.M. Alzheimer amyloid-β oligomer bound to postsynaptic prion protein activates Fyn to impair neurons. *Nat. Neurosci.* **2012**, *15*, 1227–1235. [CrossRef] [PubMed]
36. Argyropoulou, A.; Aligiannis, N.; Trougakos, I.P.; Skaltsounis, A.-L. Natural compounds with anti-ageing activity. *Nat. Prod. Rep.* **2013**, *30*, 1412–1437. [CrossRef]
37. Campisi, J.; Kapahi, P.; Lithgow, G.J.; Melov, S.; Newman, J.C.; Verdin, E. From discoveries in ageing research to therapeutics for healthy ageing. *Nature* **2019**, *571*, 183. [CrossRef]
38. Fibrich, B.D.; Lall, N. Chapter 3-Fighting the Inevitable: Skin Aging and Plants. In *Medicinal Plants for Holistic Health and Well-Being*; Lall, N., Ed.; Academic Press: Cambridge, MA, USA, 2018; pp. 77–115, ISBN 978-0-12-812475-8.
39. Parvez, S.; Kang, M.; Chung, H.-S.; Bae, H. Naturally occurring tyrosinase inhibitors: Mechanism and applications in skin health, cosmetics and agriculture industries. *Phytother. Res.* **2007**, *21*, 805–816. [CrossRef]
40. Kusumawati, I.; Indrayanto, G. Chapter 15—Natural Antioxidants in Cosmetics. In *Studies in Natural Products Chemistry*; Atta-ur-Rahman, Ed.; Elsevier: Amsterdam, The Netherlands, 2013; Volume 40, pp. 485–505.
41. Martins, A.; Vieira, H.; Gaspar, H.; Santos, S. Marketed marine natural products in the pharmaceutical and cosmeceutical industries: Tips for success. *Mar. Drugs* **2014**, *12*, 1066–1101. [CrossRef]
42. Corinaldesi, C.; Barone, G.; Marcellini, F.; Dell'Anno, A.; Danovaro, R. Marine Microbial-Derived Molecules and Their Potential Use in Cosmeceutical and Cosmetic Products. *Mar. Drugs* **2017**, *15*, 118. [CrossRef]

43. Parvez, S.; Kang, M.; Chung, H.-S.; Cho, C.; Hong, M.-C.; Shin, M.-K.; Bae, H. Survey and mechanism of skin depigmenting and lightening agents. *Phytother. Res.* **2006**, *20*, 921–934. [CrossRef]
44. Skropeta, D.; Pastro, N.; Zivanovic, A. Kinase Inhibitors from Marine Sponges. *Mar. Drugs* **2011**, *9*, 2131–2154. [CrossRef]
45. Ning, C.; Wang, H.-M.D.; Gao, R.; Chang, Y.-C.; Hu, F.; Meng, X.; Huang, S.-Y. Marine-derived protein kinase inhibitors for neuroinflammatory diseases. *Biomed. Eng. Online* **2018**, *17*, 46. [CrossRef]
46. Li, T.; Wang, N.; Zhang, T.; Zhang, B.; Sajeevan, T.P.; Joseph, V.; Armstrong, L.; He, S.; Yan, X.; Naman, C.B. A Systematic Review of Recently Reported Marine Derived Natural Product Kinase Inhibitors. *Mar. Drugs* **2019**, *17*, 493. [CrossRef]
47. Della Sala, G.; Agriesti, F.; Mazzoccoli, C.; Tataranni, T.; Costantino, V.; Piccoli, C. Clogging the Ubiquitin-Proteasome Machinery with Marine Natural Products: Last Decade Update. *Mar. Drugs* **2018**, *16*, 467. [CrossRef]
48. Fenical, W.; Jensen, P.R.; Palladino, M.A.; Lam, K.S.; Lloyd, G.K.; Potts, B.C. Discovery and development of the anticancer agent salinosporamide A (NPI-0052). *Bioorg. Med. Chem.* **2009**, *17*, 2175–2180. [CrossRef]
49. Pettit, G.R.; Tan, R.; Gao, F.; Williams, M.D.; Doubek, D.L.; Boyd, M.R.; Schmidt, J.M.; Chapuis, J.C.; Hamel, E. Isolation and structure of halistatin 1 from the eastern Indian Ocean marine sponge Phakellia carteri. *J. Org. Chem.* **1993**, *58*, 2538–2543. [CrossRef]
50. Pettit, G.R.; Gao, F.; Doubek, D.L.; Boyd, M.R.; Hamel, E.; Bai, R.; Schmidt, J.M.; Tackett, L.P.; Ruetzler, K. Antineoplastic Agents. Part 252. Isolation and Structure of Halistatin 2 from the Comoros Marine Sponge Axinella carteri. *Gazz. Chim. Ital.* **1993**, *123*, 371–377. [CrossRef]
51. Pettit, G.R.; Tan, R.; Herald, D.L.; Williams, M.D.; Cerny, R.L. Antineoplastic Agents. 277. Isolation and Structure of Phakellistatin 3 and Isophakellistatin 3 from a Republic of Comoros Marine Sponge. *J. Org. Chem.* **1994**, *59*, 1593–1595. [CrossRef]
52. Pettit, G.R.; Gao, F.; Schmidt, J.M.; Chapuis, J.-C.; Cerny, R.L. Isolation and structure of axinastatin 5 from a republic of comoros marine sponge. *Bioorg. Med. Chem. Lett.* **1994**, *4*, 2935–2940. [CrossRef]
53. Rudi, A.; Aknin, M.; Gaydou, E.M.; Kashman, Y.; Sodwanones, K.L.M. New triterpenes from the marine sponge Axinella weltneri. *J. Nat. Prod.* **1997**, *60*, 700–703. [CrossRef]
54. Gauvin, A.; Smadja, J.; Aknin, M.; Faure, R.; Gaydou, E.-M. Isolation of bioactive 5α,8α-epidioxy sterols from the marine sponge *Luffariella cf. variabilis*. *Can. J. Chem.* **2000**, *78*, 986–992. [CrossRef]
55. Campos, P.-E.; Pichon, E.; Moriou, C.; Clerc, P.; Trépos, R.; Frederich, M.; De Voogd, N.; Hellio, C.; Gauvin-Bialecki, A.; Al-Mourabit, A. New Antimalarial and Antimicrobial Tryptamine Derivatives from the Marine Sponge Fascaplysinopsis reticulata. *Mar. Drugs* **2019**, *17*, 167. [CrossRef]
56. Campos, P.-E.; Herbette, G.; Chendo, C.; Clerc, P.; Tintillier, F.; De Voogd, N.J.; Papanagnou, E.-D.; Trougakos, I.P.; Jerabek, M.; Bignon, J.; et al. Osirisynes G-I, New Long-Chain Highly Oxygenated Polyacetylenes from the Mayotte Marine *Sponge Haliclona* sp. *Mar. Drugs* **2020**, *18*, 350. [CrossRef]
57. Harwood, C.R.; Mouillon, J.-M.; Pohl, S.; Arnau, J. Secondary metabolite production and the safety of industrially important members of the *Bacillus subtilis* group. *FEMS Microbiol. Rev.* **2018**, *42*, 721–738. [CrossRef]
58. Mondol, M.A.M.; Shin, H.J.; Islam, M.T. Diversity of secondary metabolites from marine *Bacillus* species: Chemistry and biological activity. *Mar. Drugs* **2013**, *11*, 2846–2872. [CrossRef]
59. Nguyen, M.T.H.D.; Thomas, T. Diversity, host-specificity and stability of sponge-associated fungal communities of co-occurring sponges. *PeerJ* **2018**, *6*, e4965. [CrossRef]
60. Naim, M.A.; Smidt, H.; Sipkema, D. Fungi found in Mediterranean and North Sea sponges: How specific are they? *PeerJ* **2017**, *5*, e3722. [CrossRef]
61. He, L.; Liu, F.; Karuppiah, V.; Ren, Y.; Li, Z. Comparisons of the fungal and protistan communities among different marine sponge holobionts by pyrosequencing. *Microb. Ecol.* **2014**, *67*, 951–961. [CrossRef]
62. Gao, Z.; Li, B.; Zheng, C.; Wang, G. Molecular detection of fungal communities in the hawaiian marine sponges *Suberites zeteki* and *Mycale armata*. *Appl. Environ. Microbiol.* **2008**, *74*, 6091–6101. [CrossRef]
63. Kubanek, J.; Jensen, P.R.; Keifer, P.A.; Sullards, M.C.; Collins, D.O.; Fenical, W. Seaweed resistance to microbial attack: A targeted chemical defense against marine fungi. *Proc. Natl. Acad. Sci. USA* **2003**, *100*, 6916–6921. [CrossRef] [PubMed]

64. Gloeckner, V.; Wehrl, M.; Moitinho-Silva, L.; Gernert, C.; Schupp, P.; Pawlik, J.R.; Lindquist, N.L.; Erpenbeck, D.; Wörheide, G.; Hentschel, U. The HMA-LMA dichotomy revisited: An electron microscopical survey of 56 sponge species. *Biol. Bull.* **2014**, *227*, 78–88. [CrossRef] [PubMed]
65. Helber, S.B.; Steinert, G.; Wu, Y.-C.; Rohde, S.; Hentschel, U.; Muhando, C.A.; Schupp, P.J. Sponges from Zanzibar host diverse prokaryotic communities with potential for natural product synthesis. *FEMS Microbiol. Ecol.* **2019**, *95*. [CrossRef] [PubMed]
66. Voogd, N.J.; De Gauvin-Bialecki, A.; Polónia, A.R.M.; Cleary, D.F.R. Assessing the bacterial communities of sponges inhabiting the remote western Indian Ocean island of Mayotte. *Mar. Ecol.* **2018**, *39*, e12517. [CrossRef]
67. Rua, C.P.J.; Gregoracci, G.B.; Santos, E.O.; Soares, A.C.; Francini-Filho, R.B.; Thompson, F. Potential metabolic strategies of widely distributed holobionts in the oceanic archipelago of St Peter and St Paul (Brazil). *FEMS Microbiol. Ecol.* **2015**, *91*, fiv043. [CrossRef] [PubMed]
68. Esteves, A.I.S.; Amer, N.; Nguyen, M.; Thomas, T. Sample processing impacts the viability and cultivability of the sponge microbiome. *Front. Microbiol.* **2016**, *7*. [CrossRef]
69. Fan, L.; Reynolds, D.; Liu, M.; Stark, M.; Kjelleberg, S.; Webster, N.S.; Thomas, T. Functional equivalence and evolutionary convergence in complex communities of microbial sponge symbionts. *Proc. Natl. Acad. Sci. USA* **2012**, *109*, E1878–E1887. [CrossRef]
70. Moitinho-Silva, L.; Steinert, G.; Nielsen, S.; Hardoim, C.C.P.; Wu, Y.-C.; McCormack, G.P.; López-Legentil, S.; Marchant, R.; Webster, N.; Thomas, T.; et al. Predicting the HMA-LMA status in marine sponges by machine learning. *Front. Microbiol.* **2017**, *8*, 8. [CrossRef]
71. Tabares, P.; Pimentel-Elardo, S.M.; Schirmeister, T.; Hünig, T.; Hentschel, U. Anti-protease and immunomodulatory activities of bacteria associated with caribbean sponges. *Mar. Biotechnol.* **2011**, *13*, 883–892. [CrossRef]
72. Souza, D.T.; Silva, F.S.P.; Da Silva, L.J.; Da Crevelin, E.J.; Moraes, L.A.B.; Zucchi, T.D.; Melo, I.S. *Saccharopolyspora spongiae* sp. nov., a novel actinomycete isolated from the marine sponge *Scopalina ruetzleri* (Wiedenmayer, 1977). *Int. J. Syst. Evol. Microbiol.* **2017**, *67*, 2019–2025. [CrossRef]
73. Vicente, J.; Stewart, A.; Song, B.; Hill, R.T.; Wright, J.L. Biodiversity of actinomycetes associated with caribbean sponges and their potential for natural product discovery. *Mar. Biotechnol.* **2013**, *15*, 413–424. [CrossRef] [PubMed]
74. Subramani, R.; Aalbersberg, W. Culturable rare Actinomycetes: Diversity, isolation and marine natural product discovery. *Appl. Microbiol. Biotechnol.* **2013**, *97*, 9291–9321. [CrossRef] [PubMed]
75. Maldonado, L.A.; Fenical, W.; Jensen, P.R.; Kauffman, C.A.; min cer, T.J.; Ward, A.C.; Bull, A.T.; Goodfellow, M. *Salinispora arenicola* gen. nov., sp. nov. and *Salinispora tropica* sp. nov., obligate marine actinomycetes belonging to the family Micromonosporaceae. *Int. J. Syst. Evol. Microbiol.* **2005**, *55*, 1759–1766. [CrossRef] [PubMed]
76. Kim, T.K.; Garson, M.J.; Fuerst, J.A. Marine actinomycetes related to the "*Salinospora*" group from the Great Barrier Reef sponge *Pseudoceratina clavata*. *Environ. Microbiol.* **2005**, *7*, 509–518. [CrossRef]
77. Xie, Q.-Y.; Wang, C.; Wang, R.; Qu, Z.; Lin, H.-P.; Goodfellow, M.; Hong, K. *Jishengella endophytica* gen. nov., sp. nov., a new member of the family Micromonosporaceae. *Int. J. Syst. Evol. Microbiol.* **2011**, *61*, 1153–1159. [CrossRef]
78. Li, L.; Zhu, H.; Xu, Q.; Lin, H.; Lu, Y. Micromonospora craniellae sp. nov., isolated from a marine sponge, and reclassification of Jishengella endophytica as Micromonospora endophytica comb. nov. *Int. J. Syst. Evol. Microbiol.* **2019**, *69*, 715–720. [CrossRef]
79. Gontang, E.A.; Fenical, W.; Jensen, P.R. Phylogenetic diversity of gram-positive bacteria cultured from marine sediments. *Appl. Environ. Microbiol.* **2007**, *73*, 3272–3282. [CrossRef]
80. Kim, M.; Oh, H.-S.; Park, S.-C.; Chun, J. Towards a taxonomic coherence between average nucleotide identity and 16S rRNA gene sequence similarity for species demarcation of prokaryotes. *Int. J. Syst. Evol. Microbiol.* **2014**, *64*, 346–351. [CrossRef]
81. Ivanova, E.P.; Vysotskii, M.V.; Svetashev, V.I.; Nedashkovskaya, O.I.; Gorshkova, N.M.; Mikhailov, V.V.; Yumoto, N.; Shigeri, Y.; Taguchi, T.; Yoshikawa, S. Characterization of *Bacillus* strains of marine origin. *Int. Microbiol.* **1999**, *2*, 267–271.

82. Bibi, F.; Yasir, M.; Al-Sofyani, A.; Naseer, M.I.; Azhar, E.I. Antimicrobial activity of bacteria from marine sponge *Suberea mollis* and bioactive metabolites of *Vibrio* sp. EA348. *Saudi J. Biol. Sci.* **2020**, *27*, 1139–1147. [CrossRef]
83. Dunlap, C.A.; Kwon, S.-W.; Rooney, A.P.; Kim, S.-J. *Bacillus paralicheniformis* sp. nov., isolated from fermented soybean paste. *Int. J. Syst. Evol. Microbiol.* **2015**, *65*, 3487–3492. [CrossRef] [PubMed]
84. Fox, G.E.; Wisotzkey, J.D.; Jurtshuk, P. How close is close: 16S rRNA sequence identity may not be sufficient to guarantee species identity. *Int. J. Syst. Bacteriol.* **1992**, *42*, 166–170. [CrossRef]
85. Román-Ponce, B.; Millán-Aguiñaga, N.; Guillen-Matus, D.; Chase, A.B.; Ginigini, J.G.M.; Soapi, K.; Feussner, K.D.; Jensen, P.R.; Trujillo, M.E. Six novel species of the obligate marine actinobacterium Salinispora, Salinispora cortesiana sp. nov., Salinispora fenicalii sp. nov., Salinispora goodfellowii sp. nov., Salinispora mooreana sp. nov., Salinispora oceanensis sp. nov. and Salinispora vitiensis sp. nov., and emended description of the genus Salinispora. *Int. J. Syst. Evol. Microbiol.* **2020**. [CrossRef]
86. Bolaños, J.; León, L.F.D.; Ochoa, E.; Darias, J.; Raja, H.A.; Shearer, C.A.; Miller, A.N.; Vanderheyden, P.; Porras-Alfaro, A.; Caballero-George, C. Phylogenetic Diversity of Sponge-Associated Fungi from the Caribbean and the Pacific of Panama and Their In Vitro Effect on Angiotensin and Endothelin Receptors. *Mar. Biotechnol.* **2015**, *17*, 533–564. [CrossRef]
87. Höller, U.; Wright, A.D.; Matthee, G.F.; Konig, G.M.; Draeger, S.; Aust, H.-J.; Schulz, B. Fungi from marine sponges: Diversity, biological activity and secondary metabolites. *Mycol. Res.* **2000**, *104*, 1354–1365. [CrossRef]
88. Wang, G.; Li, Q.; Zhu, P. Phylogenetic diversity of culturable fungi associated with the hawaiian sponges *Suberites zeteki* and *Gelliodes fibrosa*. *Antonie Van Leeuwenhoek* **2008**, *93*, 163–174. [CrossRef]
89. Ding, B.; Yin, Y.; Zhang, F.; Li, Z. Recovery and phylogenetic diversity of culturable fungi associated with marine sponges *Clathrina luteoculcitella* and *Holoxea* sp. In the South China Sea. *Mar. Biotechnol.* **2011**, *13*, 713–721. [CrossRef]
90. Wang, M.; Liu, F.; Crous, P.W.; Cai, L. Phylogenetic reassessment of *Nigrospora*: Ubiquitous endophytes, plant and human pathogens. *Persoonia* **2017**, *39*, 118–142. [CrossRef]
91. Abdelmohsen, U.R.; Bayer, K.; Hentschel, U. Diversity, abundance and natural products of marine sponge-associated actinomycetes. *Nat. Prod. Rep.* **2014**, *31*, 381–399. [CrossRef]
92. Jensen, P.R.; Moore, B.S.; Fenical, W. The marine actinomycete genus *Salinispora*: A model organism for secondary metabolite discovery. *Nat. Prod. Rep.* **2015**, *32*, 738–751. [CrossRef]
93. Liu, X.; Liu, F.; Liu, S.; Li, H.; Ling, P.; Zhu, X. Poly-γ-glutamate from *Bacillus subtilis* inhibits tyrosinase activity and melanogenesis. *Appl. Microbiol. Biotechnol.* **2013**, *97*, 9801–9809. [CrossRef] [PubMed]
94. Kim, K.; Leutou, A.S.; Jeong, H.; Kim, D.; Seong, C.N.; Nam, S.-J.; Lim, K.-M. Anti-Pigmentary Effect of (-)-4-Hydroxysattabacin from the Marine-Derived Bacterium Bacillus sp. *Mar. Drugs* **2017**, *15*, 138. [CrossRef]
95. Dey, G.; Bharti, R.; Dhanarajan, G.; Das, S.; Dey, K.K.; Kumar, B.N.P.; Sen, R.; Mandal, M. Marine lipopeptide Iturin A inhibits Akt mediated GSK3β and FoxO3a signaling and triggers apoptosis in breast cancer. *Sci. Rep.* **2015**, *5*, 10316. [CrossRef]
96. Niggemann, J.; Bozko, P.; Bruns, N.; Wodtke, A.; Gieseler, M.T.; Thomas, K.; Jahns, C.; Nimtz, M.; Reupke, I.; Brüser, T.; et al. Baceridin, a cyclic hexapeptide from an epiphytic *Bacillus* strain, inhibits the proteasome. *ChemBioChem* **2014**, *15*, 1021–1029. [CrossRef]
97. Candela, T.; Fouet, A. Poly-gamma-glutamate in bacteria. *Mol. Microbiol.* **2006**, *60*, 1091–1098. [CrossRef]
98. Siddiquee, S. Chapter 4—Recent Advancements on the Role of Biologically Active Secondary Metabolites from Aspergillus. In *New and Future Developments in Microbial Biotechnology and Bioengineering*; Gupta, V.K., Rodriguez-Couto, S., Eds.; Elsevier: Amsterdam, The Netherlands, 2018; pp. 69–94, ISBN 978-0-444-63501-3.
99. El-Kady, I.A.; Zohri, A.N.A.; Hamed, S.R. Kojic Acid Production from Agro-Industrial By-Products Using Fungi. *Biotechnol. Res. Int.* **2014**, *2014*, 1–10. [CrossRef]
100. Abdel-Lateff, A. Chaetominedione, a new tyrosine kinase inhibitor isolated from the algicolous marine fungus *Chaetomium* sp. *Tetrahedron Lett.* **2008**, *49*, 6398–6400. [CrossRef]
101. Woodhouse, J.N.; Fan, L.; Brown, M.V.; Thomas, T.; Neilan, B.A. Deep sequencing of non-ribosomal peptide synthetases and polyketide synthases from the microbiomes of Australian marine sponges. *ISME J.* **2013**, *7*, 1842–1851. [CrossRef]

102. Vicente, J.; Stewart, A.K.; Van Wagoner, R.M.; Elliott, E.; Bourdelais, A.J.; Wright, J.L.C. Monacyclinones, new angucyclinone metabolites isolated from *Streptomyces* sp. M7_15 associated with the puerto rican sponge *Scopalina ruetzleri*. *Mar. Drugs* **2015**, *13*, 4682–4700. [CrossRef]
103. Bode, H.B.; Bethe, B.; Höfs, R.; Zeeck, A. Big effects from small changes: Possible ways to explore nature's chemical diversity. *Chembiochem* **2002**, *3*, 619–627. [CrossRef]
104. Kjer, J.; Debbab, A.; Aly, A.H.; Proksch, P. Methods for isolation of marine-derived endophytic fungi and their bioactive secondary products. *Nat. Protoc.* **2010**, *5*, 479–490. [CrossRef] [PubMed]
105. Simister, R.L.; Schmitt, S.; Taylor, M.W. Evaluating methods for the preservation and extraction of DNA and RNA for analysis of microbial communities in marine sponges. *J. Exp. Mar. Biol. Ecol.* **2011**, *397*, 38–43. [CrossRef]
106. Caporaso, J.G.; Kuczynski, J.; Stombaugh, J.; Bittinger, K.; Bushman, F.D.; Costello, E.K.; Fierer, N.; Peña, A.G.; Goodrich, J.K.; Gordon, J.I.; et al. QIIME allows analysis of high-throughput community sequencing data. *Nat. Methods* **2010**, *7*, 335–336. [CrossRef] [PubMed]
107. Magoč, T.; Salzberg, S.L. FLASH: Fast length adjustment of short reads to improve genome assemblies. *Bioinformatics* **2011**, *27*, 2957–2963. [CrossRef] [PubMed]
108. Edgar, R.C. Search and clustering orders of magnitude faster than BLAST. *Bioinformatics* **2010**, *26*, 2460–2461. [CrossRef] [PubMed]
109. Wang, Q.; Garrity, G.M.; Tiedje, J.M.; Cole, J.R. Naïve bayesian classifier for rapid assignment of rRNA sequences into the new bacterial taxonomy. *Appl. Environ. Microbiol.* **2007**, *73*, 5261–5267. [CrossRef] [PubMed]
110. Shannon, C.E. A mathematical theory of communication. *Bell Syst. Tech. J.* **1948**, *27*, 379–423. [CrossRef]
111. Chao, A. Nonparametric estimation of the number of classes in a population. *Scand. J. Stat.* **1984**, *11*, 265–270.
112. Yoon, S.-H.; Ha, S.-M.; Kwon, S.; Lim, J.; Kim, Y.; Seo, H.; Chun, J. Introducing EzBioCloud: A taxonomically united database of 16S rRNA gene sequences and whole-genome assemblies. *Int. J. Syst. Evol. Microbiol.* **2017**, *67*, 1613–1617. [CrossRef]
113. Lee, J.C.; Park, H.R.; Park, D.J.; Lee, H.B.; Kim, Y.B.; Kim, C.J. Improved production of teicoplanin using adsorbent resin in fermentations. *Lett. Appl. Microbiol.* **2003**, *37*, 196–200. [CrossRef]
114. Goff, G.L.; Adelin, E.; Cortial, S.; Servy, C.; Ouazzani, J. Application of solid-phase extraction to agar-supported fermentation. *Bioprocess Biosyst Eng* **2013**, *36*, 1285–1290. [CrossRef] [PubMed]
115. Hubert, J.; Angelis, A.; Aligiannis, N.; Rosalia, M.; Abedini, A.; Bakiri, A.; Reynaud, R.; Nuzillard, J.-M.; Gangloff, S.C.; Skaltsounis, A.-L.; et al. In vitro dermo-cosmetic evaluation of bark extracts from common temperate trees. *Planta Med.* **2016**, *82*, 1351–1358. [CrossRef] [PubMed]
116. Masuda, T.; Yamashita, D.; Takeda, Y.; Yonemori, S. Screening for tyrosinase inhibitors among extracts of seashore plants and identification of potent inhibitors from *Garcinia subelliptica*. *Biosci. Biotechnol. Biochem.* **2005**, *69*, 197–201. [CrossRef] [PubMed]

microorganisms

MDPI

Review

Combating Parasitic Nematode Infections, Newly Discovered Antinematode Compounds from Marine Epiphytic Bacteria

Nor Hawani Salikin [1,2], Jadranka Nappi [1], Marwan E. Majzoub [1] and Suhelen Egan [1,*]

1 Centre for Marine Science and Innovation, School of Biological, Earth and Environmental Sciences, UNSW, Sydney, NSW 2052, Australia; norhawani0203@gmail.com (N.H.S.); j.nappi@unsw.edu.au (J.N.); m.majzoub@unsw.edu.au (M.E.M.)

2 School of Industrial Technology, Universiti Sains Malaysia, USM, 11800 Penang, Malaysia

* Correspondence: s.egan@unsw.edu.au

Received: 4 November 2020; Accepted: 8 December 2020; Published: 11 December 2020

Abstract: Parasitic nematode infections cause debilitating diseases and impede economic productivity. Antinematode chemotherapies are fundamental to modern medicine and are also important for industries including agriculture, aquaculture and animal health. However, the lack of suitable treatments for some diseases and the rise of nematode resistance to many available therapies necessitates the discovery and development of new drugs. Here, marine epiphytic bacteria represent a promising repository of newly discovered antinematode compounds. Epiphytic bacteria are ubiquitous on marine surfaces where they are under constant pressure of grazing by bacterivorous predators (e.g., protozoans and nematodes). Studies have shown that these bacteria have developed defense strategies to prevent grazers by producing toxic bioactive compounds. Although several active metabolites against nematodes have been identified from marine bacteria, drug discovery from marine microorganisms remains underexplored. In this review, we aim to provide further insight into the need and potential for marine epiphytic bacteria to become a new source of antinematode drugs. We discuss current and emerging strategies, including culture-independent high throughput screening and the utilization of *Caenorhabditis elegans* as a model target organism, which will be required to advance antinematode drug discovery and development from marine microbial sources.

Keywords: antinematode compound; anthelminthic drugs; marine epiphytic bacteria; marine biofilm; marine environment; parasitic nematode; *Caenorhabditis elegans*

1. Introduction

Humans are vulnerable to infectious diseases caused by parasitic helminths (nematodes) resulting in morbidity and mortality within the population [1–3]. Approximately 24% of the global human population, corresponding to 1.5 billion people, suffer from parasitic helminth infections [4]. There are almost 300 nematodes associated with zoonotic diseases that are able to infect humans, including some of the most devastating parasites such as *Ascaris lumbricoides* (roundworm), *Ancylostoma duodenale* (hookworm), *Gnathostoma spinigerum*, *Halicephalobus gingivalis* and *Trichinella spiralis* (Trichina worm) [1,5,6]. While some parasitic nematodes (e.g., *Ancylostoma duodenale*, *Strongyloides stercoralis* and *Halicephalobus gingivalis*) can penetrate human skin or invade through existing skin lesions [7,8], several parasites infect humans via ingestion of food products contaminated with the embryonated eggs (e.g., *Ascaris lumbricoides* and *Trichuris trichiura*) [9,10] or from eating raw or undercooked freshwater fish, birds, frogs or reptiles contaminated with the parasitic nematode larvae (e.g., *Gnathostoma spinigerum*, *Dracunculus medinensis*, *Eustrongylides* sp. and *Trichinella spiralis*) [11–13]. Apart from causing diseases, a high nematode burden also reduces human fecundity [14] and affects children through malnutrition, stunted development and cognitive delay [15,16].

Plant parasitic nematodes (PPNs) also cause diseases [17,18] either by damaging the root system, retarding the plant development or by exposing plants to secondary bacterial, fungal or viral infections [19–22]. It is estimated that damages caused by PPNs result in a >12% loss in global crop productivity and an average annual loss of ~US$215 billion [23]. Parasitic nematodes also impede the productivity of fisheries and aquaculture industries, resulting in worldwide economic losses and health hazards (i.e., zoonotic diseases or allergies) for consumers [24,25]. For example, the global financial loss in the finfish industry due to parasitic infection is estimated to be as large as US$134 million per annum [24]. Major fish products such as Atlantic mackerel, herring, European hake, Atlantic cod and anchovy are commonly associated with parasitic nematodes e.g., *Anisakis* sp. and *Pseudoterranova* sp. [26], which are transmitted to humans via ingestion of undercooked or raw fish [27,28]. Furthermore, parasitic nematodes cause different pathophysiological symptoms in livestock, reduce meat quality and cause animal mortality [29,30]. In Australia, the economic loss due to parasite infection (e.g., *Ostertagia* sp. and *Trichostrongylus* sp.) and the cost of control management is estimated to be ~AU$1 billion [31–33] whereas in Kenya, South Africa and India, ~US$26, 46 and 103 million is spent just to control the *Haemonchus contortus* nematode infection of livestock, respectively [34,35]. While diminishing animal fecundity, nematode parasites such as *Ostertagia ostertagi* (brown stomach worm) and *Dictyocaulus viviparus* (lungworm) reduce dairy production levels (~1.25 L/day/animal or 1.6 L/day/cow, respectively) [36,37].

Currently, the most promising longer-term control strategy is dependent on pharmaceutically derived chemotherapeutic treatments to kill parasitic nematodes and/or mitigate the spread of infection [38,39]. Unfortunately, the emergence of drug resistance among parasitic nematodes, due to prolonged treatment and incorrect drug dosage with an unpredictable infection trend as a result of climate change, is alarming [40–44]. The current prevalence of drug resistance is not only limited to the older classes of antinematode drugs but also those introduced in recent years. For example, monepantel resistance in *T. circumcincta*, *Trichostrongylus colubriformis* and *H. contortus* has been reported in New Zealand and the Netherlands within just five years of launching the drug [45–47].

Drug resistance has also been reported for almost all of the currently available anthelmintic drugs and nematicide classes including piperazine, benzimidazoles, levamisole, (pyrantel and morantel), paraherquamide, ivermectin (macrocylic lactones and milbemycins), emodepside (cyclodepsipeptides, PF1022A) and nitazoxanide [47–49].

Given the impacts of resistant parasitic nematodes to human and economic growth, novel antinematode chemotherapeutic agents are urgently needed as a preventive control against parasite infestation [4,50]. For decades, microorganisms, particularly bacteria, have served as a precious source for bioactive compounds, some of which have been developed into novel drugs including those with nematicidal activity [51,52]. This review highlights the potential of marine epiphytic bacteria as a new repository for newly discovered antinematode metabolites and the underlying mechanism of antinematode compound production by marine microbial biofilms. Furthermore, the utilization of *C. elegans* as a surrogate organism for antinematode drug development will be reviewed.

2. Antinematode Drug Discovery: Transition from Terrestrial to Marine-Derived Microbial Compounds

Terrestrial plants and plant extracts have been documented as an ancient therapeutic treatment against parasitic nematodes [53] and today, extensive studies to isolate plant-derived nematicidal compounds are still ongoing [54,55]. However, the re-discovery rate of bioactive metabolites is high [4], reducing the number of newly discovered compounds in the drug discovery pipeline [56]. Moreover, external factors, e.g., specific planting season, environmental temperature and humidity, may also affect the compounds' reproducibility by the plant producers [4]. Given those inevitable challenges, microbial-derived anthelmintic compounds offer a promising solution [57].

Extensive exploration of terrestrial microbial compounds for therapeutic drug development was initiated in the 20th century [53]. Since then, more than 50,000 beneficial bioactive metabolites have been successfully identified [58,59] some of which have potent nematotoxicity via distinct modes of action (MOAs) (Table 1). Unfortunately, after almost 50 years of drug screening, few newly discovered compounds have been identified from terrestrial-borne microorganisms [59] hence the requirement for a new source of antinematode drug discovery to combat the rapidly growing nematode resistance. Here, the underexplored marine ecosystem, with highly diverse unidentified macro- and microorganisms, represents a new repository for novel nematicidal drugs [59]. In fact, marine bioactive compounds have been acknowledged as having substantial chemical novelty compared to terrestrial metabolites [60].

The marine environment represents the largest biome on the earth (70% of the earth surface; ~361 million km^2 with average depth of 3730 m) and provides a habitat for a wide diversity of life that outnumbers terrestrial environments [61–64]. Microorganisms are abundant in this ecosystem (10^5 to 10^6 of cells per milliliter), reaching an average 10^{28} to 10^{29} cells either in the open ocean, deep sea, sediment or on the subsurface [64–66]. However, the majority of marine bacteria remain unrecognized or uncultivable [67]. Given the enormous diversity and untapped bioactive potential, marine microorganisms are likely to produce newly discovered bioactive compounds with a well-defined molecular architecture and biological function including nematicidal activities [68,69].

Table 1. Examples of nematicidal compounds produced by bacteria isolated from terrestrial environments and their modes of action (MOAs) against the target nematodes.

Microbial Producer	Compound	Mode of Action	Target Nematode	Affected Nematode Region	Reference
Bacillus thuringiensis	Crystal toxin Cry5B, Cry21A	Toxin binds to nematode glycoconjugate receptor and disrupt the intestinal cells membrane integrity. This action causes fall of nematode brood size and mortality	*Ancylostoma ceylanicum*, *Ascaris suum*, *C. elegans*	Gastrointestinal system	[70–72]
Bacillus simplex, *B. subtilis*, *B. weihenstephanensis*, *Microbacterium oxydans*, *Stenotrophomonas maltophilia*, *Streptomyces lateritius* and *Serratia marcescens*	Volatile organic compound (VOC) i.e., benzaldehyde, benzeneacetaldehyde, decanal, 2-nonanone, 2-undecanone, cyclohexene and dimethyl disulfide	VOCs reduce nematode motility and death	*Panagrellus redivivus*, *Bursaphelenchus xylophilus*	Unknown	[73]
Streptomyces avermectinius	Avermectin and ivermectin (semi-synthetic)	Compound exposure resulted in pharyngeal paralysis and nematode death	*Haemonchus contortus*, *Brugia malayi*, *C. elegans*	Neuromuscular system	[48,74–76]
Serratia marcescens	Prodigiosin	Compound is toxic against juveniles larvae and inhibit egg hatching competency	*Radopholus similis*, *Meloidogyne javanica*	Unknown	[77]
Pseudomonas aeruginosa	Phenazine toxin (phenazine-1-carboxylic, pyocyanin and 1-hydroxyphenazine)	Phenazine-1-carboxylic shows fast killing activity against nematode in acidic environment whilst pyocyanin is toxic in neutral or basic pH. The toxicity of 1-hydroxyphenazine is not dependent on environmental pH. Continuous exposure to phenazine affects protein homeostasis and causes neurodegeneration	*C. elegans*	Neuromuscular system, cell mitochondria and protein folding	[78,79]

Table 1. *Cont.*

Microbial Producer	Compound	Mode of Action	Target Nematode	Affected Nematode Region	Reference
Pseudomonas aeruginosa	Exotoxin A and other undetermined effectors	Slow-killing activity against nematode is based on infection-like process thus resulting in accumulation of bacteria in the gut. Continuous exposure leads to ceased pharyngeal pumping, nematode immobility and death	*C. elegans*	Gastrointestinal system	[80]
Pseudomonas aeruginosa	Chitinase enzyme	Chitinase degrades nematode cuticle, intestine and egg shell leading to the animal death	*C. elegans*	Cuticle, eggs, gastrointestinal system	[81]
Pseudomonas plecoglossicida	Glycolipid biosurfactant	Reduction of nematode development, survival and fecundity	*C. elegans*	Unknown	[82]

3. Surface Associated Marine Bacteria: A Reservoir for Novel Antimicrobial and Antinematode Drug Discovery

Marine inhabitants, particularly microorganisms, are continuously exposed to multiple detrimental interactions imposed by competitors and predators [83,84] and different physical–chemical variables such as fluctuating temperature, pH, UV exposure, salinity, toxic compounds and desiccation, particularly in the intertidal zone [85–87]. As a survival strategy, some marine bacteria adhere to each other and/or surfaces and are embedded in enclosed matrices to form a biofilm (Figure 1) [88,89].

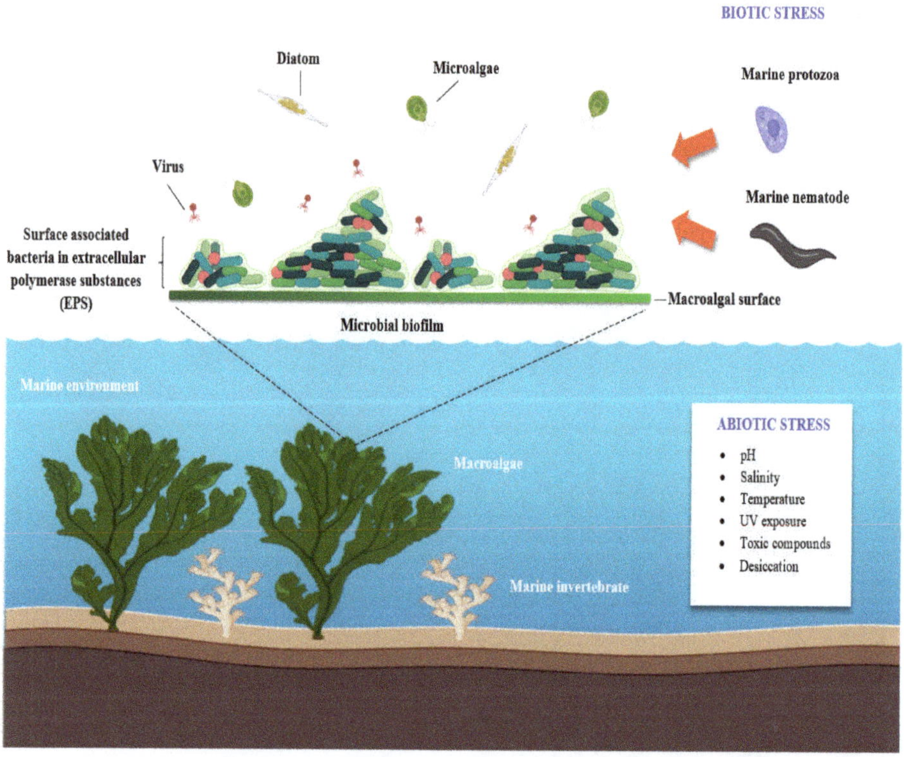

Figure 1. Surface-associated marine bacteria live on the nutrient-rich marine surfaces (for example macroalgae or cnidaria) in the form of biofilms. These marine biofilms are exposed to biotic (intra- and/or inter-species interaction with other microorganisms or predators i.e., protozoa and nematodes) and abiotic physical–chemical stressors. The predator–prey interactions lead to the production of nematicidal metabolites by the surface-associated marine bacteria, while the harsh environmental conditions enhance the chemical, molecular and functional properties of the produced microbial compounds. Image created with BioRender.com.

The continuous development of biofilm on marine surfaces leads to epibiosis, which involves multispecies biofilm formation [90,91]. Macroalgal surfaces, for example, are a hot-spot for colonization by opportunistic epibionts such as algal spores, invertebrate larvae, diatoms, fungi and other bacteria [85,92], mostly due to the accumulation of nutrients and macroalgal exudates composed of organic carbon and nitrogen particles [93–95]. Consequently, the competition among marine microorganisms to reserve a space within the biofilm community is tremendously intense and bacterial strains that are equipped with broad-spectrum inhibitory phenotypes are likely to be successful epibiotic colonizers (Figure 1) [83,96].

In addition, predation by heterotrophic protozoa and bacterivorous nematodes represents another biotic stress resulting in major mortality for both planktonic and surface-associated bacteria in the marine habitat (Figure 1) [85,97]. Protozoans, for example *Rhynchomonas nasuta* and *Cafeteria roenbergensis*, are among the most abundant ubiquitous species in the ocean and are the major controllers of the food web in the marine environment through their function as bacterial predators [98–100]. Nematodes such as *Pareudiplogaster pararmatus* are among the natural consumers of organic biomass in benthic habitats actively grazing bacterial mats and the biofilms of biotic surfaces [101,102].

The omnipresence of inter- and intra-species interactions supports the evolution of diverse defense strategies by surface-associated marine bacteria. Such defense mechanisms include the production of bioactive compounds showing antibacterial, antifungal, antitumor, antifouling, antiprotozoal and antinematode activities (Table 2) [103,104]. Interestingly, the physical–chemical properties, molecular structure and functional features of those marine microbial compounds are believed to be shaped by the naturally harsh conditions of the marine environment [105]. Moreover, it is speculated that bioactive metabolites originally attributed to marine invertebrates such as sponges, tunicates, bryozoans and molluscs are actually produced by their associated microorganisms [106,107]. For example, the antibiotic peptides andrimid and trisindoline isolated from the sponge *Hyatella* sp. and *Hyrtios altum* are believed to be produced by symbiotic *Vibrio* sp. [108,109]. An antitumor cyclic peptide leucamide A isolated from the sponge *Leucetta microraphis* is closely related to compounds produced by cyanobacterial symbionts [110]. In addition, a commercialized antitumor drug Didemnin B initially isolated from the tunicate *Trididemnum solidum* [111] was recently demonstrated to be produced by symbiotic bacteria *Tistrella mobilis* and *T. bauzanensis* [112,113]. These observations also hold true for numerous antinematode compounds that have been successfully isolated from marine eukaryotes, with production of many now attributed to host-associated microorganisms [51,114].

The potential for marine surface-associated microorganisms to be repositories of unusual gene functions and bioactivities is further corroborated by global biodiversity studies such as the Tara Oceans [69,115] and the Global Ocean Sampling (GOS) expedition [116,117]. These and other studies continue to reveal unprecedented levels of information on microbial diversity, which has subsequently led to investigations on underexplored bacterial diversity in various marine ecosystems. Furthermore, abundant and newly discovered biosynthetic gene clusters encoding rare non-ribosomal peptides (NRPS), polyketides (PKS) and NRPS–PKS hybrids have been discovered from marine biofilm samples, further strengthening the concept of the marine environment as a rich source of newly discovered bioactive compounds [68,69].

Table 2. Example of anthelmintic or nematicidal bioactivities isolated from marine bacteria.

Marine Microbial Producer	Compound	Associated Surface/Host	Mode of Action	Responsible Gene(S)	Reference
Microbulbifer sp. D250	Violacein	Algae *Delisea pulchra*	Facilitate bacterial accumulation accompanied by tissue damage and apoptosis	*VioA-VioE*	[118]
Pseudoalteromonas tunicata D2	Tambjamine	Algae *Ulva australis*	Slow-killing activity by a heat-resistant tambjamine and substantial bacterial colonization in the nematode gut	*TamA-TamT*	[119]
Pseudoalteromonas tunicata D2	Unknown	Algae *Ulva australis*	Fast-killing activity by a heat sensitive unknown compound through colonization-independent manner	Unknown	[119]
Uncultured alpha-proteobacterium, JN874385 (strain U95)	Unknown	Algae *Ulva australis*	Undetermined	Possibly NRPS gene	[120]
Aequorivita sp.	Unknown	Antarctic marine sediment	Undetermined	Unknown	[121]
Pseudovibrio sp. Pv348, 1413, HE818384 (strain D323)	Unknown	Algae *Delisea pulchra*	Undetermined	Unknown	[120]
Heterologous clone jj117 (NCBI Accession number SRX4339430)	Unknown	*Ulva australis* metagenomic library	Undetermined	ATP-grasp protein/alpha-E protein/transglutaminase protein/protease	[122]
Vibrio atlanticus strain S-16	Volatile organic compounds (VOC)	Scallop *Argopecten irradians*	Undetermined	Unknown	[123]
Virgibacillus dokdonensis MCCC 1A00493	VOC (acetaldehyde, dimethyl disulfide, ethylbenzene and 2-butanone	Polymetallic nodules in the deep sea	Direct contact killing activity, fumigation. Acetaldehyde had a fumigant activity to impede egg hatching	Unknown	[124]

Table 2. *Cont.*

Marine Microbial Producer	Compound	Associated Surface/Host	Mode of Action	Responsible Gene(S)	Reference
Pseudoalteromonas rubra	Unknown	Marine organisms (copepod or fish) or environmental samples	Undetermined	Unknown	[125,126]
Pseudoalteromonas piscicida	Unknown	Marine organisms (copepod or fish) or environmental samples	Undetermined	Unknown	[125,126]
Arthrobacter davidanieli	Unknown	Marine environmental sample	Undetermined	Unknown	[125,127]
Pseudoalteromonas luteoviolacea	Unknown	Marine organisms (copepod or fish) or environmental samples	Undetermined	Unknown	[125,126]
Photobacterium halotolerans	Unknown	Marine organisms (copepod or fish) or environmental samples	Undetermined	Unknown	[125,126]

4. *Caenorhabditis elegans* as a Model Organism for Antinematode Drug Discovery and Development

Antinematode drug research is impeded by several challenges. These include (i) similarity of biochemical reactions between parasitic nematodes and the infected host, (ii) complex parasite life cycles that involve infections in multiple hosts, (iii) different parasite geographical locations and (iv) rapid development of resistant phenotypes. Therefore, an easily maintained nematode model with the capability for rapid screening of potential antinematode compounds represents a solution to some of these challenges [128,129].

Sydney Brenner and colleagues first introduced the free-living soil nematode *C. elegans* in 1965 as an animal model for research including anti-infective and antinematode drug studies [130–135]. Owing to its small size (1–1.5 mm-adult length, 80 μm-diameter), transparency, rapid life cycle (~3 days) and diet based on a simple bacterial culture of *Escherichia coli* OP50, *C. elegans* has emerged as a valuable animal model [134,136,137]. *C. elegans* possess several features which make the organism an efficient and low-cost surrogate organism for antinematode drug discovery. Unlike parasitic nematodes which require vertebrate hosts for reproduction and maintenance [48,138], synchronized *C. elegans* can be easily propagated at the desired life stage on Nematode Growth Media (NGM) ready to be used for antinematode drug studies [139]. The earliest antinematode drug testing using synchronized *C. elegans* individuals was performed by exposing the animals to nematicidal agents incorporated into the agar media [140]. After a few years, the screening protocol evolved rapidly with the development of high throughput screening methods employing micro-fluidic systems or high-content screening (HCS) technologies, allowing for fast and large-scale drug testing (~14,000 to 360,000 of compounds) using *C. elegans* as the model organism (Figure 2) [141,142]. Today, important drug discoveries, such as benzimidazoles [143], ivermectin and its analogues, moxidectin, milbemycin oxime, doramectin, selamectin, abamectin, eprinomectin [144] and the nematicidal activity of crystal protein insecticide Cry5B, Cry21A from *Bacillus thuringiensis* [72,145], can be attributed to the use of *C. elegans* as an effective animal model [48].

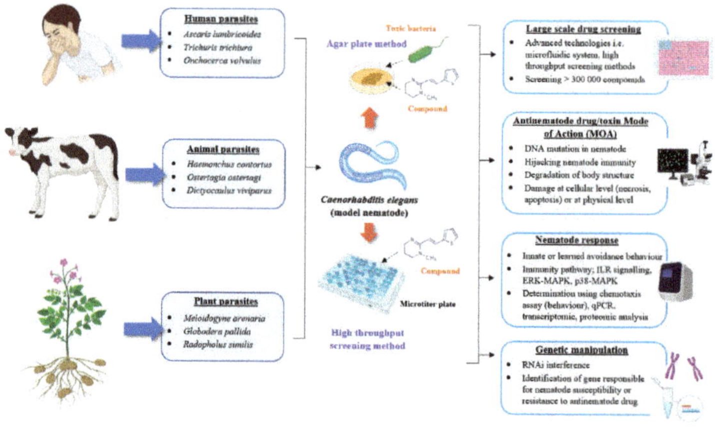

Figure 2. Schematic diagram representing the functional role of *C. elegans* as a model nematode for the development of antinematode drugs. The propagation and handling methods of human, animal or plant-parasitic nematodes in the laboratory are challenging. *C. elegans* can be used as a surrogate nematode for the initial screening against the potential compounds or microorganisms with nematotoxic properties either via the conservative agar plate method or through high throughput screening technologies. *C. elegans*-based research offers several advantages including nematode genetic manipulation, determination of drug MOA, evaluation of the resulting nematode responses and a large-scale initial drug screening due to easy nematode maintenance and propagation in the laboratory. Image created with BioRender.com.

C. elegans shares many conserved genes and protein functions with parasitic nematodes. Analysis of the intestinal parasite *Strongyloides stercoralis* genetic sequences showed 85% of protein homologs to *C. elegans* genes. The infective stage of *S. stercoralis* (L3i/dauer) also shows an increased proportion of protein homologs to *C. elegans* dauer larvae [146]. Genetic manipulation and RNA interference (RNAi) studies have been widely performed on *C. elegans* to provide a better understanding of the nematode response against the nematicidal drugs at the molecular level (Figure 2). Quantitative polymerase chain reaction (qPCR) and "omic" technologies, e.g., transcriptomic profiling and proteomics, are also being used to provide a global snapshot of the molecular response of *C. elegans* to drug exposure [147,148]. Given the conserved gene homologs and protein function among the members of phylum Nematoda, these studies provide insight into the possible mechanisms used by parasitic nematodes against similar drug exposure [48].

The majority of the current drugs used to treat parasitic nematode infections target proteins that regulate neuromuscular activity including neurotransmitter receptors and ion channels [48]. Utilization of *C. elegans* as a model organism confers a better understanding of the mode of action (MOA) of potential nematicidal drugs (Figure 2). Studies have shown that *C. elegans'* neuromuscular system, its major neurotransmitters GABA (4-aminobutyric acid) and glutamate and its enzyme choline acetyltransferase, responsible for the synthesis of the neurotransmitter acetylcholine, display strong similarities to parasitic roundworms *Ascaris suum* and *Ascaris lumbricoides* [48,149–152]. Therefore, exposure of *C. elegans* to antinematode compounds which target the neurotransmitter receptor and ion channels has enabled the discovery of the target binding molecule and the resulting toxicity to parasitic nematodes [153]. For example, observations of body muscle contraction and spastic paralysis in *C. elegans* exposed to levamisole, led Lewis and colleagues [154] to determine that binding to the muscle acetylcholine receptors was key to its activity. The initiation of amino-acetonitrile derivative (AAD) toxic activity against nematodes by binding to a nicotinic acetylcholine receptor was also revealed via forwards genetic screening using *C. elegans* mutants [155]. More recently, the MOA of paraherquamide, a broad-spectrum nicotinic nematicidal alkaloid isolated from *Penicillium paraherquei* [156], and antinematode plant-derived compounds [157] were also elucidated using *C. elegans* [157–159]. While using *C. elegans* to identify specific targets is promising, assessing side effects related to these newly discovered compounds will be important considering many of the neurotransmitter targets will also be present in the host organism [48]. Nevertheless, in the event that novel drugs negatively impact non-target organisms, derivatives may be developed with reduced side effects [160,161].

5. Conclusions

The increasing prevalence of parasitic nematode infection and the emergence of antinematode drug resistance represent critical global issues, impacting human wellbeing and economic development. Moreover, increasing temperatures and changing weather patterns, moisture and rainfall as a result of global climate change may also escalate the prevalence of parasitic nematode diseases worldwide [40–44]. Although for decades many anthelmintic/nematicidal drugs have been successfully derived from terrestrial microorganisms (see examples in Table 1), the paucity of new antinematode drug classes and the challenges now presented with drug resistance underline the importance of discovering new antinematode bioactive metabolites from natural sources [4,50]. This review highlighed the potential of marine epiphytic bacteria as a new platform for novel antinematode drug development. Marine epiphytic bacteria are highly diverse, harboring unique genes expressing newly discovered bioactive metabolites with commercially or pharmaceutically relevant biological potentials including antinematode activities [118–120]. Owing to the extraordinary molecular structure and physical/chemical properties, marine bioactive compounds are regarded as "blue gold from the ocean" and are believed to be a promising source for future novel antinematode drugs. The ability to uncover these novel marine-derived antinematode drugs will be dependent on the successful implementation of innovative cultures, culture-independent techniques and high-throughput bioassays, for which the model nematode *C. elegans* is well suited.

Author Contributions: N.H.S., J.N., M.E.M. and S.E. developed the conceptual ideas. N.H.S. wrote the initial draft. All authors have read and agreed to the published version of the manuscript.

Funding: This research received no external funding.

Acknowledgments: Authors would like to thank CMSI staff and students for helpful discussions. Nor Hawani Salikin was supported by the Ministry of Higher Education Malaysia and Universiti Sains Malaysia under the Academic Staff Training Scheme Fellowship.

Conflicts of Interest: The authors declare no conflict of interest.

References

1. Cox, F.E.G. History of human parasitic diseases. *Infect. Dis. Clin. N. Am.* **2004**, *18*, 171–188. [CrossRef] [PubMed]
2. Bañuls, A.-L.; Thomas, F.; Renaud, F. Of parasites and men. *Infect. Genet. Evol.* **2013**, *20*, 61–70. [CrossRef] [PubMed]
3. Viney, M. How can we understand the genomic basis of nematode parasitism? *Trends Parasitol.* **2017**, *33*, 444–452. [CrossRef] [PubMed]
4. Garcia-Bustos, J.F.; Sleebs, B.E.; Gasser, R.B. An appraisal of natural products active against parasitic nematodes of animals. *Parasites Vectors* **2019**, *12*, 306. [CrossRef] [PubMed]
5. Ashford, R.; Crewe, W. *Parasites of Homo Sapiens: An Annotated Checklist of the Protozoa, Helminths and Arthropods for Which We are Home*, 2nd ed.; Taylor & Francis: London, UK; New York, NY, USA, 2003; pp. 87–93.
6. Stoltzfus, D.J.; Pilgrim, A.A.; Herbert, D.B.R. Perusal of parasitic nematode 'omics in the post-genomic era. *Mol. Biochem. Parasitol.* **2017**, *215*, 11–22. [CrossRef] [PubMed]
7. Papadi, B.; Eberhard, M.E.; Boudreaux, C.; Bishop, H.; Mathison, B.; Tucker, J.A. *Halicephalobus gingivalis*: A rare cause of fatal meningoencephalomyelitis in humans. *Am. J. Trop. Med. Hyg.* **2013**, *88*, 1062–1064. [CrossRef]
8. Bryant, A.S.; Ruiz, F.; Gang, S.S.; Castelletto, M.L.; Lopez, J.B.; Hallem, E.A. A critical role for thermosensation in host seeking by skin-penetrating nematodes. *Curr. Biol.* **2018**, *28*, 2338–2347.e6. [CrossRef]
9. Bundy, D.A.; De Silva, N.; Appleby, L.J.; Brooker, S.J. Intestinal nematodes: Ascariasis. In *Hunter's Tropical Medicine and Emerging Infectious Diseases*, 10th ed.; Ryan, E.T., Hill, D.R., Solomon, T., Aronson, N.E., Endy, T.P., Eds.; Elsevier: Amsterdam, The Netherlands, 2020; pp. 840–844. [CrossRef]
10. Jourdan, P.M.; Lamberton, P.H.L.; Fenwick, A.; Addiss, D.G. Soil-transmitted helminth infections. *Lancet* **2018**, *391*, 252–265. [CrossRef]
11. Frean, J. Gnathostomiasis acquired by visitors to the Okavango Delta, Botswana. *Trop. Med. Infect. Dis.* **2020**, *5*, 39. [CrossRef]
12. Eberhard, M.L.; Cleveland, C.A.; Zirimwabagabo, H.; Yabsley, M.J.; Ouakou, P.T.; Ruiz-Tiben, E. Guinea worm (*Dracunculus medinensis*) infection in a wild-caught frog, Chad. *Emerg. Infect. Dis.* **2016**, *22*, 1961–1962. [CrossRef]
13. Magnino, S.; Colin, P.; Dei-Cas, E.; Madsen, M.; McLauchlin, J.; Nöckler, K.; Maradona, M.P.; Tsigarida, E.; Vanopdenbosch, E.; Van Peteghem, C. Biological risks associated with consumption of reptile products. *Int. J. Food Microbiol.* **2009**, *134*, 163–175. [CrossRef] [PubMed]
14. Persson, G.; Ekmann, J.R.; Hviid, T.V.F. Reflections upon immunological mechanisms involved in fertility, pregnancy and parasite infections. *J. Reprod. Immunol.* **2019**, *136*, 102610. [CrossRef] [PubMed]
15. Ezeamama, A.E.; McGarvey, S.T.; Olveda, R.M.; Acosta, L.P.; Kurtis, J.D.; Bellinger, D.C.; Langdon, G.C.; Manalo, D.L.; Friedman, J.F. Helminth infection and cognitive impairment among Filipino children. *Am. J. Trop. Med. Hyg.* **2005**, *72*, 540–548. [CrossRef] [PubMed]
16. Doni, N.Y.; Zeyrek, F.Y.; Simsek, Z.; Gurses, G.; Sahin, I. Risk factors and relationship between intestinal parasites and the growth retardation and psychomotor development delays of children in Şanlıurfa, Turkey. *Turk. Parazitolojii Derg.* **2015**, *39*, 270–276. [CrossRef] [PubMed]
17. Abad, P.; Gouzy, J.; Aury, J.; Castagnone-Sereno, P.; Danchin, E.G.J.; Deleury, E.; Perfus-Barbeoch, L.; Anthouard, V.; Artiguenave, F.; Blok, V.C.; et al. Genome sequence of the metazoan plant-parasitic nematode *Meloidogyne incognita*. *Nat. Biotechnol.* **2008**, *26*, 909–915. [CrossRef] [PubMed]

18. Bernard, G.C.; Egnin, M.; Bonsi, C. The impact of plant-parasitic nematodes on agriculture and methods of control. In *Nematology-Concepts, Diagnosis and Control*; Manjur-Shah, M., Mahamood, M., Eds.; IntechOpen: London, UK, 2017. [CrossRef]
19. Jones, J.T.; Haegeman, A.; Danchin, E.G.J.; Gaur, H.S.; Helder, J.; Jones, M.G.K.; Kikuchi, T.; Manzanilla-López, R.; Palomares-Rius, J.E.; Wesemael, W.M.L.; et al. Top 10 plant-parasitic nematodes in molecular plant pathology. *Mol. Plant Pathol.* **2013**, *14*, 946–961. [CrossRef]
20. Kassie, Y.G.; Ebrahim, A.S.; Mohamed, M.Y. Interaction effect between *Meloidogyne incognita* and *Fusarium oxysporum* f. sp. *lycopersici* on selected tomato (*Solanum lycopersicum* L.) genotypes. *Afr. J. Agric. Res.* **2020**, *15*, 330–342. [CrossRef]
21. Liu, W.; Park, S.-W. Underground mystery: Interactions between plant roots and parasitic nematodes. *Curr. Plant Biol.* **2018**, *15*, 25–29. [CrossRef]
22. Villate, L.; Morin, E.; Demangeat, G.; Van Helden, M.; Esmenjaud, D. Control of *Xiphinema index* populations by fallow plants under greenhouse and field conditions. *Phytopathology* **2012**, *102*, 627–634. [CrossRef]
23. Abd-Elgawad, M.; Askary, T.H. Impact of phytonematodes on agriculture economy. In *Biocontrol Agents of Phytonematodes*; Askary, T.H., Martinelli, P.R.P., Eds.; CAB International Publishing: Oxfordshire, UK; Boston, MA, USA, 2015; pp. 12–17.
24. Shinn, A.; Pratoomyot, J.; Bron, J.E.; Paladini, G.; Brooker, E.; Brooker, A.J. Economic costs of protistan and metazoan parasites to global mariculture. *Parasitology* **2015**, *142*, 196–270. [CrossRef]
25. Mehrdana, F.; Buchmann, K. Excretory/secretory products of anisakid nematodes: Biological and pathological roles. *Acta Vet. Scand.* **2017**, *59*. [CrossRef]
26. Levsen, A.; Svanevik, C.S.; Cipriani, P.; Mattiucci, S.; Gay, M.; Hastie, L.C.; Bušelić, I.; Mladineo, I.; Karl, H.; Ostermeyer, U.; et al. A survey of zoonotic nematodes of commercial key fish species from major European fishing grounds—Introducing the FP7 PARASITE exposure assessment study. *Fish. Res.* **2018**, *202*, 4–21. [CrossRef]
27. Aibinu, I.E.; Smooker, P.M.; Lopata, A.L. *Anisakis* nematodes in fish and shellfish- from infection to allergies. *Int. J. Parasitol. Parasites Wildl.* **2019**, *9*, 384–393. [CrossRef] [PubMed]
28. Shamsi, S. Seafood-borne parasitic diseases: A "one-health" approach is needed. *Fishes* **2019**, *4*, 9. [CrossRef]
29. Mushonga, B.; Habumugisha, D.; Kandiwa, E.; Madzingira, O.; Samkange, A.; Segwagwe, B.E.; Jaja, I.F. Prevalence of *Haemonchus contortus* infections in sheep and goats in Nyagatare District, Rwanda. *J. Vet. Med.* **2018**, *2018*, 3602081. [CrossRef] [PubMed]
30. Elseadawy, R.; Abbas, I.; Al-Araby, M.; Hildreth, M.B.; Abu-Elwafa, S. First evidence of *Teladorsagia circumcincta* infection in sheep from Egypt. *J. Parasitol.* **2019**, *105*, 484–490. [CrossRef]
31. McLeod, R. Costs of major parasites to the Australian livestock industries. *Int. J. Parasitol.* **1995**, *25*, 1363–1367. [CrossRef]
32. Roeber, F.; Jex, A.R.; Gasser, R.B. Impact of gastrointestinal parasitic nematodes of sheep, and the role of advanced molecular tools for exploring epidemiology and drug resistance—An Australian perspective. *Parasites Vectors* **2013**, *6*, 153. [CrossRef]
33. Sackett, D.; Sackett, H.; Abbott, K.; Barber, M. Assessing the Economic Cost of Endemic Disease on the Profitability of Australian Beef Cattle and Sheep Producers. *Meat & Livestock Australia Report AHW*. 2006. Available online: https://www.mla.com.au/Research-and-development/Search-RD-reports/RD-report-details/Animal-Health-and-Biosecurity/Assessing-the-economic-cost-of-endemic-disease-on-the-profitability-of-Australian-beef-cattle-and-sheep-producers/120 (accessed on 16 May 2020).
34. McLeod, R.S. Economic impact of worm infestations in small ruminants in South East Asia, India and Austrailia. In *Worm Control of Small Ruminants in Tropical Asia*; Sani, R.A., Gray, G.D., Baker, R.L., Eds.; ACIAR Monograph: Canberra, Australia, 2004; Volume 113, pp. 23–33.
35. Maqbool, I.; Wani, Z.A.; Shahardar, R.A.; Allaie, I.M.; Shah, M.M. Integrated parasite management with special reference to gastro-intestinal nematodes. *J. Parasit. Dis.* **2017**, *41*, 1–8. [CrossRef]
36. Rashid, M.; Akbar, H.; Ahmad, L.; Hassan, M.A.; Ashraf, K.; Saeed, K.; Gharbi, M. A systematic review on modelling approaches for economic losses studies caused by parasites and their associated diseases in cattle. *Parasitology* **2019**, *146*, 129–141. [CrossRef]
37. May, K.; Brügemann, K.; König, S.; Strube, C. The effect of patent *Dictyocaulus viviparus* (re)infections on individual milk yield and milk quality in pastured dairy cows and correlation with clinical signs. *Parasites Vectors* **2018**, *11*, 24. [CrossRef] [PubMed]

38. Abongwa, M.; Martin, R.J.; Robertson, A.P. A brief review on the mode of action of antinematodal drugs. *Acta Vet. Beogr.* **2017**, *67*, 137–152. [CrossRef] [PubMed]
39. Zajíčková, M.; Nguyen, L.T.; Skálová, L.; Stuchlíková, L.R.; Matoušková, P. Anthelmintics in the future: Current trends in the discovery and development of new drugs against gastrointestinal nematodes. *Drug Discov. Today* **2020**, *25*, 430–437. [CrossRef] [PubMed]
40. Fox, N.J.; Marion, G.; Davidson, R.S.; White, P.C.L.; Hutchings, M.R. Climate-driven tipping-points could lead to sudden, high-intensity parasite outbreaks. *R. Soc. Open Sci.* **2015**, *2*, 140296. [CrossRef]
41. Verschave, S.H.; Charlier, J.; Rose, H.; Claerebout, E.; Morgan, E.R. Cattle and nematodes under global change: Transmission models as an ally. *Trends Parasitol.* **2016**, *32*, 724–738. [CrossRef] [PubMed]
42. Morgan, E.R.; Aziz, N.A.A.; Blanchard, A.; Charlier, J.; Charvet, C.; Claerebout, E.; Geldhof, P.; Greer, A.W.; Hertzberg, H.; Hodgkinson, J.; et al. 100 questions in livestock helminthology research. *Trends Parasitol.* **2019**, *35*, 52–71. [CrossRef]
43. Genchi, C.; Rinaldi, L.; Mortarino, M.; Genchi, M.; Cringoli, G. Climate and *Dirofilaria* infection in Europe. *Vet. Parasitol.* **2009**, *163*, 286–292. [CrossRef]
44. McIntyre, K.M.; Setzkorn, C.; Hepworth, P.J.; Morand, S.; Morse, A.P.; Baylis, M. Systematic assessment of the climate sensitivity of important human and domestic animals pathogens in Europe. *Sci. Rep.* **2017**, *7*, 7134. [CrossRef]
45. Scott, I.; Pomroy, W.; Kenyon, P.; Smith, G.; Adlington, B.; Moss, A. Lack of efficacy of monepantel against *Teladorsagia circumcincta* and *Trichostrongylus colubriformis*. *Vet. Parasitol.* **2013**, *198*, 166–171. [CrossRef]
46. Brom, R.V.D.; Moll, L.; Kappert, C.; Vellema, P. *Haemonchus contortus* resistance to monepantel in sheep. *Vet. Parasitol.* **2015**, *209*, 278–280. [CrossRef]
47. Ploeger, H.W.; Everts, R.R. Alarming levels of anthelmintic resistance against gastrointestinal nematodes in sheep in the Netherlands. *Vet. Parasitol.* **2018**, *262*, 11–15. [CrossRef] [PubMed]
48. Holden-Dye, L.; Walker, R.J. Anthelmintic drugs and nematocides: Studies in *Caenorhabditis elegans*. In *WormBook: The Online Review of C. elegans Biology*; The *C. elegans* Research Community, Ed.; WormBook, 2014; Available online: http://www.wormbook.org/chapters/www_anthelminticdrugs.2/anthelminticdrugs.2.html (accessed on 24 June 2020).
49. Idris, A.O.; Wintola, O.A.; Afolayan, A.J. Helminthiases; prevalence, transmission, host-parasite interactions, resistance to common synthetic drugs and treatment. *Heliyon* **2019**, *5*, e01161. [CrossRef] [PubMed]
50. Elfawal, M.A.; Savinov, S.N.; Aroian, R.V. Drug screening for discovery of broad-spectrum agents for soil-transmitted nematodes. *Sci. Rep.* **2019**, *9*, 12347. [CrossRef] [PubMed]
51. Sekurova, O.N.; Schneider, O.; Zotchev, S.B. Novel bioactive natural products from bacteria via bioprospecting, genome mining and metabolic engineering. *Microb. Biotechnol.* **2019**, *12*, 828–844. [CrossRef]
52. Deng, Q.; Zhou, L.; Luo, M.; Deng, Z.; Zhao, C. Heterologous expression of Avermectins biosynthetic gene cluster by construction of a Bacterial Artificial Chromosome library of the producers. *Synth. Syst. Biotechnol.* **2017**, *2*, 59–64. [CrossRef]
53. Monaghan, R.L.; Tkacz, J.S. Bioactive microbial products: Focus upon mechanism of action. *Annu. Rev. Microbiol.* **1990**, *44*, 271–331. [CrossRef]
54. Giovanelli, F.; Mattellini, M.; Fichi, G.; Flamini, G.; Perrucci, S. In vitro anthelmintic activity of four plant-derived compounds against sheep gastrointestinal nematodes. *Vet. Sci.* **2018**, *5*, 78. [CrossRef]
55. Spiegler, V.; Liebau, E.; Hensel, A. Medicinal plant extracts and plant-derived polyphenols with anthelmintic activity against intestinal nematodes. *Nat. Prod. Rep.* **2017**, *34*, 627–643. [CrossRef]
56. Gaudêncio, S.P.; Pereira, F. Dereplication: Racing to speed up the natural products discovery process. *Nat. Prod. Rep.* **2015**, *32*, 779–810. [CrossRef]
57. Shalaby, H.; Ashry, H.; Saad, M.; Farag, T. In vitro effects of *Streptomyces* tyrosinase on the egg and dult worm of *Toxocara vitulorum*. *Iran. J. Parasitol.* **2020**, *15*, 67–75.
58. Bérdy, J. Bioactive microbial metabolites. *J. Antibiot.* **2005**, *58*, 1–26. [CrossRef] [PubMed]
59. Xiong, Z.-Q.; Wang, J.-F.; Hao, Y.-Y.; Wang, Y. Recent advances in the discovery and development of marine microbial natural products. *Mar. Drugs* **2013**, *11*, 700–717. [CrossRef] [PubMed]
60. Kong, D.-X.; Jiang, Y.-Y.; Zhang, H.-Y. Marine natural products as sources of novel scaffolds: Achievement and concern. *Dug Discov. Today* **2010**, *15*, 884–886. [CrossRef] [PubMed]

61. Costello, M.J.; Cheung, A.; De Hauwere, N. Surface area and the seabed area, volume, depth, slope, and topographic variation for the world's seas, oceans, and countries. *Environ. Sci. Technol.* **2010**, *44*, 8821–8828. [CrossRef] [PubMed]
62. Fenical, W.; Jensen, P.R. Developing a new resource for drug discovery: Marine actinomycete bacteria. *Nat. Chem. Biol.* **2006**, *2*, 666–673. [CrossRef]
63. Kanase, H.; Singh, K.N. Marine pharmacology: Potential, challenges, and future in India. *J. Med. Sci.* **2018**, *38*, 49–53. [CrossRef]
64. Flemming, H.-C.; Wuertz, S. Bacteria and archaea on earth and their abundance in biofilms. *Nat. Rev. Microbiol.* **2019**, *17*, 247–260. [CrossRef]
65. Bar-On, Y.M.; Phillips, R.; Milo, R. The biomass distribution on earth. *Proc. Natl. Acad. Sci. USA* **2018**, *115*, 6506–6511. [CrossRef]
66. Magnabosco, C.; Lin, L.-H.; Dong, H.; Bomberg, M.; Ghiorse, W.; Stan-Lotter, H.; Pedersen, K.; Kieft, T.L.; Van Heerden, E.; Onstott, T.C. The biomass and biodiversity of the continental subsurface. *Nat. Geosci.* **2018**, *11*, 707–717. [CrossRef]
67. Pascoal, F.; Magalhães, C.; Costa, R. The link between the ecology of the prokaryotic rare biosphere and its biotechnological potential. *Front. Microbiol.* **2020**, *11*, 231. [CrossRef]
68. Blockley, A.; Elliott, D.R.; Roberts, A.P.; Sweet, M.J. Symbiotic microbes from marine invertebrates: Driving a new era of natural product drug discovery. *Diversity* **2017**, *9*, 49. [CrossRef]
69. Zhang, W.; Ding, W.; Li, Y.-X.; Tam, C.; Bougouffa, S.; Wang, R.; Pei, B.; Chiang, H.; Leung, P.; Lu, Y.; et al. Marine biofilms constitute a bank of hidden microbial diversity and functional potential. *Nat. Commun.* **2019**, *10*, 1–10. [CrossRef] [PubMed]
70. Conlan, J.V.; Thompson, R.C.A.; Khamlome, B.; Pallant, L.; Fenwick, S.; Elliot, A.; Sripa, B.; Blacksell, S.D.; Vongxay, K. Soil-transmitted helminthiasis in Laos: A community-wide cross-sectional study of humans and dogs in a mass drug administration environment. *Am. J. Trop. Med. Hyg.* **2012**, *86*, 624–634. [CrossRef] [PubMed]
71. Urban, J.F., Jr.; Hu, Y.; Miller, M.M.; Scheib, U.; Yiu, Y.Y.; Aroian, R.V. *Bacillus thuringiensis*-derived Cry5B has potent anthelmintic activity against *Ascaris suum*. *PLoS Negl. Trop. Dis.* **2013**, *7*, e2263. [CrossRef] [PubMed]
72. Wei, J.-Z.; Hale, K.; Carta, L.; Platzer, E.; Wong, C.; Fang, S.-C.; Aroian, R.V. *Bacillus thuringiensis* crystal proteins that target nematodes. *Proc. Natl. Acad. Sci. USA* **2003**, *100*, 2760–2765. [CrossRef]
73. Gu, Y.-Q.; Mo, M.-H.; Zhou, J.-P.; Zou, C.-S.; Zhang, K.-Q. Evaluation and identification of potential organic nematicidal volatiles from soil bacteria. *Soil Biol. Biochem.* **2007**, *39*, 2567–2575. [CrossRef]
74. Burg, R.W.; Miller, B.M.; Baker, E.E.; Birnbaum, J.; Currie, S.A.; Hartman, R.; Kong, Y.-L.; Monaghan, R.L.; Olson, G.; Putter, I.; et al. Avermectins, new family of potent anthelmintic agents: Producing organism and fermentation. *Antimicrob. Agents Chemother.* **1979**, *15*, 361–367. [CrossRef]
75. Omura, S. Ivermectin: 25 years and still going strong. *Int. J. Antimicrob. Agents* **2008**, *31*, 91–98. [CrossRef]
76. Campbell, C.W. History of avermectin and ivermectin, with notes on the history of other macrocyclic lactone antiparasitic agents. *Curr. Pharm. Biotechnol.* **2012**, *13*, 853–865. [CrossRef]
77. Rahul, S.; Chandrashekhar, P.; Hemant, B.; Chandrakant, N.; Laxmikant, S.; Satish, P. Nematicidal activity of microbial pigment from *Serratia marcescens*. *Nat. Prod. Res.* **2014**, *28*, 1399–1404. [CrossRef]
78. Cezairliyan, B.; Vinayavekhin, N.; Grenfell-Lee, D.; Yuen, G.J.; Saghatelian, A.; Ausubel, F.M. Identification of *Pseudomonas aeruginosa* Phenazines that kill *Caenorhabditis elegans*. *PLoS Pathog.* **2013**, *9*, e1003101. [CrossRef] [PubMed]
79. Ray, A.; Rentas, C.; Caldwell, G.A.; Caldwell, K.A. Phenazine derivatives cause proteotoxicity and stress in *C. elegans*. *Neurosci. Lett.* **2015**, *584*, 23–27. [CrossRef] [PubMed]
80. Tan, M.-W.; Mahajan-Miklos, S.; Ausubel, F.M. Killing of *Caenorhabditis elegans* by *Pseudomonas aeruginosa* used to model mammalian bacterial pathogenesis. *Proc. Natl. Acad. Sci. USA* **1999**, *96*, 715–720. [CrossRef] [PubMed]
81. Chen, L.; Jiang, H.; Cheng, Q.; Chen, J.; Wu, G.; Kumar, A.; Sun, M.; Liu, Z. Enhanced nematicidal potential of the chitinase pachi from *Pseudomonas aeruginosa* in association with Cry21Aa. *Sci. Rep.* **2015**, *5*, 14395. [CrossRef] [PubMed]
82. Sabarinathan, D.; Amirthalingam, M.; Sabapathy, P.C.; Govindan, S.; Palanisamy, S.; Kathirvel, P. Anthelmintic efficacy of glycolipid biosurfactant produced by *Pseudomonas plecoglossicida*: An insight from mutant and transgenic forms of *Caenorhabditis elegans*. *Biodegradation* **2019**, *30*, 203–214. [CrossRef]

83. Rao, D.; Webb, J.S.; Kjelleberg, S. Competitive interactions in mixed-species biofilms containing the marine bacterium *Pseudoalteromonas tunicata*. *Appl. Environ. Microbiol.* **2005**, *71*, 1729–1736. [CrossRef] [PubMed]
84. Welsh, R.M.; Zaneveld, J.R.; Rosales, S.M.; Payet, J.P.; Burkepile, D.E.; Thurber, R.V. Bacterial predation in a marine host-associated microbiome. *ISME J.* **2016**, *10*, 1540–1544. [CrossRef]
85. de Carvalho, C.C. Marine biofilms: A successful microbial strategy with economic implications. *Front. Mar. Sci.* **2018**, *5*, 126. [CrossRef]
86. Chiu, J.M.Y.; Thiyagarajan, V.; Tsoi, M.M.Y.; Qian, P.Y. Qualitative and quantitative changes in marine biofilms as a function of temperature and salinity in summer and winter. *Biofilms* **2005**, *2*, 183–195. [CrossRef]
87. Ortega-Morales, B.O.; Chan-Bacab, M.J.; Rosa-García, S.D.C.D.L.; Camacho-Chab, J.C. Valuable processes and products from marine intertidal microbial communities. *Curr. Opin. Biotechnol.* **2010**, *21*, 346–352. [CrossRef]
88. Antunes, J.; Leão, P.; Vasconcelos, V. Marine biofilms: Diversity of communities and of chemical cues. *Environ. Microbiol. Rep.* **2019**, *11*, 287–305. [CrossRef]
89. Vlamakis, H.; Chai, Y.; Beauregard, P.B.; Losick, R.; Kolter, R. Sticking together: Building a biofilm the *Bacillus subtilis* way. *Nat. Rev. Microbiol.* **2013**, *11*, 157–168. [CrossRef] [PubMed]
90. Egan, S.; Harder, T.; Burke, C.; Steinberg, P.; Kjelleberg, S.; Thomas, T. The seaweed holobiont: Understanding seaweed–bacteria interactions. *FEMS Microbiol. Rev.* **2013**, *37*, 462–476. [CrossRef] [PubMed]
91. Katharios-Lanwermeyer, S.; Xi, C.; Jakubovics, N.; Rickard, A.H. Mini-review: Microbial coaggregation: Ubiquity and implications for biofilm development. *Biofouling* **2014**, *30*, 1235–1251. [CrossRef] [PubMed]
92. Steinberg, P.D.; De Nys, R. Chemical mediation of colonisation of seaweed surfaces. *J. Phycol.* **2002**, *38*, 621–629. [CrossRef]
93. Dang, H.; Lovell, C.R. Microbial surface colonization and biofilm development in marine environments. *Microbiol. Mol. Biol. Rev.* **2016**, *80*, 91–138. [CrossRef] [PubMed]
94. Armstrong, E.; Yan, L.; Boyd, K.G.; Wright, P.C.; Burgess, J.G. The symbiotic role of marine microbes on living surfaces. *Hydrobiologia* **2001**, *461*, 37–40. [CrossRef]
95. Haas, A.F.; Wild, C. Composition analysis of organic matter released by cosmopolitan coral reef-associated green algae. *Aquat. Biol.* **2010**, *10*, 131–138. [CrossRef]
96. Thomas, T.; Evans, F.F.; Schleheck, D.; Mai-Prochnow, A.; Burke, C.; Penesyan, A.; Dalisay, D.S.; Stelzer-Braid, S.; Saunders, N.F.W.; Johnson, J.; et al. Analysis of the *Pseudoalteromonas tunicata* genome reveals properties of a surface-associated life style in the marine environment. *PLoS ONE* **2008**, *3*, e3252. [CrossRef]
97. Matz, C.; Webb, J.S.; Schupp, P.J.; Phang, S.Y.; Penesyan, A.; Egan, S.; Steinberg, P.; Kjelleberg, S. Marine biofilm bacteria evade eukaryotic predation by targeted chemical defense. *PLoS ONE* **2008**, *3*, e2744. [CrossRef]
98. Patterson, D.J.; Lee, W.J. Geographic distribution and diversity of free-living heterotrophic flagellates. In *The Flagellates: Unity, Diversity and Evolution*; Leadbeater, B.S.C., Green, J.C., Eds.; Taylor & Francis: London, UK, 2000; pp. 277–283.
99. Hisatugo, K.F.; Mansano, A.S.; Seleghim, M.H. Protozoans bacterivory in a subtropical environment during a dry/cold and a rainy/warm season. *Braz. J. Microbiol.* **2014**, *45*, 145–151. [CrossRef] [PubMed]
100. De Corte, D.; Paredes, G.; Yokokawa, T.; Sintes, E.; Herndl, G.J. Differential response of *Cafeteria roenbergensis* to different bacterial and archaeal prey characteristics. *Microb. Ecol.* **2019**, *78*, 1–5. [CrossRef] [PubMed]
101. Weitere, M.; Erken, M.; Majdi, N.; Arndt, H.; Norf, H.; Reinshagen, M.; Traunspurger, W.; Walterscheid, A.; Wey, J.K. The food web perspective on aquatic biofilms. *Ecol. Monogr.* **2018**, *88*, 543–559. [CrossRef]
102. Moens, T.; Traunspurger, W.; Bergtold, M. Feeding ecology of free-living benthic nematodes. In *Freshwater Nematodes. Ecology and Taxonomy*; Abebe, E., Traunspurger, W., Andrássy, I., Eds.; CAB International Publishing: Oxfordshire, UK, 2006; pp. 105–131.
103. Adnan, M.; Alshammari, E.M.; Patel, M.; Ashraf, S.A.; Khan, S.; Hadi, S. Significance and potential of marine microbial natural bioactive compounds against biofilms/biofouling: Necessity for green chemistry. *PeerJ* **2018**, *6*, e5049. [CrossRef] [PubMed]
104. Penesyan, A.; Kjelleberg, S.; Egan, S. Development of novel drugs from marine surface associated microorganisms. *Mar. Drugs* **2010**, *8*, 438–459. [CrossRef]
105. Rocha-Martin, J.; Harrington, C.; Dobson, A.D.W.; O'Gara, F. Emerging strategies and integrated systems microbiology technologies for biodiscovery of marine bioactive compounds. *Mar. Drugs* **2014**, *12*, 3516–3559. [CrossRef]
106. Thomas, T.R.A.; Kavlekar, D.P.; LokaBharathi, P.A. Marine drugs from sponge-microbe association-a review. *Mar. Drugs* **2010**, *8*, 1417–1468. [CrossRef]

107. Proksch, P.; Edrada, R.A.; Ebel, R. Drugs from the seas—current status and microbiological implications. *Appl. Microbiol. Biotechnol.* **2002**, *59*, 125–134. [CrossRef]
108. Oclarit, J.M.; Okada, H.; Ohta, S.; Kaminura, K.; Yamaoka, Y.; Iizuka, T.; Miyashiro, S.; Ikegami, S. Anti-bacillus substance in the marine sponge, *Hyatella* species, produced by an associated *Vibrio* species bacterium. *Microbios* **1994**, *78*, 7–16.
109. Kobayashi, M.; Aoki, S.; Gato, K.; Matsunami, K.; Kurosu, M.; Kitagawa, I. Marine natural products. XXXIV. Trisindoline, a new antibiotic indole trimer, produced by a bacterium of *Vibrio* sp. separated from the marine sponge *Hyrtios altum*. *Chem. Pharm. Bull.* **1994**, *42*, 2449–2451. [CrossRef]
110. König, G.M.; Kehraus, S.; Seibert, S.F.; Abdel-Lateff, A.; Müller, D. Natural products from marine organisms and their associated microbes. *ChemBioChem* **2006**, *7*, 229–238. [CrossRef] [PubMed]
111. Rinehart, K.L.; Gloer, J.B.; Hughes, R.G.; Renis, H.E.; McGovren, J.P.; Swynenberg, E.B.; Stringfellow, D.A.; Kuentzel, S.L.; Li, L.H. Didemnins: Antiviral and antitumor depsipeptides from a Caribbean tunicate. *Science* **1981**, *212*, 933–935. [CrossRef] [PubMed]
112. Xu, Y.; Kersten, R.D.; Nam, S.-J.; Lu, L.; Al-Suwailem, A.M.; Zheng, H.; Fenical, W.H.; Dorrestein, P.C.; Moore, B.S.; Qian, P.-Y. Bacterial biosynthesis and maturation of the didemnin anti-cancer agents. *J. Am. Chem. Soc.* **2012**, *134*, 8625–8632. [CrossRef] [PubMed]
113. Tsukimoto, M.; Nagaoka, M.; Shishido, Y.; Fujimoto, J.; Nishisaka, F.; Matsumoto, S.; Harunari, E.; Imada, C.; Matsuzaki, T. Bacterial production of the tunicate-derived antitumor cyclic depsipeptide didemnin B. *J. Nat. Prod.* **2011**, *74*, 2329–2331. [CrossRef]
114. Romano, G.; Costantini, M.; Sansone, C.; Lauritano, C.; Ruocco, N.; Ianora, A. Marine microorganisms as a promising and sustainable source of bioactive molecules. *Mar. Environ. Res.* **2017**, *128*, 58–69. [CrossRef]
115. Sunagawa, S.; Coelho, L.P.; Chaffron, S.; Kultima, J.R.; Labadie, K.; Salazar, G.; Djahanschiri, B.; Zeller, G.; Mende, D.R.; Alberti, A.; et al. Structure and function of the global ocean microbiome. *Science* **2015**, *348*, 1261359. [CrossRef]
116. Venter, J.C.; Remington, K.; Heidelberg, J.F.; Halpern, A.L.; Rusch, D.; Eisen, J.A.; Wu, D.; Paulsen, I.T.; Nelson, K.E.; Nelson, W.; et al. Environmental genome shotgun sequencing of the Sargasso Sea. *Science* **2004**, *304*, 66–74. [CrossRef]
117. Rusch, D.B.; Halpern, A.L.; Sutton, G.; Heidelberg, K.B.; Williamson, S.; Yooseph, S.; Wu, D.; Eisen, J.A.; Hoffman, J.M.; Remington, K.; et al. The Sorcerer II Global Ocean Sampling Expedition: Northwest Atlantic through Eastern Tropical Pacific. *PLoS Biol.* **2007**, *5*, e77. [CrossRef]
118. Ballestriero, F.; Daim, M.; Penesyan, A.; Nappi, J.; Schleheck, D.; Bazzicalupo, P.; Di Schiavi, E.; Egan, S. Antinematode activity of violacein and the role of the insulin/IGF-1 pathway in controlling violacein sensitivity in *Caenorhabditis elegans*. *PLoS ONE* **2014**, *9*, e109201. [CrossRef]
119. Ballestriero, F.; Thomas, T.; Burke, C.; Egan, S.; Kjelleberg, S. Identification of compounds with bioactivity against the nematode *Caenorhabditis elegans* by a screen based on the functional genomics of the marine bacterium *Pseudoalteromonas tunicata* D2. *Appl. Environ. Microbiol.* **2010**, *76*, 5710–5717. [CrossRef]
120. Penesyan, A.; Ballestriero, F.; Daim, M.; Kjelleberg, S.; Thomas, T.; Egan, S. Assessing the effectiveness of functional genetic screens for the identification of bioactive metabolites. *Mar. Drugs* **2013**, *11*, 40–49. [CrossRef] [PubMed]
121. Esposito, F.P.; Ingham, C.; Hurtado-Ortiz, R.; Bizet, C.; Tasdemir, D.; De Pascale, D. Isolation by miniaturized culture chip of an Antarctic bacterium *Aequorivita* sp. with antimicrobial and anthelmintic activity. *Biotechnol. Rep.* **2018**, *20*, e00281. [CrossRef] [PubMed]
122. Nappi, J. Discovery of Novel Bioactive Metabolites from Marine Epiphytic Bacteria and Assessment of Their Ecological Role. Ph.D. Thesis, The University of New South Wales, Sydney, Australia, February 2019.
123. Yu, J.; Du, G.; Li, R.; Li, L.; Li, Z.; Zhou, C.; Chen, C.; Guo, D. Nematicidal activities of bacterial volatiles and components from two marine bacteria, *Pseudoalteromonas marina* strain H-42 and *Vibrio atlanticus* strain S-16, against the pine wood nematode, *Bursaphelenchus xylophilus*. *Nematology* **2015**, *17*, 1011–1025. [CrossRef]
124. Huang, D.; Yu, C.; Shao, Z.; Cai, M.-M.; Li, G.-Y.; Zheng, L.; Yu, Z.-N.; Zhang, J. Identification and characterization of nematicidal volatile organic compounds from deep-sea *Virgibacillus dokdonensis* MCCC 1A00493. *Molecules* **2020**, *25*, 744. [CrossRef] [PubMed]
125. Neu, A.-K.; Månsson, M.; Gram, L.; Prol-García, M.J. Toxicity of bioactive and probiotic marine bacteria and their secondary metabolites in *Artemia* sp. and *Caenorhabditis elegans* as eukaryotic model organisms. *Appl. Environ. Microbiol.* **2014**, *80*, 146–153. [CrossRef]

126. Gram, L.; Melchiorsen, J.; Bruhn, J.B. Antibacterial activity of marine culturable bacteria collected from a global sampling of ocean surface waters and surface swabs of marine organisms. *Mar. Biotechnol.* **2010**, *12*, 439–451. [CrossRef]
127. Wietz, M.; Månsson, M.; Bowman, J.S.; Blom, N.; Ng, Y.; Gram, L. Wide Distribution of closely related, antibiotic-producing *Arthrobacter* strains throughout the Arctic Ocean. *Appl. Environ. Microbiol.* **2012**, *78*, 2039–2042. [CrossRef]
128. Mathew, M.D.; Mathew, N.D.; Miller, A.; Simpson, M.; Au, V.; Garland, S.; Gestin, M.; Edgley, M.L.; Flibotte, S.; Balgi, A.; et al. Using *C. elegans* forward and reverse genetics to identify new compounds with anthelmintic activity. *PLoS Negl. Trop. Dis.* **2016**, *10*, e0005058. [CrossRef]
129. Blasco-Costa, I.; Poulin, R. Parasite life-cycle studies: A plea to resurrect an old parasitological tradition. *J. Helminthol.* **2017**, *91*, 647–656. [CrossRef]
130. Burns, A.R.; Luciani, G.M.; Musso, G.; Bagg, R.; Yeo, M.; Zhang, Y.; Rajendran, L.; Glavin, J.; Hunter, R.; Redman, E.; et al. *Caenorhabditis elegans* is a useful model for anthelmintic discovery. *Nat. Commun.* **2015**, *6*, 7485. [CrossRef]
131. Leung, M.C.K.; Williams, P.L.; Benedetto, A.; Au, C.; Helmcke, K.J.; Aschner, M.; Meyer, J.N. *Caenorhabditis elegans*: An emerging model in biomedical and environmental toxicology. *Toxicol. Sci.* **2008**, *106*, 5–28. [CrossRef] [PubMed]
132. Queirós, L.; Pereira, J.; Gonçalves, F.J.M.; Pacheco, M.; Aschner, M.; Pereira, P. *Caenorhabditis elegans* as a tool for environmental risk assessment: Emerging and promising applications for a "nobelized worm". *Crit. Rev. Toxicol.* **2019**, *49*, 411–429. [CrossRef] [PubMed]
133. Wang, H.; Park, H.; Liu, J.; Sternberg, P.W. An efficient genome editing strategy to generate putative null mutants in *Caenorhabditis elegans* using CRISPR/Cas9. *G3* **2018**, *8*, 3607–3616. [CrossRef] [PubMed]
134. Riddle, D.L.; Blumenthal, T.; Meyer, B.J.; Priess, J.R. *C elegans II*, 2nd ed.; Cold Spring Harbor Laboratory Press: New York, NY, USA, 1997; Section I, The Biological Model. Available online: https://www.ncbi.nlm.nih.gov/books/NBK20086/ (accessed on 5 July 2020).
135. Kong, C.; Eng, S.-A.; Lim, M.-P.; Nathan, S. Beyond traditional antimicrobials: A *Caenorhabditis elegans* model for discovery of novel anti-infectives. *Front. Microbiol.* **2016**, *7*, 1956. [CrossRef]
136. Frézal, L.; Félix, M.-A. The natural history of model organisms: *C. elegans* outside the Petri dish. *eLife* **2015**, *4*, e05849. [CrossRef]
137. Artal-Sanz, M.; de Jong, L.; Tavernarakis, N. *Caenorhabditis elegans*: A versatile platform for drug discovery. *Biotechnol. J.* **2006**, *1*, 1405–1418. [CrossRef]
138. Lok, J.B.; Unnasch, T.R. Transgenesis in Animal Parasitic Nematodes: *Strongyloides spp.* and *Brugia* spp. In *WormBook: The Online Review of C. elegans Biology*; The *C. elegans* Research Community, Ed.; WormBook, 2018. Available online: https://www.ncbi.nlm.nih.gov/books/NBK174830/ (accessed on 23 June 2020).
139. Stiernagle, T. Maintenance of *C. elegans*. In *WormBook: The Online Review of C. elegans Biology*; The *C. elegans* Research Community, Ed.; WormBook, 2006. Available online: http://www.wormbook.org/chapters/www_strainmaintain/strainmaintain.html (accessed on 30 May 2020).
140. Brenner, S. The genetics of *Caenorhabditis elegans*. *Genetics* **1974**, *77*, 71–94.
141. Fraietta, I.; Gasparri, F. The development of high-content screening (HCS) technology and its importance to drug discovery. *Expert Opin. Drug Discov.* **2016**, *11*, 501–514. [CrossRef]
142. Midkiff, D.; San-Miguel, A. Microfluidic technologies for high throughput screening through sorting and on-chip culture of *C. elegans*. *Molecules* **2019**, *24*, 4292. [CrossRef]
143. Driscoll, M.; Dean, E.; Reilly, E.; Bergholz, E.; Chalfie, M. Genetic and molecular analysis of a *Caenorhabditis elegans* beta-tubulin that conveys benzimidazole sensitivity. *J. Cell Biol.* **1989**, *109*, 2993–3003. [CrossRef]
144. Haber, C.L.; Heckaman, C.L.; Li, G.P.; Thompson, D.P.; Whaley, H.A.; Wiley, V.H. Development of a mechanism of action-based screen for anthelmintic microbial metabolites with avermectin like activity and isolation of milbemycin-producing *Streptomyces* strains. *Antimicrob. Agents Chemother.* **1991**, *35*, 1811–1817. [CrossRef] [PubMed]
145. Marroquin, L.D.; Elyassnia, D.; Griffitts, J.S.; Feitelson, J.S.; Aroian, R.V. *Bacillus thuringiensis* (Bt) toxin susceptibility and isolation of resistance mutants in the nematode *Caenorhabditis elegans*. *Genetics* **2000**, *155*, 1693–1699.

146. Mitreva, M.; McCarter, J.P.; Martin, J.; Dante, M.; Wylie, T.; Chiapelli, B.; Pape, D.; Clifton, S.W.; Nutman, T.B.; Waterston, R.H. Comparative genomics of gene expression in the parasitic and free-living nematodes *Strongyloides stercoralis* and *Caenorhabditis elegans*. *Genome Res.* **2004**, *14*, 209–220. [CrossRef] [PubMed]
147. Kumarasingha, R.; Young, N.D.; Manurung, R.; Lim, D.S.L.; Tu, C.-L.; Palombo, E.A.; Shaw, J.M.; Gasser, R.B.; Boag, P.R. Transcriptional alterations in *Caenorhabditis elegans* following exposure to an anthelmintic fraction of the plant *Picria fel-terrae* Lour. *Parasites Vectors* **2019**, *12*, 181. [CrossRef] [PubMed]
148. Mir, D.A.; Balamurugan, K. Global proteomic response of *Caenorhabditis elegans* against PemKSa toxin. *Front. Cell. Infect. Microbiol.* **2019**, *9*, 172. [CrossRef] [PubMed]
149. Angstadt, J.D.; Donmoyer, J.E.; Stretton, A.O. Retrovesicular ganglion of the nematode *Ascaris*. *J. Comp. Neurol.* **1989**, *284*, 374–388. [CrossRef] [PubMed]
150. Johnson, C.D.; Stretton, A.O. Localization of choline acetyltransferase within identified motoneurons of the nematode *Ascaris*. *J. Neurosci.* **1985**, *5*, 1984–1992. [CrossRef]
151. Johnson, C.D.; Stretton, A.O. GABA-immunoreactivity in inhibitory motor neurons of the nematode *Ascaris*. *J. Neurosci.* **1987**, *7*, 223–235. [CrossRef]
152. Davis, R. Neurophysiology of glutamatergic signalling and anthelmintic action in *Ascaris suum*: Pharmacological evidence for a kainate receptor. *Parasitology* **1998**, *116*, 471–486. [CrossRef]
153. Weeks, J.C.; Robinson, K.J.; Lockery, S.R.; Roberts, W.M. Anthelmintic drug actions in resistant and susceptible C. elegans revealed by electrophysiological recordings in a multichannel microfluidic device. *Int. J. Parasitol. Drugs Drug Resist.* **2018**, *8*, 607–628. [CrossRef]
154. Lewis, J.; Wu, C.-H.; Levine, J.; Berg, H. Levamisole-resitant mutants of the nematode *Caenorhabditis elegans* appear to lack pharmacological acetylcholine receptors. *Neuroscience* **1980**, *5*, 967–989. [CrossRef]
155. Kaminsky, R.; Ducray, P.; Jung, M.; Clover, R.; Rufener, L.; Bouvier, J.; Weber, S.S.; Wenger, A.; Wieland-Berghausen, S.; Goebel, T.; et al. A new class of anthelmintics effective against drug-resistant nematodes. *Nature* **2008**, *452*, 176–180. [CrossRef] [PubMed]
156. Yamazaki, M.; Okuyama, E.; Kobayashi, M.; Inoue, H. The structure of paraherquamide, a toxic metabolite from *Penicillium paraherquei*. *Tetrahedron Lett.* **1981**, *22*, 135–136. [CrossRef]
157. Hernando, G.; Turani, O.; Bouzat, C. *Caenorhabditis elegans* muscle Cys-loop receptors as novel targets of terpenoids with potential anthelmintic activity. *PLoS Negl. Trop. Dis.* **2019**, *13*, e0007895. [CrossRef] [PubMed]
158. Schaeffer, J.M.; Blizzard, T.A.; Ondeyka, J.; Goegelman, R.; Sinclair, P.J.; Mrozik, H. [^{3}H]Paraherquamide binding to *Caenorhabditis elegans*: Studies on a potent new anthelmintic agent. *Biochem. Pharmacol.* **1992**, *43*, 679–684. [CrossRef]
159. Ruiz-Lancheros, E.; Viau, C.; Walter, T.N.; Francis, A.; Geary, T. Activity of novel nicotinic anthelmintics in cut preparations of *Caenorhabditis elegans*. *Int. J. Parasitol.* **2011**, *41*, 455–461. [CrossRef] [PubMed]
160. Guo, Z. The modification of natural products for medical use. *Acta Pharm. Sin. B* **2017**, *7*, 119–136. [CrossRef]
161. Taman, A.; Azab, M. Present-day anthelmintics and perspectives on future new targets. *Parasitol. Res.* **2014**, *113*, 2425–2433. [CrossRef]

microorganisms

Review

An Overview on Industrial and Medical Applications of Bio-Pigments Synthesized by Marine Bacteria

Ali Nawaz [1], Rida Chaudhary [1], Zinnia Shah [1], Laurent Dufossé [2,*], Mireille Fouillaud [2], Hamid Mukhtar [1] and Ikram ul Haq [1]

1 Institute of Industrial Biotechnology, GC University Lahore, Lahore 54000, Pakistan; ali.nawaz@gcu.edu.pk (A.N.); ridachdry.789@gmail.com (R.C.); syedazinniashah@gmail.com (Z.S.); hamidmukhtar@gcu.edu.pk (H.M.); dr.ikramulhaq@gcu.edu.pk (I.u.H.)

2 CHEMBIOPRO Lab, ESIROI Agroalimentaire, University of Réunion Island, 97400 Saint-Denis, France; mireille.fouillaud@univ-reunion.fr

* Correspondence: laurent.dufosse@univ-reunion.fr; Tel.: +33-668-731-906

Abstract: Marine bacterial species contribute to a significant part of the oceanic population, which substantially produces biologically effectual moieties having various medical and industrial applications. The use of marine-derived bacterial pigments displays a snowballing effect in recent times, being natural, environmentally safe, and health beneficial compounds. Although isolating marine bacteria is a strenuous task, these are still a compelling subject for researchers, due to their promising avenues for numerous applications. Marine-derived bacterial pigments serve as valuable products in the food, pharmaceutical, textile, and cosmetic industries due to their beneficial attributes, including anticancer, antimicrobial, antioxidant, and cytotoxic activities. Biodegradability and higher environmental compatibility further strengthen the use of marine bio-pigments over artificially acquired colored molecules. Besides that, hazardous effects associated with the consumption of synthetic colors further substantiated the use of marine dyes as color additives in industries as well. This review sheds light on marine bacterial sources of pigmented compounds along with their industrial applicability and therapeutic insights based on the data available in the literature. It also encompasses the need for introducing bacterial bio-pigments in global pigment industry, highlighting their future potential, aiming to contribute to the worldwide economy.

Keywords: natural colors; bio-pigments; quorum sensing; marine bacteria; biosynthesis; biological activities; industrial applications; therapeutic insights; global pigment market

Citation: Nawaz, A.; Chaudhary, R.; Shah, Z.; Dufossé, L.; Fouillaud, M.; Mukhtar, H.; ul Haq, I. An Overview on Industrial and Medical Applications of Bio-Pigments Synthesized by Marine Bacteria. *Microorganisms* **2021**, 9, 11. https://doi.org/10.3390/microorganisms9010011

Received: 9 November 2020
Accepted: 18 December 2020
Published: 22 December 2020

1. Introduction

1.1. Microbial Pigments

The production of bio-pigments from bacterial species is being conducted globally with soaring interest under the research of microbial autecology. A massive array of these compounds, also referred to as "bioactive pigmented molecules", can be derived from both Gram-positive and Gram-negative bacterial species. Production of these pigments in the marine environment is mediated through the complex mechanism of "quorum sensing" [1] or also can be induced through exposure to different stress conditions in external environments. Quorum sensing is the mechanism whereby individual bacterial cells can coordinate with others in their colony to carry out constitutive functions especially involving the secretion of numerous specific chemical compounds. These compounds can help them with survival, competence, bioluminescence, biofilm formation, and even sporulation, etc. Bio-pigments can be produced by triggering regulatory quorum sensing mechanisms of these species and can be extensively used in various bio-medical and bio-industrial sectors, including textiles, food, pharmaceutical, and cosmetic industries, owing to their beneficial attributes and biological activities [2,3]. These are moreover convenient to harvest in large volumes through utilizing simple gene manipulating strategies. The rising consumer concerns regarding safety and quality of industrial products holds a significant ground as to why scientists are shifting their focus towards naturally derived, non-toxic, and eco-friendly pigment alternatives [4].

1.2. Bacterial Pigments as Natural Colorants

The use of synthetic pigments goes back to the 1850s when these were put in trend for the first time due to their supercilious coloring properties, lower prices, and easy production strategies [5], the significance of which remains empirically the same to this day. The importance of artificial/synthetic coloring agents is still based on the fact that the appearance of food items influences consumer's emotions, attitudes, and preferences. Let us say, if a carrot is not red, the consumer is most probably expected to reject it. The same can be applied in regards with the cosmetic industry, where the product apparel decides its fate. Thus, need for "synthetic pigments" cannot be overseen if client orientation is to be fulfilled [6]. The only progress made today is the shift towards naturally derived pigments rather than continuing the use of artificially synthesized ones, which have been denounced for their serious threat to consumer's well-being [7]. Cancers of skin, liver, and bladder have been found positively related to the use of artificial pigments because of their high azo-dye/heavy metal compositions. Furthermore, the precursors involved and the waste generated through their production process is environmentally hazardous as well [8,9]. The outcry against the use of synthetic colorants in many health-conscious countries has already caused the ban of several artificial colorants, including Blue NO 1, Blue NO 2, Blue FCF, and Yellow NO 6 [10].

Bio-pigments, however, are eco-friendly and proved additionally propitious as anti-toxic, antitumor, antioxidant, anticancer, and antimicrobial agents [2]. Other advantages include fast and economic extraction techniques, higher yield, and time- and cost-efficient production. Moreover, the production of microbial pigments can also be made more convenient by the optimization capacity of their growth parameters [11]. Keeping the capacity of bio-pigments into consideration, many biotech industries are now developing protocols for efficient extraction of natural pigments as a replacement to synthetic counterparts. For instance, natural pigments such as zeaxanthin, saproxanthin, myxol and many others which illicit antioxidant activities are being instigated against artificial antioxidants such as butylated hydroxyl toluene and butylated hydroxyl acids [12,13].

1.3. Marine Ecosystem as a Source of Pigment Producing Bacterial Species

The study of a likely natural ecosystem serves as the initial-most important research step needed to find an environment that can entertain the diversity of bio-pigment sources. The marine environment is a habitat for almost 80% of all life forms [14]. It serves as a rich source of aquatic microbial species that exhibit comparatively more augmented diversity than their telluric counterparts [15]. The marine environment is presently being considered as an attractive fount for bio-pigment sources [16]. Numerous bacterial isolates from such biotopes have already been tested for pigment production. At the same time, many of them are also being utilized for various industrial purposes as well [15]. The preference of pigments produced by marine microorganisms is based on their ability to persist in extremities such as highly acidic/alkaline environments (pH < 4 and >9), extreme temperatures (-2–15 °C and 60–110 °C), and under limited substrate availability [17,18]. Apart from bacterial isolates, halophilic archaea are extensively disseminated in the marine ecosystem. Pigmented compounds from marine archaea are also prioritized owing to their ability to tolerate hyper saline and basic pH environments, besides their potential to withstand osmolytes (such as 2-sulfotrehalose) or high ionic strength [19,20].

Concerns regarding environmental conservation and consumers' preferences have stimulated the interests of researchers and stake holders in exploring nontoxic, eco-friendly, and biodegradable commodities. Bacterially produced bio-pigments (bpBPs) have growing importance not only on account of their dyeing potential, but also due to their medicinal properties. Likewise, awareness regarding the carcinogenic and other pernicious effects of synthetic colorants has kindled a fresh enthusiasm towards the utilization of bacterial pigments in the food industry as safer alternatives to use as antioxidants, color intensifiers, flavor enhancers, and food additives.

Extraction of natural pigments from microorganisms populating environments exclusive of soil is a topic of current interest. Marine environment has become a captivating subject matter for microbiologists, pharmacologists, and biochemists in order to extract water based bacterial pigments. With the recent increase in awareness towards the benefits of natural over synthetic products, the bio-pigment industry is likely to increase its global market. The review aims at discussing the therapeutic and industrial significance of marine derived bacterial pigments helping to delineate the consequence of furthering the scope of these studies. It provides a comprehensive overview of potentiality and competence of marine bacteria as a source of bio-pigments by critically summarizing the scientific researches and accumulated data in the literature and the prominence of these bio-pigments in strengthening the overall pigment market by reviewing latest industry market research, reports, and statistics.

2. Marine Bacterial Species as Sources of Bio-Pigments

The marine environment has been investigated for almost 300,000 known species, which constitutes only a small fraction of the total number of explorable pigment producing bacterial species. Bacterial species isolated from marine sediments or seawater such as *Streptomyces* sp., *Pontibacter korlensis* sp., *Pseudomonas* sp., *Bacillus* sp., and *Vibrio* sp. produce an array of pigmented compounds including prodigiosin, astaxanthin, pyocyanin, melanin, and beta carotene, respectively (Table 1). These pigments belong to a range of compound classes, for instance, carotenes are a subclass of carotenoids that have unsaturated polyhydrocarbon structures, prodiginines have a pyrrolyldipyrromethene core structure, tambjamines are alkaloid molecules, while violacein compounds are indole derivatives derived from tryptophan metabolism (Figure 1) [1,2,21]. These and other such pigments, despite their class diversity, share a functional likeness due to the presence of aromatic rings in their structures.

Table 1. Marine bacterial sources of colored pigmented compounds.

Pigments	Marine Bacterial Species	References
Prodigiosin	*Hahella chejuensis* sp. *Pseudoalteromonas rubra* sp. *Streptomyces* sp. SCSIO 11594 *Vibrio* sp. (Strain MI-2) *Serratia marcescens* sp. IBRL USM 84 *Zooshikella ganghwensis* gen. nov., sp. nov.	[22–27]
Undecylprodigiosin	*Streptomyces* sp. UKMCC_PT15 *Streptomyces* sp.SCSIO 11594 Novel strain of Actinobacterium sp., *Saccharopolyspora* sp.	[24,28,29]
Heptylprodigiosin	*Spartinivicinus ruber* gen. nov., sp. nov	[30]
Cycloprodigiosin	*Pseudoalteromonas denitrificans* sp. *Pseudoalteromonas rubra* sp. ATCC 29570	[15,31]
Norprodigiosin	*Serratia* sp. WPRA3	[32]

Table 1. Marine bacterial sources of colored pigmented compounds.

Pigments	Marine Bacterial Species	References
Astaxanthin	*Brevundimonas scallop* sp. Zheng & Liu *Corynebacterium glutamicum* sp. *Brevundimonas* sp. strain N-5 *Sphingomicrobium astaxanthinifaciens* sp. nov *Rhodovulum sulfidophilum* sp. *Pontibacter korlensis* sp. AG6 *Exiguobacterium* sp. *Altererythrobacter ishigakiensis* sp. NBRC 107699 *Rhodotorula* sp. *Paracoccus haeundaensis* sp.	[33–42]
Zeaxanthin	*Sphingomonas phyllosphaerae* sp. KODA19-6 *Mesoflavibacter aestuarii* sp. nov. *Aquibacter zeaxanthinifaciens* gen. nov., sp. nov. *Zeaxanthinibacter enoshimensis* gen. nov., sp. nov *Gramella planctonica* sp. nov *Mesoflavibacter zeaxanthinifaciens* gen. nov., sp. nov *Formosa* sp. KMW *Sphingomonas phyllosphaerae* sp. $FA2^T$ *Sphingomonas (Blastomonas) natatoria* sp. DSM 3183^T *Muricauda lutaonensis* sp. CC-HSB-11T	[29,43–50]
Lycopene	*Blastochloris tepida* sp. *Salinicoccus roseus* sp.	[51,52]
Beta carotene	Cyanobacterium *Synechococcus* sp. *Micrococcus* sp. *Vibrio owensii* sp. *Flavicella marina* gen. nov., sp. nov. *Gordonia terrae* sp.TWRH01	[53–57]
Canthaxanthin	*Brevibacterium* sp. *Gordonia* sp. *Dietzia* sp.	[54]
Pyocyanin	*Pseudomonas aeruginosa* sp.	[58]
Scytonemin	*Cyanobacterial* sp. *Nostoc punctiforme* sp. ATCC	[59,60]
Violacein	*Pseudoalteromonas luteoviolacea* sp. *Pseudoalteromonas ulvae* sp. TC14 *Pseudoalteromonas* sp. 520P1 *Microbulbifer* sp. D250 *Pseudoalteromonas amylolytica* sp. nov *Chromobacterium violaceum* sp. *Collimonas* sp.	[61–67]
Melanin	*Streptomyces* sp. *Pseudomonas* sp. *Marinomonas mediterranea* sp. MMB-1T *Pseudomonas stutzeri* sp. *Bacillus* sp. BTCZ31 *Streptomyces* sp. MVCS13 *Providencia rettgeri* sp. strain BTKKS1 *Marinobacter alkaliphilus* sp. *Leclercia* sp. *Halomonas meridian* sp. *Nocardiopsis dassonvillei* sp. strain JN1 *Vibrio alginolyticus* sp. strain BTKKS3	[68–77]
Tambjamines	*Pseudoalteromonas tunicata* sp. *Pseudoalteromonas citrea* sp.	[78,79]

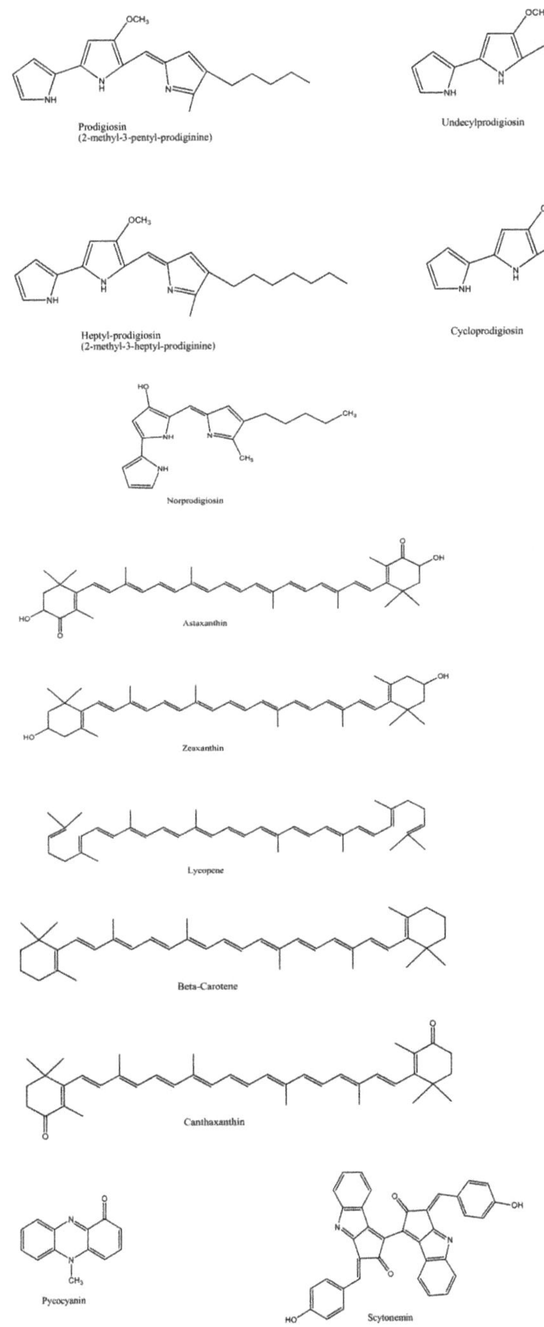

Figure 1. *Cont.*

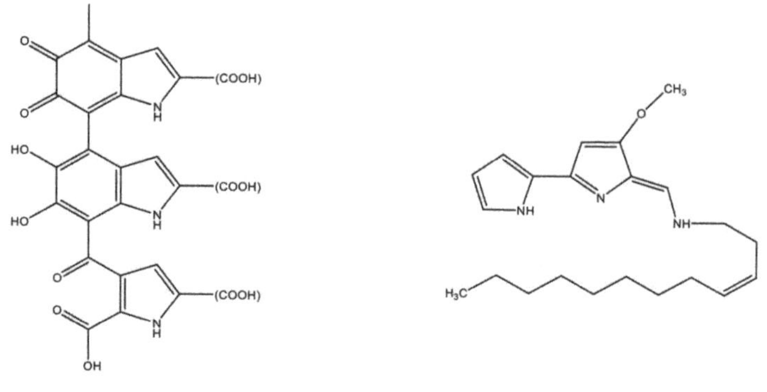

Melanin

Tambjamine

Figure 1. Chemical structures of various bacterial pigments.

3. Biosynthesis of Bacterial Pigments

The potential of marine bacterial isolates as a leading source of bio-pigments demands an extensive understanding of bio-mechanisms responsible for yielding pigmented molecules. Different studies have reported the proposed biosynthetic pathways of pigment production by marine bacterial isolates along with biochemically characterized enzymatic transformations (Figure 2). However, it is still unclear if the proposed pathways are distinct for marine or terrestrial bacterial species, or may be the same in both cases.

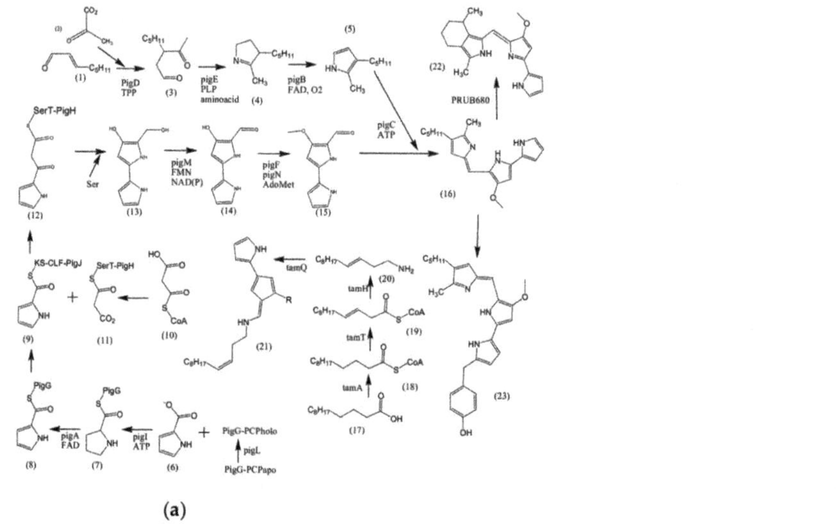

(a)

Figure 2. *Cont.*

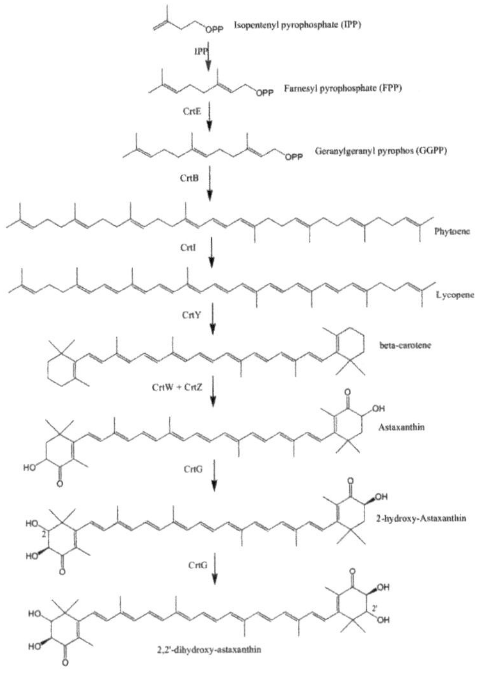

(b)

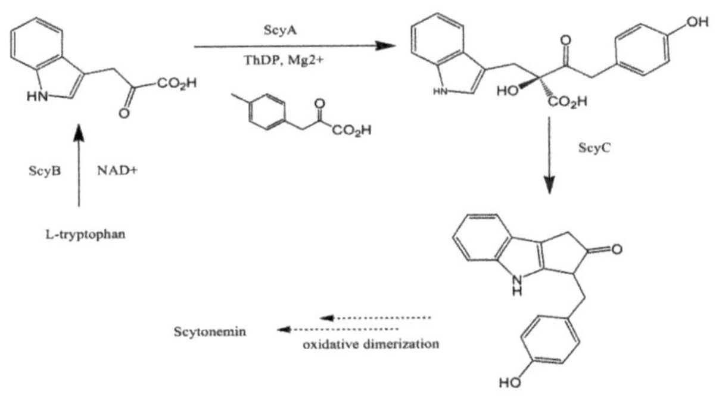

(c)

Figure 2. *Cont.*

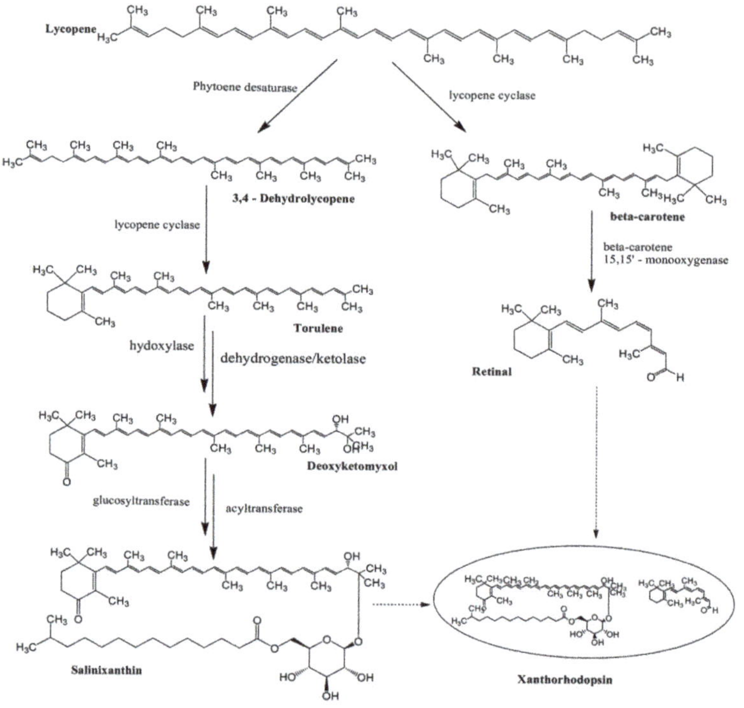

(d)

Figure 2. Proposed biosynthetic pathways of few bacterially produced bio-pigments. (**a**) Biosynthesis of Prodiginine analogs; MAP biosynthesis; MBC biosynthesis; Tambjamine biosynthesis; Cyloprodigiosin biosynthesis; 2-(p-hydroxybenzyl)prodigiosin (HBPG) biosynthesis. (**b**) Biosynthesis of carotenoids. (**c**) Biosynthesis of scytonemin. (**d**) Biosynthesis of salinixanthin and retinal pigments. (**a**) Biosynthesis of prodigioinine analogs. MAP Biosythesis (Green): (1) 2octenal, (2) Pyruvate, (3) 3-acetyloctanal, (4) H2MAP (5) MAP. MBC Biosynthesis (Blue), (6) L-proline, (7) L-prolyl-S-PCP intermediate, (8) Pyrrolyl2-carboxyl-S-PCP, (9) Pyrrole-2-carboxyl thioester, (10) Malonyl-CoA, (11) Bound malonyl, (12) pyrrolyl-β-ketothioester on PigH, (13) 4-hydroxy-2,20-bipyrrole-5methanol (HBM), (14) 4-hydroxy-2,20-bipyrrole-5-carbaldehyde (HBC), (15) MBC, (16) Prodigiosin. Tambjamine Biosynthesis, (17) Dodecenoic acid, (18) Activated fatty acid, (19) CoA-ester, (20) Enamine, (21) Tambjamine, (22) Cycloprodigiosin (cPrG) &, (23) 2-(p-hydroxybenzyl)prodigiosin(HBPG) Biosynthesis. (**b**). Biosynthesis of carotenoids: CrtE: GGPP synthase, IPP: Isopentenyl pyrophosphate, GGPP: Geranylgeranyl pyrophos, CrtB: Phytoene synthase, Crtl: Phytoene desaturase, CrtY: lycopene β-cyclase, CrtW: β-carotene ketolase, CrtZ: β-carotene hydroxylase, CrtG: Astaxanthin 2,2′-β-ionone ring hydroxylase gene. (**c**). Biosynthesis of scytonemin: Scytonemin biosynthetic enzymes: ScyA, ScyB, ScyC (ScyA: a thiamin-dependent enzyme, ScyC: enzyme annotated as a hypothetical protein), ThDP: Thiamine diphosphate, NAD: Nicotinamide adenine dinucleotide, Mg^{2+}: Magnesium ion.

3.1. Biosynthesis of Prodiginine Analogs

2-methyl-3-n-amyl-pyrrole (MAP) biosynthesis: This pathway involves three genes; *pigB*, *pigD*, and *pigE*. At first, PigD carries out the addition of pyruvate to 2-octenal in the presence of coenzyme thiamine pyrophosphate (TPP). As a result, 3-acetyloctanal formation occurs along with the release of CO_2 molecule. PigE catalyzes the transfer of an amino group to the aldehyde, followed by cyclization, resulting in the formation of H2MAP. PigB carries out further oxidation to form MAP (Figure 2a) [80,81].

4-methoxy-2,2′-bipyrrole-5-carbaldehyde (MBC) biosynthesis: This pathway involves seven genes: *pigA*, *pigF-J*, *pigL*, and *pigM*. 4′-phosphopantetheinyl transferase (PigL) carries out the activation of peptidyl carrier protein (PCP) domain of PigG by introducing 4′-phosphopantetheinyl group. Formation of L-prolyl-S-PCP intermediate occurs by the transfer of L-prolyl group of L-proline to the thiol group of phosphopantetheine, carried out by PigI and ATP. PigA further catalyzes the oxidation of the intermediate to pyrrolyl-2-carboxyl-S-PCP. Pyrrole-2-carboxyl thioester is generated by the transfer of pyrrole-2-carboxyl group of PigG to the cysteine active site at PigJ. Phosphopantetheinylated ACP domains of PigH provide binding sites for malonyl group of malonyl-CoA. Decarboxylation of bound malonyl results in condensation with pyrrole-2-carboxyl thioester and leads to the formation of pyrrolyl-β-ketothioester on PigH. Generation of 4-hydroxy-2,2′-bipyrrole-5-methanol (HBM) occurs by decarboxylation between serine and pyrrolyl-β-ketothioester, catalyzed by PigH [80,82]. 4-hydroxy-2,2′-bipyrrole-5-carbaldehyde (HBC) is formed when PigM oxidizes the alcohol group of HBM. Methyltransferase (PigF) and oxidoreductase (PigN) further carries out the methylation of HBC hydroxyl group to form MBC [81]. After the formation of MAP and MBC, PigC utilizes ATP to perform terminal condensation of these pyrroles, synthesizing prodigiosin.

Cycloprodigiosin (cPrG) biosynthesis: The cyclization of undecylprodiginine in order to form metacycloprodigiosin and butyl-meta-cycloheptylprodiginine is carried out by *mcpG* and *redG*, respectively [83]. Studies also revealed that a homologus gene (PRUB680) encodes an alkylglycerol monooxygenase-like protein away from *pig* biosynthetic gene cluster [84]. The respective enzyme demonstrates regiospecificity through C-H activation, resulting in cyclization of prodigiosin to form cPrG [78].

Tambjamine (tam) biosynthesis: Tambjamines have MBC moiety but lack MAP moiety, rather have an enamine group. Enamine biosynthetic pathway involves three genes; *tamT*, *tamH*, and *afaA* [85]. Acyl CoA synthetase (TamA) activates dodecenoic acid [86]. Dehydrogenase (TamT) carries out the oxidation of activated fatty acid, incorporating a π-bond to the fatty acyl side chain at its C-3 carbon. Further, the reduction of CoA-ester, followed by transamination to dodec-3-en-1-amine is facilitated by reductase/aminotransferase (TamH). MBC and enamine then undergoes condensation in order to form tambjamine, catalyzed by TamQ [85].

3.2. Biosynthesis of Carotenoids

Carotenoids are yellow, orange, and red colored pigmented compounds that are further subdivided into carotenes and xanthophylls. So far, 700 carotenoids have been reported, and among them beta-carotene, lutein, canthaxanthin, astaxanthin, lycopene, and zeaxanthin are the highly valued carotenoids [87]. Universal precursors for C40 and C50 carotenoid biosynthesis are two 5 C subunits: isopentenyl diphosphate (IPP) plus its isomeric form dimethylallyl diphosphate (DMAPP). IPP/DMAPP isomerase (IDI) carries out the isomerization of IPP into DMAPP. Geranylgeranyl diphosphate (GGPP) synthase further catalyzes the addition of one DMAPP molecule with three IPP molecules to generate an immediate precursor, i.e., C20 geranylgeranyl diphosphate (GGPP) [88]. Phytoene synthase carries out the first committed step of carotenoid biosynthesis i.e., condensing two GGPP molecules to form phytoene (C40), which is further desaturated by phytoene desaturase by the incorporation of four double bonds in its structure. This desaturated structure is a red-colored compound unlike its colorless parent molecule, and is called lycopene. Lycopene further undergoes several modifications to produce different carotenoids. Beta-carotene is generated by the cyclization of lycopene, carried out by lycopene beta-cyclase. It is then converted into canthaxanthin and zeaxanthin by catalytic activity of two protein classes: beta-carotene ketolase and beta-carotene hydroxylase, respectively, next to the formation of astaxanthin [30]. Beta-carotene ketolase represented by CrtW and CrtO types adds the ketone group to carbon 4/40 of the b-ionone ring. However, beta-carotene hydroxylase, encompassed by CrtR, CrtZ, and P450 types carries out the hydroxylation of carbon 3/30 of the b-ionone ring [89]. 2,2′-β-ionone ring hydroxylase introduces hydroxyl group to the β-ionone ring of astaxanthin and results in the formation of 2,2′-dihydroxy-astaxanthin (Figure 2b) [33].

3.3. Biosynthesis of Scytonemin

Biosynthesis of scytonemin involves three scytonemin biosynthetic enzymes; ScyA, ScyB, and ScyC. ScyB carries out the conversion of L-tryptophan to 3-indole pyruvic acid. ScyA (thiamin-dependent enzyme) performs the coupling of 3-indole pyruvic acid with p-hydroxyphenylpyruvic acid and results in the formation of b-ketoacid, whose cyclization is further carried out by ScyC (enzyme annotated as hypothetical protein) [90]. The resulting tricyclic ketone resembles half of the skeleton of scytonemin (Figure 2c) [91].

3.4. Biosynthesis of Salinixanthin and Retinal

Retinal: Lycopene cyclase converts lycopene into β-carotene. Breakdown of β-carotene into two retinal molecules is further catalyzed by a gene annotated as β-carotene 15,15′-monooxygenase (orf4) (Figure 2d) [92,93].

Salinixanthin: Xanthorhodopsin (orf2) (a light-driven proton pump) has two chromophores; retinal and salinixanthin [94,95]. Phytoene desaturase (CrtI) converts lycopene to 3,4-dehydrolycopene, which is further converted to torulene by lycopene cyclase [92]. Subsequently, conversion of torulene to salinixanthin is catalyzed by hydroxylase, ketolase or dehydrogenase, glucosyltransferase, and acyltransferase, having reactions involved similar to that of biosynthetic reactions of myxol and canthaxanthin [96,97].

4. Industrial and Therapeutic Applications

4.1. Therapeutic Applications

4.1.1. Antibacterial Activity

Antibacterial properties of various bacterially produced bio-pigments of marine origin have been reported against an array of bacterial species, e.g., prodigiosin, cycloprodiogisin (from *Z. rubidus* sp. S1-1), and the yellow pigment (extracted from *Micrococcus* sp. strain MP76) have shown antibacterial activity against *Staphylococcus aureus* sp. and *Escherichia coli* sp. [98,99]. Other bacterial strains that are reportedly inhibited by prodigiosin and cycloprodigiosin are *Bacillus subtilis* sp. and *Salmonella enterica* serovar Typhimurium [98]. Likewise, the yellow pigment has shown activity against *P. aeruginosa* sp. as well [99]. Nor-prodigiosin synthesized by marine *Serratia* sp. has also been reported to exhibit inhibition activity against *Vibrio paraheamolyticus* sp. and *B. subtilis* sp. [32]. These studies strengthen the utilization of bpBPs as potential alternatives to synthetic medicinal compounds.

Furthermore, inhibition activities recorded against *Citrobacter* sp. by pyocyanin and pyorubin [58] and *P. aeruginosa* sp. by violacein pigment (purified from Antarctic *Iodobacter* sp.) [100], further stretches the range of marine-derived bpBP's potential against pathogenic bacterial species to opportunistic bacterial species. There are numerous correspondingly published studies. The pigment "melanin" from marine *Streptomyces* sp., for instance, demonstrated antibacterial activity against *E. coli* sp., *S. typhi* sp., *S. paratyphi* sp., *Proteus mirabilis* sp., *Vibrio cholera* sp., *S. aureus* sp., and *Klebsiella oxytoca* sp. [68]. A bright pink-orange colored pigment extracted from *Salinicoccus* sp. (isolated from Nellore sea coast) also showed antimicrobial potential against several bacterial strains including *E.coli* sp., *Klebsiella pneumoniae* sp., *B. subtilis* sp., *Proteus vulgaris* sp., *P. aeruginosa* sp., and *S. aureus* sp. [101]. Hence, these and similar other studies all indicate the exploration of marine bacterial species as a dynamic approach to derive antibacterial compounds.

A few studies also seemingly suggest that a single pigment from different species may exhibit activities against various target microorganisms. One example is violacein, a violet colored pigment extracted from Antarctic bacterium *Janthinobacterium* sp. SMN 33.6, which showed antibacterial activity against multi-resistant bacteria: *S. aureus* sp. ATCC 25923, *E. coli* sp. ATCC 25922, *Kocuria rhyzophila* sp. ATCC 9341, and *S. typhimurium* sp. ATCC 14028 [102], and that extracted from *Collimonas* sp. showed antibacterial activity against *Micrococcus luteus* sp. [67].

4.1.2. Antifungal Activity

Studies have also been carried out to determine the antifungal potential of natural pigmented compounds. Several studies have reported the antifungal activity of marine-derived bacterial pigments, among which violacein from *Chromobacterium* sp. and prodiginine pigments (prodigiosin and cycloprodigiosin) extracted from Indonesian marine bacterium *P. rubra* sp. reported to exhibit antagonistic activity against *Candida albicans* sp. [23,103]. Violacein also inhibited several other fungal strains, including *Penicillium expansum* sp., *Fusarium oxysporum* sp., *Rhizoctonia solani* sp., and *Aspergillus flavus* sp. Studies have also reported that violacein (extracted from a pure *Chromobacterium* sp.) shows comparable antifungal activity to that of bavistin and amphotericin B, highlighting the potential of marine-derived bpBPs as effective antifungal agents over existing synthetic antifungal compounds [103].

4.1.3. Anticancer Activity

Exploring anticancer compounds from marine microbes has been considered a hot spot in natural product research. Several studies have been carried out in order to examine the antitumor ability of marine bacterial pigments. Anticancer activity of marine-derived bpBPs has been explored against several cancerous cell lines. Astaxanthin and 2-(p-hydroxybenzyl) prodigiosin (HBPG) isolated from *P. kolensis* sp. and *P. rubra* sp. displayed significant cytotoxicity against human breast cancer cell line (MCF-7) and human ovarian adenocarcinoma cell line, respectively [38,104]. PCA (Phenazine -1-carboxylic acid) pigment extracted from marine *P. aeruginosa* sp. GS-33 correspondingly showed inhibition against SK-MEL-2 (human skin melanoma cell line) [105]. Another pigment violacein extracted from Antarctic bacterium isolate, identified as a member of the genus *Janthinobacterium* (named as *Janthinobacterium* sp. strain UV13), revealed its antiproliferative activity in HeLa cells. Studies further confirmed the potential of violacein as an anticancer agent to cisplatin drug (anticancer chemotherapy drug) in cervix cell carcinoma [106]. It has also been reported that a single pigment can express anticancer activity against multiple cancerous cell lines. Synthetically derived tambjamines isolated from the marine bacterium *P. tunicata* sp. have shown significant apoptosis inducing effects against various cancer cell lines including glioblastoma cell line (SF-295), M14 melanoma cell line (MDA-MB-435), ileocecal colorectal adenocarcinoma cell line (HCT-8), and promyelocytic leukemia cells (HL-60) [107]. Carotenoid pigments extracted from marine *Arthrobacter* sp. G2O (isolated from the Caspian Sea) exhibited antitumor activity on esophageal squamous cancerous cells [108]. Likewise, prodigiosin homolog extracted from marine bacterium *Serratia proteamacula* sp. was also found to exhibit high antitumor activity [109], indicating the potential of marine-derived bpBPs in antitumor therapy.

4.1.4. Antioxidant Activity

Marine-derived bpBPs are also being explored for their antioxidant activity. 3R saproxanthin and myxol pigments (from marine bacterium belonging to genus *Flavobacteriacae*) exhibited antioxidant activity against lipid peroxidation and also showed neuroprotective activity against L-glutamate toxicity [110]. The antioxidant activities of zeaxanthin (extracted from marine bacterium of genus *Muricauda*) [111] and melanin (from marine *Pseudomonas stutzeri* sp.) [112] have also been identified. Another pigment, phycocyanin extracted from marine bacterium *Geitlerinema* sp TRV57, demonstrated appreciable antioxidant activity [113]. Crude pigment extracted from the marine bacterium *Streptomyces bellus* sp. MSA1 also displayed 82% of DPPH (2,2-diphenyl-1-picryl-hydrazyl-hydrate) activity and said to exhibit radical scavenging activity [114]. Likewise, pigment crude extract from *Zobellia laminarie* sp. 465 (isolated from sea sponge) reported to exhibit high antioxidant values for ABTS-L (capture of the 2,2-azino-bis(3-ethylbenzothiazoline)-6-sulphonic acid ($ABTS^+$) radical of the lipophilic fraction) [115], suggesting the importance of marine derived bacterial pigments in pharmaceutical and medicinal industries.

4.1.5. Antiviral Activity

The advancing viral pandemics have taken a toll over the limited pool of existing antiviral agents, which has led to a rigorous search for newer, natural compounds with better antiviral capacities. Various studies on marine bpBPs suggest them as potential candidates. Prodigiosin extracted from *Serratia rubidaea* sp. *RAM_Alex* showed antiviral activity against hepatitis C virus (HCV) upon injecting HepG2 (human liver cancer cell line) cells with 2% of HCV infected serum (Table 2) [116]. Other carotenoid pigments (from *Natrialba* sp. M6) have also displayed complete elimination of HCV and clearance of 89.42% of hepatitis B virus (HBV) [117], indicating the use of marine pigments as availing antiviral agents.

Table 2. Therapeutic applications of bio-pigments extracted from marine bacterial isolates.

Pigments	Marine Bacterial Species	Therapeutic Applications	References
Prodigiosin and cycloprodigiosin	*Zooshikella rubidus* sp. S1-1	Antibacterial	[98]
	Pseudoalteromonas rubra sp. PS1 and SB14	Antifungal	[23]
Prodigiosin	*Serratia rubidaea* sp. *RAM_Alex*	Antiviral	[116]
2-(p-hydroxybenzyl) prodigiosin	*Psuedoalteromonas rubra* sp.	Cytotoxic	[104]
Astaxanthin	*Pontibacter korlensis* sp. AG6	Anticancer	[38]
2,2 dihydroxyastaxanthin	*Brevundimonas* sp.	Antioxidant	[33]
Zeaxanthin	*Muricauda aquimarina* sp. *Muricauda olearia* sp.	Nitric oxide scavenging Inhibition of lipid peroxidation DPPH radical scavenging activities	[111]
Lycopene	*Arthrobacter* sp. G2O	Antitumor	[108]
Beta Carotene	*Cyanobacterium* sp.	Antioxidant Antidiabetic Antitumor	[118]
Pyocyanin and pyorubrin	*Psuedomonas aeroginosa* sp.	Antibacterial	[58]
Melanin	*Streptomyces* sp.	Antibacterial	[68]
	Nocardiopsis sp.	Antiquorum sensing	[76]
	Pseudomonas stutzeri sp.	Antioxidant	[112]
Poly melanin	*Leclercia* sp. BTCZ22	Antibiotic resistance	[75]
Tambjamines	*Psuedoalteromonas tunicata* sp.	Anticancer	[107]
Violacein	*Janthinobacterium* sp. SMN 33.6	Antibacterial	[102]
	Collimonas sp.	Antibacterial	[67]
	Iodobacter sp.	Antibacterial	[100]
	Chromobacterium sp.	Antifungal	[103]
	Janthinobacterium sp. UV13	Anticancer	[106]
3R Saproxanthin and Myxol	*Flavobacteriacae* sp.	Antioxidant	[110]
PCA	*Pseudomonas aeruginosa* sp. GS-33	Effectivity against melanoma cell cancer	[105]
Bright pink-orange colored pigment	*Salinicoccus* sp.	Antibacterial	[101]
Yellow pigment	*Micrococcus* sp. MP76	Antibacterial	[99]

4.2. Industrial Applications

4.2.1. Bio-Pigments as Food Colorants

Researchers have concluded that marine-derived bpBPs can be utilized to provide full-scale commercial production of food-grade pigments, owing to their little or no threats to consumer health. They also showed pleasant colors at low concentrations. Pyorubrin and pyocyanin, for example, extracted from *P. aeruginoasa* sp., when assessed for their utilization as food colorings with agar, gave pleasing colors at 25 mg mLG^{-1} [58]. The utilization of bpBPs was also suggested as a feed additive to promote growth and enhance the coloration of ornamental fishes [119]. Furthermore, prodigiosin (from marine bacterium *Zooshikella* sp.) has been reported to exhibit good staining properties and a three months shelf life [120], which hints toward a sustainable aspect of marine-derived pigmented molecules as food colorants.

4.2.2. Bio-Pigments as Dyeing Agents

The worldwide demand for clothes is rising exponentially. Newly, there is an increase in the insistence of incorporating antimicrobial properties in fabrics. Lee et al. identified a novel marine bacterium *Z. rubidus* sp. S1-1 that produced two significant pigments, i.e., prodigiosin and cycloprodigiosin. These were used to dye cotton and silk fabrics. Results revealed that the application of red-pigmented extract solution on fabrics reduced the growth rate of *S. aureus* sp. KCTC 1916 by 96.62% to 99.98% and *E. coli* sp. KCTC 1924 by 91.37% to 96.98% [98]. Furthermore, *Vibrio* sp. isolated from marine sediment produced a bright red colored prodiginine pigment that was used to dye nylon 66, silk, wool, acrylic, and modacrylic fabrics to obtain a pretty deep-colored shade. The dyed silk and wool fabrics also showed antibacterial activity against *E. coli* sp. and *S. aureus* sp. [121]. Researchers at Ulsan National Institute have also reported the synthesis of antibacterial fabric by using violacein pigment extracted from *C. violacea* sp. [122,123]. Prodigiosin pigment extracted from *Serratia* sp. BTWJ8 effectively dyed paper, PMMA (Polymethyl methacrylate sheets), and rubber latex. Rubber is commonly used in day to day life either in houses or industries. PMMA have been widely utilized for the construction of lenses for exterior lights of automobiles. Different concentrations of prodigiosin produced variable color shades that revealed its affectivity as a coloring agent [124].

4.2.3. Use in Cosmetics

The cosmetic industry is an expeditiously emerging global business market. About 2000 companies in the United States of America are cosmetic manufacturers. It is estimated that American adults use seven different skincare products per day for everyday grooming [125]. The cosmetic industry has a worth of 10.4, 10.6, and 13.01 billion euros in the UK, France, and Germany, respectively [126]. Considering the cosmetic market value worldwide, researchers have also made efforts to explore the use of marine-derived bpBPs in skincare products. The addition of the pigment PCA in a solution of commercial sunscreen enhanced its UV-B (ultraviolet B-rays) protection and increased the SPF (sun protection factor) values up to 10% to 30% [105].

Similarly, melanin incorporated cream (named cream F3) was synthesized by concentrates of seaweed (*Gelidium spinosum*) and melanin pigment (extracted from marine bacterium *Halomonas venusta* sp.). Cream F3 showed high SPF values and photoprotective activity and demonstrated great effectivity in wound healing as well. Moreover, the formulated cream also exhibited antibacterial activity against skin pathogens; *Streptococcus pyogenes* sp. (MTCC 442), and *S. aureus* sp. (MTCC 96) [127]. Another research reported the effectivity of melanin (extracted from marine bacterium *Vibrio natriegens* sp.) in protecting mammalian cells from UV irradiation. Results revealed 90% survival rate of HeLa cells in melanized cell culture [128]. In another report, Bio lip balm made from crude pigment (extracted from *S. bellus* sp. MSA1) in a mixture of coconut oil, lanolin, and shredded bee wax [114] suggested the use of melanin pigment as a significant ingredient in several beauty care products as well.

4.2.4. Antifouling Agent

Billions of dollars have been spent each year to control fouling activities on different objects placed in the marine environment. Biofouling on ships such as dreadnoughts increased the roughness of the hull, which promotes frictional resistance, ultimately leading to an increase in fuel consumption and other corresponding environmental compliances. Heavy metal-based antifoulants cause severe environmental complications, which further mandate the need for "eco-friendly" antifouling agents. Researchers have also revealed the use of marine-derived bpBPs for their role as an antifouling agent, for instance, prodigiosin extracted from *Serratia.* sp. was reported to exhibit antifouling activity against marine fouling bacterial species such as *Gallionella* sp. and *Alteromonas* sp. It also inhibited the adhesion of *Cyanobacterium* sp. on the glass surface [129]. Likewise, another pigment, poly-melanin synthesized by the marine bacterium *P. lipolytica* sp., prevented metamorphosis and decreased the invertebrate larval settlement [130], hence indicating the role of marine bacterial pigments as potential antifoulants.

4.2.5. Photosensitizers

The use of prodigiosin has also been reported as photosensitizers in solar cells. The high photostability of extracted prodigiosin demonstrated its use as a sensitizer in dye-sensitized solar cells (DSSC) (Table 3) [131]. This study suggests the viability of bpBPs in addition to that of prodigiosin for the construction of cost-effective and low tech industrially produced DSSC.

Table 3. Industrial applications of bio-pigments extracted from marine bacterial isolates.

Pigments	Marine Bacterial Species	Industrial Applications	References
Prodigiosin and cycloprodigiosin	*Zooshikella rubidus* sp. S1-1	Dyeing potential	[98]
Prodigiosin	*Vibrio* sp.	Dyeing of fabrics	[121]
	Serratia marcescens sp.11E	Photosensitizers	[131]
	Serratia sp. BTWJ8	Dyeing of paper, PMMA and rubber latex	[124]
	Zooshikella sp.	Food colorant Staining	[120]
	Serratia marcescens sp. CMST 07	Antifouling	[129]
Lycopene	*Streptomyces* sp.	Food grade pigments Feed additive Colorant	[119]
Pyocyanin and pyorubrin	*Psuedomonas aeroginosa* sp.	Food colorings	[58]
Scytonemin	*Lyngbya aestuarii* sp.	Sunscreen	[91]
Melanin	*Vibrio natriegens* sp.	Protection from UV irradiation	[128]
	Halomonas venusta sp.	Sunscreen Wound healing	[127]
	Nocardiopsis sp.	Antibiofilm	[76]
Poly melanin	*Pseudoalteromonas lipolytica* sp.	Antifouling agent	[130]
	Vibrio natriegens sp.	Removal of heavy metals and environmental pollutants	[128]
PCA	*Pseudomonas aeruginosa* sp. GS-33	Enhance SPF values UV-B protection	[105]

5. Industrial Importance and Global Market Trends of Pigmented Compounds

Pigments are already utilized in various nutritional supplements, antibiotics, skin care, and other industrial products (Table 4). The most valuable pigments in the global market are beta-carotene, lutein, and astaxanthin (Figure 3). Astaxanthin has its wide use in nutraceutical industries owing to its antioxidant properties and numerous health benefits. It has also been in wide use in cosmetic industries due to its antiaging activity. Moreover, astaxanthin is being utilized in aquaculture industries to carry out the pigmentation of shrimps, trouts, and salmons. At the industrial scale, astaxanthin production is accomplished using *Paracoccus* sp. It was predicted that the sales volume of astaxanthin by the year 2020 would be 1.1 billion US dollars [132], and the astaxanthin market is estimated to reach up to 3.4 billion US dollars with CAGR (compound annual growth rate) of 16.2% in 2027 [133].

Table 4. Different industrial products and nutritional supplements utilizing pigmented compounds along with manufacturers, product brands, suppliers and company coverage.

Pigments	Products/Nutritional Supplements	Company Coverage/Manufacturers/ Product Brands/Suppliers	References/Links
		Pigments from Bacterial Origin	
Prodigiosin	Prodigiosin *Serratia marcescens*-CAS 82-89-3-CalbiocheM	Sigma-Aldrich	[134]
Astaxanthin	*Paracoccus* Powder (Astaxanthin Powder) 2Z	Brine Shrimp Direct	[135]
	Paracoccus Powder, Natural Source of Astaxanthin, 50g	NoCoast AQUATICS	[136]
Violacein	Violacein (from *Janthinobacterium lividum*)	Sigma-Aldrich	[137]
		Synthetic Pigments and Pigments derived from other sources	
Prodigiosin	Prodigiosin, Antibiotic Prodigiosin 25c	My BioSource Leap Chem Co., Ltd	[138–140]
		Prodigiosin Suppliers	
		Hangzhou Dayangchem Co. Ltd. Puyer Bio Pharm Ltd Santa Cruz Biotechnology	
Astaxanthin	Lucantin® Pink (Astaxanthin) AstaSana™ 10% FS J-Bio™ Astaxanthin BioAstin®Hawaiian Astaxanthin	BASF Nutrition DSM GMP Global Marketing, Inc Cyanotech Corporation	[141–145]
		Astaxanthin Suppliers	
		Aecochem Corp. Simagchem Corporation Hangzhou Dayangchem Co., Ltd. Xiamen Hisunny Chemical Co., Ltd.	

Table 4. *Cont.*

Pigments	Products/Nutritional Supplements	Company Coverage/Manufacturers/ Product Brands/Suppliers	References/Links
Zeaxanthin	ZeaGold®, Zeaxanthin MacuShield®softgel capsule OPTISHARP®(Zeaxanthin) 20% FS.	Kalsec AGP Limited DSM Nutritional Products, Inc.	[146–149]
		Zeaxanthin Suppliers	
		Shanghai Worldyang Chemical Co., Ltd. Sancai Industry Co., Ltd. Carbone Scientific Co., Ltd. BLD Pharmatech Ltd.	
Lycopene	redivivo®(Lycopene) 10% FS Lyc-O-Mato®	DSM Nutritional Products, Inc. LycoRed Ltd.	[150–152]
		Lycopene Suppliers	
		Haihang Industry Co., Ltd. Shanghai Worldyang Chemical Co., Ltd. Junwee Chemical Co., Ltd. B.M.P. Bulk Medicines & Pharmaceuticals GmbH.	
Beta-Carotene	Beta-Carotene 10% DC Beta-Carotene 1% SD Beta-Carotene 30% in corn oil CaroCare®, Beta-Carotene Lyc-O-Beta 7.5% VBA	Barrington Nutritionals DSM Nutritional Products LycoRed Ltd.	[153–156]
		Beta-Carotene Suppliers	
		Puyer BioPharma Ltd. Sancai Industry Co., Ltd. United New Materials Technology SDN.BHD. Hangzhou Keying Chem Co., Ltd.	
Lutein	FloraGLO®Lutein 20% SAF	DSM Nutritional Products, Inc.	[157,158]
		Lutein Suppliers	
		Beckmann-Kenko GmbH. New Natural Biotechnology Co., Ltd. BuGuCh & Partners. Stauber Performance Ingredients, Inc. (previous Pharmline, Inc.)	

FDA accepted the use of beta-carotene as a color additive in food products in the year 1964. Additionally, in 1977, the use of beta-carotene got approved in cosmetics also. The E-Number allotted to beta-carotene is E160a. Over and above, canthaxanthin use in food and broiler chicken feed got authorized in 1969, and the E-Number assigned to it is E161g [159]. Lycopene is also being utilized for many industrial purposes, and it is reckoned that the lycopene market will grow at a rate of 5.3% CAGR, by the end of 2026 [160].

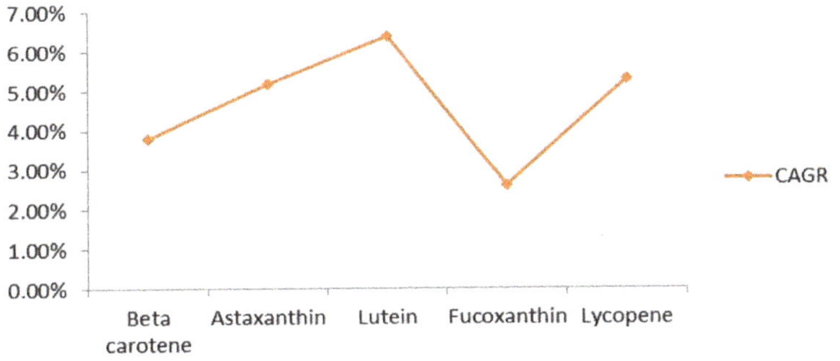

Figure 3. Prospective compound annual growth rate (CAGR) of several pigments by the year 2026 [161–165].

The global market potential of carotenoid pigments is estimated to reach up to 5.7% by 2022 [166]. Europe and USA are the key business markets for carotenoid pigments [118]. It is expected that the global carotenoid market will increase from 1397.59 million US$ in 2018 to 2124.68 million dollars by the end of 2025, at the CAGR of 6.16% [167]. For a long time, different classes of pigmented compounds have occupied the entire market due to their wide range of applications in different industries (Figure 4). It is anticipated that by the year 2022, the global market of food colorants will reach up to 3.75 billion US$, along with farming colorant market, to touch 2.03 billion US$ by the year 2022 [166]. Europe holds the forefront for cutting synthetic colorants' economy by utilizing natural dyes, which make up 85% of total dyes produced. It is evaluated that growing interests towards ready to eat and pre-packaged food items in China, India, and Middle-East countries will also drive the market of food colorants in the Asia Pacific as well [168].

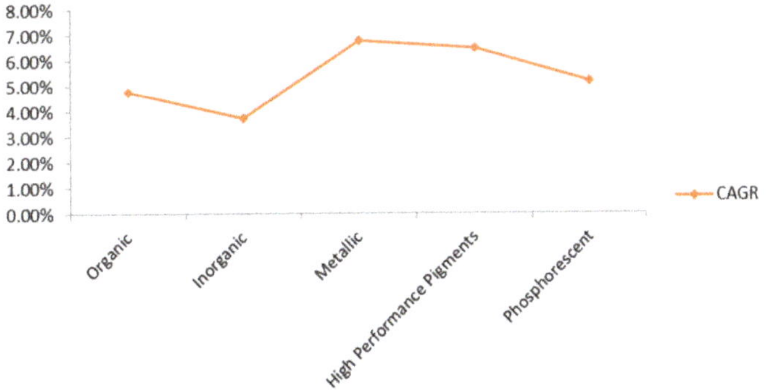

Figure 4. Expected market worth of different classes of pigments during, the forecasting period: 2019 to 2026 [169–173].

6. Conclusions

Marine bacterial pigments can be a potential substitute for synthetic products to fulfill market demand and to ensure the public well-being. Putting aside the fact that synthetic medicines combat bacterial infections, they also pose adverse effects in terms of health. Likewise, artificial colorants due to the presence of azo dyes and heavy metals can also ignite cancer and other allergies. Microbial pigments derived from marine bacteria can be a promising approach to tackle the detrimental effects of these synthetic compounds.

Furthermore, marine bacteria tolerate a vast range of environmental conditions. Due to their unique biological properties, natural pigments from the marine environment also have wide range of applications in pharmaceutical, food, cosmetics, paper, and textile industries [1]. Marine bacterial species can be cultured in vitro and are genetically modified to get the desired level of pigments. Anyhow, there are certain limitations in implementing these naturally derived marine bacterial pigments on an industrial scale significantly: less percentage annual production, technological imperfections, low stability, high need for cost investments, and health complications [11]. For replacing synthetic products, efforts are required to explore new microbial sources and finding better optimization techniques to enhance the production of pigmented compounds. Genetic engineering and other strain development techniques should also be further studied to harvest bioactive pigments from marine bacterial species.

Author Contributions: Conceptualization: I.u.H.; formal analysis: Z.S. and M.F.; investigation: A.N., R.C., and Z.S.; methodology, A.N, R.C., and M.F.; resources: M.F. and I.u.H.; supervision: L.D. and H.M.; validation: L.D. and H.M.; writing—original draft: A.N., R.C., and Z.S.; writing—review and editing: L.D., H.M., and I.u.H. All authors have read and agreed to the published version of the manuscript.

Funding: This research received no external funding.

Conflicts of Interest: The authors declare no conflict of interest.

References

1. Ramesh, C.; Vinithkumar, N.V.; Kirubagaran, R. Marine Pigmented Bacteria: A Prospective Source of Antibacterial Compounds. *J. Nat. Sci. Biol. Med.* **2019**, *10*, 104–113. [CrossRef]
2. Venil, C.K.; Zakaria, Z.A.; Ahmad, W.A. Bacterial Pigments and Their Applications. *Process Biochem.* **2013**, *48*, 1065–1079. [CrossRef]
3. Saviola, B. Pigments and Pathogenesis. *J. Mycobact. Dis.* **2014**, *4*, 5. [CrossRef]
4. Shindo, K.; Misawa, N. New and Rare Carotenoids Isolated from Marine Bacteria and Their Antioxidant Activities. *Mar. Drugs* **2014**, *12*, 1690–1698. [CrossRef] [PubMed]
5. Azman, A.-S.; Mawang, C.-I.; Abubakar, S. Bacterial Pigments: The Bioactivities and as an Alternative for Therapeutic Applications. *Nat. Prod. Commun.* **2018**, *13*, 1747–1754. [CrossRef]
6. Dufossé, L. Pigments, Microbial. In *Reference Module in Life Sciences*; Elsevier: London, UK, 2016.
7. Rodriguez-Amaya, D.B. Natural Food Pigments and Colorants. *Curr. Opin. Food Sci.* **2016**, *7*, 20–26. [CrossRef]
8. Kumar, A.; Vishwakarma, H.S.; Singh, J.; Dwivedi, S.; Kumar, M. Microbial Pigments: Production and Their Applications in Various Industries. *Int. J. Pharm. Chem. Biol. Sci.* **2015**, *5*, 203–212.
9. Numan, M.; Bashir, S.; Mumtaz, R.; Tayyab, S.; Rehman, N.; Khan, A.L.; Shinwari, Z.K.; Al-Harrasi, A. Therapeutic Applications of Bacterial Pigments: A Review of Current Status and Future Opportunities. *3Biotech.* **2018**, *8*, 207. [CrossRef]
10. The Menace of Synthetic Non-Food Colours. *Business Recorder*. Available online: https://fp.brecorder.com/2008/02/20080216695580/ (accessed on 4 June 2020).
11. Galasso, C.; Corinaldesi, C.; Sansone, C. Carotenoids from Marine Organisms: Biological Functions and Industrial Applications. *Antioxidants* **2017**, *6*, 96. [CrossRef]
12. Petruk, G.; Roxo, M.; Lise, F.D.; Mensitieri, F.; Notomista, F.; Wink, M.; Izzo, V.; Monti, D.M. The Marine Gram-Negative Bacterium *Novosphingobium* sp. PP1Y as a Potential Source of Novel Metabolites with Antioxidant Activity. *Biotechnol. Lett.* **2019**, *41*, 273–281. [CrossRef]
13. Stolz, P.; Obermayer, B. Manufacturing Microalgae for Skin Care. *Cosmet. Toilet.* **2015**, *120*, 99–106. [CrossRef]
14. Bruckner, A.W. Life-Saving Products from Coral Reefs. *Issues Sci. Technol.* **2002**, *18*, 35.
15. Soliev, A.B.; Hosokawa, K.; Enomoto, K. Bioactive Pigments from Marine Bacteria: Applications and Physiological Roles. *Evid. Based Complement. Altern. Med.* **2011**, 1–17. [CrossRef] [PubMed]
16. Pawar, R.; Mohandass, C.; Rajasabapathy, R.; Meena, R.M. Molecular Diversity of Marine Pigmented Bacteria in the Central Arabian Sea with Special Reference to Antioxidant Properties. *Cah. Biol. Mar.* **2018**, *59*, 409–420.
17. Baharum, S.N.; Beng, E.K.M.; Mokhtar, M.A.A. Marine Microorganisms: Potential Application and Challenges. *J. Biol. Sci.* **2010**, *10*, 555–564. [CrossRef]
18. Podar, M.; Reysenbach, A.-L. New Opportunities Revealed by Biotechnological Explorations of Extremophiles. *Curr. Opin. Biotechnol.* **2006**, *17*, 250–255. [CrossRef]
19. Giani, M.; Garbayo, I.; Vílchez, C.; Martínez-Espinosa, R.M. Haloarchaeal Carotenoids: Healthy Novel Compounds from Extreme Environments. *Mar. Drugs* **2019**, *17*, 524. [CrossRef]
20. Oren, A. Halophilic Microbial Communities and Their Environments. *Curr. Opin. Biotechnol.* **2015**, *33*, 119–124. [CrossRef]
21. Pierson, L.S.; Pierson, E.A. Metabolism and Function of Phenazines in Bacteria: Impact on the Behavior of Bacteria in the Environment and Biotechnological Process. *Appl. Microbiol. Biotechnol.* **2010**, *86*, 1659–1670. [CrossRef]

22. Kim, D.; Kim, J.F.; Yim, J.H.; Kwon, S.-K.; Lee, C.H.; Lee, H.K. Red to Red—The Marine Bacterium *Hahella chejuensis* and Its Product Prodigiosin for Mitigation of Harmful Algal Blooms. *J. Microbiol. Biotechnol.* **2008**, *18*, 1621–1629.
23. Setiyono, E.; Adhiwibawa, M.A.S.; Indrawati, R.; Prihastyanti, M.N.U.; Shioi, Y.; Brotosudarmo, T.H.P. An Indonesian Marine Bacterium, *Pseudoalteromonas rubra*, Produces Antimicrobial Prodiginine Pigments. *ACS Omega* **2020**, *5*, 4626–4635. [CrossRef] [PubMed]
24. Song, Y.; Liu, G.; Li, J.; Huang, H.; Zhang, X.; Zhang, H.; Ju, J. Cytotoxic and Antibacterial Angucycline- and Prodigiosin-Analogues from the Deep-Sea Derived *Streptomyces* sp. SCSIO 11594. *Mar. Drugs* **2015**, *13*, 1304–1316. [CrossRef] [PubMed]
25. Morgan, S.; Thomas, M.J.; Walstrom, K.M.; Warrick, E.C.; Gasper, B.J. Characterization of Prodiginine Compounds Produced by a *Vibrio species* Isolated from Salt Flat Sediment along the Florida Gulf Coast. *Fine Focus* **2016**, *3*, 33–51. [CrossRef]
26. Ibrahim, D.; Nazari, T.F.; Kassim, J.; Lim, S.-H. Prodigiosin—An Antibacterial Red Pigment Produced by *Serratia marcescens* IBRL USM 84 Associated with a Marine Sponge *Xestospongia testudinaria*. *J. Appl. Pharm. Sci.* **2014**, *4*, 1–6. [CrossRef]
27. Yi, H.; Chang, Y.-H.; Oh, H.W.; Bae, K.S.; Chun, J. *Zooshikella ganghwensis* gen. nov., sp. nov., Isolated from Tidal Flat Sediments. *Int. J. Syst. Evol. Microbiol.* **2003**, *53*, 1013–1018. [CrossRef]
28. Abidin, Z.A.Z.; Ahmad, A.; Latip, J.; Usup, G. Marine *Streptomyces* sp. UKMCC_PT15 Producing Undecylprodigiosin with Algicidal Activity. *J. Teknol.* **2016**, *78*, 11-2. [CrossRef]
29. Bramhachari, P.V.; Mutyala, S.; Bhatnagar, I.; Pallela, R. Novel Insights on the Symbiotic Interactions of Marine Sponge-Associated Microorganisms: Marine Microbial Biotechnology Perspective. In *Marine Sponges: Chemicobiological and Biomedical Applications*, 1st ed.; Pallela, R., Ehrlich, H., Eds.; Springer: New Delhi, India, 2016; pp. 69–95. [CrossRef]
30. Huang, Z.; Dong, L.; Lai, Q.; Liu, J. *Spartinivicinus ruber* gen. nov., sp. nov., a Novel Marine *Gamma proteobacterium* Producing Heptylprodigiosin and Cycloheptylprodigiosin as Major Red Pigments. *Front. Microbiol.* **2020**, *11*, 11. [CrossRef]
31. Xie, B.-B.; Shu, Y.-L.; Qin, Q.-L.; Rong, J.-C.; Zhang, X.-Y.; Chen, X.-L.; Zhou, B.-C.; Zhang, Y.-Z. Genome Sequence of the Cycloprodigiosin-Producing Bacterial Strain *Pseudoalteromonas rubra* ATCC 29570 T. *J. Bacteriol.* **2012**, *194*, 1637–1638. [CrossRef]
32. Jafarzade, M.; Yahya, N.A.; Shayesteh, F.; Usup, G.; Ahmad, A. Influence of Culture Conditions and Medium Composition on the Production of Antibacterial Compounds by Marine *Serratia* sp. WPRA3. *J. Microbiol.* **2013**, *51*, 373–379. [CrossRef]
33. Liu, H.; Zhang, C.; Zhang, X.; Tana, K.; Zhang, H.; Cheng, D.; Ye, T.; Li, S.; Ma, H.; Zheng, H. A Novel Carotenoids-Producing Marine Bacterium from Noble Scallop *Chlamys nobilis* and Antioxidant Activities of Its Carotenoid Compositions. *Food Chem.* **2020**, *320*, 126629. [CrossRef]
34. Henke, N.A.; Heider, S.A.E.; Peters-Wendisch, P.; Wendisch, V.F. Production of the Marine Carotenoid Astaxanthin by Metabolically Engineered *Corynebacterium glutamicum*. *Mar. Drugs* **2016**, *14*, 124. [CrossRef] [PubMed]
35. Asker, D. Isolation and Characterization of a Novel, Highly Selective Astaxanthin-Producing Marine Bacterium. *J. Agric. Food Chem.* **2017**, *65*, 9101–9109. [CrossRef] [PubMed]
36. Shahina, M.; Hameed, A.; Lin, S.-Y.; Hsu, Y.-H.; Liu, Y.-C.; Cheng, I.-C.; Lee, M.-R.; Lai, W.-A.; Lee, R.-J.; Young, C.-C. *Sphingomicrobium astaxanthinifaciens* sp. nov., an Astaxanthin-Producing Glycolipid-Rich Bacterium Isolated from Surface Seawater and Emended Description of the Genus Sphingomicrobium. *Int. J. Syst. Evol. Microbiol.* **2013**, *63*, 3415–3422. [CrossRef] [PubMed]
37. Mukoyama, D.; Takeyama, H.; Kondo, Y.; Matsunaga, T. Astaxanthin Formation in the Marine Photosynthetic Bacterium *Rhodovulum sulfidophilum* Expressing crtI, crtY, crtW and crtZ. *FEMS Microbiol. Lett.* **2006**, *265*, 69–75. [CrossRef]
38. Pachaiyappan, A.; Sadhasivam, G.; Kumar, M.; Muthuvel, A. Biomedical Potential of Astaxanthin from Novel Endophytic Pigment Producing Bacteria *Pontibacter korlensis* AG6. *Waste Biomass Valoriz.* **2020**, 1–11. [CrossRef]
39. Balraj, J.; Pannerselvam, K.; Jayaraman, A. Isolation of Pigmented Marine Bacteria *Exiguobacterium* sp. from Peninsular Region of India and a Study on Biological Activity of Purified Pigment. *Int. J. Sci. Techol. Res.* **2014**, *3*, 375–384.
40. Shi, X.-L.; Wu, Y.-H.; Cheng, H.; Zhang, X.-Q.; Wang, C.-S.; Xu, X.-W. Complete Genome Sequence of Astaxanthin-Producing Bacterium *Altererythrobacter ishigakiensis*. *Mar. Genom.* **2016**, *30*, 77–79. [CrossRef]
41. Zhao, Y.; Guo, L.; Xia, Y.; Zhuang, X.; Chu, W. Isolation, Identification of Carotenoid-Producing *Rhodotorula* sp. from Marine Environment and Optimization for Carotenoid Production. *Mar. Drugs* **2019**, *17*, 161. [CrossRef]
42. Lee, J.H.; Kim, Y.T. Cloning and Characterization of the Astaxanthin Biosynthesis gene Cluster from the Marine Bacterium *Paracoccus haeundaensis*. *Gene* **2006**, *370*, 86–95. [CrossRef]
43. Lee, J.H.; Hwang, Y.M.; Baik, K.S.; Choi, K.S.; Ka, J.-O.; Seong, C.N. *Mesoflavibacter aestuarii* sp. nov., a Zeaxanthin Producing Marine Bacterium Isolated from Seawater. *Int. J. Syst. Evol. Microbiol.* **2014**, *64*, 1932–1937. [CrossRef]
44. Hameed, A.; Shahina, M.; Lin, S.-Y.; Lai, W.-A.; Hsu, Y.-H.; Liu, Y.-C.; Young, C.-C. *Aquibacter zeaxanthinifaciens* gen. nov., sp. nov., a Zeaxanthin-Producing Bacterium of the Family *Flavobacteriaceae* Isolated from Surface Seawater, and Emended Descriptions of the Genera *Aestuariibaculum* and *Gaetbulibacter*. *Int. J. Syst. Evol. Microbiol.* **2014**, *64*, 138–145. [CrossRef] [PubMed]
45. Asker, D.; Beppu, T.; Ueda, K. *Zeaxanthinibacter enoshimensis* gen. nov., sp. nov., a Novel Zeaxanthin-Producing Marine Bacterium of the Family *Flavobacteriaceae*, Isolated from Seawater Off Enoshima Island, Japan. *Int. J. Syst. Evol. Microbiol.* **2007**, *57*, 837–843. [CrossRef] [PubMed]
46. Shahina, M.; Hameed, A.; Lin, S.-Y.; Lee, R.-J.; Lee, M.-R.; Young, C.-C. *Gramella planctonica* sp. nov., a Zeaxanthin-Producing Bacterium Isolated from Surface Seawater, and Emended Descriptions of *Gramella aestuarii* and *Gramella echinicola*. *Antonie van Leeuwenhoek* **2014**, *105*, 771–779. [CrossRef] [PubMed]

47. Asker, D.; Beppu, T.; Ueda, K. *Mesoflavibacter zeaxanthinifaciens* gen. nov., sp. nov., a Novel Zeaxanthin Producing Marine Bacterium of the Family *Flavobacteriaceae*. *Syst. Appl. Microbiol.* **2007**, *30*, 291–296. [CrossRef]
48. Sowmya, R.; Sachindra, N.M. Carotenoid Production by *Formosa* sp. KMW, Marine Bacteria of *Flavobacteriaceae* Family: Influence of Culture Conditions and Nutrient Composition. *Biocatal. Agric. Biotechnol.* **2015**, *4*, 559–567. [CrossRef]
49. Thawornwiriyanun, P.; Tanasupawat, S.; Dechsakulwatana, C.; Techkarnjanaruk, S.; Suntornsuk, W. Identification of Newly Zeaxanthin-Producing Bacteria Isolated from Sponges in the Gulf of Thailand and Their Zeaxanthin Production. *Appl. Biochem. Biotechnol.* **2012**, *167*, 2357–2368. [CrossRef]
50. Hameed, A.; Arun, A.B.; Ho, H.-P.; Chang, C.-M.J.; Rekha, P.D.; Lee, M.-R.; Young, C.-C. Supercritical Carbon Dioxide Micronization of Zeaxanthin from Moderately Thermophilic Bacteria *Muricauda lutaonensis* CC-HSB-11T. *J. Agric. Food Chem.* **2011**, *59*, 4119–4124. [CrossRef]
51. Seto, R.; Takaichi, S.; Kurihara, T.; Kishi, R.; Honda, M.; Takenaka, S.; Tsukatani, Y.; Madigan, M.T.; Wang-Otomo, Z.Y.; Kimura, Y. Lycopene-Family Carotenoids Confer Thermostability on Photocomplexes from a New Thermophilic Purple Bacterium. *Biochemistry* **2020**, *59*, 2351–2358. [CrossRef]
52. Ramanathan, G.; Ramalakshmi, P. Studies on Efficacy of Marine Bacterium *Salinicoccus roseus* Pigment for Their Bioactive Potential. *Eur. J. Biomed. Pharm. Sci.* **2017**, *4*, 330–334.
53. Montero, O.; Macías-Sánchez, M.D.; Lama, C.M.; Lubián, L.M.; Mantell, C.; Rodríguez, M.; De la Ossa, E.M. Supercritical CO_2 Extraction of â-Carotene from a Marine Strain of the Cyanobacterium *Synechococcus* Species. *J. Agric. Food Chem.* **2005**, *53*, 9701–9707. [CrossRef]
54. Hamidi, M.; Kozani, P.S.; Kozani, P.S.; Pierre, G.; Michaud, P.; Delattre, C. Marine Bacteria Versus Microalgae: Who Is the Best for Biotechnological Production of Bioactive Compounds with Antioxidant Properties and Other Biological Applications? *Mar. Drugs.* **2020**, *18*, 28. [CrossRef] [PubMed]
55. Sibero, M.T.; Bachtiarini, T.U.; Trianto, A.; Lupita, A.H.; Sari, D.P.; Igarashi, Y.; Harunari, E.; Sharma, A.R.; Radjasa, O.K.; Sabdono, A. Characterization of a Yellow Pigmented Coral-Associated Bacterium Exhibiting Anti-Bacterial Activity against Multidrug Resistant (MDR) Organism. *Egypt. J. Aquat. Res.* **2018**, *45*, 81–87. [CrossRef]
56. Teramoto, M.; Nishijima, M. *Flavicella marina* gen. nov., sp. nov., a Carotenoid-Producing Bacterium from Surface Seawater. *Int. J. Syst. Evol. Microbiol.* **2015**, *65*, 799–804. [CrossRef] [PubMed]
57. Loh, W.L.C.; Huang, K.-C.; Ng, H.S.; Lan, J.C.-W. Exploring the Fermentation Characteristics of a Newly Isolated Marine Bacteria Strain, *Gordonia terrae* TWRH01 for Carotenoids Production. *J. Biosci. Bioeng.* **2020**, *130*, 187–194. [CrossRef]
58. Saha, S.; Thavasi, T.R.; Jayalakshmi, S. Phenazine Pigments from *Pseudomonas aeruginosa* and Their Application as Antibacterial Agent and Food Colourants. *Res. J. Microbiol.* **2008**, *3*, 122–128. [CrossRef]
59. Fulton, J.M.; Arthur, M.A.; Freeman, K.H. Subboreal Aridity and Scytonemin in the Holocene Black Sea. *Org. Geochem.* **2012**, *49*, 47–55. [CrossRef]
60. Soule, T.; Palmer, K.; Gao, Q.; Potrafka, R.M.; Stout, V.; Garcia-Pichel, F. A Comparative Genomics Approach to Understanding the Biosynthesis of The sunscreen Scytonemin in Cyanobacteria. *BMC Genom.* **2009**, *10*, 336. [CrossRef]
61. Thøgersen, M.S.; Delpin, M.W.; Melchiorsen, J.; Kilstrup, M.; Månsson, M.; Bunk, B.; Spröer, C.; Overmann, J.; Nielsen, K.F.; Gram, L. Production of the Bioactive Compounds Violacein and Indolmycin Is Conditional in a *maeA* Mutant of *Pseudoalteromonas luteoviolacea* S4054 Lacking the Malic Enzyme. *Front. Microbiol.* **2016**, 7. [CrossRef]
62. Aye, A.M.; Bonnin-Jusserand, M.; Brian-Jaisson, F.; Ortalo-Magne, A.; Culioli, G.; Nevry, R.K.; Rabah, N.; Blache, Y.; Molmeret, M. Modulation of Violacein Production and Phenotypes Associated with Biofilm by Exogenous Quorum Sensing N-acylhomoserine Lactones in the Marine Bacterium *Pseudoalteromonas ulvae* TC14. *Microbiology* **2015**, *161*, 2039–2052. [CrossRef]
63. Dang, H.T.; Yotsumoto, K.; Enomoto, K. Draft Genome Sequence of Violacein-Producing Marine Bacterium *Pseudoalteromonas* sp. 520P1. *Genome Announc.* **2014**, 2. [CrossRef]
64. Ballestriero, F.; Daim, M.; Penesyan, A.; Nappi, J.; Schleheck, D.; Bazzicalupo, P.; Schiavi, E.D.; Egan, S. Antinematode Activity of Violacein and the Role of the Insulin/IGF-1 Pathway in Controlling Violacein Sensitivity in *Caenorhabditis elegans*. *PLoS ONE* **2014**, *9*, e109201. [CrossRef] [PubMed]
65. Wu, Y.-H.; Cheng, H.; Xu, L.; Jin, X.-B.; Wang, C.-S.; Xu, X.-W. Physiological and Genomic Features of a Novel Violacein-Producing Bacterium Isolated from Surface Seawater. *PLoS ONE* **2017**, *12*, e0179997. [CrossRef] [PubMed]
66. Yada, S.; Wang, Y.; Zou, Y.; Nagasaki, K.; Hosokawa, K.; Osaka, I.; Arakawa, R.; Enomoto, K. Isolation and Characterization of Two Groups of Novel Marine Bacteria Producing Violacein. *Mar. Biotechnol.* **2008**, *10*, 128–132. [CrossRef] [PubMed]
67. Hakvåg, S.; Fjærvik, E.; Klinkenberg, G.; Borgos, S.E.F.; Josefsen, K.D.; Ellingsen, T.E.; Zotchev, S.B. Violacein-Producing *Collimonas* sp. from the Sea Surface Microlayer of Coastal Waters in Trøndelag, Norway. *Mar. Drugs* **2009**, *7*, 576–588. [CrossRef]
68. Vasanthabharathi, V.; Lakshminarayanan, R.; Jayalakshmi, S. Melanin Production from Marine *Streptomyces*. *Afr. J. Biotechnol.* **2011**, *10*, 11224–11234. [CrossRef]
69. Tarangini, K.; Mishra, S. Production, Characterization and Analysis of Melanin from Isolated Marine *Pseudomonas* sp. Using Vegetable Waste. *Res. J. Eng. Sci.* **2013**, *2*, 40–46.
70. Lucas-Elío, P.; Goodwin, L.; Woyke, T.; Pitluck, S.; Nolan, M.; Kyrpides, N.C.; Detter, J.C.; Copeland, A.; Teshima, H.; Bruce, D.; et al. The Genomic Standards Consortium Complete Genome Sequence of the Melanogenic Marine Bacterium *Marinomonas mediterranea* Type Strain (MMB-1T). *Stand. Genom. Sci.* **2012**, *6*, 63–73. [CrossRef]

71. Manirethan, V.; Raval, K.; Balakrishnan, R.M. Adsorptive Removal of Trivalent and Pentavalent Arsenic from Aqueous Solutions Using Iron and Copper Impregnated Melanin Extracted from the Marine Bacterium *Pseudomonas stutzeri*. *Environ. Pollut.* **2019**, *257*, 113576. [CrossRef]
72. Kurian, N.K.; Nair, H.P.; Bhat, S.G. Evaluation of Anti-Inflammatory Property of Melanin from Marine *Bacillus* spp. BTCZ31. *Asian J. Pharm. Clin. Res.* **2015**, *8*, 251–255.
73. Sivaperumal, P.; Kamala, K.; Rajaram, R.; Mishra, S.S. Melanin from Marine *Streptomyces* sp. *(MVCS13) with Potential Effect against Ornamental Fish Pathogens of Carassius auratus. Biocatal. Agric. Biotechnol.* **2014**, *3*, 134–141. [CrossRef]
74. Kurian, N.K.; Bhat, S.G. Photoprotection and Anti-Inflammatory Properties of Non–Cytotoxic Melanin from Marine Isolate *Providencia rettgeri* Strain BTKKS1. *Biosci. Biotechnol. Res. Asia* **2017**, *14*, 1475–1484. [CrossRef]
75. Kurian, N.K.; Nair, H.P.; Bhat, S.G. Characterization of Melanin Producing Bacteria Isolated from 96m depth Arabian Sea Sediments. *Res. J. Biotechnol.* **2019**, *14*, 64–71.
76. Kamarudheen, N.; Naushad, T.; Rao, K.V.B. Biosynthesis, Characterization and Antagonistic Applications of Extracellular Melanin Pigment from Marine *Nocardiopsis Sps. Indian J. Pharm. Educ. Res.* **2019**, *53*, 112–120. [CrossRef]
77. Kurian, N.K.; Bhat, S.G. Food, Cosmetic and Biological Applications of Characterized DOPA-Melanin from *Vibrio alginolyticus* strain BTKKS3. *Appl. Biol. Chem.* **2018**, *61*, 163–171. [CrossRef]
78. Sakai-Kawada, F.E.; Ip, C.G.; Hagiwara, K.A.; Awaya, J.D. Biosynthesis and Bioactivity of Prodiginine Analogs in Marine Bacteria, *Pseudoalteromonas*: A Mini Review. *Front. Microbiol.* **2019**, *10*, 1715. [CrossRef] [PubMed]
79. Picott, K.J.; Deichert, J.A.; De Kemp, E.M.; Schatte, G.; Sauriol, F.; Ross, A.C. Isolation and Characterization of Tambjamine MYP1, a Macrocyclic Tambjamine Analogue from Marine Bacterium *Pseudoalteromonas citrea*. *MedChemComm* **2019**, *10*, 478–483. [CrossRef]
80. Harris, A.K.P.; Williamson, N.R.; Slater, H.; Cox, A.; Abbasi, S.; Foulds, I.; Simonsen, H.T.; Leeper, F.J.; Salmond, G.P.C. The *Serratia* Gene Cluster Encoding Biosynthesis of the Red Antibiotic, Prodigiosin, Shows Species- and Strain-Dependent Genome Context Variation. *Microbiology* **2004**, *150*, 3547–3560. [CrossRef]
81. Williamson, N.R.; Simonsen, H.T.; Ahmed, R.A.A.; Goldet, G.; Slater, H.; Woodley, L.; Leeper, F.J.; Salmond, G.P.C. Biosynthesis of the Red Antibiotic, Prodigiosin, in *Serratia*: Identification of a Novel 2-methyl-3-n-amyl-Pyrroie (MAP) Assembly Pathway, Definition of the Terminal Condensing Enzyme, and Implications for Undecylprodigiosin Biosynthesis in *Streptomyces*. *Mol. Microbiol.* **2005**, *56*, 971–989. [CrossRef]
82. Garneau-Tsodikova, S.; Dorrestein, P.C.; Kelleher, N.L.; Walsh, T.C. Protein Assembly Line Components in Prodigiosin Biosynthesis: Characterization of PigA, G, H, I, J. *J. Am. Chem. Soc.* **2006**, *128*, 12600–12601. [CrossRef]
83. Kimata, S.; Izawa, M.; Kawasaki, T.; Hayakawa, Y. Identification of a Prodigiosin Cyclization Gene in the Roseophilin Producer and Production of a New Cyclized Prodigiosin in a Heterologous Host. *J. Antibiot.* **2017**, *70*, 196–199. [CrossRef]
84. De Rond, T.; Stow, P.; Eigl, I.; Johnson, R.E.; Chan, L.J.G.; Goyal, G.; Baidoo, E.E.K.; Hillson, N.J.; Petzold, C.J.; Sarpong, R.; et al. Oxidative Cyclization of Prodigiosin by an Alkylglycerol Monooxygenase-Like Enzyme. *Nat. Chem. Biol.* **2017**, *13*, 1155–1159. [CrossRef] [PubMed]
85. Burke, C.; Thomas, T.; Egan, S.; Kjelleberg, S. The Use of Functional Genomics for the Identification of a Gene Cluster Encoding for the Biosynthesis of an Antifungal Tambjamine in the Marine Bacterium *Pseudoalteromonas tunicata*: Brief Report. *Environ. Microbiol.* **2007**, *9*, 814–818. [CrossRef] [PubMed]
86. Marchetti, P.M.; Kelly, V.; Simpson, J.P.; Ward, M.; Campopiano, D.J. The Carbon chain-Selective Adenylation Enzyme TamA: The Missing Link between Fatty Acid and Pyrrole Natural Product Biosynthesis. *Org. Biomol. Chem.* **2018**, *16*, 2735–2740. [CrossRef] [PubMed]
87. Asker, D.; Awad, T.S.; Beppu, T.; Ueda, K. Rapid and Selective Screening Method for Isolation and Identification of Carotenoid-Producing Bacteria. *Methods Mol. Biol.* **2018**, *1852*, 143–170. [CrossRef]
88. Yabuzaki, J. Carotenoids Database: Structures, Chemical Fingerprints and Distribution among Organisms. *Database* **2017**, *2017*. [CrossRef]
89. Scaife, M.A.; Burja, A.M.; Wright, P.C. Characterization of Cyanobacterial β-Carotene Ketolase and Hydroxylase Genes in *Escherichia coli*, and Their Application for Astaxanthin Biosynthesis. *Biotechnol. Bioeng.* **2009**, *103*, 944–955. [CrossRef]
90. Soule, T.; Stout, V.; Swingley, W.D.; Meeks, J.C.; Garcia-Pichel, F. Molecular Genetics and Genomic Analysis of Scytonemin Biosynthesis in *Nostoc punctiforme* ATCC 29133. *J. Bacteriol.* **2007**, *189*, 4465–4472. [CrossRef]
91. Balskus, E.P.; Case, R.J.; Walsh, C.T. The Biosynthesis of Cyanobacterial Sunscreen Scytonemin in Intertidal Microbial Mat Communities. *FEMS Microbiol. Ecol.* **2011**, *77*, 322–332. [CrossRef]
92. Estrada, A.F.; Maier, D.; Scherzinger, D.; Avalos, J.; Al-Babili, S. Novel Apo Carotenoid Intermediates in *Neurospora crassa* Mutants Imply a New Biosynthetic Reaction Sequence Leading to Neurosporaxanthin Formation. *Fungal Genet. Biol.* **2008**, *45*, 1497–1505. [CrossRef]
93. Liao, L.; Su, S.; Zhao, B.; Fan, C.; Zhang, J.; Li, H.; Chen, B. Biosynthetic Potential of a Novel Antarctic Actinobacterium *Marisediminicola antarctica* ZS314T Revealed by Genomic Data Mining and Pigment Characterization. *Mar. Drugs.* **2019**, *17*, 388. [CrossRef]
94. Balashov, S.P.; Imasheva, E.S.; Boichenko, V.A.; Antón, J.; Wang, J.M.; Lanyi, J.K. Xanthorhodopsin: A Proton Pump with a Light-Harvesting Carotenoid Antenna. *Science* **2005**, *309*, 2061–2064. [CrossRef] [PubMed]

95. Lanyi, J.K.; Balashov, S.P. Xanthorhodopsin: A Bacteriorhodopsin-Like Proton Pump with a Carotenoid Antenna. *Biochim. Biophys. Acta* **2008**, *1777*, 684–688. [CrossRef] [PubMed]
96. Graham, J.E.; Bryant, D.A. The Biosynthetic Pathway for Myxol-2′fucoside (Myxoxanthophyll) in the Cyanobacterium *Synechococcus* sp. Strain PCC 7002. *J. Bacteriol.* **2009**, *191*, 3292–3300. [CrossRef] [PubMed]
97. Hannibal, L.; Lorquin, J.; Dortoli, N.A.; Garcia, N.; Chaintreuil, C.; Massonboivin, C.; Dreyfus, B.; Giraud, E. Isolation and Characterization of Canthaxanthin Biosynthesis Genes from the Photosynthetic Bacterium *Bradyrhizobium* sp. Strain ORS278. *J. Bacteriol.* **2000**, *182*, 3850–3853. [CrossRef] [PubMed]
98. Lee, J.C.; Kim, Y.-S.; Park, S.; Kim, J.; Kang, S.-J.; Lee, M.-H.; Ryu, S.; Choi, J.M.; Oh, T.-K.; Yoon, J.-H. Exceptional Production of Both Prodigiosin and Cycloprodigiosin as Major Metabolic Constituents by a Novel Marine Bacterium, *Zooshikella rubidus* S1-1. *Appl. Environ. Microbiol.* **2011**, *77*, 4967–4973. [CrossRef] [PubMed]
99. Karbalaei-Heidari, H.R.; Partovifar, M.; Memarpoor-Yazdi, M. Evaluation of the Bioactive Potential of Secondary Metabolites Produced by a New Marine *Micrococcus* Species Isolated from the Persian Gulf. *Avicenna J. Med. Biotechnol.* **2020**, *12*, 61–65. [PubMed]
100. Atalah, J.; Blamey, L.; Muñoz-Ibacache, S.; Gutierrez, F.; Urzua, M.; Encinas, M.V.; Páez, M.; Sun, J.; Blamey, J.M. Isolation and Characterization of Violacein from an Antarctic *Iodobacter*: A Non-Pathogenic Psychrotolerant Microorganism. *Extremophiles* **2019**, *24*, 43–52. [CrossRef]
101. Srilekha, V.; Krishna, G.; Srinivas, V.S.; Charya, M.A.S. Antimicrobial evaluation of bioactive pigment from *Salinicoccus* sp. isolated from Nellore sea coast. *Int. J. Biotechnol. Biochem.* **2017**, *13*, 211–217.
102. Asencio, G.; Lavin, P.; Alegría, K.; Domínguez, M.; Bello, H.; González-Rocha, G.; González-Aravena, M. Antibacterial Activity of the Antarctic Bacterium *Janthinobacterium* sp. SMN 33.6 against Multi-Resistant Gram-Negative Bacteria. *Electron. J. Biotechnol.* **2014**, *17*, 1–5. [CrossRef]
103. Sasidharan, A.; Sasidharan, N.K.; Amma, D.B.N.S.; Vasu, R.K.; Nataraja, A.V.; Bhaskaran, K. Antifungal Activity of Violacein Purified from a Novel Strain of *Chromobacterium* sp. NIIST (MTCC 5522). *J. Microbiol.* **2015**, *53*, 694–701. [CrossRef]
104. Fehér, D.; Barlow, R.S.; Lorenzo, P.S.; Hemscheidt, T.K. A 2-Substituted Prodiginine, 2-(p-Hydroxybenzyl) Prodigiosin, from *Pseudoalteromonas rubra*. *J. Nat. Prod.* **2008**, *71*, 1970–1972. [CrossRef] [PubMed]
105. Patil, S.; Paradeshi, J.; Chaudhari, B. Anti-Melanoma and UV-B Protective Effect of Microbial Pigment Produced by Marine *Pseudomonas aeruginosa* GS-33. *Nat. Prod. Res.* **2016**, *30*, 2835–2839. [CrossRef] [PubMed]
106. Alem, D.; Marizcurrena, J.J.; Saravia, V.; Davyt, D.; Martinez-Lopez, W.; Castro-Sowinski, S. Production and Antiproliferative Effect of Violacein, a Purple Pigment Produced by an Antarctic Bacterial Isolate. *World J. Microbiol. Biotechnol.* **2020**, *36*, 120. [CrossRef] [PubMed]
107. Pinkerton, D.M.; Banwell, M.G.; Garson, M.J.; Kumar, N.; De Moraes, M.O.; Cavalcanti, B.C.; Pessoa, C. Antimicrobial and Cytotoxic Activities of Synthetically Derived Tambjamines C and E - J, BE-18591, and a Related Alkaloid from the Marine Bacterium *Pseudoalteromonas tunicata*. *Chem. Biodivers.* **2010**, *7*, 1311–1324. [CrossRef]
108. Afra, S.; Makhdoumi, A.; Matin, M.M.; Feizy, J. A Novel Red Pigment from Marine *Arthrobacter* sp. G20 with Specific Anticancer Activity. *J. Appl. Microbiol.* **2017**, *123*, 1228–1236. [CrossRef]
109. Miao, L.; Wang, X.; Jiang, W.; Yang, S.; Zhou, H.; Zhai, Y.; Zhou, X.; Dong, K. Optimization of the Culture Condition for an Antitumor Bacterium *Serratia proteamacula* 657 and Identification of the Active Compounds. *World J. Microbiol. Biotechnol.* **2012**, *29*, 855–863. [CrossRef]
110. Shindo, K.; Kikuta, K.; Suzuki, A.; Katsuta, A.; Kasai, H.; Yasumoto-Hirose, M.; Matsuo, Y.; Misawa, N.; Takaichi, S. Rare Carotenoids, (3R)-saproxanthin and (3R,2′S)-myxol, Isolated from Novel Marine Bacteria (*Flavobacteriaceae*) and Their Antioxidative Activities. *Appl. Microbiol. Biotechnol.* **2007**, *74*, 1350–1357. [CrossRef]
111. Prabhu, S.; Rekha, P.D.; Young, C.C.; Hameed, A.; Lin, S.-Y.; Arun, A.B. Zeaxanthin Production by Novel Marine Isolates from Coastal Sand of India and Its Antioxidant Properties. *Appl. Biochem. Biotechnol.* **2013**, *171*, 817–831. [CrossRef]
112. Kumar, G.; Sahu, N.; Reddy, G.N.; Prasad, R.B.N.; Nagesh, N.; Kamal, A. Production of Melanin Pigment from *Pseudomonas stutzeri* Isolated from Red Seaweed *Hypneamusci formis*. *Lett. Appl. Microbiol.* **2013**, *57*, 295–302. [CrossRef]
113. Renugadevi, K.; Nachiyar, C.V.; Sowmiya, P.; Sunkar, S. Antioxidant Activity of Phycocyanin Pigment Extracted from Marine Filamentous Cyanobacteria *Geitlerinema* sp TRV57. *Biocatal. Agric. Biotechnol.* **2018**, *16*, 237–242. [CrossRef]
114. Srinivasan, M.; Keziah, S.M.; Hemalatha, M.; Devi, C.S. Pigment from *Streptomyces bellus* MSA1 Isolated from Marine Sediments. *IOP Conf. Ser. Mater. Sci. Eng.* **2017**, 263. [CrossRef]
115. Silva, T.R.; Tavares, R.S.N.; Canela-Garayoa, R.; Eras, J.; Rodrigues, M.V.N.; Neri-Numa, I.A.; Pastore, G.M.; Rosa, L.H.; Schultz, J.A.A.; Debonsi, H.M.; et al. Chemical Characterization and Biotechnological Applicability of Pigments Isolated from Antarctic Bacteria. *Mar. Biotechnol.* **2019**, *21*, 416–429. [CrossRef] [PubMed]
116. Metwally, R.A.; Abeer, A.; El-Sikaily, A.; El-Sersy, N.A.; Ghozlan, H.; Sabry, S. Biological Activity of Prodigiosin from *Serratia rubidaea* RAM_Alex. *Res. J. Biotechnol.* **2019**, *14*, 100.
117. Hegazy, G.E.; Abu-Serie, M.M.; Abo-Elela, G.M.; Ghozlan, H.; Sabry, S.A.; Soliman, N.A.; Abdel-Fattah, Y.R. In Vitro Dual (Anticancer and Antiviral) Activity of the Carotenoids Produced by Haloalkaliphilic Archaeon *Natrialba* sp. M6. *Sci. Rep.* **2020**, *10*, 5986. [CrossRef]
118. Torregrosa-Crespo, J.; Montero, Z.; Fuentes, J.L.; García-Galbis, M.R.; Garbayo, I.; Vílchez, C.; Martínez-Espinosa, R.M. Exploring the Valuable Carotenoids for the Large-Scale Production by Marine Microorganisms. *Mar. Drugs* **2018**, *16*, 203. [CrossRef]

119. Dharmaraj, S.; Ashokkumar, B.; Dhevendaran, K. Food-Grade Pigments from *Streptomyces* sp. Isolated from the Marine Sponge *Callyspongia diffusa*. *Food Res. Int.* **2009**, *42*, 487–492. [CrossRef]
120. Ramesh, C.; Vinithkumar, N.V.; Kirubagaran, R.; Venil, C.K.; Dufossé, L. Applications of Prodigiosin Extracted from Marine Red Pigmented Bacteria *Zooshikella* sp. and Actinomycete *Streptomyces* sp. *Microorganisms* **2020**, *8*, 556. [CrossRef]
121. Alihosseini, F.; Ju, K.-S.; Lango, J.; Hammock, B.D.; Sun, G. Antibacterial Colorants: Characterization of Prodiginines and Their Applications on Textile Materials. *Biotechnol. Prog.* **2008**, *24*, 742–747. [CrossRef]
122. Anti-Bacterial Fabric Holds Promise for Fighting Superbug. Science Daily. *Ulsan National Institute of Science and Technology (UNIST)*. Available online: https://www.sciencedaily.com/releases/2016/03/160308091646.htm (accessed on 8 March 2020).
123. Michael, R. New Antibiotic Dye May Help Prevent Infectious Diseases. *Contagion Live*. Available online: https://www.contagionlive.com/news/new-antibiotic-dye-may-help-prevent-infectious-diseases (accessed on 4 April 2020).
124. Krishna, J.G.; Jacob, A.; Kurian, P.; Elyas, K.K.; Chandrasekaran, M. Marine Bacterial Prodigiosin as Dye for Rubber Latex, Polymethyl Methacrylate Sheets and Paper. *Afr. J. Biotechnol.* **2013**, *12*, 2266–2269. [CrossRef]
125. Derikvand, P.; Llewellyn, C.A.; Purton, S. Cyanobacterial Metabolites as a Source of Sunscreens and Moisturizers: A Comparison with Current Synthetic Compounds. *Eur. J. Phycol.* **2017**, *52*, 43–56. [CrossRef]
126. Consumption Value of Cosmetics and Personal Care in Europe in 2018, by Country (in Million Euros). Available online: https://www.statista.com/statistics/382100/european-cosmetics-market-volume-by-country/ (accessed on 24 June 2020).
127. Poulose, N.; Sajayan, A.; Ravindran, A.; Sreechitra, T.; Vardhan, V.; Selvin, J.; Kiran, G.S. Photoprotective Effect of Nanomelanin-Seaweed Concentrate in Formulated Cosmetic Cream: With Improved Antioxidant and Wound Healing Properties. *J. Photochem. Photobiol. B Biol.* **2020**, 205. [CrossRef]
128. Wang, Z.; Tschirhart, T.; Schultzhaus, Z.; Kelly, E.E.; Chen, A.; Oh, E.; Nag, O.; Glaser, E.R.; Kim, E.; Lloyd, P.F.; et al. Characterization and Application of Melanin Produced by the Fast-Growing Marine Bacterium *Vibrio natriegens* Through Heterologous Biosynthesis. *Appl. Environ. Microbiol.* **2019**, *86*. [CrossRef] [PubMed]
129. Priya, K.A.; Satheesh, S.; Balasubramaniem, A.K.; Varalakshmi, P.; Gopal, S.; Natesan, S. Antifouling Activity of Prodigiosin from Estuarine Isolate of *Serratia marcescens* CMST 07. In *Microbiological Research In Agroecosystem Management*, 1st ed.; Velu, R.K., Ed.; Springer: New Delhi, India, 2013; pp. 11–21. [CrossRef]
130. Zeng, Z.; Guo, X.-P.; Cai, X.; Wang, P.; Li, B.; Yang, J.-L.; Wang, X. Pyomelanin from *Pseudoalteromonas lipolytica* Reduces Biofouling. *Microbiol. Biotechnol.* **2017**, *10*, 1718–1731. [CrossRef] [PubMed]
131. Hernández-Velasco, P.; Morales-Atilano, I.; Rodríguez-Delgado, M.; Rodríguez-Delgado, J.M.; Luna-Moreno, D.; Ávalos-Alanís, F.G.; Villarreal-Chiu, J.F. Photoelectric Evaluation of Dye-Sensitized Solar Cells Based on Prodigiosin Pigment Derived from *Serratia marcescens* 11E. *Dyes Pigment.* **2020**, *177*, 108278. [CrossRef]
132. Global Astaxanthin Market—Sources, Technologies and Applications. Available online: https://www.marketresearch.com/Industry-Experts-v3766/Global-Astaxanthin-Sources-Technologies-Applications-8827191/ (accessed on 30 March 2020).
133. Astaxanthin Market Size, Share & Trends Analysis Report by Source, by Product, by Application and Segment Forecasts, 2020–2027. Available online: https://www.reportlinker.com/p05868776/Astaxanthin-Market-Size-Share-Trends-Analysis-Report-By-Source-By-Product-By-Application-And-Segment-Forecasts.html (accessed on 1 June 2020).
134. Prodigiosin *Serratia marcescens* - CAS 82-89-3 - Calbiochem. *Merck: A Leader in Life Sciences*. Available online: https://www.merckmillipore.com/INTL/en/product/Prodigiosin-Serratia-marcescens-CAS-82-89-3-Calbiochem,EMD_BIO-529685?ReferrerURL=https%3A%2F%2Fwww.google.com%2F&bd=1 (accessed on 6 June 2020).
135. Paracoccus Powder (Astaxanthin Powder) 2OZ. *Brine Shrimp Direct*. Available online: https://www.brineshrimpdirect.com/paracoccus-powder-astaxanthin-powder-2-ounce (accessed on 6 June 2020).
136. Paracoccus Powder, Natural Source of Astaxanthin, 50g. *NoCoast Aquatics*. Available online: https://www.nocoastaquatics.com/products/paracoccus-powder-astaxanthin (accessed on 7 June 2020).
137. Violacein (V9389)—Datasheet—Sigma-Aldrich. Available online: https://www.sigmaaldrich.com/content/dam/sigmaaldrich/docs/Sigma/Datasheet/2/v9389dat.pdf (accessed on 6 August 2020).
138. Prodigiosin antibiotic. *My Biosource*. Available online: https://www.mybiosource.com/antibiotic/prodigiosin/651317 (accessed on 6 August 2020).
139. Suppliers for Prodigiosin 25c. *BuyersGuideChem*. Available online: https://www.buyersguidechem.com/AliefAus.php?pnumm=121199138105 (accessed on 15 June 2020).
140. Suppliers for Prodigiosin. *BuyersGuideChem*. Available online: https://www.buyersguidechem.com/AliefAus.php?pnumm=553227439869&modus=einprod (accessed on 21 June 2020).
141. Lucantin® Pink (Astaxanthin). Animal Nutrition Products. *BASF*. Available online: https://nutrition.basf.com/global/en/animal-nutrition/products/lucantin-pink.html (accessed on 15 June 2020).
142. AstaSana™ 10% FS. UL Prospector. Available online: https://www.ulprospector.com/pt/eu/Food/Detail/15324/541591/AstaSana-10-FS (accessed on 15 June 2020).
143. J-Bio™ Astxanthin. GMP Global Marketing, Inc. Available online: https://www.gmpglobalmarketing.com/portfolio-items/j-bio-astaxanthin/ (accessed on 27 June 2020).
144. BioAstin® Hawaiian Astaxanthin. Cyanotech Corporation. Foods of Hawaii. Available online: https://www.foodsofhawaii.com/author/cyanotech-corporation/?user=132 (accessed on 27 June 2020).

145. Suppliers for Astaxanthin: BuyersGuideChem. Available online: https://www.buyersguidechem.com/AliefAus.php?pnumm=244231380730 (accessed on 1 July 2020).
146. ZeaGold®, Zeaxanthin. Kalsec. Available online: https://www.kalsec.com/products/zeaxanthin/ (accessed on 1 July 2020).
147. MacuShield® softgel capsule. AGP Limited. Available online: https://agp.com.pk/agp_products/macushield/ (accessed on 3 July 2020).
148. OPTISHARP® (Zeaxanthin) 20% FS by DSM Nutritional Products, Inc. Prospector. Available online: https://www.ulprospector.com/en/na/Food/Detail/6295/239896/OPTISHARP-Zeaxanthin-20-FS (accessed on 1 July 2020).
149. Suppliers for Zeaxanthin. Available online: https://www.buyersguidechem.com/AliefAus.php?pnumm=750686624198 (accessed on 1 July 2020).
150. Redivivo® (Lycopene) 10% FS by DSM Nutritional Products, Inc. UL Prospector. Available online: https://www.ulprospector.com/en/na/Food/Detail/6295/239893/redivivo-Lycopene-10-FS (accessed on 3 September 2020).
151. Lyc-O-Mato® -Packed with Powder of Tomato by LycoRed Ltd. NUTRA Ingredients. Available online: https://www.nutraingredients.com/Product-innovations/Lyc-O-Mato-R-Packed-with-the-Power-of-the-Tomato (accessed on 3 September 2020).
152. Suppliers for CAS 502-65-8. BuyersGuideChem. Available online: https://www.buyersguidechem.com/AliefAus.php?pnumm=130817126231 (accessed on 2 June 2020).
153. Food, Beverage and Nutrition. Barrington Nutritionals. Available online: https://www.ulprospector.com/en/na/Food/Suppliers/2252/Barrington-Nutritionals?st=1 (accessed on 3 September 2020).
154. CaroCare®, Beta-Carotene by DSM Nutritional Products, Inc. New Hope Network. Available online: https://www.newhope.com/ingredients-general/carocare-dsm-natural-choice-carotene (accessed on 7 September 2020).
155. Lyc-O-Beta 7.5% VBA by LycoRed Ltd. Available online: https://www.lycored.com/lyc-o-beta-7-5-vbaf/ (accessed on 7 September 2020).
156. Suppliers for CAS 7235-40-7. BuyersGuideChem. Available online: https://www.buyersguidechem.com/AliefAus.php?pnumm=180131918792 (accessed on 7 September 2020).
157. FloraGLO® Lutein 20% SAF by DSM Nutritional Products, Inc. UL Prospector. Available online: https://www.ulprospector.com/en/na/Food/Detail/6295/239891/FloraGLO-Lutein-20-SAF (accessed on 10 August 2020).
158. Suppliers for Lutein. BuyersGuideChem. Available online: https://www.buyersguidechem.com/AliefAus.php?pnumm=330278064387&modus=einprod&anzahl_produkte_cas=4 (accessed on 1 July 2020).
159. Summary of Color Additives for Use in the United States in Foods, Drugs, Cosmetics, and Medical Devices, U.S. Food and Drug Administration. Available online: https://www.fda.gov/industry/color-additive-inventories/summary-color-additives-use-united-states-foods-drugs-cosmetics-and-medical-devices (accessed on 15 November 2017).
160. Global Lycopene Market Forecast Report to 2026: Includes Company Profiling with Detailed Strategies, Financials and Recent Developments—Research and Markets.com. Available online: https://www.businesswire.com/news/home/20190430005701/en/Global-Lycopene-Market-Forecast-Report-2026-Includes (accessed on 30 November 2020).
161. Beta Carotene Market Size, Share, Trends, & Industry Analysis Report by Source (Algae, Fruits & Vegetables, & Synthetic); by Application (Food & Beverages, Dietary Supplements, Cosmetics, & Animal Feed); by Regions: Segment Forecast, 2018–2026. Available online: https://www.polarismarketresearch.com/industry-analysis/beta-carotene-market (accessed on 1 December 2020).
162. Astaxanthin Market 2020: CAGR of 5.2% with top Countries Data, Latest Trends, Market Size, Share, Global Industry Analysis & Forecast to 2026. Press Release. Available online: https://www.wfmj.com/story/42684950/astaxanthin-market-2020-cagr-of-52-with-top-countries-data-latest-trends-market-size-share-global-industry-analysis-amp-forecast-to-2026 (accessed on 27 September 2020).
163. Lutein Market to Reach USD 454.8 million by 2026. Reports and Data. Available online: https://www.prnewswire.com/news-releases/lutein-market-to-reach-usd-454-8-million-by-2026--reports-and-data-300941985.html (accessed on 1 December 2020).
164. Fucoxanthin (CAS 3351-86-8) Market 2020 Is Expected to See Magnificent Spike in CAGR with Global Industry Brief Analysis by Top Countries Data Which Includes Driving Factors by Manufacturers Growth and Forecast 2026. Press Release. Available online: https://www.marketwatch.com/press-release/fucoxanthin-cas-3351-86-8-market-2020-is-expected-to-see-magnificent-spike-in-cagr-with-global-industry-brief-analysis-by-top-countries-data-which-includes-driving-factors-by-manufacturers-growth-and-forecast-2026-2020-09-17 (accessed on 2 December 2020).
165. Lycopene—Global Market Outlook (2017–2026). Research and Markets. Available online: https://www.researchandmarkets.com/reports/4765062/lycopene-global-market-outlook-2017-2026 (accessed on 1 December 2020).
166. Saini, R.K.; Keum, Y.-S. Microbial Platforms to Produce Commercially Vital Carotenoids at Industrial Scale: An Updated Review of Critical Issues. *J. Ind. Microbiol. Biotechnol.* **2019**, *46*, 657–674. [CrossRef]
167. Global Carotenoids Market—Premium Insight, Competitive News Feed Analysis, Company Usability Profiles, Market Sizing & Forecasts to 2025. Available online: https://www.marketresearch.com/360iResearch-v4164/Global-Carotenoids-Premium-Insight-Competitive-13036149/ (accessed on 4 July 2020).
168. Venil, C.K.; Dufossé, L.; Devi, P.R. Bacterial Pigments: Sustainable Compounds with Market Potential for Pharma and Food Industry. *Front. Sustain. Food Syst.* **2020**, *4*, 100. [CrossRef]

169. Global Organic Pigments Market Is Segmented by Type (Azo Pigments, Phthalocyanine Pigments, High Performance Pigments (HPPs), Others), by Application (Printing Inks, Paints & Coatings, Plastics, Others (Textiles, Cosmetics, Food)), by Source (Natural Organic Pigments, Synthetic Organic Pigments), and by Region (North America, Latin America, Europe, Asia Pacific, Middle East, and Africa)—Share, Size, Outlook, and Opportunity Analysis, 2019–2026. Available online: https://www.datamintelligence.com/research-report/organic-pigments-market (accessed on 4 December 2020).
170. Global Inorganic Pigments Market Is Driven by Rising Demand from End Users and Expanding Paint and Coating Industry. Data Bridge Market Research. Available online: https://www.databridgemarketresearch.com/news/global-inorganic-pigments-market (accessed on 4 December 2020).
171. Global Metallic Pigments Market Is Driven by Prevailing Demand of Packaging Applications Mainly in the Food and Beverages, Tobacco Industry, Gifts Wraps among Others. Data Bridge Market Research. Available online: https://www.databridgemarketresearch.com/news/global-metallic-pigments-market (accessed on 5 December 2020).
172. Global High-Performance Pigment Market Is Segmented by Type (Organic High-Performance Pigments, In-Organic High-Performance Pigments), by Application (Coatings, Plastics, Cosmetic Products, Others (Includes Inks)), and by Region (North America, Latin America, Europe, Asia Pacific, Middle East, and Africa)—Share, Size, Outlook, and Opportunity Analysis, 2019–2026. Available online: https://www.datamintelligence.com/research-report/high-performance-pigment-market (accessed on 5 December 2020).
173. Global Phosphorescent Pigments Market—Industry Trends and Forecast to 2026. *Data Bridge Market Research*. Available online: https://www.databridgemarketresearch.com/reports/global-phosphorescent-pigments-market (accessed on 6 December 2020).

microorganisms

MDPI

Article

Pentaminomycins C–E: Cyclic Pentapeptides as Autophagy Inducers from a Mealworm Beetle Gut Bacterium

Sunghoon Hwang [1], Ly Thi Huong Luu Le [2], Shin-Il Jo [3], Jongheon Shin [1], Min Jae Lee [2,*] and Dong-Chan Oh [1,*]

1 Natural Products Research Institute, College of Pharmacy, Seoul National University, 1 Gwanak-ro, Gwanak-gu, Seoul 08826, Korea; sunghooi@snu.ac.kr (S.H.); shinj@snu.ac.kr (J.S.)
2 Department of Biochemistry and Molecular Biology, College of Medicine, Seoul National University, Seoul 03080, Korea; huongluuly94@snu.ac.kr
3 Welfare Division, Seoul Zoo, Seoul Grand Park, Gwacheon, Gyeonggi 13829, Korea; metather@seoul.go.kr
* Correspondence: minjlee@snu.ac.kr (M.J.L.); dongchanoh@snu.ac.kr (D.-C.O.); Tel.: +82-2-740-8254 (M.J.L.); +82-2-880-2491 (D.-C.O.); Fax: +82-2-744-4534 (M.J.L.); +82-2-762-8322 (D.-C.O.)

Received: 14 August 2020; Accepted: 8 September 2020; Published: 10 September 2020

Abstract: Pentaminomycins C–E (**1–3**) were isolated from the culture of the *Streptomyces* sp. GG23 strain from the guts of the mealworm beetle, *Tenebrio molitor*. The structures of the pentaminomycins were determined to be cyclic pentapeptides containing a modified amino acid, N^5-hydroxyarginine, based on 1D and 2D NMR and mass spectroscopic analyses. The absolute configurations of the amino acid residues were assigned using Marfey's method and bioinformatics analysis of their nonribosomal peptide biosynthetic gene cluster (BGC). Detailed analysis of the BGC enabled us to propose that the structural variations in **1–3** originate from the low specificity of the adenylation domain in the nonribosomal peptide synthetase (NRPS) module 1, and indicate that macrocyclization can be catalyzed noncanonically by penicillin binding protein (PBP)-type TE. Furthermore, pentaminomycins C and D (**1 and 2**) showed significant autophagy-inducing activities and were cytoprotective against oxidative stress in vitro.

Keywords: insect; mealworm; gut bacteria; OSMAC; cyclic peptides; biosynthetic pathway; autophagy inducer

1. Introduction

Microbial secondary metabolites have been utilized for the discovery and development of innumerable medicinal drugs, such as antibiotics, anticancer agents, and immunosuppressive medicines [1]. However, as more microbial compounds have been reported, the discovery of structurally unique and biologically active compounds has become more difficult. Thus, the development and application of new technologies in exploring new or less-investigated metabolites is particularly important in the field of microbial natural product chemistry. One possible approach is the chemical examination of relatively unexplored microbes [2]. For example, insect-associated bacteria, which are expected to play various roles in the life cycles of hosts due to their range of secondary metabolites, are one of the most unexplored chemical sources [3]. In particular, insects are assumed to actively associate with chemically prolific actinobacteria, which generally originate from the soil. Indeed, studying the metabolites of insect-associated actinomycetes led to the discovery of the new cyclic depsipeptide dentigerumycin [4], a geldanamycin analog natalamycin [5], the polyketide alkaloid camporidine A [6], *N*-acetylcysteamine-bearing indanone thioester formicin A [7], and chlorinated cyclic peptides, nicrophorusamides [8]. These bacterial secondary metabolites display antifungal, anti-inflammatory, anticancer, and antibacterial activities. In addition, *Streptomyces* spp. of the dung beetle *Copris tripartius* have also demonstrated the diversity of bioactive secondary metabolites of insect-associated bacteria, including a tricyclic lactam [9,10], a dichlorinated indanone [11], 2-alkenyl cinnamic acid-bearing cyclic peptides [12], and naphthoquinone-oxindoles [13].

Besides focusing on relatively understudied microbes, another method to efficiently discover new bioactive secondary metabolites is to diversify the culture conditions of one strain, which may induce the biosynthesis of various microbial metabolites that have not been produced under a previous set of conditions. This one strain many compounds (OSMAC) approach [14] is based on the fact that diverse biosynthetic gene clusters exist in the genome of a single strain.

Inspired by these two approaches for the efficient discovery of new bioactive metabolites, we collected mealworm beetles, *Tenebrio molitor*, and isolated actinobacteria from the guts. Initial chemical profiling and subsequent chemical analysis of *Streptomyces* sp. GG23 identified the production of lydiamycin A [15]. Detailed structure elucidation enabled us to revise the previously reported structure [16]. Analysis of the full genome of *Streptomyces* sp. GG23 disclosed its chemical capacity due to the fact that it possesses 31 biosynthetic gene clusters for various structural classes of secondary metabolites in its 8.7 Mb genome (Table S2). Therefore, we altered the culture conditions to induce the production of bacterial metabolites other than lydiamycin. As a result of diverse changes in the composition of the culture media, the production of a series of structurally distinct peptides, namely pentaminomycins C–E (**1–3**) (Figure 1), which were barely detected by LC-ESI-MS as minor metabolites when lydiamycin A was produced, was significantly increased. Pentaminomycins A and B, which were produced by *Streptomyces* sp. from Japanese soil have been previously reported as inhibitors of melanin synthesis [17], while pentaminomycin C, which was produced by *Streptomyces cacaoi* isolated from Nigerian cacao beans, has been identified as an antibiotic against Gram-positive bacteria [18]. Thus, we herein report our investigation into the evaluation of pentaminomycins C–E (**1–3**) as autophagy inducers and antagonizers against menadione-induced oxidative stress, which have not been previously reported for pentaminomycins. Overall, we report the structures, biological activities, and biosynthetic pathways of pentaminomycins C–E.

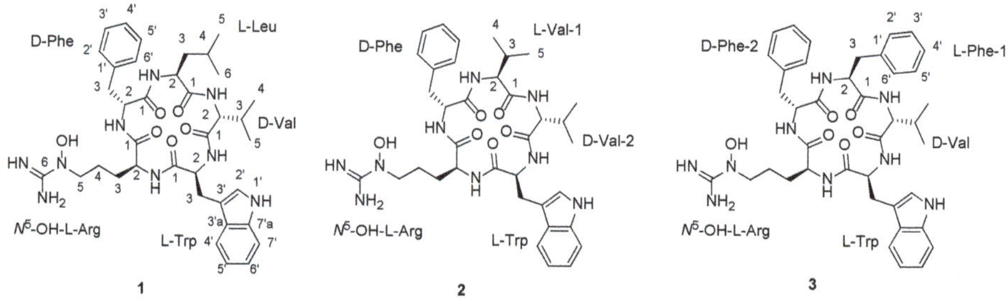

Figure 1. Structures of pentaminomycins C–E (**1–3**).

2. Materials and Methods

2.1. General Experimental Procedures

Optical rotations were measured using a JASCO P-2000 polarimeter (JASCO, Easton, MD, USA). UV spectra were recorded using a Chirascan Plus Applied Photophysics Ltd. (Applied Photophysics, Leatherhead, Surrey, UK). Infrared (IR) spectra were obtained on a Thermo NICOLET iS10 spectrometer (Thermo Fisher Scientific, Waltham, MA, USA). ^1H, ^{13}C, and two dimensional nuclear magnetic resonance (NMR) spectra were acquired using a Bruker Avance 800 MHz spectrometer (Bruker, Billerica, MA, USA) with the NMR solvent of DMSO-d_6 (reference chemical shifts: δ_C 39.5 ppm and δ_H 2.50 ppm) at the National Center for Inter-university Research Facilities (NCIRF) at Seoul National University. Electrospray ionization (ESI) low-resolution liquid chromatography-mass spectrometry

(LC/MS)data were measured with an Agilent Technologies 6130 quadrupole mass spectrometer (Agilent, Santa Clara, CA, USA) coupled to an Agilent Technologies 1200 series high-performance liquid chromatography (HPLC) instrument using a reversed-phase $C_{18}(2)$ column (Phenomenex Luna, 100 × 4.6 mm). High-resolution fast atom bombardment (HR-FAB) mass spectra were recorded using a Jeol JMS-600 W high-resolution mass spectrometer (JEOL, Akishima, Tokyo, Japan) at the NCIRF. Semi-preparative HPLC separations were achieved using a Gilson 305 pump and a Gilson UV/VIS-155 detector (Gilson, Middleton, WI, USA).

2.2. *Bacterial Isolation and Identification*

The mealworm beetle used in this study was randomly selected as the specimens of insects raised at the Seoul Grand Park, Gwacheon-si, Gyeonggi Province, Republic of Korea. The selected specimen was identified as mealworm beetles, *Tenebrio molitor* Linnaues, based on external morphological characters. The mealworm beetle-associated actinomycete strain GG23 was isolated from the gut of adult mealworm beetles by using starch-casein agar (SCA), which contained 10 g of soluble starch, 0.3 g of casein, 2 g of KNO_3, 0.05 g of $MgSO_4{\cdot}7H_2O$, 2 g of K_2HPO_4, 2 g of NaCl, 0.02 g of $CaCO_3$, 0.01 g of $FeSO_4{\cdot}7H_2O$, and 18 g of agar in 1 L of distilled water. The strain was identified as *Streptomyces* sp. (GenBank accession number MT033037), which is closest to *Streptomyces cacaoi* (identity of 99.6%) (GenBank accession number NZ_MUBL01000000) based on 16S rRNA gene sequence analysis.

2.3. *Cultivation and Extraction*

The GG23 strain was initially cultured with 50 mL of modified PST medium (containing 5 g peptone, 5 g sucrose, and 5 g tryptone in 1 L distilled water) in a 125-mL Erlenmeyer flask. After 3 d of culture on a rotary shaker at 200 rpm and 30 °C, 5 mL of the culture was transferred to 200 mL of the same medium in a 500-mL Erlenmeyer flask. The culture was maintained for 3 d under the same conditions as the 50-mL stage, and 10 mL of the culture was inoculated into 1 L of PST medium in 2.8-L Fernbach flasks (60 each × 1 L, total volume 60 L) for 4-d incubation at 170 rpm and 30 °C. The whole culture was extracted twice with 120 L of ethyl acetate, and organic layer was separated and concentrated in vacuo to yield 25 g of dry material.

2.4. *Isolation of Pentaminomycins C–E*

The dried extract was re-suspended with celite in MeOH followed by drying in vacuo. The celite-adsorbed extract was loaded onto 2 g of a pre-packed C_{18} Sepak resin. The extract was fractionated using five different compositions of aqueous MeOH (i.e., 20, 40, 60, 80, and 100 vol% MeOH). Pentaminomycins C–E (**1–3**) eluted in the 80 and 100 vol% MeOH fractions. The fractions were dried, redissolved in MeOH, and filtered to remove insoluble particles. Pentaminomycins C–E were further purified by preparative reversed-phase HPLC with a Phenomenex Luna 10 μm $C_{18}(2)$ 250 × 21.20 mm column with a gradient system of 30–50 vol% aqueous MeCN over 30 min (flow rate: 10 mL/min, detection: UV 210 nm). Pentaminomycin D (**2**) eluted at 20 min under these HPLC conditions, whereas pentaminomycins C (**1**) and E (**3**) eluted together at 23 min. The eluates were subjected to semi-preparative reversed-phase HPLC using a YMC-Pack CN 250 × 10 mm, S-5 μm, 12 nm column with an isocratic system of 35 vol% MeCN containing 0.05% trifluoroacetic acid (flow rate: 2 mL/min, detection: UV 210 nm). In the final purification, pentaminomycins C (**1**) (5 mg), D (**2**) (3 mg), and E (**3**) (4 mg) were collected at retention times of 20, 28, and 33 min, respectively.

2.4.1. Pentaminomycin C (**1**)

White powder; $[\alpha]^{20}_{D}$-50 (c 0.1, MeOH); UV (MeOH) λ_{max} (log ε) 218 (2.04), 281 (0.35) nm; IR (neat) ν_{max} 3302, 2965, 1673, 1537, 1444, 1199, 1140 cm^{-1}; ^1H and ^{13}C NMR data, see Table 1; HR-FAB-MS *m/z* 718.4041 $[M+H]^+$ (calcd. for $C_{37}H_{52}N_9O_6$ 718.4035).

Table 1. NMR data for **1–3** in DMSO-d_6.

Pentaminomycin C (1)				Pentaminomycin D (2)				Pentaminomycin E (3)			
Position		δ_C, type	δ_H, mult (J in Hz)	Position		δ_C, type	δ_H, mult (J in Hz)	Position		δ_C, type	δ_H, mult (J in Hz)
L-Leu	1	172.2, C		L-Val	1	171.4, C		L-Phe	1	170.6, C	
	2	50.5, CH	4.40, ddd (15.0,9.0,7.0)		2	57.5, CH	4.12, dd (7.5,7.5)		2	52.8, CH	4.67, dt (9.0,7.5)
	3	41.1, CH_2	1.34, m		3	30.7, CH	1.77, m		3	38.1, CH_2	2.84, dd (13.5,7.0) 2.78, dd (13.5,7.5)
	4	24.2, CH	1.43, m		4	19.0, CH_3	0.83, d (7.0)		1′	137.1, C	
	5	22.7, CH_3	0.85, d (6.5)		5	18.3, CH_3	0.76, d (7.0)		2′	129.1, CH	7.18, m
	6	22.0, CH_3	0.82, d (6.5)		NH		7.50, m		3′	128.0, CH	7.22, m
	NH		7.55, d (9.0)						4′	126.2, C	7.17, m
									NH		7.69, d (9.0)
D-Val	1	171.3, C		D-Val	1	171.4, C		D-Val	1	171.2, C	
	2	59.9, CH	3.70, dd (10.0,7.5)		2	60.2, CH	3.70, dd (10.0,7.5)		2	59.9, CH	3.70, dd (10.0,7.5)
	3	28.5, CH	1.65, m		3	28.1, CH	1.64, m		3	28.5, CH	1.65, m
	4	19.0, CH_3	0.75, d (6.5)		4	19.2, CH_3	0.77, d (7.0)		4	19.0, CH_3	0.75, d (6.5)
	5	18.5, CH_3	0.34, d (6.5)		5	18.5, CH_3	0.31, d (7.0)		5	18.5, CH_3	0.34, d (6.5)
	NH		8.41, d (7.5)		NH		8.45, d (7.5)		NH		8.41, d (7.5)
L-Trp	1	171.7, C		L-Trp	1	171.7, C		L-Trp	1	171.6, C	
	2	55.3, CH	4.29, ddd (11.0,8.0,3.5)		2	55.3, CH	4.28, ddd (11.5,8.0,3.0)		2	55.3, CH	4.27, ddd (11.5,8.0,3.5)
	3	26.9, CH_2	3.18, dd (14.5,3.0) 2.80, dd (14.5,12.0)		3	26.9, CH_2	3.19, dd (14.5,3.0) 2.91, dd (14.5,11.5)		3	26.9, CH_2	3.17, dd (14.5,3.0) 2.90, dd (14.5,11.5)
	2-NH		8.59, d (8.0)		2-NH		8.63, d (8.0)		2-NH		8.58, d (8.0)
	1′(NH)		10.78, br s		1′(NH)		10.78, br s		1′(NH)		10.76, br s
	2′	123.9, CH	7.17, m		2′	123.9, CH	7.17, m		2′	123.8, CH	7.16, m
	3′	110.2, C			3′	110.2, C			3′	110.2, C	
	3′a	126.8, C			3′a	126.8, C			3′a	126.8, C	

Table 1. *Cont.*

	Pentaminomycin C (1)				Pentaminomycin D (2)				Pentaminomycin E (3)		
	4′	117.9, CH	7.51, d (8.0)		4′	117.8, CH	7.51, m		4′	117.8, CH	7.50, d (8.0)
	5′	118.3, CH	6.98, t (7.5)		5′	118.2, CH	6.98, dd (7.5,7.5)		5′	118.2, CH	6.97, dd (7.5,7.5)
	6′	120.8, CH	7.05, t (7.5)		6′	120.8, CH	7.04, dd (7.5,7.5)		6′	120.8, CH	7.04, dd (7.5,7.5)
	7′	111.3, CH	7.31, d (8.0)		7′	111.3, CH	7.30, d (8.0)		7′	111.3, CH	7.30, d (8.0)
	7′a	136.1, C			7′a	136.1, C			7′a	136.1, C	
N^5-OH-L-Arg	1	170.4, C		N^5-OH-L-Arg	1	170.4, C		N^5-OH-L-Arg	1	170.4, C	
	2	52.8, CH	4.16, dt (7.0,7.0)		2	52.9, CH	4.16, dt (8.0,7.0)		2	53.0, CH	4.16, dt (7.0,7.0)
	3	28.1, CH_2	1.53, m		3	28.2, CH_2	1.53, m		3	28.2, CH_2	1.52, m
	4	22.1, CH_2	1.33, m, 1.18, m		4	22.0, CH_2	1.33, m, 1.15, m		4	22.1, CH_2	1.34, m, 1.17, m
	5	50.4, CH_2	3.43, m		5	50.5, CH_2	3.42, m		5	50.5, CH_2	3.42, m
	N^5-OH		10.49, br s		N^5-OH		10.55, br s		N^5-OH		10.47, s
	6	157.3, C			6	157.4, C			6	157.3, C	
	6-NH(3H)		7.45, br s		6-NH(3H)		7.50, br s		6-NH(3H)		7.43, br s
	NH		7.29, d (7.5)		NH		7.23, m		NH		7.26, d (7.5)
D-Phe	1	170.6, C		D-Phe	1	170.7, C		D-Phe	1	170.6, C	
	2	53.7, CH	4.46, ddd (9.0,9.0,6.0)		2	53.5, CH	4.52, ddd (9.5,8.5,6.0)		2	53.5, CH	4.47, ddd (9.0,9.0,5.5)
	3	34.2, CH_2	2.96, dd (14.0,5.5) 2.80, dd (14.0,9.5)		3	33.9, CH_2	2.98, dd (14.0,5.0) 2.79, dd (14.0,10.0)		3	34.0, CH_2	2.93, dd (14.0,5.5) 2.76, dd (14.0,9.5)
	1′	137.9, C			1′	138.0, C			1′	137.9, C	
	2′, 6′	129.0, CH	7.24, m		2′, 6′	129.0, CH	7.24, m		2′, 6′	128.9, CH	7.20, m
	3′, 5′	128.0, CH	7.23, m		3′, 5′	128.0, CH	7.23, m		3′, 5′	128.0, CH	7.22, m
	4′ NH	126.2, CH	7.17, m 8.85, d (8.0)		4′ NH	126.2, CH	7.17, m 8.85, d (8.0)		4′ NH	126.2, CH	7.17, m 8.89, d (8.5)

^1H and ^{13}C NMR data were recorded at 800 and 200 MHz, respectively (J: ^1H-^1H coupling constant).

2.4.2. Pentaminomycin D (**2**)

White powder; $[\alpha]_D^{20}$-8 (c 0.1, MeOH); UV (MeOH) λ_{max} (log ε) 218 (2.20), 281 (0.36) nm; IR (neat) ν_{max} 3296, 2964, 1672, 1536, 1444, 1199, 1140 cm^{-1}; ^1H and ^{13}C NMR data, see Table 1; HR-FAB-MS *m/z* 704.3876 $[M+H]^+$ (calcd. for $C_{36}H_{50}N_9O_6$ 704.3879).

2.4.3. Pentaminomycin E (**3**)

White powder; $[\alpha]_D^{20}$-31 (c 0.1, MeOH); UV (MeOH) λ_{max} (log ε) 216 (1.91), 282 (0.25) nm; IR (neat) ν_{max} 3300, 2964, 1672, 1536, 1445, 1199, 1140 cm^{-1}; ^1H and ^{13}C NMR data, see Table 1; HR-FAB-MS *m/z* 752.3879 $[M+H]^+$ (calcd. for $C_{40}H_{50}N_9O_6$ 752.3879).

*2.5. Marfey's Analysis of Pentaminomycins D and E (**2** and **3**)*

A sample (1 mg) of pentaminomycin D (**2**) was hydrolyzed in 0.5 mL of 6 N HCl at 100 °C for 1 h. After hydrolysis, the reaction vial was cooled in an ice bucket for 3 min. After this time, the reaction solvent was evaporated in vacuo, and the hydrolysate containing the free amino acids was dissolved in 100 µL of 1N $NaHCO_3$. Subsequently, 50 µL of a 10 mg/mL L-FDLA solution in acetone was added to the solution. The reaction mixture was stirred at 80 °C for 3 min, then 50 µL of 2N HCl was used to neutralize the reaction mixture, which was subsequently diluted using 300 µL of a 50 vol% aqueous MeCN solution. An aliquot (20 µL) of the reaction mixture was analyzed by LC/MS using a Phenomenex $C_{18}(2)$ column (Luna, 100 × 4.6 mm, 5 µm) under gradient solvent conditions (flow rate 0.7 mL/min; UV 340 nm detection; 10–60 vol% $MeCN/H_2O$ containing 0.1% formic acid over 50 min). LC/MS analysis indicated that during acid hydrolysis, N^5-hydroxyarginine was converted to arginine. The L-FDLA derivatives of the two valine (37.5 and 44.6 min), tryptophan (41.0 min), arginine (22.6 min), and phenylalanine (47.5 min) residues of pentaminomycin D were detected by LC/MS analysis. The same procedure was performed for authentic L-and D-Val, Trp, Arg, and Phe to compare the retention times with those of the amino acids from **2** (Table S1 and Figure S17). The absolute configurations of pentaminomycin E (**3**) were also established in the same manner.

2.6. Genome Analysis and the Biosynthetic Pathway

Whole genome sequencing of the GG23 strain was performed using PacBio RS II (Chunlab, Inc., Seocho-gu, Seoul, Korea).The sequencing data were assembled with PacBio SMRT Analysis v. 2.3.0, using a hierarchical genome assembly process (HGAP) protocol. Nucleotide sequences of the *Streptomyces* sp. GG23 genomes were generated in three contigs with a total of 8,666,993 base pairs. Gene prediction was performed using Prodigal v. 2.6.2, and sequences were annotated with EggNOG v. 4.5, Swissprot, KEGG, and SEED (Chunlab, Inc., Seocho-gu, Seoul, Republic of Korea) The biosynthetic gene clusters (BGCs) were analyzed using antiSMASH v. 5.0 [19].

2.7. Autophagic Flux Assay

To examine cellular autophagic flux after treatment of pentaminomycins, HEK293 cells (≈80% confluence) were treated with 0, 3.125, 6.25, 12.5, or 25 µM pentaminomycin C, D, or E for 8 h. Whole cell extracts were prepared using RIPA buffer (50 mM Tris-HCl (pH 8.0), 1% of NP-40, 0.5% of deoxycholate, 0.1% of sodium dodecyl sulfate (SDS), and 150 mM NaCl) supplemented with protease inhibitor cocktails. The lysates were then centrifuged at 16,000× *g* for 30 min at 4 °C. The supernatants were separated by SDS-PAGE and transferred to a polyvinylidene difluoride (PVDF) membrane (Merck Millipore, Darmstadt, Germany). Subsequently, the membranes were blocked with 5% non-fat milk and probed with the following antibodies: anti-LC3 (L7543, MilliporeSigma, St. Louis, MO, USA), anti-SQSTM1 (sc-28359, Santa Cruz Biotechnology, Dallas, TX, USA), anti-GABARAPL1 (D5R9Y, Cell Signaling Technology, Dallas, TX, USA), and anti-GAPDH (A1978, MilliporeSigma). The membranes were then incubated with a horseradish peroxidase-conjugated anti-mouse IgG antibody (81-6720, Invitrogen, Carlsbad, CA, USA) or an anti-rabbit IgG antibody (G21234, Invitrogen), and visualized using an ECL system (Thermo Fisher Scientific, Waltham, MA, USA).

2.8. Cytotoxicity Assays

The cell viability was assessed using the CellTiter-Glo Luminescent Cell Viability Assay (Promega) kit as previously described [20]. More specifically, the cells were grown in a black wall/clear-bottom 96-well plate and treated with either pentaminomycins C–E (20 μM; **1**–**3**), menadione (25 μM), or combinations of pentaminomycins and menadione for 8 h at the indicated concentrations. After the addition of luminescence substrates in the same volume as the cell culture medium, the mixtures were incubated for 2 min at room temperature on a shaker, followed by 10 min incubation at room temperature to stabilize the luminescence signal prior to measurement.

3. Results and Discussion

3.1. Structural Elucidation

Pentaminomycin C (**1**) was purified as a white powder, and its molecular formula was established to be $C_{37}H_{51}N_9O_6$ based on HRFABMS data along with ^{1}H and ^{13}C NMR data (Table 1). Further analysis of the NMR spectra (Figures S1–S5) confirmed this compound as the previously reported cyclic peptide, pentaminomycin C [18], which consists of five amino acids: leucine, valine, tryptophan, N^5-hydroxyarginine, and phenylalanine. The sequence of the amino acids was confirmed as leucine-valine-tryptophan-N^5-hydroxyarginine-phenylalanine by HMBC correlations as reported in the literature [18].

Pentaminomycin D (**2**) was isolated as a white powder. Based on HRFABMS and NMR data, the molecular formula of 2 was determined to be $C_{36}H_{49}N_9O_6$ with 17 double bond equivalents. Based on this molecular formula, pentaminomycin D (**2**) possesses one less CH_2 group than **1**. The ^{1}H NMR spectrum of **2** (Figure S6) showed the presence of five exchangeable amide NH groups (δ_H 8.85, 8.63, 8.45, 7.50, and 7.23) and five α-protons (δ_H 4.52, 4.28, 4.16, 4.12, and 3.70) in the amino acid residues, suggesting that **2** was also a pentapeptide-derived compound. The ^{13}C NMR spectrum (Figure S7) also confirmed **2** to be a peptidic metabolite through its five carbonyl signals (δ_C 171.7, 171.4, 171.4, 170.7, and 170.4) and five α-carbon signals (δ_C 60.2, 57.5, 55.3, 53.5, and 52.9), which is consistent with the ^{1}H NMR spectrum. Further analysis of the ^{13}C NMR spectrum identified 15 sp^2 carbon atoms (δ_C 157.4–110.2) and 11 aliphatic carbon atoms (δ_C 50.5–18.3). The odd number of sp^2 carbon atoms indicated the existence of an imine-type functional group. Assuming that pentaminomycin D (**2**) possessed an imine group, eight double bonds and five carbonyl groups accounted for 13 double bond equivalents out of 17, suggesting that this metabolite possessed four rings.

Analysis of the 1D (^{1}H and ^{13}C) and 2D (HSQC, COSY, and HMBC) NMR spectroscopic data (Figures S6–S10) of **2** identified the amino acid residues (Figure 2). More specifically, the conspicuous 1′-NH moiety (δ_H 10.78) was found to have a COSY correlation with H-2′ (δ_H 7.17), connecting C-1′ and C-2′. C-2′ was located adjacent to C-3′, as determined by the H-2′/C-3′ HMBC correlation. The $^3J_{HH}$ correlations of H-4′ (δ_H 7.51), H-5′ (δ_H 6.98), H-6′ (δ_H 7.04), and H-7′ (δ_H 7.30), and their ^{1}H-^{1}H coupling constants (J = 7.5 Hz), allowed the construction of an *ortho*-substituted 6-membered aromatic ring. HMBC signals from 1′-NH and H-7′ to C-7′a (δ_C 136.1) and from H-2′ and H-5′ to C-3′a (δ_C 126.8) secured the 1′-NH-C-7a′ and C-3′-C-3′a connectivity and allowed the elucidation of the indole structure. H_2-3 (δ_H 3.19 and 2.91) displayed HMBC correlations with C-3′ (δ_C 110.2), indicating C-3 methylene substitution at C-3′. In addition, 2-NH (δ_H 8.63)/H-2 (δ_H 4.28) and H-2/H_2-3 COSY and H-2/C-1 HMBC correlations confirmed the presence of a tryptophan unit, while an array of COSY correlations starting from an NH group (δ_H 7.50) to H_3-4 and H_3-5 confirmed the existence of a valine residue. In a similar manner, COSY correlations from an amide NH moiety (δ_H 8.45) to a dimethyl group allowed the elucidation of an addition valine unit. Furthermore, correlation of the α-proton at δ_H 4.52 with an amide proton (δ_H 8.85) and β-protons (δ_H 2.98 and 2.79) was also observed, as were

HMBC correlations between these β-protons with the quaternary C-1′ carbon atom of the aromatic ring (δ_C 138.0) and the C-2′ carbon atom (δ_C 129.0). Two overlapping methine carbon peaks (2 × CH) at δ_C 129.0 and 128.0, and the second-order signals observed for 5H indicated the presence of a phenylalanine residue. This unit was further assigned by HMBC correlations from H-2′ and H-5′ to C-6′ (δ_C 126.2) and from H-3′ and H-6′ to C-1′. The presence of the N^5-hydroxyarginine moiety was deciphered by consecutive COSY correlations from NH (δ_H 7.23) to H-5 (δ_H 3.42). The H_2-5 methylene protons showed COSY correlations only with H_2-4 (δ_H 1.33 and 1.15), placing the methylene unit at the terminus of this spin system. Moreover, the ^{13}C chemical shift of C-5 (δ_C 50.5) indicated that this carbon atom was bound to a nitrogen atom. This partial structure and the four elucidated amino acids (Trp, Phe, and two Val's) accounted for the $C_{35}H_{45}N_6O_5$ portion of the molecular formula $C_{36}H_{49}N_9O_6$, thereby leaving a CH_4N_3O unit for structural elucidation. Thus, this last carbon (δ_C 157.4), which was preliminarily diagnosed as an imine carbon, was correlated with H_2-5. Its chemical shift is typical for guanidine carbon, and the presence of three broad singlet protons at δ_H 7.50 confirmed the presence of a guanidine group containing the N^5-OH group (δ_H 10.55), thereby indicating that this last fragment is an N^5-hydroxyarginine residue.

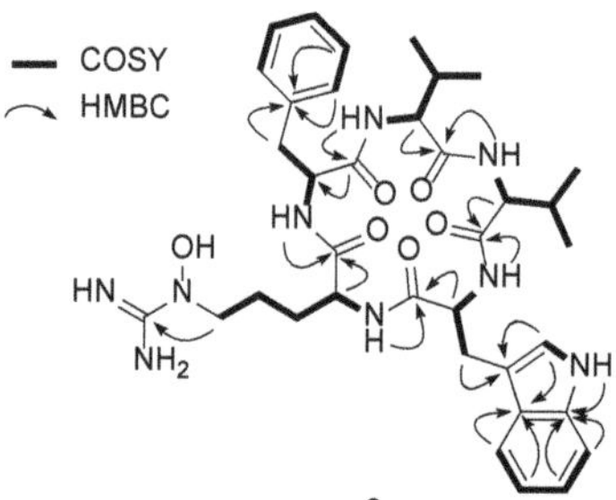

2

Figure 2. Key COSY and HMBC correlations of pentaminomycin D (**2**).

The seven double bonds, one imine group, five carbonyl groups, and three ring structures of the Phe and Trp residues accounted for 16 double bond equivalents out of 17 for pentaminomycin D (**2**). An additional ring structure was therefore confirmed in the sequence analysis of the amino acids using the HMBC spectrum (Figure S10). More specifically, the HMBC correlations from the α-proton (δ_H 4.12) of Val-1 and the amide proton (δ_H 8.45) of Val-2 to the C-1 atom (δ_C 171.4) of Val-1 established the connectivity of Val-1 to Val-2. The connection of Val-2 to Trp was supported by the heteronuclear correlation from the NH moiety (δ_H 8.63) of Trp to the C-1 atom (δ_C 171.4) of Val-2 in the HMBC spectrum. An additional HMBC correlation from the amide NH group (δ_H 7.23) of N^5-OH-Arg to the amide carbonyl carbon atom (δ_C 171.7) of Trp secured the sequence of Trp to N^5-OH-Arg. Furthermore, the NH proton (δ_H 7.23) of N^5-OH-Arg correlated with the amide carbonyl carbon (δ_C 170.7) of Phe in the HMBC spectrum, which established the linkage of arginine to phenylalanine. Lastly, the cyclized structure was completed by the confirmation of an HMBC correlation from the NH unit (δ_H 8.85) of Phe to the carbonyl carbon atom (δ_C 171.4) of Val-1, finally establishing the planar structure of **2** as a new cyclic pentapeptide.

Pentaminomycin E (**3**) was also purified as a white powder. The molecular formula of this compound was determined to be $C_{40}H_{49}N_9O_6$ based on HRFABMS and NMR data (Table 1). Comparing the NMR spectra of **3** with those of **1** and **2**, additional aromatic protons and carbons were observed, indicating the presence of an additional aromatic group in 3. Comprehensive analysis of the 1D and 2D NMR spectra (Figures S11–S15) confirmed the amino acid residues to be two phenylalanine residues, valine, tryptophan, and N^5-hydroxyarginine. The amino acid sequence of the structure was subsequently determined by analysis of the HMBC correlations, and was confirmed to be Phe-1-Val-Trp-N^5-OH-Arg-Phe-2. The new metabolites, namely pentaminomycins D and E (**2** and **3**), share cyclic pentapeptide features with pentaminomycins A and B, including Trp and N^5-OH Arg [17]. However, pentaminomycins D and E incorporate Phe instead of Leu adjacent to N^5-OH-Arg, unlike in the cases of pentaminomycins A and B [17]. In addition, pentaminomycin E was identified as the first congener bearing two Phe units in the pentaminomycin series.

To determine the absolute configurations of pentaminomycins D and E (**2** and **3**), acid hydrolysis and derivatization of the hydrolysates with Marfey's reagent (*N*-(5-fluoro-2,4-dinitrophenyl)-L-leucine amide (L-FDLA)) were carried out [21]. By comparing the retention times from LC/MS analysis of the L-FDLA derivatives with the same reaction products of authentic L and D amino acids, the absolute configurations of the α-carbons were determined. The absolute configurations of the amino acids in pentaminomycin D (**2**) were established as L-valine, D-valine, L-tryptophan, N^5-hydroxy-L-arginine, and D-phenylalanine. In a similar process, the absolute configurations of the amino acid residues present in pentaminomycin E (**3**) were determined to be D-valine, L-tryptophan, N^5-hydroxy-L-arginine, L-phenylalanine, and D-phenylalanine. Since two valine residues of the opposite configuration exist in **2**, whereas **3** contains both L- and D-phenylalanine residues, the exact assignments of the configurations were subjected to genomic analysis of the biosynthetic gene cluster for the pentaminomycins because the L- and D-Val units in **2** and the L- and D-Phe residues in **3** are not distinguishable by NMR spectroscopic analysis.

3.2. Biosynthetic Pathway

Analysis of the whole genome sequence of the *Streptomyces* sp. GG23 strain identified the putative biosynthetic gene cluster responsible for the pentaminomycins. The 8.7 Mb draft genome consisting of three contigs was analyzed using antiSMASH 5.0 [19]. In total, 31 gene clusters were involved in the biosynthesis of polyketides, nonribosomal peptides, and terpenes (Table S2). The BGC of the pentaminomycins showed a high similarity to a previous report into pentaminomycin C [18]. The total length of the BGC is approximately 83.6 kb encompassing 53 open reading frames (Table S3) including two NRPS genes, three post-modification genes, seven transport and regulatory genes, and five tryptophan biosynthesis genes (Figure 3A).

The NRPS (non-ribosomal peptide synthetase) gene for the pentaminomycins is *penN2*, which encodes five NRPS modules. Each module incorporates an amino acid to produce a pentapeptide chain. The first module without the epimerase domain flexibly recruits an amino acid of the group L-valine, L-leucine, and L-phenylalanine. The amino acid introduced by the second module is fixed as valine, whose absolute configuration is the D form because of the action of the epimerase domain in this module. This shows that the Val-2 residue introduced by module 2 has a D configuration and the other valine unit (Val-1) is in the L form in pentaminomycin D (**2**). Accordingly, L-Trp should be tethered after D-Val by module 3, and the arginine moiety is connected by module 4. The peptide chain is completed after the linkage of the last amino acid, D-Phe, by module 5 with the action of the epimerase domain. This also confirmed the absolute configuration of pentaminomycin E (**3**), in which Phe-2 is present in the D form, whereas Phe-1 possesses the L configuration. Post-modular modification by processes such as hydroxylation and cyclization finalized the biosynthesis of the pentaminomycins. *N*-hydroxylation on the arginine unit is possibly facilitated by the cytochrome P450 enzymes PenB and/or PenC. Cyclization of the pentapeptide chain was proposed to be catalyzed by serine hydrolase *penA* in the previously reported biosynthesis of pentaminomycin C [18]. However, our detailed analysis found that *penA* is the coding gene for penicillin binding protein (PBP)-type thioesterase (TE). PBP-type TE or the standalone cyclase is reported as a peptidyl cyclase included in the β-lactamase superfamily [22,23]. Cyclic peptides that use PBP-type TEs have been previously reported, including desotamide [24–26], surugamide [27], ulleungmycin [28], noursamycin [29], curacomycin [30], and mannopeptimycin [31]. These compounds share a structurally common feature in that the initial NRPS module must introduce an L-amino acid, while the terminal module recruits a D-amino acid. This is due to the fact that the structure of PBP-type TE is analogous to the penicillin-binding protein [32]. The penicillin-binding protein detects the D-alanyl-D-alanine moiety in peptidoglycan precursors to contribute to transpepdidation for cell wall construction in bacteria [33]. Similarly, PBP-type TE also detects the D-amino acid at the C-terminus of the NRPS peptide chain and catalyzes peptidyl macrocyclization. Based on the NRPS modules, the biosynthesis of the pentaminomycins starts with an L-amino acid (L-Val, L-Leu, or L-Phe) and ends with D-Phe, facilitating the cyclization by PBP-type TE (PenA) (Figure 3B).

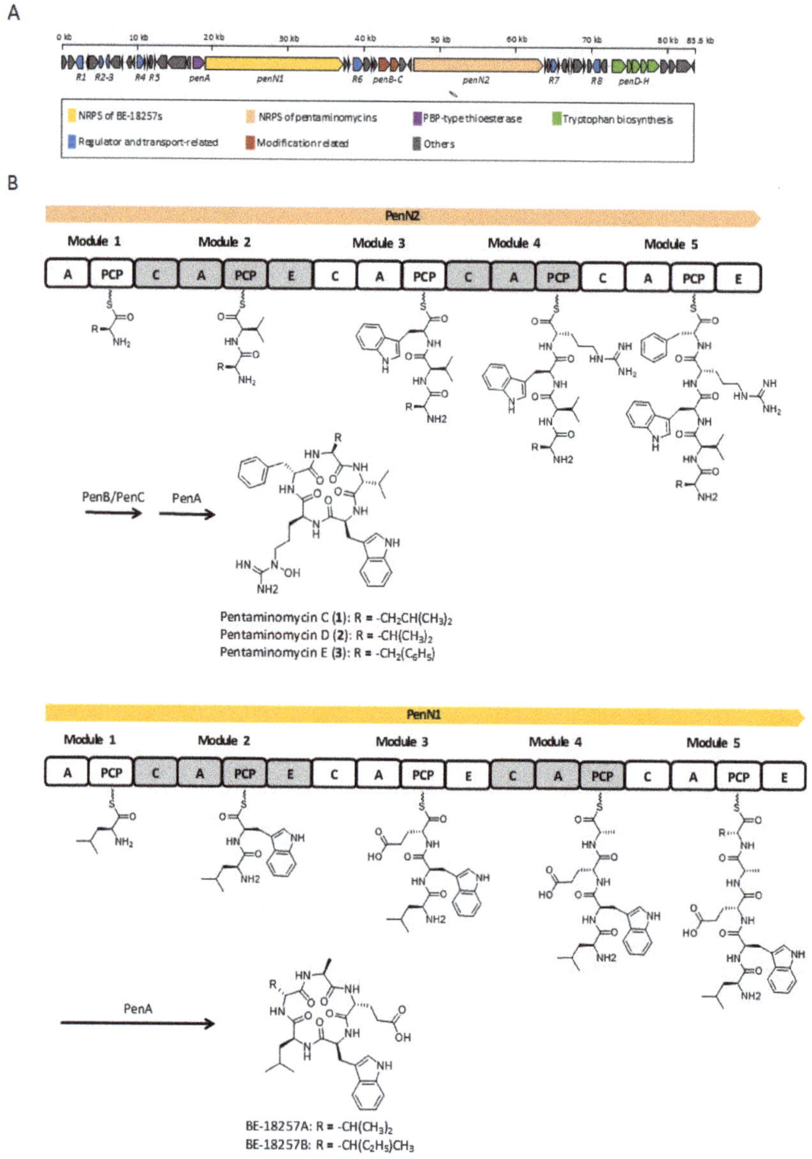

Figure 3. Proposed biosynthesis pathway for the BE-18257s and the pentaminomycins. (**A**) Genetic organization of putative biosynthetic gene cluster of the pentaminomycins. (**B**) Putative biosynthetic pathway for the pentaminomycins and the BE-18257s with the nonribosomal peptide synthetase (NRPS) modular organization. C, condensation domain; A, adenylation domain; PCP, peptidyl carrier protein; E, epimerase domain.

The BGC contains another NRPS gene, namely *penN1*, which is located close to the *penN2* gene (Figure 3B). Detailed analysis of the sequence revealed that *penN1* is also composed of five NRPS modules biosynthesizing another series of cyclic peptides, i.e., BE-18257A and B, which were reported to be endothelin-binding inhibitors [34]. In our chemical analysis of *Streptomyces* sp. GG23 based on LC/MS data, the production of these cyclic peptides was detected (Figure S16). The pentapeptide chains of BE-18257s are assumed to be cyclized by PenA because *penN1* does not possess canonical TE at the end of the NRPS module and no other PBP-type TE genes rather than *penA* were identified in the BGC. Furthermore, the NRPS gene of the BE-18267s initially introduces L-Leu and completes the biosynthesis of the pentapeptide chain with D-Val or D-Leu, which is suitable for the utilization of PBP-type TE. Additionally, tryptophan biosynthetic genes (i.e., *penD–H*) exist in the BGC [35]. The pentaminomycins and BE-18257s both contain tryptophan in their structures, and so it is hypothesized that the Trp units of these different cyclic peptides share the Trp biosynthetic gene. Even though sharing of the PBP-type TE and Trp biosynthetic genes has to be proven by a follow-up study, if confirmed, the above example could be considered an unusual case in which independent NRPSs share core genes for their biosynthesis.

Based on the determined structures for pentaminomycins C–E (**1**–**3**), the amino acid introduced by module 1 appears to be flexible. The Leu, Val, or Phe variation could be explained by the amino acid sequence of the binding pocket providing substrate specificity to the adenylation domain [36]. Eight amino acid residues exist in the binding pocket to determine the specificity, whereby the amino acid residues of the binding pocket of module 1 are Asp235-Ala236-Leu239-Trp278-Met299-Gly301-Val322-Val330. Although many variables exist in the binding pockets, one of representative sequence for aromatic amino acids such as phenylalanine and tryptophan is Asp235-Ala236-Leu239-Val278-Met299-Gly301-Ala322-Val330. Comparing the two residue sequences, the only differences were that the Val278 and Ala322 residues in the typical binding pocket were replaced with Trp278 and Val322, respectively, in module 1 of *penN2*. This means that the adenylation domain of module 1 acts to incorporate an aromatic amino acid, such as Phe or Trp. However, because of these substitutions (Val278→Trp and Ala322→Val), module 1 potentially gains promiscuity to recruit variable hydrophobic amino acids, such as Leu, Val, or Phe, as a substrate. These amino acid sequence changes result in a smaller pocket volume, thereby avoiding larger aromatic side chains, such as that of Trp, which is consistent with the determined structures of pentaminomycins C–E. Indeed, instead of Trp, the smaller pocket seems to prefer smaller hydrophobic amino acids, such as Val or Leu, thereby accounting for the production of pentaminomycins C and D.

3.3. Evaluation of the Bioactivity

The majority of amino acids constituting the pentaminomycins are non-polar residues that are expected to penetrate mammalian cells by simple diffusion and affect the membrane dynamics in the cell. In addition, similar cyclic peptides with lipophilic side chains often induce autophagy in cultured human cells [37,38]. We proceeded to examine whether pentaminomycins C–E affected the cellular autophagic flux by monitoring the conjugation of phosphatidylethanolamine (PE) to ATG8 proteins, such as microtubule-associated protein light chain 3 (LC3) and γ-aminobutyric acid receptor-associated protein (GABARAP), which is the hallmark of autophagy induction [39]. When HEK293T cells were treated with pentaminomycins C and D (**1** and **2**), the levels of lipidated forms of LC3 and GABARAPL1 (LC3-II and GABARAPL1-II, respectively) were significantly elevated in a moderate dose-dependent manner, while pentaminomycin E (**3**) did not exert a similar phenomenon (Figure 4A,B). The key autophagic receptor p62/SQSTSM1 remained virtually unchanged after compound treatment (Figure 4A).

Changes in the autophagic flux manifested as elevated levels of cellular LC3-II and GABARAPL1-II after treatment with pentaminomycins C and D may originate from either an increased overall cellular autophagy or the reduced autolysosomal degradation of LC3-II and GABARAPL1-II. To determine the molecular mechanism, the cells were treated with BafA1, which inhibits autophagy at a late stage by blocking the fusion between the autophagosome and the lysosome, prior to treatment with the pentaminomycins. We observed a modest increase in GABARAPL1-II upon exposure to pentaminomycins C (**1**) and D (**2**), but not pentaminomycin E (**3**), when the cells were cotreated with BafA1 (Figure 4C,D). Taken together, our data largely point that pentaminomycins C and D induce global autophagy instead of inhibiting cellular autophagic flux, although the underlying molecular mechanism and direct target molecules of pentaminomycins should therefore be determined.

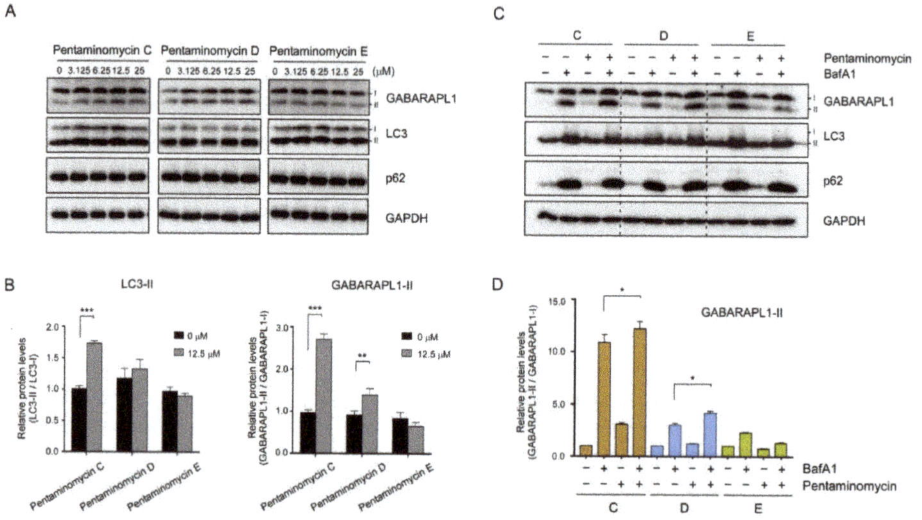

Figure 4. Effects of pentaminomycins C–E (**1**–**3**) on cellular autophagy in mammalian cells. HEK293T cells were treated with various concentrations of pentaminomycins for 8 h. (**A**) Whole cell lysates were harvested and subjected to SDS-PAGE followed by immunoblotting against the indicated antibodies. (**B**) Quantification of LC3-II and GABARAPL1-II in the presence of pentaminomycins (12.5 μM) using the multiple immunoblot images. Data were normalized to those of non-lipidated proteins. Data represent mean ± SD from three independent experiments. **, $p < 0.01$ and ***, $p < 0.001$ (one-way analysis of variance (ANOVA) with Bonferroni's multiple comparison test). (**C**) Pentaminomycins C and D, but not E, induce global cellular autophagy. HEK293T cells were cotreated with pentaminomycins C–E (20 μM) and a downstream autophagy inhibitor bafilomycin A1 (BafA1; 100 nM) for 12 h. (**D**) Quantification of GABARAPL1-II normalized to GABARAPL1-I in the presence or absence of pentaminomycins and BafA1. *, $p < 0.05$ (one-way ANOVA with Bonferroni's multiple comparison test).

Due to the fact that autophagy contributes to the degradation of oxidized proteins [40], we examined the effect of pentaminomycins C and D (**1** and **2**) on the oxidative stress induced by menadione [41]. Measurement of the cell viability based on the intracellular ATP levels revealed that the autophagy inducers, pentaminomycins C and D, potently protected HEK293 cells against menadione-induced cytotoxicity (Figure 5). Pentaminomycins C and D showed significantly reduced cell death after 4 and 2 h cotreatment with menadione, respectively. These protective effects were more prominent with longer pentaminomycin incubation times (Figure 5), thereby suggesting that autophagy induction by the pentaminomycins may accelerate oxidized protein clearance in cells and be beneficial in terms of cell protection under oxidative stress. However, it has yet to be determined whether natural compounds originating from mealworm beetle-associated bacteria can delay the pathologic process involving oxidatively damaged proteins, such as neurodegeneration and aging. Our results may therefore offer a novel strategy to modulate cellular autophagy and oxidative stress responses in cells.

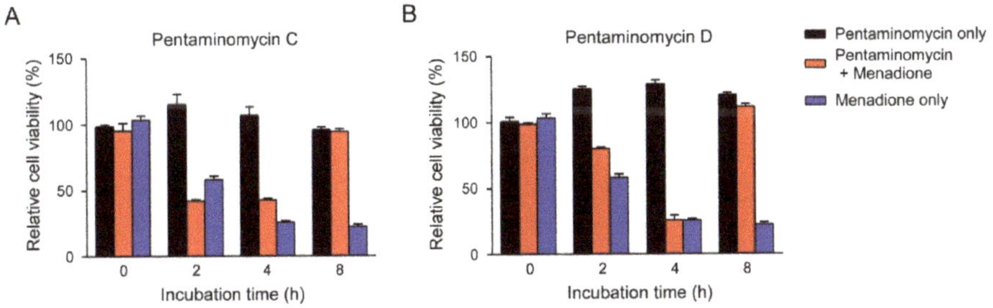

Figure 5. Alleviation of menadione-mediated cytotoxicity by (**A**) pentaminomycin C and (**B**) pentaminomycin D. Oxidative stress was induced by menadione (25 μM) for the indicated time periods in HEK93 cells, which were cotreated with either pentaminomycin C or D. The relative cell viability was assessed using the CellTiter-Glo assay and the values are represented as mean ± SD ($n = 3$).

4. Conclusions

Alteration of the culture conditions by changing the composition of culture medium, for a *Streptomyces* strain isolated from the gut of the mealworm beetle, *Tenebrio molitor*, enabled the production of cyclic pentapeptides, pentaminomycins C–E (**1–3**). The structures of **1–3** were assigned by combinational spectroscopic analysis. In addition, Marfey's analysis and bioinformatic investigations of the nonribosomal peptide synthetase (NRPS) biosynthetic gene cluster established the absolute configurations of the new metabolites, pentaminomycins D and E. Detailed sequence analysis of the adenylation domains in the NRPS modules revealed that the structural variations among **1–3** originate from the low specificity for hydrophobic amino acids in module 1. In addition, it was found that cyclization of the pentaminomycins can be catalyzed by a penicillin binding protein (PBP)-type thioesterase (TE), which is a noncanonical TE requiring L- and D-amino acids in the starting and terminal units, respectively. Pentaminomycins C and D (**1** and **2**), but not pentaminomycin E (**3**), exhibited significant autophagy-inducing activity based on LC3 and GABARAPL1 lipidation in both the presence and absence of BafA1. Importantly, cells treated with pentaminomycins C and D showed enhanced resistance to the oxidative stress induced by menadione, providing strong evidence that activation of cellular autophagic flux antagonizes the harmful effects of oxidized proteins. Although the underlying molecular mechanism requires further elucidation, our findings collectively suggest that some pentaminomycins may exhibit therapeutic potential against diseases associated with chronic oxidative stress and incompetent cellular responses. The discovery of pentaminomycins C–E therefore indicates that biotechnical investigation into relatively unexploited insect-associated bacteria may be a promising strategy to explore microbial metabolites with unique biosynthetic pathways and interesting biological activities.

Supplementary Materials: The following are available online at http://www.mdpi.com/2076-2607/8/9/1390/s1, Figures S1–S5: ^{1}H, ^{13}C, COSY, HSQC, and HMBC NMR data of **1** in DMSO-d_6, Figures S6–S10: ^{1}H, ^{13}C, COSY, HSQC, and HMBC NMR data of **2** in DMSO-d_6, Figures S11–S15: ^{1}H, ^{13}C, COSY, HSQC, and HMBC NMR data of **3** in DMSO-d_6, Table S1: LC/MS analysis of L-FDLA derivatives of **2** and **3**, Figure S16: Production of BE-18257s detected by LC/MS analysis. Figure S17: LC/MS chromatograms of L-FDLA derivatives of **2** and **3**, Table S2: antiSMASH output table of the whole genome analysis of *Streptomyces* sp. GG23, Table S3: Putative functions of ORFs in the pentaminomycin biosynthetic gene cluster.

Author Contributions: Conceptualization, S.H., J.S., M.J.L. and D.-C.O.; Data curation, S.H.; Formal analysis, L.T.H.L.L. and M.J.L.; Funding acquisition, S.H. and D.-C.O.; Investigation, S.H. and L.T.H.L.L.; Methodology, L.T.H.L.L. Resources, S.-I.J.; Supervision, M.J.L. and D.-C.O.; Writing—original draft, S.H., L.T.H.L.L., S.-I.J., M.J.L. and D.-C.O.; Writing—review and editing, S.H., J.S., M.J.L. and D.-C.O. All authors have read and agreed to the published version of the manuscript.

Funding: This work was supported by the National Research Foundation (NRF) of Korea (2018R1A4A1021703; 2020R1A2C2003518) funded by the Ministry of Science and ICT of Korea.

Acknowledgments: NMR and MS experiments were technically supported by the National Center for Inter-university Research Facilities (NCIRF) at Seoul National University.

Conflicts of Interest: The authors declare no conflict of interest.

References

1. Pettit, R.K. Small-molecule elicitation of microbial secondary metabolites. *Microb. Biotechnol.* **2011**, *4*, 471–478. [CrossRef] [PubMed]
2. Harvey, A. Strategies for discovering drugs from previously unexplored natural products. *Drug Discov. Today* **2000**, *5*, 294–300. [CrossRef]
3. Beemelmanns, C.; Guo, H.; Rischer, M.; Poulsen, M. Natural products from microbes associated with insects. *Beilstein J. Org. Chem.* **2016**, *12*, 314–327. [CrossRef] [PubMed]
4. Oh, D.-C.; Poulsen, M.; Currie, C.R.; Clardy, J. Dentigerumycin: A bacterial mediator of an ant-fungus symbiosis. *Nat. Chem. Biol.* **2009**, *5*, 391–393. [CrossRef] [PubMed]
5. Kim, K.H.; Ramadhar, T.R.; Beemelmanns, C.; Cao, S.; Poulsen, M.; Currie, C.R.; Clardy, J. Natalamycin A, an ansamycin from a termite-associated *Streptomyces* sp. *Chem. Sci.* **2014**, *5*, 4333–4338. [CrossRef] [PubMed]
6. Hong, S.-H.; Ban, Y.H.; Byun, W.S.; Kim, D.; Jang, Y.-J.; An, J.S.; Shin, B.; Lee, S.K.; Shin, J.; Yoon, Y.J.; et al. Camporidines A and B: Antimetastatic and anti-inflammatory polyketide alkaloids from a gut Bacterium of *Camponotus kiusiuensis*. *J. Nat. Prod.* **2019**, *82*, 903–910. [CrossRef]
7. Du, Y.E.; Byun, W.S.; Lee, S.B.; Hwang, S.; Shin, Y.-H.; Shin, B.; Jang, Y.-J.; Hong, S.; Shin, J.; Lee, S.K.; et al. Formicins, *N*-acetylcysteamine-bearing indenone thioesters from a wood ant-associated bacterium. *Org. Lett.* **2020**, *22*, 5337–5341. [CrossRef]
8. Shin, Y.-H.; Bae, S.; Sim, J.; Hur, J.; Jo, S.-I.; Shin, J.; Suh, Y.-G.; Oh, K.-B.; Oh, D.-C. Nicrophorusamides A and B, antibacterial chlorinated cyclic peptides from a gut bacterium of the carrion beetle *Nicrophorus concolor*. *J. Nat. Prod.* **2017**, *80*, 2962–2968. [CrossRef]
9. Park, S.-H.; Moon, K.; Bang, H.-S.; Kim, S.-H.; Kim, D.-G.; Oh, K.-B.; Shin, J.; Oh, D.-C. Tripartilactam, a cyclobutane-bearing tricyclic lactam from a *Streptomyces* sp. in a dung beetle's brood ball. *Org. Lett.* **2012**, *14*, 1258–1261. [CrossRef]
10. Hwang, S.; Kim, E.; Lee, J.; Shin, J.; Yoon, Y.J.; Oh, D.-C. Structure revision and the biosynthetic pathway of tripartilactam. *J. Nat. Prod.* **2020**, *83*, 578–583. [CrossRef]
11. Kim, S.-H.; Kwon, S.H.; Park, S.-H.; Lee, J.K.; Bang, H.-S.; Nam, S.-J.; Kwon, H.C.; Shin, J.; Oh, D.-C. Tripartin, a histone demethylase inhibitor from a bacterium associated with a dung beetle larva. *Org. Lett.* **2013**, *15*, 1834–1837. [CrossRef] [PubMed]
12. Um, S.; Park, S.H.; Kim, J.; Park, H.J.; Ko, K.; Bang, H.-S.; Lee, S.K.; Shin, J.; Oh, D.-C. Coprisamides A and B, new branched cyclic peptides from a gut bacterium of the dung beetle *Copris tripartitus*. *Org. Lett.* **2015**, *17*, 1272–1275. [CrossRef]
13. Um, S.; Bach, D.-H.; Shin, B.; Ahn, C.-H.; Kim, S.-H.; Bang, H.-S.; Oh, K.-B.; Lee, S.K.; Shin, J.; Oh, D.-C. Naphthoquinone–oxindole alkaloids, coprisidins A and B, from a gut-associated bacterium in the dung beetle, *Copris tripartitus*. *Org. Lett.* **2016**, *18*, 5792–5795. [CrossRef]
14. Bode, H.B.; Bethe, B.; Höfs, R.; Zeeck, A. Big effects from small changes: Possible ways to explore nature's chemical diversity. *ChemBioChem* **2002**, *3*, 619–627. [CrossRef]
15. Huang, X.; Roemer, E.; Sattler, I.; Moellmann, U.; Christner, A.; Grabley, S. Lyidamycins A-D: Cyclodepsipeptides with antimycobacterial properties. *Angew. Chem. Int. Ed.* **2006**, *45*, 3067–3072. [CrossRef] [PubMed]
16. Hwang, S.; Shin, D.; Kim, T.H.; An, J.S.; Jo, S.-I.; Jang, J.; Hong, S.; Shin, J.; Oh, D.-C. Structural revision of lydiamycin A by reinvestigation of the stereochemistry. *Org. Lett.* **2020**, *22*, 3855–3859. [CrossRef] [PubMed]
17. Jang, J.-P.; Hwang, G.J.; Kwon, M.C.; Ryoo, I.-J.; Jang, M.; Takahashi, S.; Ko, S.-K.; Osada, H.; Jang, J.-H.; Ahn, J.S. Pentaminomycins A and B, hydroxyarginine-containing cyclic pentapeptides from *Streptomyces* sp. RK88-1441. *J. Nat. Prod.* **2018**, *81*, 806–810. [CrossRef] [PubMed]
18. Kaweewan, I.; Hemmi, H.; Komaki, H.; Kodani, S. Isolation and structure determination of a new antibacterial peptide pentaminomycin C from *Streptomyces cacaoi* subsp. *cacaoi*. *J. Antibiot.* **2020**, *73*, 224–229. [CrossRef] [PubMed]

19. Blin, K.; Shaw, S.; Steinke, K.; Villebro, R.; Ziemert, N.; Lee, S.Y.; Medema, M.H.; Weber, T. antiSMASH 5.0: Updates to the secondary metabolite genome mining pipeline. *Nucleic Acids Res.* **2019**, *47*, W81–W87. [CrossRef]
20. Choi, W.H.; Yun, Y.; Park, S.; Jeon, J.H.; Lee, J.; Lee, J.H.; Yang, S.-A.; Kim, N.-K.; Jung, C.H.; Kwon, Y.T.; et al. Aggresomal sequestration and STUB1-mediated ubiquitylation during mammalian proteaphagy of inhibited proteasomes. *Proc. Natl. Acad. Sci. USA* **2020**, *117*, 19190–19200. [CrossRef]
21. Fujii, K.; Ikai, Y.; Oka, H.; Suzuki, M.; Harada, K. A nonempirical method using LC/MS for determination of the absolute configuration of constituent amino acids in a peptide: Combination of Marfey's method with mass spectrometry and its practical application. *Anal. Chem.* **1997**, *69*, 5146–5151. [CrossRef]
22. Kuranaga, T.; Matsuda, K.; Sano, A.; Kobayashi, M.; Ninomiya, A.; Takata, K.; Matsunaga, S.; Wakimoto, T. Total synthesis of the nonribosomal peptide surugamide B and identification of a new offloading cyclase family. *Angew. Chem. Int. Ed.* **2018**, *57*, 9447–9451. [CrossRef] [PubMed]
23. Matsuda, K.; Kobayashi, M.; Kuranaga, T.; Takada, K.; Ikeda, H.; Matsunaga, S.; Wakimoto, T. SurE is a trans-acting thioesterase cyclizing two distinct non-ribosomal peptides. *Org. Biomol. Chem.* **2019**, *17*, 1058–1061. [CrossRef] [PubMed]
24. Miao, S.; Anstee, M.R.; LaMarco, K.; Matthew, J.; Huang, L.H.T.; Brasseur, M.M. Inhibition of bacterial RNA polymerases. Peptide metabolites from the cultures of *Streptomyces* sp. *J. Nat. Prod.* **1997**, *60*, 858–861. [CrossRef]
25. Song, Y.; Li, Q.; Liu, X.; Chen, Y.; Zhang, Y.; Sun, A.; Zhang, W.; Zhang, J.; Ju, J. Cyclic hexapeptides from the deep South China Sea-derived *Streptomyces scopuliridis* SCSIO ZJ46 active against pathogenic Gram-positive bacteria. *J. Nat. Prod.* **2014**, *77*, 1937–1941. [CrossRef]
26. Fazal, A.; Webb, M.E.; Seipke, R.F. The desotamide family of antibiotics. *Antibiotics* **2020**, *9*, 452. [CrossRef]
27. Takada, K.; Ninomiya, A.; Naruse, M.; Sun, Y.; Miyazaki, M.; Nogi, Y.; Okada, S.; Matsunaga, S. Surugamides A–E, cyclic octapeptides with four D-Amino acid residues, from a marine *Streptomyces* sp.: LC–MS-aided inspection of partial hydrolysates for the distinction of D- and L-amino acid residues in the sequence. *J. Org. Chem.* **2013**, *78*, 6746–6750. [CrossRef]
28. Son, S.; Hong, Y.-S.; Jang, M.; Heo, K.T.; Lee, B.; Jang, J.-P.; Kim, J.-W.; Ryoo, I.-J.; Kim, W.-G.; Ko, S.-K.; et al. Genomics-driven discovery of chlorinated cyclic hexapeptides ulleungmycins A and B from a *Streptomyces* species. *J. Nat. Prod.* **2017**, *80*, 3025–3031. [CrossRef]
29. Mudalungu, C.M.; von Törne, W.J.; Voigt, K.; Rückert, C.; Schmitz, S.; Sekurova, O.N.; Zotchev, S.B.; Süssmuth, R.D. Noursamycins, chlorinated cyclohexapeptides identified from molecular networking of *Streptomyces noursei* NTR-SR4. *J. Nat. Prod.* **2019**, *82*, 1478–1486. [CrossRef]
30. Kaweewan, I.; Komaki, H.; Hemmi, H.; Kodani, S. Isolation and structure determination of new antibacterial peptide curacomycin based on genome mining. *Asian J. Org. Chem.* **2017**, *6*, 1838–1844. [CrossRef]
31. Magarvey, N.A.; Haltli, B.; He, M.; Greenstein, M.; Hucul, J.A. Biosynthetic pathway for mannopeptimycins, lipoglycopeptide antibiotics active against drug-resistant Gram-positive pathogens. *Antimicrob. Agents Chemother.* **2006**, *50*, 2167–2177. [CrossRef] [PubMed]
32. Matsuda, K.; Zhai, R.; Mori, T.; Kobayashi, M.; Sano, A.; Abe, I.; Wakimoto, T. Heterochiral coupling in non-ribosomal peptide macrolactamization. *Nat. Catal.* **2020**, *3*, 507–515. [CrossRef]
33. Goffin, C.; Ghuysen, J.M. Biochemistry and comparative genomics of SxxK superfamily acyltransferases offer a clue to the mycobacterial paradox: Presence of penicillin-susceptible target proteins versus lack of efficiency of penicillin as therapeutic agent. *Microbiol. Mol. Biol. Rev.* **2002**, *66*, 702–738. [CrossRef]
34. Kojiri, K.; Ihara, M.; Nakajima, S.; Kawamura, K.; Funaishi, K.; Yano, M.; Suda, H. Endothelin-binding inhibitors, BE-18257A and BE-18257B. *J. Antibiot.* **1991**, *44*, 1342–1347. [CrossRef] [PubMed]
35. Radwanski, E.R.; Last, R.L. Tryptophan biosynthesis and metabolism: Biochemical and molecular genetics. *Plant. Cell* **1995**, *7*, 921–934. [CrossRef] [PubMed]
36. Challis, G.L.; Ravel, J.; Townsend, C.A. Predictive, structure-based model of amino acid recognition by nonribosomal peptide synthetase adenylation domains. *Chem. Biol.* **2000**, *7*, 211–224. [CrossRef]
37. Peraro, L.; Zou, Z.; Makwana, K.M.; Cummings, A.E.; Ball, H.L.; Yu, H.; Lin, Y.-S.; Levine, B.; Kritzer, J.A. Diversity-oriented stapling yields intrinsically cell-penetrant inducers of autophagy. *J. Am. Chem. Soc.* **2017**, *139*, 7792–7802. [CrossRef] [PubMed]

38. Wang, L.; Li, M.-D.; Cao, P.-P.; Zhang, C.-F.; Huang, F.; Xu, X.-H.; Liu, B.-L.; Zhang, M. Astin B, a cyclic pentapeptide from *Aster tataricus*, induces apoptosis and autophagy in human hepatic L-02 cells. *Chem. Biol. Interact.* **2014**, *223*, 1–9. [CrossRef] [PubMed]
39. Kim, E.; Park, S.; Lee, J.H.; Mun, J.Y.; Choi, W.H.; Yun, Y.; Lee, J.; Kim, J.H.; Kang, M.-J.; Lee, M.J. Dual function of USP14 deubiquitinase in cellular proteasomal activity and autophagic flux. *Cell Rep.* **2018**, *24*, 732–743. [CrossRef] [PubMed]
40. Filomeni, G.; De Zio, D.; Cecconi, F. Oxidative stress and autophagy: The clash between damage and metabolic needs. *Cell Death Differ.* **2015**, *22*, 377–388. [CrossRef]
41. Choi, W.H.; De Poot, S.A.; Lee, J.H.; Kim, J.H.; Han, D.H.; Kim, Y.K.; Finley, D.; Lee, M.J. Open-gate mutants of the mammalian proteasome show enhanced ubiquitin-conjugate degradation. *Nat. Commun.* **2016**, *7*, 1–12. [CrossRef] [PubMed]

microorganisms

MDPI

Article

Actinomycetes from the Red Sea Sponge *Coscinoderma mathewsi*: Isolation, Diversity, and Potential for Bioactive Compounds Discovery

Yara I. Shamikh [1,2,†], Aliaa A. El Shamy [3,†], Yasser Gaber [4,5], Usama Ramadan Abdelmohsen [6,7,8], Hashem A. Madkour [9], Hannes Horn [10], Hossam M. Hassan [11], Abeer H. Elmaidomy [11], Dalal Hussien M. Alkhalifah [12] and Wael N. Hozzein [13,14,*]

1 Department of Microbiology and Immunology, Nahda University in Beni Suef, Beni-Suef 65211, Egypt; yara.shamikh@nub.edu.eg
2 Consultant, Virology Department, Egypt Center for Research and Regenerative Medicine (ECRRM), Cairo 11517, Egypt
3 Department of Microbiology and Public Health, Faculty of Pharmacy, Heliopolis University for Sustainable Development, Cairo 11785, Egypt; aliaa.ali@hu.edu.eg
4 Department of Microbiology and Immunology, Faculty of Pharmacy, Beni-Suef University, Beni-Suef 62511, Egypt; yasser.gaber@pharm.bsu.edu.eg
5 Department of Pharmaceutics and Pharmaceutical Technology, College of Pharmacy, Mutah University, Karak 61710, Jordan
6 Department of Pharmacognosy, Faculty of Pharmacy, Minia University, Minia 61519, Egypt; usama.ramadan@mu.edu.eg
7 Department of Pharmacognosy, Faculty of Pharmacy, Deraya University, 7 Universities Zone, New Minia 61111, Egypt
8 Department of Pharmacognosy, College of Pharmacy, King Khalid University, Abha 61441, Saudi Arabia
9 Department of Marine and Environmental Geology, National Institute of Oceanography and Fisheries, Red Sea Branch, Hurghada 84511, Egypt; madkour_hashem@yahoo.com
10 Independent Researcher, 69126 Heidelberg, Germany; hannesdhorn@gmail.com
11 Department of Pharmacognosy, Faculty of Pharmacy, Beni-Suef University, Beni-Suef 62511, Egypt; abuh20050@yahoo.com (H.M.H.); abeerabdelhakium@yahoo.com (A.H.E.)
12 Biology Department, College of Science, Princess Nourah Bint Abdulrahman University, Riyadh 11451, Saudi Arabia; DHALkalifah@pnu.edu.sa
13 Bioproducts Research Chair, Zoology Department, College of Science, King Saud University, Riyadh 11451, Saudi Arabia
14 Botany and Microbiology Department, Faculty of Science, Beni-Suef University, Beni-Suef 62521, Egypt
* Correspondence: whozzein@ksu.edu.sa
† These authors contributed equally to this work.

Received: 15 April 2020; Accepted: 21 May 2020; Published: 23 May 2020

Abstract: The diversity of actinomycetes associated with the marine sponge *Coscinoderma mathewsi* collected from Hurghada (Egypt) was studied. Twenty-three actinomycetes were separated and identified based on the 16S rDNA gene sequence analysis. Out of them, three isolates were classified as novel species of the genera *Micromonospora*, *Nocardia*, and *Gordonia*. Genome sequencing of actinomycete strains has revealed many silent biosynthetic gene clusters and has shown their exceptional capacity for the production of secondary metabolites, not observed under classical cultivation conditions. Therefore, the effect of mycolic-acid-containing bacteria or mycolic acid on the biosynthesis of cryptic natural products was investigated. Sponge-derived actinomycete *Micromonospora* sp. UA17 was co-cultured using liquid fermentation with two mycolic acid-containing actinomycetes (*Gordonia* sp. UA19 and *Nocardia* sp. UA 23), or supplemented with pure mycolic acid. LC-HRESIMS data were analyzed to compare natural production across all crude extracts. *Micromonospora* sp. UA17 was rich with isotetracenone, indolocarbazole, and anthracycline analogs. Some co-culture extracts showed metabolites such as a chlorocardicin, neocopiamycin A, and chicamycin B that were not found in the respective monocultures, suggesting a mycolic acid effect on the induction of cryptic natural product biosynthetic pathways. The antibacterial, antifungal, and antiparasitic activities for the different cultures extracts were also tested.

Keywords: sponges; actinomycetes; cryptic; *Micromonospora; Nocardia; Gordonia;* mycolic acid; LC-HRESIMS

---

1. Introduction

Actinomycetes are Gram-positive bacteria living in a wide range of aquatic, terrestrial environments and produce a variety of diverse bioactive compounds [1–4]. This phylum also has been found in a range of marine organisms such as corals, sponges, and jellyfish [5–7]. Actinomycetes from the marine environment have been reported to produce most of the bioactive compounds identified from the marine ecosystems [8,9]. These compounds belong to a variety of classes including polyketides, alkaloids, fatty acids, peptides, and terpenes [10–14]. There are many potential bioactivities of these compounds ranging from antibacterial, antifungal and antiparasitic to antioxidant and immunomodulatory activities [15–17]. With advances in sequencing technologies, actinomycete genomes have revealed many biosynthetic genes that encode for natural products not observed under standard fermentation conditions [18–22]. Previous methods were used to induce cryptic metabolites including chemical, molecular, and biological elicitation [22–29]. Altering the fermentation conditions (pH, media composition, and temperature) using the "one strain many compounds" (OSMAC) approach has been used to induce silent or poor expressed metabolic pathways [30–33]. Co-cultivation of microbial strains is a widely known approach to induce significant changes in the microbial metabolomes [20].

Mycolic acids are high-molecular-weight α-branched, β-hydroxyl fatty acids, which are located in the cell wall of certain bacterial genera such as *Corynebacterium, Mycobacterium Nocardia, Rhodococcus,* and *Segniliparus* [34–36]. They play a major role in shaping the cell wall and protect against chemical substances [37,38]. The structure of each mycolic acid is thought to be genus-specific and differs in the length of the carbon chain [34,39]. For example, members of the genus *Corynebacterium* have C_{50}-C_{56} and the genus *Rhodococcus* has been found to contain C_{34}-C_{52}. Onaka et al., 2011, reported the induction of a red pigment by *Streptomyces lividans TK23* after co-cultivation with living cells of the mycolic acid-containing bacterium *Tsukamurella pulmonis* TP-B0596 [40]. It was shown that the metabolite profiles of several *Streptomyces* strains were changed after co-cultivation with the mycolic acid-containing bacterium *T. pulmonis.* Combined culture of *S. endus* S-522 with *T. pulmonis* resulted in the identification of a novel antibiotic, alchivemycin A [40]. Recently, Hoshino *et al.*, 2015, isolated the di- and tri-cyclic macrolactams niizalactams A–C from the co-culture of *Streptomyces* sp. NZ-6 and the mycolic acid-containing bacterium *Tsukamurella pulmonis* TP-B0596 [41]. Arcyriaflavin E, a new cytotoxic indolocarbazole alkaloid, was isolated by co-cultivation of mycolic acid-containing bacteria and *Streptomyces cinnamoneus* NBRC 13823 [42]. Chojalactones A–C, cytotoxic butanolides were isolated from the co-culture of *Streptomyces* sp. cultivated with mycolic acid-containing bacterium *Tsukamurella pulmonis* TP-B0596 [43]. These studies highlight the efficacy of the co-cultivation strategy with mycolic acid-containing bacteria, for the discovery of cryptic natural products. Interestingly, all of those studies co-cultivate mycolic acid-containing bacteria with terrestrial actinomycetes however the effect on marine actinomycetes to our knowledge is yet to be investigated.

In this study, isolates of novel species belonging to the genera *Micromonospora*, *Nocardia*, and *Gordonia* were identified, and the effect of pure mycolic acid and mycolic acid-containing bacteria actinomycete *Gordonia* sp. UA19 and *Nocardia* sp. UA 23 on the secondary metabolite production of sponge-derived actinomycete *Micromonospora* sp. UA17 was examined by analyzing using LC-HRMS/MS data via metabolomes tools. The antibacterial, antifungal, and antiparasitic activities for the different cultures extracts were also tested.

2. Materials and Methods

2.1. Description of the Area for Sponge Collection

The study area lies about 5 km to the north of Hurghada at latitudes 27°17′01.0″ N, and longitudes 33°46′21.0″ E (Figure 1). This site is characterized by a long patchy reef, representing the front edge of a wide and shallow reef flat with many depressions and lagoons. The depth ranged from about 3 m at the reef front with a gentle slope towards deep water. The area was exposed to strong waves, and the currents follow the prevailing current direction in the Red Sea from north to south. A medium development undergoes along the coast of this area. The bottom topography of this area is characterized by seagrasses and algae in intertidal and subtidal areas in addition to coral. The samples collected from this area, namely Ahia Reefs. *Coscinoderma mathewsi* was identified by El-Sayd Abed El-Aziz (Department of Invertebrates Lab., National Institute of Oceanography and Fisheries, Red Sea Branch, 84511 Hurghada, Egypt).

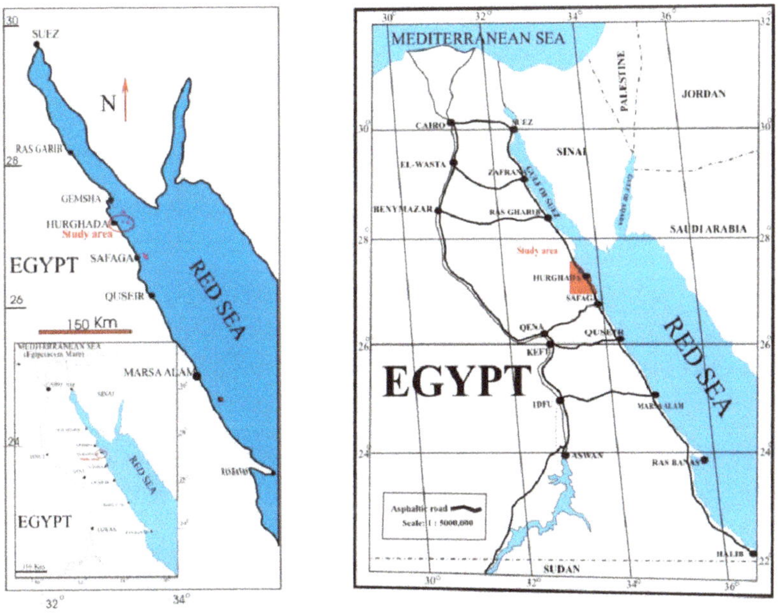

Figure 1. Location map of the study area along the Egyptian Red Sea coast.

2.2. Chemicals and Reagents

All chemicals were of high analytical grade, purchased from Sigma Chemical Co Ltd. (St. Louis, MO, USA).

2.3. Actinomycetes Isolation

The sponge biomass was transferred to the laboratory in a plastic bag containing seawater. Sponge specimens were washed with sterile seawater, cut into pieces of ~1 cm^3, and then thoroughly homogenized in a sterile mortar with 10 volumes of sterile seawater. The supernatant was serially diluted (10^{-1}, 10^{-2}, 10^{-3}) and subsequently plated onto agar plates. Three different media (M1, ISP2, and Marine Agar (MA)) were used for the isolation of actinomycetes. All media were supplemented with 0.2 µm pore size filtered cycloheximide (100 µg/mL), nystatin (25 µg/mL) and nalidixic acid (25 µg/mL) to facilitate the isolation of slow-growing actinomycetes. All media contained Difco Bacto agar (18 g/L) and were prepared in 1 L artificial seawater (NaCl 234.7 g, $MgCl_2.6\ H_2O$ 106.4 g, Na_2SO_4

39.2 g, $CaCl_2$ 11.0 g, $NaHCO_3$ 1.92 g, KCl 6.64 g, KBr 0.96 g, H_3BO_3 0.26 g, $SrCl_2$ 0,24 g, NaF 0.03 g and ddH_2O to 10.0 L). The inoculated plates were incubated at 30 °C for 6–8 weeks. Distinct colony morphotypes were picked and re-streaked until visually free of contaminants. *Micromonospora* sp. UA17, *Gordonia* sp. UA19 and *Nocardia* sp. UA 23 was cultivated on ISP2 medium. The isolates were maintained on plates at the fridge and in 20% glycerol at −80 °C.

2.4. Molecular Identification and Phylogenetic Analysis

The systematic position of the 16S rDNA sequences was analyzed with the SINA web aligner and the search and class option [44]. Closest relatives and type strains were obtained from GenBank using nucleotide Blast against nt and refseq_rna databases, respectively [45]. Alignments were calculated using again the SINA web aligner v1.2.11 (variability profile: bacteria). For maximum-likelihood tree construction RAxML v8.2.12 (-f a -m GTRGAMMA) was used with 100 bootstrap replicates [46]. Trees were visualized with interactive Tree of Life (iTol) v5.5 [47]. The 16S rDNA sequences of *Micromonospora* sp. UA17, *Nocardia* sp. UA23, and *Gordonia* sp. UA19 were deposited in GenBank under the accession numbers MT271359, MT271360, and MT271361.

2.5. Co-cultivation and Extract Preparation

Three actinomycetes were subjected to liquid fermentation as follows; each strain was fermented in 2 L Erlenmeyer flasks each containing 1.5 L ISP2 medium. After incubation of monocultures and co-cultures, the liquid cultures were grown for 10 days at 30 °C while shaking at 150 rpm. The culture was then filtered and the supernatant was extracted with ethyl acetate. The ethyl acetate extracts were stored at 4 °C. Mycolic acid was used at the concentration (5 μg/mL).

2.6. Metabolic Profiling

Ethyl acetate extracts from samples were prepared at 1 mg/mL for mass spectrometry analysis. The recovered ethyl acetate extract was subjected to metabolic analysis using LC-HR-ESI-MS according to Abdelmohsen et al. [33]. An Acquity Ultra Performance Liquid Chromatography system connected to a Synapt G2 HDMS quadrupole time-of-flight hybrid mass spectrometer (Waters, Milford, CT, USA) was used. Positive and negative ESI ionization modes were utilized to carry out the high-resolution mass spectrometry coupled with a spray voltage at 4.5 kV, the capillary temperature at 320 °C, and mass range from *m/z* 150–1500. The MS dataset was processed and data were extracted using MZmine 2.20 based on the established parameters [48]. Mass ion peaks were detected and accompanied by chromatogram builder and chromatogram deconvolution. The local minimum search algorithm was addressed and isotopes were also distinguished via the isotopic peaks of grouper. Missing peaks were displayed using the gap-filling peak finder. An adduct search along with a complex search was carried out. The processed data set was next subjected to molecular formula prediction and peak identification. The positive and negative ionization mode data sets from the respective extract were dereplicated against the DNP (Dictionary of Natural Products) databases.

2.7. Mycolic Acid Detection

The existence of mycolic acid in the bacterial strains (*Gordonia* sp. UA19 and *Nocardia* sp. UA 23) was investigated following the protocol by Onaka et al., 2011 [40]. After 5–7 days fermentation, a broth culture (50 mL) was harvested and centrifuged (5000 rpm for 15 min), the resulting pellet was resuspended in 20 mL of 10% KOH-MeOH and then hydrolyzed by heating at 100 °C for 2 h. The solution was cooled to room temperature, and the hydrolyzed residues were acidified with 6 N HCl and then extracted using *n*-hexane (30 mL). The hexane phase was collected and evaporated in vacuo. The residue was re-suspended in 20 mL of benzene-MeOH-H_2SO_4 (10:20:1) solution and incubated for 2 h at 100 °C. The solution was cooled to room temperature, and the esterified residue was extracted using 20 mL of water and *n*-hexane (1:1). Mycolic acid was obtained by concentrating on the *n*-hexane phase. To confirm the extraction procedure, a hexane phase aliquot was subjected to thin-layer chromatography (TLC) (silica gel 60 F254; Merck); using an *n*-hexane-diethyl ether (4:1) mobile phase, and then dipped in 50% H_2SO_4. The plates were heated at 150 °C, and the methyl ester derivatives of mycolic acid were detected as brown colored spots.

2.8. Antibacterial Activity

Antibacterial activity was tested against *Staphylococcus aureus* NCTC 8325, *Enterococcus faecalis*, *Escherichia coli* and *Pseudomonas aeruginosa* (Culture Collections Public Health England, Porton Down, UK) [49]. After 24 h incubation at 37 °C, broth cultures were diluted in Müller-Hinton broth (1:100) and cultivated again until the cells reached the exponential growth phase. Cells (10^5 cells/mL) were incubated in the presence of various concentrations of the tested extracts in DMSO to the last volume of 200 µL in a 96-well plate at 37 °C. The final concentration of DMSO was 0.8% in each well. After 18 h of incubation, the optical density of the cultures was determined at 550 nm using an ELISA microplate reader (Dynatech Engineering Ltd., Willenhall, UK). The lowest concentration of the compound that inhibits bacterial growth was defined as the minimal inhibitory concentration (MIC), where chloramphenicol was used as a positive control (0.3 µg/mL).

2.9. Antifungal Activity

Antifungal activity was done by re-suspending a colony of *Candida albicans* 5314 (ATCC 90028) (Culture Collections Public Health England, Porton Down, UK) [50], in 2 mL of 0.9% NaCl. Four microliters of this suspension were transferred to 2 mL of HR medium. Various concentrations of the test extracts were diluted in 100 µL of a medium in a 96-well microplate with a final DMSO concentration of 0.4%. One hundred microliters of the *Candida* suspension were added to each well then incubated at 30 °C for 48 h. Optical density was measured at 530 nm for control well without *Candida* cells, and the MIC was detected. Amphotericin B was used as a positive control (MIC 0.4 µg/mL).

2.10. Anti-Trypanosomal Activity

The anti-trypanosomal activity was carried out according to the protocol of Huber and Koella using 104 trypanosomes per mL of *Trypanosoma brucei* brucei strain TC 221, which were cultivated in Complete Baltz Medium. Trypanosomes were tested in 96-well plate against different concentrations of test extracts at 10–200 µg/mL in 1% DMSO to a final volume of 200 µL. As a control, 1% DMSO and the parasite without the extract was used in each plate to show no effect of 1% DMSO. The plates were then incubated at 37 °C in an atmosphere of 5% CO_2 for 24 h. After the addition of 20 µL of Alamar Blue, the activity was measured after 48 and 72 h by light absorption using an MR 700 Microplate Reader at a wavelength of 550 nm with a reference wavelength of 650 nm. The MIC values of the test extracts were quantified in by linear interpolation of three independent measurements. Suramin was used as a positive control (MIC 0.23 µg/mL).

2.11. Statistical Analysis

All experiments were carried out in triplicate. The data were presented as the means ± standard error of the mean (SEM) of at least three independent experiments. The differences among various treatment groups were determined by ANOVA followed by Dunnett's test using PASW Statistics® version18 (Quarry Bay, Hong Kong). The difference of $p < 0.05$ considered statistically significant compared with a vehicle-treated control group and showed by a * symbol. The MIC values were determined using a nonlinear regression curve fitting analysis using GraphPad Prism software version 6 (La Jolla, CA, USA).

3. Results and Discussion

3.1. Molecular Identification and Phylogenetic Analysis

The actinomycete diversity of the Red Sea sponge *Coscinoderma mathewsi* was investigated. Twenty-three isolates were selected based on their cultural characteristics appearance. The 16S rDNA genes were sequenced, and the resulted sequences were blasted against the GenBank database. The isolates were found to belong to six different genera, *Gordonia, Kocuria, Nocardia, Micrococcus, Micromonospora,* and *Microbacterium.* Three new species (*Micromonospora* sp. UA17, *Gordonia* sp. UA19, and *Nocardia* sp. UA 23) were identified based on sequence similarities < 98.2%. The sequence similarities of the three isolates against the type strains ranging from 95.39% to 96.97% (Tables 1–3).

Table 1. List of validly published strains of genus Micromonospora. Identity calculated against strain Micromonospora sp. UA17.

Isolate	Accession ID	Identity [%]	Source	Ref
Micromonospora terminaliae DSM 101760	CP045309.1	96.678	Surface sterilized stem of Thai medicinal plant *Terminalia mucronata*	[51]
Micromonospora inositola DSM 43819	LT607754.1	96.263	forest soil	[52]
Micromonospora cremea CR30	NR_108478.1	96.258	rhizosphere of *Pisum sativum*	[53]
Micromonospora rosaria DSM 803	NR_026282.1	96.125	unknown	[54]
Micromonospora palomenae NEAU-CX1	NR_136848.1	96.055	Nymphs of stinkbug (*Palomena viridissima* Poda)	[55]

Table 2. List of validly published strains of genus Gordonia. Identity calculated against strain Gordonia sp. UA19.

Isolate	Accession ID	Identity [%]	Source	Ref
Gordonia hongkongensis HKU50	NR_152023.1	95.386	human blood culture	[56]
Gordonia terrae 3612	CP016594.1	95.320	soil	[57]
Gordonia bronchialis DSM 43247	NR_074529.1	94.470	human sputum	[58]
Gordonia desulfuricans 213E	NR_028734.1	94.412	soil	[59]
Gordonia rubripertincta DSM 43248	NR_043330.1	94.345	soil	[60]

Table 3. List of validly published strains of genus Nocardia. Identity calculated against strain Nocardia sp. UA23.

Isolate	Accession ID	Identity [%]	Source	Ref
Nocardia xestospongiae ST01-07	NR_156866.1	96.972	*Xestospongia sp.*	[61]
Nocardia amikacinitolerans NBRC 108937	NR_117564.1	96.972	human eye (clinical isolate)	[62]
Nocardia arthritidis DSM 44731	NR_115824.1	96.898	human sputum	[63]
Nocardia araoensis NBRC 100135	NR_118199.1	96.677	human	[64]
Nocardia beijingensis DSM 44636	NR_118618.1	96.529	mud from a sewage ditch	NA

The phylogenetic tree for *Micormonospora* reveals the type strain *Micromonospora terminaliae* DSM 101760 to be the closest to *Micromonospora* sp. UA27 but did not show a specific cluster (Figure 2). The isolate *Gordonia* sp. UA19 was shown to be closest to the strain *Gordonia* sp. EG50, originally isolated from a marine sponge in the Red Sea and both seem to form an own cluster next to the obtained type strains (Figure 3). Isolate *Nocardia* sp. UA23 was placed with three strains, also isolated from marine sponges in either the Red Sea or the South China Sea (Figure 4).

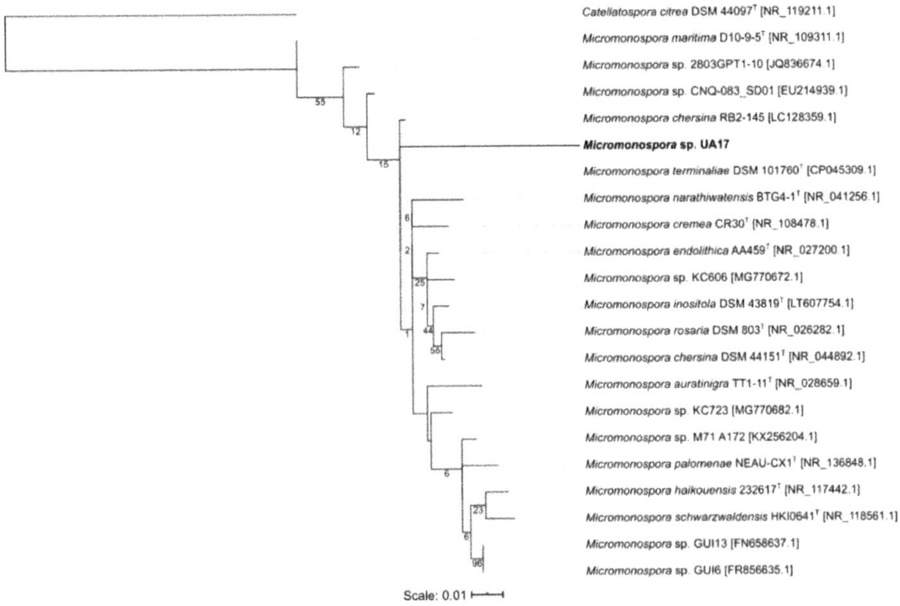

Figure 2. Maximum-likelihood tree of 21 *Micromonospora* representatives and one *Catellatospora* strain as an outgroup. Bootstrap values (100 resamples) are given in percent at the nodes of the tree. The isolate *Micromonospora* sp. UA17 obtained in this study is presented in bold.

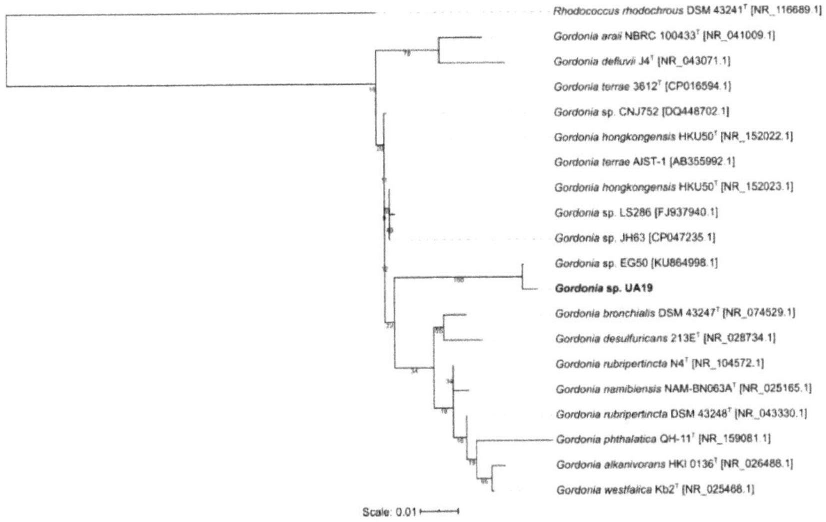

Figure 3. Maximum-likelihood tree of 19 *Gordonia* representatives and one *Rhodococcus* strain as an outgroup. Bootstrap values (100 resamples) are given in percent at the nodes of the tree. The isolate *Gordonia* sp. UA19 obtained in this study is presented in bold.

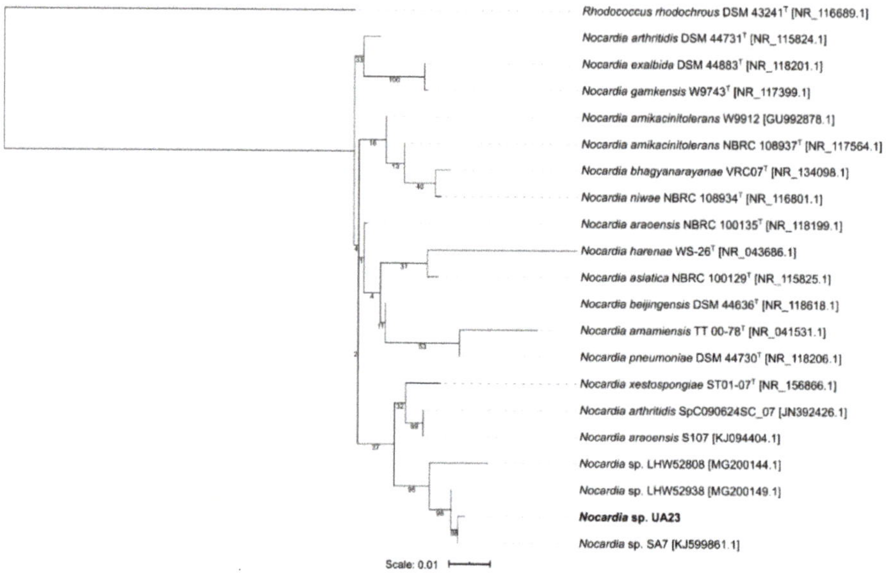

Figure 4. Maximum-likelihood tree of 20 *Nocardia* representatives and one *Rhodococcus* strain as an outgroup. Bootstrap values (100 resamples) are given in percent at the nodes of the tree. The isolate *Nocardia* sp. UA23 obtained in this study is presented in bold.

It has been noticed that the discovery and isolation of new secondary metabolites are becoming hard tasks, as many of the gene clusters encoding for proteins involved in the production of these compounds are normally silenced under lab cultivation conditions [22,65]. To activate these biosynthetic pathways, several strategies have been developed [66,67]. One of the most effective ones consists of the co-cultivation of different microorganisms. The fermentation of both microorganisms in a common environment creates a competitive interaction between them. This fight for survival may induce the synthesis of secondary metabolites that have a defending function against the other microorganism present in the culture medium, thus resulting in the silent activation of gene clusters. Another efficient method for inducing the production of cryptic secondary metabolites is the use of elicitors [48,67,68]. Elicitors are molecules that unregulate the expression of gene clusters involved in the biosynthesis of secondary metabolites in bacteria and fungi.

Sponge-derived actinomycete *Micromonospora* sp. UA17 was co-cultured using liquid fermentation with two mycolic acid-containing actinomycetes (*Gordonia* sp. UA19 and *Nocardia* sp. UA 23) or supplemented with pure mycolic acid. The crude extracts were tested against bacteria, fungi, and the human parasite *Trypanosoma brucei.*

3.2. *Metabolomic Profiling of Monoculture and Co-Culture Crude Extracts*

Metabolite profiles from crude extracts of the actinobacterial monoculture extracts (*Micromonospora* sp. UA17, *Gordonia* sp. UA19 and *Nocardia* sp. UA 23), besides co-cultures with two strains of mycolic acid-containing bacteria (*Gordonia* sp. UA19 and *Nocardia* sp. UA 23) and monocultures amended with mycolic acid were analyzed. The existence of mycolic acid in the selected strains was confirmed as described by Onaka et al., 2011 [40]. The richest metabolites (in terms of several metabolites produced) were observed when the strain *Micromonospora* sp. UA17 was co-cultured with *Nocardia* sp. UA 23 (which contains mycolic acid) or when supplemented with mycolic acid.

3.2.1. Chemical Dereplication of Micromonospora sp. UA17

Analyzing the *Micromonospora* sp. UA17, several hits were proposed (Supplementary Table S1, Supplementary Figure S1a). The molecular ion mass peaks at *m/z* 467.1350021, and 451.1401825 $[M-H]^+$, for the predicted molecular formulas $C_{25}H_{24}O_9$ and $C_{25}H_{24}O_8$ gave hits of the isotetracenone type antitumor antibiotics atramycin A (**1**), and B (**2**) (Supplementary Figure S1b) that were previously isolated from *Streptomyces atratus* [69]. The mass ion peak at *m/z* 465.1557922, corresponding to the suggested molecular formula $C_{27}H_{22}N_4O_4$ $[M-H]^+$ fit an antibiotic indolocarbazole derivative compound TAN-1030A (**3**) that was previously isolated from *Streptomyces longisporoflavus* R-19 [70]. The ion mass peak at *m/z* 323.091309 $[M+H]^+$ for the predicted molecular formulas $C_{19}H_{14}O_5$ gave hits of the anthracyclinone antibiotic fujianmycin A (**4**) which was isolated from *Streptomyces* sp. GW71/2497 [71]. Two major ion peaks with the *m/z* values of 529.171692 and 543.187439 $[M-H]^+$ with molecular formulas $C_{27}H_{30}O_{11}$ and $C_{28}H_{32}O_{11}$ were detected and dereplicated as anthracycline antibiotic mutactimycin C (**5**) and A (**6**), respectively, which were isolated earlier from *Streptomyces* sp. 1254 [72]. In addition, the mass ion peaks at *m/z* 479.171387 $[M-H]^+$, for the predicted molecular formula $C_{28}H_{24}N_4O_4$ was dereplicated another antibiotic indolocarbazole derivative 7-Oxostaurosporine (**7**), which was previously detected in *Streptomyces platensis*, and reported as an inhibitor of protein kinase C (Figure S1b) [73].

3.2.2. Chemical Dereplication of Gordonia sp. UA19

Analyzing the *Gordonia* sp. UA19, few hits were proposed (Supplementary Table S2, Supplementary Figure S2a). The molecular ion mass peak at *m/z* 199.0866394 $[M+H]^+$ for the predicted molecular formulas $C_{12}H_{10}N_2O_2$ gave hits of the 5-(3-indolyl)oxazole type antiviral pimprinine (**8**) (Supplementary Figure S2b) that were previously isolated from *Streptomyces pimprina* [74] and reported as inhibitors against the replication of EV71 and ADV-7 [75]. Other than indolocarbazole derivative founded in *Micromonospora* sp. UA17, which had promising activities, heroin the mass ion peak at *m/z* 404.123993 which corresponded to the suggested molecular formula $C_{22}H_{17}N_3O_5$ $[M+H]^+$ fit an indenotryptoline compound-cladoniamide C (**9**) that was previously isolated from *Streptomyces uncialis* [76], and yet showed no activity. Additionally, atramycin A (**1**), B (**2**), and fujianmycin A (**4**) metabolites were also dereplicated based on the mass ion peaks and in agreement with the molecular formulas.

3.2.3. Chemical Dereplication of Nocardia sp. UA 23

When analyzing the *Nocardia* sp. UA 23, several hits were proposed (Supplementary Table S3, Supplementary Figure S3a,b). The molecular ion mass peak at *m/z* 161.08075 $[M+H]^+$ for the predicted molecular formula $C_7H_{12}O_4$ gave hits of the carbasugar gabosine-B (**10**) that were previously isolated from *Streptomyces albus* [77] and reported as having DNA-binding properties [78]. The molecular ion mass peaks at *m/z* 397.093338 $[M-H]^+$ for the predicted molecular formula $C_{22}H_{14}N_4O_4$ gave hits of the quinoline-5,8-diones type antitumor antibiotic lavendamycin (**11**) that was previously isolated from *Streptomyces lavendulae* C-22030 S [79]. The mass ion peak at *m/z* 300.049118, corresponding to the suggested molecular formula $C_{10}H_{12}ClN_5O_4$ $[M-H]^+$ fits the purine derivative 2-chloroadenosine (**12**) that was previously isolated from *Streptomyces rishiriensis* 265, Sp-265 (FERM-p 5921) [80] and reported as having a suppression effect in seizure [81]. The molecular ion mass peaks at *m/z* 249.107941 $[M+H]^+$ for the predicted molecular formula $C_9H_{16}N_2O_6$ gave hits of the amino acid antibiotic malioxamycin (**13**) that was previously isolated from *Streptomyces lydicus* [82] and reported as having a role in the management of *Streptococcus pneumoniae* infection [83]. In addition, the mass ion peaks at *m/z* 407.167511$[M-H]^+$ for the predicted molecular formula $C_{16}H_{28}N_2O_{10}$ was dereplicated as Enkastine I (**14**), which was a glycopeptide derivative, and was previously detected in *Streptomyces albus* ATCC 21838 [84]. According to the literature, Enkastine I was reported as a potent inhibitor of the endopeptidase 24.11, with an IC_{50} of 1.8×10^{-9} M [84]. Likewise, the molecular formula $C_{14}H_{18}N_2O_5$ was characterized as the antitumor antibiotic benzodiazepine derivative chicamycin A (**15**), from the mass ion peak at *m/z* 293.1146317 [M-H]$^+$, which was previously obtained from *Streptomyces albus* [85,86]. The mass ion peaks at *m/z* 279.0975189 [M+H]$^+$ and 215.1289215 [M-H]$^+$ in agreement with the predicted molecular formulas $C_{13}H_{14}N_2O_5$, and $C_{11}H_{20}O_4$, and were dereplicated as benzodiazepine antibiotic RK-1441B (**16**), and the aliphatic alcohol antidepressant ketalin (**17**), respectively. These metabolites have been isolated earlier from *Streptomyces griseus* [87], and *Streptomyces lavendulae* Tue 1668 [88,89], respectively. In addition, the mass ion peak at *m/z* 1058.672668 [M+H]$^+$ for the predicted molecular formula $C_{54}H_{95}N_3O_{17}$ was dereplicated as macrolide derivative antibiotic copiamycin which has in vitro activity against *Candida albicans, Torulopsis glabrata,* and *Trichomonas vaginalis.* Lo (**18**) and was previously detected in *Streptomyces hygroscopicus var. crystallogenes* [90,91].

3.2.4. Chemical Dereplication of Strains UA17 + UA19

Analyzing co-culture *Micromonospora* sp. UA17 with mycolic acid-containing *Gordonia* sp. UA19 strain, interestingly dereplicated several hits (Supplementary Table S4, Supplementary Figure S4a, S4b). The molecular ion mass peak at *m/z* 252.124252 [M-H]$^+$ for the predicted molecular formula $C_{13}H_{19}NO_4$ gave hits of the piperidine derivative MY 336a (**19**) that was previously isolated from *Streptomyces gabonae* [92]. The molecular ion mass peaks at *m/z* 533.1067098 [M-H]$^+$ for the predicted molecular formula $C_{23}H_{23}ClN_4O_9$ gave hits of the *β*-lactam derivative antibiotic chlorocardicin (20) that was previously isolated from *Streptomyces spp.* [93], and reported as an inhibitor to peptidoglycan biosynthesis [93]. The molecular ion mass peaks at *m/z* 335.056519 [M-H]$^+$ for the predicted molecular formula $C_{19}H_{12}O_6$ gave hits of the benzo[a]anthracene derivative antibiotic WS 5995-A (**21**) that was previously isolated from *Streptomyces auranticolor* 5995 (FERM-p 5365) [94]. The molecular ion mass peaks at *m/z* 1044.656982 [M+H]$^+$ for the predicted molecular formula $C_{53}H_{93}N_3O_{17}$ gave hits of the macrolide derivative Neocopiamycin A (**22**) that was previously isolated from *Streptomyces hygroscopicus var. crystallogenes* [95] and reported to be more active against Gram-positive bacteria and fungi but less toxic than copiamycin (**18**) [90]. The previously identified metabolites atramycin A (**1**), B (**2**), pimprinine (**8**), and copiamycin (**18**), were also dereplicated based on the mass ion peaks and in agree with the molecular formulas.

3.2.5. Chemical Dereplication of Strains UA17 + UA23

Analyzing co-culture *Micromonospora* sp. UA17 with mycolic acid-containing *Nocardia* sp. UA 23 strain, dereplicated several hits (Supplementary Table S5, Supplementary Figure S5a,b). A compound at *m/z* 215.1390686 [M+H]$^+$, corresponding to the suggested molecular formula $C_{10}H_{18}N_2O_3$ was dereplicated as Alkaloid derivative LL-BH-872*α* (**23**), which was formerly reported from *Streptomyces hinnulinus* [96]. The mass ion peaks at *m/z* 247.1076736, 276.170578, 280.082809, 224.0916748, and 261.0881348 for the predicted molecular formulas $C_{13}H_{14}N_2O_3$, $C_{15}H_{21}N_3O_2$, $C_{13}H_{15}NO_6$, $C_{11}H_{13}NO_4$, and $C_{13}H_{14}N_2O_4$ were dereplicated as benzodiazepine DC 81 (**24**), eserine Alkaloid (**25**), amino derivative 13-hydroxy-streptazolin (**26**), pyridine-piperidine derivative A 58365B (**27**), and also benzodiazepine antitumor antibiotic chicamycin B (**28**), respectively, which were previously detected in *Streptomyces roseiscleroticus* do-81 (FERM-p 6502) [97], *Streptomyces griseofuscus* [98], *Streptomyces chromofuscus* [99], *Streptomyces sp.* A1 [100], and *Streptomyces albus* [86], respectively. Whereas that at *m/z* 229.1546555, corresponding to the suggested molecular formula $C_{11}H_{20}N_2O_3$ was dereplicated as imidazolidine derivative Libramycin-A (**29**), which was formerly reported from *Streptomyces sp.* [101]. Likewise, the molecular formulas $C_{16}H_{27}N_3O_6$, and $C_{11}H_{11}NO_5$ was characterized as *β*-propiolactone amino acid belactosin C (**30**), and dehydrodioxolide B alkaloids (**31**), from the mass ion peaks at *m/z* 356.183197, and 236.0565987, which was previously obtained from *Streptomyces sp.* KY11780 [102], and *Streptomyces tendae* [103], respectively. Moreover, the characteristic metabolites atramycin A (**1**), chicamycin A (**15**), and chlorocardicin (**20**) were also dereplicated based on the mass ion peaks and in agreement with the molecular formulas.

3.2.6. Chemical Dereplication of Strain UA17 with Mycolic Acid

Analyzing co-culture *Micromonospora* sp. UA17 with mycolic acid, dereplicated several hits (Supplementary Table S6, Supplementary Figure S6a,b). A compound at *m/z* 213.102264, corresponding to the suggested molecular formula $C_{13}H_{12}N_2O$ was dereplicated as 5-(3-indolyl)oxazole derivative antiviral pimprinethine (**32**), which was formerly reported from *Streptoverticillium olivoreticuli* [75,104]. The mass ion peaks at *m/z* 259.1089478, 190.049843, 227.1178055, and 238.0722198 for the predicted molecular formulas $C_{14}H_{16}N_2O_3$, $C_{10}H_7NO_3$, $C_{14}H_{14}N_2O$, and $C_{11}H_{13}NO_5$ were dereplicated as diketopiperazine derivative maculosin (**33**), antioxidant 3-Hydroxyquinoline-2-carboxylic acid (**34**), benzodiazepine derivative antitumor antibiotic prothracarcin (**35**), and amino acid N-acetyl-3,4-dihydroxy-L-phenylalanine (**36**), respectively which were previously detected in *Streptomyces rochei 87051-3* [105], *Streptomyces cyaneofuscatus* M-157 [106,107], *Streptomyces umbrosus* [108,109], and *Streptomyces akiyoshiensis* ATCC13480 L127 mutants [110], respectively. Whereas that at *m/z* 303.1353149, corresponding to the suggested molecular formula $C_{16}H_{20}N_2O_4$ was dereplicated as benzodiazepine derivative antitumor antibiotic tomaymycin (**37**), which was formerly reported from *Streptomyces achromogenes-tomaymyceticus* [111,112]. Likewise, the molecular formulas $C_{16}H_{25}N_7O_4S$, $C_{25}H_{29}N_3O_{10}S$ were characterized as purine derivative antibiotic cystocin (**38**), and β-lactam derivative deoxycephamycin B (**39**) from the mass ion peaks at *m/z* 410.1608276, and 564.1655884 which were previously obtained from *Streptomyces* sp. GCA0001 [113] and *Streptomyces olivaceus* SANK 60384 (NRRL 3851) [114,115], respectively. Moreover, the characteristic metabolites, ketalin (**17**), chlorocardicin (**20**), DC 81 (**24**), eserine (**25**), 13-hydroxy-streptazolin (**26**), chicamycin B (**28**), libramycin-A (**29**), belactosin C (**30**), dehydrodioxolide B (**31**), were also dereplicated based on the mass ion peaks and in agreement with the molecular formulas.

3.3. *Antibacterial, Antifungal, and Anti-Trypanosomal Activities*

In this investigation, the crude extracts of the actinobacterial monoculture extracts (*Micromonospora* sp. UA17, *Gordonia* sp. UA19, and *Nocardia* sp. UA 23), beside co-cultures with two strains of mycolic acid-containing bacteria (*Gordonia* sp. UA19, and *Nocardia* sp. UA 23) and monocultures amended with mycolic acid were evaluated for their antibacterial, antifungal, and anti-trypanosomal activities against *Staphylococcus aureus* NCTC 8325, *Escherichia coli, Pseudomonas aeruginosa, Candida albicans* 5314, and *Trypanosoma brucei TC 221,* respectively. The results showed that *Micromonospora* sp. UA17 co-cultured with the two strains of mycolic acid-containing bacteria (*Gordonia* sp. UA19, and *Nocardia* sp. UA 23) and monocultures amended with mycolic acid were more active against *Staphylococcus aureus* NCTC 8325, *Enterococcus faecalis,* and *Candida albicans* 5314 compared with monoculture extracts, where UA17 + UA23 had recorded the highest inhibition activities with MIC value of 4.2, 3.9, and 3.8 μg/mL, respectively (Table 4). However, no activities were detected against *Escherichia coli* and *Pseudomonas aeruginosa.* These results suggest the mycolic acid affected the induction of bacterial natural product biosynthetic pathways. All tested extracts showed low activity against *Trypanosoma brucei* TC 221 (MIC > 100 μg/mL), except *Nocardia* sp. UA 23 which recorded the highest inhibition activities with MIC value of 7.2 μg/mL (Table 4).

Table 4. Results of the crude extracts of the actinobacterial monoculture extracts (*Micromonospora* sp. UA17, *Gordonia* sp. UA19, and *Nocardia* sp. UA 23), beside co-cultures with two strains of mycolic acid-containing bacteria (*Gordonia* sp. UA19, and *Nocardia* sp. UA 23) and monocultures amended with mycolic acid against *Staphylococcus aureus* NCTC 8325, *Candida albicans* 5314, and *Trypanosoma brucei* TC 221.

Sample Code	MIC (μg/mL)			MIC (μg/mL, 72 h.)
	Staphylococcus aureus NCTC 8325	*Enterococcus faecalis*	*Candida albicans* 5314	*Trypanosoma brucei* TC 221
Micromonospora sp. UA17	15.6	14.3	13.2	>100
Gordonia sp. UA19	35.7	31.9	16.8	>100
Nocardia sp. UA 23	38.9	39.2	25.7	7.2 *
UA17 + UA19	8.6 *	7.4 *	6.4 *	>100
UA17 + UA23	4.2 *	3.9 *	3.8 *	>100
UA17 + Myc	4.7 *	3.8 *	5.9 *	>100

MIC value of compounds against tested the microorganism, which was defined as minimal inhibitory concentration (MIC). Data were expressed as mean ± 212 S.E.M ($n = 3$). One-way analysis of variance (ANOVA) followed by Dunnett's test was applied. Graph Pad Prism 5 was used for statistical calculations (Graph pad Software, San Diego, CA, USA). * Significant ($p < 0.05$).

4. Conclusions

The rapidly growing number of actinomycete genome sequences highlighted their potential for biosynthesizing a plethora of natural products that are much higher than expected during classical laboratory conditions. Biological elicitation (co-cultivation) of actinomycetes is an effective strategy to provoke the expression of unexpressed or poorly expressed secondary metabolites and further increasing their chemical diversity. This study highlighted the effect of co-culture with mycolic acid-producing microorganisms or mycolic acid itself in the induction of the biosynthesis of many metabolites; although they are known or previously isolated, it was first highlighted by this species because of the effect of co-culturing or the elicitor mycolic acid. On the other hand, some peaks showed no hits during dereplication which suggests they may be new metabolites and need further investigation in scale-up fermentation. The induction of these metabolites qualitatively and/or quantitatively may be the attributed to the difference in biological activities. As in *Micromonospora* sp. UA17 co-cultures with the two strains of mycolic acid-containing bacteria (*Gordonia* sp. UA19, and *Nocardia* sp. UA 23), monocultures amended with mycolic acid were more active against *Staphylococcus aureus* NCTC 8325, *Enterococcus faecalis*, and *Candida albicans* 5314 compared with monoculture extracts, where UA17 + UA23 had recorded the highest inhibition activities with MIC value of 4.2, 3.9, and 3.8 μg/mL, respectively. These results suggest that mycolic acid affected the induction of bacterial natural product biosynthetic pathways. On the other hand, all tested extracts showed low activity against *Trypanosoma brucei* TC 221, except *Nocardia* sp. UA 23 which recorded the highest inhibition activity with an MIC value of 7.2 μg/mL.

Supplementary Materials: The following are available online at http://www.mdpi.com/2076-2607/8/5/783/s1.

Author Contributions: Conceptualization: U.R.A. and H.M.H.; methodology: Y.I.S., U.R.A. and W.N.H.; software: U.R.A. and A.H.E.; formal analysis: A.H.E., A.A.E.S. and H.A.M.; investigation: U.R.A. and H.M.H.; resources: U.R.A. and H.H.; data curation: Y.G. and U.R.A.; writing—original draft: U.R.A. and A.H.E.; writing—review and editing: U.R.A., A.H.E., H.M.H. and W.N.H.; supervision: D.H.M.A. and W.N.H.; project administration: U.R.A.; funding acquisition: D.H.M.A. and W.N.H. All authors have read and agreed to the published version of the manuscript.

Funding: This research received no external funding.

Acknowledgments: This research was funded by the Deanship of Scientific Research at Princess Nourah bint Abdulrahman University through the Fast-track Research Funding Program. We thank M. Müller and M. Krischke (University of Würzburg) for LC-MS measurement.

Conflicts of Interest: The authors declare that no conflict of interest exist.

References

1. Hameş-Kocabaş, E.E.; Ataç, U.Z.E.L. Isolation strategies of marine-derived *actinomycetes* from sponge and sediment samples. *J. Microbiol. Methods* **2012**, *88*, 342–347.
2. Adegboye, M.F.; Babalola, O.O. Taxonomy and ecology of antibiotic producing actinomycetes. *Afr. J. Agric. Res.* **2012**, *15*, 2255–2261.
3. Zhao, K.; Penttinen, P.; Guan, T.W.; Xiao, J.; Chen, Q.A.; Xu, J.; Lindstrom, K.; Zhang, L.L.; Zhang, X.P.; Strobel, G.A. The diversity and antimicrobial activity of endophytic actinomycetes isolated from medicinal plants in *Panxi Plateau* China. *Curr. Microbiol.* **2011**, *62*, 182–190. [CrossRef] [PubMed]
4. Ukhari, M.; Thomas, A.; Wong, N. Culture Conditions for Optimal Growth of Actinomycetes from Marine Sponges. In *Developments in Sustainable Chemical and Bioprocess Technology*; Springer: New York, NY, USA, 2013; pp. 203–210. [CrossRef]
5. Vicente, J.; Stewart, A.; Song, B.; Hill, R.T.; Wright, J.L. Biodiversity of Actinomycetes associated with Caribbean sponges and their potential for natural product discovery. *Mar. Biotechnol.* **2013**, *15*, 413–424. [CrossRef]
6. Sun, W.; Peng, C.S.; Zhao, Y.Y.; Li, Z.Y. Functional gene–guided discovery of type II polyketides from culturable actinomycetes associated with soft coral *Scleronephthya* sp. *PLoS ONE* **2012**, *7*, e42847. [CrossRef]
7. Cheng, C.; MacIntyre, L.; Abdelmohsen, U.R.; Horn, H.; Polymenakou, P.; Edrada-Ebel, R.; Hentschel, U. Biodiversity, anti–trypanosomal activity screening, and metabolomics profiling of actinomycetes isolated from Mediterranean sponges. *PLoS ONE* **2015**, *10*, e0138528. [CrossRef]
8. Abdelmohsen, U.R.; Bayer, K.; Hentschel, U. Diversity, abundance, and natural products of marine sponge–associated actinomycetes. *Nat. Prod. Rep.* **2014**, *31*, 381–399. [CrossRef]
9. Muller, R.; Wink, J. Future potential for anti–infectives from bacteria–how to exploit biodiversity and genomic potential. *Int. J. Med. Microbiol.* **2014**, *304*, 3–13. [CrossRef]
10. Dalisay, D.S.; Williams, D.E.; Wang, X.L.; Centko, R.; Chen, J.; Andersen, R.J. Marine sediment–derived Streptomyces bacteria from British Columbia, Canada are a promising microbiota resource for the discovery of antimicrobial natural products. *PLoS ONE* **2013**, *8*, e77078. [CrossRef]
11. Eltamany, E.E.; Abdelmohsen, U.R.; Ibrahim, A.K.; Hassanean, H.A.; Hentschel, U.; Ahmed, S.A. New antibacterial xanthone from the marine sponge–derived Micrococcus sp. EG45. *Bioorg. Med. Chem. Lett.* **2014**, *24*, 4939–4942. [CrossRef]
12. Abdelmohsen, U.R.; Zhang, G.L.; Philippe, A.; Schmitz, W.; Pimentel-Elardo, S.M.; Hertlein-Amslinger, B.; Hentschel, U.; Bringmann, G. Cyclodysidins A–D, cyclic lipopeptides from the marine sponge–derived *Streptomyces* strain RV15. *Tetrahedron Lett.* **2012**, *53*, 23–29. [CrossRef]
13. Abdelmohsen, U.R.; Szesny, M.; Othman, E.M.; Schirmeister, T.; Grond, S.; Stopper, H.; Hentschel, U. Antioxidant and anti–Protease activities of diazepinomicin from the sponge–associated *Micromonospora* strain RV115. *Mar. Drugs* **2012**, *10*, 2208–2221. [CrossRef] [PubMed]
14. Subramani, R.; Aalbersberg, W. Marine actinomycetes: An ongoing source of novel bioactive metabolites. *Microbiol. Res.* **2012**, *167*, 571–580. [CrossRef] [PubMed]
15. Solanki, R.; Khanna, M.; Lal, R. Bioactive compounds from marine actinomycetes. *Indian J. Microbiol.* **2008**, *48*, 410–431. [CrossRef]
16. Abdelmohsen, U.R.; Yang, C.; Horn, H.; Hajjar, D.; Ravasi, T.; Hentschel, U. Actinomycetes from Red Sea sponges: Sources for chemical and phylogenetic diversity. *Mar. Drugs* **2014**, *12*, 2771–2789. [CrossRef]
17. Grkovic, T.; Abdelmohsen, U.R.; Othman, E.M.; Stopper, H.; Edrada-Ebel, R.; Hentschel, U.; Quinn, R.J. Two new antioxidant actinosporin analogues from the calcium alginate beads culture of sponge associated Actinokineospora sp. strain EG49. *Bioorg. Med. Chem. Lett.* **2014**, *24*, 5089–5092. [CrossRef]
18. Ziemert, N.; Lechner, A.; Wietz, M.; Millan-Aguinaga, N.; Chavarria, K.L.; Jensen, P.R. Diversity, and evolution of secondary metabolism in the marine actinomycete genus Salinispora. *Proc. Natl. Acad. Sci. USA* **2014**, *111*, 1130–1139. [CrossRef]
19. Udwary, D.W.; Zeigler, L.; Asolkar, R.N.; Singan, V.; Lapidus, A.; Fenical, W.; Jensen, P.R.; Moore, B.S. Genome sequencing reveals complex secondary metabolome in the marine actinomycete *Salinispora tropica*. *Proc. Natl. Acad. Sci. USA* **2007**, *104*, 10376–10381. [CrossRef]
20. Marmann, A.; Aly, A.H.; Lin, W.H.; Wang, B.G.; Proksch, P. Co–Cultivation–A Powerful Emerging Tool for Enhancing the Chemical Diversity of Microorganisms. *Mar. Drugs* **2014**, *12*, 1043–1065. [CrossRef]

21. Cimermancic, P.; Medema, M.H.; Claesen, J.; Kurita, K.; Brown, L.C.W.; Mavrommatis, K.; Pati, A.; Godfrey, P.A.; Koehrsen, M.; Clardy, J.; et al. Insights into secondary metabolism from a global analysis of prokaryotic biosynthetic gene clusters. *Cell* **2014**, *158*, 412–421. [CrossRef]
22. Abdelmohsen, U.R.; Grkovic, T.; Balasubramanian, S.; Kamel, M.S.; Quinn, R.J.; Hentschel, U. Elicitation of secondary metabolism in actinomycetes. *Biotechnol. Adv.* **2015**, *33*, 798–811. [CrossRef] [PubMed]
23. Liu, G.; Chater, K.F.; Chandra, G.; Niu, G.Q.; Tan, H.R. Molecular regulation of antibiotic biosynthesis in *streptomyces*. *Microbiol. Mol. Biol. Rev.* **2013**, *77*, 112–143. [CrossRef] [PubMed]
24. Ochi, K.; Hosaka, T. New strategies for drug discovery: Activation of silent or weakly expressed microbial gene clusters. *Appl. Microbiol. Biotechnol.* **2013**, *97*, 87–98. [CrossRef] [PubMed]
25. Brakhage, A.A. Regulation of fungal secondary metabolism. *Nat. Rev. Microbiol.* **2013**, *11*, 21–32. [CrossRef]
26. Rutledge, P.J.; Challis, G.L. Discovery of microbial natural products by activation of silent biosynthetic gene clusters. *Nat. Rev. Microbiol.* **2015**, *13*, 509–523. [CrossRef]
27. Letzel, A.C.; Pidot, S.J.; Hertweck, C. A genomic approach to the cryptic secondary metabolome of the anaerobic world. *Nat. Prod. Rep.* **2013**, *30*, 392–428. [CrossRef]
28. Luo, Y.; Huang, H.; Liang, J.; Wang, M.; Lu, L.; Shao, Z.; Cobb, R.E.; Zhao, H. Activation, and characterization of a cryptic polycyclic tetramate macrolactam biosynthetic gene cluster. *Nat. Commun.* **2013**, *4*, 2894. [CrossRef]
29. Zhu, H.; Sandiford, S.K.; van Wezel, G.P. Triggers and cues that activate antibiotic production by actinomycetes. *J. Ind. Microbiol. Biotechnol.* **2014**, *41*, 371–386. [CrossRef]
30. Bode, H.B.; Bethe, B.; Hofs, R.; Zeeck, A. Big effects from small changes: Possible ways to explore nature's chemical diversity. *ChemBioChem* **2002**, *3*, 619–627. [CrossRef]
31. Paranagama, P.A.; Wijeratne, E.M.K.; Gunatilaka, A.A.L. Uncovering biosynthetic potential of plant–associated fungi: Effect of culture conditions on metabolite production by *Paraphaeosphaeria quadriseptata and Chaetomium chiversii*. *J. Nat. Prod.* **2007**, *70*, 1939–1945. [CrossRef]
32. Wei, H.; Lin, Z.; Li, D.; Gu, Q.; Zhu, T. OSMAC (One Strain Many Compounds) approach in the research of microbial metabolites a review. *Wei Sheng Wu Xue Bao* **2010**, *50*, 701–709. [PubMed]
33. Abdelmohsen, U.R.; Cheng, C.; Viegelmann, C.; Zhang, T.; Grkovic, T.; Ahmed, S.; Quinn, R.J.; Hentschel, U.; Edrada-Ebel, R. Dereplication strategies for targeted isolation of new antitrypanosomal actinosporins A and B from a marine sponge associated–Actinokineospora sp. EG49. *Mar. Drugs* **2014**, *12*, 1220–1244. [CrossRef] [PubMed]
34. Butler, W.R.; Guthertz, L.S. Mycolic acid analysis by high–performance liquid chromatography for identification of Mycobacterium species. *Clin. Microbiol. Rev.* **2001**, *14*, 704–726. [CrossRef] [PubMed]
35. Rivera-Betancourt, O.E.; Karls, R.; Grosse-Siestrup, B.; Helms, S.; Quinn, F.; Dluhy, R.A. Identification of mycobacteria based on spectroscopic analyses of mycolic acid profiles. *Analyst* **2013**, *138*, 6774–6785. [CrossRef] [PubMed]
36. Butler, W.R.; Floyd, M.M.; Brown, J.M.; Toney, S.R.; Daneshvar, M.I.; Cooksey, R.C.; Carr, J.; Steigerwalt, A.G.; Charles, N. Novel mycolic acid–containing bacteria in the family Segniliparaceae fam. nov., including the genus *Segniliparus* gen. nov., with descriptions of *Segniliparus rotundus* sp nov and *Segniliparus rugosus* sp. nov. *Int. J. Syst. Evol. Microbiol.* **2005**, *55*, 1615–1624. [CrossRef] [PubMed]
37. Marrakchi, H.; Laneelle, M.A.; Daffe, M. Mycolic Acids: Structures, Biosynthesis, and Beyond. *Chem. Biolog.* **2014**, *21*, 67–85. [CrossRef] [PubMed]
38. Jamet, S.; Slama, N.; Domingues, J.; Laval, F.; Texier, P.; Eynard, N.; Quemard, A.; Peixoto, A.; Lemassu, A.; Daffe, M.; et al. The Non–Essential Mycolic Acid Biosynthesis Genes hadA and hadC Contribute to the Physiology and Fitness of Mycobacterium smegmatis. *PLoS ONE* **2015**, *10*, e0145883. [CrossRef]
39. Glickman, M.S.; Cox, J.S.; Jacobs, W.R. A novel mycolic acid cyclopropane synthetase is required for cording, persistence, and virulence of Mycobacterium tuberculosis. *Mol. Cell* **2000**, *5*, 717–727. [CrossRef]
40. Onaka, H.; Mori, Y.; Igarashi, Y.; Furumai, T. Mycolic Acid–Containing Bacteria Induce Natural–Product Biosynthesis in Streptomyces Species. *Appl. Environ. Microbiol.* **2011**, *77*, 400–406. [CrossRef]
41. Hoshino, S.; Okada, M.; Wakimoto, T.; Zhang, H.; Hayashi, F.; Onaka, H.; Abe, I. Niizalactams A–C, Multicyclic Macrolactams Isolated from Combined Culture of Streptomyces with Mycolic Acid–Containing Bacterium. *J. Nat. Prod.* **2015**, *78*, 3011–3017. [CrossRef]

42. Hoshino, S.; Zhang, L.; Awakawa, T.; Wakimoto, T.; Onaka, H.; Abe, I. Arcyriaflavin E, a new cytotoxic indolocarbazole alkaloid isolated by combined–culture of mycolic acid–containing bacteria and *Streptomyces cinnamoneus* Nbrc 13823. *J. Antibiotics* **2015**, *68*, 342–344. [CrossRef] [PubMed]
43. Hoshino, S.; Wakimoto, T.; Onaka, H.; Abe, I. Chojalactones A–C, cytotoxic butanolides isolated from Streptomyces sp. cultivated with mycolic acid–containing bacterium. *Org. Lett.* **2015**, *17*, 1501–1504. [CrossRef] [PubMed]
44. Pruesse, E.; Peplies, J.; Glöckner, F.O. SINA: Accurate high–throughput multiple sequence alignment of ribosomal RNA genes. *Bioinformatics* **2012**, *28*, 1823–1829. [CrossRef] [PubMed]
45. Altschul, S.F.; Gish, W.; Miller, W.; Myers, E.W.; Lipman, D.J. Basic local alignment search tool. *J. Mol. Biol.* **1990**, *215*, 403–410. [CrossRef]
46. Stamatakis, A. RAxML version 8: A tool for phylogenetic analysis and post–analysis of large phylogenies. *Bioinformatics* **2014**, *30*, 1312–1313. [CrossRef]
47. Letunic, I.; Bork, P. Interactive Tree of Life (iTOL) v4: Recent updates and new developments. *Nucleic Acids Res.* **2019**, *47*, 256–259. [CrossRef]
48. Tawfike, A.; Attia, E.Z.; Desoukey, S.Y.; Hajjar, D.; Makki, A.A.; Schupp, P.J.; Edrada-Ebel, R.; Abdelmohsen, U.R. New bioactive metabolites from the elicited marine sponge–derived bacterium *Actinokineospora spheciospongiae* sp. nov. *AMB Express* **2019**, *9*, 12. [CrossRef]
49. Liu, M.; Lu, J.; Müller, P.; Turnbull, L.; Burke, C.M.; Schlothauer, R.C.; Carter, D.A.; Whitchurch, C.B.; Harry, E.J. Antibiotic–specific differences in the response of Staphylococcus aureus to treatment with antimicrobials combined with manuka honey. *Fron. Microbiol.* **2015**, *5*, 779. [CrossRef]
50. Lum, K.Y.; Tay, S.T.; Le, C.F.; Lee, V.S.; Sabri, N.H.; Velayuthan, R.D.; Hassan, H.; Sekaran, S.D. Activity of novel synthetic peptides against *Candida albicans*. *Sci. Rep.* **2015**, *5*, 9657. [CrossRef]
51. Kaewkla, O.; Thamchaipinet, A.; Franco, C.M.M. Micromonospora terminaliae sp. nov., an endophytic actinobacterium isolated from the surface–sterilized stem of the medicinal plant *Terminalia mucronata*. *Int. J. Syst. Evol. Microbiol.* **2017**, *67*, 225–230. [CrossRef]
52. Carro, L.; Nouioui, I.; Sangal, V.; Meier-Kolthoff, J.P.; Trujillo, M.E.; del Carmen Montero-Calasanz, M.; Sahin, N.; Smith, D.L.; Kim, K.E.; Peluso, P. Genome–based classification of micromonosporae with a focus on their biotechnological and ecological potential. *Sci. Rep.* **2018**, *8*, 1–23. [CrossRef] [PubMed]
53. Carro, L.; Pukall, R.; Spröer, C.; Kroppenstedt, R.M.; Trujillo, M.E. Micromonospora cremea sp. nov. and Micromonospora zamorensis sp. nov., isolated from the rhizosphere of *Pisum sativum*. *Int. J. Syst. Evol. Microbiol.* **2012**, *62*, 2971–2977. [CrossRef] [PubMed]
54. Kasai, H.; Tamura, T.; Harayama, S. Intrageneric relationships among Micromonospora species deduced from gyrB–based phylogeny and DNA relatedness. *Int. J. Syst. Evol. Microbiol.* **2000**, *50*, 127–134. [CrossRef] [PubMed]
55. Fang, B.; Liu, C.; Guan, X.; Song, J.; Zhao, J.; Liu, H.; Li, C.; Ning, W.; Wang, X.; Xiang, W. Two new species of the genus Micromonospora: *Micromonospora palomenae* sp. nov. and *Micromonospora harpali* sp. nov. isolated from the insects. *Antonie Van Leeuwenhoek* **2015**, *108*, 141–150. [CrossRef] [PubMed]
56. Tsang, C.-C.; Xiong, L.; Poon, R.W.; Chen, J.H.; Leung, K.-W.; Lam, J.Y.; Wu, A.K.; Chan, J.F.; Lau, S.K.; Woo, P.C. Gordonia hongkongensis sp. nov., isolated from blood culture and peritoneal dialysis effluent of patients in Hong Kong. *Int. J. Syst. Evol. Microbiol.* **2016**, *66*, 3942–3950. [CrossRef]
57. Russell, D.A.; Bustamante, C.A.G.; Garlena, R.A.; Hatfull, G.F. Complete genome sequence of Gordonia terrae 3612. *Genome Announc.* **2016**, *4*, e01058-16. [CrossRef]
58. Ivanova, N.; Sikorski, J.; Jando, M.; Lapidus, A.; Nolan, M.; Lucas, S.; Del Rio, T.G.; Tice, H.; Copeland, A.; Cheng, J.-F. Complete genome sequence of *Gordonia bronchialis* type strain (3410 T). *Stand. Genomic Sci.* **2010**, *2*, 19–28. [CrossRef]
59. Kim, S.B.; Brown, R.; Oldfield, C.; Gilbert, S.C.; Iliarionov, S.; Goodfellow, M. Gordonia amicalis sp. nov., a novel dibenzothiophene–desulphurizing actinomycete. *Int. J. Syst. Evol. Microbiol.* **2000**, *50*, 2031–2036. [CrossRef]
60. Shen, F.-T.; Lu, H.-L.; Lin, J.-L.; Huang, W.-S.; Arun, A.; Young, C.-C. Phylogenetic analysis of members of the metabolically diverse genus Gordonia based on proteins encoding the gyrB gene. *Res. Microbiol.* **2006**, *157*, 367–375. [CrossRef]
61. Thawai, C.; Rungjindamai, N.; Klanbut, K.; Tanasupawat, S. Nocardia xestospongiae sp. nov., isolated from a marine sponge in the Andaman Sea. *Int. J. Syst. Evol. Microbiol.* **2017**, *67*, 1451–1456. [CrossRef]

62. Ezeoke, I.; Klenk, H.-P.; Pötter, G.; Schumann, P.; Moser, B.D.; Lasker, B.A.; Nicholson, A.; Brown, J.M. Nocardia amikacinitolerans sp. nov., an amikacin–resistant human pathogen. *Int. J. Syst. Evol. Microbiol.* **2013**, *63*, 1056–1061. [CrossRef] [PubMed]
63. Conville, P.S.; Zelazny, A.M.; Witebsky, F.G. Analysis of secA1 gene sequences for identification of Nocardia species. *J. Clin. Microbiol.* **2006**, *44*, 2760–2766. [CrossRef] [PubMed]
64. Conville, P.S.; Murray, P.R.; Zelazny, A.M. Evaluation of the Integrated Database Network System (IDNS) SmartGene software for analysis of 16S rRNA gene sequences for identification of Nocardia species. *J. Clin. Microbiol.* **2010**, *48*, 2995–2998. [CrossRef] [PubMed]
65. Nett, M.; Ikeda, H.; Moore, B.S. Genomic basis for natural product biosynthetic diversity in the actinomycetes. *Nat. Prod. Rep.* **2009**, *26*, 1362–1384. [CrossRef]
66. El-Hawary, S.S.; Sayed, A.M.; Mohammed, R.; Hassan, H.M.; Zaki, M.A.; Rateb, M.E.; Mohammed, T.A.; Amin, E.; Abdelmohsen, U.R. Epigenetic Modifiers Induce Bioactive Phenolic Metabolites in the Marine–Derived Fungus *Penicillium brevicompactum*. *Mar. Drugs* **2018**, *16*, 253. [CrossRef]
67. Dashti, Y.; Grkovic, T.; Abdelmohsen, U.R.; Hentschel, U.; Quinn, R.J. Actinomycete Metabolome Induction/Suppression with N–Acetylglucosamine. *J. Nat. Prod.* **2017**, *80*, 828–836. [CrossRef]
68. Dinesh, R.; Srinivasan, V.; Sheeja, T.E.; Anandaraj, M.; Srambikkal, H. Endophytic actinobacteria: Diversity, secondary metabolism, and mechanisms to unsilence biosynthetic gene clusters. *Crit. Rev. Microbiol.* **2017**, *43*, 546–566. [CrossRef]
69. Fujioka, K.; Furihata, K.; Shimazu, A.; Hayakawa, Y.; Seto, H. Isolation and characterization of atramycin A and atramycin B, new isotetracenone type antitumor antibiotics. *J. Antibiot.* **1991**, *44*, 1025–1028. [CrossRef]
70. Cai, Y.; Fredenhagen, A.; Hug, P.; Peter, H.H. A nitro analogue of staurosporine and other minor metabolites produced by a *Streptomyces longisporoflavus* strain. *J. Antibiot.* **1995**, *48*, 143–148. [CrossRef]
71. Maskey, R.P.; Grün-Wollny, I.; Laatsch, H. Resomycins AC: New Anthracyclinone Antibiotics Formed by a Terrestrial Streptomyces sp. *J. Antibiot.* **2003**, *56*, 795–800. [CrossRef]
72. Jin, W. Isolation and structure determination of mutactimycin A, a new anthracycline antibiotic. *Kangshengsu* **1990**, *15*, 399–406.
73. Osada, H.; Koshino, H.; Kudo, T.; Onose, R.; Isono, K. A new inhibitor of protein kinase C, Rk–1409 (7–oxostaurosporine). *J. Antibiot.* **1992**, *45*, 189–194. [CrossRef] [PubMed]
74. Naik, S.; Harindran, J.; Varde, A. Pimprinine, an extracellular alkaloid produced by Streptomyces CDRIL–312: Fermentation, isolation, and pharmacological activity. *J. Biotechnol.* **2001**, *88*, 1–10. [CrossRef]
75. Wei, Y.; Fang, W.; Wan, Z.; Wang, K.; Yang, Q.; Cai, X.; Shi, L.; Yang, Z. Antiviral effects against EV71 of pimprinine and its derivatives isolated from Streptomyces sp. *Virol. J.* **2014**, *11*, 195. [CrossRef]
76. Williams, D.E.; Davies, J.; Patrick, B.O.; Bottriell, H.; Tarling, T.; Roberge, M.; Andersen, R.J. Cladoniamides A– G, tryptophan–derived alkaloids produced in culture by Streptomyces uncialis. *Org. Lett.* **2008**, *10*, 3501–3504. [CrossRef] [PubMed]
77. Roscales, S.; Plumet, J. Biosynthesis and biological activity of carbasugars. *Int. J. Carbohydr. Chem.* **2016**, *2016*. [CrossRef]
78. Tang, Y.Q.; Maul, C.; Höfs, R.; Sattler, I.; Grabley, S.; Feng, X.Z.; Zeeck, A.; Thiericke, R. Gabosines L, N and O: New Carba-Sugars from Streptomyces with DNA-Binding Properties. *Eur. J. Org. Chem.* **2000**, 149–153. [CrossRef]
79. Balitz, D.; Bush, J.; Bradner, W.; Doyle, T.; O'herron, F.; Nettleton, D. Isolation of lavendamycin a new antibiotic from *Streptomyces lavendulae*. *J. Antibiot.* **1982**, *35*, 259–265. [CrossRef]
80. Takahashi, E.; Beppu, T. Anew nucleosidic antibiotic AT–265. *J. Antibiot.* **1982**, *35*, 939–947. [CrossRef]
81. Ates, N.; Ilbay, G.; Sahin, D. Suppression of generalized seizures activity by intrathalamic 2–chloroadenosine application. *Exp. Biol. Med.* **2005**, *230*, 501–505. [CrossRef]
82. Takeuchi, M.; Inukai, M.; Enokita, R.; Iwado, S.; Takahashi, S.; Arai, M. Malioxamycin, a new antibiotic with spheroplast–forming activity. *J. Antibiot.* **1980**, *33*, 1213–1219. [CrossRef] [PubMed]
83. Macgowan, A.P.; Bowker, K.E.; Wootton, M.; Holt, H.A. Activity of moxifloxacin, administered once a day, against *Streptococcus pneumoniae* in an in vitro pharmacodynamic model of infection. *Antimicrob. Agents Chemother.* **1999**, *43*, 1560–1564. [CrossRef] [PubMed]
84. Vértesy, L.; Fehlhaber, H.W.; Kogler, H.; Schindler, P.W. Enkastines: Amadori Products with a Specific Inhibiting Action against Endopeptidase–24.11–from *Streptomyces albus* and by Synthesis. *Liebigs Ann.* **1996**, 121–126. [CrossRef]

85. Tsunakawa, M.; Kamei, H.; Konishi, M.; Miyaki, T.; Oki, T.; Kawaguchi, H. Porothramycin, a new antibiotic of the anthramycin group: Production, isolation, structure, and biological activity. *J. Antibiot.* **1988**, *41*, 1366–1373. [CrossRef]
86. Konishi, M.; Hatori, M.; Tomita, K.; Sugawara, M.; Ikeda, C.; Nishiyama, Y.; Imanishi, H.; Miyaki, T.; Kawaguchi, H. Chicamycin, a new antitumor antibiotic. *J. Antibiot.* **1984**, *37*, 191–199. [CrossRef]
87. Osada, H.; Ishinabe, K.; Yano, T.; Kajikawa, K.; Isono, K. New Pyrrolobenzodiazepine Antibiotics, RK-1441A and B I. Biological Properties. *Agric. Biol. Chem.* **1990**, *54*, 2875–2881. [CrossRef]
88. Huang, X.; He, J.; Niu, X.; Menzel, K.D.; Dahse, H.M.; Grabley, S.; Fiedler, H.P.; Sattler, I.; Hertweck, C. Benzopyrenomycin, a Cytotoxic Bacterial Polyketide Metabolite with a Benzo a pyrene-Type Carbocyclic Ring System. *Angew. Chem. Int. Ed.* **2008**, *47*, 3995–3998. [CrossRef]
89. Pham, T.H.; Gardier, A.M. Fast–acting antidepressant activity of ketamine: Highlights on brain serotonin, glutamate, and GABA neurotransmission in preclinical studies. *Pharmacol. Therapeut.* **2019**, 58–90. [CrossRef]
90. Arai, T.; Uno, J.; Horimi, I.; Fukushima, K. Isolation of neocopiamycin A from *Streptomyces hygroscopicus* var. crystallogenes, the copiamycin source. *J. Antibiot.* **1984**, *37*, 103–109. [CrossRef]
91. Seiga, K.; Yamaji, K. Microbiological study of copiamycin. *Appl. Environ. Microbiol.* **1971**, *21*, 986–989. [CrossRef]
92. Butnariu, M.; Buțu, A. Functions of collateral metabolites produced by some actinomycetes. In *Microbial Pathogens and Strategies for Combating them: Science, Technology, and Education*; Formatex Research Center: Badajoz, Spain, 2013; pp. 1419–1425.
93. Nisbet, L.J.; Mehta, R.J.; Oh, Y.; Pan, C.H.; Phelen, C.G.; Polansky, M.J.; Shearer, M.C.; Giovenella, A.J.; Grappel, S.F. Chlorocardicin, a monocyclic β–lactam from a Streptomyces sp. *J. Antibiot.* **1985**, *38*, 133–138. [CrossRef] [PubMed]
94. Ikushima, H.; Iguchi, E.; Kohsaka, M.; Aoki, H.; Imanaka, H. *Streptomyces auranticolor* sp. nov., a new anticoccidial antibiotics producer. *J. Antibiot.* **1980**, *33*, 1103–1106. [CrossRef] [PubMed]
95. Song, X.; Yuan, G.; Li, P.; Cao, S. Guanidine–containing polyhydroxyl macrolides: Chemistry, biology, and structure–activity relationship. *Molecules* **2019**, *24*, 3913. [CrossRef] [PubMed]
96. Sugawara, A.; Kubo, M.; Hirose, T.; Yahagi, K.; Tsunoda, N.; Noguchi, Y.; Nakashima, T.; Takahashi, Y.; Welz, C.; Mueller, D. Jietacins, azoxy antibiotics with potent nematocidal activity: Design, synthesis, and biological evaluation against parasitic nematodes. *Eur. J. Med. Chem.* **2018**, *145*, 524–538. [CrossRef] [PubMed]
97. Kamal, A.; Reddy, P.; Reddy, D.R. The effect of C2–fluoro group on the biological activity of DC–81 and its dimers. *Bioorg. Med. Chem. Lett.* **2004**, *14*, 2669–2672. [CrossRef] [PubMed]
98. Russotti, G.; Göklen, K.E.; Wilson, J.J. Development of a pilot–scale microfiltration harvest for the isolation of physostigmine from *Streptomyces griseofuscus* broth. *J. Chem. Technol. Biotechnol.* **1995**, *63*, 37–47. [CrossRef]
99. Mynderse, J.S.; O'Connor, S.C. Quiuolizine and indolizine enzyme inhibitors. U.S. Patent 4,508,901, 2 April 1985.
100. Mayer, M.; Thiericke, R. Biosynthesis of streptazolin. *J. Org. Chem.* **1993**, *58*, 3486–3489. [CrossRef]
101. Chaudhary, H.S.; Soni, B.; Shrivastava, A.R.; Shrivastava, S. Diversity and versatility of actinomycetes and its role in antibiotic production. *J. Appl. Pharm. Sci.* **2013**, *3*, S83–S94.
102. Beck, J.; Guminski, Y.; Long, C.; Marcourt, L.; Derguini, F.; Plisson, F.; Grondin, A.; Vandenberghe, I.; Vispé, S.; Brel, V. Semisynthetic neoboutomellerone derivatives as ubiquitin–proteasome pathway inhibitors. *Bioorg. Med. Chem.* **2012**, *20*, 819–831. [CrossRef]
103. Blum, S.; Groth, I.; Rohr, J.; Fiedler, H.P. Biosynthetic capacities of actinomycetes. 5. Dioxolides, novel secondary metabolites from Streptomyces tendae. *J. Basic Microbiol.* **1996**, *36*, 19–25. [CrossRef]
104. Koyama, Y.; Yokose, K.; Dolby, L.J. Isolation, characterization, and synthesis of pimprinine, pimprinethine and pimprinaphine, metabolites of Streptoverticillium olivoreticuli. *Agric. Biol. Chem.* **1981**, *45*, 1285–1287. [CrossRef]
105. Lee, H.-B.; Choi, Y.-C.; Kim, S.-U. Isolation and identification of maculosins from *Streptomyces rochei* 87051–3. *Appl. Biol. Chem.* **1994**, *37*, 339–342.
106. Ortiz-López, F.; Alcalde, E.; Sarmiento-Vizcaíno, A.; Díaz, C.; Cautain, B.; García, L.; Blanco, G.; Reyes, F. New 3–Hydroxyquinaldic Acid Derivatives from Cultures of the Marine Derived Actinomycete Streptomyces cyaneofuscatus M–157. *Mar. Drugs* **2018**, *16*, 371. [CrossRef] [PubMed]

107. Massoud, M.A.; el Bialy, S.A.; Bayoumi, W.A.; el Husseiny, W.M. Synthesis of new 2–and 3–hydroxyquinoline–4–carboxylic acid derivatives as potential antioxidants. *Heterocycl. Commun.* **2014**, *20*, 81–88. [CrossRef]
108. Mori, M.; Uozumi, Y.; Kimura, M.; Ban, Y. Total syntheses of prothracarcin and tomaymycin by use of palladium catalyzed carbonylation. *Tetrahedron* **1986**, *42*, 3793–3806. [CrossRef]
109. Shimizu, K.-i.; Kawamoto, I.; Tomita, F.; Morimoto, M.; Fujimoto, K. Prothracarcin, a novel antitumor antibiotic. *J. Antibiot.* **1982**, *35*, 972–978. [CrossRef]
110. Smith, K.C.; White, R.L.; Le, Y.; Vining, L.C. Isolation of N–acetyl–3, 4–dihydroxy–L–phenylalanine from Streptomyces akiyoshiensis. *J. Nat. Prod.* **1995**, *58*, 1274–1277. [CrossRef]
111. Kannan, S. Screening for antiviral activity of Actinomycetes isolated from soil sediments. Ph.D. Thesis, Nandha College of Pharmacy, Erode, India, 2009.
112. Hurley, L.H. Pyrrolo (1, 4) benzodiazepine antitumor antibiotics. Comparative aspects of anthramycin, tomaymycin and sibiromycin. *J. Antibiot.* **1977**, *30*, 349–370. [CrossRef]
113. Deepika, L.; Kannabiran, K. Antagonistic activity of streptomyces vitddk1 spp.(gu223091) isolated from the coastal region of tamil nadu, India. *Pharmacologyoline* **2010**, *1*, 17–29.
114. Harada, S. New Cephalosporins. *J. Chromatogr. Libr.* **1989**, *43*, 233–257.
115. Lee, H.C.; Liou, K.; Kim, D.H.; Kang, S.-Y.; Woo, J.-S.; Sohng, J.K. Cystocin, a novel antibiotic, produced byStreptomyces sp. GCA0001: Biological activities. *Arch. Pharm. Res.* **2003**, *26*, 446–448. [CrossRef] [PubMed]

microorganisms

MDPI

Article

Bioactivity of Serratiochelin A, a Siderophore Isolated from a Co-Culture of *Serratia* sp. and *Shewanella* sp.

Yannik Schneider [1,*,†], Marte Jenssen [1,*,†], Johan Isaksson [2], Kine Østnes Hansen [1], Jeanette Hammer Andersen [1] and Espen H. Hansen [1]

1 Marbio, Faculty for Fisheries, Biosciences and Economy, UiT—The Arctic University of Norway, Breivika, N-9037 Tromsø, Norway; kine.o.hanssen@uit.no (K.Ø.H.); jeanette.h.andersen@uit.no (J.H.A.); espen.hansen@uit.no (E.H.H.)

2 Department of Chemistry, Faculty of Natural Sciences, UiT—The Arctic University of Norway, Breivika, N-9037 Tromsø, Norway; johan.isaksson@uit.no

* Correspondence: yannik.k.schneider@uit.no (Y.S.); marte.jenssen@uit.no (M.J.); Tel.: +47-7764-9267 (Y.S.); +47-7764-9275 (M.J.)

† These authors contributed equally to the work.

Received: 15 June 2020; Accepted: 10 July 2020; Published: 14 July 2020

Abstract: Siderophores are compounds with high affinity for ferric iron. Bacteria produce these compounds to acquire iron in iron-limiting conditions. Iron is one of the most abundant metals on earth, and its presence is necessary for many vital life processes. Bacteria from the genus *Serratia* contribute to the iron respiration in their environments, and previously several siderophores have been isolated from this genus. As part of our ongoing search for medicinally relevant compounds produced by marine microbes, a co-culture of a *Shewanella* sp. isolate and a *Serratia* sp. isolate, grown in iron-limited conditions, was investigated, and the rare siderophore serratiochelin A (**1**) was isolated with high yields. Compound **1** has previously been isolated exclusively from *Serratia* sp., and to our knowledge, there is no bioactivity data available for this siderophore to date. During the isolation process, we observed the degradation product serratiochelin C (**2**) after exposure to formic acid. Both **1** and **2** were verified by 1-D and 2-D NMR and high-resolution MS/MS. Here, we present the isolation of **1** from an iron-depleted co-culture of *Shewanella* sp. and *Serratia* sp., its proposed mechanism of degradation into **2**, and the chemical and biological characterization of both compounds. The effects of **1** and **2** on eukaryotic and prokaryotic cells were evaluated, as well as their effect on biofilm formation by *Staphylococcus epidermidis*. While **2** did not show bioactivity in the given assays, **1** inhibited the growth of the eukaryotic cells and *Staphylococcus aureus*.

Keywords: Serratiochelin A; Serratiochelin C; *Serratia* sp.; siderophore; iron; anticancer; natural products; microbial biotechnology; degradation; antibacterial; *S. aureus*

1. Introduction

Iron is the fourth most abundant metal in the Earth's crust and is an absolute requirement for life [1]. Iron is an essential nutrient vital for several biological processes, such as respiration, gene regulation, and DNA biosynthesis [1,2]. Despite its abundance, iron is a growth-limiting factor for organisms in many environments [1]. To tackle this, microorganisms produce a vast range of iron-chelating compounds, called siderophores. Siderophores are compounds of low molecular weight (<1000 Da) that have high affinity and selectivity for ferric iron (iron(III)) [1], with the function of mediating iron uptake by microbial cells [3]. Siderophore production is commonly regulated by the iron concentration in the surroundings [4]. The siderophores are accumulated by membrane-bound iron receptors and brought inside the cell by active transport. Subsequently, the iron is normally reduced from iron(III) to iron(II). Since the affinity towards iron(II) is much lower than to iron(III), the iron is released from the iron-siderophore complex and can be utilized by the microorganism [4]. One of the major functional groups of siderophores is catecholate. Many siderophores of the catecholate type contain building blocks consisting of dihydroxybenzoic acid coupled to an amino acid [3]. The first catecholate-type siderophore, a glycine conjugate of 2,3-dihydroxybenzoic acid, was identified in 1958. The compound was produced by *Bacillus subtilis* under iron-poor conditions [5].

The genus *Serratia* is part of the family *Enterobacteriaceae*, whose type species is *Serratia marcescens* [6,7]. Species of the genus *Serratia* have been detected in diverse habitats, such as soil, humans, invertebrates, and water. For *Serratia plymuthica*, water appears to be the principal habitat [6]. Bacteria from this genus are also problematic in health care, as *Serratia marcescens* is an opportunistic pathogen causing infections in immunocompromised patients. One of the pathogenicity factors of the bacterium is its production of potent siderophores. Several different siderophores are produced by bacteria of this genus [8], one example being the serratiochelins produced by *Serratia* sp. V4 [9,10].

The serratiochelins are catecholate siderophores produced by *Serratia* sp. [9,10]. In a paper from 2012 by Seyedsayamdost and co-authors, a new siderophore biosynthetic pathway was proposed for the production of the serratiochelins [10]. The new pathway consisted of genes originating and recombined from two known siderophore biosynthetic clusters: The clusters for enterobactin (*Escherichia coli*) and vibriobactin (*Vibrio cholera*). The study mentions three different serratiochelins, serratiochelin A (**1**), B, and C (**2**); the structures of **1** and **2** can be seen in Figure 1. In the study from 2012, only two of the three compounds, **1** and serratiochelin B, were found in the untreated culture extracts, while **2** is a hydrolysis product of **1**, which was produced in the presence of formic acid. Sayedsayamdost et al. indicated that **1** and serratiochelin B were the native compounds produced by the bacterium [10].

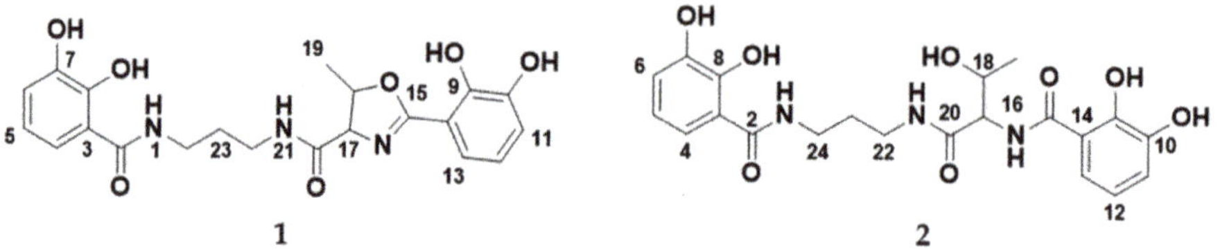

Figure 1. The structures of serratiochelin A (**1**) and C (**2**).

Siderophores are of pharmaceutical interest. They can be used in their native form to treat iron overload diseases, like sickle cell disease. Desferal® (Deferoxamine) is a siderophore-based drug used to treat iron poisoning and thalassemia major, a disease that leads to iron overload, which can lead to severe organ damage [11,12]. Siderophores can furthermore be used to facilitate active uptake of antibiotics by bacteria, and by the production of siderophore-antibiotic drug conjugates (SADCs). For some antibiotics, this strategy can reduce the minimal inhibitory concentration (MIC) by 100-fold, compared to an unbound antibiotic that enters the bacterial cell by passive diffusion [13]. The sideromycins are one example of SADCs. Albomycin, which belongs to this group, enters via the ferrichrome transporter, and has broad-spectrum antibiotic activity and is active against different Gram-negative bacteria [4,14]. The main problem with the use of SADCs is that most pathogenic bacteria have different routes for iron uptake, which could lead to higher frequency in resistance [4].

Due to the important role of *Shewanella* sp. and *Serratia* sp. in the environmental iron cycle, we were intrigued by observing a compound in high yields in an iron-limited co-culture of the two bacteria, which was not found in cultures supplemented with iron nor in axenic cultures of the bacteria. Here, we report the isolation of **1** from a co-culture of *Shewanella* sp. and *Serratia* sp. The degradation of **1** into **2** in the presence of acid was confirmed. To our knowledge, there is no published data regarding the bioactivity of these compounds. In this study, **1** and **2** were tested against a panel of bacterial and human cells, and for their ability to inhibit biofilm formation of the biofilm-producing bacterium *S. epidermidis*.

2. Materials and Methods

2.1. Bacterial Isolates

Compound **1** was isolated from bacterial cultures started from a non-axenic glycerol stock. The bacterial glycerol stock originally contained a *Leifsonia* sp. isolate. The stock was found to be contaminated with both *Shewanella* sp. and *Serratia* sp. after several steps of cultivation and production of new glycerol stock solutions. The non-axenic glycerol stock was inoculated onto three different agar plates, in order to gather information of the different isolates present. Originally, the *Leifsonia* sp. isolate was provided as an axenic culture by The Norwegian Marine Biobank (Marbank, Tromsø, Norway) (Reference number: M10B719). The bacterium was isolated from the intestine/stomach of an Atlantic hagfish (*Myxine glutinosa*) collected by benthic trawl in Hadselfjorden (Norwegian Sea, 16th of April, 2010). The bacterium was grown in liquid FMAP medium (15 g Difco Marine Broth (Becton Dickinson and Company, Franklin Lakes, NJ, USA), 5 g peptone from casein, enzymatic digest (Sigma, St. Louis, MS, USA), 700 mL ddH_2O, and 300 mL filtrated sea water) until sufficient turbidity, and cryo-conserved at −80 °C with 30% glycerol (Sigma). Filtration of sea water was done through a Millidisk® 40 Cartridge with a Durapore® 0.22-µm filter membrane (Millipore, Burlington, MA, USA).

2.2. PCR and Identification of the Strains

The glycerol stock was plated onto three different types of agar: FMAP agar (FMAP medium with 15 g/L agar), DVR1 agar (6.7 g malt extract (Sigma), 11.1 g peptone from casein, enzymatic digest (Sigma), 6.7 g yeast extract (Sigma), 0.5 L filtered sea water, 0.5 L ddH_2O), and potato glucose agar (Sigma). The plates with bacteria were incubated at 10 °C until sufficient growth, and transferred to 4 °C for temporary storage. This plating experiment resulted in the discovery of three different bacterial isolates, based on bacterial morphology and sequencing of the 16S rRNA gene. Clear colonies were picked from the plates, and inoculated into 100 µL of autoclaved ddH_2O. The samples were stored at −20 °C until PCR amplification. The characterization of the bacterial strains was done with sequencing of the 16S rRNA gene through colony PCR and Sanger sequencing. The primer set used for amplification of the gene was the 27F primer (forward primer; 5′-AGAGTTTGATCMTGGCTCAG) and the 1429R primer (reverse primer; 5′-TACCTTGTTACGACTT), both from Sigma. Prior to the amplification PCR, the bacterial samples were vortexed and diluted 1:100 and 1:1000 in UltraPure Water (BioChrom GmbH, Berlin, Germany). For PCR, 1 µL of the diluted bacterial sample was combined in a 25-µL PCR reaction, together with 12.5 µL DreamTaq Green PCR Master Mix (2×) (Thermo Scientific, Vilnius, Lithuania), 10.5 µL ultrapure water, and 0.5 µL of the forward and reverse primers (10 µM) mentioned above. The amplification was done using a Mastercycler ep gradient S (Eppendorf AG, Hamburg, Germany) with the following program: 95 °C initial denaturation for 3 min, followed by 35 cycles of 95 °C for 30 s, 47 °C for 30 s, and 72 °C for 1 min. Final extension was 72 °C for 10 min. The success and purity of the PCR reaction was analyzed on a 1.0% agarose gel (Ultrapure™ Agarose, Invitrogen, Paisley, UK) with Gel-Red® Nucleic Acid Gel Stain (Biotium, Fremont, CA, USA), and the results were documented using a Syngene Bioimaging system (Syngene, Cambridge, UK). Successfully amplified samples were purified by the A'SAP PCR clean up kit (ArcticZymes, Tromsø, Norway). The purified PCR product was used for sequencing PCR, using 1 µL PCR product, 2 µL BigDye™ 3.1 (Applied Biosystems, Foster City, CA, USA), 2 µL 5× sequencing buffer (Applied Biosystems, Foster City, CA, USA), 4 µL of UltraPure water, and 1 µL of primer (1 µM of 27F primer or 1429R primer). The program for the sequencing PCR was as follows: 96 °C initial denaturation for 1 min, followed by 30 cycles of 96 °C for 10 s, 47 °C for 5 s, and 60 °C for 2 min. The PCR product was sequenced at the University Hospital of North Norway (Tromsø, Norway).

The forward and reverse sequences obtained were assembled using the Geneious Prime® 2020.0.5 software (https://www.geneious.com). The sequences were assembled by using the built-in Geneious assembler. Prior to assembly, the sequences were trimmed using a 0.05 error probability limit. Sequence homology comparison was conducted using the built-in Basic Local Alignment Search

Tool (BLAST) [15] in Geneious, excluding environmental samples, metagenomes, and uncultured microorganisms, for phylogenetic identification of the strains.

To identify which strain was responsible for the production of **1**, the three bacterial strains were isolated on separate agar plates and inoculated in small cultures of DVR1 medium (for media contents, see below). The bacteria were pelleted by centrifugation, and the supernatant was diluted 1:1 in methanol and ran on the UHPLC-HR-MS for identification of the compound.

2.3. Fermentation and Extraction of Bacterial Cultures

For extraction of compounds, the bacteria were cultivated in 1000-mL flasks containing 300 mL DVR1 medium (6.7 g malt extract (Sigma), 11.1 g peptone from casein, enzymatic digest (Sigma), 6.7 g yeast extract (Sigma), 0.5 L filtered sea water, and 0.5 L ddH_2O) cultures for 16 days, at 10 °C and 130 rpm. A total of 12 flasks were inoculated, giving 3.6 L of culture. The medium was autoclaved for 30 min at 120 °C prior to inoculation. Cultures were started by loop inoculation from the non-axenic glycerol stock solution.

Extraction of metabolites from the liquid media was done with Diaion® HP-20 resin (Supelco, Bellefonte, PA, USA). The resin was activated by incubation in methanol for 30 min, followed by washing with ddH_2O for 15 min, and added to the cultures (40 g/L). The cultures were incubated with resin for 3 days prior to compound extraction. For extraction, the resin beads were separated from the liquid by vacuum filtration through a cheesecloth mesh (Dansk Hjemmeproduktion, Ejstrupholm, Denmark), the resin was washed with ddH_2O, and finally extracted two times with methanol. The extract was vacuum filtered through a Whatman No. 3 filter paper (Whatman plc, Maidstone, UK), and dried under reduced pressure at 40 °C.

2.4. Fractionation by FLASH Chromatography

Due to the degradation of **1** in the presence of acid, the culture extract was fractionated for bioactivity testing and structure verification, using FLASH chromatography (Biotage SP4™ system, Uppsala, SE), removing the use of acid in the purification process. The extract (3667.9 mg) was re-dissolved in 90% methanol, before adding Diaion® HP20-SS resin (Supelco) in a ratio of 1:1.5 (resin:dry extract, *w*/*w*) and drying under reduced pressure at 40 °C. Due to the high amount of the extract, it was fractionated in two rounds. FLASH columns were prepared with 6.5 g activated Diaion® HP-20SS resin per column. The dried extract was applied to the column, and ran with a water: methanol gradient from 5–100% methanol over 36 min at a flow rate of 12 mL/min. This resulted in 15 fractions per run. The fractions eluting at 100% methanol were analyzed on the UHPLC-HR-MS, and the purest fraction (fraction no 13, >95% pure based on UV/Vis) was used and dried under reduced pressure at 40 °C. The fraction yielded 50.9 mg and was used for the bioactivity testing.

2.5. UHPLC-HR-MS and Dereplication

UHPLC-HR-MS data for dereplication and to analyze the various experiments was recorded using an Acquity I-class UPLC (Waters, Milford, MA, USA) coupled to a PDA detector and a Vion IMS QToF (Waters). The chromatographic separation was performed using an Acquity C-18 UPLC column (1.7 µm, 2.1 mm × 100 mm) (Waters). Mobile phases consisted of acetonitrile (HiPerSolv, VWR, Radnor, PA, USA) for mobile phase B and ddH_2O produced by the in-house Milli-Q® system (Millipore, Burlington, MA, USA) as mobile phase A, both containing 1% formic acid (*v*/*v*) (33015, Sigma). The gradient was run from 10% to 90% B in 12 min at a flow rate of 0.45 mL/min. Samples were run in ESI+ and ESI- ionization mode. The data was processed and analyzed using UNIFI 1.9.4 (Waters). Exact masses and isotopic distributions were calculated using ChemCalc (https://www.chemcalc.org).

2.6. Purification by Preparative HPLC

Initially, the purification of **1** and **2** was done by preparative HPLC-MS using a 600 HPLC pump, a 3100 mass spectrometer, a 2996 photo diode array detector, and a 2767 sample manager (Waters).

For infusion of the eluents into the ESI-quadrupole-MS, a 515 HPLC pump (Waters) and a flow splitter were used and 80% methanol in ddH_2O (*v/v*) acidified with 0.2% formic acid (Sigma) as make-up solution at a flow rate of 0.7 mL/min. The columns used for isolation were a Sunfire RP-18 preparative column (10 µm, 10 mm × 250 mm) and XSelect CSH preparative fluoro-phenyl column (5 µm, 10 mm × 250mm), both columns were purchased from Waters. The mobile phases for the gradients were A (ddH_2O with 0.1% (*v/v*) formic acid) and B (acetonitrile with 0.1% (*v/v*) formic acid), flow rate was set to 6 mL/min. Acetonitrile (Prepsolv®, Merck, Darmstad, Germany) and formic acid (33015, Sigma) were purchased in appropriate quality, ddH_2O was produced with the in-house Milli-Q® system. The collected fractions were reduced to dryness at 40 °C in vacuo and freeze drying using an 8L laboratory freeze dryer (Labconco, Fort Scott, KS, USA).

2.7. NMR analysis

NMR spectra were acquired in DMSO-d_6 on a Bruker Avance III HD spectrometer (Bruker, Billerica, MA, USA) operating at 600 MHz for protons, equipped with an inverse TCI cryo probe enhanced for ^1H, ^{13}C, and ^2H. All NMR spectra were acquired at 298 K, in 3-mm solvent-matched Shigemi tubes using standard pulse programs for proton, carbon, HSQC, HMBC, COSY, and ROESY with gradient selection and adiabatic versions where applicable. $^1H/^{13}C$ chemical shifts were referenced to the residual solvent peak (DMSO-d_6: $\delta H = 2.50$, $\delta C = 39.51$).

2.8. Cultivation Study

Due to the hypothesis that the compound had iron-chelating properties for the bacteria, a cultivation study with and without the addition of iron to the medium was conducted. To investigate if the production was temperature specific, the bacteria were also grown at two different temperatures. The bacteria were grown in DVR1 medium and DVR2 medium (DVR1 with added 5.5 mL $FeSO_4$ 7 H_2O (8 g/L stock, ≙ 28.8 mM Fe)), at room temperature and at 10 °C with 130 rpm shaking. Samples were taken from the cultures, under sterile conditions, after 7, 14, and 21 days, for chemical analysis by UHPLC-HR-MS. From the cultures, 5 mL of sample were taken and centrifuged to pellet the bacteria, 1 mL of the supernatant was transferred to a new tube and centrifuged again, before sterile filtration using an Acrodisc syringe filter 0.2 µm, supor membrane (Pall Corp., East Hills, NY, USA) The filtered sample was mixed 1:1 with methanol prior to injecting on the UHPLC-HR-MS for investigation.

2.9. Marfey's Amino Acid Analysis

A small quantity of **1** was dissolved in 1 mL of 6N HCL and incubated for 6 h at 110 °C using 1.5-mL reaction tubes and a thermoblock. After cooling down to room temperature, the reaction was reduced to dryness by vacuum centrifugation at 40 °C. The dry sample after hydrolysis was re-dissolved in 100 µL of H_2O. The derivatization was carried out by mixing the re-dissolved hydrolystate with 180 µL FDAA in acetone (Marfey's reagent, Sigma), N_α-(2,4-Dinitro-5-fluorophenyl)-L-alaninamide), and 20 µL 1N $NaHCO_3$. The reaction was incubated at 40 °C using a thermoblock. After incubation, the reaction was acidified with 30 µL of 1N HCl and diluted with 2.5 mL of methanol. Then, 0.1 mg of L-threonine and D-threonine dissolved in 100 µL water were used to prepare standards of the amino acids using the same derivatization procedure as described for the sample hydrolysate. The standards and sample diluted in methanol were analyzed using UHPLC-MS/MS as described above.

2.10. Iron Chelation Experiment

For testing the capability of **1** and **2** to chelate iron, a chelation assay was performed. The molecule was dissolved in water (0.2 mg/mL) and 75 µL of the molecule were mixed with 25 µL of 10 mg/mL $FeCl_3 \times 6\ H_2O$. The preparation was done in HPLC vials, the reaction was thoroughly mixed by vortexing, centrifuged, and subsequently analyzed by UHPLC-MS/MS.

2.11. Hydrolyzation with Formic Acid

For testing the liability for hydrolyzation, a 1-mg sample of **1** was dissolved in 1 mL 10% (*v/v*) DMSO *aq*. and incubated for 24 h at room temperature with formic acid concentrations of 0% (control), 0.1%, 1.0%, 5.0%, and 10% (*v/v*). The reaction product was analyzed by UHPLC-MS/MS.

2.12. Production of Serratiochelin C

For testing the bioactivity of **2** in comparison to **1**, a sample of non-degraded **1** was hydrolyzed by adding 10% (*v/v*) formic acid and incubation over 24 h at room temperature. The formic acid was removed by vacuum centrifugation at 40 °C and subsequent freeze drying using a laboratory freeze dryer (Labconco).

2.13. Bioactivity Testing

2.13.1. Growth Inhibition Assay

To determine antimicrobial activity, a bacterial growth inhibition assay was executed. Compounds **1** and **2** were tested against *Staphylococcus aureus* (ATCC 25923), *Escherichia coli* (ATCC 259233), *Enterococcus faecialis* (ATCC 29122), *Pseudomonas aeruginosa* (ATCC 27853), *Streptococcus agalactiae* (ATCC 12386), and Methicillin-resistant *Staphylococcus aureus* (MRSA) (ATCC 33591), all strains from LGC Standards (Teddington, London, UK). *S. aureus*, MRSA, *E. coli*, and *P. aeruginosa* were grown in Muller Hinton broth (275730, Becton). *E. faecalis* and *S. agalactiae* were cultured in brain hearth infusion broth (53286, Sigma). Fresh bacterial colonies were transferred to the respective medium and incubated at 37 °C overnight. The bacterial cultures were diluted to a culture density representing the log phase and 50 μL/well were pipetted into a 96-well microtiter plate (734-2097, Nunclon™, Thermo Scientific, Waltham, MA, USA). The final cell density was 1500–15,000 colony forming units/well. The compound was diluted in 2% (*v/v*) DMSO (Dimethyl sulfoxide) in ddH_2O, and the final assay concentration was 50% of the prepared sample, since 50 μL of sample in DMSO/water were added to 50 μL of bacterial culture. After adding the samples to the plates, they were incubated over night at 37 °C and the growth was determined by measuring the optical density at $\lambda = 600$ nm (OD_{600}) with a 1420 Multilabel Counter VICTOR3™ (Perkin Elmer, Waltham, MA, USA). A water sample was used as the reference control, growth medium without bacteria as a negative control, and a dilution series of gentamycin (32 to 0.01 μg/mL, A2712, Merck) as the positive control and visually inspected for bacterial growth. The positive control was used as a system suitability test and the results of the antimicrobial assay were only considered valid when the positive control was passed. The final concentration of DMSO in the assays was ≤2% (*v/v*), known to have no effect in the tested bacteria. The data was processed using GraphPad Prism 8 (GraphPad, San Diego, CA, USA).

2.13.2. Cell Proliferation Assay

The inhibitory effect **1** and **2** was tested using an MTS in vitro cell proliferation assay against two cell lines: The human melanoma cell line A2058 (ATCC, CLR-1147™), and for general cytotoxicity assessment, the non-malignant MRC5 lung fibroblast cells (ATCC CCL-171™) were employed. The cells were cultured and assayed in Roswell Park Memorial Institute medium (RPMI-16040, FG1383, Merck) containing 10% (*v/v*) fetal bovine serum (FBS, 50115, Biochrom, Holliston, MA, USA). The cell concentration was 4000 cells/well for the lung fibroblast cells and 2000 cells/well for the cancer cells. After seeding, the cells were incubated for 24 h at 37 °C and 5% CO_2. The medium was then replaced with fresh RPMI-1640 medium supplemented with 10% (*v/v*) FBS and gentamycin (10 μg/mL, A2712, Merck). After adding 10 μL of sample diluted in 2% (*v/v*) DMSO in ddH_2O, the cells were incubated for 72 h at 37 °C and 5% CO_2. For assaying the viability of the cells, 10 μL of CellTiter 96® AQueous One Solution Reagent (G3581, Promega, Madison, WI, USA) containing tetrazolium [3-(4,5-dimethylthiazol-2-yl)-5-(3-carboxymethoxyphenyl)-2-(4-sulfophenyl)-2*H*-tetrazolium, inner salt] and phenazine ethosulfate was added to each well and incubated for one hour. The tests were executed with three technical replicates. The plates were read using a DTX 880 plate reader (Beckman Coulter, CA, USA) by measuring the absorbance at λ = 485 nm. The cell viability was calculated using the media control. As a negative control, RPMI-1640 with 10% (*v/v*) FBS and 10% (*v/v*) DMSO (Sigma) was used as a positive control. The data was processed and visualized using GraphPad Prism 8.

2.13.3. Biofilm Inhibition Assay

For testing the inhibition of biofilm formation, the biofilm-producing *Staphylococcus epidermidis* (ATCC 35984) was grown in Tryptic Soy Broth (TSB, 105459, Merck, Kenilworth, NJ, USA) overnight at 37 °C. The overnight culture was diluted in fresh medium with 1% glucose (D9434, Sigma) before being transferred to a 96-well microtiter plate; 50 µL/well were incubated overnight with 50 µL of the test compound dissolved in 2% (*v/v*) DMSO aq. added in duplicates. The bacterial culture was removed from the plate and the plate was washed with tap water. The biofilm was fixed at 65 °C for 1 h before 70 µL of 0.1% crystal violet (115940, Millipore) were added to the wells for 10 min of incubation. Excess crystal violet solution was then removed and the plate dried for 1 h at 65 °C. Seventy microliters of 70% ethanol were then added to each well and the plate incubated on a shaker for 5–10 min. Biofilm formation inhibition were assessed by the presence of violet color and was measured at 600-nm absorbance using a 1420 Multilabel Counter VICTOR3™. Fifty microliters of a non-biofilm-forming *Staphylococcus haemolyticus* (clinical isolate 8-7A, University Hospital of North Norway Tromsø, Norway) mixed in 50 µL of autoclaved Milli-Q water was used as a control; 50 µL of *S. epidermidis* mixed in 50 µL of autoclaved Milli-Q water was used as the control for biofilm formation; and 50 µL of TSB with 50 µL of autoclaved Milli-Q water was used as a medium blank control.

3. Results

Compound **1** was isolated from a co-culture of *Serratia* sp. and *Shewanella* sp. when cultivated in an iron-limited medium. The bacteria were also cultivated in iron-supplemented media, where **1** was not detected. Compound **1** was only produced in co-cultures started directly from the glycerol stock by loop inoculation, and not found in any axenic cultures. The cultures were extracted, and the extracts were fractionated using FLASH chromatography to isolate serratiochelin A (**1**), a siderophore previously isolated exclusively from a *Serratia* sp., also when grown under iron-limited conditions [10]. During preparative HPLC-MS isolation, it was observed that the compound was degraded, and the degradation product was found to be serratiochelin C (**2**), which corresponds to previous observations [10]. A study of the iron binding of the compounds and a degradation study with formic acid was conducted. The structures of the compounds were verified by 1-D and 2-D NMR and MS experiments, and Marfey's analysis was used to find the configuration of the threonine moiety of the molecule. Compound **1** and **2** were tested for their antibacterial activities, their abilities to inhibit the formation of biofilm, and their toxicity towards human cells. This is the first study on the bioactivity of **1** since its original discovery in 1994 [9].

3.1. Identification of Co-Culture and Serratiochelin A Production Strain

When streaking out the glycerol stock onto three different agar plates, three morphologically different bacterial colonies were observed (Figure S14). The 16S rRNA gene of these bacteria was amplified and sequenced by Sanger sequencing, showing that the stock solution contained *Leifsonia* sp. (original isolate in stock), *Shewanella* sp., and *Serratia* sp. The 16S rRNA sequences for the three isolates can be found in the Supplementary Material (Texts S15–S17). *Shewanella* sp. and *Serratia* sp. are assumed to be of marine origin, as strains of the same genera have been cultivated at the same time as the *Leifsonia* sp. isolate, and the 16S rRNA sequences are similar to two strains of the Marbank strain collection (*Shewanella* sp. M10B851 and *Serratia* sp. M10B861, Marbank ID). In order to investigate if all bacteria were able to co-exist in the liquid culture started from the glycerol stock, a 450-mL culture of DVR1 was inoculated with the glycerol stock (identically as was done with the culture from which **1** was isolated) and the culture was streaked out on agar after 3 and 10 days of cultivation.

After three days of cultivation, the colony forming units (CFUs) of both *Shewanella* sp. and *Serratia* sp. were observed on the plates (Figure S14), proven by morphological identification and sequencing of the 16S rRNA gene. After 10 days of culturing, no CFUs of *Shewanella* sp. were observed from the culture, and the experiment detected exclusively CFUs of *Serratia* sp. No *Leifsonia* colonies were observed from the liquid cultures, not after 3 nor 10 days of cultivation. This indicates that the *Serratia* sp. isolate outgrow the other two isolates in the cultivation done in this study. After re-streaking the three bacterial isolates present in the glycerol stock to obtain pure cultures, the different isolates were cultivated separately in 50-mL cultures in DVR1 to identify the actual producer of **1**. Compound **1** was only produced in co-cultures started directly from the glycerol stock, and not by any of the cultures started from axenic colonies from agar plates.

3.2. Dereplication and Isolation

Serratiochelin A (**1**) was obtained as a brown powder. The bacterial extracts and fractions were analyzed using UHPLC-IMS-MS and **1** was detected at *m/z* 430.1594 ($[M+H]^+$) in ESI+ eluting at 4.45 min. The calculated elemental composition was $C_{21}H_{23}N_3O_7$ (Calc. *m/z* 430.1614 $[M+H]^+$), corresponding to 12 degrees of unsaturation. The elemental composition gave several hits for natural products in available databases, including serratiochelin A (**1**). As **1** had been previously isolated from *Serratia* sp., we saw it as a clear possibility that we had a positive identification of the compound. However, to confirm this, isolation and structure elucidation was necessary. After the first round of isolation using preparative HPLC, we detected two species of the product, one at RT= 4.45 min (**1**) and another at RT = 2.07 min (**2**), both having the same *m/z* and elemental composition in ESI+. We later confirmed that the *m/z* of **2** in ESI+ was not the *m/z* of the molecular ion due to neutral water loss in the ion source. The masses of **1** and **2** are thus not equivalent, which was later confirmed by ESI-ionization, which confirmed the mass of **2** to be equal to that of **1**+H_2O.

It was not possible to obtain **1** as a pure compound after the purification, as it was always accompanied by **2**, indicating a possible degradation of **1**. Compound **2**, on the other hand, was obtained as a pure compound after using preparative HPLC for isolation. To distinguish between the two molecules, the collision cross section (CCS) and drift time of the compounds were compared, and the samples were also investigated in ESI- (see Table 2 for the respective values, the high- and low-energy MS spectra, as well as UV/Vis spectra for **1** and **2** that are given in Figures S11 and S12). For isolating **1**, FLASH chromatography was used, since there was no **2** detected using this protocol, where no acid was employed. The collected fractions were assayed individually using UHPLC-MS and the first fraction eluting at 100% methanol was found to be sufficiently pure for structure elucidation via NMR and further bioactivity testing (results of the purity assay are given in Figure 2), yielding 50.9 mg **1** from 3667.9 mg of extract. Compound **1** was not readily dissolved in water and methanol but it dissolved in DMSO. Solutions of **1** were prepared in 100% DMSO and further diluted in water. The same was done with **2**, which also dissolved in methanol.

Serratiochelin C (**2**) was obtained as a brown powder, after acid-catalyzed degradation of **1**. From the ESI-, it was possible to elucidate the elemental composition of **2**. Compound **2** was detected, with *m/z* 446.1568 ($[M-H]^-$) in ESI- eluting at 2.07 min. The calculated elemental composition was $C_{21}H_{25}N_3O_8$ (Calc. *m/z* 446.1563 $[M-H]^-$), corresponding to 11 degrees of unsaturation.

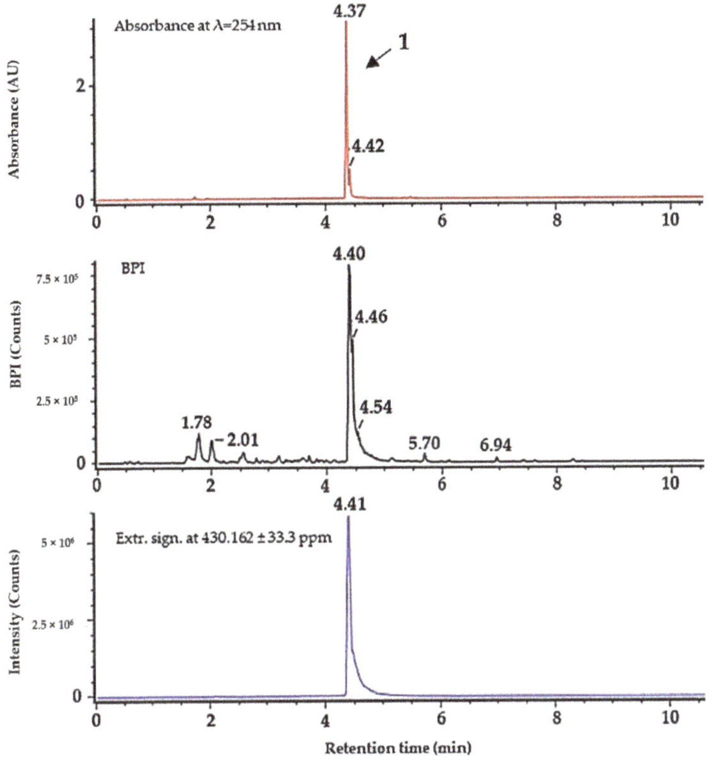

Figure 2. Purity of serratiochelin A (**1**) after isolation using FLASH chromatography, analyzed by UHPLC-MS. Top (in red) absorbance at 254 nm, middle (black) BPI chromatogram, bottom (blue) extracted signal for m/z = 430.162 (±33.3 ppm). ΔRT for UV/Vis detector is ~−0.05 min.

3.3. Structure Elucidation

Close inspection of 1-D (^{1}H and ^{13}C, Table 1) and 2-D (HSQC, HMBC, COSEY, and ROESY) NMR data of **1** confirmed that we isolated the previously reported compound serratiochelin A (**1**). All NMR spectra can be seen in the Supplementary Material (Figures S1–S5). Key COSY and HMBC correlations used to assign the structure of **1** can be seen in Figure 3.

In preparations treated with formic acid, we detected a third molecule eluting at 2.60 min. According to its signal, fragments, and retention time, we concluded it was serratiochelin B [10]. Serratiochelin B was not isolated or verified by NMR. Serratiochelin B and **2** were not present within the crude extract or within fractions obtained by FLASH chromatography but were detected after treatment with acid. The conformation of threonine was found to be L by Marfey's method, which is in compliance what has been published previously [10]. Results are given within the Supplementary Material (Figure S13).

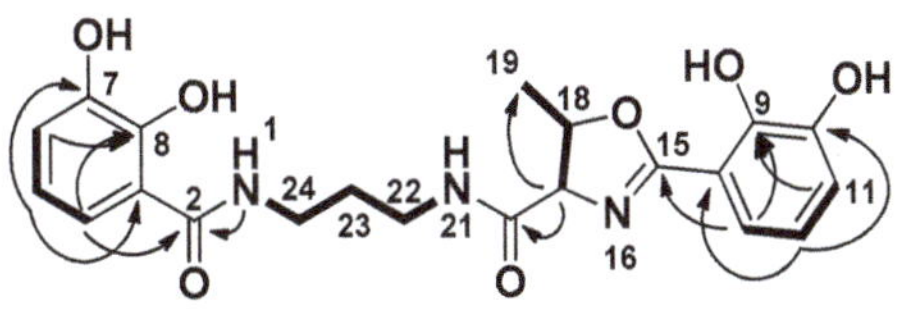

Figure 3. Key COSY (bold) and HMBC (arrow) correlations for serratiochelin A (**1**).

Table 1. ^{1}H- and ^{13}C-NMR data for serratiochelin A (**1**) and C (**2**) in DMSO-d_6.

NMR Data	Serratiochelin A (1)		Serratiochelin C (2)	
Position	δ_C, Type	δ_H (J in Hz)	δ_C, Type	δ_H (J in Hz)
1		8.30, t (5.9)		8.01, t (5.9)
2	169.6, C		169.73, C	
3	114.9, C		114.92, C	
4	117.1, CH	7.25, dd (8.1, 1.5)	117.04, CH	7.25, dd (8.2, 1.5)
5	117.8, CH	6.67, t (7.9)	117.84, CH	6.66, t (8.0)
6	118.7, CH	6.9, dd (7.8, 1.4)	118.65, CH	6.89, dd (7.8, 1.5)
7	146.3, C		146.27, C	
8	149.8, C		149.81, C	
9	148.3, C		146.12, C	
10	145.7, C		148.27, C	
11	119.4, CH	6.96, dd (7.8, 1.6)	118.18, CH	6.92, dd (7.7, 1.5)
12	118.7, CH	6.73, t (7.9)	117.77, CH	6.69, t (7.9)
13	117.9, CH	7.07, dd (7.9, 1.6)	118.92, CH	7.37, dd (8.1, 1.6)
14	110.3, C		116.77, C	
15	165.7, C		168.01, C	
16				8.66, s
17	73.7, CH	4.45, d (87.3)	59.18, CH	4.34, dd (8.0, 4.4)
18	78.8, CH	4.86, p (6.4)	66.38, CH	4.10, qd (6.1, 4.7)
19	20.7, CH_3	1.45, d (6.3)	20.30, CH_3	1.09, d (6.4)
20	169.8, C		169.99, CH	
21		8.81, s		8.78, t (5.3)
22	36.7, CH_2	3.30, m	36.58, CH_2	3.29, q (6.7)
23	28.9, CH_2	1.72, p (7.0)	28.96, CH_2	1.67, p (7.0)
24	36.6, CH_2	3.18, m	36.41, CH_2	3.20–3.08, m

The structure of serratiochelin C (**2**) was confirmed in a similar manner to that of **1**. All NMR spectra can be seen in the Supplementary Material (Figures S6–S10).

3.4. Detection of Iron Chelation

Compounds **1** and **2** were mixed with aqueous $FeCl_3$ solution to investigate if the compounds were able to chelate iron. Both **1** and **2** chelated iron, and the mass spectrometric data given in Table 2 indicate chelation of iron by the loss of three protons through coordination, as published previously [10]. The calculated exact mass for chelation of **1** was *m/z* 483.0729 ([M+Fe-2H]$^+$) and for **2** and serratiochelin B *m/z* 501.0835 ([M+Fe-2H]$^+$). In ESI-, the calculated *m/z* ratios were *m/z* 481.0572 ([M+Fe-4H]$^+$) for **1** and *m/z* 499.0678 ([M+Fe-4H]$^+$) for **2**.

Table 2. IMS and MS data for the apo- and ferrylspecies of serratiochelin A (**1**), serratiochelin B, and serratiochelin C (**2**).

Compounds	Form	Ionization	RT * [min]	*m/z*	CCS ** [A^2]	Drift Time [ms]
Serratiochelin C (earliest eluting)	apo	ESI+	2.07	430.1610 ***	202.88	7.00
	apo	ESI-	2.05	446.1568 ***	198.35	6.94
	ferri	ESI+	2.09	501.0844	208.06	6.75
	ferri	ESI-	2.11	499.0680	203.54	7.12
Serratiochelin B (middle eluting)	apo	ESI+	2.64	448.1714	210.99	6.84
	apo	ESI-	2.60	446.1573	199.52	6.98
	ferri	ESI+	2.61	501.0822	211.91	6.89
	ferri	ESI-	2.62	499.0673	201.16	7.04
Serratiochelin A (late eluting)	apo	ESI+	4.45	430.1611	202.85	6.99
	apo	ESI-	4.37	428.1466	201.47	7.05
	ferri	ESI+	4.46	483.0723	206.88	6.70
	ferri	ESI-	4.39	481.5058	208.50	7.30

* Retention time, ** Collision cross section, *** Loss of water of apo-serratiochelin C (**2**) in ESI+, not in ESI-, and not for the ferri-siderophores.

3.5. Degradation Study with Formic Acid

The study confirmed that the degradation was triggered by formic acid. In order to obtain a pure sample of **1**, we used FLASH fraction no. 13, which predominantly contained **1** since during the extraction process and the FLASH chromatography, no formic acid or acidic solution is used that could induce degradation. This sample was used for the degradation study. Formic acid at concentrations of 0.1%, 1.0%, 5.0%, and 10% (*v/v*) were tested and compared to the control (no acid), as can be seen in Figure 4. It was found that the degradation correlates with the concentration of formic acid. The degradation takes place not only in the presence of formic acid. When incubated with 1% (*v/v*) hydrochloric acid or acetic acid, we observed degradation to approximately the same extent (data not shown). The acid-catalyzed degradation mechanism turning **1** into **2** via intermediates **1a–e** can be seen in Figure 5.

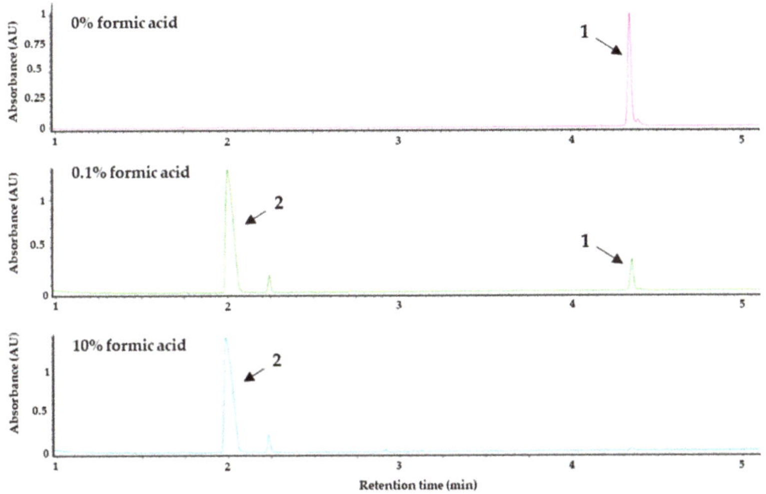

Figure 4. UV-max plot chromatogram. Degradation study showing the effect of formic acid on serratiochelin A (**1**). The purified sample of **1** was treated with different concentrations of formic acid (% (*v/v*)) for 24 h at room temperature and subsequently analyzed via UHPLC-PDA-MS. The chromatograms of the control (0% formic acid), 0.1% formic acid, and 10% formic acid are given above. The degradation of **1** (RT = 4.45 min) into serratiochelin C (**2**) (RT = 2.07 min) corresponds to the amount of formic acid used.

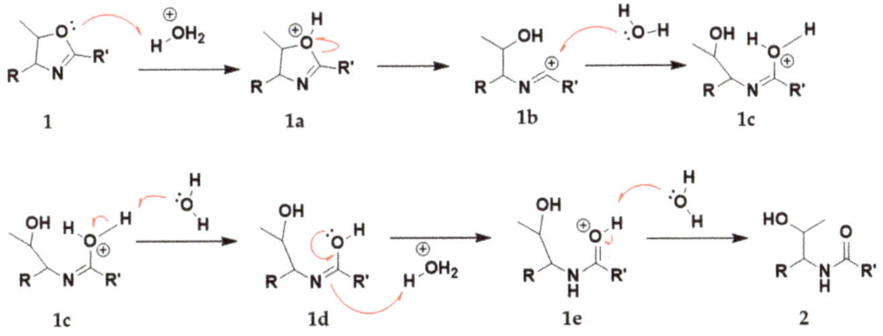

Figure 5. The proposed acid-catalyzed degradation reaction of the central methylated oxazoline ring of **1**, turning **1** into **2** via intermediates **1a** to **1e**.

3.6. Cultivation Study

The cultivation study revealed that **1** was only produced in the iron-deficient co-cultures, as can be seen in Figure 6. Cultures grown in media supplemented with 160 µM $FeSO_4$ did not produce **1** after 7, 14, and 21 days when grown at room temperature nor when grown at 10 °C (Figure 6). Within the iron-deficient cultures, **1** was detected after 7, 14, and 21 days cultivation at 10 °C as well as when cultivated at room temperature. Additionally, when extracting two cultures grown for 14 days at 10 °C using solid-phase extraction, there was no **1** present within the iron-supplemented media while it was a major component in the extract of the iron-deficient culture. Serratiochelin B and **2** were not detected in the cultures, nor in crude extracts after solid-phase extraction.

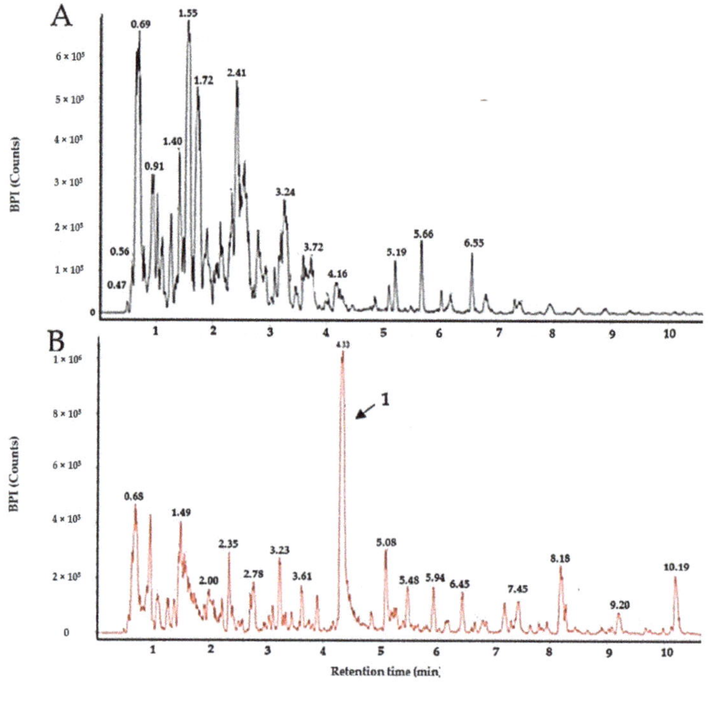

Figure 6. BPI chromatograms of the extracts of two co-cultures. (**A**) The extract of a 14-day culture (10 °C) supplemented with 160 µM Fe(III). (**B**) The extract of a 14-day culture (10 °C) grown in iron-deficient media. The peak of serratiochelin A (**1**) is indicated by the black arrow.

3.7. Bioassays

The growth-inhibiting properties of **1** and **2** were tested against several Gram-positive and Gram-negative strains. The antimicrobial assay detected an effect of **1** on *S. aureus*. Interestingly, there was no effect of **2** on *S. aureus* detected in the assay. There was no antimicrobial effect of **1** and **2** against *S. agalactiae*, *P. aeruginosa*, *E. coli*, *E. faecalis*, and MRSA observed. The results against all the test strains can be seen in Figure 7. The antimicrobial assay with *S. aureus* was repeated to verify the effect of **1**. Among the tested concentrations, 25 µM was the lowest concentration of **1**, which completely inhibited the growth of *S. aureus*, as displayed in Figure 8. Compound **1** and **2** were also tested for their ability to inhibit biofilm formation by *S.epidermidis* in concentrations up to 200 µM. Compound **1** showed some weak effects (assay result of ~ 40%, meaning 60% inhibition, normal cut-off used for further investigation is minimum 70% inhibition) at 200 µM. Compound **2** showed no visible effect up to 200 µM.

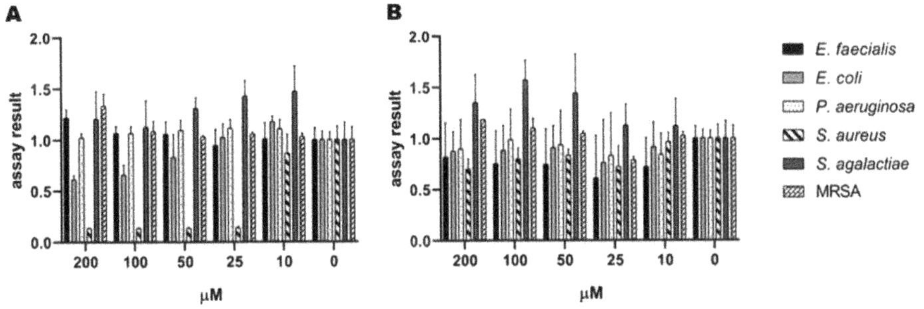

Figure 7. Initial screen of antibacterial activity of (**A**) serratiochelin A (**1**) and (**B**) serratiochelin C (**2**) on *E. faecialis*, *E. coli*, *P. aeruginosa*, *S. agalactiae*, and MRSA, normalized assay results. The experiment was executed twice with two technical replicates each.

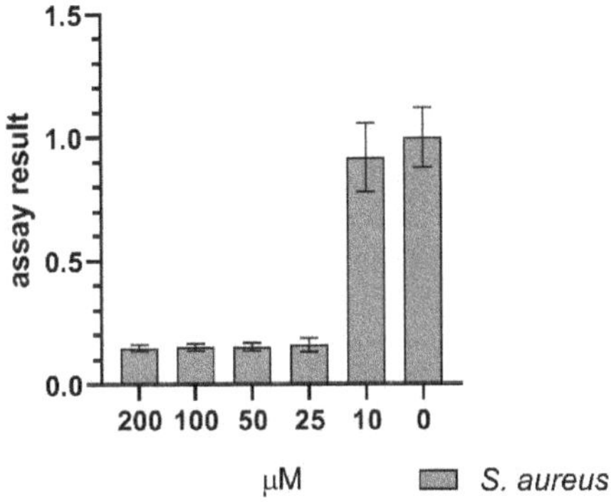

Figure 8. Effect of serratiochelin A (**1**) on *S. aureus* showing inhibition of growth down to 25 µM, normalized assay results. The assay was executed in four experiments with 3 × 2 and 1 × 3 technical replicates.

The effects of the compounds on eukaryotic cells was evaluated using the human melanoma cell line A2058 and the non-malign lung fibroblast cell line MRC5, see Figure 9. The effect of **2** on both cell lines is insufficient, while **1** reduces the cell proliferation of both MRC5 and A2058 cells. The effect of **1** is stronger against MRC5 cells than against A2058.

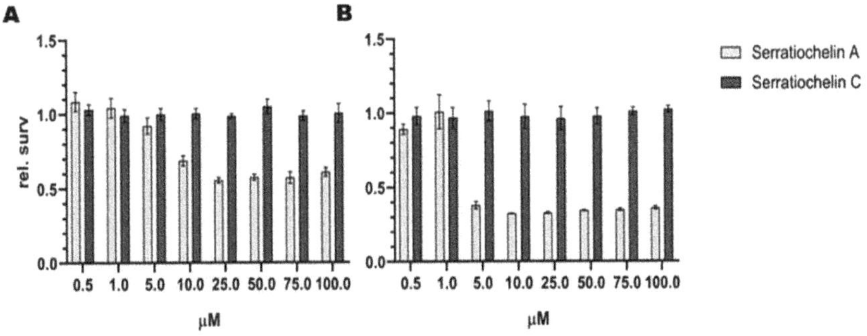

Figure 9. Antiproliferative effect of serratiochelin A (**1**) and C (**2**) on (**A**) A2058 (melanoma) and (**B**) MRC-5 (non-malignant lung fibroblasts) cell lines. The experiments were repeated twice with three technical replicates.

4. Discussion

In this study, a siderophore was isolated from a co-culture of a *Shewanella* sp. and *Serratia* sp. bacteria, both of which come from bacterial genera that are important for environmental iron metabolism. Bacteria from the genus *Shewanella* are known for their important role in iron metabolism, especially in aquatic environments. Previously, several siderophores have been isolated from bacteria of the genus *Serratia*, among these serratiochelin A (**1**).

Shewanella is a genus of Gram-negative rod-shaped γ-proteobacteria, within the order Alteromonadales, found mostly in aquatic habitats [16]. Bacteria from this genus have been isolated from several aquatic sources, both marine and freshwater [17–20]. The genus was established in 1985 [21], after a reconstruction of the *Vibrionaceae* family. *Shewanella* is part of the monogeneric family Shewanellaceae [16], which consists only of this one genera. The genus has high respiratory diversity, with the capability to respire approximately 20 different compounds, including toxic compounds and insoluble metals, one example being reducing Fe(III) chelate and Fe(III) oxide to produce soluble Fe(II) [22]. Bacteria from this genus are often involved in the iron metabolism in their environment, and several iron chelators (siderophores) have been isolated from this genus. Putrebactin is a cyclic dihydroxamate siderophore, produced and isolated from *S. putrefaciens* [23].

To investigate if the three bacterial isolates present in the glycerol stock co-exist in the liquid DVR1 cultures, the culture was streaked out on several agar plates after 3 and 10 days of incubation. The *Shewanella* colonies appeared first, followed by *Serratia* forming colonies on top of the *Shewanella* sp. colonies (Figure S14). After 10 days, there were only colony forming units of *Serratia* sp. present from the liquid co-culture, and the *Shewanella* could not be detected when streaked out on agar. Serratiochelins have previously only been isolated from the *Serratia* genus, and are considered to be rare siderophores [10]. As *Serratia* completely dominates the *Serratia-Shewanella* co-culture after 10 days, and based on data reported regarding previous isolation of **1** [**9**,**10**], it seems to be reasonable to hypothesize that *Serratia* is the true producer of **1** in this co-culture and that it is outcompeting *Shewanella* because of its specific iron acquisition. As **1** was not observed in axenic cultures of *Shewanella* or *Serratia*, we assume that the co-culturing is inducing the production of the compound, possibly due to the competition for iron in the culture.

Compound **1** was isolated from the co-culture after modifying the purification protocol. The degradation of **1** to **2** was triggered by formic acid used in the mobile phase during chromatographic isolation of the compounds. We confirmed that the degradation correlates with the concentration of formic acid as previously published [10]. In addition, the same acidic hydrolyzation of an oxazoline ring was also observed for the compound agrobactin after exposure to hydrochloric acid [24]. We also confirmed the chelation of iron in a hexadecanoate coordination indicated by the loss of three protons, observed in HR-MS experiments [10]. Compound **1** was only produced when no additional iron was added to the co-culture. In the presence of iron, **1** was not detected in the bacterial culture media. We did not detect serratiochelin B and **2** in the culture media, extract, or FLASH fractions (where no acid was used). Previously, it was reported that **1** and serratiochelin B are the initial biosynthetic products of *Serratia* [10]. For our isolate, the results strongly indicate that **1** is the only biosynthetic product, while serratiochelin B and **2** are degradation products of **1**. To obtain **1**, its liability for acid degradation is a significant disadvantage. The FLASH liquid chromatography represents a rather inefficient method for isolation since we were taking only the fraction with the highest purity. Thus, a considerable amount of compound eluted before and after together with other impurities, which diminished the yield of pure **1** significantly, and the purification protocol was not optimized regarding yields but for obtaining **1** without its degradation product. We assume that within the producer isolates' natural environment, **1** is, however, most likely not degrading into **2** due to the rather alkaline pH of seawater [25].

The acid-free isolation enabled us to isolate **1** for bioactivity testing. Since there is no bioactivity data present for **1** and **2**, and the purpose of our investigation was to find new bioactive molecules, it was prioritized for isolation. The testing of both compounds revealed some interesting insights into their bioactivity. Compound **2** displayed no activity in the tested assays and at the tested concentrations, while **1** had antibacterial activity against *S. aureus* and toxic effects on both eukaryotic cell lines tested. Its antibacterial effect was specific towards *S. aureus*, while not having an effect on the other bacteria, including MRSA. Its cytotoxic effect was evaluated against the melanoma cell line (A2058), as we frequently observed that it is the most sensitive cancer cell line in our screening of extracts and compounds. The non-malignant lung fibroblasts (MRC5) was included as a general control of toxicity. The observed effect was stronger on lung fibroblasts than melanoma cells. Of interest to us was the observed difference in activity between **1** and **2** despite the fact that the two structures are closely related. It is questionable if the antiproliferative effect of **1** is caused by iron deprivation as observed for other siderophores [26] or by another effect. The same applies for the observed antibacterial effect on *S.aureus*, while the lack of effect on the other bacteria might indicate a specific target. Both molecules are capable of chelating iron, so either **1** has a higher affinity to iron than **2**, or it has another mode of action. The species-specific antibacterial effect indicates the latter. Gokarn and co-authors investigated the effect of iron chelation by exochelin-MS, mycobactin S, and deferoxiamine B on mammalian cancer cell lines and an antiproliferative effect was observed at concentrations between 0.1 to 1.0 mg/mL. Only HEPG2 cells have shown 23% cell survival at 20 μg/mL for mycobactin S. They observed a different sensitivity among the tested cell lines and siderophores [26]. Compound **1** had an effect at concentrations of <43 μg/mL (40% cell survival was detected at 2.15 μg/mL of **1** against MRC5). Therefore, testing of **1** against more cell lines and testing of **2** at higher concentrations would be an approach for further studies on the antiproliferative effects of **1** and **2**. Some siderophores are known to have additional functions, such as a virulence factor and modulation of the host of a pathogen [27]. Assuming another mode of action than iron chelation, the most relevant structural difference would be the 5-methyl-2-oxazoline heterocycle in **1**, which is hydrolyzed in **2**. Oxazole and oxazoline moieties are structural motives present in molecules with an antibacterial and antiproliferative effect [28,29]. They are ligands to a number of different protein targets and can be regarded as "privileged structures" [29,30]. Further bioactivity elucidation of the two serratiochelins and the mode of action studies of **1** will be the subject of further investigation.

5. Conclusions

We proved the production of **1** in high yields by a co-culture of *Serratia* sp. and *Shewanella* sp., while the compound was not observed in axenic cultures. We confirmed the iron chelation, as well as the degradation of **1** to **2**. We did not observe the production of any compound that could be related to serratiochelin B in the bacterial cultures nor in the extract, but we observed its generation in traces during acid-induced degradation, which gives rise to the assumption that serratiochelin B and **2** are both hydrolyzation products of **1** in this study.

While **1** showed antiproliferative activity on human cancer cells but also on non-malignant lung fibroblasts, and a specific antimicrobial effect on *S. aureus*, **2** did not show any bioactivity in the assays conducted in this study. Since **1** and **2** differ in the presence of a structural motif that can be seen as a privileged structure, we hypothesize that the hydrolyzation of the 5-methyl-2-oxazoline explains the difference in bioactivity. The liability for hydrolyzation, however, represents a strong disadvantage for developing this candidate further as a drug lead.

Supplementary Materials: The following are available online at http://www.mdpi.com/2076-2607/8/7/1042/s1, Figure S1: ^{1}H NMR (600 MHz, DMSO-d_6) spectrum of **1**; Figure S2: ^{13}C (151 MHz, DMSO-d_6) spectrum of **1**; Figure S3: HSQC + HMBC (600 MHz, DMSO-d_6) spectrum of serratiochelin A; Figure S4: COSY (600 MHz, DMSO-d_6) spectrum of **1**; Figure S5: ROESY (600 MHz, DMSO-d_6) spectrum of **1**; Figure S6: ^{1}H NMR (600 MHz, DMSO-d_6) spectrum of **2**; Figure S7: ^{13}C (151 MHz, DMSO-d_6) spectrum of **2**; Figure S8: HSQC + HMBC (600 MHz, DMSO-d_6) spectrum of **2**; S9: HSQC + HMBC (600 MHz, DMSO-d_6) spectrum of **2** (**2**), zoomed in crowded area; Figure S10: COSY (600 MHz, DMSO-d_6) spectrum of **2**; Figure S11: Mass spectra of **1** and **2**; Figure S12: UV/Vis spectra of **1**and **2**; Figure S13: Chromatograms of Marfey's analysis of **1**; Figure S14: Isolation of the bacteria and co-culture of *Serratia* sp. and *Shewanella* sp., Text S15: Consensus sequence of *Shewanella* sp.; Text S16: Consensus sequence of *Serratia* sp.; Text S17: Consensus sequence of *Leifsonia* sp.

Author Contributions: Conceptualization, J.H.A., Y.S. and M.J.; investigation, M.J. and Y.S.; structure elucidation J.I.; writing—original draft preparation, M.J., K.Ø.H. and Y.S.; writing—review and editing, J.I., J.H.A. and E.H.H. All authors have read and agreed to the published version of the manuscript.

Funding: This project received funding from the following projects: The Marie Skłodowska-Curie Action MarPipe of the European Union and from UiT-The Arctic University of Norway (grant agreement GA 721421 H2020-MSCA-ITN-2016), The DigiBiotics project of the Research Counsil of Norway (project iD 269425) and the AntiBioSpec project of UiT the Arctic University of Norway (Cristin iD 20161326). The publication charges for this article have been funded by a grant from the publication fund of UiT-The Arctic University of Norway.

Acknowledgments: Yannik Schneider has been supported by the MarPipe Project, and Marte Jenssen by the DigiBiotics project and the AntiBioSpec project. Kirsti Helland and Marte Albrigtsen are gratefully acknowledged for executing the bioactivity assays and Chun Li for his help with identifying the strains. We want to acknowledge our colleagues of The Norwegian Marine Biobank (Marbank) for sampling and isolation of the bacterial strains.

Conflicts of Interest: The authors declare no conflict of interest.

References

1. Andrews, S.C.; Robinson, A.K.; Rodríguez-Quiñones, F. Bacterial iron homeostasis. *FEMS Microbiol. Rev.* **2003**, *27*, 215–237. [CrossRef]
2. Sandy, M.; Butler, A. Microbial iron acquisition: marine and terrestrial siderophores. *Chem. Rev.* **2009**, *109*, 4580–4595. [CrossRef]
3. Winkelmann, G. Microbial siderophore-mediated transport. *Biochem. Soc. Trans.* **2002**, *30*, 691–696. [CrossRef]
4. Hider, R.C.; Kong, X. Chemistry and biology of siderophores. *Nat. Prod. Rep.* **2010**, *27*, 637–657. [CrossRef] [PubMed]
5. Ito, T.; Neilands, J.B. Products of "low-iron fermentation" with bacillus subtilis: Isolation, characterization and synthesis of 2, 3-dihydroxybenzoylglycine. *J. Am. Chem. Soc.* **1958**, *80*, 4645–4647. [CrossRef]
6. Dworkin, M.; Falkow, S.; Rosenberg, E.; Schleifer, K.-H.; Stackebrandt, E. *The Genus Serratia*; Springer: New York, NY, USA, 2006; pp. 219–244. [CrossRef]
7. Martinec, T.; Kocur, M. The taxonomic status of Serratia marcescens Bizio. *Int. J. Syst. Evol. Microbiol.* **1961**, *11*, 7–12. [CrossRef]
8. Khilyas, I.; Shirshikova, T.; Matrosova, L.; Sorokina, A.; Sharipova, M.; Bogomolnaya, L. Production of siderophores by Serratia marcescens and the role of MacAB efflux pump in siderophores secretion. *BioNanoScience* **2016**, *6*, 480–482. [CrossRef]
9. Ehlert, G.; Taraz, K.; Budzikiewicz, H. Serratiochelin, a new catecholate siderophore from Serratia marcescens. *Z. Nat.* **1994**, *49*, 11–17. [CrossRef]
10. Seyedsayamdost, M.R.; Cleto, S.; Carr, G.; Vlamakis, H.; João Vieira, M.; Kolter, R.; Clardy, J. Mixing and matching siderophore clusters: Structure and biosynthesis of serratiochelins from *Serratia* sp. V4. *J. Am. Chem. Soc.* **2012**, *134*, 13550–13553. [CrossRef]
11. Modell, B.; Letsky, E.A.; Flynn, D.M.; Peto, R.; Weatherall, D.J. Survival and desferrioxamine in thalassaemia major. *Br. Med. J.* **1982**, *284*, 1081–1084. [CrossRef] [PubMed]
12. Saha, M.; Sarkar, S.; Sarkar, B.; Sharma, B.; Bhattacharjee, S.; Tribedi, P. Microbial siderophores and their potential applications: A review. *Environ. Sci. Pollut. Res.* **2015**, *23*, 3984–3999. [CrossRef] [PubMed]
13. Braun, V.; Pramanik, A.; Gwinner, T.; Koberle, M.; Bohn, E. Sideromycins: Tools and antibiotics. *Biometals* **2009**, *22*, 3–13. [CrossRef]
14. Pramanik, A.; Stroeher, U.H.; Krejci, J.; Standish, A.J.; Bohn, E.; Paton, J.C.; Autenrieth, I.B.; Braun, V. Albomycin is an effective antibiotic, as exemplified with Yersinia enterocolitica and Streptococcus pneumoniae. *Int. J. Med. Microbiol.* **2007**, *297*, 459–469. [CrossRef]
15. Altschul, S.F.; Gish, W.; Miller, W.; Myers, E.W.; Lipman, D.J. Basic local alignment search tool. *J. Mol. Biol.* **1990**, *215*, 403–410. [CrossRef]
16. Ivanova, E.P.; Flavier, S.; Christen, R. Phylogenetic relationships among marine Alteromonas-like proteobacteria: Emended description of the family Alteromonadaceae and proposal of Pseudoalteromonadaceae fam. nov., Colwelliaceae fam. nov., Shewanellaceae fam. nov., Moritellaceae fam. nov., Ferrimonadaceae fam. nov., Idiomarinaceae fam. nov. and Psychromonadaceae fam. nov. *Int. J. Syst. Evol. Microbiol.* **2004**, *54*, 1773–1788. [CrossRef]

17. Lee, O.O.; Lau, S.C.; Tsoi, M.M.; Li, X.; Plakhotnikova, I.; Dobretsov, S.; Wu, M.C.; Wong, P.K.; Weinbauer, M.; Qian, P.Y. *Shewanella irciniae* sp. nov., a novel member of the family Shewanellaceae, isolated from the marine sponge Ircinia dendroides in the Bay of Villefranche, Mediterranean Sea. *Int. J. Syst. Evol. Microbiol.* **2006**, *56*, 2871–2877. [CrossRef] [PubMed]
18. Ivanova, E.P.; Nedashkovskaya, O.I.; Sawabe, T.; Zhukova, N.V.; Frolova, G.M.; Nicolau, D.V.; Mikhailov, V.V.; Bowman, J.P. *Shewanella affinis* sp. nov., isolated from marine invertebrates. *Int. J. Syst. Evol. Microbiol.* **2004**, *54*, 1089–1093. [CrossRef]
19. Kizhakkekalam, K.V.; Chakraborty, K.; Joy, M. Antibacterial and antioxidant aryl-enclosed macrocyclic polyketide from intertidal macroalgae associated heterotrophic bacterium Shewanella algae. *Med. Chem. Res.* **2019**, *29*, 145–155. [CrossRef]
20. Bowman, J.P.; McCammon, S.A.; Nichols, D.S.; Skerratt, J.H.; Rea, S.M.; Nichols, P.D.; McMeekin, T.A. *Shewanella gelidimarina* sp. nov. and *Shewanella frigidimarina* sp. nov., novel Antarctic species with the ability to produce eicosapentaenoic acid (20:5 omega 3) and grow anaerobically by dissimilatory Fe(III) reduction. *Int. J. Syst. Evol. Microbiol.* **1997**, *47*, 1040–1047. [CrossRef]
21. MacDonell, M.T.; Colwell, R.R. Phylogeny of the Vibrionaceae, and recommendation for two new genera, Listonella and Shewanella. *Syst. Appl. Microbiol.* **1985**, *6*, 171–182. [CrossRef]
22. Hau, H.H.; Gralnick, J.A. Ecology and biotechnology of the genus Shewanella. *Annu. Rev. Microbiol.* **2007**, *61*, 237–258. [CrossRef] [PubMed]
23. Ledyard, K.M.; Butler, A. Structure of putrebactin, a new dihydroxamate siderophore produced by Shewanella putrefaciens. *J. Biol. Inorg. Chem.* **1997**, *2*, 93–97. [CrossRef]
24. Ong, S.A.; Peterson, T.; Neilands, J.B.; Ong, S.A. Agrobactin, a siderophore from Agrobacterium Tumefaciens. *J. Biol. Chem.* **1979**, *254*, 1860–1865. [PubMed]
25. Halevy, I.; Bachan, A. The geologic history of seawater pH. *Science* **2017**, *355*, 1069–1071. [CrossRef]
26. Gokarn, K.; Sarangdhar, V.; Pal, R.B. Effect of microbial siderophores on mammalian non-malignant and malignant cell lines. *BMC Complement. Altern. Med.* **2017**, *17*, 145. [CrossRef]
27. Behnsen, J.; Raffatellu, M. Siderophores: More than stealing iron. *MBio* **2016**, *7*, e01906–e01916. [CrossRef]
28. Chiacchio, M.A.; Lanza, G.; Chiacchio, U.; Giofrè, S.V.; Romeo, R.; Iannazzo, D.; Legnani, L. Oxazole-based compounds as anticancer agents. *Curr. Med. Chem.* **2019**, *26*, 7337–7371. [CrossRef]
29. Zhang, H.-Z.; Zhao, Z.-L.; Zhou, C.-H. Recent advance in oxazole-based medicinal chemistry. *Eur. J. Med. Chem.* **2018**, *144*, 444–492. [CrossRef]
30. Kim, J.; Kim, H.; Park, S.B. Privileged structures: Efficient chemical "navigators" toward unexplored biologically relevant chemical spaces. *J. Am. Chem. Soc.* **2014**, *136*, 14629–14638. [CrossRef]

microorganisms

MDPI

Article

Baikalomycins A-C, New Aquayamycin-Type Angucyclines Isolated from Lake Baikal Derived *Streptomyces* sp. IB201691-2A

Irina Voitsekhovskaia [1,†,‡], Constanze Paulus [2,†], Charlotte Dahlem [3], Yuriy Rebets [4], Suvd Nadmid [4,§], Josef Zapp [3], Denis Axenov-Gribanov [1,5], Christian Rückert [6], Maxim Timofeyev [1,5], Jörn Kalinowski [6], Alexandra K. Kiemer [3] and Andriy Luzhetskyy [2,4,*]

1 Institute of Biology, Irkutsk State University, 664003 Irkutsk, Russia; irina.voytsekhovskaya@gmail.com (I.V.); denis.axengri@gmail.com (D.A.-G.); m.a.timofeyev@gmail.com (M.T.)

2 Helmholtz Institute for Pharmaceutical Research Saarland, 66123 Saarbrücken, Germany; Constanze.Paulus@helmholtz-hzi.de

3 Pharmaceutical Biology, Saarland University, 66123 Saarbrücken, Germany; charlotte.dahlem@uni-saarland.de (C.D.); j.zapp@mx.uni-saarland.de (J.Z.); pharm.bio.kiemer@mx.uni-saarland.de (A.K.K.)

4 Pharmaceutical Biotechnology, Saarland University, 66123 Saarbrücken, Germany; y.rebets@mx.uni-saarland.de (Y.R.); suvdn@yahoo.com (S.N.)

5 Baikal Research Centre, 664003 Irkutsk, Russia

6 Technology Platform Genomics, Center for Biotechnology (CeBiTec), Bielefeld University, 33615 Bielefeld, Germany; Christian.Rueckert@cebitec.uni-bielefeld.de (C.R.); joern@cebitec.uni-bielefeld.de (J.K.)

* Correspondence: a.luzhetskyy@mx.uni-saarland.de; Tel.: +49-681-302-70200

† These authors contributed equally to this work.

‡ Current address: Interfaculty Institute of Microbiology and Infection Medicine (IMIT), Microbiology/Biotechnology, Eberhard Karls University of Tübingen, 72074 Tübingen, Germany; irina.voitsekhovskaia@uni-tuebingen.de.

§ Current address: School of Pharmacy, Mongolian National University of Medical Sciences, 14210 Ulaanbaatar, Mongolia.

Received: 7 April 2020; Accepted: 5 May 2020; Published: 7 May 2020

Abstract: Natural products produced by bacteria found in unusual and poorly studied ecosystems, such as Lake Baikal, represent a promising source of new valuable drug leads. Here we report the isolation of a new *Streptomyces* sp. strain IB201691-2A from the Lake Baikal endemic mollusk *Benedictia baicalensis*. In the course of an activity guided screening three new angucyclines, named baikalomycins A–C, were isolated and characterized, highlighting the potential of poorly investigated ecological niches. Besides that, the strain was found to accumulate large quantities of rabelomycin and 5-hydroxy-rabelomycin, known shunt products in angucyclines biosynthesis. Baikalomycins A–C demonstrated varying degrees of anticancer activity. Rabelomycin and 5-hydroxy-rabelomycin further demonstrated antiproliferative activities. The structure elucidation showed that baikalomycin A is a modified aquayamycin with β-D-amicetose and two additional hydroxyl groups at unusual positions (6a and 12a) of aglycone. Baikalomycins B and C have alternating second sugars attached, α-L-amicetose and α-L-aculose, respectively. The gene cluster for baikalomycins biosynthesis was identified by genome mining, cloned using a transformation-associated recombination technique and successfully expressed in *S. albus* J1074. It contains a typical set of genes responsible for an angucycline core assembly, all necessary genes for the deoxy sugars biosynthesis, and three genes coding for the glycosyltransferase enzymes. Heterologous expression and deletion experiments allowed to assign the function of glycosyltransferases involved in the decoration of baikalomycins aglycone.

Keywords: natural products; angucycline; aquayamycin; glycosyltransferase; Lake Baikal; *Streptomyces*

1. Introduction

Angucyclines are by far the largest group of aromatic polyketides solely produced by actinobacteria [1,2]. These natural products have diverse biological activities, including antibacterial, anticancer, antiviral, enzyme inhibitory, fungicidal, and others. Angucyclines are assembled by the repetitive condensation of malonyl-CoA to produce a common benz[a]anthracene intermediate [2]. This reaction is performed by type II polyketide synthase enzymes. Two pathways to form the benz[a]anthracene aglycone of angucyclines exist. In most cases, the initial decaketide chain undergoes a series of cyclizations and aromatizations dictated by cyclases/aromatases to produce an angular core structure [3]. This major route takes place in the biosynthesis of vineomycins [4], jadomycins [5], and landomycins [6], for example. However, at least in two cases, BE-7585A and PD116198, the polyketide chain is folded into an anthracyclinone intermediate that further undergoes oxidative A-ring opening and rearrangement into an angular benz[a]anthracene structure [7,8]. The first common benz[a]anthracene intermediate for both routes, UWM6, can be preserved or further modified by a series of oxidation and reduction events and decorated by extensive glycosylation or other modifications [2]. Glycosylation is a distinctive feature of many angucyclines sharing a common aglycone structure. The length and composition of oligosaccharide chains have a strong impact on biological activities [9].

Saquayamycins are a large group of angucyclines that, together with urdamycins, derive from the aquayamycin-type aglycone (Figure 1) [10]. More than ten derivatives of saquayamycins are known that mainly differ by the glycosylation pattern [4,11–15]. The largest representative of this group of natural products, saquayamycin Z, contains nine sugars [16]. Vineomycins and grincamycins as well as recently discovered Sch 47554–47555 and saprolmycins A–E also possess an aquayamycin-type aglycone [17,18]. Several other natural products have a modified aquayamycin-like aglycone structure, namely moromycins A and B and N05WA963 A–C, that lack the angular hydroxyl groups at positions 4a and 12b, resulting in a fully aromatic ring B [19,20]. N05WA963 A–C possesses an additional methoxy group at C-5. Lastly, several natural products with an opened ring A are known to derive directly from aquayamycin-type angucyclines. These include fridamycins A–E, himalomycins A and B, vineomycin B2, amicenomycins A, and the recently discovered vineomycin D [21–23]. It is still under discussion if these compounds are true naturally occurring products of biosynthetic pathways or are derived from the acid hydrolysis of respective angucyclines [21,24–27].

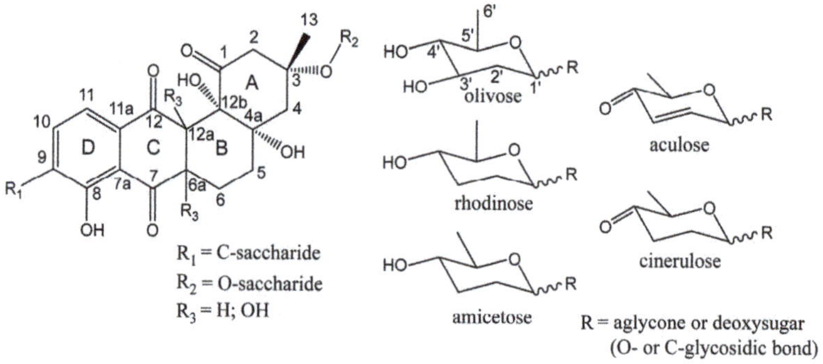

Figure 1. Structures of an aglycone and deoxy sugars typical for the aquayamycin-type angucyclines. Carbon atoms of the aglycone and sugars (with ′) are labeled according to IUPAC rules. The rings of the aglycone are indicated as A–D. The compound with R_1 as a D-olivose is historically considered as a common "aquayamycin-type aglycone".

The distinct feature of the aquayamycin-type angucyclines is the presence of oligosaccharide chains attached at positions C-9 and C-3 (Figure 1). Typically, the first sugar at the C-9 position is C-linked D-olivose as in the case of saquayamycins, vineomycins, moromycins, grincamycins and N05WA963 A–C. At the same time, saquayamycins, vineomycins, moromycins, and grincamycins at C-3 have *O*-glycosidically attached L-rhodinose. N05WA963 A–C do not have a saccharide chain at C-3 [20]. Differences in the length and composition of the attached oligosaccharide chains determine the variety of aquayamycin-type angucyclines. Saquayamycins (except saquayamycin Z) and moromycins have diverse disaccharides at C-3 and C-9 [11,15,19]. Grincamycins and vineomycins have a trisaccharide at C-9 consisting of D-olivose, L-rhodinose and the ketosugar L-aculose or L-cinerulose (aculose derivative) [14,21]. A similar glycosylation pattern was observed in saprolmycins A–E with either trisaccharide or disaccharide at C-9 and L-aculose or L-cinerulose as a first and the only sugar at the C-3 position [18]. Sch 47554 and Sch 47555 also contain L-aculose at C-3 but carry C-bound D-amicetose (stereoisomer of rhodinose) at C-9 extended with either L-amicetose or L-aculose [17]. Lastly, amicenomycins A and B have the trisaccharide L-amicetose-L-amicetose-L-rhodinose at C-3 and a single D-olivose at C-9 [22]. As can be seen, despite the common aglycone, aquayamycin-type angucyclines represent a diverse group of natural products due to differences in the glycosylation pattern.

The glycosylation events are well studied in the case of saquayamycin Z and saquayamycins G-K [16,28]. The saquayamycins G–K biosynthetic gene cluster (*sqn*) contains three genes *sqnGT1–3* encoding glycosyltransferases. However, only SqnGT2 was shown to be essential for decorating the aglycone, while the other two, SqnGT1 and SqnGT3, are proposed to act as chaperons, modulating the SqnGT2 activity [28]. Similarly, the Sch 47554 and Sch 47555 biosynthetic gene cluster also encode three glycosyltransferases (*schS7*, *schS9* and *schS10*) [29]. Genetic studies have shown that SchS7 attaches D-amicetose at C-9 and SchS9 further extends the saccharide chain when SchS10 attaches L-aculose at C-3 position [30]. Gene clusters for saprolmycins (*spr*) and grincamycins (*gcn*) biosynthesis have been recently cloned and were found to contain three genes encoding glycosyltransferases [27,31]. This implies that the different decoration pattern of these angucyclines results from the differences in functional properties of the glycosylation enzymes.

Here we report the characterization of the new aquayamycin-type angucycline antibiotics baikalomycins A–C produced by *Streptomyces* sp. IB201691-2A. The strain was isolated from the Lake Baikal endemic gastropod *Benedictia baicalensis*. Baikalomycins demonstrated moderate anticancer and antibacterial activities. The genome sequencing and mining led to identification of a gene cluster responsible for the biosynthesis of baikalomycins. Heterologous expression and gene deletion experiments supported this finding and provided hints on the glycosylation steps in baikalomycins biosynthesis.

2. Materials and Methods

2.1. Bacterial Strains, Culture Conditions and Routine Procedures

Streptomyces sp. IB201691-2A and *Rhodococcus* sp. IB201691-2A2 were isolated during this work. *Streptomyces albus* J1074 was used as a host for the heterologous expression of the baikalomycin biosynthetic gene cluster [32]. For the routine cloning, *Escherichia coli* XL1Blue (Agilent, Santa Clara, CA, USA) has been used and intergenic conjugation was carried out with *E. coli* ET12567 (pUB307) [33]. *S. cerevisiae* BY4742 was used for transformation-associated recombination cloning [34]. *E. coli* strains were grown in Luria–Bertani (LB) broth. Actinobacteria strains were cultured on soya flour mannitol agar (MS) medium and in liquid tryptic soy broth medium (TSB; Sigma-Aldrich, St. Louis, MO, USA). If necessary, the following antibiotics have been added: apramycin (50 $\mu g \cdot mL^{-1}$), spectinomycin (100 $\mu g \cdot mL^{-1}$), phosphomycin (100 $\mu g \cdot mL^{-1}$) and carbenicillin (100 $\mu g \cdot mL^{-1}$) (Sigma-Aldrich, St. Louis, MO, USA; Roth, Karlsruhe, Germany). The chromogenic substrate X-gluc with 100 $\mu g \cdot mL^{-1}$ concentration was used to detect the GUS (β-glucuronidase) activity.

Plasmid and total DNA isolation, *E. coli* transformation and *E. coli/Streptomyces* intergeneric conjugation were performed according to standard protocols [35,36]. *S. cerevisiae* BY4742 was transformed with the standard LiAc protocol [36]. Enzymes, including restriction endonucleases, ligase, Taq DNA polymerase, Klenow fragment of DNA polymerase I, were used according to manufacturer's recommendations (New England Biolabs, Ipswich, MA, USA; Thermo Fischer Scientific, Waltham, MA, USA; Agilent, Santa Clara, CA, USA).

2.2. Sampling and Actinobacteria Isolation

Endemic mollusks *Benedictia baicalensis* (Gerstfeldt, 1859) [37] were collected from Lake Baikal near Bolshiye Koty village (51°54′19″ N 105°4′31″ E, western shore of Lake Baikal) at depths of 50 and 100 m in February 2016 using deep-water traps. Each mollusks' sample included up to 5 specimens. Mollusks were surface-washed with sterile water, 70% ethanol, and again with sterile water to eliminate transient microorganisms. Afterwards, prepared samples were homogenized in 20% sterile glycerol and stored at −20 °C. Homogenates were thawed on ice and plated on MS plates supplemented with phosphomycin (50 μg/mL) and cycloheximide (100 μg/mL). Plates were incubated at 28 °C for 14 days. Colonies with typical for actinobacteria morphology were picked on a fresh MS plate and further characterized.

2.3. 16S rRNA Gene Sequencing and Phylogenetic Analysis

Strains were grown in 10 mL of TSB medium at 28 °C for 3 days and 180 rpm and total DNA was isolated using standard method [35]. The *16S* rRNA gene was amplified by PCR with the modified universal 8F and 1492R primers (Supplementary Table S1) [38]. PCR was carried out with initial denaturation at 95 °C for 3 min, followed by 30 cycles of 95 °C for 35 s, 51 °C for 40 s and 72 °C for 110 s, with an end extension at 72 °C for 7 min. The PCR products were purified using the Wizard SV Gel and PCR Clean-Up System (Promega, Madison, WI, USA) and sequenced using 8F and 1492R primers (Supplementary Table S1) [38]. The forward and reverse sequences were assembled with Bioedit software (version 7.2.5, Tom A. Hall, Department of Microbiology, North Carolina State University, North Carolina, USA, freeware). Evolutionary analyses were conducted in MEGA7 using *16S* rRNA gene sequences of related strains (Supplementary Table S2) [39]. The evolutionary history was inferred using the neighbor-joining method [40].

2.4. Screening the Culture Conditions for Biological Activity of Streptomyces sp. IB2016I91-2A

The following media were used for metabolites production by *Streptomyces* sp. IB2016I91-2A: SM1 (soy flour 10 g, glucose 18 g, Na_2SO_4 1 g, $CaCO_3$ 0.2 g, pH 7.0, 1 L tap water), SM17 (soy flour 5 g, glucose 2 g, glycerol 40 g, soluble starch 2 g, peptone 5 g, yeast extract 5 g, NaCl 5 g, $CaCO_3$ 2 g, pH 6.4, 1 L tap water), SM12 (soy flour 10 g, glucose 50 g, peptone 4 g, meat extract 4 g, yeast extract 1 g, NaCl 2.5 g, $CaCO_3$ 5 g, pH 7.6, 1 L tap water), SM20 (maltose 20 g, peptone 5 g, meat extract 5 g, yeast 3 g, $MgSO_4 \times 7\,H_2O$ 1 g, NaCl 3 g, pH 7.2, 1 L tap water), SM24 (yeast extract 9 g, peptone 1.8 g, glucose 20 g, KH_2PO_4 1 g, $MgSO_4 \times 7\,H_2O$ 0.5 g, pH 6.2, 1 L distilled water), SM25 (peptone 10 g, malt extract 21 g, glycerol 40 g, pH 6.5, 1 L distilled water), SM27Ac (soy flour 10 g, glucose 50 g, peptone 4 g, meat extract 4 g, yeast extract 1 g, NaCl 2.5 g, $CaCO_3$ 5 g, soluble starch 5 g, pH 4.5 with HCl, 1 L tap water), SM27N and SM27A1 (SM27Ac at pH 7.0 and pH 8.7 with NaOH respectively), R2 (malt extract 10 g, yeast extract 4 g, glucose 4 g, artificial sea water 0.5 L, pH 7.8, 0.5 L tap water), and Hopwood minimal medium [35]. The strain was inoculated into 50 mL of TSB medium in 500 mL flasks and grown for 3 days at 28 °C on a rotary shaker at 180 rpm. An inoculation was carried out of 5 mL of seed culture into 50 mL of production media and incubated for 8 days at 28 °C on a rotary shaker at 180 rpm. The metabolites from cultural broth were extracted with equal volume of ethyl acetate. Organic solvent was evaporated and the obtained extracts were dissolved in 500 μL of methanol.

2.5. LC-MS and LC-HRMS Analysis

LC-MS (Liquid chromatography–mass spectrometry) measurements were performed on a Dionex Ultimate 3000 RSLC (Thermo Fischer Scientific, Waltham, MA, USA) system using a BEH C18, 100 × 2.1 mm, 1.7 μm d_p column (Waters, Eschborn, Germany). Injection volume amounts to 1 μL and elution was achieved by a linear gradient (5–95% over 18 min) of solvent B (distilled acetonitrile with 0.1% of formic acid) against solvent A (bi-distilled water with 0.1% of formic acid). The column thermostat was set to 45 °C and a flow rate of 600 μL/min was used. UV spectra were recorded using DAD detector in the range of 200–600 nm and mass spectrometry data were collected on amazon SL speed (Bruker, Billerica, MA, USA) with an Apollo II ESI source in a range of 200–2000 *m*/*z*. High-resolution mass spectroscopic data (HRMS) with LC were acquired on a Dionex Ultimate 3000 RSLC system (Thermo Fischer Scientific, Waltham, MA, USA) using a BEH C18, 100 × 2.1 mm, 1.7 μm d_p column (Waters, Eschborn, Germany). A linear gradient from 5–95% solvent B (distilled acetonitrile + 0.1% formic acid) against solvent A (bi-distilled water + 0.1% formic acid) at a flow rate of 450 μL/min and 45 °C column temperature was used to separate 1 μL sample. UV spectroscopic data were collected by a DAD detector in the range of 200–600 nm. Mass spectroscopic data were acquired with an LTQ Orbitrap mass spectrometer (Thermo Fischer Scientific, Waltham, MA, USA). LC-MS data were collected, processed, and analyzed with Bruker Compass Data Analysis software, version 4.2 (Bruker, Billerica, MA, USA) and the Thermo Xcalibur software, version 3.0 (Thermo Fischer Scientific, Waltham, MA, USA). Dereplication was carried out by means of the Dictionary of Natural Products Database, version 10.0 (CRC Press, Baca Raton, FL, USA) with accurate mass, UV absorption maxima, and biological source as parameters [41].

2.6. Isolation and Purification of Compounds 1–5

For purification of the angucyclines, the strain *Streptomyces* sp. IB2016I91-2A was cultivated in 10 L of SM27N medium (pH 7.0), as described above. The metabolites were extracted with ethyl acetate from the cultural liquid and solvent was removed under reduced pressure. The obtained extract, 2.04 g, was dissolved in 13 mL of methanol and subjected to size-exclusion chromatography. The crude extract was loaded on a glass column (1 m) packed with Sephadex® LH 20 (total volume ~ 700 mL; Sigma-Aldrich, St. Louis, MO, USA). Methanol was used as eluent and fractions were collected every 15 min with a speed of 1–2 drops per second. Fractions were analyzed on LC-MS and targeted fractions further purified through preparative and semipreparative high performance liquid chromatography (HPLC) with the following equipment. A preparative HPLC system the Dionex Ultimate 3000 (Thermo Fischer Scientific, Waltham, MA, USA) equipped with a Nucleodur C18 HTEC column (150 × 21 mm, 5 μm) and linear gradient from 5–95% solvent B (acetonitrile + 0.1 formic acid) against solvent A (water + 0.1 % formic acid) over 28 min with a flow rate of 17 mL/min was used for initial purification. Obtained fractions containing the targeted compounds were further purified on a semipreparative HPLC system Agilent 1260 Series (Agilent Technologies, Santa Clara, CA, USA). Compounds 1, 2, and 5 were purified using Jupiter proteo C12 column (250 × 10 mm, 4 μm; Phenomenex, Madrid Ave, Torrance, CA, USA). Compounds 1 and 2 were obtained using a gradient starting from 30% of solvent B (acetonitrile + 0.1 formic acid, A: water + 0.1 % formic acid) and an increase of solvent B to 95% over 25 min. For compound 5, a multistep gradient from 5–75% B over 8 min and an increase to 85% B over 24 min was used. Compounds 3 and 4 were separated on the same system with Synergy Fusion RP column (250 × 10 mm, 4 μm; Phenomenex, Madrid Ave, Torrance, CA, USA) using a multistep gradient starting from 5% B (acetonitrile + 0.1% formic acid, A: water + 0.1% formic acid) to 45% over 17 min and a further increase to 95% B over 8 min. In a second step, compounds 3 and 4 were purified once more using the same column and a multistep gradient from 5–50% B over 10 min and an increase to 70% over 15 min. In case of compounds 1, 2, and 5, a flow rate of 5.0 mL/min and for compounds 3 and 4, a flow rate of 4.0 mL/min were used. Column thermostat was set to 45 °C and UV spectra were recorded in the range of 200–600 nm with DAD (Diode-Array Detector) detector.

NMR (Nuclear magnetic resonance) spectra were recorded in deuterated methanol (CD_3OD) and deuterated dimethyl sulfoxide (DMSO-d_6) at 298 K on a Bruker Avance III spectrometers (700 and 500 MHz; Bruker, MA, USA), both equipped with a 5 mm TXI cryoprobe. NMR data were analyzed using Topspin, version 3.5 pl7 (Bruker, Billerica, MA, USA).

2.7. Genome Sequencing and Bioinformatics

For isolation of total DNA, *Streptomyces* sp. IB2016I91-2A was grown in R5A medium [42] at 28 °C on a rotary shaker (180 rpm) for four days and a salting out procedure was used to obtain DNA [35]. For genome sequencing, an Illumina paired-end sequencing library (TruSeq sample preparation kit; Illumina, USA) was constructed as recommended by the manufacturer. The draft genome sequence was achieved on an Illumina MySeq system in rapid run mode (2 × 250 nt) with a pair distance of 500 bp. Subsequent to sequencing, the processed data were subjected to *de novo* assembly using SPAdes (version 3.8.1) [43] with default settings. Genome annotation was carried out using prokka v1.11 and GenDB 2.0 platform [44,45]. Secondary metabolism gene clusters were analyzed by the genome mining tool antiSMASH [46]. The genome sequence of *Streptomyces* sp. IB2016I91-2A was deposited under accession number SPQF00000000 in GenBank database.

2.8. Gene Disruption of the Glycosyltransferase Genes baiGT2 and baiGT3

For deletion of *baiGT2*, the fragment 3L (2.485 kb) and the fragment 4R (2.366 kb) were amplified by PCR with primer pairs 3L-FHindIII and 3L-REcoRV, 4R-FEcoRV and 4R-RXbaI, respectively (Supplementary Table S1). For deletion of *baiGT3*, the fragment 5L (2.366 kb) and 6R (2.574 kb) were amplified with primer pairs 5L-FHindIII and 5L-REcoRV, 6R-FEcoRV and 6R-RXbaI, respectively. Obtained PCR fragments were cloned into a pJET1.2/blunt cloning vector (Thermo Fischer Scientific, Waltham, MA, USA) resulting in the plasmids pJET3L and pJET4R, and pJET5L and pJET6R. The 4R fragment was retrieved with *Eco*RV and *Xho*I and ligated into pJET3L digested with the same enzymes, giving pJET34. The 5L fragment was retrieved with *Eco*RV and *Xba*I and ligated into *Eco*RV-*Xba*I digested pJET6R, resulting in pJET56. The spectinomycin resistance gene *aadA* was obtained from pHP45Ω as *Eco*RV fragment and cloned into *Eco*RV digested pJET34 to yield pJET34sp and pJET56 to yield pJET56sp. The resulting plasmids were digested with *Hind*III, fragments that corresponded to deletion constructs 34 sp and 56 sp were gel-purified and treated with Klenow fragment of *E. coli* DNA polymerase I (New England Biolabs, Ipswich, MA, USA) and sub-cloned into pKG1132 vector digested with *Eco*RV creating the final constructs pKG1132sp-34 and pKG1132sp-56. The plasmids were introduced into *Streptomyces* sp. IB201691-2A using intergeneric conjugation. Exconjugants were screened for white spectinomycin-resistant colonies on MS plates supplemented with X-gluc and spectinomycin. The deletion of *baiGT2* and *baiGT3* genes was confirmed by PCR using the CheckGT2F and CheckGT2R, and the CheckGT3F and CheckGT3R primer pairs (Supplementary Table S1)

2.9. Cloning of the bai Gene Cluster Using Transformation-Associated Recombination (TAR) Technique

The two 2.5 kb (TAR1) and 2.4 kb (TAR2) DNA fragments flanking the 44 kb *bai* biosynthetic gene cluster were amplified using primer pairs 91-2aTAR1-FNotI /91-2aTAR1-RNheI and 91-2aTAR2-FNheI/91-2aTAR2-RHindIII (Supplementary Table S1), respectively, and cloned into pJET1.2/blunt cloning vector (Thermo Fischer Scientific, USA). The TAR1 was retrieved with *Hind*III and *Nhe*I and ligated into *Nhe*I/*Hind*III digested pJetTAR2. The resulting construct was digested by *Hind*III/*Not*I and sub-cloned into a pCLY10 vector [47]. The final construct was linearized with *Nhe*I and co-transformed into *S. cerevisiae* BY4742 with *Streptomyces* sp. IB201691-2A chromosomal DNA in ratios 1:1, 1:2, 1:3, 1:5. Transformants were grown onto the selection medium YNB supplemented with yeast synthetic drop-out medium supplements without leucine containing 1% of glucose for 4 days (Sigma-Aldrich, St. Louis, MO, USA). Yeast colonies were screened by PCR using the primers 91-2aCheck-FNotI and 91-2aCheck-RHindIII and pCLY10-RNotV (Supplementary Table S1). Total DNA was isolated from the positive clone pCLY8.13-10bai and transformed into *E. coli* XL1Blue. The plasmid containing the cloned *bai* cluster was named p8-13bai and verified by digesting with restriction enzymes *Eco*RI, *Xho*I, *Not*I and *Kpn*I. p8-13bai was introduced into *S. albus* J1074 via intergeneric conjugation. The recombinant *S. albus* J1074/p8-13bai was cultivated in SM27N and extracts were analyzed for production of baikalomycins as described above.

2.10. Biological Activity Assays

The biological activity of the crude extracts from small-scale cultivation of *Streptomyces* sp. IB2016I91-2A was screened using a disk diffusion assay. The 6 mm paper disks were loaded with 40 μL of each extract and dried. *Bacillus subtilis* ATCC 6633, *Pseudomonas putida* KT 2440, *Escherichia coli* K 12 were grown in liquid LB medium and *Saccharomyces cerevisiae* BY4742 in YPD. Test cultures were spread on solid LB and YPD medium and dried paper discs were placed on top. The plates were incubated at 37 °C for 12 h and 30 °C for 2 days (in case of *Saccharomyces cerevisiae*) and zones of inhibition were measured manually.

The minimal inhibitory concentration (MIC) was determined against the Gram-positive bacteria *Staphylococcus carnosus* DSM 20501 and *Mycobacterium smegmatis* DSM 43286, against the Gram-negative bacteria *Erwinia persicina* DSM 19328 and *Pseudomonas putida* KT2440, and against the yeast *Candida glabrata* DSM 11226. The minimal inhibitory concentrations were estimated by a standard serial dilutions protocol in 200 μL in 96-well plates using DMSO as solvent. Kanamycin was used as a positive control, and DMSO was used as negative control. 190 μL of bacterial test cultures in appropriate media (1:500 dilution of overnight culture) were added to each well containing 10 μL of compound solution. Plates were shaken at 30 °C for 16–20 h. To each well, 5 μL of thiazolyl blue tetrazolium bromide (10 mg/mL; Sigma-Aldrich, St. Louis, MO, USA) solution was added, and the plates were incubated at 30 °C for an additional 10–30 min. MICs were determined as the concentration of antibiotic in the well where the color of thiazolyl blue tetrazolium bromide was not changed from yellow to dark blue.

2.11. Anticancer Activities of Isolated Compounds

Assays were performed using the human tumor cell lines A549 (lung carcinoma), Huh7.5 (hepatocellular carcinoma), MCF7 (breast adenocarcinoma), and SW620 (colorectal adenocarcinoma). All cell lines were cultured in RPMI-1640 (A549, HuH7.5) or DMEM (MCF-7, SW620) supplemented with 10% fetal bovine serum, 100 U/mL penicillin, 100 mg/mL streptomycin, and 2 mM glutamine. The cells were maintained at 37 °C in a humidified atmosphere of 5% CO_2.

A cell viability assay (MTT assay) was performed as described below. Cells were seeded in appropriate numbers to reach confluency the next day and were then treated with the respective compounds in different concentrations for 48 h. Stock solutions of the compounds were prepared in DMSO, and solvent controls were tested concurrently. The viability of adherent cells was determined by replacing the supernatants with 0.5 mg/mL MTT (3-(4,5-dimethylthiazole-2-yl)-2,5 diphenyltetrazolium bromide; Sigma-Aldrich, St. Louis, MO, USA) solution in respective culture media. After a 30 min incubation, the formazan crystals were dissolved in DMSO, and the absorbance was measured at 560 nm in a microplate reader (GloMax Discover, Promega, Madison, WI, USA). IC_{50} values were calculated by non-linear regression using OriginPro (OriginLab Corporation, Northampton, MA, USA).

A proliferation assay (ECIS) was performed as follow. Cell proliferation was measured using the electric cell-substrate impedance sensing (ECIS®) system (Applied BioPhysics, Road in Troy, NY, USA). On the day before cell seeding, the arrays were pre-incubated with full cell culture medium at 37 °C. A549 cells were grown on 96-well ECIS arrays (96W10E+, with 10 electrodes per well), and the impedance measurement (every 15 min for 100 h, 16,000 Hz) was started directly after cell seeding (8000 cells per well). The cells were left to attach for 5 h before the compounds were added to the cells at the indicated concentrations. Control cells were treated with the diluted solvent DMSO. Impedance was normalized to the value at 7 h after inoculation.

3. Results and Discussion

3.1. Isolation and Characterization of *Streptomyces* sp. IB201691-2A

The Lake Baikal endemic mollusk *Benedictia baicalensis* [37] was sampled (12× samples with five specimens, each) at a depth of 50 and 100 m in February 2016 at Lake Baikal in Bolshiye Koty village (51°54′19″ N 105°4′31″ E, western shore of Lake Baikal) and actinobacteria were isolated as described in the materials and methods section. Several actinobacteria-like colonies with similar morphology were found in three samples from 50 m and two samples from 100 m depth. After re-plating on fresh MS medium, all five isolates were noticed to contain two different species (Figure 2A). One of them demonstrated surface growth with pale orange colonies and did not form aerial mycelium and spores (Figure 2B). Based on *16S* rRNA gene sequence analysis, this bacterium was identified as a *Rhodococcus* sp. designated as IB201691-2A2 (Supplementary Figure S1).

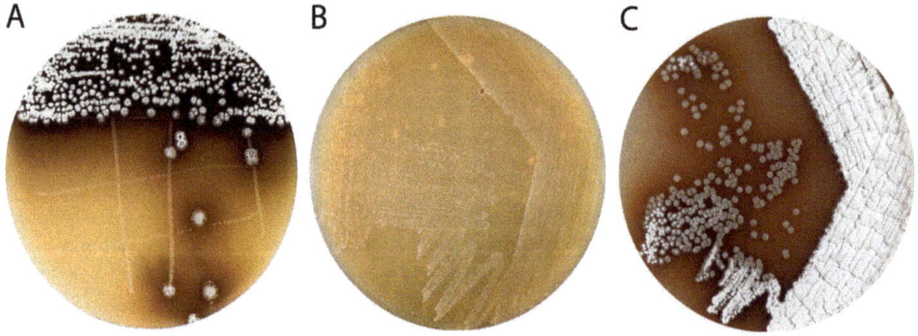

Figure 2. Actinobacteria species co-isolated from *Benedictia baicalensis* grown on soya flour mannitol (MS) agar. (**A**) Original mixture of *Streptomyces* sp. IB201691-2A and *Rhodococcus* sp. IB201691-2A2 obtained from a single colony grown from a plated homogenate of *B. baicalensis*; (**B**) Pure culture of *Rhodococcus* sp. IB201691-2A2; (**C**) Pure culture of *Streptomyces* sp. IB201691-2A.

The second bacterium has substrate growth typical for streptomycetes, forms white spores, and produces dark-brown pigment (Figure 2C). The 16S rRNA gene phylogeny analysis placed the strain into the genus *Streptomyces* and it was named *Streptomyces* sp. IB201691-2A (Supplementary Figure S2). Gene sequences of *16S* rRNA of *Rhodococcus* sp. IB201691-2A2 and *Streptomyces* sp. IB201691-2A specimens isolated from different samples were identical (data not shown). *Streptomyces* sp. IB201691-2A is closely related to *S. ederensis* NBRC 15410, producing moenomycins and *S. umbrinus* NBRC 13091 producing phaeochromycin and diumycins (moenomycin derivatives). Both strains were isolated from soil and are heterotypic synonyms of phaeochromycin-producing *S. phaeochromogenes* (Supplementary Figure S2) [48,49]. The *S. phaeochromogenes* strain was originally discovered as a producer of angucycline PD116198 [7]. In this study, we focused on *Streptomyces* sp. IB201691-2A strain.

3.2. Production and Isolation of Baikalomycins

For biological activity screening *Streptomyces* sp. IB201691-2A was cultured in nine different liquid media at three temperatures (13, 28, and 37 °C) and extracts were tested against four test-cultures, including Gram-positive and Gram-negative bacteria and yeast (Supplementary Table S3). The strain was found to produce compounds active only against *B. subtilis*. SM27N medium and 28 °C conditions were preferable for accumulation of these bioactive metabolites, as seen from the largest inhibition zone (Supplementary Table S3).

The crude extract of *Streptomyces* sp. IB201691-2A grown in SM27N was analyzed by high-resolution LC-MS and dereplicated using the Dictionary of Natural Products database. Two known compounds rabelomycin (**1**) and 5-hydroxy-rabelomycin (**2**) were identified, with the former being the major product of the strain (Figure 3A,B; Supplementary Figures S3 and S4). Rabelomycin, a well-known shunt product in the biosynthesis of angucyclines, was also co-isolated together with vineomycins from extracts of *S. matensis* subsp. *vineus* and himalomycins A and B from *Streptomyces* sp. B6921 [4,25,50]. Besides these two compounds, several peaks with characteristics for angucyclines absorption spectra and *m/z* ranging from 330 to 620 (in negative mode) are present in the extract of *Streptomyces* sp. IB201691-2A. However, we were not able to identify these compounds within the Dictionary of Natural Products.

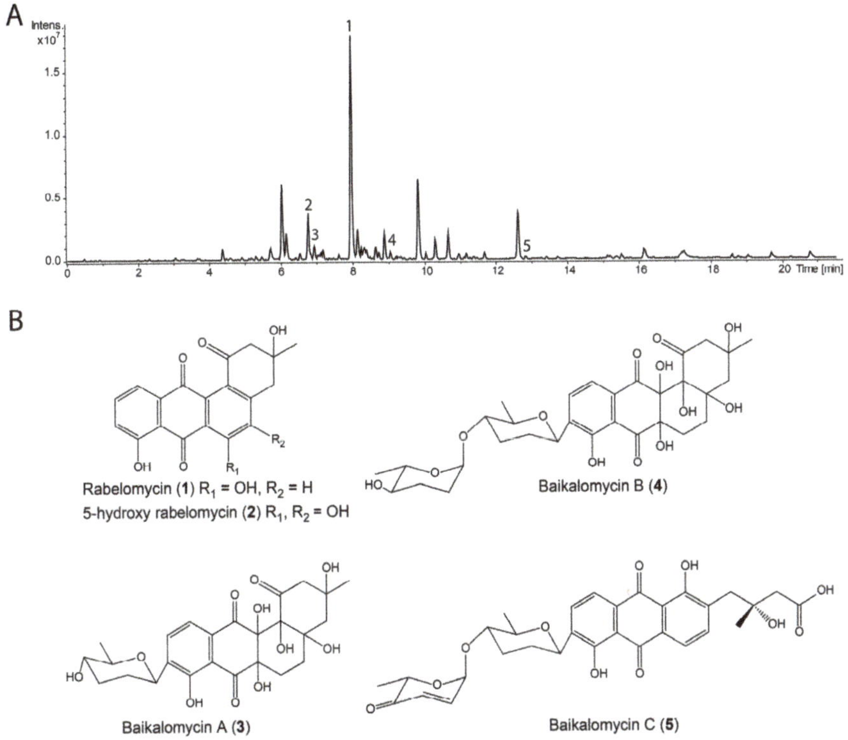

Figure 3. (**A**) LC chromatogram of crude extract of *Streptomyces* sp. IB201691-2A; (**B**) structure of compounds isolated from the extract of *Streptomyces* sp. IB201691-2A: rabelomycin (**1**) and 5-hydroxy-rabelomycin (**2**) and baikalomycins A-C (**3–5**).

Streptomyces sp. IB201691-2A was cultivated in 10 L of the SM27N medium and metabolites were extracted, giving 2.04 g of a crude extract. Metabolites were fractionated by size exclusion chromatography and the targeted compounds were purified by the preparative HPLC. Five pure compounds were obtained (with 80–85% purity): Rabelomycin (**1**) (6.5 mg), 5-Hydroxy-rabelomycin (**2**) (1.8 mg), baikalomycin A (**3**) (RT 6.9 min, 0.9 mg), baikalomycin B (**4**) (RT 9.1 min, 1.1 mg) and baikalomycin C (**5**) (RT 12.8 min, 0.7 mg) (Figure 3A,B; Supplementary Figures S3–S5).

3.3. Structure Elucidation of Baikalomycins

The structures of rabelomycin (**1**) and 5-hydroxy-rabelomycin (**2**) were confirmed by the comparison of the ^{1}H and ^{13}C NMR data to the previously reported data (Figure 3B; Supplementary Table S4) [51,52].

Baikalomycin A (**3**) was obtained as a pale-yellow solid (purity > 80 mol%, according to ^{1}H NMR, Supplementary Figure S6). The molecular formula of $C_{25}H_{30}O_{11}$ and *m/z* 487.1634 [*M*-H_2O-H]$^-$ (calc. *m/z* 487.1604 [*M*-H_2O-H]$^-$) was determined on the basis of HRESIMS (High-resolution electrospray ionisation mass spectrometry) data, indicating 11 degrees of unsaturation (Figure 3B; Supplementary Figure S5). The compound exhibited UV absorption at 240, 285, and 355 nm. The analysis of ^{1}H and 2D HSQC (Heteronuclear single quantum coherence spectroscopy) and HMBC (Heteronuclear Multiple Bond Correlation) spectra revealed the presence of 25 carbon atoms, 12 quaternary carbons, five methine, six methylene, two methyl carbons and two *ortho*-coupled aromatic protons at δ_H 7.84 (dd, 8 Hz, 0.6 Hz, H-10) and δ_H 7.58 (d, 8 Hz, H-11) which indicate a tetrasubstituted aromatic ring. Key correlations in the HMBC spectrum from H-11 to C-7, C-7a, C-8, C-9, C-12, from H-6 to C-6a, C-12a, C-7, C-12, from H-5 to C-4a, C-12b, C-6a and from H-4 to C-4a, C-12b revealed the presence of the well-known aquayamycin-type core structure (ring A–D) substituted at position C-9 (Figure 4A; Supplementary Table S5; Supplementary Figures S6–S10). However, ring B is found to be fully saturated, which is only known for a few cases, e.g., moromycins A and B, grecocyclines, and N05WA963 A–C [19,20,53]. The ^{13}C values, ranging from 70–80 ppm for C-4a, C-6a, C-12a, and C-12b indicated the presence of hydroxy groups whereas C-5 (δ_C 30.0, δ_H 1.65 and 2.21) and C-6 (δ_C 26.0, δ_H 2.18 and 2.47) were identified as CH_2 groups. As known for the aquayamycin compounds, ring A bears a hydroxy and methyl group at C-3. The absolute configuration at this position has been determined as *R* in several previously isolated angucyclines with a similar biosynthetic origin, e.g., in urdamycin or saquayamycin [11,54]. Therefore, the configuration in baikalomycins A as well as in B is likely to be the same. In addition, one anomeric proton at δ_H 4.77 (δ_C 74.0), one methine proton at δ_H 3.22 (δ_C 72.5), two methylene signals at δ_H 1.43/2.20 (δ_C 33.0) and δ_H 1.63/2.11 (δ_C 33.5), and one methyl signal at δ_H 1.32 (δ_C 18.5) confirm the presence of a 2,3,6-tridesoxy-hexose unit. As known for other aquayamycin natural products, the sugar was attached to the aglycone at C-9, which was supported by HMBC correlations from H-1′ to C-8, C-9, C-10, as well as from H-2′ to C-9 (Figure 4A). Due to the large coupling constants of H-1′ (11 Hz), H-4′ (11, 9 Hz), and H-5′ (9 Hz), the respective protons must be in axial positions and therefore the sugar should be β-amicetose. The respective ROESY (Rotating frame Overhause Effect Spectroscopy) crosspeaks provided further proof for the given structure (Figure 4B). We assume that the sugar is *D* configured, as only the β-D-form of amicetose has been found among the C-glycosides in aquayamycin-like compounds so far [9]. Due to the fact that all substances could only be isolated in very small quantities and should still be subjected to biologic testing, a hydrolysis of the glycosides to determine the absolute configuration of the sugar components by means of optical rotation was not possible.

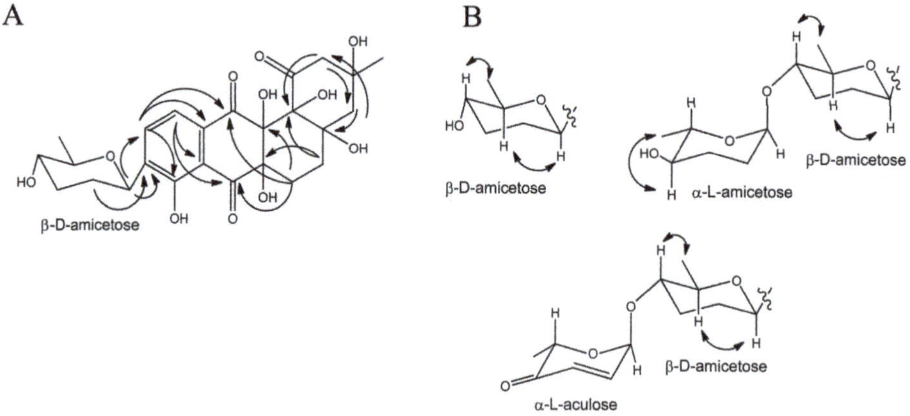

Figure 4. (**A**) Selected HMBC correlations within baikalomycin A; (**B**) Selected ROESY correlations which support the designated configuration of attached sugars.

Baikalomycin B (**4**) was isolated as a pale-yellow solid (purity > 85 mol% according to ^{1}H NMR, Supplementary Figure S11) with the molecular formula $C_{31}H_{40}O_{13}$ and *m/z* 601.2332 [*M*-H_2O-H]$^-$ (calc. *m/z* 601.2285 [*M*-H_2O-H]$^-$), suggesting 12 degrees of unsaturation (Figure 3B; Supplementary Figure S5). Similar to (**3**), this compound showed absorption at 240, 285, and 355 nm. NMR data of ^{1}H and ^{13}C largely resembles the data of (**3**), however, with one major difference. The presence of an additional anomeric proton at δ_H 4.80 (δ_C 99.5), together with one methine proton at δ_H 3.15, two methylene groups at δ_H 1.76/1.84 and δ_H 1.74/1.78, and one methyl group at δ_H 1.18 indicate the presence of a second 2,3,6-tridesoxy-hexose moiety, which is *O*-glycosidically linked to the first sugar. HMBC correlations from H-1′′ to C-4′ confirm the connectivity of both sugar units (Figure 4B; Supplementary Table S5; Supplementary Figures S11–S15). The second sugar moiety was as well identified as amicetose due to the large diaxial coupling $J_{H4''H5''}$ 9 Hz. Although the small coupling constant of H-1′′ (J_{HH} = 2.5 Hz) indicates the α-anomer in this case. Most probably, the α-amicetose is *L* configured as it was found for most of the terminal, O-glycosidically bound sugars in aquayamycin-like natural products [9].

Baikalomycin C (**5**) was obtained as a yellow solid (purity > 80 mol% according to ^{1}H NMR, Supplementary Figure S16) and showed *m/z* 579.1887 [*M*-H]$^-$ (calc. *m/z* 579.187184 [*M*-H]$^-$) that corresponds to molecular formula $C_{31}H_{32}O_{11}$ indicating 16 degrees of unsaturation (Figure 3B; Supplementary Figure S5). The UV absorption at 228, 258, 294, and 441 nm suggests an increased conjugated system compared to (**3**) and (**4**). Analysis of ^{1}H NMR, HSQC, and HMBC spectra revealed an anthraquinone core formed by ring B, C, and D (Figure 4B; Supplementary Table S5; Supplementary Figures S16–S21). This finding was concluded from the *ortho*-coupled aromatic proton signals at δ_H 7.91 (d, 8 Hz, H-10) and δ_H 7.86 (d, 8 Hz, H-11) and HMBC correlations from H-10 to C-8, C-11a and from H-11 to C-7a, C-9, and C-12. A second *ortho*-coupled pair of protons at δ_H 7.76 (d, 8 Hz, H-5) and δ_H 7.80 (d, 8 Hz, H-6) confirm the presence of another aromatic ring. HMBC correlations from H-6 to C-12a and C-7 and from H-5 to C6a establish the connection of this ring to the quinone ring C. In contrast to (**3**) and (**4**), the former ring A was opened between C-1 and C-12b leading to a phenolic hydroxyl group at C-12b (δ_C 162.56) and a 3-hydroxy-3-methyl butanoic acid side chain attached to C-4a (Figure 3B). Similar to baikalomycin A and B, the C-glycosidic sugar attached to the aglycone at C-9 was identified as β-D-amicetose due to comparable chemical shift values and coupling constants for H-1′, H-4′ and H-5′ (Supplementary Table S5). An additional anomeric signal at δ_H 5.40 and δ_C 96.21 indicates a second O-glycosidic bonded sugar which is supported by corresponding HMBC correlations from H-1′′ to C-4′ (Supplementary Table S5). This sugar consists of two olefinic protons δ_H 6.98 (H-2′′), 6.05 (H-3′′), one methine proton δ_H 4.61, a methyl group at δ_H 1.32 and a carbonyl signal at δ_C 198.82 (C-4′′) and was thus identified as aculose. The small coupling constant of H-1′′ (3.5 Hz) is consistent with the α-anomer (Supplementary Table S5). Until today, only α-L-aculose was found in aquayamycin-type natural products, which let us assume that we have the sugar moiety [9]. The aglycone of baikalomycin C is the same as in amicenomycin B and himalomycin A [22,25]. Thus, the absolute configuration at C-3 can be assigned as *R*.

Despite the similarity to aquayamycin-type compounds, baikalomycins A and B possess an uncommon aglycone (hydroxy groups at C-6a and C-12a) that has been found in only one other angucycline, namely a derivative of PD 116198 isolated from *S. phaeochromogenes* WP 3688 [7]. As opposed to this, several angucyclinones and glycosylated angucyclines with epoxide function at these positions are known, including simocyclinones and grecocycline A [53,55]. Compounds with one hydroxyl group at C-6a or C-12a, like panglimycins A-B, saccharothrixmicine B, and kiamycin are also described [56–58]. The standalone case, however, is grecocycline B with a thiol group at C-6a and hydroxyl group at C-12a. Importantly, we were not able to find tri- and tetra-saccharides of baikalomycins in the extracts of *Streptomyces* sp. IB201691-2A.

The stereochemistry of the chiral centers in ring B of baikalomycins A and B remains uncertain. The spatial positions of the hydroxy groups at C-4a and C-12b were concluded to be *cis* as it has been described for all comparable angucyclines beforehand. However, the chiral centers C-6a and C-12a of the new aglycone could not be elucidated due to the lack of reliable signals for the hydroxyl protons even in DMSO-d6. Therefore, respective correlations involving the hydroxyl protons could not be found in the ROESY spectra.

3.4. Biological Activities of Baikalomycins

Rabelomycins and baikalomycins were tested for antibacterial and anticancer activities. The compounds **1**, **2**, and **5** showed moderate and weak activity against *Staphylococcus carnosus* DSMZ 20501 and *Mycobacterium smegmatis* DSMZ 43286 (Table 1). Also, **1**, **2**, and **4** were moderately active against *Erwinia persicina*.

Table 1. Activity tests of baikalomycins A–C (**3–5**), rabelomycin (**1**), and 5-hydroxy-rabelomycin (**2**).

Test Strain	MIC, μM				
	3	**4**	**5**	**1**	**2**
Erwinia persicina DSMZ 19328	>500	250	n.t.	31	125
Pseudomonas putida KT2440	>500	>500	>500	>500	>500
Candida glabrata DSMZ 11226	>500	>500	>500	>500	>500
Staphylococcus carnosus DSMZ 20501	>500	>500	62	62	125
Mycobacteriaum smegmatis DSMZ 43286	>500	>500	250	31	125

Anticancer activities were determined against the human cancer cell lines A549 (lung carcinoma), Huh7.5 (hepatocellular carcinoma), MCF7 (breast adenocarcinoma), and SW620 (colorectal adenocarcinoma) by MTT assay [59]. Baikalomycins A and B showed moderate to weak effects on A549 and MCF7 cell viability (Table 2). The compounds **1**, **2**, and **5** exerted a more potent activity with IC_{50} values in the low micromolar range on all four cell lines (Table 2, Supplementary Figure S22A).

Table 2. IC_{50} values [μM] of baikalomycins A–C (**3–5**), rabelomycin (**1**), and 5-hydroxy-rabelomycin (**2**) ± SEM against human tumor cell lines, treated for 48 h.

Compound	A549	Huh7.5	MCF7	SW620
3	58.51 ± 5.15	inactive	53.19 ± 3.36	inactive
4	46.26 ± 0.52	inactive	inactive	inactive
5	42.43 ± 3.71	7.62 ± 0.47	13.35 ± 1.33	3.87 ± 0.69
1	9.78 ± 0.49	7.21 ± 0.70	21.94 ± 1.59	7.82 ± 0.40
2	9.11 ± 0.59	11.91 ± 2.94	27.39 ± 2.17	13.43 ± 0.72

A549 cells were chosen to further analyze the potential antiproliferative actions in an impedance-based assay. Antiproliferative effects correlated with activities observed in A549 cells in the MTT assay: while **3**, **4**, and **5** showed no antiproliferative effects (data not shown), **1** and, to a greater extent, **2** reduced cell proliferation in concentrations showing no effect in the MTT assays (cell viability > 80% after 48 h treatment) (Supplementary Figure S22B).

3.5. Streptomyces sp. IB201691-2A Genome Sequencing and Analysis

The genome of *Streptomyces* sp. IB201691-2A has been sequenced and assembled into 109 contigs with an overall size of 11,410,308 bp (including 61,648 undefined nucleotides). The G+C content was found to be 70% that is typical for streptomycetes. The largest contig is 1155 kbp. The genome of *Streptomyces* sp. IB201691-2A consists of a single chromosome, based on the sequence coverage, and contains 10,023 predicted CDSs, 5 rRNA gene clusters, 86 tRNA, and one transfer-messenger RNA genes. AntiSMASH analysis revealed the presence of 38 gene clusters potentially involved in secondary metabolites biosynthesis (Supplementary Table S6). The genome is enriched with the genes for siderophores production, including those typical for actinobacteria desferrioxamine and two aerobactin-like siderophores biosynthetic gene clusters, as well as a more unusual scabichelin biosynthetic gene cluster, originally found in *S. scabies* [60]. The enrichment of *Streptomyces* sp. IB201691-2A with siderophores encoding gene clusters is not unusual, since the iron content of the Lake Baikal water is relatively low [61]. Biosynthetic products could also be clearly predicted for three terpene gene clusters, such as hopene (cluster 7), albaflavenone (cluster 13), and earthy-musty odor-causing 2-methylisoborneol (cluster 22) [62–64].

3.6. Identification of Baikalomycins Biosynthetic Gene Cluster

The genome of *Streptomyces* sp. IB201691-2A contains only one gene cluster encoding type II PKS and oligosaccharides biosynthesis enzymes that could be putatively involved in biosynthesis of baikalomycins (Figure 5A; Supplementary Table S6). This cluster, further designated as the *bai* gene cluster, is predicted to be 34 kbp in length and is located at the 3′ edge of scaffold00031. It encodes 27 complete and 1 incomplete CDSs. The BLAST analysis of the incomplete open reading frame (ORF) revealed that it is coding for putative acyl-CoA carboxylase with the closest homologue being SchP1 from Sch47554/47555 biosynthesis [29].

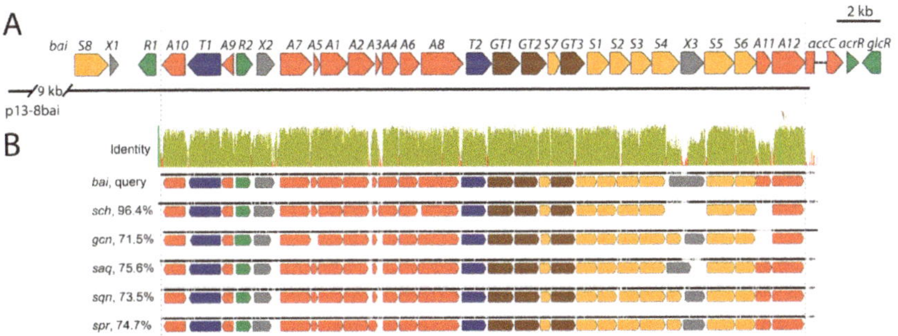

Figure 5. (**A**) Genetic organization of baikalomycin biosynthetic gene cluster (*bai*) from *Streptomyces* sp. IB 201691-2A. Region cloned in p13-8bai is shown below; (**B**) alignment of gene clusters responsible for biosynthesis of saquayamycin-type angucyclines from different actinobacteria. Mean pairwise identity over all pairs in the column: green—100% identity; green-brown—at least 30% and under 100% identity; red—below 30% identity. *bai*—baikalomycin gene cluster from *Streptomyces* sp. IB201691-2A; *sch*—Sch47554/47555 gene cluster from *Streptomyces* sp. SCC-2136; *gcn*—grincamycin gene cluster from *S. lusitanus* SCSIO LR32; *saq*—saquayamycin gene cluster from *S. nodosus* ATCC4899; *sqn*—saquayamycin gene cluster from *Streptomyces* sp. KY40-1; *spr*—saprolomycin gene cluster from *Streptomyces* sp. TK08046.

The search within the genome of *Streptomyces* sp. IB201691-2A identified a missing part of the acyl-CoA carboxylase gene at the 5′ edge of scaffold00038 (Supplementary Figure S23). We performed multiple sequence alignment of *bai*, *sch* (Sch47554/47555 from *Streptomyces* sp. SCC-2136), *saq* (saquayamycin from *S. nodosus* ATCC4899), *sqn* (saquayamycin from *Streptomyces* sp. KY40-1), *spr* (saprolomycin from *Streptomyces* sp. TK08046) and *gcn* (grincamycin from *S. lusitanus* SCSIO LR32) gene clusters (Figure 5B). This allowed determining the core of the *bai* cluster defined by genes *baiA10* encoding putative NADP-oxidase and *baiA12* methylmalonyl-CoA carboxylase. The genes within this region are present in all six closely related type II PKS-encoding clusters. The core of the *bai* gene cluster shares a 96.4% nucleotide sequence identity with the *sch* cluster in pairwise alignment (Figure 5B; Supplementary Figure S23). At the same time, this number is lower for the other four related clusters. The high degree of similarity shows a close evolutional relationship between the mentioned biosynthetic genes that is reflected in the high similarity of produced compounds (Supplementary Figure S24).

3.6.1. Genes Putatively Involved in Biosynthesis of Aglycone Core

We attempted to predict the function of Bai enzymes in the assembly of baikalomycins. Genes *baiA1*, *baiA2*, and *baiA3* show high sequence similarity to *saqA*, *B*, *C*, *sqn H*, *I*, *J* and *SchP6*, 7, 8, encoding components of the "minimal PKS" complex (Supplementary Table S7): BaiA1 ketoacyl synthase α, BaiA2 ketoacyl synthase β (chain length factor CLF), and BaiA3 acyl carrier protein (ACP). These enzymes catalyze the synthesis of the nascent decaketide chain by repetitive condensation of one acetyl-CoA and nine malonyl-CoA units. BaiA4, showed high similarity to SchP5 (98%) and UrdD (86%, urdamycin ketoreductase from *S. fradiae* Tü2717) ketoreductases, responsible for the first reduction of the nascent polyketide chain at the C-9 position [29,65]. Two genes *baiA5/A6* code for putative cyclases/aromatases performing cyclization of the polyketide chain into the benz[a]anthracene structure. BaiA5/A6 resembles SchP4/P9 (99%/99%) and UrdF/L (73%/87%). *baiA7* encodes a close homologue of SrpB (77%) and UrdE (78%) proteins. UrdE is an oxygenase catalyzing hydroxylation of an angucycline aglycone at the C-12 position [66]. Lastly, *bai*A8 is encodes a second oxygenase enzyme with high identity to UrdM (85%) and SprI (88%) and seems to be involved in the hydroxylation in C-4a and C-12b [67]. Additionally, baikalomycins A and B bear hydroxy groups at positions C-6a and C-12a and lack C-5/C-6 double bond. A complete non-aromatic ring B has been rarely found in angucycline structures and thus there is no experimental data on how it is formed. It can be only assumed that hydroxylation of C-6a/C-12a and reduction of the double bond is catalyzed by oxygenase-reductase BaiA7, as it was hypothesized for kiamycin biosynthesis [68]. However, another scenario cannot be excluded. Two genes on the left edge of the *bai* cluster, *baiA9* and *baiA10*, code for putative flavin-oxidoreductase and NADPH-dependent reductase, respectively. The latter enzyme might provide the activity needed to reduce the double bond at C-5/C-6, when the former can be involved in processing of keto groups at C-6a/C-12a. Lastly, *baiA11* codes for putative 4′-phosphopantetheinyl transferase, possibly involved in activation of ACP, and *baiA12* encodes putative methylmalonyl-CoA carboxyltransferase, that might participate in precursors supply.

3.6.2. Genes Putatively Involved in Deoxysugars Biosynthesis and Attachment

Two deoxysugars are present in the structure of baikalomycins: amicetose, in D- and L-configuration, and L-aculose. The *bai* gene cluster contains ten genes predicted to be involved in the biosynthesis of deoxysugars. Four enzymes, required for biosynthesis of the common precursor NDP-4-keto-2,6-dideoxy-D-glucose, are encoded by *baiS1* (glucose-1-phosphate thymidylyltransferase), *baiS2* (dTDP-glucose 4,6-dehydratase), *baiS5* (NDP-hexose 2,3-dehydratase), and *baiS6* (glucose-fructose oxidoreductase) (Figure 5A; Supplementary Table S7). Subsequently, the generation of D-amicetose is accomplished through the action of BaiS4 (dTDP-4-amino-4,6-dideoxy-D-glucose transaminase) and BaiS3 (dTDP-6-deoxy-L-talose-4-dehydrogenase). Lastly, epimerization at C-5′ would lead to formation of L-amicetose. BaiS7 has a high homology to UrdZ1 (71%) dTDP-4- dehydrorhamnose-3-epimerase.

The biosynthetic steps in the L-aculose formation were previously described for aclacinomycins and grincamycins [27,69]. In both cases L-aculose derives from rhodinose by the action of flavin-dependent oxidoreductases AknOx and GcnQ. Catalysis is performed in two steps: first, rhodinose is converted to cinerulose by oxidation at the C-4″ position, followed by dehydrogenation to form double bond between C2″ and C3″ [69]. Unlike in the AknOx performed reaction, in the case of GcnQ no cinerulose intermediate has been observed [27]. Based on similarity with GcnQ and AknOx, it can be assumed that the flavin-dependent oxidoreductases SprY and SqnQ convert rhodinose to acculose in the case of saprolmycins and saquayamycins biosynthesis [28,31]. Corresponding genes are located at the 3′ edge of respective gene clusters downstream of the putative methylmalonyl-CoA carboxylase gene. In both, the *sch* and *bai* clusters, a gene encoding flavin-dependent oxidoreductase is missing in this region (Figure 5A). A BLAST search against *Streptomyces* sp. IB201691-2A genome using AknOx and GcnQ sequences as query resulted in one positive hit with high amino acid sequence identity (57.3% and 67.9%, respectively). The corresponding gene, 91_A2_44060 (*baiS8*), is located just upstream of initially defined borders of the *bai* cluster (Figure 5A; Supplementary Table S7). The *baiS8* orthologue in *sch* cluster *schA26*, also encodes flavin-dependent oxidoreductase. This makes BaiS8 and SchA26 the best candidates for enzymes catalyzing the conversion of amicetose to aculose.

Three genes, *baiGT1*, *baiGT2*, and *baiGT3*, are predicted to be involved in glycosylation steps in the assembly of baikalomycins (Figure 5A; Supplementary Table S7). BaiGT3 shows high homology to the C-glycosyltransferases SchS7 (98%), SprGT3 (79%), and UrdGT2 (73%), which catalyze a transfer of the first sugar to the C-9 position of the aglycone. Two other enzymes, BaiGT1 and BaiGT2, are predicted to be *O*-glycosyltransferases based on their homology to SchS10 (97%) and SchS9 (98%), respectively [30].

3.6.3. Genes Involved in Regulation, Resistance, and with Unknown Functions

Two genes within the *bai* cluster might be involved in the regulation of baikalomycins production and control of its transport (Figure 5A; Supplementary Table S7). The *baiR1* gene product shows similarity to the LndI (66% identity) transcriptional regulator controlling landomycin E biosynthesis [70]. *baiR2* codes for TetR family transcriptional regulators that are widely distributed in bacteria, including *Streptomyces* genera, and typically control the expression of antibiotics transporter genes [71]. Two genes, bai*T1* and *baiT2*, which encode proteins putatively participating in the transport of baikalomycins and strain self-resistance, are found within the *bai* gene cluster. The *baiT1* gene product showed similarity to a major facilitator superfamily of the DHA2 group, typically implicated in multidrug resistance [72]. *baiT2* codes for a protein with 65% identity to PgaJ, a putative transmembrane efflux protein from the gaudimycin biosynthesis pathway [73]. Three genes *baiX1*, *X2*, and *X3* code for conserved hypothetical proteins with orthologues in many actinobacteria genomes but without functions assigned.

3.7. Inactivation of Genes Encoding Glycosyltransferases in bai Cluster

In order to prove that the identified gene cluster is responsible for the production of baikalomycins and to investigate the specificity of glycosylation steps during biosynthesis we aimed to delete the genes *baiGT1*, *baiGT2*, and *baiGT3* within the chromosome of *Streptomyces* sp. IB201691-2A. For this, the suicide vector-based strategy was employed. However, the strain was found to be poorly genetically tractable. It took more than 20 attempts with each construct to obtain a few transconjugants. Unfortunately, we failed to introduce the *baiGT1* deletion construct into *Streptomyces* sp. IB201691-2A. In the case of two other genes, *baiGT2* and *baiGT3*, mutants were obtained by replacing the corresponding regions of the chromosome of *Streptomyces* sp. IB201691-2A with the spectinomycin resistance cassette. Mutant strains were cultivated in the production medium and extracted metabolites were analyzed by LC-MS (Figure 6). As expected, *Streptomyces* sp. IB201691-2AΔGT3, lacking *baiGT3*, accumulated rabelomycin (**1**) and 5-hydroxy- rabelomycin (**2**), but not glycosylated angucyclines (Figure 6). Furthermore, we found an additional peak (**X1**) in the extract of the mutant strain with RT of 7.56 and *m*/*z* 373.0945 [*M*-H2O-H]$^-$ that corresponds to the mass calculated for aglycone of baikalomycins A and B (exact mass 392.11073, calculated *m*/*z* 373.092340 [*M*-H2O-H]$^-$) (Figure 6; Supplementary Figure S25). This compound is also present in the extract of parental strain but in much smaller quantities.

Deletion of *baiGT2* abolished production of baikalomycins B and C. However, *Streptomyces* sp. IB201691-2AΔGT2 demonstrated increased accumulation of baikalomycin A that is barely detectable in the parental strain (Figure 6). These data lead us believe that BaiGT3 catalyzes the attachment of first amicetose at the C-9 of baikalomycins' aglycone, followed by addition of a second amicetose by BaiGT2.

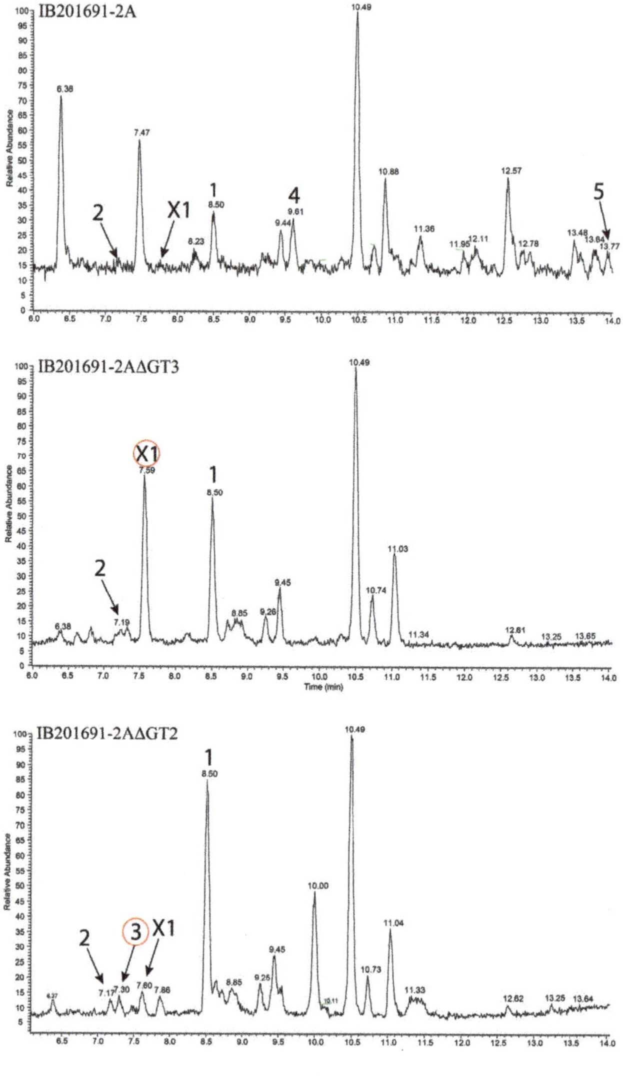

Figure 6. Analysis of LC-HRMS chromatograms of extracts of parental *Streptomyces* sp. IB201691-2A and mutant strains IB201691-2AΔGT3 and IB201691-2AΔGT2, lacking genes *baiGT3* and *baiGT2*, respectively, encoding baikalomycin glycosyltransferases. Compounds identified in the extracts: **1**—rabelomycin; **2**—5-hydroxy-rabelomycin; **3**—baikalomycin A; **4**—baikalomycin B; **5**—baikalomycin C; **X1**—baikalomycin aglycone.

The second sugar might stay unprocessed or is converted to aculose by action of BaiS8, similar to how it occurs in the case of grincamycins and Sch 47554/47555 biosynthesis [27,30]. At the same time, the function of BaiGT1 remained unclear. In the case of Sch 47554/47555 biosynthesis SchS10, the BaiGT1 orthologue is responsible for introduction of a sugar (aculose or amicetose) at C-3 position. However, *Streptomyces* sp. IB201691-2A did not produce baikalomycin trisaccharides under multiple tested conditions. We were also not able to inactivate the *baiGT1* gene due to technical difficulties.

We have cloned the entire *bai* gene cluster using transformation-associated recombination technique. As result, a plasmid p8-13bai, carrying 44.01 kb fragment of *Streptomyces* sp. IB201691-2A genomic

DNA with the *bai* gene cluster, however, lacking *aacC* and other downstream genes, (Figure 5A) was expressed in *S. albus* J1074. As a result, the recombinant strain failed to produce baikalomycins A–C (Figure 7). At the same time, three peaks with *m/z* of 697.2882 [*M*-H]$^-$ (RT 12.64, 12.89 and 13.11 min) and two peaks with *m/z* of 695.2734 [*M*-H]$^-$ (RT 14.46 and 14.83 min) and characteristic for angucylines absorption spectrum could be found in the extract of *S. albus* J1074/p8-13bai (Figure 7, Supplementary Figure S26). These compounds are absent in the extract of *Streptomyces* sp. IB201691-2A. Based on the detected mass and UV spectra these metabolites are proposed to be baikalomycins trisaccharides closely related to Sch-47554/47555. MS2 data support this idea. All five new metabolites have typical angucycline type fragmentation pattern with the characteristic loss of one sugar, most probably at C-3 position (Supplementary Figure S27). Most plausible, that the variety of masses of the detected compounds originate from incomplete conversion of amicetose to aculose, but rather to cineruloses intermediate in one or both saccharides at C-3 and C-9 positions, as it was shown for grincamycins [27].

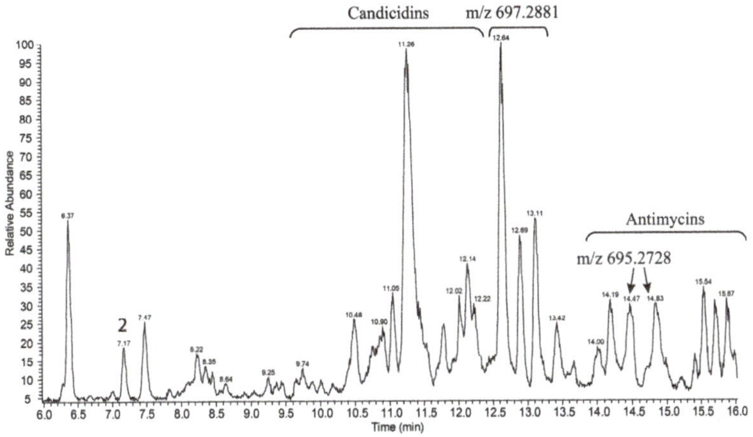

Figure 7. LC-MS chromatogram of extracts of *S. albus* J1074 carrying plasmid p8-13bai, with baikalomycin gene cluster from *Streptomyces* sp. IB201691-2A. Host exogenous compounds and new metabolites arisen from the expression of the *bai* gene cluster are highlighted.

In the latter case, expression of the grincamycin biosynthesis genes in *S. coelicolor* yielded vineomycin A1 with two aculose moieties instead of cineruloses, present in grincamycins. It is obvious that the baikalomycins biosynthetic enzymes behave differently in the natural producer and the heterologous host. In fact, *Streptomyces* sp. IB201691-2A did not produce angucycline trisaccharides under any of tested cultivation conditions but accumulated large amounts of shunt product rabelomycin and its derivatives. Vice versa, *S. albus* carrying the *bai* gene cluster accumulated the angucycline trisaccharides. In conclusion, six gene clusters encoding the aquayamycin-type angucyclines biosynthesis have a high degree of sequence identity (Figure 5B) and, at the same time, are responsible for the production of a great variety of related, but still different, natural products. Such variety, obviously, cannot be deduced from the nucleotide sequence analysis and thus leaves high chances to discover new natural products from the highly similar biosynthetic gene clusters.

Supplementary Materials: The following are available online at http://www.mdpi.com/2076-2607/8/5/680/s1.

Author Contributions: I.V., C.P., C.D., S.N., Y.R. performed the experiments and analyzed the data. I.V. isolated the strains. D.A.-G. collected samples from Lake Baikal and performed the experiments. C.R. performed the genome sequencing and analysis. S.N., C.P. performed structure elucidation. J.Z. discussed structure elucidation. M.T., J.K., A.K.K., and A.L. planned the experiments, analyzed data, and edited the manuscript. All authors have read and agreed to the published version of the manuscript.

Funding: This research was funded by Deutscher Akademischer Austauschdienst (DAAD), grant number 57299291 (personal grant for I.V.). Further, this research has received funding from BMBF grant "MyBio" 031B0344B

(University of Saarland), projects RFBR (project N 16-34-00686) (IV) (Irkutsk State University), RSF (project N 18-74-00018) (D.A.-G.) (Irkutsk State University), LFFP (project 17/08) and project of Russian Ministry of Science and Education N FZZE-2020-0026 (Irkutsk State University).

Acknowledgments: We thank Sergey Zotchev for pCLY10 vector.

Conflicts of Interest: The authors declare no conflict of interest. The funders had no role in the design of the study; in the collection, analyses, or interpretation of data; in the writing of the manuscript, or in the decision to publish the results.

References

1. Rohr, J.; Thiericke, R. Angucycline group antibiotics. *Nat. Prod. Rep.* **1992**, *9*, 103–137. [CrossRef] [PubMed]
2. Kharel, M.K.; Pahari, P.; Shepherd, M.D.; Tibrewal, N.; Nybo, S.E.; Shaaban, K.A.; Rohr, J. Angucyclines: Biosynthesis, mode-of-action, new natural products, and synthesis. *Nat. Prod. Rep.* **2012**, *29*, 264–325. [CrossRef] [PubMed]
3. Zhou, H.; Li, Y.; Tang, Y. Cyclization of aromatic polyketides from bacteria and fungi. *Nat. Prod. Rep.* **2010**, *27*, 839–868. [CrossRef] [PubMed]
4. Imamura, N.; Kakinuma, K.; Ikekawa, N.; Tanaka, H.; Omura, S. Biosynthesis of vineomycins A1 and B2. *J. Antibiot.* **1982**, *35*, 602–608. [CrossRef] [PubMed]
5. Kulowski, K.; Wendt-Pienkowski, E.; Han, L.; Yang, K.; Vining, L.C.; Hutchinson, C.R. Functional Characterization of the *jadI* gene as a cyclase forming angucyclinones. *J. Am. Chem. Soc.* **1999**, *121*, 1786–1794. [CrossRef]
6. Ostash, B.; Rebets, Y.; Yuskevich, V.; Luzhetskyy, A.; Tkachenko, V.; Fedorenko, V. Targeted disruption of *Streptomyces globisporus lndF* and *lndL* cyclase genes involved in landomycin E biosynthesis. *Folia Microbiol.* **2003**, *48*, 484–488. [CrossRef]
7. Gould, S.J.; Halley, K.A. Biosynthesis of the benz[a]anthraquinone antibiotic PD 116198: Evidence for a rearranged skeleton. *J. Am. Chem. Soc.* **1991**, *113*, 5092–5093. [CrossRef]
8. Sasaki, E.; Ogasawara, Y.; Liu, H.-W. A biosynthetic pathway for BE-7585A, a 2-thiosugar-containing angucycline-type natural product. *J. Am. Chem. Soc.* **2010**, *132*, 7405–7417. [CrossRef]
9. Elshahawi, S.I.; Shaaban, K.A.; Kharel, M.K.; Thorson, J.S. A comprehensive review of glycosylated bacterial natural products. *Chem. Soc. Rev.* **2015**, *44*, 7591–7697. [CrossRef]
10. Sezaki, M.; Kondo, S.; Maeda, K.; Umezawa, H.; Ohno, M. The structure of aquayamycin. *Tetrahedron* **1970**, *26*, 5171–5190. [CrossRef]
11. Uchida, T.; Imoto, M.; Watanabe, Y.; Miura, K.; Dobashi, T.; Matsuda, N.; Sawa, T.; Naganawa, H.; Hamada, M.; Takeuchi, T.; et al. Saquaymycins, new aquayamycin-group antibiotics. *J. Antibiot.* **1985**, *38*, 1171–1181. [CrossRef] [PubMed]
12. Sekizawa, R.; Iinuma, H.; Naganawa, H.; Hamada, M.; Takeuchi, T.; Yamaizumi, J.; Umezawa, K. Isolation of novel saquayamycins as inhibitors of farnesyl-protein transferase. *J. Antibiot.* **1996**, *49*, 487–490. [CrossRef] [PubMed]
13. Antal, N.; Fiedler, H.-P.; Stackebrandt, E.; Beil, W.; Ströch, K.; Zeeck, A. Retymicin, galtamycin B, saquayamycin Z and ribofuranosyllumichrome, novel secondary metabolites from *Micromonospora* sp. Tü 6368. I. Taxonomy, fermentation, isolation and biological activities. *J. Antibiot.* **2005**, *58*, 95–102. [CrossRef] [PubMed]
14. Huang, H.; Yang, T.; Ren, X.; Liu, J.; Song, Y.; Sun, A.; Ma, J.; Wang, B.; Zhang, Y.; Huang, C.; et al. Cytotoxic angucycline class glycosides from the deep sea actinomycete *Streptomyces lusitanus* SCSIO LR32. *J. Nat. Prod.* **2012**, *75*, 202–208. [CrossRef] [PubMed]
15. Shaaban, K.A.; Ahmed, T.A.; Leggas, M.; Rohr, J. Saquayamycins G-K, cytotoxic angucyclines from *Streptomyces* sp. Including two analogues bearing the aminosugar rednose. *J. Nat. Prod.* **2012**, *75*, 1383–1392. [CrossRef]
16. Erb, A.; Luzhetskyy, A.; Hardter, U.; Bechthold, A. Cloning and sequencing of the biosynthetic gene cluster for saquayamycin Z and galtamycin B and the elucidation of the assembly of their saccharide chains. *Chembiochem Eur. J. Chem. Biol.* **2009**, *10*, 1392–1401. [CrossRef]
17. Chu, M.; Yarborough, R.; Schwartz, J.; Patel, M.G.; Horan, A.C.; Gullo, V.P.; Das, P.R.; Puar, M.S. Sch 47554 and Sch 47555, two novel antifungal antibiotics produced from a *Streptomyces* sp. *J. Antibiot.* **1993**, *46*, 861–865. [CrossRef]

18. Nakagawa, K.; Hara, C.; Tokuyama, S.; Takada, K.; Imamura, N. Saprolmycins A-E, new angucycline antibiotics active against *Saprolegnia parasitica*. *J. Antibiot.* **2012**, *65*, 599–607. [CrossRef]
19. Abdelfattah, M.S.; Kharel, M.K.; Hitron, J.A.; Baig, I.; Rohr, J. Moromycins A and B, isolation and structure elucidation of C-glycosylangucycline-type antibiotics from *Streptomyces* sp. KY002. *J. Nat. Prod.* **2008**, *71*, 1569–1573. [CrossRef]
20. Ren, X.; Lu, X.; Ke, A.; Zheng, Z.; Lin, J.; Hao, W.; Zhu, J.; Fan, Y.; Ding, Y.; Jiang, Q.; et al. Three novel members of angucycline group from *Streptomyces* sp. N05WA963. *J. Antibiot.* **2011**, *64*, 339–343. [CrossRef]
21. Imamura, N.; Kakinuma, K.; Ikekawa, N.; Tanaka, H.; Omura, S. The structure of vineomycin B2. *J. Antibiot.* **1981**, *34*, 1517–1518. [CrossRef] [PubMed]
22. Kawamura, N.; Sawa, R.; Takahashi, Y.; Sawa, T.; Kinoshita, N.; Naganawa, H.; Hamada, M.; Takeuchi, T. Amicenomycins A and B, New Antibiotics from *Streptomyces* sp. MJ384-46F6. *J. Antibiot.* **1995**, *48*, 1521–1524. [CrossRef] [PubMed]
23. Peng, A.; Qu, X.; Liu, F.; Li, X.; Li, E.; Xie, W. Angucycline glycosides from an intertidal sediments strain *Streptomyces* sp. and their cytotoxic activity against hepatoma carcinoma cells. *Mar. Drugs* **2018**, *16*, 470. [CrossRef] [PubMed]
24. Henkel, T.; Zeeck, A. Derivatives of saquayamycins A and B. Regio- and diastereoselective addition of alcohols to the L-aculose moiety. *J. Antibiot.* **1990**, *43*, 830–837. [CrossRef] [PubMed]
25. Maskey, R.P.; Helmke, E.; Laatsch, H. Himalomycin A and B: Isolation and structure elucidation of new fridamycin type antibiotics from a marine streptomyces isolate. *J. Antibiot.* **2003**, *56*, 942–949. [CrossRef]
26. Chen, Q.; Mulzer, M.; Shi, P.; Beuning, P.J.; Coates, G.W.; O'Doherty, G.A. *De novo* asymmetric synthesis of fridamycin E. *Org. Lett.* **2011**, *13*, 6592–6595. [CrossRef]
27. Zhang, Y.; Huang, H.; Chen, Q.; Luo, M.; Sun, A.; Song, Y.; Ma, J.; Ju, J. Identification of the grincamycin gene cluster unveils divergent roles for GcnQ in different hosts, tailoring the L-rhodinose moiety. *Org. Lett.* **2013**, *15*, 3254–3257. [CrossRef]
28. Salem, S.M.; Weidenbach, S.; Rohr, J. Two cooperative glycosyltransferases are responsible for the sugar diversity of saquayamycins isolated from *Streptomyces* sp. KY 40-1. *ACS Chem. Biol.* **2017**, *12*, 2529–2534. [CrossRef]
29. Basnet, D.B.; Oh, T.-J.; Vu, T.T.H.; Sthapit, B.; Liou, K.; Lee, H.C.; Yoo, J.-C.; Sohng, J.K. Angucyclines Sch 47554 and Sch 47555 from *Streptomyces* sp. SCC-2136: Cloning, sequencing, and characterization. *Mol. Cells* **2006**, *22*, 154–162.
30. Fidan, O.; Yan, R.; Gladstone, G.; Zhou, T.; Zhu, D.; Zhan, J. New insights into the glycosylation steps in the biosynthesis of sch47554 and sch47555. *Chembiochem Eur. J. Chem. Biol.* **2018**, *19*, 1424–1432.
31. Kawasaki, T.; Moriyama, A.; Nakagawa, K.; Imamura, N. Cloning and identification of saprolmycin biosynthetic gene cluster from *Streptomyces* sp. TK08046. *Biosci. Biotechnol. Biochem.* **2016**, *80*, 2144–2150. [CrossRef]
32. Chater, K.F.; Wilde, L.C. *Streptomyces albus* G mutants defective in the *Sal*GI restriction-modification system. *J. Gen. Microbiol.* **1980**, *116*, 323–334. [CrossRef] [PubMed]
33. Flett, F.; Mersinias, V.; Smith, C.P. High efficiency intergeneric conjugal transfer of plasmid DNA from *Escherichia coli* to methyl DNA-restricting streptomycetes. *Fems Microbiol. Lett.* **1997**, *155*, 223–229. [CrossRef] [PubMed]
34. Kouprina, N.; Larionov, V. Transformation-associated recombination (TAR) cloning for genomics studies and synthetic biology. *Chromosoma* **2016**, *125*, 621–632. [CrossRef] [PubMed]
35. Kieser, T.; Bibb, M.J.; Buttner, M.J.; Chater, K.F.; Hopwood, D.A. *Practical Streptomyces Genetics*; John Innes Foundation: Norwich, UK, 2000; p. 613.
36. Sambrook, J.; Fritsch, E.F.; Maniatis, T. *Molecular Cloning: A Laboratory Manual*; Cold Spring Harbor Laboratory Press: Cold Spring Harbor, NY, USA, 1989; p. 1659.
37. Hausdorf, B.; Röpstorf, P.; Riedel, F. Relationships and origin of endemic Lake Baikal gastropods (*Caenogastropoda: Rissooidea*) based on mitochondrial DNA sequences. *Mol. Phylogenetics Evol.* **2003**, *26*, 435–443. [CrossRef]
38. Turner, S.; Pryer, K.M.; Miao, V.P.W.; Palmer, J.D. Investigating deep phylogenetic relationships among cyanobacteria and plastids by small subunit rRNA sequence analysis. *J. Eukaryot. Microbiol.* **1999**, *46*, 327–338. [CrossRef] [PubMed]

39. Kumar, S.; Stecher, G.; Tamura, K. MEGA7: Molecular evolutionary genetics analysis version 7.0 for bigger datasets. *Mol. Biol. Evol.* **2016**, *33*, 1870–1874. [CrossRef]
40. Saitou, N.; Nei, M. The neighbor-joining method: A new method for reconstructing phylogenetic trees. *Mol. Biol. Evol.* **1987**, *4*, 406–425.
41. Running, W. Computer Software Reviews. Chapman and Hall Dictionary of Natural Products on CD-ROM. *J. Chem. Inf. Model.* **1993**, *33*, 934–935. [CrossRef]
42. Fernández, E.; Weißbach, U.; Reillo, C.S.; Braña, A.F.; Méndez, C.; Rohr, J.; Salas, J.A. Identification of two genes from *Streptomyces argillaceus* encoding glycosyltransferases involved in transfer of a disaccharide during biosynthesis of the antitumor drug mithramycin. *J. Bacteriol.* **1998**, *180*, 4929–4937. [CrossRef]
43. Bankevich, A.; Nurk, S.; Antipov, D.; Gurevich, A.A.; Dvorkin, M.; Kulikov, A.S.; Pyshkin, A.V. SPAdes: A new genome assembly algorithm and its applications to single-cell sequencing. *J. Comput. Biol.* **2012**, *19*, 455–477. [CrossRef]
44. Meyer, F. GenDB-an open source genome annotation system for Prokaryote genomes. *Nucleic Acids Res.* **2003**, *31*, 2187–2195. [CrossRef] [PubMed]
45. Seemann, T. Prokka: Rapid prokaryotic genome annotation. *Bioinform* **2014**, *30*, 2068–2069. [CrossRef] [PubMed]
46. Weber, T.; Blin, K.; Duddela, S.; Krug, D.; Kim, H.U.; Bruccoleri, R.; Lee, S.Y.; Fischbach, M.A.; Müller, R.; Wohlleben, W.; et al. antiSMASH 3.0-a comprehensive resource for the genome mining of biosynthetic gene clusters. *Nucleic Acids Res.* **2015**, *43*, 237–243. [CrossRef]
47. Bilyk, O.; Sekurova, O.; Zotchev, S.B.; Luzhetskyy, A. Cloning and heterologous expression of the grecocycline biosynthetic gene cluster. *PLoS ONE* **2016**, *11*, 1–17. [CrossRef] [PubMed]
48. Ritacco, F.V.; Eveleigh, D.E. Molecular and phenotypic comparison of phaeochromycin-producing strains of *Streptomyces phaeochromogenes* and *Streptomyces ederensis*. *J. Ind. Microbiol. Biotechnol.* **2008**, *35*, 931–945. [CrossRef] [PubMed]
49. Williams, S.T.; Goodfellow, M.; Alderson, G.; Wellington, E.M.; Sneath, P.H.; Sackin, M.J. Numerical classification of *Streptomyces* and related genera. *J. Gen. Microbiol.* **1983**, *129*, 1743–1813. [CrossRef]
50. Olano, C.; Méndez, C.; Salas, J.A. Antitumor compounds from marine actinomycetes. *Mar. Drugs* **2009**, *7*, 210–248. [CrossRef]
51. Gould, S.J.; Cheng, X.C.; Halley, K.A. Biosynthesis of dehydrorabelomycin and PD 116740: Prearomatic deoxygenation as evidence for different polyketide synthases in the formation of benz[a]anthraquinones. *J. Am. Chem. Soc.* **1992**, *114*, 10066–10068. [CrossRef]
52. Liu, W.-C.; Parker, W.L.; Slusarchyk, D.S.; Greenwood, G.L.; Graham, S.F.; Meyers, E. Isolation, characterization, and structure of rabelomycin, a new antibiotic. *J. Antibiot.* **1970**, *23*, 437–441. [CrossRef]
53. Paululat, T.; Kulik, A.; Hausmann, H.; Karagouni, A.D.; Zinecker, H.; Imhoff, J.F.; Fiedler, H.-P. Grecocyclines: New Angucyclines from *Streptomyces* sp. Acta 1362. *Eur. J. Org. Chem.* **2010**, *2010*, 2344–2350. [CrossRef]
54. Rohr, J.; Zeeck, A. Metabolic products of microorganisms. 240 Urdamycins, new angucycline antibiotics from *Streptomyces fradiae*. II Structural studies of urdamycins B to F. *J. Antibiot.* **1987**, *40*, 459–467. [CrossRef] [PubMed]
55. Holzenkämpfer, M.; Walker, M.; Zeeck, A.; Schimana, J.; Fiedler, H.-P. Simocyclinones, novel cytostatic angucyclinone antibiotics produced by *Streptomyces antibioticus* Tü 6040 II. Structure elucidation and biosynthesis. *J. Antibiot.* **2002**, *55*, 301–307. [CrossRef] [PubMed]
56. Fotso, S.; Mahmud, T.; Zabriskie, T.M.; Santosa, D.A.; Sulastri; Proteau, P.J. Angucyclinones from an Indonesian *Streptomyces* sp. *J. Nat. Prod.* **2008**, *71*, 61–65. [CrossRef] [PubMed]
57. Kalinovskaya, N.I.; Kalinovsky, A.I.; Romanenko, L.A.; Dmitrenok, P.S.; Kuznetsova, T.A. New angucyclines and antimicrobial diketopiperazines from the marine mollusk-derived actinomycete *Saccharothrix espanaensis* An 113. *Nat. Prod. Commun.* **2010**, *5*, 597–602. [CrossRef] [PubMed]
58. Xie, Z.; Liu, B.; Wang, H.; Yang, S.; Zhang, H.; Wang, Y.; Ji, N.; Qin, S.; Laatsch, H. Kiamycin, a unique cytotoxic angucyclinone derivative from a marine *Streptomyces* sp. *Mar. Drugs* **2012**, *10*, 551–558. [CrossRef]
59. Seif, M.; Hoppstädter, J.; Breinig, F.; Kiemer, A.K. Yeast-mediated mRNA delivery polarizes immuno-suppressive macrophages towards an immuno-stimulatory phenotype. *Eur. J. Pharm. Biopharm.* **2017**, *117*, 1–13. [CrossRef]

60. Kodani, S.; Bicz, J.; Song, L.; Deeth, R.J.; Ohnishi-Kameyama, M.; Yoshida, M.; Ochi, K.; Challis, G.L. Structure and biosynthesis of scabichelin, a novel tris-hydroxamate siderophore produced by the plant pathogen *Streptomyces scabies* 87.22. *Org. Biomol. Chem.* **2013**, *11*, 4686–4694. [CrossRef]
61. Kulikova, N.N.; Mekhanikova, I.V.; Chebykin, E.P.; Vodneva, E.V.; Timoshkin, O.A.; Suturin, A.N. Chemical element composition and amphipod concentration function in Baikal littoral zone. *Water Resour.* **2017**, *44*, 497–511. [CrossRef]
62. Ghimire, G.P.; Oh, T.-J.; Lee, H.C.; Sohng, J.K. Squalene-hopene cyclase (*Spterp25*) from *Streptomyces peucetius*: Sequence analysis, expression and functional characterization. *Biotechnol. Lett.* **2009**, *31*, 565–569. [CrossRef]
63. Gürtler, H.; Pedersen, R.; Anthoni, U.; Christophersen, C.; Nielsen, P.H.; Wellington, E.M.; Pedersen, C.; Bock, K. Albaflavenone, a sesquiterpene ketone with a zizaene skeleton produced by a streptomycete with a new rope morphology. *J. Antibiot.* **1994**, *47*, 434–439. [CrossRef]
64. Komatsu, M.; Tsuda, M.; Omura, S.; Oikawa, H.; Ikeda, H. Identification and functional analysis of genes controlling biosynthesis of 2-methylisoborneol. *Proc. Natl. Acad. Sci. USA* **2008**, *105*, 7422–7427. [CrossRef] [PubMed]
65. Decker, H.; Haag, S. Cloning and characterization of a polyketide synthase gene from *Streptomyces fradiae* Tü2717, which carries the genes for biosynthesis of the angucycline antibiotic urdamycin A and a gene probably involved in its oxygenation. *J. Bacteriol.* **1995**, *177*, 6126–6136. [CrossRef] [PubMed]
66. Patrikainen, P.; Kallio, P.; Fan, K.; Klika, K.D.; Shaaban, K.A.; Mäntsälä, P.; Rohr, J.; Yang, K.; Niemi, J.; Metsä-Ketelä, M. Tailoring enzymes involved in the biosynthesis of angucyclines contain latent context-dependent catalytic activities. *Chem. Biol.* **2012**, *19*, 647–655. [CrossRef]
67. Faust, B.; Hoffmeister, D.; Weitnauer, G.; Westrich, L.; Haag, S.; Schneider, P.; Decker, H.; Künzel, E.; Rohr, J.; Bechthold, A. Two new tailoring enzymes, a glycosyltransferase and an oxygenase, involved in biosynthesis of the angucycline antibiotic urdamycin A in *Streptomyces fradiae* Tü2717. *Microbiology* **2000**, *146*, 147–154. [CrossRef]
68. Zhang, H.; Wang, H.; Wang, Y.; Cui, H.; Xie, Z.; Pu, Y.; Pei, S.; Li, F.; Qin, S. Genomic sequence-based discovery of novel angucyclinone antibiotics from marine *Streptomyces* sp. W007. *Fems Microbiol. Lett.* **2012**, *332*, 105–112. [CrossRef] [PubMed]
69. Alexeev, I.; Sultana, A.; Mäntsälä, P.; Niemi, J.; Schneider, G. Aclacinomycin oxidoreductase (AknOx) from the biosynthetic pathway of the antibiotic aclacinomycin is an unusual flavoenzyme with a dual active site. *Proc. Natl. Acad. Sci. USA* **2007**, *104*, 6170–6175. [CrossRef] [PubMed]
70. Rebets, Y.; Ostash, B.; Luzhetskyy, A.; Hoffmeister, D.; Braña, A.; Mendez, C.; Salas, J.A.; Bechthold, A.; Fedorenko, V. Production of landomycins in *Streptomyces globisporus* 1912 and *S. cyanogenus* S136 is regulated by genes encoding putative transcriptional activators. *Fems Microbiol. Lett.* **2003**, *222*, 149–153. [CrossRef]
71. Cuthbertson, L.; Nodwell, J.R. The TetR family of regulators. *Microbiol. Mol. Biol. Rev. Mmbr* **2013**, *77*, 440–475. [CrossRef]
72. Hassan, K.A.; Brzoska, A.J.; Wilson, N.L.; Eijkelkamp, B.A.; Brown, M.H.; Paulsen, I.T. Roles of DHA2 family transporters in drug resistance and iron homeostasis in *Acinetobacter* spp. *J. Mol. Microbiol. Biotechnol.* **2011**, *20*, 116–124. [CrossRef]
73. Palmu, K.; Ishida, K.; Mäntsälä, P.; Hertweck, C.; Metsä-Ketelä, M. Artificial reconstruction of two cryptic angucycline antibiotic biosynthetic pathways. *Chembiochem Eur. J. Chem. Biol.* **2007**, *8*, 1577–1584. [CrossRef]

microorganisms

MDPI

Article

Evaluation of Antiviral, Antibacterial and Antiproliferative Activities of the Endophytic Fungus *Curvularia papendorfii*, and Isolation of a New Polyhydroxyacid †

Afra Khiralla [1,2], Rosella Spina [1,*], Mihayl Varbanov [1], Stéphanie Philippot [1], Pascal Lemiere [1], Sophie Slezack-Deschaumes [3], Philippe André [4], Ietidal Mohamed [5], Sakina Mohamed Yagi [5] and Dominique Laurain-Mattar [1,*]

1 Université de Lorraine, CNRS, L2CM, F-54000 Nancy, France; aafraa21@hotmail.com (A.K.); mihayl.varbanov@univ-lorraine.fr (M.V.); stephanie.philippot@univ-lorraine.fr (S.P.); pascal.lemiere@univ-lorraine.fr (P.L.)

2 Botany Department, Faculty of Sciences and Technologies, Shendi University, P.O. Box 142 Shendi, Sudan

3 Université de Lorraine, Inra, LAE, F-54000 Nancy, France; sophie.deschaumes@univ-lorraine.fr

4 Université de Strasbourg, UMR 7021 CNRS, 67401 Illkirch, France; philippe.andre@unistra.fr

5 Department of Botany, Faculty of Science, University of Khartoum, 11115 Khartoum, Sudan; ietidalem11@gmail.com (I.M.); sakinayagi@gmail.com (S.M.Y.)

* Correspondence: rosella.spina@univ-lorraine.fr (R.S.); dominique.mattar@univ-lorraine.fr (D.L.-M.); Tel.: +33-3-7274-5226 (R.S.); +33-3-7274-5675 (D.L.-M.)

† This article is dedicated in memory of Pr. Annelise Lobstein, professor in University of Strasbourg, France.

Received: 9 July 2020; Accepted: 31 August 2020; Published: 4 September 2020

Abstract: An endophytic fungus isolated from *Vernonia amygdalina,* a medicinal plant from Sudan, was taxonomically characterized as *Curvularia papendorfii.* Ethyl acetate crude extract of *C. papendorfii* revealed an important antiviral effect against two viral pathogens, the human coronavirus HCoV 229E and a norovirus surrogate, the feline coronavirus FCV F9. For the last one, 40% of the reduction of the virus-induced cytopathogenic effect at lower multiplicity of infection (MOI) 0.0001 was observed. Selective antibacterial activity was obtained against *Staphylococcus* sp. (312 μg/mL), and interesting antiproliferative activity with half maximal inhibitory concentration (IC_{50}) value of 21.5 ± 5.9 μg/mL was observed against human breast carcinoma MCF7 cell line. Therefore, *C. papendorfii* crude extract was further investigated and fractionated. Twenty-two metabolites were identified by gas chromatography coupled to mass spectrometry (GC–MS), and two pure compounds, mannitol and a new polyhydroxyacid, called kheiric acid, were characterized. A combination of spectroscopic methods was used to elucidate the structure of the new aliphatic carboxylic acid: kheiric acid (3,7,11,15-tetrahydroxy-18-hydroxymethyl-14,16,20,22,24-pentamethyl-hexacosa-4E,8E,12E,16,18-pentaenoic acid). Kheiric acid showed an interesting result with a minimum inhibitory concentration (MIC) value of 62.5 μg/mL against meticillin-resistant *Staphylococcus aureus* (MRSA). Hence, endophytes associated with medicinal plants from Sudan merit more attention, as they could be a treasure of new bioactive compounds.

Keywords: *Curvularia papendorfii*; endophytic fungi; human coronavirus HCoV 229E; *Staphylococcus* sp.; MRSA; antiproliferative activity; polyhydroxyacid; kheiric acid

1. Introduction

Currently, the priority in research is the discovery of alternative treatments for viral and bacterial infections and cancer diseases. For human coronavirus (HCoV) and noroviruses, there is no vaccine or effective antivirals to prevent or control infections. Human coronaviruses (HCoVs) are a set of viruses that induce respiratory disease of varying severity, including common cold and pneumonia [1,2]. This wide-ranging family of viruses infects many species of mammals, including humans [3]. Their ability for interspecies transmission has led to the emergence of the Severe Acute Respiratory Syndrome (SARS) [4] and the Middle East Respiratory Syndrome (MERS) [5], both associated with high mortality and morbidity.

Noroviruses frequently cause acute gastroenteritis outbreaks globally, which is associated with heavy economic burden [6], as norovirus-induced gastroenteritis is particularly acute in the elderly and in young children. It is a highly resistant and infectious virus, with an infectious dose close to 20 virions and is easily transmitted through person-to-person contact [7]. The feline calicivirus strain F9 (FCV F9) is used to study the biology of norovirus [8], given the difficulties of growing human noroviruses in laboratory conditions.

The emerging of antimicrobial resistance is a problem for society [9]. Methicillin-resistant *S. aureus* (MRSA) is one of the pathogen strains causing the majority of hospital infections and effectively escapes the effects of antibacterial drugs [10]. MRSA causes different infections in the blood, heart, skin, soft tissue and bones. MRSA is also responsible for nosocomial infections. Treatment is often difficult, and currently, there is a need to develop new antimicrobials [11].

A bioprospecting of the bioactive molecules is reported from endophytic fungi, and the genus Curvularia and Bipolaris revealed interesting biological activities [12]. Some crude extracts of endophytes, isolated from medicinal plants of Sudan, have proven to be very rich and promising resources of bioactive compounds. For example, *Aspergillus* sp. associated with *Trigonella foenum-graecum* revealed powerful antioxidant activity [13,14]. *Byssochlamys spectabilis* and *Alternaria* sp. isolated from *Euphorbia prostrata* showed potent antiproliferative and antibacterial activities, respectively [15].

The genus Curvularia belongs to the family Pleosporaceae. Members of this genus are of widespread distribution in tropical and subtropical regions and are commonly isolated from a wide range of plant species as well as from soil. Many species are known to be plant pathogens [16]. Some researchers have demonstrated that crude extracts of some *Curvularia* species have interesting properties including antimicrobial, antioxidant, phytotoxicity and leishmanicidal activities [12,17–19].

Several natural products from different chemical classes were purified from the genus Curvularia: alkaloids such as curvulamine and curindolizine [20,21]; polyketides such as apralactone A, curvulide A and cochliomycin A [22–24]; quinones such as cynodontin and lunatin [25,26]; and terpenes such as zaragozic acid A [27].

In this study, an endophytic fungus was isolated from *Vernonia amygdalina*, and then the taxonomic characterization of the isolate was established. The ethyl acetate crude extract of the isolated endophyte was investigated for antiviral, antibacterial and antiproliferative activities. The crude extract was fractionated to afford pure compounds. Each fraction was analyzed by gas chromatography coupled to mass spectrometry (GC–MS). Structure elucidation of pure compounds was done using different spectroscopy techniques, and the biological activities were evaluated.

2. Materials and Methods

2.1. Chemicals

Ethyl acetate (≥99.9%), cyclohexane (≥99.9%), methanol (≥99.9%), acetic acid (≥99.9%), dichloromethane (≥99.9%), formic acid (≥99.9%), dextrose (≥99.9%), agar (≥99.9%), silica gel, tetramethylsilane (TMS), deuterated methanol, deuterated pyridine and deuterated dimethyl sulfoxide were purchased from Sigma-Aldrich Co. LLC, Steinheim am Albuch, Germany. All solvents used were LC analytical grade. The mixture of alkanes standard from C10 to C40 was purchased from Merck KGaA, Darmstadt, Germany.

2.2. Endophytic Fungus: Isolation and Taxonomic Characterization

Leaf and stem samples of *V. amygdalina* were collected from a plant species growing wild in Khartoum State, Sudan. Voucher specimen TN4010 was deposited at the herbarium of the Botany Department, Faculty of Sciences, University of Khartoum, Khartoum, Sudan. The protocol for the isolation of endophytic fungus from *V. amygdalina* plant materials was the same procedure described in the article of A. Khiralla et al. [15].

For the taxonomic characterization of the isolated fungus, the protocol is the same described by A. Khiralla et al. [15].

2.3. Cultivation of the Fungus and Extraction of the Metabolites

The fungus was cultivated and extracted according to the previous protocol [18] with minor modifications. Briefly, fungal strain was cultured on 750 Petri dishes (15 L). The Petri dishes were incubated at 28 ± 2 °C for 14 days. Then, the cultured plates were macerated using ethyl acetate for 24 h. The extraction was repeated three times [15]. The organic extract was stored at 4 °C after filtration and evaporation.

2.4. Biological Assays

2.4.1. Cytotoxicity Tests: Cells, Media and Protocols

L132 (ATCC® CCL5™) and CRFK (CCL-94™) cell lines were cultured in antibiotic-free Minimum Essential Medium Eagle (MEM, M4655, Sigma-Aldrich, St. Quentin Fallavier, France) complemented with 10% fetal bovine serum (FBS) (CVFSV F00-0U, Eurobio, Les Ulis, France).

Cytotoxicity of the crude extract was assessed in 96-well tissue culture plates. For this purpose, cells were seeded at 10^4 cells per well in 96-well plates. The dry extract was dissolved in dimethylsulfoxide (DMSO), called DMSO-solubilized extract, or in sterile water called water-solubilized extract, then diluted in MEM medium complemented with 2% FBS. One hundred microliters of the diluted extract at increasing concentrations (2 to 256 μg/mL) were added to the cells monolayers 24 h following seeding. For the organic extract test, DMSO was used at 1% final concentration in order to avoid decrease in cell viability. The plates were incubated for 72 h at 37 °C in a 5% CO_2 atmosphere. Viability of cells was evaluated with the MTT assay based on the reduction of the MTT by cellular metabolism into purple formazan in living cells [28].

After 72 h, the supernatants were replaced by 100 μL of MTT (0.5 mg/mL) (M2128, Sigma-Aldrich, St. Quentin Fallavier, France), prepared in MEM medium complemented with 2% FBS, and added to each well. The plates were incubated for 2 h at 37 °C. Finally, the wells were washed and formazan crystals were solubilized by the addition of 100 μL of DMSO (04474701, Biosolve, Dieuze, France). The plates were agitated until complete dissolution, and then the absorbance was read at 540 nm using a 96-well plate spectrophotometer (Multiskan GO, Thermo Scientific, Saint Herblain, France). Percentages of survival compared to control cells were calculated and the maximal concentration with no cytotoxic effect was determined using Microsoft Excel 2010 (Microsoft Corp., Redmond, WA, USA) and GraphPAD Prism v. 5 software (GraphPAD, San Diego, CA, USA).

2.4.2. Antiviral Assay: Media, Viruses and Protocols

The infection medium, used for the antiviral assays, was the same as the growth medium, but 2% FBS was added instead. The human coronavirus HCoV 229E strain was propagated and quantified in L132 cells. CRFK cells were used for infection with the feline calicivirus FCV strain F9. Virus quantification was performed according to Reed and Muench's method [29]. Briefly, the cells (10^4 cells/well) were grown in 96-well tissue culture plates and incubated for 72 h in the presence of the HCoV 229E or FCV F9, at 33 °C and 37 °C, respectively, in a 5% CO_2 atmosphere with serial 10-fold diluted virus suspensions in order to test multiplicity of infection (MOI), defined as the ratio of infectious virions to cells in a culture, between 0.0001 and 1. This protocol is similar to the article [30]

The virus-induced cytopathogenic effect (CPE) was determined after 72 h of infection. By the method of Reed and Muench, the titers were counted as 50% Cell Culture Infectious Doses ($CCID_{50}$)/mL. All virus stocks were stored at −80 °C until used.

The antiviral activity was evaluated by the reduction of the virus-induced cytopathogenic effect (CPE), characterized by different parameters such as rounding, vacuolation, syncytia formation and cell death of the cell monolayer. Different treatments were evaluated against each virus in 96-well tissue culture plates. One day before infection, cells were seeded at a concentration of 10^4 cells/well. The next day, medium was removed and replaced by a mix consisting of diluted virus suspension and appropriate concentration of the extract. For each viral isolate (titers at least 10^6 (CCID50)/mL), two sets of 1:10 serial dilutions were used from 1:10 (MOI 1) to 1:100,000 (MOI 0.0001), the first set without antiviral agent, and the second with a non-cytotoxic concentration of the extract. A blank without any cell, virus or extract was added. Eight wells were tested the same way for each assay, control or blank ($n = 8$). Plates were then incubated with HCoV 229E or FCV F9 at 33 °C and 37 °C, respectively, and were consequently checked for virus-induced CPE on days 1, 2 and 3 post-infection, using an inverted light microscope. The tests were read for determination of viral CPE when cell destruction in infected untreated cultures was at its maximum post-infection.

Consequently, estimation of the cytopathogenic effect was determined by the crystal violet (CV) assay, according to a previously described protocol with same adaptation [31]. Removal of cell culture medium, washing of cells with 1 × PBS ant then fixing with 3.7% formaldehyde (533,998, Sigma-Aldrich, St. Quentin Fallavier, France) for 5 min are doing in the CV uptake assay. Next, the cell monolayers were stained with 0.1% CV in PBS (C3886, Sigma-Aldrich, St. Quentin Fallavier, France) added to the same set of plates used to obtain the visual scores. After 30 min incubation at room temperature, the dye was removed, all the wells were washed two times with PBS, and uptaken CV was then solubilized with 100 µL methanol (525,102, Carlo Erba, Val-de-Reuil, France) per well and left for 5 min at room temperature. The color intensity of the dye uptake by the cells was measured with a 96-well plate spectrophotometer (Multiskan GO, Thermo Scientific, Saint Herblain, France) by reading optical density (OD) at 540 nm. The percentage of cytopathogenic effect (% CPE) was calculated for treated and non-treated infected wells according to the formula: $((OD_{sample} - \text{mean } OD_{blank})/\text{mean } OD_{control}) \times 100$ where control was non-treated and non-infected cells. Antiviral activity was expressed as decrease of virus-induced CPE due to the treatment: % CPE (non-treated) − % CPE (extract-treated). Results are presented as the mean values obtained from at least two independent experiments.

Immunofluorescence Analysis (IFA)

An immunofluorescence analysis (IFA) was used to detect HCoV 229E and FCV F9 protein expression in infected host cells. Briefly, L132 and CRFK cells seeded on 96-well cell culture plates were grown for 24 h at 37 °C in 5% CO_2 [30]. The cells were then infected with HCoV 229E and FCV F9 virus at MOI 1 and MOI 0.0001 and incubated for 24 h at 33 °C and 37 °C, respectively. At 24 h post-infections, 2% of paraformaldehyde was used to fix the cells and blocked in 5% bovine serum albumin (BSA) in 1% Triton-X-100 PBS. The infected cells were incubated with anti-HCoV 229E (FIPV3-70, St Cruz Biotechnology, Heidelberg, Germany) or anti-FCV F9 (FCV1-43, St Cruz Biotechnology) mouse primary antibody (1:500) for 1 h, washed three times with PBS and then incubated with 1:400-diluted FITC-labeled goat anti-mouse IgG (sc-2010, St Cruz Biotechnologies, Heidelberg, Germany) for 30 min. The identification of positive foci was done using fluorescence microscopy under an inverted fluorescence microscope (Zeiss, Marly-Le-Roi, France) after DAPI duplicate staining.

2.4.3. Antibacterial Assay: Cells, Media and Protocols

In this study, eighteen standard strains of bacteria were used. Gram-negative bacteria: *Pseudomonas aeruginosa* (CIP82118), *Salmonella enterica* subspecies *enterica* sérovar Abony, *Escherichia coli* (ATCC 8739). Gram-positive bacteria: *Staphylococcus aureus* (ATCC 6538), MRSA, *S. arlettae*, *S. capitis*, *S. hominis*,

S. auricularis, S. epidermidis, S. haemolyticus, S. xylosus, S. lugdunensis, S. sciuri, Enterococcus faecalis, E. faecium, Bacillus cereus, Kytococcus sedentarius [14].

Information for the media and protocols are in the article by A. Khiralla et al. [15].

2.4.4. Antiproliferative Activity: Cells, Media and Protocol

The cells used for the determination of antiproliferative activities were the human colon adenocarcinoma (HT29 and HCT116) and human breast adenocarcinoma (MCF7). The media used for the cultivation and protocols are described in the article by Khiralla A. et al. [15]. For antiproliferative activities, the respective IC_{50} value was calculated from results obtained from quadruplicate determination of two independent experiments (n = 8). IC_{50} value was expressed as µg/mL of extract diluted in DMSO.

2.5. Analytical and Spectroscopic Analysis

For identification and characterization of metabolites, the following were used: a thin-layer chromatograph (TLC GF_{254} plates (Merck), an infrared (IR) spectrometer (Perkin-Elmer model 1650 FTIR), a Bruker Avance III 400 MHz spectrometer, and a gas chromatography system coupled with a mass spectrometer (GC–MS, QP2010-Shimadzu equipment). The column used was an SLB5 column DB-5 ms, the procedure of which was the same as that described in our previous works [32,33]. The following were also used: a liquid chromatography system (U3000-Dionex apparatus) coupled with a mass spectrometer (Bruker Daltonics micrOTOF-Q^{TM}); a high-resolution liquid chromatography system (HPLC, Merck Hitachi Lachrom) for analytical analysis, in which the analytical column used was a C18 ODS Hypersil^{TM} 5 µM 250 × 4.6 mm column (Thermo Scientific, USA); a preparative HPLC (Gilson), for which the semi-preparative column used was ODS Hypersil^{TM} C18 250x10mm (ThermoScientific, USA). For all technical characteristics of equipment, refer to Elmi et al.'s [32] and A. Khiralla's work [14].

2.6. Purification and Identification of Metabolites from C. papendorfii Crude Extract

The dark brown ethyl acetate crude extract of *C. papendorfii* contained a white precipitate. The precipitate was physically separated, washed with ethyl acetate and then subjected to fractionation and analyzed. At the end, we recovered 52 mg. The first analysis was realized by TLC. The mobile phase for TLC was made of 7/3/0.1 (*v/v/v*) ethyl acetate/cyclohexane/glacial acetic acid. The plates were observed under UV254 and UV365 nm and then sprayed with sulfuric acid reagent. The precipitate was injected to GC–MS, LC–MS and HPLC. The precipitate (25 mg) was dissolved in 1 mL of a mixture of methanol and water (0.9 mL and 0.1 mL respectively) and purified with semipreparative HPLC, using a linear gradient consisting to a mixture of methanol with 2% formic acid and water with 2% formic acid to give compound **1** (7 mg).

The ethyl acetate fraction was purified with an open column of silica gel, with a gradient of cyclohexane/ethyl acetate, (9:1 to 2:8 (*v/v*)) and then ethyl acetate/methanol/acetic acid (8:1:1 (*v/v/v*)). At the end, the column was washed with methanol. The tubes were collected using TLC profile. Ten fractions (F1, F2, F3, F4, F5, F6, F7, F8, F9 and F10) were obtained, and all were evaporated under a vacuum. In the fraction F6, a crystalline solid could be observed, compound **2** (12 mg). Fraction F10 was purified using an open column (using a mobile phase constituted by EtOAc/MeOH/acetic acid). The purification led to six subfractions called F10.A, F10.B, F10.C, F10.D, F10.E and F10.F. All subfractions were analyzed by TLC and GC–MS. A total of 23 compounds were determined by GC–MS. The identification of the chemical compounds was obtained by comparison of mass spectra with the spectra present in the NIST (National Institute of Standards and Technology) library. The retention index was calculated using alkane standard mixture (C10–C40) under the same operating conditions.

3. Results and Discussion

3.1. Isolation and Taxonomic Characterization of the Fungal Strain

The endophytic fungus was isolated from both leaf and stem samples of *V. amygdalina* plant, which showed a colonization frequency (CF) of 90%. High CF was also recorded on previous studies on fungal endophytes communities such as in Puerto Rico and Sudan [15,34]. ITS sequences were deposited in GenBank and then compared using a BLAST search [32]. The isolated fungus was identified as *C. papendorfii* with 99% identity (Genbank number KR673909) (Figure 1).

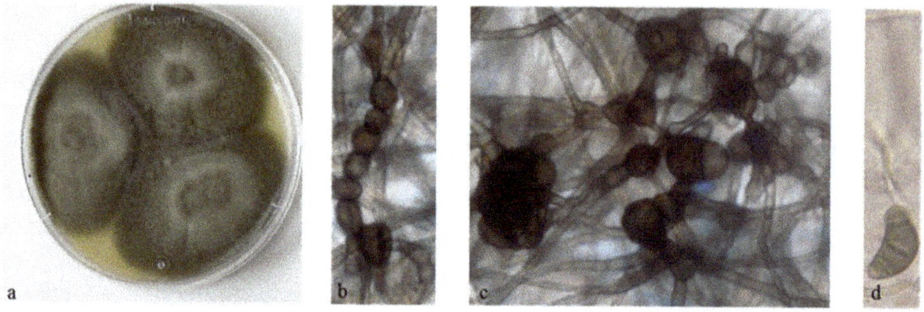

Figure 1. *Curvularia papendorfii* fungus; (**a**): culture on Potato dextrose agar (PDA) plate; (**b**,**c**): chlamydospores ×400; (**d**): conidia ×400 [14].

In this study, for the first time, *C. papendorfii* was reported as a fungal endophyte in *V. amygdalina*, collected in Sudan, although most *Curvularia* species occur as tropical and subtropical plant pathogens [12,35]. The genus Curvularia was also recovered from several plants as endophytes [36,37], as endolichens [38] and as marine-derived fungus [22,23,39].

3.2. Screening of Biological Activities of C. papendorfii Crude Extract

3.2.1. Cytotoxic Effects

In order to exclude non-specific activities of the extract, cytotoxic effects had to be evaluated on cell lines L132 (A) and CRFK (B) cells, and the maximum non-toxic concentration for the cells had been determined (Figure 2).

The ethyl acetate crude extract of *C. papendorfii* was dissolved at 25 mg/mL in dimethylsulfoxide (DMSO). This extract is called DMSO-solubilized extract. Water was also tested to solubilize the extract and evaluate the effect of extract dissolution on cytotoxic effect. This extract is called water-solubilized extract. L132 and CRFK cell lines were treated with different concentrations of extract. The extract seemed to be more toxic when dissolved in DMSO for both cell lines. The CRFK cell line was more sensitive to the DMSO-solubilized extract than the L132 cell line. According to these results, the concentration 16 μg/mL was the dose chosen for antiviral tests. The low toxicity of water-solubilized extract (10.5 ± 6.9%) allowed application in the antiviral treatment at 128 μg/mL.

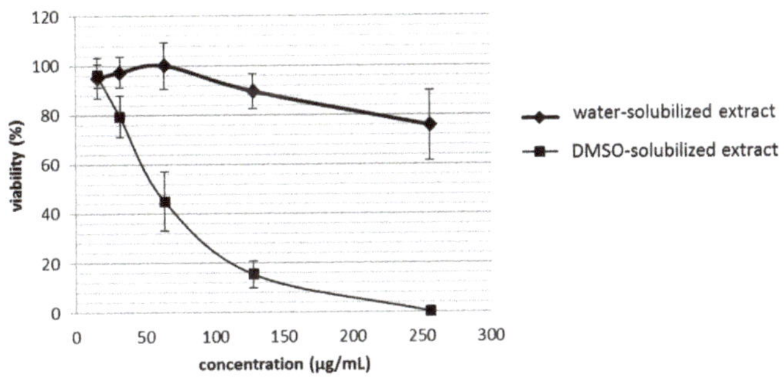

A.

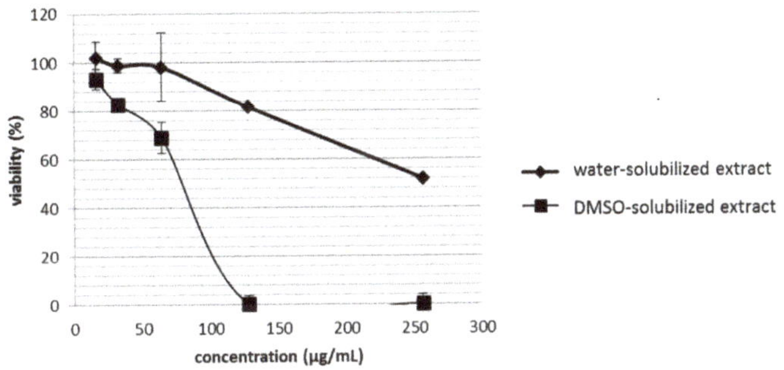

B.

Figure 2. Cytotoxicity of *C. papendorfii* DMSO-solubilized extract and water-solubilized extract after 72 h treatment on L132 (**A**) and CRFK (**B**) cells. Data show mean for $n = 3$ independently performed experiments. Bars indicate standard deviations.

3.2.2. Antiviral Activity of Crude Extract

For both viruses, HCoV and noroviruses, there is no vaccine or antivirals drugs for the prevention and the treatment of infection. In the absence of curative antiviral strategies, the bioactive molecules of fungal origin are particularly important as novel drug candidates.

The impact of the crude extract on viral infection was evaluated by the reduction of virus-induced cytopathogenic effect. L132 and CRFK were infected by the human coronavirus and the feline calicivirus, respectively, in the presence or absence of the extract for 72 h. The use of L132 cell line allowed producing a high level of viral titers and obtaining high sensitivity in antiviral evaluation, as previously demonstrated [40], notwithstanding the fact that there has been a contamination with HeLa cells as described recently by ATCC. This cell line is still made available by ATCC as a reference cell line, and it is currently used as a host cell line for HCoV 229E, as it allows one to obtain reproducible results in terms of antiviral tests. Given that this study does not involve or require specific organ or tissue of presumptive origin and that human coronaviruses have wide tissue and cellular tropism, the results of the virus production and assays remain uncompromised in these conditions.

Cell monolayers were then stained with CV, and the percentage of CPE was calculated. A decrease of CPE was reported for both viruses at low multiplicity of infection ratio (Figures 3 and 4).

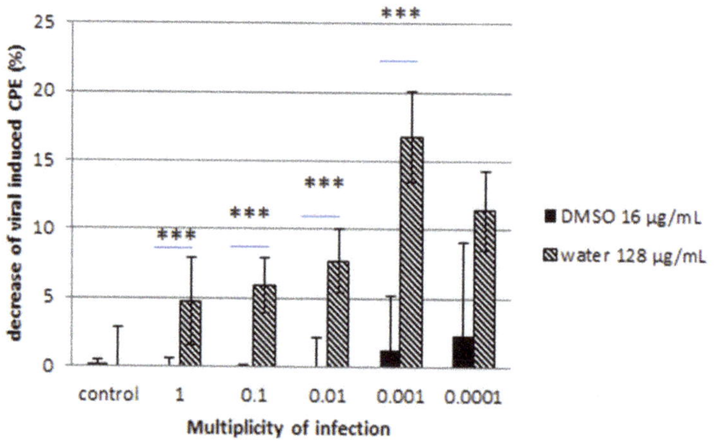

Figure 3. Effect of *C. papendorfii* DMSO-solubilized extract and water-solubilized extract on the infection of L132 cells by the human coronavirus 229E. Results are presented as a decrease of virus-induced cytopathogenic effect (CPE) at 72 h post-infection, calculated as % CPE (non-treated) − % CPE (extract-treated). Data are shown for at least two independent experiments. *T-test* statistical analysis was performed with GraphPAD Prism 5 software (*** means $p < 0.005$).

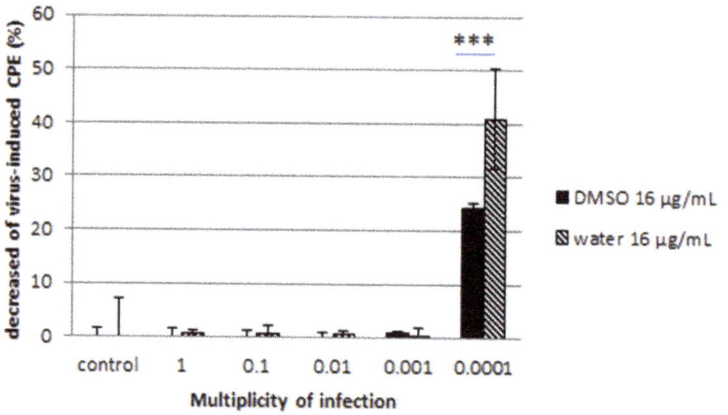

Figure 4. Effect of *C. papendorfii* DMSO-solubilized extract and water-solubilized extract on the infection of CRFK cells by the feline calicivirus FCV F9. Results are presented as a decrease of virus-induced CPE, at 72 h post-infection, calculated as % CPE (non-treated) − % CPE (extract-treated). Data are shown for at least two independent experiments. Bars indicate standard deviations. *T-test* statistical analysis was performed with GraphPAD Prism 5 software (*** means $p < 0.005$).

The water-solubilized extract was more effective than the DMSO-solubilized extract. The reduction reached 40% for the water-solubilized extract at lower MOI for the feline calicivirus while it was more moderate for the coronavirus. These observations were confirmed by the immunofluorescence assay, where a clear reduction of HCoV- and FCV F9-positive host cells was noted (Figure 5).

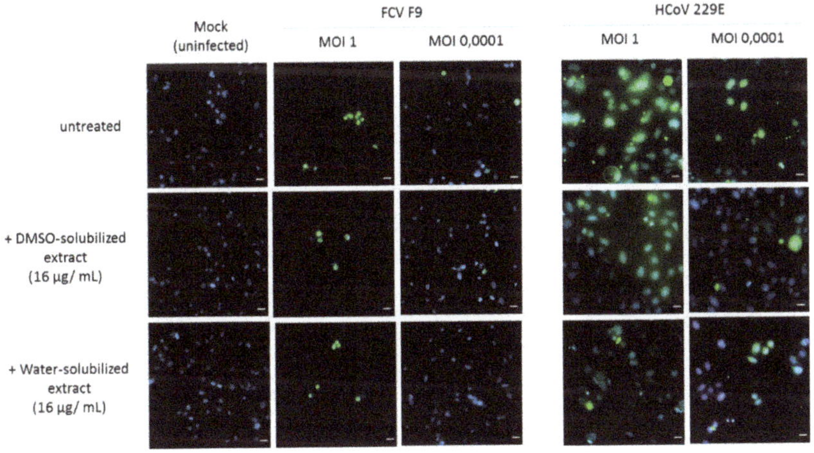

Figure 5. Immunofluorescence staining of FCV F9- and human coronavirus (HCoV) 229E-infected cells. CRFK and L132 cells were infected with FCV F9 and HCoV 229E, respectively, for 24 h and treated with DMSO-solubilized extract and water-solubilized extract (16 µg/mL) prior to fixation with methanol. The infected cells were then incubated with anti-FCV F9 and anti-HCoV 229E antibodies, then stained with goat anti-mouse FITC secondary antibody and treated with DAPI as counterstain. The photos were obtained at 40× magnification power. Scale bars: 20 µm.

Here we report an important antiviral effect of *C. papendorfii* extracts on two viral pathogens—the human coronavirus HCoV 229E and a norovirus surrogate, the feline coronavirus FCV F9.

Our exploration of the antiviral properties of DMSO-solubilized extract and water-solubilized extract of the fungal endophyte *C. papendorfii* indicated an important antiviral effect on enveloped viruses such as HCoV 229E at all MOI tested, even at MOI as high as MOI 1. This effect is particularly interesting in the case of water-solubilized extract with a reduction of the virus-induced cytopathogenic effect over 15% at MOI 0.001. The impact on virus-induced CPE by water-solubilized extract was also important in the case of non-enveloped viruses, represented in our study by FCV F9, with close to 40% reduction. This effect was observed exclusively at lower MOI, such as MOI 0.0001, which can be explained by the important infectivity and resistance of the virions. Indeed, non-enveloped viruses are much more stable and may stay active in wastewaters and on environmental surfaces for several months [41,42]. On the other hand, enveloped viruses are less stable and more prone to degradation. Enveloped viruses of the respiratory tract like influenza virus and coronavirus can persist on surfaces only for several days [43]. We observed that incubation with the plant extract during infection impaired the productive replication of both viruses in a MOI-dependent manner. The results obtained suggest that *C. papendorfii* antiviral activity might be partially due to a direct interaction of the compounds in the extract with the viral envelope, given the effect on HCoV. However, the reduction of the FCV F9 infection at lower MOI is a piece of evidence in favor of extract-induced interference with intracellular virus-induced macromolecular synthesis, thus hampering the viral replication. The predominant antiviral effect of water-solubilized extract indicated that polar biomolecules may be in the center of the antiviral activity, which is in accordance with the existing literature [44].

Our results allowed us to describe for the first time the capacity of *C. papendorfii* extracts to inhibit viral infection. Previous study has shown that crude extracts of one endophytic *Curvularia* species isolated from Garcinia plants [45] present antimycobacterial properties but no antiviral activity using herpes simplex virus (HSV-1) infection, even at the highest concentration (50 µg/mL) tested [45].

3.2.3. Antibacterial Activity of Crude Extract

A preliminary antibacterial screening was performed for ethyl acetate crude extract of *C. papendorfii* using agar disk diffusion method against 18 Gram-positive and Gram-negative bacterial strains. The results obtained indicated that the ethyl acetate crude extract of *C. papendorfii* had an effective antimicrobial activity against most Gram-positive bacteria, and no effect was observed against three Gram-negative bacterial strains [14].

The crude extract exhibited a maximum inhibition zone of 13 mm against MRSA and *S. aureus*. The inhibition zones were compared with the positive control, nitrofurantoin (27 mm), while 12 mm inhibition zone was reported against *S. epidermidis* and *S. capitis*; 10 mm against *S. lentus*, *S. warneri*, *S. sciuri*, *S. xylosus*, *S. haemolyticus* and *S. lugdunensis*; and 9 mm against *E. faecalis*, *K. sedentarius* and *S. arlettae*. In contrast, no inhibitory effect against *E. faecium*, *B. cereus*, *P. aeruginosa*, *E. coli* or *S. abony* was observed (Table 1).

Table 1. Antimicrobial activity of crude extract of *C. papendorfii* isolated from *V. amygdalina* against several Gram-positive and Gram-negative bacterial strains in an agar diffusion assay.

Bacterial Strains	Inhibition Zone (mm) with Ethyl Acetate Extract
P. aeruginosa	6
E. coli	6
Salmonella Abony	6
Staphylococcus aureus	13
MRSA	13
S. arlettae	9
S. lentus	10
S. epidermidis	12
S. haemolyticus	10
S. xylosus	10
S. sciuri	10
S. warneri	10
S. capitis	12
S. lugdunensis	10
E. faecalis	9
E. faecium	6
B. cereus	6
K. sedentarius	9

The inhibition zone was measured in mm and derived from experiments in triplicates. Standard deviation value is ±0.58 for all bacteria.

For *Staphylococcus aureus* and MRSA, the inhibition zone of 27 mm was determined with Nitrofurantoin (100 µg). Nitrofurantoin was used as a positive control when the inhibition zone was ≥13 mm [46].

The antibacterial activities of crude extract of *C. papendorfii* were confirmed by the broth dilution method, and the MIC values were 312 µg/mL for *S. aureus* as well as for MRSA.

However, several studies revealed that *Curvularia* sp. crude extracts have promising antibacterial activities. The extract of *C. lunata* showed antibacterial activities against *S. aureus* and *S. typhi* [17], and the ethyl acetate crude extract of *C. tuberculata* has shown antibacterial activities against *S. aureus*, *E. coli* and *P. aeruginosa* [39]. Moderate activities of extract of *Curvularia* B34 against *Bacillus subtilis*, *Listeria monocytogenes* and *Salmonella* bacteria have been observed [37]. *C. hawaiiensis*, an endophytic fungus isolated from *Calotropis procera*, showed antibacterial activity against *Serratia marcescens* [47]. The extract of endophytic fungus *Curvularia* sp. T12 isolated from *Rauwolfia macrophylla* had antibacterial activity against *E. coli*, *Micrococcus luteus*, *Pseudomonas agarici* and *S. warneri* [48].

3.2.4. Antiproliferative Activity of Crude Extract

Antiproliferative activity of ethyl acetate crude extract of *C. papendorfii* was evaluated using human cancer cell lines. The results obtained showed that the IC_{50} value was 21.5 ± 5.9 µg/mL for human breast adenocarcinoma (MCF7), and the IC_{50} values were higher than 100 µg/mL for the two cell lines of human colon adenocarcinoma (HT29 and HCT116). Response–dose curves established from results obtained by MTT assays after MCF7, HT29 and HCT116 cells exposure of *C. papendorfii* are presented in Figure S1.

In our previous work [15], the stem extract of *V. amygdalina*, the host plant of endophytic *C. papendorfii*, also showed cytotoxic effect against HT29 and MCF7 (IC_{50} values of 15.3 ± 3.6 and 49.6 ± 4.4 µg/mL respectively), while leaf extract had strong cytotoxicity (IC_{50} value of 5.6 ± 0.4 µg/mL) against HCT116 and moderate cytotoxicity (IC_{50} value of 27.5 ± 5.7 and 31.9±5.1 µg/mL) against MCF7 and HT29, respectively.

The cytotoxicity of ethyl acetate crude extract of *C. papendorfii* for MCF7 cell line may be in relationship with the chemical constituents as has been shown with polyketides extracted from marine-derived fungus *Curvularia* sp. [23].

3.3. Phytochemical Analysis

After purification, two pure compounds were identified, using mono and bidimensional Nuclear Magnetic Resonance (NMR), Mass Spectrometry (MS) and Infrared (IR) analysis. One is a new compound, never described in the literature and given the trial name kheiric acid (compound **1**), and the second one is a known structure, mannitol (compound **2**). The other compounds were identified by GC–MS analysis and compared with literature bibliography.

For the characterization of compound **1**, an absorption band on the IR spectrum at 2914 cm^{-1} is typical of a carboxylic group. An absorption band at 1714 cm^{-1} is predictive of a C=O group. The weak absorption band at 1666 cm^{-1} indicates alkenes functionality. A strong absorption band at 3383 cm^{-1} indicates OH groups (Figure S3).

The high-resolution electrospray ionization mass spectrometry (HRESIMS), positive mode, gave molecular ion peak of *m/z* 573.3781 $[M + Na]^+$ (Figure S4). The HRESIMS, negative mode, gave a molecular ion peak of *m/z* 549.3793 $[M - H]^-$. The possible molecular formula is $C_{32}H_{54}O_7$. The index of hydrogen deficiency (IHD) reveals an unsaturation index of 6.

Referring to the ^{13}C NMR spectrum and the ^{1}H NMR spectrum, 32 carbon signals and 48 protons are evident. By *J*-modulated spin-echo (*J*-mod) spectrum, it is possible to observe six CH_3, six CH_2, and seventeen CH signals in addition to three quaternary carbon atoms. By ^{1}H and ^{13}C NMR spectrum in methanol d_4 and consequently by observation of COSY and HMBC spectra, it may be deduced that six signals are indicative of methyl groups—22.9 (C-30), 20.7 (C-32), 20.3 (C-31), 17.9 (C-27), 13.5 (C-28), 11.7 (C-26) ppm; ten carbon signals are indicative of olefinic carbons—140.0 (C-19), 139.0 (C-16), 136.3 (C-12), 136.0 (C-9), 135.9 (C-18), 135.6 (C-4), 134.1 (C-8), 129.9 (C-17), 128.9 (C-5), 128.8 (C-13) ppm—and five peaks signify carbons with a carbon–oxygen single bond: 83.7 (C-15), 73.4 (C-11), 73.7 (C-7), 70.3 (C-3), 61.2 (C-29) ppm. The last one is characteristic of primary alcohol. The others are secondary hydroxyl functions. The deshielding of the carbon C-1 at 175.3 ppm suggests the presence of carbonyl function of carboxylic acid. The methylene group H-2 and H-2′ at 2.47 ppm is correlated in HMBC with acid carbon C-1 (175.3 ppm). The methylene group at 2.47 ppm is also correlated with two other signals in HMBC: C-3 (70.3 ppm) and C-4 (135.6 ppm). An H-C coupling was observed between H-4 and C-5, corresponding to one double bond. In the ^{1}H NMR spectra, we observed a section comprising two repeated 4 carbon units: a methylene (H-6 and H-10), a hydroxymethine (H-7 and H-11) and a disubstituted double bond (H-8/H-9 and H-12/H-13). In this section a set of twinned NMR signals are present. The data suggest the presence of 3 double-bonded carbons with a coupling constant of 15 Hz (H-4/H-5), (H-8/H-9), (H-12/H-13) that confirm the configuration trans. Two trisubstituted unsaturated bonds were observed, C-16/C-17 and C-18/C-19. One of the methyl groups is located on the second trisubstituted unsaturated bonds (d, 1.77 ppm, CH3-28 at C-16). By COSY, it was possible to observe the correlation between CH3-28 and H-17. The primary alcohol CH_2OH (AB system for H-29 and H-29′ at C-18) is located on one of the two trisubstituted unsaturated bonds. The NMR characterization was also reported in A. Khiralla's work [14].

The NMR characterization was also realized using pyridine d_5 as solvent due to signal overlap and the high degree of functionalization. By the analysis, the presence may be deduced of 32 signals of carbons in the ^{13}C NMR spectrum in pyridine d_5 and 53 protons in the ^1H NMR. From ^1H NMR spectrum of compound **1** in pyridine d_5, it was possible to observe a good separation in the region of double bonds (6.4 ppm–5.9 ppm). We can clearly determine the presence of five methyl groups (doublets respectively a 2.17, 1.16, 1.06, 0.91, 0.84 ppm) and one other CH_3 (triplet at 0.85 ppm). Four methylene protons were distinguished at 2.58 ppm and 2.64 ppm. All the data are shown in Table 2 as reported in [14].

Table 2. NMR data of pure compound **1**, kheiric acid.

Position	**Methanol d_4**		**Pyridine d_5**	
	^{13}C δ_C	**^1H δ_H (*J* in Hz)**	**^{13}C δ_C**	**^1H δ_H (*J* in Hz)**
C-1	175.3	-	175.0	-
C-2	43.5	2.47 m (2H, H-2, H-2′)	44.6	3.05 dd (1H, *J* = 15 Hz, *J* = 8.5 Hz, H-2′) 2.96 dd (1H, *J* = 15 Hz, *J* = 5 Hz, H-2)
C-3	70.3	4.47 m (1H, H-3)	69.9	5.18 m (1H, H-3)
C-4	135.6	5.59 dd (1H, H-4, *J* = 6.6 Hz, 18 Hz)	136.5	6.10 dd (1H, H-4, *J* = 6 Hz, 15 Hz)
C-5	128.9	5.73 dd (1H, H-5, *J* = 7 Hz, 15 Hz)	128.0	6.32 dd (1H, H-5, *J* = 7 Hz, 15 Hz)
C-6	41.4	2.26 m (2H, H-6, H-6′)	42.0	2.58 m (2H, H-6, H-6′)
C-7	73.7	4.05 m (1H, H-7)	72.7	4.52 m (1H, H-7)
C-8	134.1	5.48 dd (1H, H-8, *J* = 7 Hz, 15 Hz)	137.0	5.97 dd (1H, H-8, *J* = 6 Hz, 15 Hz)
C-9	136.0	5.64 dd (1H, H-9, *J* = 7 Hz, 15 Hz)	127.9	6.21 dd (1H, H-9, *J* =7.6 Hz, 15 Hz)
C-10	41.5	2.23 m (2H, H-10, H-10′)	42.1	2.64 m (2H, H-10, H-10′)
C-11	73.4	4.03 m (1H, H-11)	72.9	4.51 m (1H, H-11)
C-12	136.3	5.54 dd (1H, H-12, *J* = 7 Hz, 18 Hz)	134.8	5.98 dd (1H, H-12, *J* = 6 Hz, 15 Hz)
C-13	128.8	5.70 dd (1H, H-13, *J* = 7 Hz, 15 Hz)	134.5	6.27 dd (1H, H-13, *J* = 7.6 Hz, 15 Hz)
C-14	41.6	2.35 m (1H, H-14)	41.35	2.72 m (1H, H-14)
C-15	83.7	3.72 d (1H, H-15, *J* = 8.6 Hz)	82.8	4.18 d (1H, *J* = 8 Hz, H-15)
C-16	139.0	-	139.5	-
C-17	129.9	5.92 s (1H, H-17)	128.9	6.51 bs (1H, H-17)
C-18	135.9	-	137.1	-
C-19	140.0	5.14 d (1H, H-19, *J* = 10 Hz)	138.3	5.40 d (1H, *J* = 10 Hz, H-19)
C-20	31.4	2.70 m (1H, H-20)	30.7	2.91 m (1H, H-20)

Table 2. *Cont.*

Position	Methanol d_4		Pyridine d_5	
	^{13}C δ_C	^{1}H δ_H (*J* in Hz)	^{13}C δ_C	^{1}H δ_H (*J* in Hz)
C-21	46.7	1.30 m (1H, H-21) 1.05 m (1H, H-21′)	45.9	1.42 m (1H, H-21′) 1.07 m (1H, H-21)
C-22	32.9	1.43 m (1H, H-22)	32.0	1.43 m (1H, H-22)
C-23	46.8	1.20 m (1H, H-23) 0.95 m (1H, H-23′)	45.9	1.22 m (1H, H-23′) 0.95 m (1H, H-23)
C-24	29.5	1.52 m (1H, H-24)	28.7	1.66 m (1H, H-24)
C-25	30.5	1.09 m (1H, H-25) 1.39 m (1H, H-25′)	29.9	1.29 m (1H, H-25′) 1.07 m (1H, H-25)
C-26	11.7	0.86 t (3H, CH3, H-26, *J* = 7 Hz)	11.7	0.85 t (3H, CH_3, *J* = 7 Hz, C-26)
C-27	17.9	0.89 d (3H, CH3, H-27, *J* = 7 Hz)	18.4	1.16 d (3H, CH_3, *J* = 6.6 Hz, C-27)
C-28	13.5	1.77 d (3H, CH3, H-28, *J* = 1.3 Hz)	14.2	2.17 d (3H, CH_3, *J* = 0.9 Hz, C-28)
C-29	61.2	4.16 AB system, d (1H, H-29, *J* = 12 Hz) 4.16 AB system, d (1H, H-29′, *J* = 12 Hz)	61.0	4.60 d (1H, *J* AB = 12 Hz, H-29′) 4.58 d (1H, *J* AB = 12 Hz, H-29)
C-30	22.9	0.99 d (3H, CH3, H-30, *J* = 6.6 Hz)	23.1	1.06 d (3H, CH_3, *J* = 6.6 Hz, C-30)
C-31	20.3	0.84 d (3H, CH3, H-31, *J* = 6.6 Hz)	20.1	0.84 d (3H, CH_3, *J* = 6 Hz, C-31)
C-32	20.7	0.87 d (3H, CH3, H-32, *J* = 6.6 Hz)	20.8	0.91 d (3H, CH_3, *J* = 6.6 Hz, C-32)
			-	4.98 broad m (5H, 5-OH)

Chemical shifts (δ) are in parts per million (ppm); coupling constants (*J*) are in hertz (Hz).

The analysis and the original NMR spectra are provided in supporting information (Figures S5–S13).

The presence of five double bonds and one insaturation of carboxylic acid suggests that the structure is a long chain. The compound **1** is a polyfunctionalized long-chain carboxylic acid.

The isolated compound was identified as 3,7,11,15-tetrahydroxy-18-hydroxymethyl-14,16,20,22,24-pentamethyl-hexacosa 4E,8E,12E,16,18-pentaenoic acid [14], which is named kheiric acid (Figure 6). Taking into account all the spectroscopic results, the structure is a new compound and has never been described in the literature.

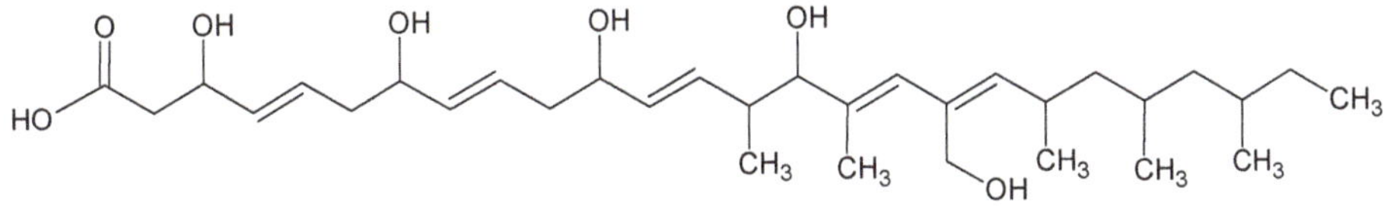

Figure 6. Chemical structure of the new polyhydroxyacid, kheiric acid (compound **1**).

The analysis of the fraction 6 of *C. papendorfii* endophytic fungus extract revealed the presence of mannitol (compound **2**). A clear crystalline solid (5 mg) was obtained. HRESIMS (positive mode) *m/z* 205.0691 $[M+Na]^+$ (calcd. for C6H14O6Na, 205.0688) (Figure S15). ^{1}H NMR (400 MHz, DMSO d_6):

δ (ppm) 4.42 (d, J = 5.4 Hz), 4.34 (t, J = 5.8 Hz), 4.15 (d, J = 7.1 Hz), 3.63 (m) 3.58 (m), 3.49 (m), 3.39 (m). ^{13}C NMR (100 MHz, DMSO d_6): δ (ppm) 71.4, 69.8, 63.9. The ^{1}H and ^{13}C spectra and bidimensional NMR were compared with original standard and with the literature (see supporting information, Figures S16–S20). The chemical structure is presented in Figure 7. This compound is a polyol extracted naturally from plants, bacteria and fungi [49,50]. Mannitol is a chemical metabolite used currently in the field of food and drugs [51].

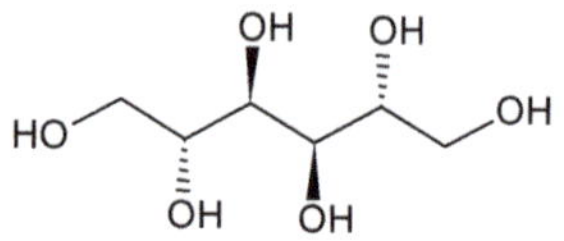

Figure 7. Chemical structure of mannitol (compound **2**).

The GC–MS analysis of the fractions and subfractions led to the identification of twenty-two compounds (Table 3).

Table 3. Metabolites identified using GC–MS analysis.

Fraction Name	Compound Number	Compound Name	Formula	Molecular Weight	Calculated Retention Index	Retention Time (min)	Similarity (%)
F1	1	2,4-Di-tert-butylphenol	C14H22O	206	1447	14.21	96
F1	2	1-Octadecene	C18H36	252	1730	17.610	96
F1	3	Undec-10-ynoic acid	C11H18O2	182	2262	22.797	80
F1	4	Benzenepropanoic acid, 3,5-bis(1,1-dimethylethyl)-4-hydroxy-, octadecyl ester	C35H62O3	530	3598	32.063	87
F3	5	Methyl palmitoleate	C17H34O2	270	1986	20.257	94
F3	6	Methyl linolelaidate	C19H34O2	294	2159	21.887	94
F3	7	Methyl petroselinate	C19H36O2	296	2166	21.947	85
F3	8	Methyl stearate	C19H38O2	298	2194	22.187	92
F10.C	9	1b,5,5,6a-Tetramethyl-octahydro-1-oxa-cyclopropa[a]inden-6-one	C13H20O2	208	2253	22.718	82
F10.C	10	4-Oxo-β-isodamascol	C13H20O2	208	2290	23.030	80
F10.C	11	Oleamide	C18H35NO	281	3188	29.548	92
F10.C	12	1,2-Octadecanediol	C18H38O2	286	3556	31.742	87
F10.C	13	Cyclopentadecanol	C15H30O	226	3796	33.665	94
F10.D	14	Dicyclohexane	C12H22	166	1269	11.733	94
F10.D	1	2,4-Di-tert-butylphenol	C14H22O	206	1450	14.300	89
F10.D	15	3-Dodecyl-2,5-furandione	C16H26O3	266	1825	18.630	86
F10.D	16	Methyl elaidate	C19H36O2	296	2031	20.680	94
F10.E	1	2,4-Di-tert-butylphenol	C14H22O	206	1450	14.297	91
F10.E	2	1-Octadecene	C18H36	252	1730	17.610	95
F10.E	17	n-Dodecenylsuccinic anhydride	C16H26O3	266	1832	18.697	86
F10.E	18	9-Eicosene	C20H40	280	2016	20.543	89
F10.E	19	11,14-Eicosadienoic acid, methyl ester	C21H38O2	322	2025	20.623	88
F10.E	20	Methyl oleate	C19H36O2	296	2031	20.683	93
Precipitate	21	Ascaridole	C10H16O2	168	2177	22.042	82
Precipitate	9	1b,5,5,6a-Tetramethyl-octahydro-1-oxa-cyclopropa[a]inden-6-one	C13H20O2	208	2253	22.718	82
Precipitate	22	1-Eicosanol	C20H42O	298	3557	31.750	90

After identification by GC–MS and comparison with NIST Mass Spectral Library, the chemical names are confirmed using PubChem [52].

The GC–MS analysis of the fraction F1 reveals the presence of four compounds: 2,4-di-tert-butylphenol, 1-octadecene, undec-10-ynoic acid and benzenepropanoic acid, 3,5-bis(1,1-dimethylethyl)-4-hydroxy-, octadecyl ester.

Fraction F3 contained a mixture of four fatty acids identified as methyl palmitoleate, methyl petroselinate, methyl stearate and methyl linolelaidate.

The identifications of five compounds were observed in fraction F10.C: 1b,5,5,6a-tetramethyl-octahydro-1-oxa-cyclopropa[a]inden-6-one, 4-oxo-β-isodamascol, oleic acid amide, 1,2-octadecanediol and cyclopentadecanol.

In fraction F10.D were identified four compounds: dicyclohexane, 3-dodecyl-2,5-furandione, methyl elaidate and 2,4-di-tert-butylphenol. This last compound was also found in the fraction F1.

In fraction F10.E, the following compounds were identified: n-dodecenylsuccinic anhydride, 9-eicosene, the methyl ester of 11,14-eicosadienoic acid, the methyl ester oleic acid and two compounds previously observed, the 2,4-di-tert-butylphenol present in the fractions F1 and F10.D and 1-octadecene, also present in the fraction F1.

Analysis of the fraction precipitate reveals three compounds: ascaridole, 1b,5,5,6a-tetramethyl-octahydro-1-oxa-cyclopropa[a]inden-6-one and 1-eicosanol.

The chemical structures are presented in Figure 8.

3.4. Biological Activities of Identified Compounds and Kheiric Acid

Some chemical compounds identified by GC–MS have shown some interesting biological properties. For example, undec-10-ynoic acid is an inhibitor of cytochrome P450 4A1 [53], and fatty acids such as methyl palmitoleate, methyl linolelaidate and methyl stearate showed antimicrobial activities against *Streptococcus mutans* and some human fungi pathogens [54]. The methyl petroselinate has antioxidant activity [55]. Ascaridol, a bicyclic monoterpenoid showed cytotoxicity against MDA MB-231 breast cancer [56]. The presence of these bioactive compounds in the crude extract of *Curvularia papendorfii* can justify its interesting biological activities as antiviral, antibacterial and antiproliferative properties.

In Table 4, the biological activities of endophytic fungus crude extract and pure kheiric acid are presented. Kheiric acid revealed moderate activity (MIC = 62.5 μg/mL) against Gram-positive MRSA and *S. aureus*. This result is promising because the isolated compound **1** has a better value than the crude extract, which showed an MIC of 312 μg/mL. No antiviral and antiproliferative activities of this compound against MCF7, HT29 or HCT116 cell lines were observed. This loss of activity could be related to the fractionation.

Some analogs of polyhydroxyacid kheiric acid were isolated from different fungal genera. Phomenoic acid ($C_{34}H_{58}O_8$) was purified from the mycelium of *Phoma lingam*; this compound showed moderate antifungal and antibacterial in vitro, in particular against *Candida albicans* [57,58]. Arthrinic acid ($C_{32}H_{54}O_9$) was isolated from the crude extract of the fungus *Arthrinium phaeospermum*. Arthrinic acid showed antifungal activity against *Botrytis cinerea*, *Rhizopus stolonifera* and *Diplodia pinea* as pathogens in horticulture [59].

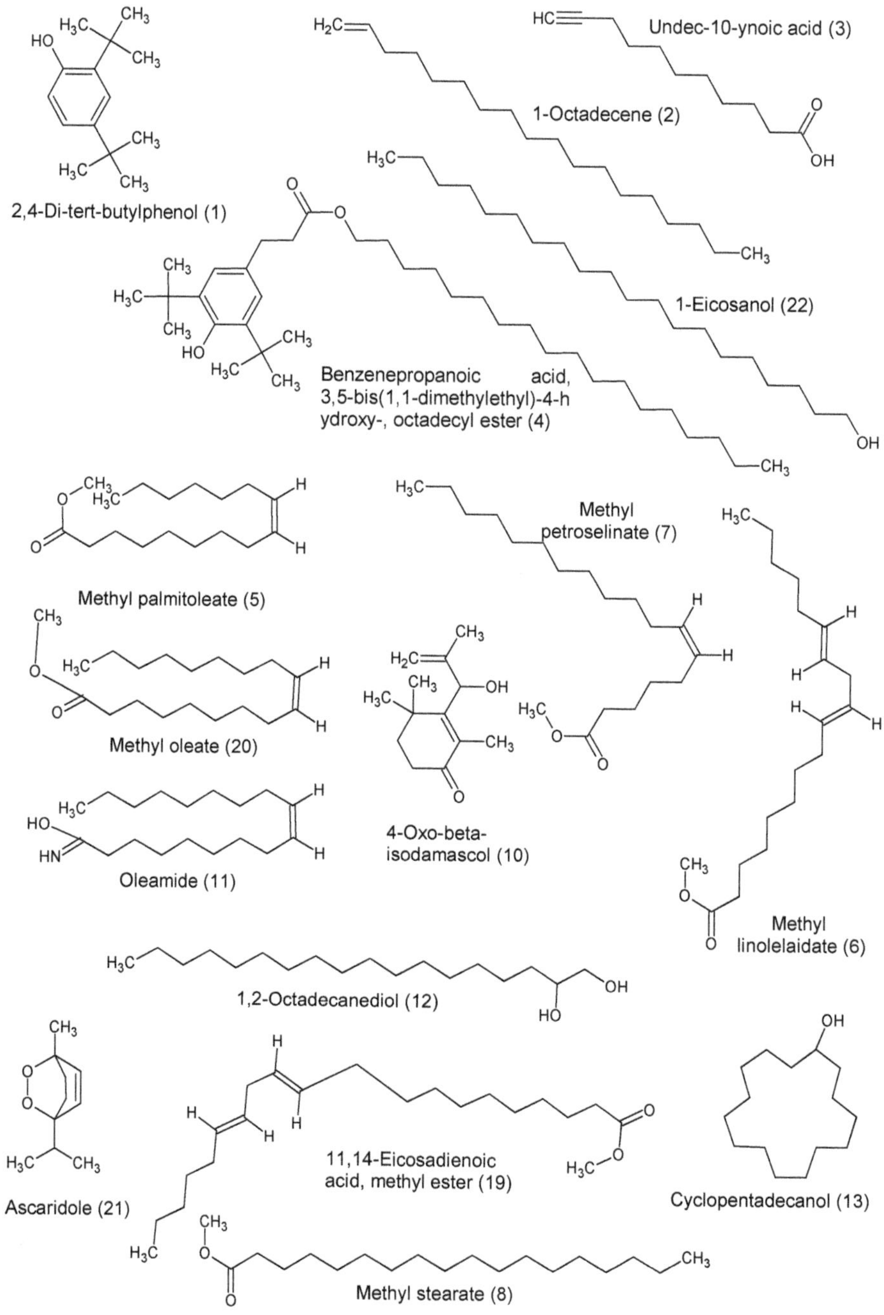

Figure 8. *Cont.*

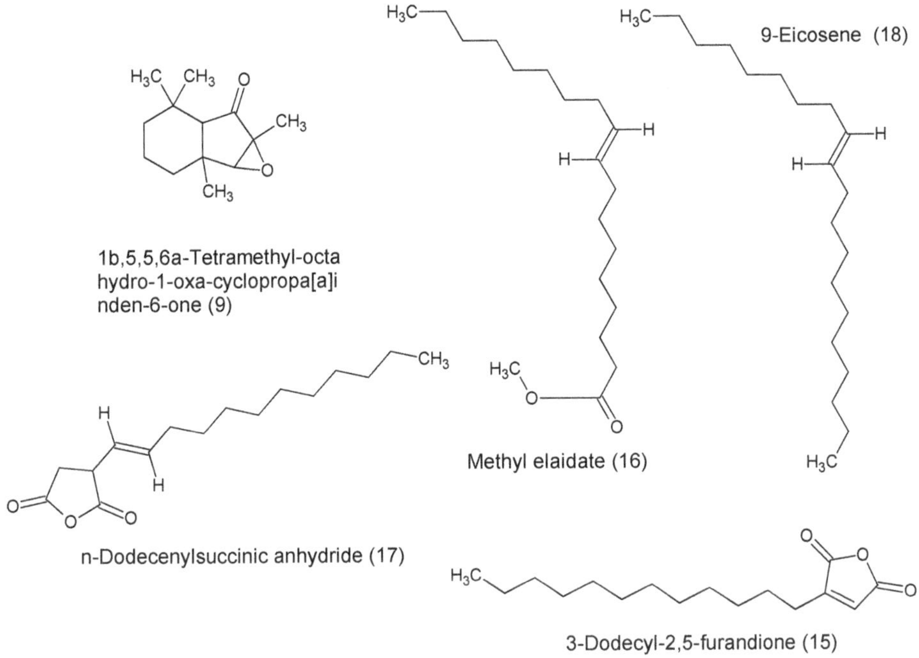

Figure 8. Chemical structures identified by GC–MS.

Table 4. Biological comparison data of *C. papendorfii* crude extract and pure kheiric acid.

Strains	Crude Extract of *C. papendorfii*	Kheiric Acid
Staphylococcus aureus	MIC = 312 μg/mL	MIC = 62.5 μg/mL
MRSA	MIC = 312 μg/mL	MIC = 62.5 μg/mL
HCoV 229E	15% of the reduction of the virus-induced cytopathogenic effect at MOI 0.001	No reduction of the virus-induced cytopathogenic effect
FCV F9	40% of the reduction of the virus-induced cytopathogenic effect at lower MOI 0.0001	No reduction of the virus-induced cytopathogenic effect
MCF7	IC_{50} = 21.5 ± 5.9 μg/mL	IC_{50} > 100 μg/mL
HT29	IC_{50} > 100 μg/mL	IC_{50} > 100 μg/mL
HCT116	IC_{50} > 100 μg/mL	IC_{50} > 100 μg/mL

4. Conclusions

C. papendorfii is an endophytic fungus associated with *V. amygdalina*, a Sudanese medicinal plant. In this study, the ethyl acetate crude extract of *C. papendorfii* was extensively studied. The reduction of 40% for water-solubilized extracts at a lower MOI of 0.0001 is an important result against the feline calicivirus FCV F9 because caliciviruses are generally much more resistant than coronaviruses. Until now, no treatment is present for this virus. In addition, this extract revealed interesting antiproliferative effects against human breast carcinoma MCF7 cell line (IC_{50} = 21.5 ± 5.9 μg/mL) and an MIC value of 312 μg/mL against meticillin-resistant *Staphylococcus aureus* (MRSA). A total of twenty-four chemical structures were identified. The major compound isolated from the ethyl acetate crude extract of *C. papendorfii* was a new compound, a polyhydroxyacid, kheiric acid (3,7,11,15-tetrahydroxy-18-hydroxymethyl-14,16,20,22,24-pentamethyl-hexacosa 4E,8E,12E,16,18-pentaenoic acid). This compound revealed moderate antibacterial activity against MRSA and *S. aureus* with MIC value 62.5 µg/mL. Once more, these results confirm the importance of endophytic fungi as a source of new biomolecules.

Supplementary Materials: The following are available online at http://www.mdpi.com/2076-2607/8/9/1353/s1, Figure S1: Response-dose curves established from results obtained by MTT assays after MCF7, HT29 and HCT116 cells exposure of *C. papendorfii*; Figure S2: Chemical structure of compound **1**; Figure S3: IR spectrum of compound **1**; Figure S4: HR-ESI-MS spectrum of compound **1**; Figure S5: ^{1}H-NMR spectrum of compound **1** in CD_3OD (400 MHz); Figure S6: ^{13}C-NMR spectrum of compound **1** in CD_3OD (100 MHz); Figure S7: ^{1}H-^{1}H COSY spectrum of compound **1** in CD_3OD; Figure S8: HSQC spectrum of compound **1** in CD_3OD; Figure S9: HMBC spectrum of compound **1** in CD_3OD; Figure S10: Zoom of HMBC spectrum of compound **1** in CD_3OD; Figure S11: Zoom of HMBC spectrum of compound **1** in CD_3OD; Figure S12: ^{1}H-NMR spectrum of compound **1** in pyridine d_5 (400 MHz); Figure S13: ^{13}C-NMR spectrum of compound **1** in pyridine d_5 (100 MHz); Figure S14: Chemical structure of compound **2**; Figure S15: HR-ESI-MS spectrum of compound **2**; Figure S16: ^{1}H-NMR spectrum of compound **2** in DMSO d_6 (400 MHz); Figure S17: ^{13}C-NMR spectrum of compound **2** in DMSO d_6 (100 MHz); Figure S18: ^{1}H-^{1}H COSY spectrum of compound **2**; Figure S19: HSQC spectrum of compound **2**; Figure S20: HMBC spectrum of compound **2**.

Author Contributions: Conceptualization, A.K., R.S., I.M., S.M.Y. and D.L.-M.; methodology, A.K., R.S., M.V., S.P., P.L., P.A., S.S.-D. and D.L.-M.; formal analysis, A.K., R.S., M.V., S.P., P.L., S.S.-D. and P.A.; investigation, A.K., R.S., M.V., S.P. and D.L.-M.; resources, A.K., R.S., D.L.-M. and S.M.Y.; data curation, A.K., R.S., M.V., S.P. and D.L.-M.; writing—original draft preparation, A.K., R.S., M.V., S.P. and D.L.-M.; writing—review and editing, A.K., R.S., M.V., S.P., P.L., S.S.-D., P.A., I.M., S.M.Y. and D.L.-M.; supervision, R.S., D.L.-M. and S.M.Y.; project administration, A.K., R.S., D.L.-M. and S.M.Y.; funding acquisition, A.K., R.S., D.L.-M. and S.M.Y. All authors have read and agreed to the published version of the manuscript.

Funding: This research was funded by L2CM laboratory and Institut Jean Barriol in University of Lorraine. The APC was funded by L2CM.

Acknowledgments: The authors acknowledge: the French Embassy in Khartoum and the Foreign Scholarships office at the Ministry of Higher Education and Scientific Research in Sudan for the financial support of scholarship of A.K.; the support of "Laboratoire Lorrain de Chimie Moléculaire L2CM UMR 7053" by the "Impact Biomolecules" project of the "Lorraine Université d'Excellence" (Investissements d'avenir—ANR); this research was also supported by the MEAE and MESRI PHC NAPATA project (N° 43443YL). François Dupire from Platform of Mass Spectrometry in UMR 7053-L2CM team, Hervé Schohn and Tzvetomira Tzanova. R.S. acknowledges Francoise Chrétien and Brigitte Fernette for their help in identification of compound **1**.

Conflicts of Interest: The authors declare no conflict of interest.

References

1. Pene, F.; Merlat, A.; Vabret, A.; Rozenberg, F.; Buzyn, A.; Dreyfus, F.; Cariou, A.; Freymuth, F.; Lebon, P. Coronavirus 229E-related pneumonia in immunocompromised patients. *Clin. Infect. Dis.* **2003**, *37*, 929–932. [CrossRef] [PubMed]
2. Lim, Y.X.; Ng, Y.L.; Tam, J.P.; Liu, D.X. Human Coronaviruses: A Review of Virus-Host Interactions. *Diseases* **2016**, *4*, 26. [CrossRef] [PubMed]
3. Woo, P.C.Y.; Lau, S.K.P.; Huang, Y.; Yuen, K.-Y. Coronavirus diversity, phylogeny and interspecies jumping. *Exp. Biol. Med.* **2009**, *234*, 1117–1127. [CrossRef] [PubMed]
4. Peiris, J.S.M.; Guan, Y.; Yuen, K.Y. Severe acute respiratory syndrome. *Nat. Med.* **2004**, *10*, S88–S97. [CrossRef]
5. Zaki, A.M.; van Boheemen, S.; Bestebroer, T.M.; Osterhaus, A.D.M.E.; Fouchier, R.A.M. Isolation of a novel coronavirus from a man with pneumonia in Saudi Arabia. *N. Engl. J. Med.* **2012**, *367*, 1814–1820. [CrossRef]
6. Bartsch, S.; Lopman, B.A.; Ozawa, S.; Hall, A.J.; Lee, B. Global economic burden of norovirus gastroenteritis. *PLoS ONE* **2016**, *11*, e0151219. [CrossRef]
7. Glass, R.I.; Parashar, U.D.; Estes, M.K. Norovirus gastroenteritis. *N. Engl. J. Med.* **2009**, *361*, 1776–1785. [CrossRef]
8. Karst, S.M.; Wobus, C.E.; Goodfellow, I.G.; Green, K.Y.; Virgin, H.W. Advances in norovirus biology. *Cell Host Microbe* **2014**, *15*, 668–680. [CrossRef]
9. WHO. General Documents on Antimicrobial Resistance. Available online: http://www.who.int/antimicrobial-resistance/publications/general-documents/en/ (accessed on 22 April 2020).

10. Kirienko, N.V.; Rahme, L.; Cho, Y.-H. *Beyond Antimicrobials: Non-Traditional Approaches to Combating Multidrug-Resistant Bacteria*; Frontiers Media SA: Lausanne, Switzerland, 2019; ISBN 9782889632565.
11. Turner, N.A.; Sharma-Kuinkel, B.K.; Maskarinec, S.A.; Eichenberger, E.M.; Shah, P.P.; Carugati, M.; Holland, T.L.; Fowler, V.G. Methicillin-resistant *Staphylococcus aureus*: An overview of basic and clinical research. *Nat. Rev. Microbiol.* **2019**, *17*, 203–218. [CrossRef]
12. Khiralla, A.; Spina, R.; Saliba, S.; Laurain-Mattar, D. Diversity of natural products of the genera Curvularia and Bipolaris. *Fungal Biol. Rev.* **2019**, *33*, 101–122. [CrossRef]
13. Khiralla, A.; Mohamed, I.; Thomas, J.; Mignard, B.; Spina, R.; Yagi, S.; Laurain-Mattar, D. A pilot study of antioxidant potential of endophytic fungi from some Sudanese medicinal plants. *Asian Pac. J. Trop. Med.* **2015**, *8*, 701–704. [CrossRef] [PubMed]
14. Khiralla, A. Phytochemical Study, Cytotoxic and Antibacterial Potentialities of Endophytic Fungi from Medicinal Plants from Sudan. Ph.D. Thesis, University of Lorraine, Nancy, France, 2015.
15. Khiralla, A.; Mohamed, I.E.; Tzanova, T.; Schohn, H.; Slezack-Deschaumes, S.; Hehn, A.; André, P.; Carre, G.; Spina, R.; Lobstein, A.; et al. Endophytic fungi associated with Sudanese medicinal plants show cytotoxic and antibiotic potential. *FEMS Microbiol. Lett.* **2016**, *363*. [CrossRef] [PubMed]
16. Hyde, K.D.; Nilsson, R.H.; Alias, S.A.; Ariyawansa, H.A.; Blair, J.E.; Cai, L.; de Cock, A.W.A.M.; Dissanayake, A.J.; Glockling, S.L.; Goonasekara, I.D.; et al. One stop shop: Backbones trees for important phytopathogenic genera: I (2014). *Fungal Divers.* **2014**, *67*, 21–125. [CrossRef]
17. Avinash, K.S.; Ashwini, H.S.; Babu, H.N.R.; Krishnamurthy, Y.L. Antimicrobial Potential of Crude Extract of *Curvularia lunata*, an Endophytic Fungi Isolated from *Cymbopogon caesius*. Available online: https://www.hindawi.com/journals/jmy/2015/185821/ (accessed on 18 April 2020).
18. Campos, F.F.; Rosa, L.H.; Cota, B.B.; Caligiorne, R.B.; Rabello, A.L.T.; Alves, T.M.A.; Rosa, C.A.; Zani, C.L. Leishmanicidal metabolites from *Cochliobolus* sp., an endophytic fungus isolated from *Piptadenia adiantoides* (Fabaceae). *PLoS Negl. Trop. Dis.* **2008**, *2*, e348. [CrossRef]
19. Wells, J.M.; Cole, R.J.; Cutler, H.C.; Spalding, D.H. *Curvularia lunata*, a New Source of Cytochalasin B. *Appl. Environ. Microbiol.* **1981**, *41*, 967–971. [CrossRef]
20. Han, W.B.; Lu, Y.H.; Zhang, A.H.; Zhang, G.F.; Mei, Y.N.; Jiang, N.; Lei, X.; Song, Y.C.; Ng, S.W.; Tan, R.X. Curvulamine, a new antibacterial alkaloid incorporating two undescribed units from a *Curvularia* species. *Org. Lett.* **2014**, *16*, 5366–5369. [CrossRef]
21. Han, W.B.; Zhang, A.H.; Deng, X.Z.; Lei, X.; Tan, R.X. Curindolizine, an Anti-Inflammatory Agent Assembled via Michael Addition of Pyrrole Alkaloids Inside Fungal Cells. *Org. Lett.* **2016**, *18*, 1816–1819. [CrossRef] [PubMed]
22. Greve, H.; Schupp, P.J.; Eguereva, E.; Kehraus, S.; König, G.M. Ten-Membered Lactones from the Marine-Derived Fungus *Curvularia* sp. *J. Nat. Prod.* **2008**, *71*, 1651–1653. [CrossRef] [PubMed]
23. Greve, H.; Schupp, P.J.; Eguereva, E.; Kehraus, S.; Kelter, G.; Maier, A.; Fiebig, H.-H.; König, G.M. Apralactone A and a New Stereochemical Class of *Curvularins* from the Marine-Derived Fungus *Curvularia* sp. *Eur. J. Org. Chem.* **2008**, *2008*. [CrossRef] [PubMed]
24. Liu, Q.-A.; Shao, C.-L.; Gu, Y.-C.; Blum, M.; Gan, L.-S.; Wang, K.-L.; Chen, M.; Wang, C.-Y. Antifouling and Fungicidal Resorcylic Acid Lactones from the Sea Anemone-Derived Fungus *Cochliobolus lunatus*. *J. Agric. Food Chem.* **2014**, *62*, 3183–3191. [CrossRef]
25. Jadulco, R.; Brauers, G.; Edrada, R.A.; Ebel, R.; Wray, V.; Sudarsono; Proksch, P. New Metabolites from Sponge-Derived Fungi *Curvularia lunata* and *Cladosporium herbarum*. *J. Nat. Prod.* **2002**, *65*, 730–733. [CrossRef] [PubMed]
26. Van Eijk, G.W.; Roeymans, H.J. Cynodontin, the tetrahydroxyanthraquinone of *Curvularia* and *Drechslera* species. *Experientia* **1977**, *33*, 1283–1284. [CrossRef]
27. Bills, G.F.; Peláez, F.; Polishook, J.D.; Diez-Matas, M.T.; Harris, G.H.; Clapp, W.H.; Dufresne, C.; Byrne, K.M.; Nallin-Omstead, M.; Jenkins, R.G.; et al. Distribution of zaragozic acids (squalestatins) among filamentous ascomycetes. *Mycol. Res.* **1994**, *98*, 733–739. [CrossRef]
28. Mosmann, T. Rapid colorimetric assay for cellular growth and survival: Application to proliferation and cytotoxicity assays. *J. Immunol. Methods* **1983**, *65*, 55–63. [CrossRef]
29. Reed, L.J.; Muench, H.; Muench, H.A.; Reed, L.I.; Reed, L.I.; Reed, L.J. A simple method of estimating fifty percent endpoints. *Am. J. Hyg.* **1938**, *27*, 493–497.

30. Thabti, I.; Albert, Q.; Philippot, S.; Dupire, F.; Westerhuis, B.; Fontanay, S.; Risler, A.; Kassab, T.; Elfalleh, W.; Aferchichi, A.; et al. Advances on Antiviral Activity of *Morus spp.* Plant Extracts: Human Coronavirus and Virus-Related Respiratory Tract Infections in the Spotlight. *Molecules* **2020**, *25*, 1876. [CrossRef]
31. Feoktistova, M.; Geserick, P.; Leverkus, M. Crystal Violet Assay for Determining Viability of Cultured Cells. *Cold Spring Harb. Protoc.* **2016**. [CrossRef]
32. Elmi, A.; Spina, R.; Abdoul-Latif, F.; Yagi, S.; Fontanay, S.; Risler, A.; Duval, R.E.; Laurain-Mattar, D. Rapid screening for bioactive natural compounds in *Indigofera caerulea* Rox fruits. *Ind. Crops Prod.* **2018**, *125*, 123–130. [CrossRef]
33. Elmi, A.; Spina, R.; Risler, A.; Philippot, S.; Mérito, A.; Duval, R.E.; Abdoul-latif, F.M.; Laurain-Mattar, D. Evaluation of Antioxidant and Antibacterial Activities, Cytotoxicity of *Acacia seyal* Del Bark Extracts and Isolated Compounds. *Molecules* **2020**, *25*, 2392. [CrossRef]
34. Gamboa, M.A.; Bayman, P. Communities of Endophytic Fungi in Leaves of a Tropical Timber Tree (*Guarea guidonia*: Meliaceae). *Biotropica* **2001**, *33*, 352–360. [CrossRef]
35. Busi, S.; Peddikotla, P.; Yenamandra, V. Secondary Metabolites of *Curvularia oryzae* MTCC 2605. *Rec. Nat. Prod.* **2009**, *3*, 204–208.
36. Kharwar, R.N.; Verma, V.C.; Strobel, G.; Ezra, D. The endophytic fungal complex of *Catharanthus roseus* (L.) G. Don. *Curr. Sci.* **2008**, *95*, 228–233.
37. Shen, X.; Cheng, Y.-L.; Cai, C.; Fan, L.; Gao, J.; Hou, C.-L. Diversity and Antimicrobial Activity of Culturable Endophytic Fungi Isolated from Moso Bamboo Seeds. *PLoS ONE* **2014**. [CrossRef] [PubMed]
38. Samanthi, K.A.U.; Wickramaarachchi, S.; Wijeratne, E.M.K.; Paranagama, P.A. Two new antioxidant active polyketides from *Penicillium citrinum*, an endolichenic fungus isolated from *Parmotreama* species in Sri Lanka. *J. Natl. Sci. Found. Sri Lanka* **2015**, *43*, 119–126. [CrossRef]
39. Geetha, V.; Venkatachalam, A.; Suryanarayanan, T.S.; Doble, M. Isolation and Characterization of New Antioxidant and Antibacterial Compounds from Algicolous Marine Fungus *Curvularia Tuberculata*. In Proceedings of the International Conference on Bioscience, Biochemistry and Bioinformatics (IPCBEE), Hyderabad, India, 16–18 December 2011; IACSIT Press: Singapore, 2011; Volume 5.
40. Bradburne, A.F. An investigation of the replication of coronaviruses in suspension cultures of L132 cells. *Arch. Gesamte Virusforsch.* **1972**, *37*, 297–307. [CrossRef]
41. Robilotti, E.; Deresinski, S.; Pinsky, B.A. Norovirus. *Clin. Microbiol. Rev.* **2015**, *28*, 134–164. [CrossRef] [PubMed]
42. Linnakoski, R.; Reshamwala, D.; Veteli, P.; Cortina-Escribano, M.; Vanhanen, H.; Marjomäki, V. Antiviral Agents from Fungi: Diversity, Mechanisms and Potential Applications. *Front. Microbiol.* **2018**, *9*, 2325. [CrossRef]
43. Kramer, A.; Schwebke, I.; Kampf, G. How long do nosocomial pathogens persist on inanimate surfaces? A systematic review. *BMC Infect. Dis.* **2006**, *6*, 130. [CrossRef]
44. Marín, L.; Miguélez, E.M.; Villar, C.J.; Lombó, F. Bioavailability of Dietary Polyphenols and Gut Microbiota Metabolism: Antimicrobial Properties. *BioMed Res. Int.* **2015**, *2015*. [CrossRef]
45. Phongpaichit, S.; Nikom, J.; Rungjindamai, N.; Sakayaroj, J.; Hutadilok-Towatana, N.; Rukachaisirikul, V.; Kirtikara, K. Biological activities of extracts from endophytic fungi isolated from Garcinia plants. *FEMS Immunol. Med. Microbiol.* **2007**, *51*, 517–525. [CrossRef]
46. CASFM. *EUCAST 2019*; La Société Française de Microbiologie: Paris, France, 2019.
47. Rani, R.; Sharma, D.; Chaturvedi, M.; Parkash Yadav, J. Antibacterial Activity of Twenty Different Endophytic Fungi Isolated from *Calotropis procera* and Time Kill Assay. *Clin. Microbiol. Open Access* **2017**, *6*. [CrossRef]
48. Kaaniche, F.; Hamed, A.; Abdel-Razek, A.S.; Wibberg, D.; Abdissa, N.; Euch, I.Z.E.; Allouche, N.; Mellouli, L.; Shaaban, M.; Sewald, N. Bioactive secondary metabolites from new endophytic fungus *Curvularia*. sp isolated from *Rauwolfia macrophylla*. *PLoS ONE* **2019**, *14*, e0217627. [CrossRef]
49. Vélëz, H.; Glassbrook, N.J.; Daub, M.E. Mannitol metabolism in the phytopathogenic fungus *Alternaria alternata*. *Fungal Genet. Biol.* **2007**, *44*, 258–268. [CrossRef]
50. Patel, T.K.; Williamson, J.D. Mannitol in Plants, Fungi, and Plant-Fungal Interactions. *Trends Plant Sci.* **2016**, *21*, 486–497. [CrossRef]
51. Song, S.H.; Vieille, C. Recent advances in the biological production of mannitol. *Appl. Microbiol. Biotechnol.* **2009**, *84*, 55–62. [CrossRef]
52. PubChem. Available online: https://pubchem.ncbi.nlm.nih.gov/ (accessed on 17 August 2020).

53. Lenart, J.; Pikuła, S. 10-Undecynoic acid, an inhibitor of cytochrome P450 4A1, inhibits ethanolamine-specific phospholipid base exchange reaction in rat liver microsomes. *Acta Biochim. Pol.* **1999**, *46*, 203–210. [CrossRef]
54. Huang, C.B.; George, B.; Ebersole, J.L. Antimicrobial activity of n-6, n-7 and n-9 fatty acids and their esters for oral microorganisms. *Arch. Oral Biol.* **2010**, *55*, 555–560. [CrossRef]
55. Dadwal, V.; Agrawal, H.; Sonkhla, K.; Joshi, R.; Gupta, M. Characterization of phenolics, amino acids, fatty acids and antioxidant activity in pulp and seeds of high altitude Himalayan crab apple fruits (*Malus baccata*). *J. Food Sci. Technol.* **2018**, *55*, 2160–2169. [CrossRef]
56. Degenhardt, R.T.; Farias, I.V.; Grassi, L.T.; Franchi, G.C.; Nowill, A.E.; Bittencourt, C.M.D.S.; Wagner, T.M.; de Souza, M.M.; Cruz, A.B.; Malheiros, A. Characterization and evaluation of the cytotoxic potential of the essential oil of *Chenopodium ambrosioides*. *Rev. Bras. Farmacogn.* **2016**, *26*, 56–61. [CrossRef]
57. Devys, M.; Férézou, J.-P.; Topgi, R.S.; Barbier, M.; Bousquet, J.-F.; Kollmann, A. Structure and biosynthesis of phomenoic acid, an antifungal compound isolated from Phoma lingam Tode. *J. Chem. Soc. Perkin Trans.* **1984**, *1*, 2133–2137. [CrossRef]
58. Topgi, R.S.; Devys, M.; Bousquet, J.F.; Kollmann, A.; Barbier, M. Phomenoic acid and phomenolactone, antifungal substances from *Phoma lingam* (Tode) Desm.: Kinetics of their biosynthesis, with an optimization of the isolation procedures. *Appl. Environ. Microbiol.* **1987**, *53*, 966–968. [CrossRef]
59. Bloor, S. Arthrinic Acid, a Novel Antifungal Polyhydroxyacid from *Arthrinium phaeospermum*. *J. Antibiot.* **2008**, *61*, 515–517. [CrossRef]

microorganisms

MDPI

Article

Comparative Transcriptomic Analysis Uncovers Genes Responsible for the DHA Enhancement in the Mutant *Aurantiochytrium* sp.

Liangxu Liu [1,2,3], Zhangli Hu [1,2,3], Shuangfei Li [1,2,3], Hao Yang [1,2,3], Siting Li [1,2,3], Chuhan Lv [1,2,3], Madiha Zaynab [1,2,3], Christopher H. K. Cheng [4], Huapu Chen [5] and Xuewei Yang [1,2,3,*]

1 Guangdong Technology Research Center for Marine Algal Bioengineering, Guangdong Key Laboratory of Plant Epigenetics, College of Life Sciences and Oceanography, Shenzhen University, Shenzhen 518060, China; liangxuliu@icloud.com (L.L.); huzl@szu.edu.cn (Z.H.); sfli@szu.edu.cn (S.L.); 3718aaa@163.com (H.Y.); siting.li@aut.ac.nz (S.L.); LVCHU0510@163.com (C.L.); 15820761674@163.com (M.Z.)

2 Shenzhen Key Laboratory of Marine Biological Resources and Ecology Environment, Shenzhen Key Laboratory of Microbial Genetic Engineering, College of Life Sciences and Oceanography, Shenzhen University, Shenzhen 518055, China

3 Longhua Innovation Institute for Biotechnology, Shenzhen University, Shenzhen 518060, China

4 School of Biomedical Sciences, The Chinese University of Hong Kong, Hong Kong 999077, China; chkcheng@cuhk.edu.hk

5 Guangdong Research Center on Reproductive Control and Breeding Technology of Indigenous Valuable Fish Species, Fisheries College, Guangdong Ocean University, Zhanjiang 524088, China; chpsysu@hotmail.com

* Correspondence: yangxw@szu.edu.cn

Received: 26 February 2020; Accepted: 5 April 2020; Published: 7 April 2020

Abstract: Docosahexaenoic acid (DHA), a *n*-3 long-chain polyunsaturated fatty acid, is critical for physiological activities of the human body. Marine eukaryote *Aurantiochytrium* sp. is considered a promising source for DHA production. Mutational studies have shown that ultraviolet (UV) irradiation (50 W, 30 s) could be utilized as a breeding strategy for obtaining high-yield DHA-producing *Aurantiochytrium* sp. After UV irradiation (50 W, 30 s), the mutant strain X2 which shows enhanced lipid (1.79-fold, 1417.37 mg/L) and DHA (1.90-fold, 624.93 mg/L) production, was selected from the wild *Aurantiochytrium* sp. Instead of eicosapentaenoic acid (EPA), 9.07% of docosapentaenoic acid (DPA) was observed in the mutant strain X2. The comparative transcriptomic analysis showed that in both wild type and mutant strain, the fatty acid synthesis (FAS) pathway was incomplete with key desaturases, but genes related to the polyketide synthase (PKS) pathway were observed. Results presented that mRNA expression levels of *CoAT, AT, ER, DH,* and *MT* down-regulated in wild type but up-regulated in mutant strain X2, corresponding to the increased intercellular DHA accumulation. These findings indicated that *CoAT*, *AT*, *ER*, *DH*, and *MT* can be exploited for high DHA yields in *Aurantiochytrium*.

Keywords: Docosahexaenoic acid (DHA); polyunsaturated fatty acids (PUFAs); mutant strain; *Aurantiochytrium* sp.; transcriptome

1. Introduction

Owing to the importance of cell membrane function and numerous cellular processes for maintaining health, long-chain polyunsaturated fatty acids (LC-PUFAs) have attracted increasing attention for human health. LC-PUFAs can be classified into two principal families, namely, omega-3 (*n*-3) and omega-6 (*n*-6) fatty acids (FAs) [1]. The typical *n*-3 LC-PUFAs are docosahexaenoic acid (DHA) and eicosapentaenoic acid (EPA), which can strongly influence monocyte physiology. Previous studies have reported that DHA could potently inhibit platelet aggregation [2], reduce hemoglobin formation [3], treat cardiovascular diseases [4], and prevent osteoporosis [5]. Currently, fatty fish including sardines [6], *Oncorhynchus keta* [7], *Thunnus* [8], etc. is being used as the primary global supply for DHA [9]. However, the industry is severely limited by the original low levels and the instability of *n*-3 LC-PUFAs, which is caused by the fish variation, the climate, and high concentrations of the cholesterol [10]. Eggs naturally contain small amounts of DHA, but new DHA enriched eggs can contain up to 258.2 mg of DHA per egg [11]. Furthermore, marine microalgae including *chrysophyta* [12], *dinoflagellate* [13], and diatom [14], are also regarded as a promising alternative as the primary producer of the EPA and DHA in marine food webs.

Marine eukaryotes, such as *Thraustochytrids* [15] and *Schizochytrium*, with abundant FA contents, have emerged as promising producers of *n*-3 LC-PUFAs [16]. The FAs content required in the industry is currently 40–45 g/L, and the biomass required is 200 g/L [17]. The fermentation process of the *Schizochytrium* sp. SR 21 was optimized with bioreactor cultivation so that the DHA content doubled up to 66.72 ± 0.31% w/w total lipids (10.15 g/L of DHA concentration) [18]. Maximum DHA yield ($Y_{p/x}$) of 21.0% and 18.9% and productivity of 27.6 mg/L-h and 31.9 mg/L-h were obtained, respectively, in a 5 L bioreactor fermentation operated with optimal conditions and dual oxygen control strategy in *Schizochytrium* sp. [19]. Nevertheless, it is difficult for the wild-type (WT) strain to meet the requirements of industrial production due to the low biomass and *n*-3 LC-PUFA content, which accounts for the high cost of the downstream process [20]. Artificial mutagenesis has been applied to obtain high-yield DHA-producing strains for industrial fermentation. Ultraviolet (UV) radiation, a kind of non-ionizing radiation, causes gene mutation via maximum absorption by purines and pyrimidines present in DNA [21]. With UV irradiation, DHA percentage of the total fatty acids up to 43.65% was achieved using the mutant *Schizochytrium* sp. [22]. Therefore, UV radiation was used as a method for mutagenesis to obtain a *Schizochytrium* strain with a high yield of DHA. There is abundant research on the effects of salinity, pH, temperature, and media optimization on the DHA production. Nevertheless, the genome and transcriptome research of *Thraustochytrid* is still rarely reported. Transcriptome sequencing and comparative analysis of *Schizochytrium* mangrovei PQ6 at different cultivation times were presented by Hoang et al. [23]. Transcriptome analysis reveals that the up-regulation of the fatty acid synthase gene promotes the accumulation of DHA in *Schizochytrium* sp. S056 when glycerol is used [24]. Transcriptome and gene expression analysis of DHA producer *Aurantiochytrium* under low-temperature conditions were conducted by Ma et al. [25]. Zhu et al. Revealed the genome information of *Thraustochytrim* sp. [26].

De novo assembly of RNA-seq data serves as an important tool for studying the transcriptomes of "non-model" organisms without existing genome sequences [27]. Recently, transcriptome analysis has emerged as an essential method for the identification of genes involved in the secondary metabolites biosynthesis [28], such as the accumulation of fatty acids in the microalgae *Nannochloropsis* sp. [29], *Schizochytrium mangrovei* PQ6 [30], *Neochloris oleoabundans* [31], *Euglena gracilis* [32], and *Rhodomonas* sp. [33]. Recent research has indicated that DHA is synthesized by two distinct pathways in *Thraustochytrids*: The polyketide synthase (PKS) pathway and the fatty acid synthase (FAS) pathway [34]. Fatty acids are synthesized through the PKS pathway via highly repetitive cycles of four reactions, including condensation by ketoacyl synthase (KS), ketoreduction (KR), dehydration, and enoyl reduction (ER) [35]. Three large subunits of a type I PKS-like PUFA synthase in *Thraustochytrium* sp. 26185 have been identified [36]. According to the FAS pathway, small molecular carbon units can be polymerized to form chain fatty acids by fatty acids desaturases and elongases [37]. There are two families of desaturases, which are fatty acid desaturases (FADs) and stearoyl-coA desaturases (SCDs). Genomic and transcriptomic analysis revealed that both the FAS and PKS pathways of PUFA production were incomplete in *Thraustochytrids* strains [38]. The dehydratase and isomerase enzymes were not detected in the *Thraustochytrids* strain SZU445 [26]. Although FAD12, FAD4, and FAD5 have been reported in *Thraustochytrids*, some *Thraustochytrids* only contains the desaturase not belonging to the FAS pathway, such as FAD6 [39]. Previous research has illustrated that the DHA synthesis pathway in *Thraustochytrids* is different from the classic fatty acid metabolism pathway and remains ambiguous [40]. By comparing the transcriptome of wild type and the mutant, it could help us to elucidate the genes involved in the fatty acid enhancement and provide valuable information for clarifying the DHA synthesis pathway.

In this study, UV mutagenesis was utilized to obtain competitive *Aurantiochytrium* sp. strain with enhanced biomass and DHA production. The key genes related to the increasing DHA accumulation were explored by comparing the transcriptome between the mutant and the parent strain.

2. Materials and Methods

2.1. Microbial Cultivation

Aurantiochytrium sp. PKU#Mn16 were previously isolated from mangrove (22°31′13.044″ N, 113°56′56.560″ E) from coastal waters in Southern China, and then maintained in the China General Microbiological Culture Collection Center (CGMCC). *Aurantiochytrium* sp. PKU#Mn16 was inoculated into M4 liquid medium made with 100% filtered natural seawater (from Mirs Bay in Shenzhen, China) containing glucose (2.00%), yeast extract (0.10%), peptone (0.15%), and KH_2PO_4 (0.025%) [41]. The seed inoculum of *Aurantiochytrium* sp. PKU#Mn16 was cultured in a shaking incubator (LYZ-123CD, Shanghai Longyue Equipment Co., Shanghai, China) at 23 °C and 180 rpm for 48 h. One hundred milliliters of medium in a 250 mL flask was inoculated with 5 mL (5% (*v*/*v*) inoculation ratio) of the above culture. Three biological replicates of each sample were examined.

2.2. UV-Mediated Mutagenesis

The microorganism solution was diluted 10^5 times and applied to the plate. Then the microorganisms on the plate were mutagenized after 24 h of incubation in a constant-temperature incubator (LR-250, Shanghai Yiheng Technology Co., Ltd., Shanghai, China) at 23 °C. Before UV mutagenesis, the UV crosslinker (SZ03-2, Shanghai Netcom Business Development Co., Ltd., Shanghai, China) was turned on for 30 min to stabilize the light waves. The plates were placed in 0 W, 10 W, 20 W, 30 W, 40 W, 50 W, 60 W, 70 W, 80 W, and 90 W UV crosslinkers and irradiated for 0 s, 6 s, 9 s, 12 s, 15 s, 18 s, 21 s, 24 s, 27 s, 30 s, 33 s, and 36 s. After mutagenesis, the plates were incubated for 48 h in the dark, and then, the number of colonies was counted, and the lethality was calculated. Three biological replicates for each sample were examined.

2.3. Biomass Determination

The mutagenized strain was cultivated as described in Section 2.1 for 48 h. The culture was then centrifuged (Z366K, HERMLE, Germany) at 10,000 rpm for 5 min to obtain the cell precipitate. After washing three times with deionized water, the cell precipitate was collected as the biomass and then lyophilized in a freeze dryer (Triad 2.51, Labconco, Kansas City, MO, USA) for 72 h. Three biological replicates for each sample were examined in the experiment.

2.4. Fatty Acid Extraction

Before the experiment, filter paper bags (Civil Administration Filter Paper Factory, Liaoning Province, China) were pretreated with a solvent mixture (chloroform:methanol ratio of 2:1 (*v*/*v*)) for 48 h and dried at 50 °C. Five hundred milligrams of freeze-dried cells were placed in a pretreated filter paper bag as a filter paper package and extracted in a Soxhlet extractor at 70 °C for 48 h (solvent as described above) [42]. Then, the filter paper package was dried and weighed. The difference between the weights before and after was the weight of the FAs. The remaining liquid was evaporated to dryness at 70 °C by a rotary evaporator. The FAs were rinsed completely with 5 mL of n-hexane and placed in a 10 mL glass tube [43]. Three biological replicates for each sample were examined.

2.5. Fatty Acid Structure and Composition Analysis

2.5.1. Fourier Transform Infrared (FTIR) Spectrometer Analysis

KBr powder was uniformly mixed with the dried cells and compressed into a sheet (KBr to dried cell ratio of approximately 100:1). KBr was used as a background and detected using a Fourier transform infrared (FTIR) spectrometer (Thermo Fisher Scientific, Waltham, MA, USA). The infrared spectrometer had a spectral range of 7800–350 cm^{-1}, and its scanning frequency was 65 spectra (16 cm^{-1} resolution). Three biological replicates of each sample were prepared [44].

2.5.2. Gas Chromatography and Mass Spectrometry (GC/MS) Analysis

The FAs obtained in Section 2.4 were added to 5 mL of a 4% sulfuric acid–methanol solution (*v*/*v*), and 100 μL of a nonanecene–methylene chloride solution (500 μg mL^{-1}) was used as an internal standard. After the tube was allowed to stand in a 65 °C water bath for 1 h, 2 mL of n-hexane and 2 mL of deionized water were added, and the mixture was shaken for 30 s. Three biological replicates for the extraction were examined. Then, the upper organic layer was transferred to a new test tube, and the organic solvent was thoroughly dried with nitrogen. Finally, 1 mL of dichloromethane was added to each tube to dissolve the FAs, and the solution was then transferred to a chromatography bottle [45].

The FAs in the chromatography bottle were diluted 100-fold and analyzed by gas chromatography mass spectrometry (GC-MS, 7890-5975 Agilent, Santa Clara, CA, USA). The GC column for FA determination was HP-5MS (19091S-433) with a stationary phase of (5%)-diphenyl (95%)-dimethylpolysiloxane, constituting a weakly polar capillary column. The column had a maximum temperature of 350 °C and dimensions of 30.0 m × 250 μm × 0.25 μm. The GC inlet temperature was 250 °C, the carrier gas was high purity He, constant pressure mode was used, the head pressure was 1.2 psi, the split ratio was 10:1, and the injection volume was 1 μL. The column temperature rise program was determined by 37 FA mixing standards. The steps for selecting the peaks for the separation of the 37 FAs in the sequence were as follows. First, the temperature was raised to 180 °C at a rate of 25 °C min^{-1} from 60 °C, increased to 240 °C at a rate of 3 °C min^{-1}, maintained for 1 min, and then heated to 250 °C at a rate of 5 °C min^{-1}. The GC-MS transfer line temperature was 250 °C, and the mass spectrometer detector selected the full scan mode [38]. Three physical replicates for each sample were prepared.

2.6. Comparative Transcriptomic Analysis

2.6.1. RNA Extraction and cDNA Library Construction

After the total RNA was extracted with TRIzol (Life Technologies, Thermo Fisher Scientific Inc.), an Illumina HiSeq 4000 system was used to construct the cDNA library. mRNA sequences were then selected and the library was prepared [46]. To assess the integrity of the total extracted RNA, an Agilent 2100 bioanalyzer was used. The preparation of two libraries of cDNA constructs and transcriptome sequencing was conducted by Huada Gene Technology Co., Ltd. (Shenzhen, China). Oligo (dt) magnetic beads were utilized for enrichment and purification of mRNAs from the total RNA of each sample. The purified mRNAs enriched were short fragments, which were reverse transcribed for first-strand synthesis, and the second strand was used for cDNA synthesis. Then, these obtained double-stranded fragments were ligated with adapters, and appropriate DNA fragments were used as PCR amplification templates.

2.6.2. Illumina Sequencing, Assembly, and Annotation

cDNA library sequencing was carried out by an Illumina HiSeq^{TM} 4000, with 100 nt paired-end reads generated. The obtained reads were then filtered based on quality parameters of GC content, sequence duplication level, Q20, and Q30. High-quality clean reads were chosen from raw reads and reads with adapters and poly-N sequences were eliminated. De novo transcriptome assembly of clean reads was implemented through the Trinity assembly database program using default parameters [47]. Trinity software consists of three modules, namely, Chrysalis, Inchworm, and Butterfly (http://trinityrnaseq.sourceforge.net/). Initially, the Inchworm module formed a k-mer dictionary by breaking sequence reads (k-mer fixed-length sequence of k nucleotides, in repetition k = 25 bp). For contig assembly, the most recurrent k-mers were selected by removing low-complexity, error-containing, and singleton k-mers. Contigs were obtained until both side sequences could not protract with k-1 overlap. Then, the Chrysalis module was used to make the de Bruijn graph and gather the linear contigs. Finally, the Butterfly module was constructed to analyze de Bruijn graphs and produce transcript sequences. Transcript assembly was performed by using all generated contigs. The main transcripts that contained more than 200 bp were selected as uni-genes. BLASTX (Altschul et al., 1997) alignment was performed against public protein databases such as the non-redundant (Nr) protein (Deng et al., 2006), Kyoto Encyclopedia of Genes and Genomes (KEGG; Kanehisa et al., 2004), Clusters of Orthologous Groups (COG), Gene Ontology (GO), and Swiss-Prot (Ashburner et al., 2000) databases, and uni-sequences such as National Center for Biotechnology Information (NCBI) Taxonomy. KEGG is a database of metabolic pathways that is used to identify the gene products and functions associated with a cellular process. This pathway analysis provides a logical understanding of the complex biological performance of different genes in a network, and the analysis is performed by using BLAST software against the KEGG database. The cDNA sequence of mutant X2 was uploaded to GenBank (Accession number: MT232522).

2.6.3. qRT-PCR Analysis

Total RNA was extracted using the TRIzol (Life Technologies, Thermo Fisher Scientific Inc.) method. For quantitative real-time PCR (qRT-PCR), primers were first designed according to transcriptomic sequence data using Primer Premier 5 software (Supplementary Table S1). Then, the SYBR Taq^{TM} Ex Premix (Tli RNaseH Plus) Kit (TaKaRa Japan) was used with the following thermocycler protocol: 95 °C for 30 s, followed by 40 cycles of 95 °C for 5 s and 60 °C for 30 s. The entire process was performed in a CFX96 BioRad RT-PCR detection system. Actin was used as a housekeeping gene, which helped us check for standard and normal gene expression. qRT-PCR was performed in 3 replicates, and relative gene expression was quantified using the $2^{-\Delta\Delta Ct}$ method [48].

2.6.4. Statistical Analysis

Analysis of variance (ANOVA) was utilized for the statistical analysis of the data. The biomass yield and DHA productions of the wild type and mutant strains were analyzed by IMB SPSS Statistics 26.0 through a one-way ANOVA. The least significant difference (LSD) test was applied to determine the significant differences among the group means at $p < 0.05$.

3. Results

3.1. Cell Mutagenesis

To obtain a DHA-rich mutant with relatively high biomass yield, *Aurantiochytrium* sp. was subjected to random mutagenesis with UV irradiation. As shown in Figure 1, the fatality rate of the cells was sensitive to both UV treatment time and power. With a UV treatment time of 30 s, as the UV power increased from 10 W to 50 W, the survival rate decreased from 92.15% to 8.29%. The mutant treated with UV power of 50 W showed a rapid decrease in survival rate after 15 s (survival rate of 83.67%). The survival rate was 2.67% when the cells were exposed to UV (power of 50 W) for 33 s. The results showed that both the UV exposure time and UV power contributed to the severity of DNA damage in the cells. At present, UV radiation has been widely used in the breeding of microbial species [49], but rarely used in the production of DHA by *Aurantiochytrium* sp. Currently, researchers either chemically mutagenize strains or optimize fermentation conditions to increase DHA production. In 2014, Choi et al. optimized the extraction method of DHA to increase DHA production. In this study, acid catalyzed hot-water extraction of docosahexaenoic acid (DHA)-rich lipids from *Aurantiochytrium* sp. [50]. Cheng Yurong et al. (2016) performed mutagenesis of *Aurantiochytrium* sp. through cold stress (4 °C and FAS inhibitors (triclosan and isoniazid) to enhance DHA enrichment [51]. Shariffah et al. (2018) optimized the levels of fructose, monosodium glutamate, and sea salt through monosodium glutamate (MSG) experiments, predicting that DHA production by *Aurantiochytrium* sp. SW1 would reach 8.82 g/L [52]. Under the UV irradiation, the DHA content (0.20 g/g dry biomass) of *Schizochytrium* sp. increased by 38.88% compared with the parent strain [53]. Thus, based on the results, UV irradiation at 50 W for 30 s was chosen as the mutagenesis condition for breeding the DHA-producing mutant strain.

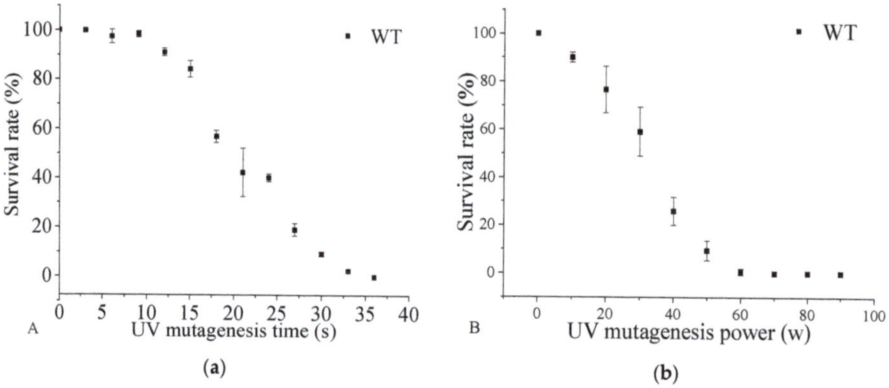

Figure 1. The survival rate of *Aurantiochytrium* sp. PKU#Mn 16 (wild type (WT)) under various ultraviolet (UV) mutagenesis time (**a**) and UV mutagenesis power (**b**). UV irradiation for 30 s (**a**) at 50 W (**b**) was selected as the WT mutation condition. All data were collected from three independent experiments. Error bars were the standard deviation.

3.2. Screening of the Mutant *Aurantiochytrium sp.*

After UV irradiation, 135 colonies were obtained from the surviving cells. The first round of mutant screening was based on dry cell weight (DCW) enhancement. As shown in Figure 2a, 14 colonies (X1 to X14) exhibited significantly enhanced cell growth compared with the parent strain. Notably, the biomass yield of mutants X2 and X4 increased 1.53 ± 0.025 and 1.52 ± 0.053-fold, respectively. The lipid and DHA contents of the mutants were also analyzed (Figure 2a). Among the 14 mutants, eight independent colonies (X1, X2, X3, X4, X5, X9, X11, and X14) exhibited increased fatty acid yield (per g of DCW; 1.09 ± 0.056-fold, 1.79 ± 0.041-fold, 1.26 ± 0.043-fold, 1.25 ± 0.064-fold, 1.23 ± 0.024-fold, 1.35 ± 0.012-fold, 1.08 ± 0.033-fold, and 1.64 ± 0.059-fold, respectively). Four mutants (X2, X3, X5, X14) showed an improved ability of DHA production. In particular, mutant strain X2 showed a marked improvement of 1.90-fold compared with the WT strain. According to the cell dry mass, lipid, and DHA content, mutant strain X2 was chosen as the preferable DHA-producing candidate for the following experiments.

To verify the hereditary stability of mutant X2, the strain was cultivated continuously in a shake flask for ten generations (Table 1). There was no significant difference for the DHA production observed among the ten generations. The DHA, lipid, and biomass yields of the tenth generation were 624.93 mg/L, 1417.37 mg/L, and 2920.60 mg/L, respectively. The results showed that UV irradiation (50 W, 30 s) could be utilized as a breeding strategy to screen for high-yield DHA-producing *Aurantiochytrium* sp.

Table 1. The total fatty acids (TFAs) and docosahexaenoic acid (DHA) contents of mutant strain *Aurantiochytrium* sp. X2 during the ten-generations subculture.

Generation	TFAs (%DCW)	DHA (% TFAs)	DHA Content (mg/L)
WT	41.07 ± 2.05	40.55 ± 2.05	321.37 ± 14.98
X2-2nd	50.00 ± 3.87	44.39 ± 3.87	624.34 ± 2.95
X2-6th	48.10 ± 0.74	44.98 ± 0.74	623.40 ± 7.14
X2-10th	48.53 ± 1.40	44.09 ± 1.40	624.93 ± 13.32

The fermentation conditions during continuous subculture remained the same and were set as: Inoculum size 10% (*v*/*v*), culture temperature 23 °C, initial pH 6.5, and fermentation medium volume 1000 mL. DCW = dry cell weight, TFAs = total fatty acids, WT = wild type strain, X2-2nd = the second-generation subculture of mutant strain X2, X2-6th = the sixth-generation subculture of mutant strain X2, X2-10th = the tenth-generation subculture of mutant strain X2. All data were collected from three independent experiments.

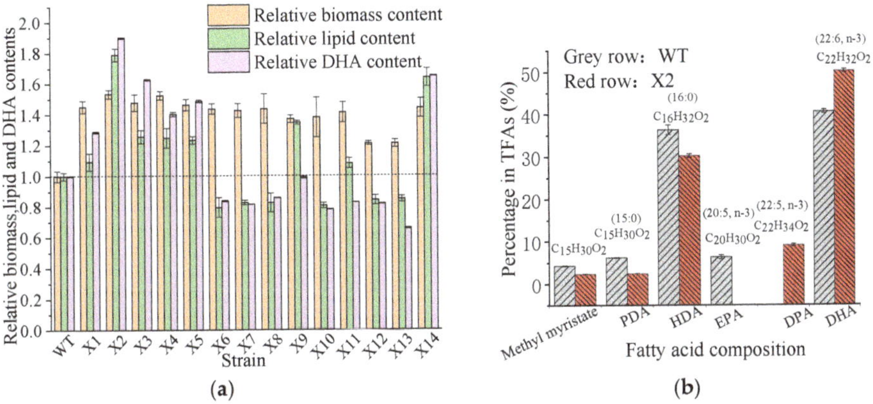

Figure 2. (**a**) The biomass, lipid and docosahexaenoic acid (DHA) contents of wild-type (WT) and 14 isolated mutant strains (X1~X14), and (**b**) the fatty acid component in total fatty acids (TFAs) of *Aurantiochytrium* sp. PKU#Mn16 (WT) and mutant strain *Aurantiochytrium* sp. X2 (X2). All data were collected from three independent experiments. Error bars represent the standard deviation. The values of biomass, lipid, and DHA contents of the wild-type strain were set to 1.0. PDA = Pentadecanoic acid, HAD = Hexadecanoic acid, EPA = Eicosapentaenoic acid, DPA = Docosapentaenoic acid, DHA = Docosahexaenoic acid.

3.3. *PUFAs Production by the Mutant Aurantiochytrium sp.*

Significant differences in FAs production were observed between the mutant and WT. As shown in Figure 2b, the amounts of LC-PUFAs (DHA (22:6, *n*-3) and EPA (20:5, *n*-3)) and saturated fatty acids (SFAs; hexadecanoic acid (HDA, 16:0) and pentadecanoic acid (PDA, 15:0)) were markedly different after UV mutation. The HDA and PDA levels decreased from 36.28% to 30.21% and 6.18% to 2.44%, respectively, whereas the DHA levels increased from 40.55% to 50.19%. The production of DHA in the mutant strain X2 increased by 23.77% compared with the WT. Interestingly, 9.07% of DPA was observed instead of EPA in the mutant X2. The results indicate that the mutation led to the transformation of SFAs to PUFAs, reflecting the mutated genes responsible for FAs carbon chain lengthening and unsaturation. Previous studies have confirmed that the accumulation of LC-PUFAs can be improved by increasing the SFA levels in the substrate, and that the long-chain saturated FAs of either C16 or C18 [54,55] could be further transferred to LC-PUFA by desaturase and elongase [56]. The culture conditions of mutant X2 were also studied to explore the appropriate conditions for DHA production. Figure 3 shows that pH 6.5, a fermentation volume of 200 mL, a culture temperature of 27 °C, and an inoculum size of 5% were suitable conditions for DHA accumulation in mutant X2.

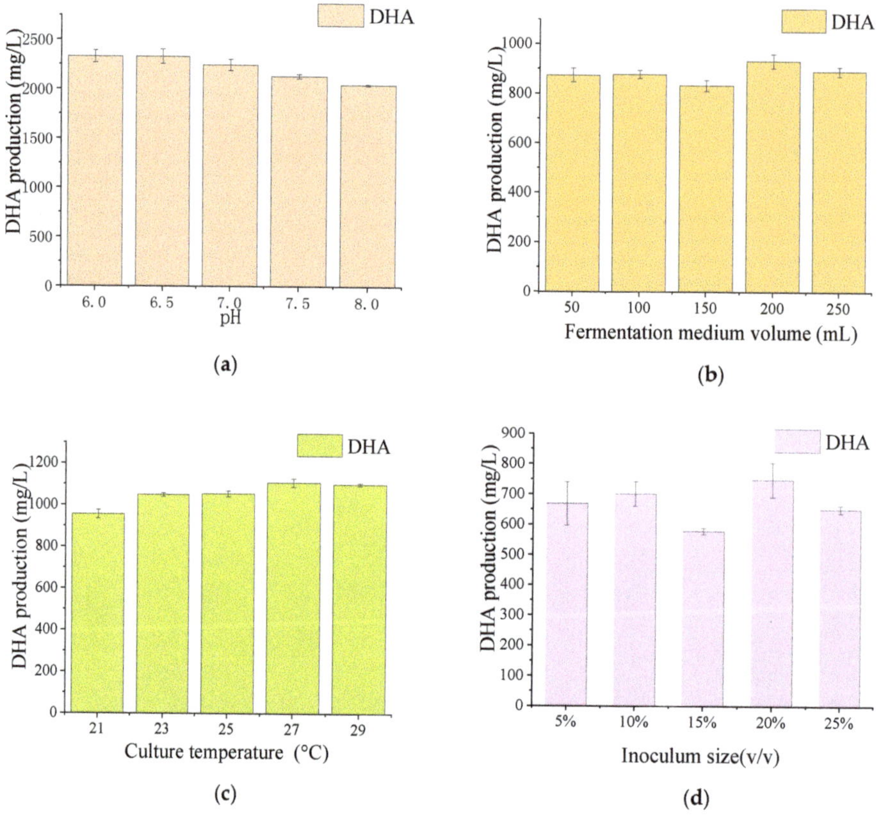

Figure 3. The effect of pH, cultivation temperature, fermentation medium volume, and inoculum size on the DHA production of mutant strain X2. The cultivation conditions were described as followed: (**a**) initial pH 6.0–8.0, culture temperature 23 °C, fermentation medium 100 mL, inoculum size 5%; (**b**) initial pH 6.5, culture temperature 23–29 °C, fermentation medium 100 mL, inoculum size 5%; (**c**) initial pH 6.5, culture temperature 23 °C, fermentation medium 50–250 mL, inoculum size 5%; (**d**) initial pH 6.5, culture temperature 23 °C, fermentation medium 100 mL, inoculum size 5%–25%. All data were collected from three independent experiments. Error bars were the standard deviation.

3.4. Sequence Analysis and Assembly

For a comprehensive understanding of the molecular mechanism underlying FA improvement in the collection, a comparative transcriptomic study was conducted between the WT and the mutant X2. The Q20 base value with a base quality greater than 20 and an error rate ≤0.01% made up more than 96.82% and 96.67% of the WT and X2 reads, respectively, indicating that the raw sequence reads were very reliable and of high quality (Table 2).

Table 2. Summary of sequencing data for wild type *Aurantiochytrium* sp. PKU#Mn16 and mutant *Aurantiochytrium* sp. X2.

Sample	Total Raw	Total Clean		Clean Reads		
	Reads (Mb) [1]	Reads (Mb) [2]	Bases (Gb) [3]	Q20 (%) [4]	Q30 (%) [5]	Ratio (%) [6]
Mn16	50.94	44.69	6.70	96.82	88.53	87.72
X2	47.43	43.14	6.47	96.67	87.97	90.94

[1] The reads amount before filtering; [2] The reads amount after filtering; [3] The total base amount after filtering; [4] The rate of bases in which quality was greater than 20 in clean reads; [5] The rate of bases in which quality was greater than 30 in clean reads; [6] The ratio of the amount of clean reads.

After trimming the adapter sequences, ambiguous nucleotides, and low-quality sequences, the qualified mRNA-based sequenced reads were subjected to transcriptome de novo assembly (Table 3). For wild type Mn 16 and mutant X2, the transcriptome assembly generated 20,874 and 18,952 uni-genes with a N50 of 1880 and 2032 bp, respectively. The analysis of N50 indicated that 50% of the assembled reads were incorporated into transcripts more than 1880 and 2032 bp. The mean length of the transcripts was 1149 and 1205 bp.

Table 3. Quality metrics of transcriptome and uni-genes assembly for wild type *Aurantiochytrium* sp. PKU#Mn16 and mutant *Aurantiochytrium* sp. X2.

Quality Metrics	Uni-genes	
	Mn16	X2
Total Number	20,874	18,952
Total Length	23,991,758	22,844,185
Mean Length	1149	1205
N50 [1]	1880	2032
N70 [2]	1264	1328
N90 [3]	480	504
GC (%) [4]	48.36	48.75

[1] The N50 length is used to determine the assembly continuity. N50 is a weighted median statistic that 50% of the total length is contained in transcripts that are equal to or larger than this value. [2] N70 is a weighted median statistic that 70% of the total length is contained in transcripts that are equal to or larger than this value. [3] N90 is a weighted median statistic that 90% of the total length is contained in transcripts that are equal to or larger than this value. [4] GC (%): the percentage of G and C bases.

3.5. *Differentially Expressed Gene Analysis by RNA-Seq*

In the comparison between WT and mutant X2, a total of 39,826 differentially expressed genes (DEGs) existed. Of these total DEGs, 1350 were downregulated and 1945 were upregulated in *Aurantiochytrium* sp. WT and mutant X2, respectively. Further elucidation of DEGs with different expression arrays was performed with hierarchical DEG clustering through Euclidean distance associated with complete linkage (Figure 4).

GO analysis of the transcriptome was based on three main categories: Biological processes, cellular components, and molecular functions (Figure 5a). In many cases, several GO terms were assigned to the same uni-gene. Biological processes, molecular functions, and cellular components related to functional subgroups were used to categorize all DEGs. DEGs in both WT and mutant X2 were implicated in cellular and metabolic processes, which are plentiful in biological processes. One of the proteins involved in catalytic and binding processes was the foremost protein in molecular function, and other cellular components included cell and cellular parts.

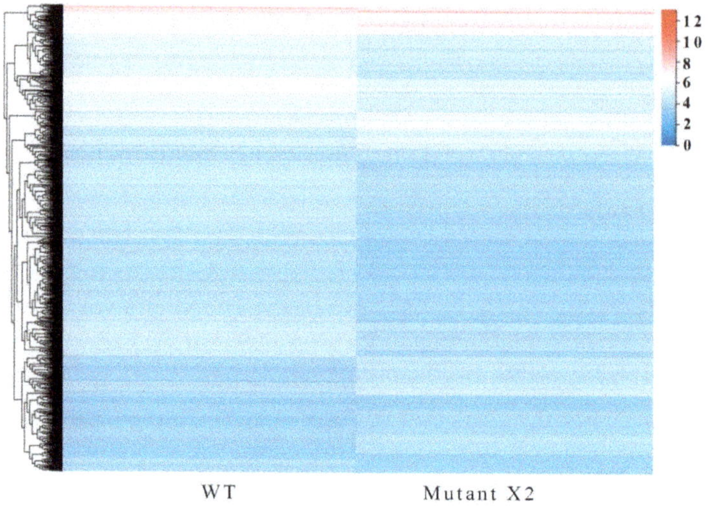

Figure 4. Heat map of the differentially expressed genes identified in *Aurantiochytrium* sp. PKU#Mn 16 (wild type) and *Aurantiochytrium* sp. X2 (mutant).

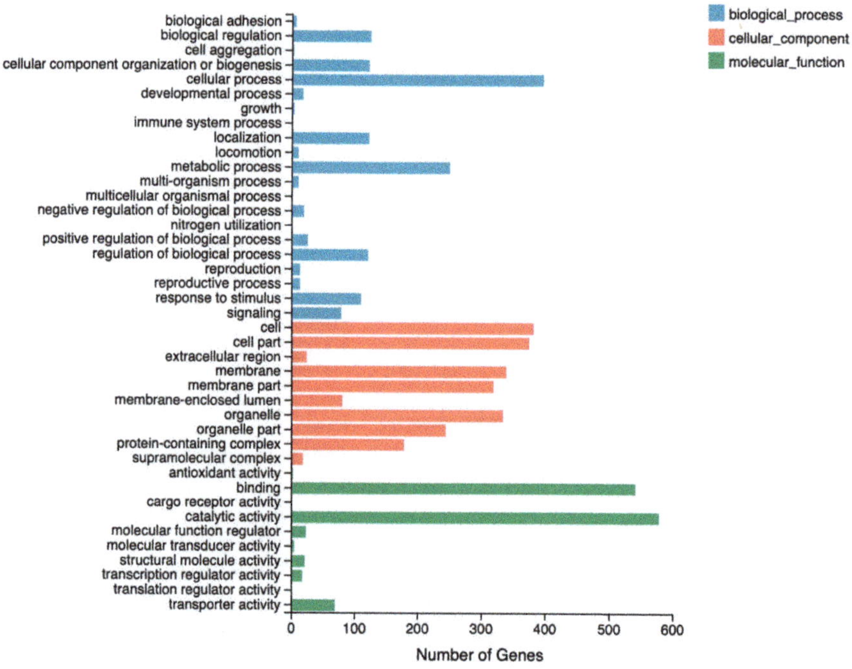

(**a**)

Figure 5. *Cont.*

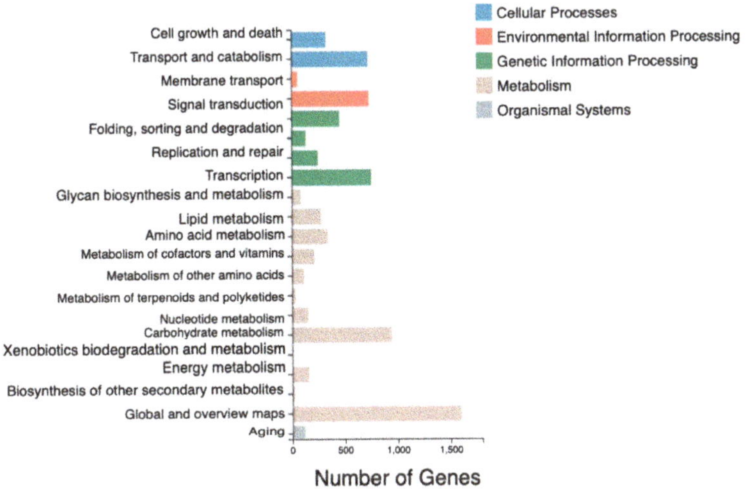

(**b**)

Figure 5. The differentially expressed genes (DEGs) of comparative transcriptomic analysis between wild type and mutant *Aurantiochytrium* sp.: (**a**) Ontology (GO); (**b**) Enriched Kyoto Encyclopedia of Genes and Genomes pathway (KEGG).

DEG-associated pathways were analyzed by using KEGG pathway tools in *Aurantiochytrium* sp., with $p < 0.05$ as the significance threshold. Nineteen significantly enriched pathways were found in WT and mutant X2 (Figure 5b). These enriched pathways were associated with lipid metabolism, carbohydrate metabolism, biosynthesis and translation, and secondary metabolite biosynthesis. KEGG pathways involved in lipid metabolism include fatty acid metabolism, glycerophospholipid metabolism, glycerolipid metabolism, fatty acid biosynthesis and secondary metabolite biosynthesis, which probably play an important role in PUFA biosynthesis.

3.6. Identification and Characterization of Genes Involved in DHA Biosynthesis

The expressions of the key genes in the PKS pathway have been determined for mutant X2, including CoA-transferase (*CoAT*), acyltransferase (*AT*), enoyl reductase (*ER*), dehydratase (*DH*), and methyltransferase (*MT*), shown as Table 4. Using transcriptomic sequencing, we identified only one fatty acid synthesis (FAS) desaturase encoded by a uni-gene. In addition, RNA sequence analysis was used to investigate key biosynthetic enzymes of PKS pathway genes (Table 4). This analysis reported that Unigene4591_All, Unigene2419_All, Unigene10491_All, CL663.Contig2_All, and CL555.Contig2_All were involved in synthesizing DHA, which was downregulated in Mn16 and upregulated in mutant X2.

Table 4. The key candidate genes related to polyketide synthase (PKS) pathway wild type *Aurantiochytrium* sp. PKU#Mn16 and mutant *Aurantiochytrium* sp. X2.

Gene ID	Protein Name	Gene Name	Control FPKM	Treat FPKM	log2
CL555.Contig2_All	Enoyl-(Acyl carrier protein) reductase	*ER*	3.85	10.1	1.400242079
CL663.Contig2_All	CoA-transferase	*CoAT*	1.14	3.04	1.394908036
Unigene10491_All	Acyltransferase	*AT*	74.10	149.53	1.006525578
Unigene2419_All	Methyltransferase domain	*MT*	8.46	18.11	1.083865306
Unigene4591_All	Dehydratase family	*DH*	0.01	1.66	7.211944308
CL94.Contig2_All	Fatty acid desaturase	*FAD*	2.02	4.34	1.101442064

3.7. *mRNA Expression Level of the Mutant X2 and WT*

qRT-PCR was used to check the DEG expression profiles associated with PKS pathways. mRNA expression levels were checked for different genes, such as *CoAT, ER, DH, MT*, and *AT*, in comparison with the levels in Mn16, showing downregulation in Mn16 but upregulation in X2 samples. The qRT-PCR results were consistent with the RNA-seq results (Figure 6) and validated the DEG expression profile.

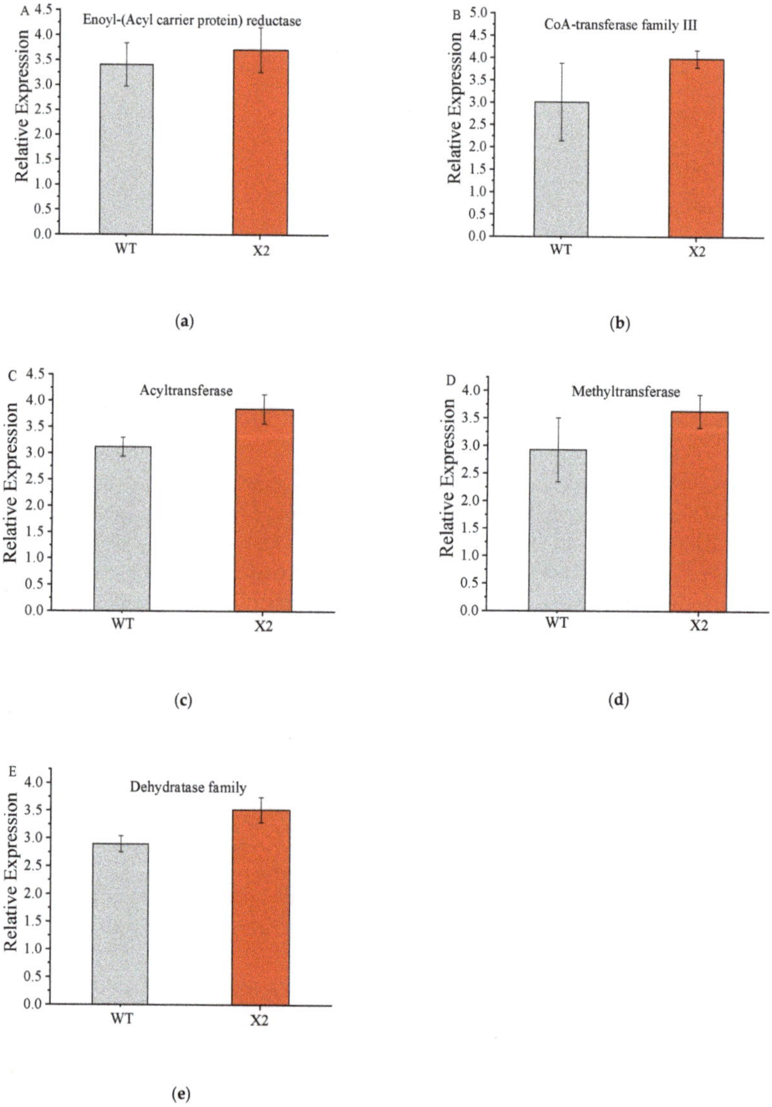

Figure 6. qPCR relative expression of genes annotated for enoyl-(acyl carrier protein) reductase (**a**), coA-transferase family II (**b**), acyltransferase (**c**), methyltransferase (**d**), and dehydratase family (**e**) of wile type (WT; *Aurantiochytrium* sp. PKU#Mn16) and mutant strain (X2; *Aurantiochytrium* sp. X2). All data were collected from three independent experiments. Error bars were the standard deviation.

4. Discussion

4.1. DHA Synthesis Enhancement

Recently, the DHA produced by microorganisms in the ocean has received increased attention [57]. Dietary supplements constitute the largest market share of 55% for *n*-3 products, followed by functional food and beverages, and pharmaceuticals [58]. The *n*-3-PUFA market is projected to show an annual growth rate of 12.8% between 2014 and 2019 and is expected to be worth USD 4300 million by 2019 (www.marketsandmarkets.com). DHA-rich oils from *thraustochytrids* are currently on the market as dietary supplements [59]. The main source of the marine *n*-3 fatty acids EPA and DHA are fish oils [60]. Approximately 200,000 tons of fish oils are used in products for the human markets. Meanwhile, the production of microbial *n*-3-rich oils constituted only 5000 tons in 2011, with marine protist *Thraustochytrids* and the heterotrophic microalgae *Chrypthecodinium cohnii* as production organisms [34]. Although industrial DHA production has been accomplished, certain approaches have been applied to enhance the synthesis of DHA via intrinsic [61,62] or extrinsic parameters [63]. Conversely, several shared features, including low adaptability, degeneration, and low production, continue to hinder significant production by strains [64]. An effective approach such as mutagenesis is broadly useful for selecting high-yield strains [65]. Alonso reported that microalgae produced high yields of DHA and EPA after mutagenesis, and increased EPA content was also observed in *Phaeodactylum tricornutum* mutated by UV light [66]. It was also reported that treating marine microalgae *Nannochloropsis salina* with UV mutagenesis, lipid accumulation of the mutant cultures was elevated to more than 3-fold that of the wild type strain. However, reduced growth rates resulted in a reduction in overall productivity [67]. Forjan et al. (2014) showed that UV-A increased the saturated:unsaturated fatty acids ratio, and hence an increase in storage lipids in *Nannochloropsis. gaditana* [68]. Liu et al. (2015) applied UV irradiation to microalgae *Chlorella sp.* and found that the biomass for the UV mutation strain was 7.6% higher and the lipid content reached the maximum value of 28.1% on day 15. Our research showed that in the mutant X2, the TFAs and DHA contents reached 81.73% and 35.24% of DCW, respectively. Results also prove that UV irradiation can be used as a breeding strategy for obtaining potential DHA-producing *Thraustochytrids* strain.

4.2. PKS Pathway

The DHA biosynthesis pathway in *Thraustochytriu* has not been fully elucidated. It has been reported that the two pathways, i.e., The PKS system and FAS pathway, they are likely to be present [69,70]. In general, eukaryotes biosynthesize the polyunsaturated fatty acids through a series of desaturation and elongation reactions catalyzed by membrane-bound enzymes such as desaturases and elongases, known as the fatty acid synthetase (FAS) system. A small amount of label in 22:6 was detected when the alga was grown in the presence of ^{14}C labeled 18:0 or 18:1 [71]. The addition of ^{13}C acetate or ^{13}C butyrate in the growth medium resulted in 22:6, with only the odd carbon atoms enriched. The commonly found extended products of FAS in nearly all organisms are long-chain saturated FAs of either C16 or C18 [54,55]. The FAS pathway comprised seven or more kinds of desaturases, including $\Delta12$, $\Delta9$, $\Delta8$, $\Delta6$, $\Delta5$, $\Delta4$, and n-3 (e.g., $\Delta17$ and $\Delta15$). However, the activity of desaturase and elongase was not detected in *Schizochytrium* by ^{14}C labeling, which implies the existence of a different DHA biosynthesis mechanism in the genus *Schizochytrium* [56]. The transcriptomic study of the mutant X2 and WT revealed that only one gene encoding desaturase was involved in the FAS pathway; however, the $\Delta4$, $\Delta6$, and $\Delta12$ desaturase genes, which are important for DHA production, were not observed in the present study. Similar to previous studies, expressed sequence tag (EST) sequencing or PCR-based detection failed to identify the probable desaturases in the FAS pathway [72]. The modification of FAs is performed to produce long-chain DHA (C22:6) by an enzyme-dependent continuous process [73]. On the other side, it has been reported that marine bacteria could produce polyunsaturated fatty acid via the polyketide synthases (PKS) pathway [74]. PKS pathway domains are likely involved in the production of DHA, such as *AT, DH, MT* and *ER* [75,76]. Genomic and transcriptomic analysis have shown that *Thraustochytrids* contained some key enzymes of PKS system such as 3-ketoacyl-synthase (KS), ketoreductase (KR), and enoyl reductase (ER) [26]. As a probable source of high-value DHA, the PKS pathway is important in DHA biosynthesis, as genes related to the PKS pathway were mined in the transcriptome study of wild *Aurantiochytrium* sp. and mutant X2 (Table 4). These uni-genes were homologs to *MT, AT, ER* and *DH*, which are crucial in polyketide synthesis. These findings suggest that DHA synthesis is likely to occur via the PKS pathway in WT and mutant X2. Currently, no evidence supports the hypothesis that DHA biosynthesis occurs via either of the two hypothetical pathways [40]. Furthermore, the formation of PUFAs of >C22 (e.g., 28:8*n*-3 and 28:7*n*-6) also occurs via the PKS pathway, which has been described in some species of oceanic dinoflagellates by Mansour M.P. (1999) [77].

4.3. Transcriptional Responses of the PKS Pathway

The omega-3 PUFAs, including EPA, DPA, and DHA, are produced by certain strains, e.g., thraustochytrids [78]. Biochemical studies have been performed to characterize the distinct enzymes from the standard PKS pathways, which is ultimately helpful for understanding the underlying biosynthetic mechanisms [79]. These findings revealed that the FAS pathway does not participate in the biosynthesis of DHA in the *Aurantiochytrium* sp. strain. PUFA synthesis is carried out by ACP (acyl carrier protein) in the PKS pathway. ACP acts as a covalent joint for chain elongation during many cycles. The synthesis of lengthy (unsaturated) fatty acids includes several enzymes in the PKS system, e.g., *AT, MT, ER*, and *DH*. A vital role is played by *AT* domains and their allies, i.e., ACPs. *AT* loads the building units onto ACP (substrate acceptor). Therefore, *AT* decides which building blocks will be incorporated into the polyketide assembly [80]. *AT* plays significant roles in the PKS pathway; here, *AT* showed down-regulation in WT and up-regulation in X2 at the transcription level. The data showed increased gene expression of the PKS pathway in the mutant. PKS-linked genes contained *DH, AT, ER*, and *MT* domains, as revealed by the transcriptomic study of *Aurantiochytrium* sp. X2 (Table 4). The mutation led to the formation of ORFC (open reading frame control), which contains two *DH* domains and one *ER* domain, with up-regulation similar to that observed by Zhi-Qian Bi (2018) [81]. The mutation also improved the expression of *DH* and *ER* in *Thraustochytriu*, which suggests increased production of DHA. It is believed that mutagenesis is a valuable strategy for the enhancement of PUFA biosynthesis. The PKS anchor gene up-regulation suggests that the PKS system is actively involved in PUFA biosynthesis, which is supported by [40].

5. Conclusions

UV mutagenesis enhanced the ability of DHA production and led to the generation of potential DHA-producing strain from *Aurantiochytrium*. The transcriptome of the WT and the mutant strain were compared to investigate the vital genes responsible for the DHA enrichment. Results showed that in both WT and mutant strain X2, FAS was incomplete and key desaturases, but genes related to the PKS pathway were observed. The qPCR revealed that the upregulation of key PKS pathway genes (*CoAT, DH, AT, ER, MT*) involved in the high yield of DHA for the mutant strain X2. The research provides valuable information for constructing a genetic engineering strain with rational design for the fatty acid composition in future work.

Supplementary Materials: The following are available online at http://www.mdpi.com/2076-2607/8/4/529/s1, Table S1: List of mRNA primers.

Author Contributions: Conceptualization, L.L., Z.H., S.L. (Shuangfei Li), and X.Y.; Data curation, L.L., Z.H., and H.Y.; Formal analysis, S.L. (Shuangfei Li) and H.Y.; Funding acquisition, S.L. (Shuangfei Li) and X.Y.; Investigation, H.Y., S.L. (Siting Li), C.L., M.Z., and X.Y.; Methodology, L.L. and Z.H.; Project administration, L.L., Z.H., S.L. (Shuangfei Li), and X.Y.; Software, C.H.K.C.; Supervision, L.L, Z.H., and X.Y.; Validation, H.C.; Visualization, C.H.K.C.; Writing—original draft, L.L., M.Z. and Z.H.; Writing—review and editing, L.L., Z.H., and X.Y. All authors have read and agreed to the published version of the manuscript.

Funding: This study was supported by National Key Research and Development Project (Grant No. 2018YFA0902500), the Natural Science Foundation of Guangdong Province (Grant No. 2018A030313139),

Joint R&D Project of Shenzhen-Hong Kong Innovation (Grant No. SGLH20180622152010394), Hong Kong Innovation and Technology Commission TCFS (GHP/087/18SZ), and sponsored by the Shenzhen Taifeng East Marine Biotechnology Co. Ltd, Natural Science Foundation of Shenzhen (Grant No. KQJSCX20180328093806045), Shenzhen science and technology application demonstration project (Grant No. KJYY20180201180253571), and Shenzhen peacock plan (Grant No. 827-000192).

Conflicts of Interest: The authors declare no conflict of interest.

References

1. Simopoulos, A.P. Evolutionary aspects of diet, the omega-6/omega-3 ratio and genetic variation: Nutritional implications for chronic diseases. *Biomed. Pharmacotherapy* **2006**, *60*, 502–507. [CrossRef] [PubMed]
2. Bormann, J.L.; Meyer, E.S.; Voigt, E.M.; Whiteheart, S.W.; Larson, M.K. Attenuation of EPA and DHA-Mediated Platelet Inhibition Effects Following Heterogeneous Agonist Treatment. *FASEB J.* **2016**, *30*, 722.
3. Malan, L.; Baumgartner, J.; Zandberg, L.; Calder, P.C.; Smuts, C.M. Iron and a mixture of DHA and EPA supplementation, alone and in combination, affect bioactive lipid signalling and morbidity of iron deficient south African school children in a two-by-two randomised controlled trial. *Prostaglandins Leukot. Essent. Fat. Acids* **2016**, *105*, 15–25. [CrossRef] [PubMed]
4. Cottin, S.C.; Alsaleh, A.; Sanders, T.A.B.; Hall, W.L. Lack of effect of supplementation with EPA or DHA on platelet-monocyte aggregates and vascular function in healthy men. *Nutr. Metab. Cardiovasc Dis.* **2016**, *26*, 743–751. [CrossRef] [PubMed]
5. Zhou, L.; Liu, Q.; Yang, M.; Wang, T.; Yao, J.; Cheng, J.; Yuan, J.; Lin, X.; Zhao, J.; Tickner, J.; et al. *Dihydroartemisinin*, an anti-malaria drug, suppresses estrogen deficiency-induced osteoporosis, osteoclast formation, and RANKL-induced signaling pathways. *J. Bone Miner Res.* **2016**, *31*, 964–974. [CrossRef]
6. Létisse, M.; Rozières, M.; Hiol, A.; Sergent, M.; Comeau, L. Enrichment of EPA and DHA from sardine by supercritical fluid extraction without organic modifier: I. Optimization of extraction conditions. *J. Supercritical Fluids* **2006**, *38*, 27–36. [CrossRef]
7. Codabaccus, M.B.; Ng, W.K.; Nichols, P.D.; Carter, C.G. Restoration of EPA and DHA in rainbow trout (*Oncorhynchus mykiss*) using a finishing fish oil diet at two different water temperatures. *Food Chem.* **2013**, *141*, 236–244. [CrossRef]
8. Scholefield, A.M.; Lipids, K.A. Cell Proliferation and Long Chain Polyunsaturated Fatty Acid Metabolism in a Cell Line from Southern Bluefin Tuna (*Thunnus maccoyii*). *Lipids* **2014**, *49*, 703–714. [CrossRef]
9. Regulska-Ilow, B.; Ilow, R.; Konikowska, K.; Kawicka, A.; Bochińska, A. FATTY ACIDS PROFILE OF THE FAT IN SELECTED SMOKED MARINE FISH. *Rocz Panstw Zakl Hig.* **2013**, *64*, 299–307.
10. Kitessa, S.M.; Abeywardena, M.; Wijesundera, C.; Nichols, P.D. DHA-containing oilseed: A timely solution for the sustainability issues surrounding fish oil sources of the health-benefitting long-chain omega-3 oils. *Nutrients* **2014**, *6*, 2035–2058. [CrossRef]
11. Shapira, N.; Weill, P.; Loewenbach, R. Egg fortification with n-3 polyunsaturated fatty acids (PUFA): Nutritional benefits versus high n-6 PUFA western diets, and consumer acceptance. *Isr. Med Assoc J.* **2008**, *10*, 262–265. [PubMed]
12. Feng, D.; Chen, Z.; Xue, S.; Zhang, W. Increased lipid production of the marine oleaginous microalgae *Isochrysis zhangjiangensis* (*Chrysophyta*) by nitrogen supplement. *Bioresour. Technol.* **2011**, *102*, 6710–6716. [CrossRef] [PubMed]
13. Tsirigoti, A.; Tzovenis, I.; Koutsaviti, A.; Economou-Amilli, A.; Ioannou, E.; Melkonian, M. Biofilm cultivation of marine dinoflagellates under different temperatures and nitrogen regimes enhances DHA productivity. *J. Appl. Phycol.* **2020**. [CrossRef]
14. Nanjappa, D.; d'Ippolito, G.; Gallo, C.; Zingone, A.; Fontana, A.J. Oxylipin diversity in the diatom family *Leptocylindraceae* reveals DHA derivatives in marine diatoms. *Mar. Drugs* **2014**, *12*, 368–384. [CrossRef] [PubMed]
15. Liu, Y.; Singh, P.; Sun, Y.; Luan, S.; Wang, G. Culturable diversity and biochemical features of *thraustochytrids* from coastal waters of Southern China. *Appl. Microbiol. Biotechnol.* **2013**, *98*, 3241–3255. [CrossRef] [PubMed]
16. Sun, X.-M.; Geng, L.-J.; Ren, L.-J.; Ji, X.-J.; Hao, N.; Chen, K.-Q.; Huang, H. Influence of oxygen on the biosynthesis of polyunsaturated fatty acids in microalgae. *Bioresour. Technol.* **2018**, *250*, 868–876. [CrossRef] [PubMed]

17. Bailey, R.B.; DiMasi, D.; Hansen, J.M.; Mirrasoul, P.J.; Ruecker, C.M.; Veeder III, G.T.; Kaneko, T.; Barclay, W.R. Enhanced production of lipids containing polyenoic fatty acid by very high density cultures of eukaryotic microbes in fermentors. U.S. Patent 6,607,900, 19 August 2012.
18. Patel, A.; Liefeldt, S.; Rova, U.; Christakopoulos, P.; Matsakas, L. Co-production of DHA and squalene by *thraustochytrid* from forest biomass. *Sci. Rep.* **2020**, *10*, 1–12. [CrossRef]
19. Wang, Q.; Ye, H.; Sen, B.; Xie, Y.; He, Y.; Park, S.; Wang, G.J.S.; Biotechnology, S. Improved production of docosahexaenoic acid in batch fermentation by newly-isolated strains of *Schizochytrium* sp. and *Thraustochytriidae* sp. through bioprocess optimization. *Synth Syst Biotechnol.* **2018**, *3*, 121–129. [CrossRef]
20. Li, D.; Zhang, K.; Chen, L.; Ding, M.; Zhao, M.; Chen, S. Selection of *Schizochytrium limacinum* mutants based on butanol tolerance. *Electron. J. Biotechnol.* **2017**, *30*, 58–63. [CrossRef]
21. Marques, T.S.; Pires, F.; Magalhães-Mota, G.; Ribeiro, P.A.; Raposo, M.; Mason, N. Development of a DNA Biodosimeter for UV Radiation. In Proceedings of the 6th International Conference on Photonics, Optics and Laser Technology, Funchal, Madeira, Portugal, 25–27 January 2018; pp. 328–333.
22. Shi-Xiong, X.U.; Jiang, Y.; Zhan, X.B.; Zheng, Z.Y.; Jian-Rong, W.U.J.I.M. Breeding of *Schizochytrium* sp.by diethyl sulfate-UV mutagenesis for high docosahexaenoic acids production. *Ind. Microbiol.* **2013**, *43*, 64–70.
23. Hoang, M.H.; Nguyen, C.; Pham, H.Q.; Van Nguyen, L.; Hoang Duc, L.; Van Son, L.; Hai, T.N.; Ha, C.H.; Nhan, L.D.; Anh, H.T.L.; et al. Transcriptome sequencing and comparative analysis of *Schizochytrium* mangrovei PQ6 at different cultivation times. *Biotechnol. Lett.* **2016**, *38*, 1781–1789. [CrossRef] [PubMed]
24. Chen, W.; Zhou, P.-p.; Zhang, M.; Zhu, Y.-m.; Wang, X.-p.; Luo, X.-a.; Bao, Z.-d.; Yu, L.-j. Transcriptome analysis reveals that up-regulation of the fatty acid synthase gene promotes the accumulation of docosahexaenoic acid in *Schizochytrium* sp. S056 when glycerol is used. *Algal Res.* **2016**, *15*, 83–92. [CrossRef]
25. Ma, Z.; Tan, Y.; Cui, G.; Feng, Y.; Cui, Q.; Song, X. Transcriptome and gene expression analysis of DHA producer *Aurantiochytrium* under low temperature conditions. *Sci. Rep.* **2015**, *5*, 14446. [CrossRef] [PubMed]
26. Zhu, X.; Li, S.; Liu, L.; Li, S.; Luo, Y.; Lv, C.; Wang, B.; Cheng, C.H.; Chen, H.; Yang, X. Genome Sequencing and Analysis of *Thraustochytriidae* sp. SZU445 Provides Novel Insights into the Polyunsaturated Fatty Acid Biosynthesis Pathway. *Mar. Drugs* **2020**, *18*, 118.
27. Haas, B.J.; Papanicolaou, A.; Yassour, M.; Grabherr, M.; Blood, P.D.; Bowden, J.; Couger, M.B.; Eccles, D.; Li, B.; Lieber, M. De novo transcript sequence reconstruction from RNA-seq using the Trinity platform for reference generation and analysis. *Nat. Protoc.* **2013**, *8*, 1494. [CrossRef] [PubMed]
28. Wang, Y.; Pan, Y.; Liu, Z.; Zhu, X.; Zhai, L.; Xu, L.; Yu, R.; Gong, Y.; Liu, L. De novo transcriptome sequencing of radish (*Raphanus sativus L.*) and analysis of major genes involved in glucosinolate metabolism. *BMC Genom.* **2013**, *14*, 836.
29. Zheng, M.; Tian, J.; Yang, G.; Zheng, L.; Chen, G.; Chen, J.; Wang, B. Transcriptome sequencing, annotation and expression analysis of *Nannochloropsis* sp. at different growth phases. *Gene* **2013**, *523*, 117–121. [CrossRef] [PubMed]
30. Hoang, M.H.; Ha, N.C.; Thom, L.T.; Tam, L.T.; Anh, H.T.L.; Thu, N.T.H.; Hong, D.D. Extraction of squalene as value-added product from the residual biomass of Schizochytrium mangrovei PQ6 during biodiesel producing process. *J. Biosci. Bioeng.* **2014**, *118*, 632–639. [CrossRef] [PubMed]
31. Rismani-Yazdi, H.; Haznedaroglu, B.Z.; Hsin, C.; Peccia, J. Transcriptomic analysis of the oleaginous microalga *Neochloris oleoabundans* reveals metabolic insights into triacylglyceride accumulation. *Biotechnol. Biofuels* **2012**, *5*, 74. [CrossRef]
32. O'neill, E.C.; Trick, M.; Hill, L.; Rejzek, M.; Dusi, R.G.; Hamilton, C.J.; Zimba, P.V.; Henrissat, B.; Field, R.A. The transcriptome of *Euglena gracilis* reveals unexpected metabolic capabilities for carbohydrate and natural product biochemistry. *Mol. Biosyst.* **2015**, *11*, 2808–2820. [CrossRef]
33. Li, H.; Wang, W.; Wang, Z.; Lin, X.; Zhang, F.; Yang, L. De novo transcriptome analysis of carotenoid and polyunsaturated fatty acid metabolism in *Rhodomonas* sp. *J. Appl. Phycol.* **2016**, *28*, 1649–1656. [CrossRef]
34. Aasen, I.M.; Ertesvåg, H.; Heggeset, T.M.B.; Liu, B.; Brautaset, T.; Vadstein, O.; Ellingsen, T.E. *Thraustochytrids* as production organisms for docosahexaenoic acid (DHA), squalene, and carotenoids. *Appl Microbiol. Biotechnol.* **2016**, *100*, 4309–4321. [CrossRef] [PubMed]
35. Hopwood, D.A.; Sherman, D.H. Molecular genetics of polyketides and its comparison to fatty acid biosynthesis. *Annu. Rev. Genet.* **1990**, *24*, 37–62. [CrossRef] [PubMed]
36. Meesapyodsuk, D.; Qiu, X.J. Biosynthetic mechanism of very long chain polyunsaturated fatty acids in *Thraustochytrium* sp. 26185. *J. Lipid Res.* **2016**, *57*, 1854–1864. [CrossRef] [PubMed]

37. Ratledge, C. Fatty acid biosynthesis in microorganisms being used for single cell oil production. *Biochimie* **2004**, *86*, 807–815. [CrossRef] [PubMed]
38. Yang, X.; Li, S.; Li, S.; Liu, L.; Hu, Z. De Novo Transcriptome Analysis of Polyunsaturated Fatty Acid Metabolism in Marine Protist *Thraustochytriidae* sp. PKU# Mn16. *Trends Food Sci. Technol.* **2020**, *97*, 35–48.
39. Patell, V.M. Delta 6 Desaturase From Thraustochytrid and its Uses Thereof. U.S. Patent 20,090,118,371, 18 January 2009.
40. Song, Z.; Stajich, J.E.; Xie, Y.; Liu, X.; He, Y.; Chen, J.; Hicks, G.R.; Wang, G. Comparative analysis reveals unexpected genome features of newly isolated *Thraustochytrids* strains: On ecological function and PUFAs biosynthesis. *BMC Genomics* **2018**, *19*, 541. [CrossRef] [PubMed]
41. Yaguchi, T.; Tanaka, S.; Yokochi, T.; Nakahara, T.; Higashihara, T. Production of high yields of docosahexaenoic acid by *Schizochytrium* sp. strain SR 21. *J. Am. Oil Chem. Soc.* **1997**, *74*, 1431–1434. [CrossRef]
42. Chesler, S.N.; Emery, A.P.; Duewer, D.L. Recovery of diesel fuel from soil by supercritical fluid extraction–gas *chromatography. J. Chromatogr. A* **1997**, *790*, 125–130. [CrossRef]
43. Kumari, P.; Reddy, C.R.K.; Jha, B. Comparative evaluation and selection of a method for lipid and fatty acid extraction from macroalgae. *Anal. Biochem.* **2011**, *415*, 134–144. [CrossRef]
44. Lun, L.W.; Gunny, A.A.N.; Kasim, F.H.; Arbain, D. Fourier transform infrared spectroscopy (FTIR) analysis of paddy straw pulp treated using deep eutectic solvent. Proceedings of ADVANCED MATERIALS ENGINEERING AND TECHNOLOGY V: International Conference on Advanced Material Engineering and Technology, Jeju Island, South Korea, 18–20 November 2016.
45. Wang, H.Q.; Liang, F.; Qiao, N.; Dong, J.X.; Zhang, L.Y.; Guo, Y.F.; Wang, H.Q.; Liang, F.; Qiao, N.; Dong, J.X. Chemical Composition of Volatile Oil from Two Emergent Plants and Their Algae Inhibition Activity. *Pol. J. Environ. Stud.* **2014**, *23*, 2371–2374.
46. Gasic, K.; Hernandez, A.; Reporter, S.S. RNA extraction from different apple tissues rich in polyphenols and polysaccharides for cDNA library construction. *Plant Mol. Biol. Rep.* **2012**, *22*, 437–438. [CrossRef]
47. Grabherr, M.G.; Haas, B.J.; Yassour, M.; Levin, J.Z.; Thompson, D.A.; Amit, I.; Adiconis, X.; Fan, L.; Raychowdhury, R.; Zeng, Q. Full-length transcriptome assembly from RNA-Seq data without a reference genome. *Nat. Biotechnol.* **2011**, *29*, 644–652. [CrossRef]
48. Livak, K.J.; Schmittgen, T.D. Analysis of relative gene expression data using real-time quantitative PCR and the 2(-Delta Delta C(T)) Method. *Methods* **2001**, *25*, 402. [CrossRef]
49. Moeller, R.; Douki, T.; Rettberg, P.; Reitz, G.; Cadet, J.; Nicholson, W.L.; Horneck, G. Genomic bipyrimidine nucleotide frequency and microbial reactions to germicidal UV radiation. *Arch. Microbiol.* **2010**, *192*, 521–529. [CrossRef] [PubMed]
50. Choi, S.A.; Jung, J.Y.; Kim, K.; Lee, J.S.; Kwon, J.H.; Kim, S.W.; Yang, J.W.; Park, J.Y. Acid-catalyzed hot-water extraction of docosahexaenoic acid (DHA)-rich lipids from *Aurantiochytrium* sp. KRS101. *Bioresour. Technol.* **2014**, *161*, 469–472. [CrossRef] [PubMed]
51. Cheng, Y.; Sun, Z.; Cui, G.; Song, X.; Qiu, C.; Cheng, Y.; Sun, Z.; Cui, G.; Song, X.; Qiu, C.J.E.; et al. A new strategy for strain improvement of *Aurantiochytrium* sp. based on heavy-ions mutagenesis and synergistic effects of cold stress and inhibitors of enoyl-ACP reductase. *Enzyme Microb. Technol.* **2016**, *93–94*, 182–190. [CrossRef]
52. Rahman, S.N.S.A.; Kalil, M.S.; Hamid, A.A. The significance of fructose and MSG in affecting lipid and docosahexaenoic acid (DHA) production of *Aurantiochytrium* sp. SW1. In *The 2017 UKM FST Postgraduate Colloquium: Proceedings of the Universiti Kebangsaan Malaysia, Faculty of Science and Technology 2017 Postgraduate Colloquium*; AIP Publishing: Melville, NY, USA.
53. Lian, M.; Huang, H.; Ren, L.; Ji, X.; Zhu, J.; Jin, L. Increase of Docosahexaenoic Acid Production by *Schizochytrium* sp. Through Mutagenesis and Enzyme Assay. *Appl. Biochem. Biotechnol.* **2010**, *162*, 935–941. [CrossRef] [PubMed]
54. Christophe, G.; Fontanille, P.; Larroche, C. Research and Production of Microbial Polyunsaturated Fatty Acids. *Bioprocess. Biomol. Prod.* **2019**. [CrossRef]
55. Gago, G.; Diacovich, L.; Arabolaza, A.; Tsai, S.-C.; Gramajo, H. Fatty acid biosynthesis in *actinomycetes. FEMS Microbiol. Rev.* **2011**, *35*, 475–497. [CrossRef] [PubMed]
56. Metz, J.G.; Roessler, P.; Facciotti, D.; Levering, C.; Dittrich, F.; Lassner, M.; Valentine, R.; Lardizabal, K.; Domergue, F.; Yamada, A.; et al. Production of Polyunsaturated Fatty Acids by Polyketide Synthases in Both Prokaryotes and Eukaryotes. *Science* **2001**, *293*, 290–293. [CrossRef] [PubMed]

57. Gabriel, D.; Isabel, M. Lipidomic methodologies for biomarkers of chronic inflammation in nutritional research: ω-3 and ω-6 lipid mediators. *Free. Radic. Biol. Med.* **2019**, *144*, 90–109.
58. Rippe, J.K. Functional Food in the Marketplace: New Products, Availability, and Implications for the Consumer. *Nutritional Health* **2012**, 451–475.
59. Gupta, A.; Barrow, C.J.; Puri, M. Omega-3 biotechnology: *Thraustochytrids* as a novel source of omega-3 oils. *Biotechnol Adv.* **2012**, *30*, 1733–1745. [CrossRef] [PubMed]
60. Engström, M.K.; Saldeen, A.-S.; Yang, B.; Mehta, J.L.; Saldeen, T. Effect of fish oils containing different amounts of EPA, DHA, and antioxidants on plasma and brain fatty acids and brain nitric oxide synthase activity in rats. *Ups J. Med. Sci.* **2009**, *114*, 206–213. [CrossRef]
61. Wang, Q.; Sen, B.; Liu, X.; He, Y.; Xie, Y.; Wang, G. Enhanced saturated fatty acids accumulation in cultures of newly-isolated strains of *Schizochytrium* sp. and *Thraustochytriidae* sp. for large-scale biodiesel production. *Sci. Total Environ.* **2018**, *631*, 994–1004. [CrossRef]
62. Meireles, L.A.; Guedes, A.C.; Malcata, F.X. Lipid class composition of the microalga *Pavlova lutheri*: Eicosapentaenoic and docosahexaenoic acids. *J. Agric. Food Chem.* **2003**, *51*, 2237–2241. [CrossRef] [PubMed]
63. Chi, Z.; Liu, Y.; Frear, C.; Chen, S. Study of a two-stage growth of DHA-producing marine algae *Schizochytrium limacinum* SR21 with shifting dissolved oxygen level. *Appl. Microbiol. Biotechnol.* **2009**, *81*, 1141–1148. [CrossRef] [PubMed]
64. Gravius, B.; Bezmalinović, T.; Hranueli, D.; Cullum, J. Genetic instability and strain degeneration in *Streptomyces rimosus*. *Appl. Environ. Microbiol.* **1993**, *59*, 2220–2228. [CrossRef] [PubMed]
65. Meireles, L.A.; Guedes, A.C.; Malcata, F.X. Increase of the yields of eicosapentaenoic and docosahexaenoic acids by the microalga *Pavlova lutheri* following random mutagenesis. *Biotechnol. Bioeng.* **2003**, *81*, 50–55. [CrossRef] [PubMed]
66. Alonso, D.L.; Belarbi, E.-H.; Fernández-Sevilla, J.M.; Rodríguez-Ruiz, J.; Grima, E.M. Acyl lipid composition variation related to culture age and nitrogen concentration in continuous culture of the microalga Phaeodactylum tricornutum. *Phytochemistry* **2000**, *54*, 461–471. [CrossRef]
67. Beacham, T.A.; Macia, V.M.; Rooks, P.; White, D.A.; Ali, S.T. Altered lipid accumulation in *Nannochloropsis salina* CCAP849/3 following EMS and UV induced mutagenesis. *Biotechnol. Rep.* **2015**, *16*, 87–94. [CrossRef]
68. Forján, E.; Garbayo, I.; Henriques, M.; Rocha, J.; Vega, J.M.; Vílchez, C. UV-A Mediated Modulation of Photosynthetic Efficiency, Xanthophyll Cycle and Fatty Acid Production of *Nannochloropsis*. *Mar. Biotechnol.* **2010**, *13*, 366–375. [CrossRef] [PubMed]
69. Chan, Y.A.; Podevels, A.M.; Kevany, B.M.; Thomas, M.G. Biosynthesis of polyketide synthase extender units. *Nat. Prod. Rep.* **2009**, *26*, 90–114. [CrossRef]
70. Ye, H.; He, Y.; Xie, Y.; Sen, B.; Wang, G. Fed-batch fermentation of mixed carbon source significantly enhances the production of docosahexaenoic acid in *Thraustochytriidae* sp. PKU# Mn16 by differentially regulating fatty acids biosynthetic pathways. *Bioresour. Technol.* **2020**, *297*, 122402.
71. Henderson, R.J.; Phytochemistry, E.E. Polyunsaturated fatty acid metabolism in the marine dinoflagellate *Crypthecodinium cohnii*. *Phytochemistry* **1991**, *30*, 1781–1787. [CrossRef]
72. Jeon, E.; Lee, S.; Won, J.I.; Han, S.O.; Kim, J.; Lee, J. Development of *Escherichia coli* MG1655 strains to produce long chain fatty acids by engineering fatty acid synthesis (FAS) metabolism. *Enzym. Microb. Technol.* **2011**, *49*, 44–51. [CrossRef]
73. Hayashi, S.; Satoh, Y.; Ujihara, T.; Takata, Y.; Dairi, T. Enhanced production of polyunsaturated fatty acids by enzyme engineering of tandem acyl carrier proteins. *Sci. Rep.* **2016**, *6*, 35441. [CrossRef]
74. Napier, J.A. Plumbing the depths of PUFA biosynthesis: A novel polyketide synthase-like pathway from marine organisms. *Trends Plant Sci.* **2002**, *7*, 0–54. [CrossRef]
75. Santín, O.; Moncalián, G. Loading of malonyl-CoA onto tandem acyl carrier protein domains of polyunsaturated fatty acid synthases. *J. Biol. Chem.* **2018**, *293*, 12491–12501. [CrossRef] [PubMed]
76. Zhang, Y.; Adams, I.P.; Ratledge, C. Malic enzyme: The controlling activity for lipid production? Overexpression of malic enzyme in *Mucor circinelloides* leads to a 2.5-fold increase in lipid accumulation. *Microbiology* **2007**, *153*, 2013–2025. [CrossRef] [PubMed]
77. Mansour, M.P.; Volkman, J.K.; Jackson, A.E.; Blackburn, S.I. The fatty acid and sterol composition of five marine dinoflagellates. *J. Phycol.* **1999**, *35*, 710–720. [CrossRef]

78. Chang, K.J.L.; Nichols, C.M.; Blackburn, S.I.; Dunstan, G.A.; Koutoulis, A.; Nichols, P.D. Comparison of *thraustochytrids Aurantiochytrium* sp., *Schizochytrium* sp., *Thraustochytrium* sp., and *Ulkenia* sp. for production of biodiesel, long-chain omega-3 oils, and exopolysaccharide. *Mar. Biotechnol.* **2014**, *16*, 396–411. [CrossRef] [PubMed]
79. Xie, Y.; Wang, G. Mechanisms of fatty acid synthesis in marine fungus-like protists. *Appl. Microbiol. Biotechnol.* **2015**, *99*, 8363–8375. [CrossRef] [PubMed]
80. Robertsen, H.L.; Musiol-Kroll, E.M.; Ding, L.; Laiple, K.J.; Hofeditz, T.; Wohlleben, W.; Lee, S.Y.; Grond, S.; Weber, T. Filling the gaps in the kirromycin biosynthesis: Deciphering the role of genes involved in ethylmalonyl-CoA supply and tailoring reactions. *Sci. Rep.* **2018**, *8*, 3230. [CrossRef]
81. Bi, Z.-Q.; Ren, L.-J.; Hu, X.-C.; Sun, X.-M.; Zhu, S.-Y.; Ji, X.-J.; Huang, H. Transcriptome and gene expression analysis of docosahexaenoic acid producer *Schizochytrium* sp. under different oxygen supply conditions. *Biotechnol. Biofuels* **2018**, *11*, 249. [CrossRef]

microorganisms

MDPI

Article

Novel Bi-Factorial Strategy against *Candida albicans* Viability Using Carnosic Acid and Propolis: Synergistic Antifungal Action

Alejandra Argüelles [1], Ruth Sánchez-Fresneda [1,2], José P. Guirao-Abad [1,2], Cristóbal Belda [3], José Antonio Lozano [4], Francisco Solano [4] and Juan-Carlos Argüelles [2,*]

1 Vitalgaia España, S.L. 30005 Murcia, Spain; a.arguellesprieto@um.es (A.A.); ruth.sanchez1@um.es (R.S.-F.); josepedro.guirao@um.es (J.P.G.-A.)
2 Área de Microbiología, Facultad de Biología, Universidad de Murcia, E-30071 Murcia, Spain
3 Hospital Universitario Sanchinarro, 28050 Madrid, Spain; cbelda@cmcbiotech.xyz
4 Departamento de Bioquímica y Biología Molecular B e Inmunología, Facultad de Medicina, Universidad de Murcia, E-30100 Murcia, Spain; jalozate@um.es (J.A.L.); psolano@um.es (F.S.)
* Correspondence: arguelle@um.es; Tel.: +34-868-88-71-31; Fax: +34-868-88-39-63

Received: 25 April 2020; Accepted: 14 May 2020; Published: 16 May 2020

Abstract: The potential fungicidal action of the natural extracts, carnosic acid (obtained from rosemary) and propolis (from honeybees' panels) against the highly prevalent yeast *Candida albicans*, used herein as an archetype of pathogenic fungi, was tested. The separate addition of carnosic acid and propolis on exponential cultures of the standard SC5314 *C. albicans* strain caused a moderate degree of cell death at relatively high concentrations. However, the combination of both extracts, especially in a 1:4 ratio, induced a potent synergistic pattern, leading to a drastic reduction in cell survival even at much lower concentrations. The result of a mathematical analysis by isobologram was consistent with synergistic action of the combined extracts rather than a merely additive effect. In turn, the capacity of SC5314 cells to form in vitro biofilms was also impaired by the simultaneous presence of both agents, supporting the potential application of carnosic acid and propolis mixtures in the prevention and treatment of clinical infections as an alternative to antibiotics and other antifungal agents endowed with reduced toxic side effects.

Keywords: carnosic acid; propolis; antifungal action; synergy; *Candida albicans*; biofilms

1. Introduction

Natural compounds obtained from plants and other sources have been successfully used for healthcare, cosmetics or food since ancient times, and they represent an important biotechnological target in today's world. Rosemary (*Rosmarinus officinalis*) is one of the most widely studied plants for this goal, because it contains several terpenoids and other bioactive compounds, of which carnosic acid (CA) is one of the most prominent since it is endowed with both antioxidant and antimicrobial activities [1,2]. Due to these properties, CA has been used successfully in several biotechnological applications and as a food additive to improve the stability of meat [3].

In turn, propolis (PP) is a generic name for a resinous and assorted material produced by honeybees (*Apis mellifera*) from a large number of substances collected from different parts of plants, including buds and exudates, with broad clinical applications [4]. The composition of PP depends on the geographic area and the flora, climate and region covered by honeybees [5,6]. Indeed, more than 300 constituents have been identified and new ones are still being recognized. Therefore, the chemical composition of propolis is not only complex but also highly variable with serious standardization problems [4]. However, the most common constituents of raw PP are typically 50% plant resins, 30% waxes,

10% essential oils, 5% pollen and 5% of other components of a diverse origin and nature. Among the large list of beneficial applications of propolis, the most well-described include anti-inflammatory, antioxidant, immunostimulant, cytostatic, antimicrobial, antitumor and hepatoprotective activities [7,8].

The present work focuses on the application of these two natural extracts, CA and PP, in the development of new antifungal chemotherapies. This field has traditionally received less attention and research effort than an investigation into their bacterial action because (i) a low number of fungal pathogens were thought to exist, most of them of an opportunistic nature; (ii) mycosis were usually superficial infections with good prognosis, except in the case of complications. Nevertheless, this scenario has changed in recent years, during which a dramatic increase in morbidity and mortality caused by septicemic mycosis has been recorded. This septicemia particularly affects patients subjected to invasive surgery, medical implants, or lengthy hospitalization, as well as the immunocompromised chronic population suffering AIDS and other viral diseases [9–11].

Moreover, there is growing concern over several fungal species that were classically considered as innocuous, and which are responsible for infectious outbreaks in hospitals [10,12]. In fact, the opportunistic ascomycetous yeast *C. albicans* now represents the fourth most common cause of nosocomial diseases worldwide [11]. Therefore, the search for new, safer, and more potent antifungal compounds is an urgent clinical need [13]. Apart from using known natural or chemically synthesized antifungal agents, which frequently cause unwanted side effects after prolonged treatment, a complementary strategy focuses on the search for novel natural products with effective fungicidal activity. Here, we report on an unexpected potent antifungal effect shown by mixtures of CA and PP. Both are natural agents with certain antioxidant properties, and they are commonly employed and commercialized as food additives or as prodrugs because of their beneficial effects on the human metabolism, without undesirable side effects.

2. Materials and Methods

2.1. Yeast Strains and Culture Conditions

The *C. albicans* SC5314 standard strain was used throughout this study. This same strain has been employed in recent studies on susceptibility to the antifungal compounds polyenes and echinocandins [14]. Yeast cell cultures (blastoconidia) of this opportunistic pathogen were grown at 37 °C by shaking in YPD medium consisting of 2% peptone, 1% yeast extract and 2% glucose.

2.2. Natural Extracts

Carnosic acid is a natural benzenediol abietane diterpene found in rosemary (*Rosmarinus officinalis*) and common sage (*Salvia officinalis*). Dried leaves of rosemary contain around 2% CA. The CA used for this work was provided by Nutrafur S.A. (Murcia, Spain). This preparation was obtained according to previously described methods [3,15,16] with slight modifications to get the highest purification. Purity was assayed by HPLC–UV-DAD using conditions described below. According to that, the content of diterpenes is enriched to reach around 80%, with CA being the most abundant one (around 72%, see Table 1 for composition in other components). This preparation was referred to as CA throughout the current manuscript and dissolved in 98% ethanol until the required concentration for the antifungal assays. However, the possibility of some other components (minor diterpenes or others) could be involved in the antifungal action cannot be totally ruled out.

Concerning propolis, several extracts from different geographic regions were initially tested. For the biotechnological and therapeutic purposes of this study, the best antimicrobial outcomes and availability corresponded to a propolis of Chinese origin, which could be provided in great amounts by Monteloeder local supplier (Parque Empresarial, Alicante, Spain). The raw PP stuffs were ground to a fine powder in a mortar and then dissolved by gentle shaking in prewarmed ethanol (98%, at 55 °C) at the required concentrations for antifungal assays. A control with 98% ethanol was run without any significant antimicrobial effect. Other samples were dissolved in dimethyl sulfoxide for chromatographic analysis (see below). According to these determinations, the total polyphenols content was estimated in the 70%–90% range. Particular details about the complex composition are out of the scope of this study. The main goal of this focused on the use of an antifungal mixture of raw PP and carnosic acid rather than the fractionation and characterization of the PP active compounds. Further studies are underway in order for fractionation and identification of the compounds or fractions responsible for the antifungal action, as well as the effect on other yeast strains.

Table 1. Composition of the semipurified carnosic acid (CA) used in this study.

Component	%
Carnosic acid	70%–75% (mean 72%)
Minor diterpenes (carnosol, rosmanol, epirosmanol and 12-methyl carnosate)	3%–7%
Other plant lipids	5%–10%
Plant carbohydrates	2%–5%
Proteins	0%–0.5%
Water	2%–4%
Mineral salts	1%–2%
Other plant materials	2%–4%

2.3. HPLC

Conditions for analytical HPLC were as described by Benavente-García et al. [17] with slight modifications. Samples of CA or PP (5 mg/mL) were dissolved in DMSO, filtered through a nylon membrane 0.45 mm pore size. 20 μL of solution were injected in a LiChrospher 100-C18 reverse-phase column (250 × 4.0 mm inner diameter) thermostatized at 30 °C. Mobile phase consisted of a gradient from acetonitrile/2.5% acetic acid aqueous solution starting with 5%/95% and finishing with 95%/5% acetonitrile/acidic water. The flux was 1 mL/min. Detection was followed by a Diodo-Array Agilent UV-Vis detector. Usually, chromatographic profiles were monitored at 280, 340 and 370 nm for differentiated identification of phenols and flavonoids.

2.4. Determination of Cell Viability

C. albicans cultures were grown at 37 °C in YPD until they reached exponential phase (OD_{600nm} = 0.8–1.0) and were then divided into several identical aliquots, which were treated with the concentrations of CA and PP indicated in the Results section. In all series of experiments, control samples with neither extract added were incubated simultaneously. Cell viability was determined in samples diluted appropriately with sterile water by plating in triplicate on solid YPD after incubation for 1–2 days at 37 °C. Between 30 and 300 colonies were counted per plate. Survival percentages were normalized to control samples (100% viability). Colony growth in solid medium was tested by spotting 5 μL from the respective 10-fold dilutions onto YPD agar. Then, the plates were incubated at 30 °C and scored after 24 or 48 h. The polyene Amphotericin B has been included as a positive control of antifungal activity.

2.5. Morphological Analysis

After exposure to the different antifungals, cell morphology was recorded with a Leica DMRB microscope using the Nomarsky interference contrast technique. The microscope was equipped with a Leica DC500 camera connected to a PC containing the Leica Application Suite V 2.5.0 R1 software.

2.6. Biofilm Formation

In vitro biofilm formation was analyzed on the surface of polystyrene 96-well microtiter plates using previously described methods (Pierce et al. 2014). Briefly, 100 μL of the analyzed *C. albicans* suspension (1.0 × 10^6 blastoconidia/mL) in RPMI 1640 was allowed to adhere and form a biofilm at 37 °C for 24 h. Following biofilm formation, the medium was aspirated, and non-adherent cells were removed by washing three times with sterile phosphate saline buffer (PBS). Then, CA and PP were added immediately as separate extracts or as mixtures and the biofilms were further incubated for 24 h. The viability of yeast cells within the biofilms was quantified by means of 2,3-bis-(2-methoxy-4-nitro-5-sulfophenyl)-2H-tetrazolium-5-carboxalinide reduction assay (XTT) reduction assay. XTT (Sigma Chemicals) was prepared as a saturated solution at 0.5 g L^{-1} in PBS, filter-sterilized through a 0.22 μm pore size filter, dispensed and stored at −70 °C. An aliquot of the stock solution of XTT (100 μL) was thawed prior to each assay and 10 mM menadione (Sigma Chemicals) in ethanol was added to obtain a final concentration of 25 μM. A 100 μL aliquot of the XTT–menadione solution was added to each well and the plates were incubated for 2 h al 37 °C. The metabolic activity of sessile *C. albicans* cells was assessed quantitatively by measuring the absorbance in a microtiter plate reader (Asys Jupiter) at 490 nm. The tetrazolium salt that accumulated following the reduction of XTT by fungal dehydrogenases was proportional to the number of viable cells present in the biofilm. A base line at 490 nm was run for background subtraction before the measurement of each microplate. Data were expressed as the percentage of metabolic activity in the treated biofilm samples with respect to the 100% (untreated controls).

3. Results and Discussion

3.1. Single Antifungal Action of CA and PP on C. albicans

Several reports have suggested the existence of noticeable antimicrobial activity in CA and PP extracts against a set of pathogenic bacteria and fungi [3,18–20]. Accordingly, we carried out a comparative analysis of this potential antifungal effect against the standard SC5314 strain of *C. albicans* using a range of concentrations similar to that previously chosen in other studies on antimicrobial tests with the same compounds [21–23]. In this regard, the concentrations of CA applied were lower than those used for PP. As can be seen in Figure 1A, the fungicidal action of CA in liquid YPD medium was substantial, a large fall in cell viability being observed on exponential *C. albicans* cells after 1 h of exposure to the diterpene. This action was proportional to the applied dose and clearly noticeable even at the lower concentration tested (50 μg/mL), and very pronounced at the highest concentration (500 μg/mL). In turn, the effect of PP was rather weak, and the extract only induced a small but statistically significant degree of cell death after addition of 500 μg/mL (Figure 1A). The parallel assays to measure colony growth on solid YPD plates showed a good correlation with the data obtained in liquid medium (Figure 1B).

The MIC_{50} values for the *C. albicans* SC5314 standard strain in the presence of CA and PP were calculated following the European Committee on Antimicrobial Susceptibility Testing (EUCAST) protocol established for this parameter [24]. The obtained values were 125 μg/mL for CA and 250 μg/mL for PP. These concentrations are within the range previously reported for other *C. albicans* genetic backgrounds of clinical or laboratory origin (EUCAST) [24].

3.2. Antifungal Activity of CA and PP Mixtures on C. albicans

We also performed a more in-depth study regarding the action of both components either as an isolated application or in combination. Different CA:PP proportions were assayed in preliminary experiments (results not shown). Figure 2 shows the degree of *C. albicans* cell death as a function of time (1–5 h) in liquid medium (Figure 2A), as well as a parallel assay of colony growth on solid YPD plates (Figure 2B). Taking into account that CA always displayed higher fungicidal action than PP at identical doses, we chose the lowest concentration of CA (50 μg/mL) and 200 μg/mL of PP to maintain a 1:4 ratio.

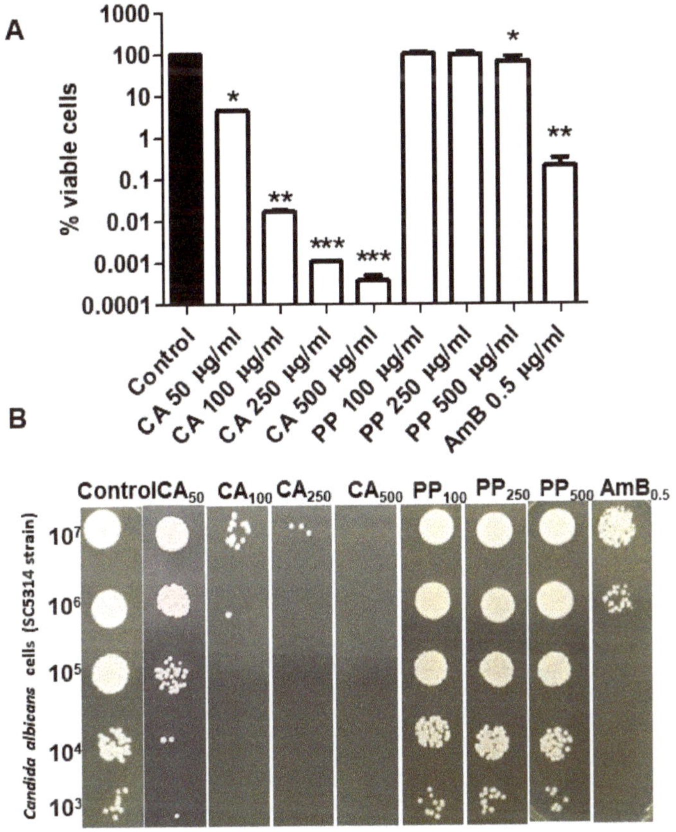

Figure 1. Fungicidal effect caused by the addition of CA or propolis (PP) on the standard strain SC5314 of *C. albicans*. Exponential cells were aliquoted and treated for 1 h at 37 °C with the indicated concentrations of CA and PP. Samples for each treatment and a control (black bar) were spread on YPD plates and the viability was determined by CFU counting (**A**). The macroscopic colonial growth in the different conditions was also recorded in solid media. 10^7 cells/mL were diluted in YPD and 10-fold dilutions thereof, were spotted in 5 µL onto YPD agar. The plates were further incubated at 37 °C and scored after 24 h (**B**). Amphotericin B (AmB) was included as a positive antifungal control. The data shown are representative of three independent experiments. Statistically significant differences (* = $P < 0.05$; ** = $P < 0.01$; ***, $P = 0.001$) with respect to an untreated control according to Mann–Whitney U test.

The experimental results presented in Figure 2A confirm that CA had greater candidacidal activity than PP, measured in terms of viable cells (CFUs, Figure 2A), the difference being clearly perceptible after only 1 h of treatment, albeit no further reduction was observed at longer times. However, the exposure of *C. albicans* cultures to a combination of CA and PP (1:4) led to a much larger loss (up to five log units) of cell viability compared with the effect of the agents added individually. This pattern was maintained throughout the experiment (Figure 2A), although a modest recovery of cells was evident after 5 h of treatment. The data from the liquid medium showed a very good correlation with the macroscopic growth recorded on solid plates (Figure 2B). Overall, we provide consistent evidence supporting the fact that the combination of CA plus PP induces a fast and strong fungicidal action, which is undoubtedly more effective than the mere addition of one of the substances alone.

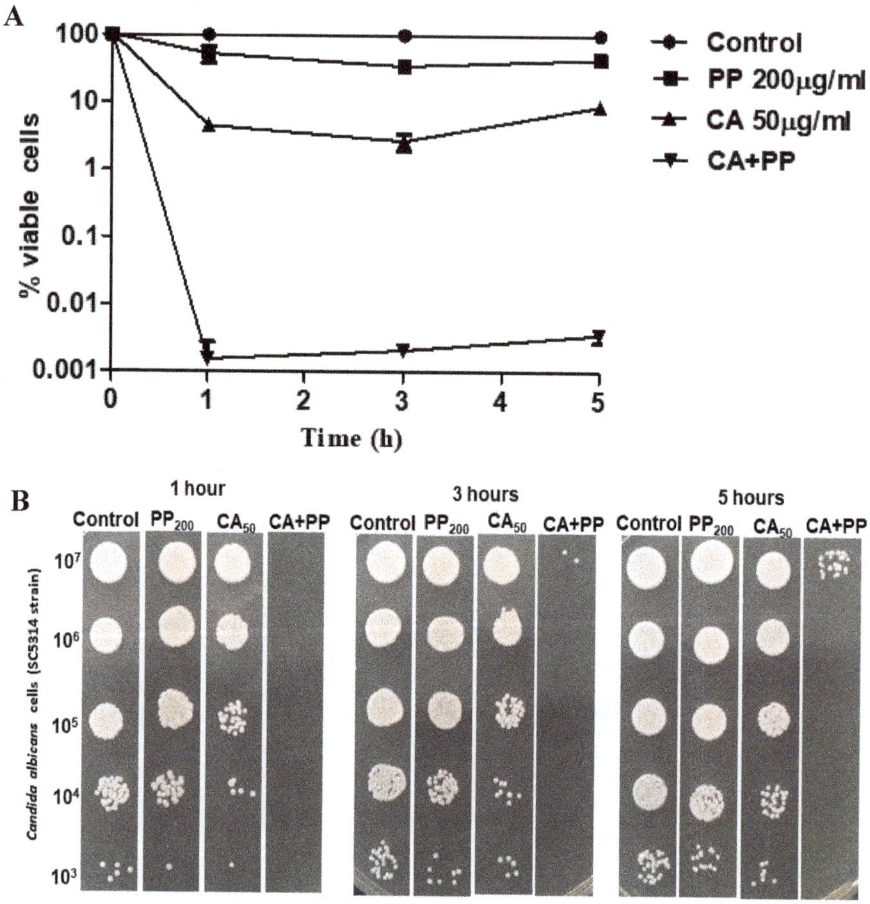

Figure 2. Time-course evolution of cell viability (%) after the addition of 50 µg/mL CA and 200 µg/mL PP, either separately or combined, to cultures of SC5314 strain growing on YPD. After incubation for 1 h at 37 °C, identical samples were harvested and washed at the indicated periods and the percentage of surviving cells in liquid medium (**A**) or the formation of colonies on solid plates (**B**) was determined. The experiments were repeated three times with consistent results. For other details, see Figure 1.

If the incubation time is lengthened from 1 h to 5 h, this loss of cell viability appears to be mitigated and a slight but consistent resumption of active growth can be observed rather than the complete elimination of the cultures (Figure 2A,B). It seems that the small number of cells surviving after 1 h was sufficient to start new cycles of bud division. Although a plausible explanation for this residual fraction of viable blastoconidia may come out from an insufficient dose of extracts applied, this should not be the case, because a five-fold increase in the concentration of CA was not dose-response (Figure 1), whereas the weak antifungal effect of PP at 100 µg/mL and 500 µg/mL was rather similar (Figure 1).

As an alternative, the presence of minority cells that possess intrinsic resistance (persister cells) might be operative [25] or the possible inactivation of at least one extract. Whatever the case, further studies are required in order to throw more light on this pattern. Bearing in mind the therapeutic potential of applying a combination of both agents, it is important to get the best formulation for this mixture and identify the intracellular effects on fungal cells during the time of exposure [2,26].

3.3. Morphological Changes Induced by Exposure to CA and PP

The hypothetical morphological changes recorded in growing SC5314 cells upon exposure to both extracts for the intermediate time of 3 h, were monitored by optical microscopy (Figure 3). Similar observations were also visible at longer times (5 h, not shown). For this purpose, identical cell samples were fixed with formalin and kept at 4 °C until their microscopical examination. The Nomarsky interferential contrast confirmed the reduction in the number of viable cells, particularly after the combined action of CA and PP (Figure 3). In this analysis, a concentration of 100 μg/mL was used, since a lower dose (50 μg/mL) did not cause alterations in the external yeast shape, despite it inducing a significant degree of cell killing (Figure 1). A morphological inspection indicated an increase of inner granularity in the samples treated with PP respect to the control. However, and somehow surprisingly, no appreciable cell lysis could be observed, as it occurs in the same strain exposed to 0.5 μg/mL Amphotericin B or 0.1 μg/mL Micafungin [14], indicating different mechanisms of action of the antibiotic and the natural agents herein studied. In turn, the addition of CA, alone or combined with PP, promoted a clear diminution of the cell volume (Figure 3).

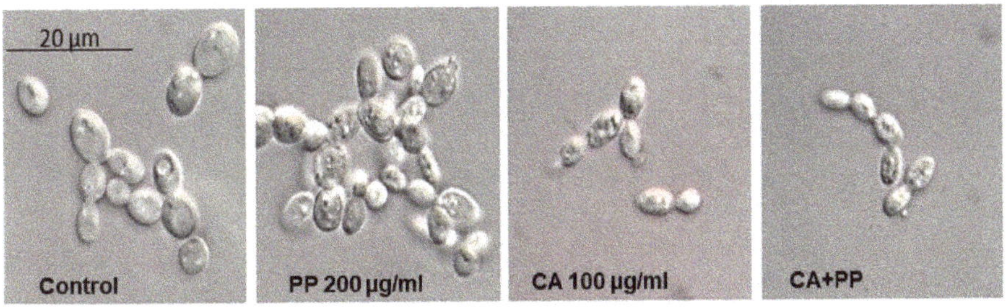

Figure 3. Morphological changes in C. albicans SC5314 blastoconidia induced by the individual addition of CA and PP or in combination. YPD-grown exponential cells (OD_{600} = 0.8) were exposed for 3 h at 37 °C to the indicated concentrations of CA and PP. An untreated sample was maintained as a control. Two similar representative images from each treatment were taken by means of Nomarsky interferential contrast.

3.4. Isobologram Plot: Mathematical Demonstration of Synergism

The analysis of the results obtained by single treatments with CA or PP in comparison to combined mixtures reveals a remarkable reduction of the required doses of both compounds to get a similar degree of cell death in cultures of *C. albicans* (Figures 1 and 2). Thus, the collated data from cell survival clearly points out the existence of a reciprocal stimulation of the fungicidal activity due to the combination of CA and PP.

Nevertheless, this outcome, which could be of importance for the potential therapeutic use of the agents, still needed to be demonstrated by a quantitative mathematical approach. The synergistic cooperation between two compounds endowed with antimicrobial activity has been carefully considered previously [27], utilizing isobolograms to confirm (or not) the strength of the effect. According to this, the MICs for each individual compound should be plotted on the corresponding Cartesian axes for their respective concentrations (Figure 4). The combined MIC value corresponds to the intersection point recorded after plotting. Thus, the action of two compounds would be additive if this point lies on the line that joins the individual MICs. However, the action would be synergistic when that point fits within the right triangle obtained, and the effect would be antagonistic if the point falls outside this triangle. In our case, the plot took as reference the corresponding MIC_{50} values calculated previously for the separate additions of CA and PP represented on the respective Cartesian axes (Figure 4). The point obtained after the supply of a mixture of both compounds corresponds to a concentration of 31.25 μg/mL (CA) and 125 μg/mL (PP) acting on growing SC5314 *C. albicans* cultures

(Figure 4). This clearly fits inside the corresponding right triangle, confirming that the combined action of CA plus PP at the stated ratio of 1:4 conveys a strong synergic fungicidal action.

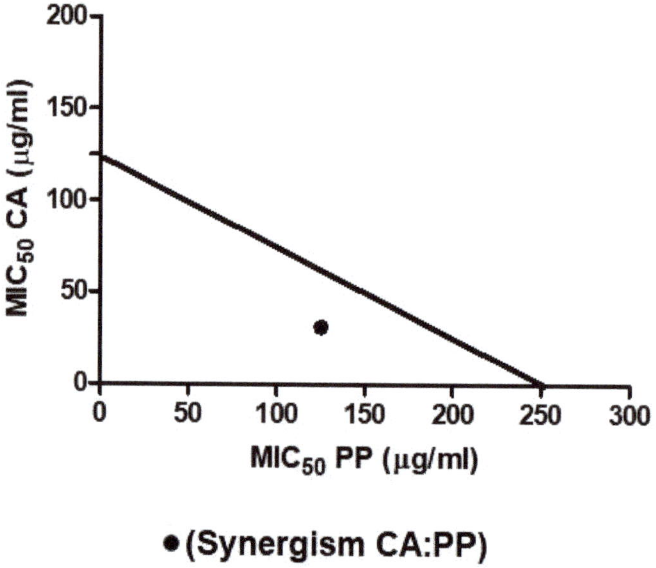

Figure 4. Isobologram plot showing the synergistic action of CA and PP on the *C. albicans* SC5314 strain. The individual MIC_{50} values calculated for CA and PP are represented on the Cartesian axes. The combination of both compounds gives rise to a point that falls inside in the right triangle obtained by drawing the line that joins the MICs of the individual agents.

3.5. The Capacity to Form Biofilms Is Impaired by the Addition of CA and PP

The infections provoked by *Candida albicans* that give rise to the formation of structured biofilms have important clinical repercussions, particularly among the immunocompromised patients [28,29]. However, the therapeutic options available for the successful treatment of drug-resistant biofilms are scarce and new strategies need to be designed to address this medical problem [28,30]. We have previously examined the efficacy of the antifungals Amphotericin B and Micafungin in connection with the trehalose biosynthetic pathway in *C. albicans* and *C. parapsilosis*, with a moderate degree of therapeutic success [14].

The action of CA and PP added individually or in combination, on the percentage of cell viability inside the preformed biofilms was assessed using the XTT reduction assay described in the Methods section. In these experiments, three concentrations of CA were used with the concomitant addition of increasing doses of PP. In these particular series, the CA:PP relation varied from 1:8 to 4:1 due to the particular characteristics of the biofilms. As can be seen in Figure 5, incubation of SC5314 sessile cells for 24 h in the presence of increasing CA:PP ratios induced a notable reduction in the metabolic activity and impaired the degree of biofilm production. Of the different formulations tested, a similar maximum degree of inhibition was achieved with the CA:PP ratios of 62.5:500 and 125:250 (Figure 5). These data demonstrate that the synergistic action of CA and PP has a direct and progressive inhibitory effect on the capacity of *C. albicans* to form biofilms. They also reinforce the potential therapeutic application proposed for the assayed formula.

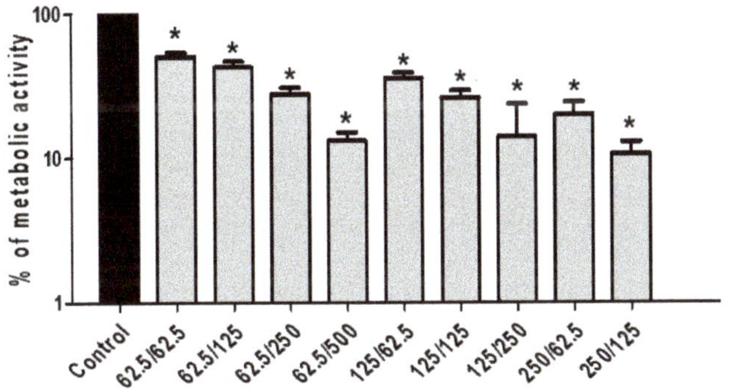

Figure 5. Biofilm formation capability recorded by different combinations of CA and PP in *C. albicans* SC5314 strain. The metabolic activity of the formed biofilms was quantified by the XTT reduction assay (see Methods). The results are expressed as the mean + standard deviation of two experiments with six replications for each group. The concentrations of CA and PP are expressed as μg/mL. Statistically significant differences (* = $P < 0.05$) were recorded with respect to an untreated control according to the Mann–Whitney U test.

4. Conclusions

Herein, we present preliminary but conclusive evidence consistent with a synergistic fungicidal action induced by the combination of two natural compounds, CA and PP. These agents are described elsewhere as antioxidants [3,18,21] and they have been used with some beneficial effects on animal cells due to their antioxidant activity and the capacity to scavenge ROS. However, the redox balance inside cells is complex, and some well-founded opinion reports are putting in doubt the widely assumed scientific belief that antioxidants agents are always beneficial for living cells [31]. Recently, some reports suggest that under specific conditions, the simultaneous presence of two antioxidants could result in an opposite cytotoxic action due to the induction of a high intracellular oxidative stress impairing mitochondrial function and aerobic metabolism [32,33]. Presumably, the synergistic action herein reported would be related to intracellular impairment of the antioxidant–prooxidant balance caused by both agents. In this regard, quinone reductases are crucial enzymes in the maintenance of the cellular redox balance, especially quinone reductase 2 [34].

Taken together, our results support the continued search for more natural bioactive molecules with the purpose of their application in clinical therapies against *C. albicans* and possibly other harmful pathogenic fungi. This strategy would enable us to avoid the undesirable noxious side effects triggered by some conventional antifungals and could surmount their low selective toxicity due to the eukaryotic nature of fungal cells. Indeed, our previous approaches testing accredited compounds, like validamycin A or resveratrol, yielded quite modest antimicrobial outcomes [35,36]. Further studies are currently underway in order to unravel the mechanisms involved in the synergistic fungicidal activity of CA and PP mixtures.

5. Patents

A patent resulted from the work here reported. Title: Synergistic composition comprising propolis and carnosic acid for use in the prevention and treatment of candidiasis. Authors. J.A. Lozano, J.C. Argüelles, A. Argüelles, R. Sánchez-Fresneda and J.P. Guirao-Abad. Reference: Europe (EP 3272344), USA (10,272,120 B2), Canada (2,975,047). Date of concession: 01/31/2019.

Author Contributions: Conceived and designed the experimental research: F.S. and J.-C.A.; Performed the experiments: A.A., R.S.-F., J.P.G.-A.; Explored the therapeutic options: C.B.; Analyzed the data: J.A.L., F.S., J.-C.A.; Contributed reagents/materials/analysis tools: C.B., J.A.L., F.S., J.-C.A.; Wrote the paper: F.S., J.C.A. All authors have read and agreed to the published version of the manuscript.

Funding: This research received no external funding. A.A., R.S.-F. and J.P.G.-A. were partially supported by a fellowship from Vitalgaia España, S.L. The APC was funded by Cespa, Servicios Públicos de Murcia, S.A. (Murcia, Spain).

Acknowledgments: We are indebted to Manuel Abadía and Angel Tomás, co-presidents of Vitalgaia España, S.L., for their continuous encouragement and warm support. J.C.A. is also grateful for the financial contract provided by Cespa, Servicios Públicos de Murcia, S.A. (Murcia, Spain). Ethical Approval was not required.

Conflicts of Interest: The authors declare no conflict of interest. The funders had no role in the design of the study; in the collection, analyses, or interpretation of data; in the writing of the manuscript, or in the decision to publish the results.

References

1. Bernardes, W.A.; Lucarini, R.; Tozatti, M.G.; Souza, M.G.; Andrade Silva, M.L.; da Silva Filho, A.A.; Martins, C.H.G.; Miller Crotti, A.E.; Pauletti, P.M.; Groppo, M.; et al. Antimicrobial activity of *Rosmarinus officinalis* against oral pathogens: Relevance of carnosic acid and carnosol. *Chem. Biodiver.* **2010**, *7*, 1835–1840. [CrossRef] [PubMed]
2. Birtić, S.; Dussort, P.; Pierre, F.-X.; Bily, A.C.; Roller, M. Carnosic acid. *Phytochemistry* **2015**, *115*, 9–19.
3. Jordán, M.J.; Castillo, J.; Bañón, S.; Martínez-Conesa, C.; Sotomayor, J.A. Relevance of the carnosic acid/carnosol ratio for the level of rosemary diterpene transfer and for improving lamb meat antioxidant status. *Food Chem.* **2014**, *151*, 212–218. [CrossRef] [PubMed]
4. Bankova, V. Chemical diversity of propolis and the problem of standardization. *J. Ethnopharm.* **2005**, *100*, 114–117. [CrossRef] [PubMed]
5. Miguel, M.G.; Antunes, M.D. Is propolis safe as an alternative medicine? *J. Pharm. Bioal. Sci.* **2011**, *3*, 479. [CrossRef]
6. Huang, S.; Zhang, C.-P.; Wang, K.; Li, G.Q.; Hu, F.L. Recent advances in the chemical composition of Propolis. *Molecules* **2014**, *19*, 19610–19632. [CrossRef]
7. Cigut, T.; Polak, T.Z.; Ga, L.; Raspor, P.; Jamnik, P. Antioxidative activity of Propolis extract in yeast cells. *J. Agric. Food Chem.* **2011**, *59*, 11449–11455. [CrossRef]
8. De Castro, P.A.; Bom, V.L.; Brown, N.A.; de Almeida, R.S.C.; Ramalho, L.N.Z.; Savoldi, M.; Goldman, M.H.S.; Berretta, A.A.; Goldman, G.H. Identification of the cell targets important for propolis-induced cell death in *Candida albicans*. *Fungal Genet. Biol.* **2013**, *60*, 74–86. [CrossRef]
9. Sardi, J.C.; Giannini, M.J.; Bernardi, T.; Fusco-Almeida, A.M.; Giannini, M.M. *Candida* species: Current epidemiology, pathogenicity, biofilm formation, natural antifungal products and new therapeutic options. *J. Med. Microbiol.* **2013**, *62*, 10–24. [CrossRef]
10. Pfaller, M.A.; Jones, R.N.; Castanheira, M. Regional data analysis of *Candida* non-albicans strains collected in United States medical sites over a 6-year period, 2006–2011. *Mycoses* **2014**, *57*, 602–611. [CrossRef]
11. Pfaller, M.A.; Castanheira, M. Nosocomial candidiasis: Antifungal stewardship and the importance of rapid diagnosis. *Med. Mycol.* **2016**, *54*, 1–22. [CrossRef] [PubMed]
12. Yapar, N. Epidemiology and risk factors for invasive candidiasis. *Therap. Clin. Risk Manag.* **2014**, *10*, 95–105. [CrossRef] [PubMed]
13. Campoy, S.; Adrio, J.L. Antifungals. *Biochem. Pharmacol.* **2017**, *133*, 86–96. [CrossRef] [PubMed]
14. Guirao-Abad, J.P.; Sánchez-Fresneda, R.; Alburquerque, B.; Hernández, J.A.; Argüelles, J.C. ROS formation is a differential contributory factor to the fungicidal action of Amphotericin B and Micafungin in *Candida albicans*. *Internatl. J. Med. Microbiol.* **2017**, *307*, 241–248. [CrossRef]
15. Kontogianni, V.G.; Tomic, G.; Nikolic, I.; Nerantzaki, A.A.; Sayyad, N.; Stosic-Grujicic, S.; Stojanovic, I.; Gerothanassis, I.P.; Tzakos, A.G. Phytochemical profile of *Rosmarinus officinalis* and *Salvia officinalis* extracts and correlation to their antioxidant and anti-proliferative activity. *Food Chem.* **2013**, *136*, 120–129. [CrossRef] [PubMed]
16. Erkan, N.; Ayranci, G.; Ayranci, E. Antioxidant activities of rosemary (*Rosmarinus Officinalis L.*) extract, blackseed (*Nigella sativa* L.) essential oil, carnosic acid, rosmarinic acid and sesamol. *Food Chem.* **2008**, *110*, 76–82. [CrossRef]
17. Benavente-García, O.; Castillo, J.; Lorente, J.; Ortuño, A.D.R.J.; Del Rio, J.A. Antioxidant activity of phenolics extracted from *Olea europaea* L. leaves. *Food Chem.* **2000**, *68*, 457–462. [CrossRef]

18. Badiee, P.; Reza Nasirzadeh, A.; Motaffaf, M. Comparison of *Salvia officinalis* L. essential oil and antifungal agents against *Candida* species. *J. Pharm. Technol. Drug Res.* **2012**, *1*, 1–5. [CrossRef]
19. Scazzocchio, F.; D'auria, F.D.; Alessandrini, D.; Pantanella, F. Multifactorial aspects of antimicrobial activity of propolis. *Microbiol. Res.* **2006**, *161*, 327–333. [CrossRef]
20. Freires, I.A.; Queiroz, V.C.; Furletti, V.F.; Ikegaki, M.; de Alencar, S.M.; Duarte, M.C.T.; Rosalen, P.L. Chemical composition and antifungal potential of Brazilian propolis against *Candida* spp. *J. Mycol. Medical.* **2016**, *26*, 122–132. [CrossRef]
21. Stepanović, S.; Antić, N.; Dakić, I.; Švabić-Vlahović, M. In vitro antimicrobial activity of propolis and synergism between propolis and antimicrobial drugs. *Microbiol. Res.* **2003**, *158*, 353–357. [CrossRef] [PubMed]
22. Pertino, M.W.; Theoduloz, C.; Butassi, E.; Zacchino, S.; Schmeda-Hirschmann, G. Synthesis, antiproliferative and antifungal activities of 1,2,3-Triazole-substituted carnosic acid and carnosol derivatives. *Molecules* **2015**, *20*, 8666–8686. [CrossRef] [PubMed]
23. Siqueira, A.B.; Rodriguez, L.R.; Santos, R.K.; Marinho, R.R.B.; Abreu, S.; Peixoto, R.F.; Gurgel, B.C.D.V. Antifungal activity of propolis against *Candida* species isolated from cases of chronic periodontitis. *Braz. Oral Res.* **2015**, *29*, 1–6. [CrossRef] [PubMed]
24. European Committee on Antimicrobial Susceptibility Testing (EUCAST). Antifungal Agents Breakpoint Tables for Interpretation of MICs European Committee on Antimicrobial Susceptibility Testing Antifungal Agents Breakpoint Tables for Interpretation of MICs. Available online: http://www.eucast.org/mic_distributions_and_ecoffs/ (accessed on 17 April 2020).
25. Wuyts, J.; Van Dijck, P.; Holtappels, M. Fungal persister cells: The basis for recalcitrant infections? *PLoS Path.* **2018**, *14*, 1–14. [CrossRef] [PubMed]
26. Suleman, T.; van Vuuren, S.; Sandasi, M.; Viljoen, A.M. Antimicrobial activity and chemometric modelling of South African propolis. *J. Appl. Microbiol.* **2015**, *119*, 981–990. [CrossRef]
27. Tallarida, R.J. Quantitative Methods for Assessing Drug Synergism. *Genes Cancer.* **2011**, *2*, 1003–1008. [CrossRef]
28. Pierce, C.G.; Saville, S.P.; Lopez-Ribot, J.L. High-content phenotypic screenings to identify inhibitors of *Candida albicans* biofilm formation and filamentation. *Pathogens Dis.* **2014**, *70*, 423–431. [CrossRef]
29. Manke, M.B.; Raut, J.S.; Dhawale, S.C.; Karuppayil, S.M. Antifungal activity of *Helicteres isora* Linn. Fruit extracts against planktonic and biofilm growth of *Candida albicans*. *J. Biol. Active Prod. Nat.* **2015**, *5*, 357–364.
30. Mukherjee, P.K.; Chandra, J. *Candida* Biofilms: Development, architecture, and resistance. *Microbiol. Spectr.* **2015**, *3*, 1–14.
31. Scudellari, M. The science myths that will not die. *Nature* **2015**, *528*, 322–325. [CrossRef]
32. Metodiewa, D.; Jaiswal, A.K.; Cenas, N.; Dickancaité, E.; Segura-Aguilar, J. Quercetin may act as a cytotoxic prooxidant after its metabolic activation to semiquinone and quinoidal product. *Free Radic. Biol. Med.* **1999**, *26*, 107–116. [CrossRef]
33. Halliwell, B. Are polyphenols antioxidants or pro-oxidants? What do we learn from cell culture and in vivo studies? *Arch. Biochem. Biophys.* **2008**, *476*, 107–112. [CrossRef]
34. Vella, F.; Ferry, G.; Delagrange, P.; Boutin, J.A. NRH:quinone reductase 2: An enzyme of surprises and mysteries. *Biochem. Pharmacol.* **2005**, *71*, 1–12. [CrossRef] [PubMed]
35. Collado-González, M.; Guirao-Abad, J.P.; Sánchez-Fresneda, R.; Belchí-Navarro, S.; Argüelles, J.C. Resveratrol lacks antifungal activity against *Candida albicans*. *World J. Microbiol. Biotechnol.* **2012**, *28*, 2441–2446.
36. Guirao-Abad, J.P.; Sánchez-Fresneda, R.; Valentín, E.; Martínez-Esparza, M.; Argüelles, J.C. Analysis of validamycin as a potential antifungal compound against *Candida albicans*. *Internatl. Microbiol.* **2013**, *16*, 217–225.

microorganisms

MDPI

Communication

Antibiofilm Activity of a *Trichoderma* Metabolite against *Xanthomonas campestris* pv. *campestris*, Alone and in Association with a Phage

Marina Papaianni [1], Annarita Ricciardelli [2], Andrea Fulgione [3], Giada d'Errico [1], Astolfo Zoina [1], Matteo Lorito [1,4], Sheridan L. Woo [4,5], Francesco Vinale [4,6,*] and Rosanna Capparelli [1]

1 Department of Agricultural Sciences, University of Naples Federico II, 80055 Portici (NA), Italy; marina.papaianni@unina.it (M.P.); giada.derrico@unina.it (G.d.); zoina@unina.it (A.Z.); lorito@unina.it (M.L.); capparel@unina.it (R.C.)

2 Department of Chemical Sciences, University of Naples Federico II, 80125 Naples, Italy; ananrita.ricciardelli@unina.it

3 Istituto Zooprofilattico Sperimentale del Mezzogiorno (IZSM), 80055 Portici (NA), Italy; Andrea.Fulgione@izsm.it

4 Institute for Sustainable Plant Protection, National Research Council, 80055 Portici (NA), Italy; woo@unina.it

5 Department of Pharmacy, University of Naples Federico II, 80131 Naples, Italy

6 Department of Veterinary Medicine and Animal Productions, University of Naples Federico II, 80137 Naples, Italy

* Correspondence: frvinale@unina.it; Tel.: +390812539338

Received: 16 March 2020; Accepted: 22 April 2020; Published: 25 April 2020

Abstract: Biofilm protects bacteria against the host's immune system and adverse environmental conditions. Several studies highlight the efficacy of lytic phages in the prevention and eradication of bacterial biofilms. In this study, the lytic activity of Xccφ1 (*Xanthomonas campestris* pv. *campestris*-specific phage) was evaluated in combination with 6-pentyl-α-pyrone (a secondary metabolite produced by *Trichoderma atroviride* P1) and the mineral hydroxyapatite. Then, the antibiofilm activity of this interaction, called a φHA6PP complex, was investigated using confocal laser microscopy under static and dynamic conditions. Additionally, the mechanism used by the complex to modulate the genes (*rpf, gumB, clp* and *manA*) involved in the biofilm formation and stability was also studied. Our results demonstrated that Xccφ1, alone or in combination with 6PP and HA, interfered with the gene pathways involved in the formation of biofilm. This approach can be used as a model for other biofilm-producing bacteria.

Keywords: *Xanthomonas campestris* pv. *campestris*; 6-pentyl-α-pyrone; antibiosis; *Trichoderma*; secondary metabolites; antibiofilm; Gram-negative bacterium; antibiotic resistance

1. Introduction

The Gram-negative bacterium *Xanthomonas campestris* pv. *campestris* (*Xcc*) is the causal agent of black-rot disease in crucifers, responsible for serious yield losses worldwide [1]. The bacterial secretion composed of exopolysaccharides can obstruct the xylem vessels, causing tissue necrosis and leaf wilting [2]. The *Xcc* produces a variety of substances, such as enzymes, that may be used by this bacterium to parasitize the host [3].

At present, the ongoing spread of antibiotic-resistant bacteria is a major public health concern that is further exacerbated by some agricultural management practices [4]. The bacterial contamination of food surfaces is the main cause of food-borne illnesses [5]. The mechanism mainly used by bacteria to improve its chance of survival, especially in weakly resistant isolates, is the formation of biofilm [6], an extracellular polymeric matrix composed of proteins, polysaccharides and DNA [7]. In this context, emerging issues in the management of infections caused by antibiotic-resistant and biofilm-forming bacteria have encouraged the development of alternative therapeutic approaches [8,9].

Several studies highlight the efficacy of lytic phages against bacteria [10,11]; however, antiphage resistance mechanisms may limit their activity [12]. We believe that this strategy, combined with the application of metabolites obtained from a specific (or given) fungus, might be a potential alternative in the prevention and control of biofilm-related infections. Thus, *Trichoderma*, a well-known beneficial fungus, was selected for its plant growth promotion [13] and biocontrol abilities [14]. The species *T. atroviride* produces a natural compound, known as 6-pentyl-α-pyrone (6PP), with interesting properties (i.e., antibiotic, plant metabolome interference) [13,15].

Interactions between biomolecules and different kinds of inorganic surfaces, like hydroxyapatite (HA) nanocrystals, biogenic silica, carbonates and phosphates, are important in numerous biological applications, such as drug and gene delivery, a possible way to selectively release prodrug in tumor tissues, and antimicrobial molecule carriers [16–18]. Previous studies have demonstrated that HA nanocrystals, a mineral rich in $Ca_{10}(PO_4)_6(OH)_2$, can be used for bacteriophage delivery and as an enhancer of its biological activities and stability [16]. Therefore, HA may act as carrier of antimicrobial compounds [17] and as a factor to improve lytic activity of bacteriophages [16], adsorbing molecules as well as particles [18].

This is the first study to describe the antibiofilm activity of 6PP, alone and in association with a phage and hydroxyapatite (a biocompatible mineral) against *Xcc*. This association may overcome the problems arising from the formation of bacterial biofilm during infection. In particular, the antibiofilm activity of 6PP helps the phage to better interact with and lyse the bacteria, and the HA can be considered as a potential carrier for the active principles.

2. Materials and Methods

2.1. Isolation and Growth of Xcc Phages

Ten grams of rhizospheric soil collected from different kohlrabi (*Brassica oleracea* var. *gongylodes*) plants showing black rot symptoms was suspended in 15 mL of nutrient broth (Sigma Aldrich) and agitated for 30 min at 20 °C. Soil sediment was removed by centrifugation (5.000 rpm for 20 min), and the supernatants were transferred to sterile flasks. *Xcc* was incubated overnight at 24 °C under shaking conditions. Cultures were centrifuged and filtered through a 0.22 μm pore-size membrane filter (MF-Millipore, Darmstadt, Germany). The filtrates were assayed for the presence of *Xcc*-infecting phage by plating a 10 μL aliquot on soft agar. After 48 h, plaque-forming units (PFU) of phage were picked and resuspended in distilled water, incubated at 37 °C for 4 h, centrifuged at 5000 rpm for 30 min and filtered as mentioned above [19]. The experiment was repeated five times.

2.2. Complex φHA6PP

The φHA6PP complex was prepared by mixing 1 mL of HA (100 mg/mL) with 1 mL of Xccφ1 (10^8 PFU/mL) and varying concentrations of 6PP (0.02, 0.01, 0.004, 0.002, 0.001, 0.0004 and 0.0002 μg/mL solubilized in dimethyl sulfoxide (DMSO)). A control sample with the solvents alone was also prepared. The mixture was incubated at room temperature, under shaking conditions, for time periods of 0, 30, 90, 180 and 300 min and 24 h. Later, samples were centrifuged, and the pellet was suspended in distilled water. A further centrifugation step was carried out, and the supernatant was screened for the presence and concentration of phage particles using the double-layer agar (DLA) method [16]. After overnight incubation at 24 °C, the optimal incubation time was selected.

2.3. Isolation of 6-Pentyl-α-pyrone

The biocontrol fungus *T. atroviride* strain P1 was maintained on potato dextrose agar (PDA, Sigma St. Louis, MO, USA) Petri plates at room temperature and sub-cultured bimonthly. Six 5 mm diameter plugs were collected from the actively growing margins of the PDA cultures and inoculated into 5 L conical flasks containing 1 L of sterile potato dextrose broth (PDB, Sigma). The stationary cultures were incubated at 25 °C for 21 days. The cultures were vacuum filtered through filter paper (Whatman No. 4; Brentford, UK). The culture broth of *T. atroviride* strain P1 was extracted exhaustively with ethyl acetate (EtOAc). The red-brown residue was subjected to flash column chromatography (Si gel; 50 g) by eluting with a gradient of EtOAc:Petroleum ether (8:2 to 10:0). Analytical thin-layer chromatography (TLC) was performed, and fractions were further purified by using silica-gel flash chromatography (Kieselgel 60, GF_{254}, 0.25 and 0.5 mm, Merck Darmstadt, Germany). Compounds were detected with UV radiation (254 or 366 nm) and by dipping the TLC plates in a 10% (*w*/*v*) aqueous solution of $CeSO_4$ or in a 5% (*v*/*v*) ethanol solution of H_2SO_4 and heating at 110 °C for 10 min. Pure metabolite was characterized by LC/MS q-TOF analysis recorded with an Agilent system (HPLC 1260 Infinity Series) coupled to a q-TOF mass spectrometer Model G6540B with a dual electrospray ionization source and equipped with a DAD system (Agilent Technologies, Santa Clara, CA, USA). This compound showed the same spectrometric data of an authentic standard previously isolated in our laboratories [15].

2.4. Antibiofilm Activity of 6-Pentyl-α-pyrone

The biofilm formation was measured using crystal violet staining. The experiment was performed to characterize the antibiofilm activity of different concentrations of 6PP on mature biofilm. A 200 µL of *Xcc* was added to each well using a 96-well plate (Falcon), and incubated at 24 °C for 72 h under static conditions to allow bacterial attachment and the biofilm formation. Then, 6PP was added at different concentrations to the complex. After 4 h, the samples were analyzed as previously reported [20].

2.5. CLSM Analysis for Static Biofilm Evaluation

Confocal laser scanning microscopy (CLSM) of *Xcc* biofilms was performed on Nunc Lab-Tek eight-well Chamber Slides (No. 177445; Thermo Scientific, Ottawa, ON, Canada). The overnight cultures of *Xcc* were diluted to a cell concentration of about 0.001 $(OD)_{600nm}$. The bacterial culture was incubated at 24 °C for 96 h to allow the *Xcc* biofilm to form. In order to assess the antibiofilm activity and the influence on cell viability of treatments, the mature biofilms were incubated for 4 h and treated as follows: (1) with 6PP (0.001 µg/mL); (2) without 6PP; (3) with the phage alone (10^8 PFU/mL); (4) with the complex of phage (10^8 PFU/mL) plus HA (10 mg/mL) and 6PP (0.001 µg/mL). The lowest concentration of 6PP showing the maximum efficacy in combination was selected for the experiments. The biofilm cell and the microscopic observations were performed as previously reported [21].

2.6. CLSM Analysis for Dynamic Biofilm Evaluation

The analysis of *Xcc* biofilms was performed using a three-channel flow cell chamber (IBI Scientific, Peosta, IA). A solution of phosphate-buffered saline (PBS, pH 7) was introduced into each channel of the cell at a controlled flow rate of 160 µL/min using an Ismatec IPC 4 peristaltic pump (Cole-Parmer GmbH, Germany). The flow system was kept free of air bubbles using a bubble trap, which created low positive pressure under the PBS flow. Then, a bacterial suspension of *Xcc* at optical density of 0.5 mL^{-1} was left in circulation through the system for 2 h, and the non-adhering cells were removed using sterile PBS for 15 min. Finally, fresh medium (nutrient broth 50% *v*/*v* in PBS) was put through the system for 48 h to allow biofilm formation. After incubation, treatments of fresh medium (NT), 6PP (0.001 µg/mL) or the φHA6PP complex was circulated separately for 3 h into each cell channel. The biofilm formation was evaluated by CLSM. The biofilm cell viability and microscopic observations were determined as previously reported [22].

2.7. RNA Extraction and Expression Profiling by qPCR

Fifty milliliters of *Xcc* was incubated at 24 °C for 72 h in a narrow-mouth glass Erlenmeyer flask under static conditions. Then, the phage (10^8 PFU/mL), 6PP (0.001 µg/mL) and φHA6PP were added to each sample. Ten milliliters of biofilm were collected after 30, 60, 90 and 120 min. Total RNA was extracted using the TRIzol protocol [23]. A NanoDrop ND-1000 (Thermo Fisher Scientific Inc.) was used to assess total RNA quantity. One microgram of purified total RNA was used as a template for first-strand cDNA synthesis using SuperScript III Reverse Transcriptase (Invitrogen). The primers were designed using the tool provided at https://www.eurofinsgenomics.eu/en/ecom/tools/qpcr-assay-design/ for all genes (Supplementary Table). Gene transcript levels were measured using Power SYBR Green PCR Master Mix as previously reported [24]. Thermocycler conditions were as follows: initial step at 95 °C for 10 min, 40 cycles of 95 °C for 15 s, (*clp* 57.1 °C; *manA* 55 °C, *rpf* 59.9 °C, *gumB* 63.7 °C) for 40 s and 72 °C for 1 min. All samples were normalized to HcrC as the reference housekeeping gene. The relative quantitative expression was determined using the $2^{-\Delta\Delta CT}$ method [25].

3. Results

3.1. *Antibiofilm Activity of 6-Pentyl-α-pyrone*

The antibiofilm activity of 6PP alone and the complex φHA6PP were evaluated against *Xcc*. The complex φHA6PP limited the biofilm formation more efficiently than 6PP used alone. The 6PP at a concentration of 0.001 μg/mL disrupted the biofilm by about 70%. The DMSO (used as solvent of 6PP at 0.1% concentration) did not interfere with biofilm integrity (Figure 1).

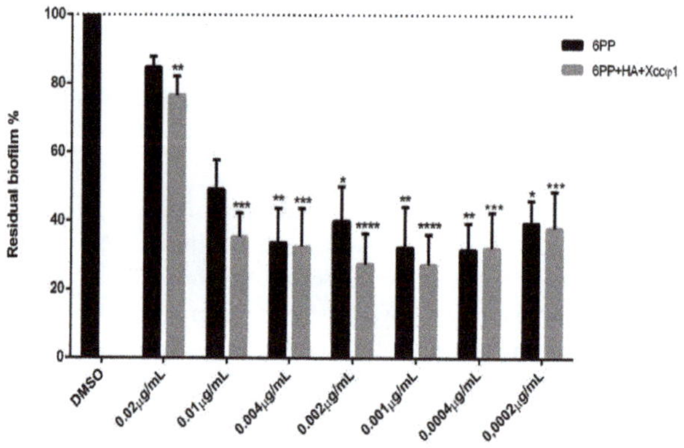

Figure 1. Antibiofilm activities of different concentrations of 6-pentyl-α-pyrone (6PP) alone, and in the complex of 6PP+hydroxyapatite and *Xanthomonas campestris* pv. *campestris* phage (Xccφ1). Biofilm was assessed after 72 h of incubation at 25 °C using a crystal violet assay. The data are expressed as percentages of residual biofilm. Values represent the mean ± SD of the three independent experiments. Absorbance, compared to the untreated control, was considered statistically significant with $p < 0.05$ (* $p < 0.05$, ** $p < 0.01$, *** $p < 0.001$, **** $p < 0.0001$) according to two-way ANOVA multiple comparisons.

3.2. *CLSM Analysis for Static Biofilm Evaluation*

CLSM analysis confirmed the results obtained from the crystal violet test. As shown in Figure 2, the 6PP (0.001 μg/mL) and the phage, when used alone, caused a small reduction in the biofilm mass. However, when 6PP was combined with phage and HA, there was a consistent reduction in biofilm formation.

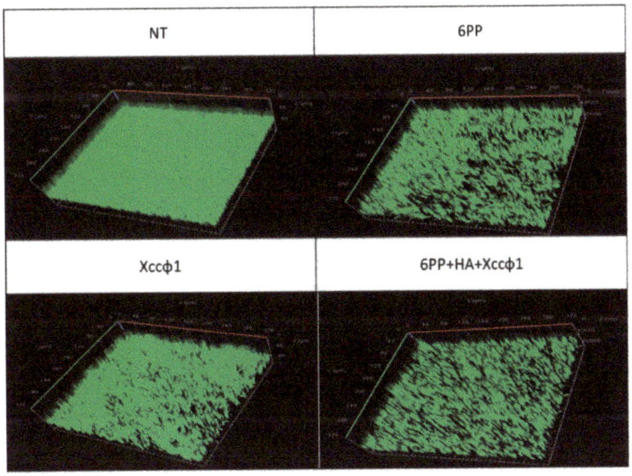

Figure 2. Confocal laser scanning microscopy (CLSM) observation of biofilm formation of *Xanthomonas campestris* pv. *campestris* under static conditions. The treatments were fresh medium, no treatment (NT); 6-pentyl-α-pyrone (6PP, 0.001 μg/mL); *Xanthomonas campestris* pv. *campestris* phage (Xccφ1); and the complex of 6PP, hydroxyapatite (HA) plus Xccφ1. Biofilm analysis was carried out on mature biofilm after 72 h of incubation at 24 °C. The three-dimensional biofilm structure was demonstrated using the LIVE/DEAD Biofilm Viability kit.

3.3. CLSM Analysis for Dynamic Biofilm Evaluation

In order to reproduce the plant xylem, the *Xcc* biofilm formation was investigated using a three-channel flow cell system. The phage alone did not show significant disruption of the biofilm structure. The 6PP (0.001 μg/mL) used alone partially disrupted the biofilm structure. However, the biofilm mass was almost totally unstructured when 6PP was combined with the phage and HA (Figure 3). The CLSM fluidic system confirmed the antibiofilm activity of φHA6PP.

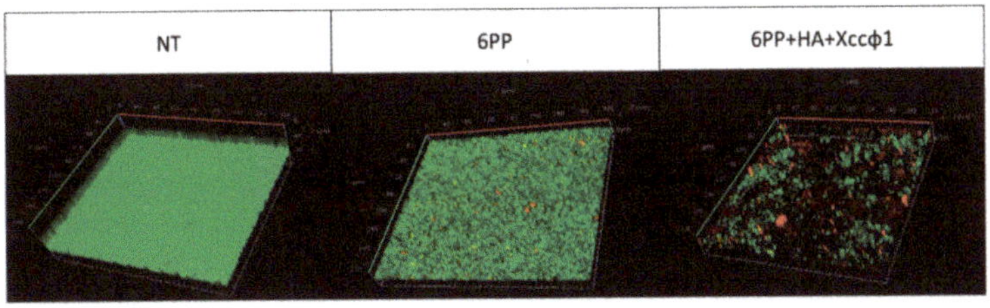

Figure 3. Confocal laser scanning microscopy analysis demonstrating biofilm formation of *Xanthomonas campestris* pv. *campestris* under dynamic conditions. The three treatments were fresh medium–no treatment (NT); 6-pentyl-α-pyrone (6PP, 0.001 μg/mL); and the complex of 6PP, hydroxyapatite (HA) plus *Xanthomonas campestris* pv. *campestris* phage (Xccφ1). Biofilm formation was performed using a three-channel flow cell. The three-dimensional biofilm structures were obtained using the LIVE/DEAD Biofilm Viability kit.

3.4. RNA Extraction and Expression Profiling by qPCR

The expression levels were evaluated of four principal genes (*rpf*, *gumB*, *clp* and *manA*) involved in the biofilm production pathways. The gene *rpf* positively regulates the synthesis of extracellular enzymes and extracellular polymeric substance (EPS) as well as several essential factors of pathogenicity and diffusible signal factor (DSF) [26]. The gene *clp* is essential in DSF production, and it is highly conserved in *Xcc* [27]. The gene *gumB* is involved in the assemblage of pentasaccharides, transfer of nonglycosidic constituents and export of xanthan [26]. Finally, the gene *manA* encodes several mannase enzymes, including mannan endo-1,4-β-mannosidase, which is a dispersing biofilm enzyme of *Xcc*. Analysis by qPCR showed a significant upregulation of *rpf*, *gumB* and *clp* genes 30 min after treatment, and the complex φHA6PP induced the highest level of upregulation ($p < 0.05$). Interestingly, the upregulation of *manA* was cyclical (every 30 min) (Figure 4).

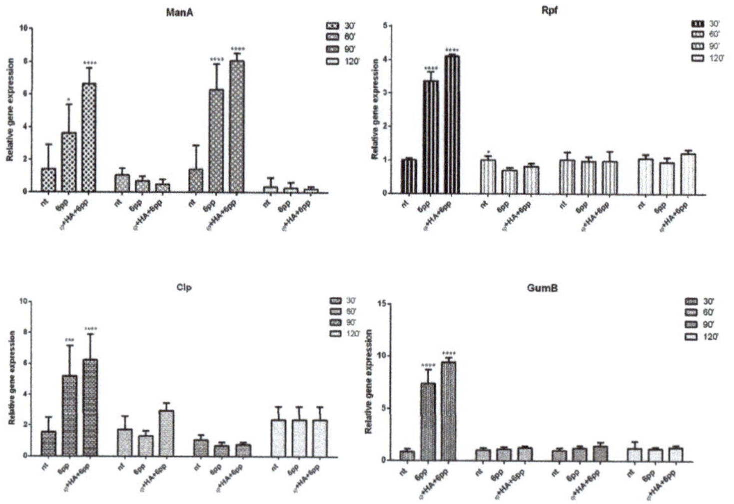

Figure 4. The expression profiles of the four genes (*rpf* at top right; *gumB* at bottom right; *clp* at bottom left; *manA* at top left) of *Xanthomonas campestris* pv. *campestris* by quantitative real-time PCR (qPCR). Biofilm samples untreated (NT), treated with 6-pentyl-α-pyrone (6PP); and the complex of hydroxyapatite, 6PP and *Xanthomonas campestris* pv. *campestris* phage (φHA6PP) were analyzed after 30, 60, 90 and 120 min. Statistical analysis was performed using two-way ANOVA multiple comparison test (* $p < 0.05$, ** $p < 0.01$, *** $p < 0.001$, **** $p < 0.0001$). The gene *HcrC* was used as reference.

4. Discussion

Worldwide research on bacterial biofilms is proceeding on several diverse fronts, with particular emphasis on the investigation of specifically expressed genes, the role of biofilms in antimicrobial resistance, the evaluation of control measures and the development of innovative strategies.

Many molecules have antibiofilm properties as well as antimicrobial activities [28]. Specifically, we investigated the biofilm formation during *Xcc* growth that plays a key role in its infective process [27]. Our results suggested that 6PP did not show as much significant antibacterial activity against free-living *Xcc* as other bacteria in the biofilm matrix (data not shown). However, this molecule displayed an effective biofilm-dissolving ability. Herein, we are reporting for the first time on the antibiofilm activity of 6PP, a secondary metabolite isolated from the culture filtrate of *T. atroviride* P1. Several species belonging to *Trichoderma* are widely known and recognized as biocontrol and biostimulant agents [29–31].

Additionally, the mineral HA is known to chemically interact not only with molecules but also with biological structures like bacteriophages [16]. In fact, the low degree of crystallinity and the presence of carbonate ions in the crystal structure make HA extremely reactive in biological systems and suitable for interaction with, and transportation of, bacteriophages. In this study, HA seems to work well as a carrier for 6PP and Xccφ1 in order to increase their activities against the pathogenic bacteria.

The compound 6PP, alone or in combination with Xccφ1 and HA, showed lower antibiofilm activity when applied at higher concentration (Figure 1, 0.02 mg/mL). It is likely that higher doses of 6PP may interfere with the bacterial endosmosis processes of damaging the cell membranes [32].

Otherwise, it is possibly the formation of aggregates that makes the compound unable to penetrate the biofilm structure [33].

Generally, the complex φHA6PP was more efficient on biofilm under the dynamic rather than under the static conditions. This combination showed potentially valid properties for further in vivo tests.

The DSF biosynthesis is modulated by a novel post-translational mechanism involving protein–protein interaction between the two DSF clusters, synthase RpfF (*rpf*) and sensor RpfC (*gumB*) [34]. Further, the quorum sensing (QS) signal is coupled with several intracellular regulatory networks through the second messenger cyclic dimeric GMP and the global regulator *clp* (gene investigated). Genomic analysis shows that the DSF-QS signaling pathway regulates diverse biological functions including virulence, biofilm dispersal and ecological competence. Future investigations could be made into the DSF-QS system in plant and human bacterial pathogens [35].

The RpfC/RpfG two-component system plays a key role in DSF signal transduction and modulates downstream DSF regulon by changing intracellular content of cyclic dimeric GMP. The increase in the content of cyclic dimeric GMP can positively influence both *manA*, involved in biofilm dispersal, and *clp*, associated with xanthan production by upregulating the gum genes (*gumB*) [34]. Figure 4 shows that, 30 min after φHA6PP treatment, all the genes are upregulated (*rpf*, *clp*, *gumB* and *manA*); this may be related to the promoter *rpf*. Interestingly, after 90 min, only *manA* is upregulated, in accordance to recent work demonstrating that the one-step growth curve of the phage completes its lytic cycle over the exact same period of time [35]. Moreover, *manA* is involved in mannose metabolism, for which the metabolic alteration activity of the phage on *Xcc* biofilm has been previously demonstrated [36]. Specifically, 30 min after treatment with the complex a stress signal, released from the bacteria activates all the pathways (studied) involved in biofilm formation. Subsequently, the activity of the phage, improved by HA and 6PP, may influence the *manA* synthesis in order to assist the biofilm dispersion. This hypothesis is supported by our results in which the treatment with the phage shows high levels of galactomannan that reduce the viscosity of the biofilm. This study highlights the ability of the complex to dysregulate the metabolic pathways via *manA*, a gene involved in processes able to promote the mature biofilm dispersion, thus making the bacterial aggregate more penetrable. For these reasons, the φHA6PP complex seems to work as a DSF molecule, able to activate *manA* expression. Furthermore, the interference on metabolic pathways involved in biofilm formation due to Xccφ1, alone or in combination with HA and eicosanoic acid, has been demonstrated [36]. Our results have provided evidence that the φHA6PP complex modifies the biofilm structure and production, thus probably interacts by biofilm solubilisation.

In conclusion, this study demonstrates that the φHX6PP complex interferes with biofilm production and promotes the mature biofilm dispersion of *Xcc*. This approach may represent a novel strategy for biofilm prevention and control of plant pathogenic bacteria.

Author Contributions: Conceptualization, F.V. and R.C.; methodology, M.P., A.R. and A.Z.; validation, S.L.W. and M.L.; formal analysis, M.P., A.F. and A.R.; investigation, M.P.; data curation, G.d. and M.P.; writing—original draft preparation, M.P., G.d. and F.V.; writing—review and editing, F.V., M.L., S.L.W. and R.C.; supervision, R.C. and F.V.; project administration, F.V. and M.L. All authors have read and agreed to the published version of the manuscript.

Funding: The study was supported by MIURPON grant Marea 03PE_00106, POR FESR CAMPANIA 2014/2020-O.S. 1.1 grant Bioagro 559, MISE CRESO Protection n. F/050421/01-03/ X32, PSR Veneto 2014/2020 Misura 16—Tipologia di intervento 16.1—Azione 2 "Sostegno ai Progetti Operativi di Innovazione (POI)"—Divine n. 3589659, PSR Campania 2014/2020 Misura 16—Tipologia di intervento 16.1—Azione 2 "Sostegno ai Progetti Operativi di Innovazione (POI)"— Progetto "DI.O.N.IS.O.", C.U.P. B98H19005010009 and MIURPON grant Linfa 03PE_00026_1.

Conflicts of Interest: The authors declare no conflict of interest.

References

1. Qian, W.; Jia, Y.; Ren, S.X.; He, Y.Q.; Feng, J.X.; Lu, L.F.; Wu, W. Comparative and functional genomic analyses of the pathogenicity of phytopathogen *Xanthomonas campestris* pv. *campestris*. *Genome Res.* **2005**, *15*, 757–767. [CrossRef] [PubMed]
2. Liao, C.T.; Chiang, Y.C.; Hsiao, Y.M. Functional characterization and proteomic analysis of lolA in *Xanthomonas campestris* pv. *campestris*. *BMC Microbiol.* **2019**, *19*, 20. [CrossRef] [PubMed]
3. Dow, J.M.; Daniels, M.J. Pathogenicity determinants and global regulation of pathogenicity of *Xanthomonas campestris* pv. *campestris*. *Curr. Top. Microbiol. Immunol.* **1994**, *192*, 29–41. [PubMed]
4. Li, B.; Webster, T.J. Bacteria antibiotic resistance: New challenges and opportunities for implant-associated orthopedic infections. *J. Orthop. Res.* **2018**, *36*, 22–32. [CrossRef]
5. Windler, M.; Leinweber, K.; Bartulos, C.R.; Philipp, B.; Kroth, P.G. Biofilm and capsule formation of the diatom *Achnanthidium minutissimum* are affected by a bacterium. *J. Phycol.* **2015**, *51*, 343–355. [CrossRef]
6. Cepas, V.; López, Y.; Munoz, E.; Rolo, D.; Ardanuy, C.; Martí, S.; Soto, S.M. Relationship between biofilm formation and antimicrobial resistance in gram-negative bacteria. *Microb. Drug Resist.* **2019**, *25*, 72–79. [CrossRef]
7. Donlan, R.M. Preventing biofilms of clinically relevant organisms using bacteriophage. *Trends Microbiol.* **2009**, *17*, 66–72. [CrossRef]
8. Stewart, P.S.; William Costerton, J. Antibiotic resistance of bacteria in biofilms. *Lancet* **2001**, *358*, 135–138. [CrossRef]
9. Del Pozo, J.L.; Patel, R. The challenge of treating biofilm-associated bacterial infections. *Clin. Pharmacol. Ther.* **2007**, *82*, 204–209. [CrossRef]
10. Harper, D.R.; Parracho, H.M.; Walker, J.; Sharp, R.; Hughes, G.; Werthén, M.; Morales, S. Bacteriophages and biofilms. *Antibiotics* **2014**, *3*, 270–284. [CrossRef]
11. Gutiérrez, D.; Vandenheuvel, D.; Martínez, B.; Rodríguez, A.; Lavigne, R.; García, P. Two phages, philPLA-RODI and philPLA-C1C, lyse mono-and dual-species staphylococcal biofilms. *Appl. Environ. Microbiol.* **2015**, *81*, 3336–3348. [CrossRef] [PubMed]
12. Pires, D.P.; Melo, L.D.R.; Vilas Boas, D.; Sillankorva, S.; Azeredo, J. Phage therapy as an alternative or complementary strategy to prevent and control biofilm-related infections. *Curr. Opin. Microbiol.* **2017**, *39*, 48–56. [CrossRef] [PubMed]
13. Vinale, F.; Sivasithamparam, K.; Ghisalberti, E.L.; Marra, R.; Barbetti, M.J.; Li, H.; Woo, S.L.; Lorito, M. A novel role for *Trichoderma* secondary metabolites in the interactions with plants. *Physiol. Mol. Plant Pathol.* **2008**, *72*, 80–86. [CrossRef]
14. Marra, R.; Ambrosino, P.; Carbone, V.; Vinale, F.; Woo, S.L.; Ruocco, M.; Gigante, S. Study of the three-way interaction between *Trichoderma atroviride*, plant and fungal pathogens by using a proteomic approach. *Curr. Genet.* **2006**, *50*, 307–321. [CrossRef]
15. Pascale, A.; Vinale, F.; Manganiello, G.; Nigro, M.; Lanzuise, S.; Ruocco, M.; Lorito, M. *Trichoderma* and its secondary metabolites improve yield and quality of grapes. *Crop Prot.* **2017**, *92*, S11–S12. [CrossRef]
16. Fulgione, A.; Ianniello, F.; Papaianni, M.; Contaldi, F.; Sgamma, T.; Giannini, C.; Lelli, M. Biomimetic hydroxyapatite nanocrystals are an active carrier for *Salmonella* bacteriophages. *Int. J. Nanomed.* **2019**, *14*, 2219–2232. [CrossRef]
17. Nocerino, N.; Fulgione, A.; Iannaccone, M.; Tomasetta, L.; Ianniello, F.; Martora, F.; Capparelli, R. Biological activity of lactoferrin-functionalized biomimetic hydroxyapatite nanocrystals. *Int. J. Nanomed.* **2014**, *9*, 1175–1184.
18. Fulgione, A.; Nocerino, N.; Iannaccone, M.; Roperto, S.; Capuano, F.; Roveri, N.; Capparelli, R. Lactoferrin adsorbed onto biomimetic hydroxyapatite nanocrystals controlling-in vivo-the *Helicobacter pylori* infection. *PLoS ONE* **2016**, *11*. [CrossRef]
19. Cross, T.; Schoff, C.; Chudoff, D.; Graves, L.; Broomell, H.; Terry, K.; Dunbar, D. An optimized enrichment technique for the isolation of *Arthrobacter* bacteriophage species from soil sample isolates. *J. Vis. Exp.* **2015**. [CrossRef]
20. Papaianni, M.; Contaldi, F.; Fulgione, A.; Woo, S.L.; Casillo, A.; Corsaro, M.M.; Garonzi, M. Role of phage ϕ1 in two strains of *Salmonella* Rissen, sensitive and resistant to phage ϕ1. *BMC Microbiol.* **2018**, *18*, 208. [CrossRef]

21. Casillo, A.; Papa, R.; Ricciardelli, A.; Sannino, F.; Ziaco, M.; Tilotta, M.; Artini, M. Anti-Biofilm activity of a long-chain fatty aldehyde from antarctic *Pseudoalteromonas haloplanktis* TAC125 against *Staphylococcus epidermidis* biofilm. *Front. Cell. Infect. Microbiol.* **2017**, *7*, 46. [CrossRef]
22. Casillo, A.; Ziaco, M.; Lindner, B.; Parrilli, E.; Schwudke, D.; Holgado, A.; Tutino, M.L. Unusual Lipid A from a cold-adapted Bacterium: Detailed structural characterization. *ChemBioChem* **2017**, *18*, 1845–1854. [CrossRef]
23. Rio, D.C.; Ares, M.; Hannon, G.J.; Nilsen, T.W. Purification of RNA using TRIzol (TRI reagent). *Cold Spring Harb. Protoc.* **2010**, *2010*, pdb.prot5439. [CrossRef]
24. Allam, A.F.; Farag, H.F.; Zaki, A.; Kader, O.A.; Abdul-Ghani, R.; Shehab, A.Y. Detection of low-intensity *Schistosoma mansoni* infection by Percoll sedimentation and real-time PCR techniques in a low-endemicity Egyptian setting. *Trop. Med. Int. Heal.* **2015**, *20*, 658–664. [CrossRef]
25. Livak, K.J.; Schmittgen, T.D. Analysis of relative gene expression data using real-time quantitative PCR and the $2-\Delta\Delta CT$ method. *Methods* **2001**, *25*, 402–408. [CrossRef]
26. Dow, J.M.; Crossman, L.; Findlay, K.; He, Y.Q.; Feng, J.X.; Tang, J.L. Biofilm dispersal in *Xanthomonas campestris* is controlled by cell-cell signaling and is required for full virulence to plants. *Proc. Natl. Acad. Sci. USA* **2003**, *100*, 10995–11000. [CrossRef]
27. Ryan, R.P.; Dow, J.M. Communication with a growing family: Diffusible signal factor (DSF) signaling in bacteria. *Trends Microbiol.* **2011**, *19*, 145–152. [CrossRef]
28. Chung, P.Y.; Khanum, R. Antimicrobial peptides as potential anti-biofilm agents against multidrug-resistant bacteria. *J. Microbiol. Immunol. Infect.* **2017**, *50*, 405–410. [CrossRef]
29. Harman, G.E. Multifunctional fungal plant symbionts: New tools to enhance plant growth and productivity. *New Phytol.* **2011**, *189*, 647–649. [CrossRef]
30. Harman, G.E.; Howell, C.R.; Viterbo, A.; Chet, I.; Lorito, M. *Trichoderma* species-Opportunistic, avirulent plant symbionts. *Nat. Rev. Microbiol.* **2004**, *2*, 43–56. [CrossRef]
31. Mukherjee, P.K.; Horwitz, B.A.; Herrera-Estrella, A.; Schmoll, M.; Kenerley, C.M. *Trichoderma* research in the genome era. *Annu. Rev. Phytopathol.* **2013**, *51*, 105–129. [CrossRef]
32. Ozkan, A.; Erdogan, A. A comparative study of the antioxidant/prooxidant effects of carvacrol and thymol at various concentrations on membrane and DNA of parental and drug resistant H1299 cells. *Nat. Prod. Commun.* **2012**, *7*, 1557–1560. [CrossRef]
33. Lewis, K. Riddle of biofilm resistance. *Antimicrob. Agents Chemother.* **2001**, *45*, 999–1007. [CrossRef]
34. Torres, P.S.; Malamud, F.; Rigano, L.A.; Russo, D.M.; Marano, M.R.; Castagnaro, A.P.; Vojnov, A.A. Controlled synthesis of the DSF cell-cell signal is required for biofilm formation and virulence in *Xanthomonas campestris*. *Environ. Microbiol.* **2007**, *9*, 2101–2109. [CrossRef]
35. Papaianni, M.; Paris, D.; Woo, S.L.; Fulgione, A.; Rigano, M.M.; Parrilli, E.; Limone, A. Plant dynamic metabolic response to bacteriophage treatment after *Xanthomonas campestris* pv. *campestris* infection. *Front. Microbiol.* **2020**, *11*, 732. [CrossRef]
36. Papaianni, M.; Cuomo, P.; Fulgione, A.; Albanese, D.; Gallo, M.; Paris, D.; Motta, A.; Iannelli, D.; Capparelli, R. Bacteriophages promote metabolic changes in bacteria biofilm. *Microorganisms* **2020**, *8*, 480. [CrossRef]

microorganisms

MDPI

Article

Screening Fungal Endophytes Derived from Under-Explored Egyptian Marine Habitats for Antimicrobial and Antioxidant Properties in Factionalised Textiles

Ahmed A. Hamed [1,2,†], Sylvia Soldatou [2,†], M. Mallique Qader [3,4], Subha Arjunan [2], Kevin Jace Miranda [2,5], Federica Casolari [2], Coralie Pavesi [2], Oluwatofunmilay A. Diyaolu [2], Bathini Thissera [3], Manal Eshelli [3,6], Lassaad Belbahri [7], Lenka Luptakova [8], Nabil A. Ibrahim [9], Mohamed S. Abdel-Aziz [1], Basma M. Eid [9], Mosad A. Ghareeb [10], Mostafa E. Rateb [3,*] and Rainer Ebel [2,*]

1 Microbial Chemistry Department, National Research Centre, 33 El-Buhouth Street, Dokki, Giza 12622, Egypt; ahmedshalbio@gmail.com (A.A.H.); mohabomerna@yahoo.ca (M.S.A.-A.)
2 Marine Biodiscovery Centre, Department of Chemistry, University of Aberdeen, Aberdeen AB24 3UE, UK; s.soldatou@rgu.ac.uk (S.S.); r02sa17@abdn.ac.uk (S.A.); r02kjm17@abdn.ac.uk (K.J.M.); f.casolari.19@abdn.ac.uk (F.C.); coralie.pavesi@edu.mnhn.fr (C.P.); r01oad17@abdn.ac.uk (O.A.D.)
3 School of Computing, Engineering & Physical Sciences, University of the West of Scotland, Paisley PA1 2BE, UK; mallique.qader@gmail.com (M.M.Q.); bathini.thissera@uws.ac.uk (B.T.); m.eshelli@hotmail.com (M.E.)
4 National Institute of Fundamental Studies, Hantana Road, Kandy 20000, Sri Lanka
5 College of Pharmacy, Adamson University, 900 San Marcelino Street, Manila 1000, Philippines
6 Food Science & Technology Department, Faculty of Agriculture, University of Tripoli, Tripoli 13538, Libya
7 Laboratory of Soil Biology, University of Neuchatel, 2000 Neuchatel, Switzerland; lassaad.belbahri@unine.ch
8 Department of Biology and Genetics, Institute of Biology, Zoology and Radiobiology, University of Veterinary Medicine and Pharmacy, 04181 Kosice, Slovakia; lenka.luptakova@uvlf.sk
9 Textile Research Division, National Research Centre, Scopus Affiliation ID 60014618, 33 EL Buhouth St., Dokki, Giza 12622, Egypt; nabibrahim49@yahoo.co.uk (N.A.I.); basmaeid@yahoo.com (B.M.E.)
10 Medicinal Chemistry Department, Theodor Bilharz Research Institute, Kornaish El Nile, Warrak El-Hadar, Imbaba, Giza 12411, Egypt; mosad_tbri@hotmail.com
* Correspondence: Mostafa.Rateb@uws.ac.uk (M.E.R.); r.ebel@abdn.ac.uk (R.E.); Tel.: +44-141-8483072 (M.E.R.); +44-1224-272930 (R.E.)
† Both authors contributed equally to the work.

Received: 24 September 2020; Accepted: 19 October 2020; Published: 21 October 2020

Abstract: Marine endophytic fungi from under-explored locations are a promising source for the discovery of new bioactivities. Different endophytic fungi were isolated from plants and marine organisms collected from Wadi El-Natrun saline lakes and the Red Sea near Hurghada, Egypt. The isolated strains were grown on three different media, and their ethyl acetate crude extracts were evaluated for their antimicrobial activity against a panel of pathogenic bacteria and fungi as well as their antioxidant properties. Results showed that most of the 32 fungal isolates initially obtained possessed antimicrobial and antioxidant activities. The most potent antimicrobial extracts were applied to three different cellulose containing fabrics to add new multifunctional properties such as ultraviolet protection and antimicrobial functionality. For textile safety, the toxicity profile of the selected fungal extract was evaluated on human fibroblasts. The 21 strains displaying bioactivity were identified on molecular basis and selected for chemical screening and dereplication, which was carried out by analysis of the MS/MS data using the Global Natural Products Social Molecular Networking (GNPS) platform. The obtained molecular network revealed molecular families of compounds commonly produced by fungal strains, and in combination with manual dereplication, further previously reported metabolites were identified as well as potentially new derivatives.

Keywords: endophytic fungi; antimicrobial; antioxidant; GNPS; textiles

1. Introduction

The emergence of antimicrobial resistance (AMR) is one of the current global health challenges, and it refers to the ability of microorganisms to stop the action of antimicrobial agents, thereby increasing the prevalence and the associated risks of infections by pathogenic bacteria, fungi, viruses, and parasites [1]. Several reports highlighted the potential dangers of AMR to global public health and economics. One of these reports estimated that, if the current rate of continuous increase of AMR should prevail, 300 million people are expected to die prematurely over the next 35 years [2]. This report also predicted that, by 2050, the world will lose between 60 and 100 trillion USD if no action is taken toward AMR. In developing countries such as Egypt, the continuous misuse and overuse of antibiotics without prescription and medical supervision play a significant role in the development of antimicrobial resistance, especially in hospital-acquired infections [3]. Recent reports showed that textile can play a significant role in reducing and preventing nosocomial infections. Textiles under appropriate temperature and moisture conditions are an excellent substrate for microbial growth. These contaminated textiles act as a rich source for microbial transmission to susceptible patients. Studies hypothesize that the use of antimicrobial textiles may significantly reduce the risk of nosocomial infections [4].

Despite the undeniable achievements and success of chemical synthesis of antimicrobial compounds, nature is still considered a treasure trove and a highly attractive renewable source of a structurally diverse array of natural products that have served as lead structures for novel drug development and continue to do so [5]. There is an urgent need to discover new drug candidates with novel mechanisms of action to counter the continuous increase of AMR [1]. Naturally occurring bioactive compounds derived from endophytic fungal extracts can attenuate these harmful effects [6]. Endophytic fungi are considered a vital source of bioactive secondary metabolites with numerous biological applications [7–9]. Many endophytic fungal strains from different environmental and geographical areas were isolated and screened to assess their ability to produce novel molecules belonging to various compounds classes, including macrolides, terpenoids, alkaloids, or peptides, and displaying antimicrobial, antioxidant, antiviral and anticancer activity [10–15].

As the number of natural products reported in the literature is steadily increasing, dereplication strategies are essential in early stages of the fractionation-guided process to avoid isolating known metabolites. The Global Natural Products Social Molecular Networking (GNPS) is a platform which allows rapid and automated comparison of fragmentation patterns based on high-resolution MS/MS data that leads to effective dereplication [16]. GNPS clusters compounds with similar structural features which translate into similar fragmentation patterns into groups of molecular families. A molecular network consists of nodes which correspond to parent ions and are linked into groups with edges which represent a cosine similarity score. Molecular networking provides effective and rapid dereplication of large and complex MS/MS datasets and has been successfully implemented in natural product research to accelerate targeted isolation of potentially new metabolites [17–19].

In late 2017, we initiated a collaborative project between Egypt and the UK aiming at the isolation of new endophytic fungal strains from under-explored marine habitats in different locations in Egypt to be screened for their antimicrobial effects, with the ultimate aim of incorporating their bioactive extracts or metabolites in textiles used in Egyptian hospitals to reduce nosocomial infections. In our efforts to assess the biological activities and the chemical diversity of such fungal strains, we isolated 32 fungal strains from marine organisms and plants collected from Hurghada, Red Sea, and Wadi El-Natrun. Based on initial biological assays and LC/MS analysis of their crude extracts, 21 fungal isolates were prioritised for further analysis and identified using a combination of 18S rRNA, ITS rRNA, β-tubulin, and calmodulin gene sequencing. The most potent antimicrobial extracts were selected for the functionalization of cellulose containing fabrics to produce textile with UV-protection and antimicrobial functional properties. Moreover, chemical investigation of the most active extracts was carried out by the analysis of MS/MS data obtained from their crude extracts through the GNPS platform, and the results revealed several already reported fungal natural products which clustered with unidentified parent ions, suggesting the presence of potentially new secondary metabolites.

2. Materials and Methods

2.1. Sample Collection

Plant and marine samples were collected from two different locations in Egypt, Wadi El-Natrun depression (El-Beheira Governorate) and Hurghada (Red Sea Governorate). In total, 10 plant samples were collected from two saline lakes of Wadi El-Natrun (Al-Hamra and Al-Beida), and 9 marine samples (6 sponges, 1 soft coral, 1 sea grass, and 1 alga) were collected using SCUBA from two reefs in Hurghada, site (1) Abu Monqar island at N 27°12′53.7″, E 33°51′11.15″, and site (2) Makady Bay South at N 26°59′42.87″, E 33°54′4.02″, at a depth between 5 and 10 m. Plant and marine samples were transferred in ice boxes to the Microbial Chemistry Department, National Research Centre (NRC), Egypt, where each specimen was given a unique code, photographed, and stored at 4 °C and −20 °C, respectively, and was kept for further investigation and analysis.

2.2. Isolation of Endophytic Fungi

Isolation of endophytic fungi from plants and marine organisms was carried out using two different media: malt agar (MA; malt extract 15 g, sea salt 24.4 g, agar 20 g, distilled water up to 1 L, pH 6) and potato dextrose agar (PDA; potato extract 4 g, dextrose 20 g, sea salt 24.4 g, agar 20 g, distilled water up to 1 L, pH 6). For plants, healthy leaves were washed thoroughly with tap water and surface sterilised in 70% ethanol for 1 min, followed by subsecutive rinsing in sterile distilled water for 1 min, 2% sodium hypochlorite for 1 min, and finally 3 times in sterile distilled water. Sterilised leaves were dried under sterile conditions, and small segments were incubated on PDA supplemented with filter-sterilised nalidixic acid (50 mg/L) and chloramphenicol (200 mg/L) to suppress the growth of bacteria. For marine sponges and algae, approximately 1 cm^3 of the inner tissue of each sample was excised under sterile conditions and directly placed onto MA or PDA following a previously reported protocol [20]. The plates were incubated at 28 °C until growth of endophytic fungi was observed. Individual colonies were picked and repeatedly sub-cultured until pure colonies were obtained. The pure fungal isolates were preserved as glycerol stocks and stored at −20 °C in the Microbial Chemistry Department, National Research Centre (NRC), Egypt.

2.3. Production and Preparation of Fungal Extracts

Fungal isolates were cultivated on rice media (100 g commercial rice in 100 mL artificial sea water, adjusted to 50% natural salinity). The cultures were incubated for 15 days at 28 °C under static conditions. After incubation, the cultures were extracted with ethyl acetate (EtOAc), and the organic phases were dried under vacuum to obtain the crude extracts.

2.4. DNA Extraction, Amplification and Sequencing

The molecular identification of selected fungal strains was carried out by extraction of the genomic DNA using Qiagen DNeasy Mini Kit following the manufacturer's instructions. The PCR reaction mixture was as follows: 1 μg genomic DNA, 1 μL (20 μM of each primer), 10 mM dNTPs mixture, 2 units of Taq DNA polymerase enzyme, and 10 μL 5× reaction buffer. Initially, amplification reactions for 18S and ITS genes, respectively, were performed using 3 primer pairs; NS3 (5′-GCAAGTCTGGTGCCAGCAGCC-3′)/NS4 (5′-CTTCCGTCAATTCCTTTAAG-3′), NS1 (5′-GTAGTCATATGCTTGTCTC-3′)/NS8 (5′-TCCGCAGGTTCACCTACGGA-3′), and ITS1

(5′-TCCGTAGGTGAACCTGCG-3′)/ITS4 (5′-TCCTCCGCTTATTGATATGC-3′ [21], and the following PCR thermal profile: denaturation step at 94 °C for 5 min, followed by 35 cycles of 94 °C for 30 s, 55 °C for 30 s, 72 °C for 90 s, and a final extension step at 72 °C for 5 min. For amplification of -tubulin genes, primer pair Bt2a (5′-TTCCCCCGTCTCCACTTCTTCATG-3′)/Bt2b (5-GACGAGATCGTTCATGTTGAACTC-3′) was used with the following PCR thermal profile: denaturation step at 95 °C for 5 min, followed by 35 cycles of 95 °C for 30 s, 58 °C for 30 s, 72 °C for 60 s, and a final extension step at 72 °C for 7 min [22]. For amplification of CaM genes, primer pair cmd5 (5′-CCGAGTACAAGGAGGCCTTC-3′)/cmd6 (5-CCGATAGAGGTCATAACGTGG-3′) was used with the following PCR thermal profile: denaturation step at 95 °C for 10 min, followed by 35 cycles of 95 °C for 30 s, 55 °C for 30 s, 72 °C for 60 s, and a final extension step at 72 °C for 7 min [23]. The amplified products were subjected to agarose gel electrophoresis, and bands of the expected sizes were excised and purified using Montage PCR Clean up kit (Millipore) or JeneJET purification kit (ThermoFisher Scientific, Basel, Switzerland) and shipped for sequencing by two commercial services, SolGent and Macrogen, South Korea. The resulting sequences were analysed by BLASTN to study their similarity and homology with the respective target gene sequences contained in the NCBI database. The phylogenetic tree was constructed based on the maximum-likelihood (ML) algorithm [24] using MEGA6 [25], with evolutionary distances computed using the Kimura 2-parameter model [26]. Validity of branches in the resulting trees was evaluated by bootstrap resampling support of the data sets with 1000 replications. Isolate 13A was not included, as only ITS sequence information was available, while strain M13 did not provide DNA of sufficient quality during the second round of sequencing.

2.5. *Antimicrobial Activity*

Antimicrobial screening of the fungal crude extracts was carried out against a set of test microbes comprising the penicillin-resistant Gram-positive bacterium *Staphylococcus aureus* ATCC 6538-P, the Gram-negative bacterium *Pseudomonas aeruginosa* ATCC 27853, the yeast *Candida albicans* ATCC 10231, and the fungus *Aspergillus niger* NRRLA-326. Initial screening and selection of strains for further chemical analysis was based on the agar diffusion method [27] (data not shown). For active fungal extracts, minimal inhibitory concentrations (MIC) were determined using the microplate dilution method [28]. In total, 10 µL of extracts at different concentrations were added to 180 µL of culture medium, i.e., lysogeny broth for bacteria or potato dextrose broth for fungi, followed by addition of 10 µL of bacterial or fungal suspension at the log phase. The plates were incubated overnight at 37 °C, and the absorbance was measured at OD600 using a Spectrostar Nano Microplate Reader (BMG Labtech GmbH, Allmendgrun, Germany). Initially, eight serial dilutions (250, 125, 62.5, 31.25, 15.63, 7.81, 3.90, 1.95 µg/mL) were prepared, and, based on the results obtained, further dilutions bracketing the lowest concentration with no observable growth of bacteria or fungi were tested in steps of 1 µg/mL. MICs are reported as the average of the lowest concentrations with no observable growth of bacteria or fungi determined in three independent experiments. Ciprofloxacin and nystatin were used as positive controls.

2.6. *Total Antioxidant Capacity (TAC)*

The total antioxidant capacity (TAC) of each fungal extract was determined following the phosphomolybdenum method [29] using ascorbic acid as standard. This assay is based on formation of a green coloured Mo(V) phosphate complex if analytes present in the sample are capable of reducing Mo(VI) to Mo(V). In total, 0.5 mL of each extract (at a concentration of 100 µg /mL in methanol) was combined in dried vials with 5 mL of reagent solution (0.6 M sulfuric acid, 28 mM sodium phosphate and 4 mM ammonium molybdate). The reaction mixture was incubated in a thermal block at 95 °C for 90 min. After the samples had cooled at room temperature, the absorbance was measured at 695 nm against a blank consisting of all reagents and solvents without the sample, which was incubated under the same conditions. All experiments were carried out in triplicate. The antioxidant activity of the sample was expressed as the number of equivalents of ascorbic acid (AAE), and the total antioxidant capacity (TAC) was calculated as follows:

$$TAC = \left(\frac{Absorbance\ sample}{Absorbance\ ascorbic\ acid}\right) * 1000$$

2.7. Grafting of MCT-βCD onto Cellulosic Substrates

Pre-chemical modification of the different cellulosic substrates (size 3 × 6 cm), namely knitted cotton jersey (30/1) (100g/m^2), knitted viscose jersey (30/1) (100 g/m^2) and plain (1/1) woven cotton (120 g/m^2), was carried out using the pad-dry-cure method. Cellulosic substrates were immersed for 20 min with 25 g/L MCT-βCD and 8 g/L Na_2CO_3 as a catalyst, followed by padding to a wet pick-up at 80% wetness, drying at 100 °C for 3 min, and curing at 160 °C for 3 min. Before testing, a final washing step was used to remove excess, partially hydrolysed and unfixed reactant, followed by drying.

2.8. Post-Hosting of Bio-Active Extracts into Hydrophobic Cavities to Impart the Demanded Functionalities

Portions of pre-modified cellulosic substrates containing the hosting cavities were post-loaded with the selected active extracts, then suspended in 15 mL methanol by exhaustion technique using an IR-dyeing machine at 40 °C for 1 h, followed by thoroughly rinsing and air drying.

2.9. Methods of Analysis

Nitrogen content of the pre-modified cellulosic substrates was assessed by Kjeldhal method. Antimicrobial effects of bioactive extract-loaded substrates against *S. aureus*, *E. coli*, *C. albicans*, and *A. niger* were evaluated qualitatively according to a published method [27]. The UV protection factor (UPF) of untreated, pre-modified, and functionalized fabric samples was evaluated according to the Australian/New Zealand Standard (AS/NZS 4399/1996) and ranked as follows: good (UPF:15–24), very good (UPF: 25–39), and excellent (UPF > 40).

2.10. In Vitro Cytotoxicity

Cytotoxicity against normal human diploid fibroblasts (WI-38) was tested using the MTT assay [30]. The cell lines were obtained from American Type Culture Collection (ATCC) via Holding company for biological products and vaccines (VACSERA), Cairo, Egypt.

2.11. Mass Spectral Data Acquisition

Extracts were dissolved in methanol at a final concentration of 0.1 mg/mL, centrifuged, and injected onto a Bruker MAXIS II Q-ToF mass spectrometer coupled to an Agilent 1290 UHPLC system. Separation was achieved using a Phenomenex Kinetex XB-C18 (2.6 mM, 100 × 2.1 mm) column and the following LC gradient profile: 5% MeCN + 0.1% formic acid to 100% MeCN + 0.1% formic acid in 15 min. MS parameters were: mass range *m/z* 100–2000, capillary voltage 4.5 kV, nebulizer gas 5.0 bar, dry gas 12.0 L/min, and dry temperature of 220 °C. MS/MS experiments were conducted under Auto MS/MS scan mode with no step collision.

2.12. Molecular Networking

The MS/MS data obtained for 21 bioactive fungal isolates were converted from Bruker DataAnalysis (.d) to .mzXML file format using MSConvert (Available online: http://proteowizard.sourceforge.net/index.html). A molecular network was created using the online workflow at GNPS (Available online: https://gnps.ucsd.edu/). The data were filtered by removing all MS/MS peaks within ±17 Da of the precursor *m/z*. MS/MS spectra were window-filtered by choosing only the top 6 peaks in the ±50 Da window throughout the spectrum. The data were clustered with MS-Cluster with a parent mass tolerance of ±2 Da and a MS/MS fragment ion tolerance of ±0.5 Da to create consensus spectra.

Further, consensus spectra that contained less than 1 spectrum were discarded. A network was created where edges were filtered to have a cosine score above 0.5 and more than 6 matched peaks. Further edges between two nodes were kept in the network only if each of the nodes appeared in each other's respective top 10 most similar nodes. The spectra in the network were then searched against GNPS' spectral libraries. The library spectra were filtered in the same manner as the input data. All matches kept between network spectra and library spectra were required to have a score above 0.6 and at least 6 matched peaks. Cytoscape version 3.6.1 was used to visually display the data as a network of nodes and edges [31].

3. Results

3.1. *Fungal Isolation and Cultivation*

A total of 32 pure fungal strains were isolated, out of which 18 were isolated from marine samples of Hurghada, and 14 were obtained from plant samples of Wadi El-Natrun. Small scale fermentation of the isolated fungal strains was carried out on solid rice medium, and the resulting extracts were subjected to biological screening for anti-bacterial, anti-fungal, and antioxidant activities.

3.2. *Molecular Identification*

Based on a combination of biological screening results as well as LC/MS analysis, 21 fungal isolates were selected and genetically identified by extraction and sequencing of DNA at SolGent and Macrogen Companies, South Korea. The resulting sequences were aligned with closely related known sequences in GenBank. Based on their ITS and 18S rDNA sequences, isolates were identified and found to belong to eight different genera, i.e., *Alternaria*, *Aspergillus*, *Byssochlamys*, *Cladosporium*, *Epicoccum*, *Penicillium*, *Sarocladium*, and *Talaromyces*. As only isolate 13A could be identified as *Alternaria alternata* with this approach, the remaining isolates were further assessed by phylogenetic analysis of concatenated ITS rDNA, β-tubulin, and calmodulin regions, respectively (for details including GenBank accession numbers, see Table 1).

Our molecular analysis revealed interesting strains that have not been studied earlier. For example, initial DNA analysis allowed us to ascertain that AS14 fungal isolate matches the species *Epicoccum nigrum*. Given that *Epicoccum nigrum* is known to englobe two genotypes, we opted for phylogenetic analysis using type strain CBS 161.73 and other *Epicoccum nigrum* characterised by Favaro et al. [20]. Phylogenetic analysis using either ITS rDNA and β-tubulin markers separately or concatenated allowed us to unambiguously identify AS14 fungal isolate as *Epicoccum nigrum* group 2 designated by Favaro et al. [20] as *Epicoccum* sp., a new species awaiting formal description. This putative new species is understudied chemically and biologically.

Table 1. Origin and identification of fungal strains by DNA sequencing.

Strain ID	Target Gene and GenBank Accession Number				ID Assigned in GenBank	Sample Origin	Type	Location
	18S	ITS	β-Tubulin	Calmodulin				
A3	-	MN114540	MT184332	MT184352	*Aspergillus terreus* strain A3	*Crella cyathophora*	sponge	Makady Bay South, Hurghada
M113	-	MK262919	MT184334	MT184354	*Aspergillus calidoustus* strain M113	*Thalassia hemprichii*	sea grass	Makady Bay South, Hurghada
M13	-	MK953943	MT184348	-	*Epicoccum nigrum* strain FAS-14	*Thalassia hemprichii*	sea grass	Makady Bay South, Hurghada
M2S3	MN328309	MT152324	MT184340	MT184360	*Byssochlamis spectabilis* strain M2S3	*Crella cyathophora*	sponge	Makady Bay South, Hurghada
M2S4	MN328341	MT152319	MT184335	MT184355	*Aspergillus niger* strain M2S4	*Crella cyathophora*	sponge	Makady Bay South, Hurghada
M35	-	MK953944	MT184347	MT184367	*Penicillium crustosum* strain FAS-28	*Siphonochalina siphonella*	sponge	Makady Bay South, Hurghada
M4	-	MK953942	MT184345	MT184365	*Talaromyces verruculosus* strain FAS-10	*Latrunculia magnifica*	sponge	Makady Bay South, Hurghada
M42	MN328356	MT152318	MT184331	MT184351	*Aspergillus terreus* strain M42	*Latrunculia magnifica*	sponge	Makady Bay South, Hurghada
SHP3	-	MN114156	MT184343	MT184363	*Cladosporium spinulosum* strain SHP3	*Thalassia hemprichii*	sea grass	Makady Bay South, Hurghada
SHP18	-	MN114621	MT184342	MT184362	*Cladosporium spinulosum* strain SHP18	*Thalassia hemprichii*	sea grass	Makady Bay South, Hurghada
13A	-	MK248606	-	-	*Alternaria alternata* strain 13A	*Phragmites australis*	plant	Lake El-Bida, Wadi El-Natrun
15F6	-	MN328763	MT184330	MT184350	*Aspergillus terreus* strain 15F6	*Hyoscyamus muticus*	plant	Lake El-Bida, Wadi El-Natrun
15F14	MN328048	MT152320	MT184336	MT184356	*Aspergillus fumigatus* strain 15F14	*Hyoscyamus muticus*	plant	Lake El-Bida, Wadi El-Natrun
2S4	-	MN115554	MT184341	MT184361	*Alternaria alternata* strain 2S4	*Juncus acutus*	plant	Lake El-Hamra, Wadi El-Natrun
2S6	-	MN330611	MT184338	MT184358	*Aspergillus fumigatus* strain 2S6	*Juncus acutus*	plant	Lake El-Hamra Wadi El-Natrun
5S1	-	MN110110	MT184339	MT184359	*Aspergillus ochraceopetaliformis* strain 5S1	*Panicum turgidum*	plant	Lake El-Hamra, Wadi El-Natrun
7S1	-	MK953941	MT184346	MT184366	*Talaromyces verruculosus* strain FAS-02	*Tamarix nilotica*	plant	Lake El-Hamra, Wadi El-Natrun
7S4	-	MN114521	MT184333	MT184353	*Aspergillus terreus* strain 7S4	*Tamarix nilotica*	plant	Lake El-Hamra, Wadi El-Natrun
7S6	MN326853	MT152321	MT184337	MT184357	*Aspergillus fumigatus* strain 7S6	*Tamarix nilotica*	plant	Lake El-Hamra, Wadi El-Natrun
7S9	-	MN114217	MT184344	MT184364	*Penicillium rubens* strain 7S9	*Tamarix nilotica*	plant	Lake El-Hamra, Wadi El-Natrun
9AS1	MN327968	MT152323	MT184349	MT184368	*Sarocladium kiliense* strain 9AS1	*Panicum turgidum*	plant	Lake El-Hamra, Wadi El-Natrun

3.3. Biological Screening

3.3.1. Antimicrobial Activity

The antimicrobial results showed that some fungal extracts exhibited promising antimicrobial activity against some tested microorganisms (Table 2). Among tested extracts, M113, M13, 7S4, 7S5 and 7S9 showed significant antibacterial activity against *Staphylococcus aureus* with MIC values between 12.3 and 31.25 µg/mL, followed by M2S3, M2S4, M35, M42, 2S4, 2S6, 5S1, 7S1, 7S6, 9AS1, SHP18, and 13A with moderate activity, i.e., MIC values between 43.0 and 117.7 µg/mL. Additionally, M113, M13, 7S4, and 7S5 extracts displayed significant antibacterial activity against *Pseudomonas aeruginosa* with MIC values between 13.7 and 31.25 µg/mL. Moreover, all extracts were screened for their antifungal activity towards *Candida albicans* and *Aspergillus niger*. The extracts of the fungal isolates M113, M13, M42, 2S4, 5S1, 7S1, 7S4, 7S5, and SHP3 showed moderate inhibitory activity against *C. albicans* with MIC values between 35.0 and 125 µg/mL, whereas none of the extracts showed significant activity toward *A. niger* (Table 2).

Table 2. Antimicrobial activity of the fungal extracts.

Sample Code	*Staphylococcus aureus* MIC (µg/mL) *	*Pseudomonas aeruginosa* MIC (µg/mL) *	*Candida albicans* MIC (µg/mL) *	*Aspergillus niger* MIC (µg/mL) *
M1 °	250	250	250	-
M113	25.0 ± 1.7	22.3 ± 1.5	52.3 ± 0.6	192.7 ± 3.8
M13	20.0 ± 2.0	42.0 ± 1.0	111.0 ± 1.7	122.7 ± 4.6
M23 °	250	250	250	-
M2S3	100.3 ± 4.0	195.0 ± 6.2	177.3 ± 4.5	-
M2S4	115.0 ± 2.6	102.7 ± 3.8	163.3 ± 2.3	-
M32 °	250	250	250	-
M35	85.0 ± 2.6	101.0 ± 1.7	180.7 ± 5.0	237.7 ± 2.1
M4	172.3 ± 4.0	185.0 ± 3.6	158.7 ± 4.7	210.3 ± 4.2
M42	43.0 ± 1.0	86.0 ± 2.6	92.7 ± 3.8	-
SHP3 °	-	-	35.0 ± 1.7	-
SHP18	105.0 ± 5.6	123.7 ± 0.6	203.0 ± 2.6	-
13A	45.0 ± 1.0	42.0 ± 1.0	102.7 ± 3.1	-
2S4	48.0 ± 2.6	47.3 ± 3.2	87.7 ± 3.8	85.0 ± 4.4
2S6	97.7 ± 2.1	103.7 ± 2.9	209.0 ± 2.0	-
5S1	50.0 ± 2.6	46.3 ± 3.5	85.3 ± 3.2	125.0 ± 0.0
7S1	54.0 ± 1.7	51.0 ± 2.0	102.7 ± 2.9	183.7 ± 5.0
7S4	12.3 ± 1.5	13.7 ± 1.2	52.3 ± 1.5	-
7S5 °	31.25	62.5	125	-
7S6	113.7 ± 0.6	55.7 ± 1.2	175.0 ± 2.6	-
7S9	47.3 ± 3.2	41.0 ± 2.6	171.0 ± 1.5	-
9AS1	117.7 ± 4.2	43.7 ± 3.1	-	-
Cip	0.078	0.156	-	-
Nys	-	-	5	10

* Based on three independent replicates; ° only tested in the initial serial dilution, see Experimental section. Cip: ciprofloxacin; Nys: nystatin; -: not active; MIC: minimum inhibitory concentration.

3.3.2. Total Antioxidant Capacity (TAC)

The fungal crude extracts were evaluated for their total antioxidant capacity (TAC) using the phosphomolybdenum method. The results presented in Figure 1 revealed that the fungal extract M13 showed the most potent antioxidant activity with a TAC value of 813.5 mg AAE/g extract (mg of ascorbic acid equivalents/g). Two extracts (M113 and M32) exhibited TAC values between 500 and 600 mg AAE/g extract, while the activity of a further three (M2S4, 5S1, and 7S1) ranged between 400 and 500 mg AAE/g extract. Four fungal extracts showed TAC values ranging between 150 and 400 mg AAE/g extract, while the remaining extracts showed weak or no activity.

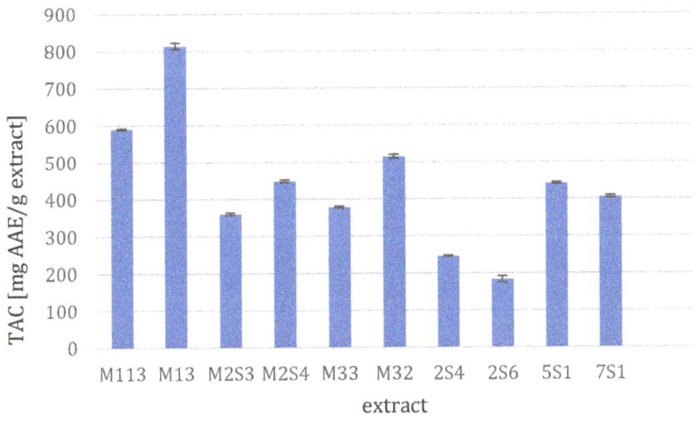

Figure 1. Total antioxidant capacity (TAC) of selected fungal extracts. Results are presented as means ± SD, $n = 3$) and are expressed as mg ascorbic acid equivalent (AAE)/g extract.

Overproduction of reactive species such as reactive oxygen species (ROS) and reactive nitrogen species (RNS) leads to generation of a phenomenon known as oxidative stress. Oxidative stress represents the imbalance between the production rate of free radicals and the antioxidants in the human body, which may be followed by several health disorders, including cancer, cardiovascular diseases, inflammation, and Alzheimer's disease [32,33]. Previous studies revealed a positive linear correlation between the antioxidant activities of the tested endophytic fungal extracts and the presence of certain chemical classes in these extracts, including phenolic acid and quinone derivatives [34], indole derivatives [6], coumarins [35], and butyrolactones [36].

3.4. Functionalization of the Nominated Cellulosic Substrates

Functionalization of mill-scoured and bleached cellulosic fabrics was carried out by grafting of monochlorotriazinyl β-cyclodextrin (MCT-βCD), as an environmentally friendly encapsulating and hosting reactive βCD, onto cellulose structure to create core-shaped hydrophobic cavities as follows:

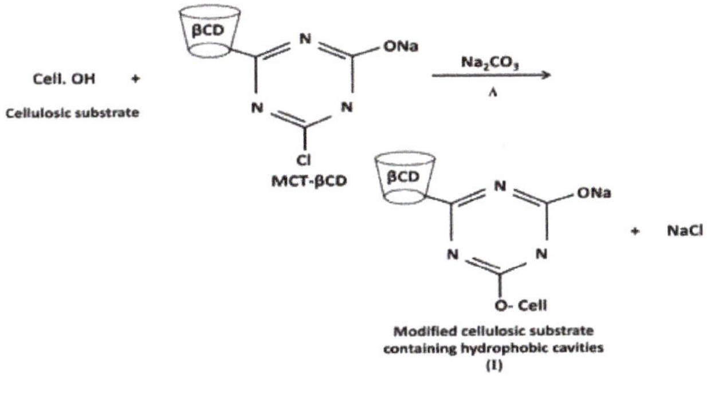

This was followed by subsequent inclusion of the nominated bioactive extracts into the host internal cavities of the grafted reactive βCD via formation of host–guest inclusion complexes [37] as follows:

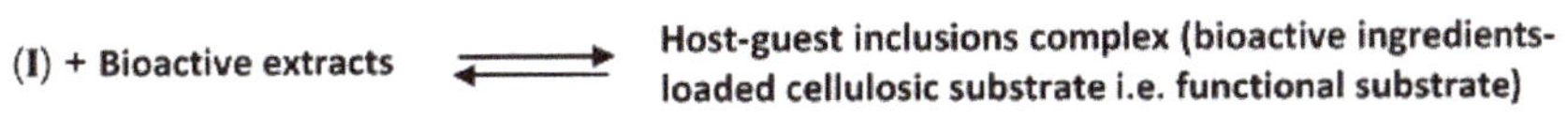

3.5. Antimicrobial Properties of the Treated Cellulosic Textiles

The imparted antimicrobial activity of the pre-modified/post-treated substrates with the selected bioactive extracts as potentially eco-friendly alternatives to synthetic antimicrobial agents are presented in Table 3. For a given set of treatment conditions, it is clear that pre-modification of the nominated cellulosic substrates with MCT-βCD had practically no inhibitory effect against the tested bacterial and fungal strains. Inclusion of any of the selected bioactive extracts into the hydrophobic cavities at treated fabrics surface brought about an improvement in their antimicrobial functionality which followed the decreasing order: M113 > 7S4 > 13A >> non-treated. It could be presumed that the extent of improvement in the imparted antimicrobial functionality of bioactive extract-loaded substrates was governed by their concentration, chemical composition, active ingredients, degree of inclusion into, and release out of the hydrophobic cavities at the finished fabrics surface as well as mode and degree of microbial inhibition. On the other hand, the efficacy of the imparted antimicrobial activity, expressed as the size of inhibition zone, against the targeted microorganisms was determined by the type of target microorganism, e.g., bacteria or fungi, their cell wall structure, their amenability to damage, and their capability to inactivate bioactive ingredients present in the bioactive extracts tested. The encapsulated bioactive extracts had practically no (in the case of M113 or 7S4) or only moderate inhibitory effects (as in the case of 13A) against the *E. coli* strain [38–42].

Table 3. Antimicrobial properties of the treated cellulosic textiles.

Extract (Concentration Tested)	Fabric Type [a]	Zone of Inhibition (mm)			
		S. aureus	*E. coli*	*C. albicans*	*A. niger*
M113 (55 μg/mL)	1	16	-	17	15
	2	18	-	18	15
	3	19	-	19	17
7S4 (70 μg/mL)	1	13	-	14	11
	2	14	-	15	12
	3	16	-	16	14
13A (62 μg/mL)	1	12	7	11	10
	2	13	8	12	11
	3	14	9	13	12
modified	1	-	-	-	-
	2	-	-	-	-
	3	-	-	-	-
untreated	1	-	-	-	-
	2	-	-	-	-
	3	-	-	-	-

[a] 1; woven; 2, C. knitted; 3, V. knitted.

Moreover, the extent of pre-modification and subsequent post-loading and releasing of the nominated bioactive ingredients, which in turn affected the antimicrobial functionality, were governed by type of cellulosic substrates and followed the decreasing order: knitted viscose > knitted cotton > woven cotton fabric, keeping other parameters constant. This reflects the differences among these substrates in weight, thickness, fabric construction amorphous/crystalline area, number, and location of accessible active site, i.e., –OH groups, extent of pre-modification, post-loading, and slow releasing of the active antimicrobial ingredients into the surrounding zone to inhibit the growth or to kill the pathogenic microorganisms [42,43].

3.6. Ultraviolet Protection

From Table 4, it can be seen that inclusion of the selected bioactive extracts into the hydrophobic cavities of the modified cellulosic substrates resulted in a remarkable enhancement in their UV protection functionality, expressed as UPF values, regardless of the substrates used and the active ingredients included. According to Table 4, the extent of improvement in UPF values depended on type of substrate, i.e., knitted cotton > woven cotton fabric > knitted viscose fabric, reflecting the differences among them in thickness, porosity, extent of modification, and subsequent hosting of the bioactive extracts ingredients. On the other hand, post-loading of the selected extracts onto the finished fabrics surface to impart UV protection capability was accompanied by a significant increase in their UPF values with the following sequence, 13A > M113 > 7S4 >> modified ≈ untreated. In line with previous studies, we would assume that the variation in UPF values of the finished substrates upon using these ingredients reflects the differences among them in extent of loading and coating the treated substrates as well as the positive role of the hosted bioactive extracts and their ingredients in absorbing, reflecting, scattering, or blocking harmful UV radiation, especially UVA and UVB, irrespective of the substrate used [44,45].

Table 4. UV protective properties of different textile fabrics treated with selected fungal extracts.

Extract (Concentration Tested)	Fabric Type [a]	UPF	Classification
M113 (55 μg/mL)	1	70.8	excellent
	2	162.2	excellent
	3	30.7	Very good
7S4 (70 μg/mL)	1	33.0	very good
	2	77.6	excellent
	3	26.9	very good
13A (62 μg/mL)	1	113.9	excellent
	2	202.8	excellent
	3	38.3	very good
modified	1	10.5	no protection
	2	15.8	good
	3	8.0	no protection
untreated	1	7.0	no protection
	2	11.2	no protection
	3	5.0	no protection

[a] 1; woven; 2, C. knitted; 3, V. knitted. UPF: UV protection factor.

3.7. Fungal Extracts Toxicity Study

Because of global concern of functionalized textile toxicity, it was very important to test the selected fungal extracts for their toxicity. Accordingly, the three crude extracts used in textile functionalisation (M113, 7S4, and 13A) were investigated for their cytotoxicity on normal human diploid fibroblasts (WI-38) in comparison with docetaxel as reference control (IC_{50} 22.2 μg/mL). The results obtained demonstrated that M113 extract has a very low toxic effect with an IC_{50} value of 88.6 μg/mL, followed by 13A extract which exhibited weak to moderate toxicity at an IC_{50} value 68.4 μg/mL, while significant toxicity was observed for strain 7S4 with an IC_{50} value 19.3 μg/mL.

3.8. Chemical Investigation Using a Molecular Network Approach

The same 21 fungal extracts selected for molecular identification, as described above (Table 1), were subjected for further chemical investigation based on a comparative untargeted metabolomics study using the Global Natural Products Social Molecular Networking (GNPS) platform (https://gnps.ucsd.edu). Analysis of the MS/MS data of the crude extracts was carried out by combining manual dereplication with the aid of various databases such as Natural Products Atlas (NPAtlas, [46] https://www.npatlas.org), AntiBase [47] (https://application.wiley-vch.de/stmdata/antibase.php), and Reaxys (https://www.reaxys.com) as well as automated dereplication using the GNPS platform. The generated molecular network (MN) (Figure 2A) consisted of 2765 nodes in total, out of which 257 corresponded to parent ions present in the media and the solvent blanks, which were excluded from the data analysis. It is worth mentioning that some nodes correspond to different quasi-molecular ions pertaining to the same molecular formula; therefore, not all observed nodes represent a single molecule. The distribution of nodes according to each fungal strain varied across all 21 selected isolates and is given in Figure 2B. The isolates for which the largest number of nodes were observed include *Aspergillus calidoustus* strain M113 (178 nodes) and *Cladosporium spinulosum* strain SHP3 (140 nodes), whereas *Aspergillus fumigatus* strain 15F14 and *Aspergillus terreus* strain 15F6 produced the lowest number of nodes with 34 and 60, respectively. In the MN, different molecular families were observed which correspond to metabolites that are commonly produced by fungi. It is important to take into consideration that manual and automated dereplication cannot provide conclusive results for all 2765 nodes that were present in the molecular network, therefore, herein we report and describe previously known fungal metabolites as well as potentially new derivatives based on the generated molecular network. Additionally, it should be noted that a definite identification of any known compounds by MS/Ms alone is not possible, as it is not possible to deduct the absolute configuration, and the potential presence of isomers cannot ultimately be ruled out. However, from a practical point of view, in our view, the likelihood of the identifications presented in the following being correct would be considered high enough (especially if more than one known compound has been pinpointed within the same cluster, and also taking into account taxonomic information, e.g., previous report from related species) that subsequent isolation efforts would be deemed not worthwhile unless the analysis of the GNPS clusters in question also suggests the presence of potentially unknown compounds.

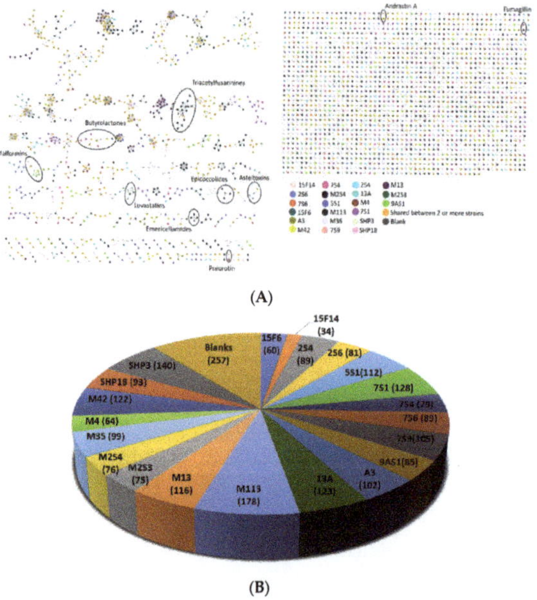

Figure 2. (**A**) Molecular network of the organic extracts of 21 biologically active Egyptian fungal strains. Nodes are colour-coded based on fungal strain ID, as shown in Table 1. The reverse triangle-shaped nodes represent previously reported metabolites identified by Global Natural Products Social Molecular Networking (GNPS)' libraries. Molecular families are shown in circles. (**B**) Distribution of unique nodes for each fungal strain observed in the molecular network.

More specifically, the previously reported cyclic depsipeptide emericellamide A (*m/z* 632.402, $[M+Na]^+$) was identified by GNPS and was produced by *Aspergillus calidoustus* strain M113, which exhibited the best antimicrobial properties using the treated cellulosic textiles and a low toxicity profile. Manual dereplication allowed the identification of emericellamide B (*m/z* 674.449 $[M+Na]^+$), which is indeed linked with emericellamide A in the same molecular cluster (Figure 3A). Emericellamides A and B were originally isolated from a marine-derived fungus *Emericella* sp. during co-cultivation experiments with the actinomycete *Salinispora arenicola* [48]. Dereplication with AntiBase of the additional parent ions *m/z* 660.433 ($[M+Na]^+$) and *m/z* 702.480 ($[M+Na]^+$) in the emericellamide cluster identified by GNPS provided no hits, which strongly suggests the presence of new emericellamide homologues, with the candidate molecular formulae suggesting an additional two or five methylene groups compared to emericellamide A, respectively. Even though there are reported lipodepsipetides with the same molecular formulae, such as scopularides and oryzamides [49], considering the structures and the relatively high cosine scores in the GNPS cluster, it is very likely that the potentially new derivatives share the same peptide backbone, and that these modifications are located in the polyketide-derived side chain. Taken together, it is highly likely that *Aspergillus calidoustus* strain M113, which was among the most promising in all of our biological assays, may produce undescribed representatives of the emericellamide family of cyclic peptides, warranting future efforts at their targeted isolation and structural characterisation.

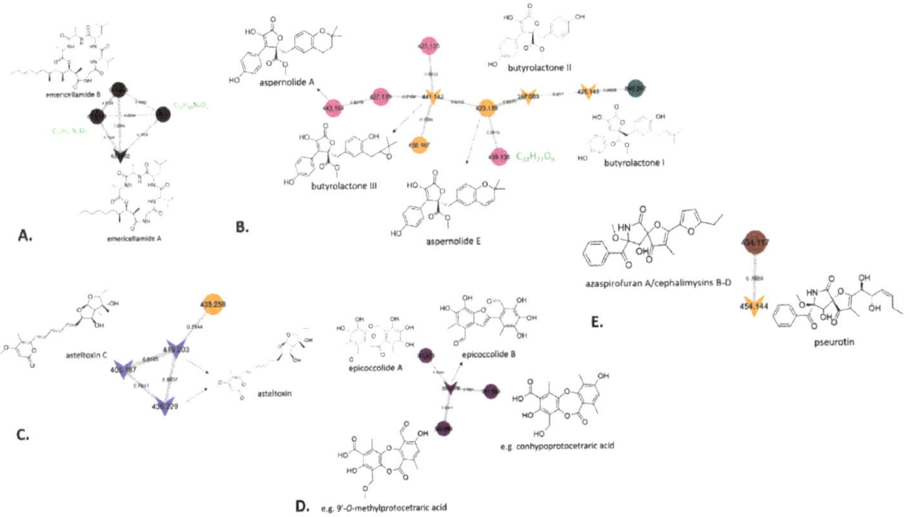

Figure 3. Selected GNPS clusters representing molecular families: (**A**) emericellamide cluster, (**B**) butyrolactone cluster, (**C**) asteltoxin cluster, (**D**) epicoccolide cluster, and (**E**) pseurotin cluster. Nodes depicted by inverted triangular shapes represent hits in the GNPS compound library. Candidate molecular formulae for potential new derivatives highlighted in green.

Moreover, GNPS identified a cluster containing butyrolactone I (*m/z* 425.145, $[M+H]^+$), II (*m/z* 357.083, $[M+H]^+$) and III (*m/z* 441.142, $[M+H]^+$), which were observed in two strains of *Aspergillus terreus* (7S4, 15F6) and in one isolate of *Penicillium crustosum*, M35 (Figure 3B). Butyrolactones are commonly found in fungal strains and have exhibited a wide range of biological activities [50,51]. Our identifications were based on matching spectra in the GNPS library, although the mass error of the pertaining molecular formulae was relatively high (data not shown). Therefore, manual analysis of the MS/MS of the three fungal extracts was carried out, which confirmed the presence of the butyrolactone derivatives, and, subsequently, butyrolactones I and II were isolated from strain 7S4, and their identity was established by NMR spectroscopy (data not shown). On this basis, while not contained in the GNPS library, further nodes in Figure 3B in all likelihood represent the known aspernolides A [52] and E [53], which are closely related to butyrolactones. It is worth noting that the node with *m/z* 439.134, corresponding to the putative molecular formula, $C_{24}H_{22}O_8$, was only produced by strain 7S4 and gave no hits when it was dereplicated manually, which suggests it may correspond to a new butyrolactone (or aspernolide) derivative. It is worth mentioning that strain 7S4 showed moderate toxicity against normal human diploid fibroblasts, although there is no indication in the literature of the toxic nature of butyrolactones, which warrants further investigation for non-traced metabolites which could be responsible for such toxicity. Additionally, further chemical investigation of this strain would allow the isolation of this potentially new metabolite, which highlights the power of sample prioritisation among large datasets using molecular networking.

Strain 5S1, which, based on the sequencing results should be closely related to *Aspergillus ochraceopetaliformis*, was the only producer of asteltoxin and asteltoxin C which clustered in the same molecular family (Figure 3C). Manual dereplication revealed that a further parent ion in this cluster with *m*/*z* 435.258 ($[M+H]^+$) was also observed in the extract of *Alternaria alternata* strain 13A, which exhibited good antibacterial property in the textile assay but was associated with moderate toxicity, which in all likelihood could be attributed to the presence of asteltoxin. While its nominal mass matches that of the known asteltoxin B, the mass error would suggest that the molecular formula may not, thus not fully ruling out the presence of a new asteltoxin derivative, which could only be clarified by further chemical investigation. The mycotoxin asteltoxin was isolated in 1979 from *Aspergillus stellatus* [54], whereas asteltoxin B was obtained from the liquid cultures of an *Aspergillus* sp. strain isolated from the octocoral *Melitodes squamata* [55]. Its structure was later revised in the course of a chemical study of the entomopathogenic fungus *Pochonia bulbillosa*, during which it was obtained along with asteltoxin C [56]. The toxicity associated with strain 13A could be attributed to the presence of such metabolites.

In the epicoccolide cluster, epicoccolide B (*m*/*z* 359.078, $[M+H]^+$) was identified by GNPS, whereas epicoccolide A (*m*/*z* 375.073, $[M+H]^+$) was confirmed through manual analysis (Figure 3D). Both polyoxygenated polyketides were previously described from an endophytic fungus *Epicoccum* sp. [57]. Interestingly, two further nodes with *m*/*z* 361.093 ($[M+H]^+$) and *m*/*z* 389.089 ($[M+H]^+$) gave no hits in the GNPS library, while manual dereplication of their candidate molecular formulae, $C_{18}H_{16}O_8$ and $C_{19}H_{16}O_9$, respectively, could not rule out that they may represent known depsidones, for example, conhypoprotocetraric acid [58] and 9′-O-methylprotocetraric acid [59], previously described from terrestrial lichens (besides a large number of other, chemically less closely related compounds with matching molecular formulae). Ultimately, it is not possible to come to a definite conclusion as to whether these compounds are known or may represent new compounds on the basis of LC/MS and GNPS data alone, thus further chemical investigation is required.

Pseurotin was originally isolated in the late 1970s from *Pseudeurotium ovalis* [60] and in the present study was identified in the obtained molecular network as a product of *Aspergillus fumigatus* strain 7S6 and *Talaromyces verruculosis* strain FAS-10 (M4). Pseurotin clustered with the node *m*/*z* 434.117 ($[M+Na]^+$) present only in the extract of the latter fungus, and manual dereplication suggested the presence of one out four fungal compounds with a molecular formula of $C_{22}H_{21}NO_7$ (Figure 3E). Azaspirofuran A [61] and cephalimysins B, C, and D [62] are all structurally related to pseurotin, however, they cannot be differentiated in the molecular network because they are stereoisomers, which highlights the limitation of MS/MS data and, therefore, GNPS in identifying isomers. In order to identify which of those four fungal metabolites corresponds to the observed node, chromatographic isolation and structure elucidation are required, but it should be noted that such an effort is not worthwhile, as it will very likely result in the re-isolation of known compounds.

4. Discussion

The rates of nosocomial infections, especially by those caused by the newly emerged antibiotic resistant bacteria, are increasing and doing so more alarmingly in the developing countries [63]. The rate of such infections has recently increased, especially in some teaching hospitals in Egypt. As textiles are an excellent substrate for bacterial growth under appropriate moisture and temperature conditions, it is not surprising that several studies have confirmed that personal contact with contaminated textiles is considered the main source of transmission of the microbial infections to susceptible patients, while indirect contact or aerosol transmission of nosocomial pathogens may also be contributing factors [64]. Few studies have addressed the use of antimicrobial textiles, especially for those staff that are in close contact with the patients. A recent study has proven that treating textiles with ionic silver after washing resulted in a significant decrease in microbial contamination, especially *S. aureus* and MRSA contamination [21].

In 2017, we initiated a collaborative project between Egypt and the UK aiming at the isolation of new endophytic fungal strains from under-explored marine habitats in different locations in Egypt to be screened for their antimicrobial effect and the incorporation of the most bioactive extracts in textiles used in Egyptian hospitals to reduce nosocomial infections. We managed to isolate 32 fungal strains from different under-explored marine habitats in Egypt. Based on antimicrobial screening against different microbial pathogens together with the total antioxidant capacity (TAC) of the fungal extracts, 21 endosymbiotic fungal isolates were selected and identified using 18S, ITS, β-tubulin, or calmodulin gene sequencing and were prioritised for further analysis. Further, inclusion of the three most active fungal extracts into the hydrophobic cavities of grafted MCT-βCD onto the cellulosic fabrics surface indicated that *Aspergillus calidoustus* strain M113 exhibited the most promising improvement in the antimicrobial textile functionality and the second best in the UV protection of functionalised cellulosic fabrics while showing only low/weak toxicity against normal human skin fibroblasts. Large scale production of this bioactive extracts along with their industrial application to develop eco-friendly multifunctional cellulosic fabrics will be carried out in our forthcoming study to determine the total cost and the feasibility.

To highlight the possible chemistry behind such antimicrobial and TAC effects, chemical investigation was carried out by the analysis of their LC-MS/MS data through the GNPS platform. The results revealed several already reported antimicrobial compounds which clustered with unidentified parent ions, suggesting the presence of new secondary metabolites. *Aspergillus calidoustus* strain M113 produced previously reported emericellamide derivatives [48] but also two potentially new congeners. The second strain with promising antimicrobial textile functionality and the third for UV protection was *Aspergillus terreus* strain 7S4, which produces a suite of known butyrolactones and aspernolides [49–53] besides one potentially new derivative. However, its skin toxicity precluded further analysis. The third recognised strain for its antimicrobial textile functionality and the best among the best in the UV-protection was *Alternaria alternata* strain 13A, but this was associated with mild skin toxicity, which could be attributed to the presence of asteltoxin based on its GNPS analysis.

5. Conclusions

In conclusion, during our Newton funding institutional links project between UK and Egypt, we managed to isolate and characterise different marine-derived endophytes from under-explored habitats in Egypt. Biological screening, textile functionalisation, and GNPS analysis allowed us to prioritise three fungal strains for future in-depth studies due to their promising chemical and biologica results, which would lead to the discovery of a convenient approach to reduce the spread of nosocomial pathogens through contaminated textiles in susceptible patients by offering safe and environmentally friendly hospital textiles.

Author Contributions: Conceptualization; A.A.H.; M.E.R.; and R.E.; methodology; A.A.H.; S.S.; M.M.Q.; S.A.; K.J.M.; F.C.; C.P.; O.A.D.; B.T.; M.E.; L.B.; L.L.; N.A.I.; M.S.A.-A.; B.M.E.; M.A.G.; M.E.R.; and R.E.; software; A.A.H.; S.S.; M.M.Q.; validation; A.A.H.; S.S.; M.M.Q.; M.E.R.; and R.E.; formal analysis; A.A.H.; S.S.; M.M.Q.; S.A.; K.J.M.; F.C.; C.P.; O.A.D.; B.T.; M.E.; L.B.; L.L.; N.A.I.; M.S.A.-A.; B.M.E.; M.A.G.; writing—original draft preparation; A.A.H.; S.S.; M.M.Q.; N.A.I.; M.E.R.; and R.E.; writing—review and editing; all co-authors; supervision; A.A.H.; M.E.R.; S.S.; and R.E.; project administration; A.A.H.; M.E.R.; and R.E. All authors have read and agreed to the published version of the manuscript.

Funding: This research was funded by the British Council Newton Fund, Institutional Links project No. 261781172 and The Science and Technology Development Fund (STDF) project No. 27701.

Acknowledgments: We would like to acknowledge the College of Physical Sciences, University of Aberdeen, for provision of infrastructure and facilities in the Marine Biodiscovery Centre. We are grateful to Teppo Rämä,

The Artic University of Norway, Tromsø, for advice relating to the identification of the fungal strains. The authors would also like to thank Mohamed A Ghani, Environmental Researcher, Red Sea Protectorates, Egypt, for his help in collecting the marine samples.

Conflicts of Interest: The authors declared no conflict of interest.

References

1. Aslam, B.; Wang, W.; Arshad, M.I.; Khurshid, M.; Muzammil, S.; Rasool, M.H.; Nisar, M.A.; Alvi, R.F.; Aslam, M.A.; Qamar, M.U.; et al. Antibiotic resistance: A rundown of a global crisis. *Infect. Drug Resist.* **2018**, *11*, 1645–1658. [CrossRef]
2. O'Neil, J. *Review on Antibiotic Resisitance. Antimicrobial Resistance: Tackling a Crisis for the Health and Wealth of Nations.* 2014. Available online: https://amr-review.org/sites/default/files/AMR%20Review%20Paper%20-%20Tackling%20a%20crisis%20for%20the%20health%20and%20wealth%20of%20nations_1.pdf (accessed on 20 September 2020).
3. Llor, C.; Bjerrum, L. Antimicrobial resistance: Risk associated with antibiotic overuse and initiatives to reduce the problem. *Ther. Adv. drug Saf.* **2014**, *5*, 229–241. [CrossRef]
4. Salvador, J.A.R.; Moreira, V.M. Natural products action on prostate cancer cell growth, signaling and survival. In *Handbook of Prostate Cancer Cell Research: Growth, Signalling, and Survival*; Meridith, A.T., Ed.; Nova Science Publishers: New York, NY, USA, 2009; pp. 1–77.
5. Dzoyem, J.P.; Melong, R.; Tsamo, A.T.; Maffo, T.; Kapche, D.G.W.F.; Ngadjui, B.T.; McGaw, L.J.; Eloff, J.N. Cytotoxicity, antioxidant and antibacterial activity of four compounds produced by an endophytic fungus *Epicoccum nigrum* associated with *Entada abyssinica*. *Braz. J. Pharmacogn.* **2017**, *27*, 251–253. [CrossRef]
6. Abdel-Aziz, M.S.; Ghareeb, M.A.; Saad, A.M.; Refahy, L.A.; Hamed, A.A. Chromatographic isolation and structural elucidation of secondary metabolites from the soil-inhabiting fungus *Aspergillus fumigatus* 3T-EGY. *Acta Chromatogr.* **2018**, *30*, 243–249. [CrossRef]
7. Ghareeb, M.; Hamed, M.; Saad, A.; Abdel-Aziz, M.; Hamed, A.; Refahy, L. Bioactive secondary metabolites from the locally isolated terrestrial fungus, *Penicillium* sp. SAM16-EGY. *Pharmacogn. Res.* **2019**, *11*, 162–170. [CrossRef]
8. Strobel, G.A. Endophytes as sources of bioactive products. *Microbes Infect.* **2003**, *5*, 535–544. [CrossRef]
9. Rateb, M.E.; Ebel, R. Secondary metabolites of fungi from marine habitats. *Nat. Prod. Rep.* **2011**, *28*, 290–344. [CrossRef] [PubMed]
10. Atiphasaworn, P.; Monggoot, S.; Gentekaki, E.; Brooks, S.; Pripdeevech, P. Antibacterial and Antioxidant Constituents of Extracts of Endophytic Fungi Isolated from *Ocimum basilicum* var. thyrsiflora Leaves. *Curr. Microbiol.* **2017**, *74*, 1185–1193. [CrossRef] [PubMed]
11. Bhadury, P.; Mohammad, B.T.; Wright, P.C. The current status of natural products from marine fungi and their potential as anti-infective agents. *J. Ind. Microbiol. Biotechnol.* **2006**, *33*, 325–337. [CrossRef]
12. Bugni, T.S.; Ireland, C.M. Marine-derived fungi: A chemically and biologically diverse group of microorganisms. *Nat. Prod. Rep.* **2004**, *21*, 143–163. [CrossRef] [PubMed]
13. Danagoudar, A.; Joshi, C.; Ravi, S.; Rohit Kumar, H.; Ramesh, B. Antioxidant and cytotoxic potential of endophytic fungi isolated from medicinal plant *Tragia involucrata* L. *Pharmacogn. Res.* **2018**, *10*, 188–194.
14. Uzma, F.; Chowdappa, S. Antimicrobial and antioxidant potential of endophytic fungi isolated from ethnomedicinal plants of Western Ghats, Karnataka. *J. Pure Appl. Microbiol.* **2017**, *11*, 1009–1025. [CrossRef]
15. Yadav, M.; Yadav, A.; Yadav, J.P. In vitro antioxidant activity and total phenolic content of endophytic fungi isolated from *Eugenia jambolana* Lam. *Asian Pac. J. Trop. Med.* **2014**, *7*, S256–S261. [CrossRef]
16. Wang, M.; Carver, J.J.; Phelan, V.V.; Sanchez, L.M.; Garg, N.; Peng, Y.; Nguyen, D.D.; Watrous, J.; Kapono, C.A.; Luzzatto-Knaan, T.; et al. Sharing and community curation of mass spectrometry data with Global Natural Products Social Molecular Networking. *Nat. Biotechnol.* **2016**, *34*, 828–837. [CrossRef]
17. Ding, C.Y.G.; Pang, L.M.; Liang, Z.X.; Goh, K.K.K.; Glukhov, E.; Gerwick, W.H.; Tan, L.T. MS/MS-Based molecular networking approach for the detection of aplysiatoxin-Related compounds in environmental marine cyanobacteria. *Mar. Drugs* **2018**, *16*, 505. [CrossRef]
18. Yang, J.Y.; Sanchez, L.M.; Rath, C.M.; Liu, X.; Boudreau, P.D.; Bruns, N.; Glukhov, E.; Wodtke, A.; De Felicio, R.; Fenner, A.; et al. Molecular networking as a dereplication strategy. *J. Nat. Prod.* **2013**, *76*, 1686–1699. [CrossRef]

19. Li, F.; Janussen, D.; Peifer, C.; Pérez-Victoria, I.; Tasdemir, D. Targeted isolation of tsitsikammamines from the antarctic deep-sea sponge *Latrunculia biformis* by molecular networking and anticancer activity. *Mar. Drugs* **2018**, *16*, 268. [CrossRef]
20. de Lima Fávaro, L.C.; de Melo, F.L.; Aguilar-Vildoso, C.I.; Araújo, W.L. Polyphasic Analysis of Intraspecific Diversity in *Epicoccum nigrum* Warrants Reclassification into Separate Species. *PLoS ONE* **2011**, *6*, e14828. [CrossRef]
21. Henríquez, M.; Vergara, K.; Norambuena, J.; Beiza, A.; Maza, F.; Ubilla, P.; Araya, I.; Chávez, R.; San-Martín, A.; Darias, J.; et al. Diversity of cultivable fungi associated with Antarctic marine sponges and screening for their antimicrobial, antitumoral and antioxidant potential. *World J. Microbiol. Biotechnol.* **2014**, *30*, 65–76. [CrossRef]
22. Visagie, C.M.; Houbraken, J.; Frisvad, J.C.; Hong, S.B.; Klaassen, C.H.W.; Perrone, G.; Seifert, K.A.; Varga, J.; Yaguchi, T.; Samson, R.A. Identification and nomenclature of the genus *Penicillium*. *Stud. Mycol.* **2014**, *78*, 343–371. [CrossRef]
23. Belbahri, L.; Moralejo, E.; Calmin, G.; Oszako, T.; García, J.A.; Descals, E.; Lefort, F. *Phytophthora polonica*, a new species isolated from declining *Alnus glutinosa* stands in Poland. *FEMS Microbiol. Lett.* **2006**, *261*, 165–174. [CrossRef] [PubMed]
24. Felsenstein, J. Evolutionary trees from DNA sequences: A maximum likelihood approach. *J. Mol. Evol.* **1981**, *17*, 368–376. [CrossRef]
25. Tamura, K.; Stecher, G.; Peterson, D.; Filipski, A.; Kumar, S. MEGA6: Molecular Evolutionary Genetics Analysis Version 6.0. *Mol. Biol. Evol.* **2013**, *30*, 2725–2729. [CrossRef] [PubMed]
26. Kimura, M. A simple method for estimating evolutionary rates of base substitutions through comparative studies of nucleotide sequences. *J. Mol. Evol.* **1980**, *16*, 111–120. [CrossRef]
27. Hamed, A.A.; Abdel-Aziz, M.S.; Fadel, M.; Ghali, M.F. Antimicrobial, antidermatophytic, and cytotoxic activities from *Streptomyces* sp. MER4 isolated from Egyptian local environment. *Bull. Natl. Res. Cent.* **2018**, *42*, 22. [CrossRef]
28. Buwa, L.V.; van Staden, J. Antibacterial and antifungal activity of traditional medicinal plants used against venereal diseases in South Africa. *J. Ethnopharmacol.* **2006**, *103*, 139–142. [CrossRef] [PubMed]
29. Prieto, P.; Pineda, M.; Aguilar, M. Spectrophotometric Quantitation of Antioxidant Capacity through the Formation of a Phosphomolybdenum Complex: Specific Application to the Determination of Vitamin E. *Anal. Biochem.* **1999**, *269*, 337–341. [CrossRef]
30. Mosmann, T. Rapid colorimetric assay for cellular growth and survival: Application to proliferation and cytotoxicity assays. *J. Immunol. Methods* **1983**, *65*, 55–63. [CrossRef]
31. Cline, M.S.; Smoot, M.; Cerami, E.; Kuchinsky, A.; Landys, N.; Workman, C.; Christmas, R.; Avila-Campilo, I.; Creech, M.; Gross, B.; et al. Integration of biological networks and gene expression data using cytoscape. *Nat. Protoc.* **2007**, *2*, 2366–2382. [CrossRef]
32. Sobeh, M.; Mahmoud, M.F.; Hasan, R.A.; Abdelfattah, M.A.O.; Sabry, O.M.; Ghareeb, M.A.; El-Shazly, A.M.; Wink, M. Tannin-rich extracts from *Lannea stuhlmannii* and *Lannea humilis* (Anacardiaceae) exhibit hepatoprotective activities in vivo via enhancement of the anti-apoptotic protein Bcl-2. *Sci. Rep.* **2018**, *8*, 9343. [CrossRef]
33. Bakchiche, B.; Gherib, A.; Bronze, M.R.; Ghareeb, M.A. Identification, quantification, and antioxidant activity of hydroalcoholic extract of *Artemisia campestris* from Algeria. *Turkish J. Pharm. Sci.* **2019**, *16*, 234–239. [CrossRef] [PubMed]
34. Prihantini, A.I.; Tachibana, S. Antioxidant compounds produced by *Pseudocercospora* sp. ESL 02, an endophytic fungus isolated from *Elaeocarpus sylvestris*. *Asian Pac. J. Trop. Biomed.* **2017**, *7*, 110–115. [CrossRef]
35. Marson Ascêncio, P.G.; Ascêncio, S.D.; Aguiar, A.A.; Fiorini, A.; Pimenta, R.S. Chemical Assessment and Antimicrobial and Antioxidant Activities of Endophytic Fungi Extracts Isolated from *Costus spiralis* (Jacq.) Roscoe (Costaceae). *Evid. Based Complement. Altern. Med.* **2014**. [CrossRef] [PubMed]
36. Runeberg, P.A.; Brusentsev, Y.; Rendon, S.M.K.; Eklund, P.C. Oxidative Transformations of Lignans. *Molecules* **2019**, *24*, 300. [CrossRef]
37. Ibrahim, N.A.; Abd El-Ghany, N.A.; Eid, B.M.; Mabrouk, E.M. Green options for imparting antibacterial functionality to cotton fabrics. *Int. J. Biol. Macromol.* **2018**, *111*, 526–553. [CrossRef]

38. Chen, M.; Wang, Y.; Xie, X.; Jiang, H.; Chen, F.; Li, W. Inclusion complex of monochlorotriazine-beta-cyclodextrin and wormwood oil: Preparation, characterization, and finishing on cotton fabric. *J. Text. Inst.* **2015**, *106*, 31–38. [CrossRef]
39. Cusola, O.; Tabary, N.; Belgacem, M.N.; Bras, J. Cyclodextrin functionalization of several cellulosic substrates for prolonged release of antibacterial agents. *J. Appl. Polym. Sci.* **2013**, *129*, 604–613. [CrossRef]
40. Ibrahim, N.A.; El-Zairy, E.M.R.; Eid, B.M. Eco-friendly modification and antibacterial functionalization of viscose fabric. *J. Text. Inst.* **2017**, *108*, 1406–1411. [CrossRef]
41. Ibrahim, N.A.; Abdalla, W.A.; El-Zairy, E.M.R.; Khalil, H.M. Utilization of monochloro-triazine β-cyclodextrin for enhancing printability and functionality of wool. *Carbohydr. Polym.* **2013**, *92*, 1520–1529. [CrossRef]
42. Ibrahim, N.A.; Khalil, H.M.; Eid, B.M.; Tawfik, T.M. Application of MCT-βCD to Modify Cellulose/Wool Blended Fabrics for Upgrading Their Reactive Printability and Antibacterial Functionality. *Fibers Polym.* **2018**, *19*, 1655–1662. [CrossRef]
43. Ibrahim, N.A.; Eid, B.M.; El-Zairy, E.R. Antibacterial functionalization of reactive-cellulosic prints via inclusion of bioactive Neem oil/βCD complex. *Carbohydr. Polym.* **2011**, *86*, 1313–1319. [CrossRef]
44. Ibrahim, N.A.; El-Zairy, E.M.R. Union disperse printing and UV-protecting of wool/polyester blend using a reactive β-cyclodextrin. *Carbohydr. Polym.* **2009**, *76*, 244–249. [CrossRef]
45. Ibrahim, N.A.; E-Zairy, W.R.; Eid, B.M. Novel approach for improving disperse dyeing and UV-protective function of cotton-containing fabrics using MCT-β-CD. *Carbohydr. Polym.* **2010**, *79*, 839–846. [CrossRef]
46. Van Santen, J.A.; Jacob, G.; Singh, A.L.; Aniebok, V.; Balunas, M.J.; Bunsko, D.; Neto, F.C.; Castaño-Espriu, L.; Chang, C.; Clark, T.N.; et al. The Natural Products Atlas: An Open Access Knowledge Base for Microbial Natural Products Discovery. *ACS Cent. Sci.* **2019**, *5*, 1824–1833. [CrossRef] [PubMed]
47. Laatsch, H. *AntiBase: The Natural Compound Identifier*; Wiley-Vch: Weinheim, Germany, 2017.
48. Oh, D.C.; Kauffman, C.A.; Jensen, P.R.; Fenical, W. Induced production of emericellamides A and B from the marine-derived fungus *Emericella* sp. in competing co-culture. *J. Nat. Prod.* **2007**, *70*, 515–520. [CrossRef] [PubMed]
49. Elbanna, A.H.; Khalil, Z.G.; Bernhardt, P.V.; Capon, R.J. Scopularides Revisited: Molecular Networking Guided Exploration of Lipodepsipeptides in Australian Marine Fish Gastrointestinal Tract-Derived Fungi. *Mar. Drugs* **2019**, *17*, 475. [CrossRef]
50. Cazar, M.E.; Schmeda-Hirschmann, G.; Astudillo, L. Antimicrobial butyrolactone I derivatives from the Ecuadorian soil fungus *Aspergillus terreus* Thorn. var terreus. *World J. Microbiol. Biotechnol.* **2005**, *21*, 1067–1075. [CrossRef]
51. Chen, M.; Wang, K.L.; Liu, M.; She, Z.G.; Wang, C.Y. Bioactive Steroid Derivatives and Butyrolactone Derivatives from a Gorgonian-Derived *Aspergillus* sp. *Chem. Biodivers.* **2015**, *12*, 1398–1406. [CrossRef]
52. Parvatkar, R.R.; D'Souza, C.; Tripathi, A.; Naik, C.G. Aspernolides A and B, butenolides from a marine-derived fungus *Aspergillus terreus*. *Phytochemistry* **2009**, *70*, 128–132. [CrossRef]
53. He, F.; Bao, J.; Zhang, X.-Y.; Tu, Z.-C.; Shi, Y.-M.; Qi, S.-H. Asperterrestide A, a Cytotoxic Cyclic Tetrapeptide from the Marine-Derived Fungus *Aspergillus terreus* SCSGAF0162. *J. Nat. Prod.* **2013**, *76*, 1182–1186. [CrossRef]
54. Kruger, G.J.; Steyn, P.S.; Vleggaar, R.; Rabie, C.J. X-Ray crystal structure of asteltoxin, a novel mycotoxin from *Aspergillus stellatus* curzi. *J. Chem. Soc. Chem. Commun.* **1979**, 441–442. [CrossRef]
55. Bao, J.; Zhang, X.Y.; Xu, X.Y.; He, F.; Nong, X.H.; Qi, S.H. New cyclic tetrapeptides and asteltoxins from gorgonian-derived fungus *Aspergillus* sp. SCSGAF 0076. *Tetrahedron* **2013**, *69*, 2113–2117. [CrossRef]
56. Adachi, H.; Doi, H.; Kasahara, Y.; Sawa, R.; Nakajima, K.; Kubota, Y.; Hosokawa, N.; Tateishi, K.; Nomoto, A. Asteltoxins from the Entomopathogenic Fungus *Pochonia bulbillosa* 8-H-28. *J. Nat. Prod.* **2015**, *78*, 1730–1734. [CrossRef] [PubMed]
57. Talontsi, F.M.; Dittrich, B.; Schüffler, A.; Sun, H.; Laatsch, H. Epicoccolides: Antimicrobial and antifungal polyketides from an endophytic fungus *Epicoccum* sp. associated with *Theobroma cacao*. *European J. Org. Chem.* **2013**, 3174–3180. [CrossRef]
58. Elix, J.A.; Lumbsch, H.T.; Wardlaw, J.H. Conhypoprotocetraric Acid, a New Lichen β-Orcinol Depsidone. *Aust. J. Chem.* **1995**, *48*, 1479–1483. [CrossRef]
59. Bézivin, C.; Tomasi, S.; Rouaud, I.; Delcros, J.G.; Boustie, J. Cytotoxic activity of compounds from the lichen: *Cladonia convoluta*. *Planta Med.* **2004**, *70*, 874–877. [CrossRef]

60. Bloch, P.; Tamm, C.; Bollinger, P.; Petcher, T.J.; Weber, H.P. Pseurotin, a New Metabolite of Pseudeurotium ovalis STOLK Having an Unusual Hetero-Spirocyclic System (Preliminary Communication). *Helv. Chim. Acta* **1976**, *59*, 133–137. [CrossRef]
61. Ren, H.; Liu, R.; Chen, L.; Zhu, T.; Zhu, W.M.; Gu, Q.Q. Two new hetero-spirocyclic γ -lactam derivatives from marine sediment-derived fungus *Aspergillus sydowi* D2-6. *Arch. Pharm. Res.* **2010**, *33*, 499. [CrossRef]
62. Yamada, T.; Kitada, H.; Kajimoto, T.; Numata, A.; Tanaka, R. The relationship between the CD cotton effect and the absolute configuration of FD-838 and its seven stereoisomers. *J. Org. Chem.* **2010**, *75*, 4146–4153. [CrossRef]
63. Borkow, G.; Gabbay, J. Biocidal textiles can help fight nosocomial infections. *Med. Hypotheses* **2008**, *70*, 990–994. [CrossRef]
64. Openshaw, J.J.; Morris, W.M.; Lowry, G.V.; Nazmi, A. Reduction in bacterial contamination of hospital textiles by a novel silver-based laundry treatment. *Am. J. Infect. Control* **2016**, *44*, 1705–1708. [CrossRef] [PubMed]

microorganisms

MDPI

Article

Microbial Natural Products as Potential Inhibitors of SARS-CoV-2 Main Protease (M^{pro})

Ahmed M. Sayed [1,†], Hani A. Alhadrami [2,3,†], Ahmed O. El-Gendy [4], Yara I. Shamikh [5,6], Lassaad Belbahri [7], Hossam M. Hassan [8], Usama Ramadan Abdelmohsen [9,10,*] and Mostafa E. Rateb [11,*]

1 Department of Pharmacognosy, Faculty of Pharmacy, Nahda University, 62513 Beni-Suef, Egypt; ahmedpharma8530@gmail.com
2 Department of Medical Laboratory Technology, Faculty of Applied Medical Sciences, King Abdulaziz University, Jeddah 21589, Saudi Arabia; hanialhadrami@kau.edu.sa
3 Special Infectious Agent Unit, King Fahd Medical Research Centre, King Abdulaziz University, Jeddah 21589, Saudi Arabia
4 Department of Microbiology, Faculty of Pharmacy, Beni-Suef University, 62514Beni-Suef, Egypt; ahmed.elgendy@pharm.bsu.edu.eg
5 Department of Microbiology & Immunology, Nahda University, Beni-Suef 62513, Egypt; yara.shamikh@nub.edu.eg
6 Department of Virology, Egypt Center for Research and Regenerative Medicine (ECRRM), 11517 Cairo, Egypt
7 Laboratory of Soil Biology, Department of Biology, University of Neuchatel, 2000 Neuchatel, Switzerland; lassaad.belbahri@unine.ch
8 Department of Pharmacognosy, Faculty of Pharmacy, Beni-Suef University, 62514 Beni-Suef, Egypt; abuh20050@yahoo.com
9 Department of Pharmacognosy, Faculty of Pharmacy, Minia University, 61519 Minia, Egypt
10 Department of Pharmacognosy, Faculty of Pharmacy, Deraya University, 61111 New Minia, Egypt
11 School of Computing, Engineering & Physical Sciences, University of the West of Scotland, Paisley PA1 2BE, UK
* Correspondence: usama.ramadan@mu.edu.eg (U.R.A.); Mostafa.Rateb@uws.ac.uk (M.E.R.)
† These authors contributed equally to this work.

Received: 18 June 2020; Accepted: 27 June 2020; Published: 29 June 2020

Abstract: The main protease (M^{pro}) of the newly emerged severe acute respiratory syndrome coronavirus 2 (SARS-CoV-2) was subjected to hyphenated pharmacophoric-based and structural-based virtual screenings using a library of microbial natural products (>24,000 compounds). Subsequent filtering of the resulted hits according to the Lipinski's rules was applied to select only the drug-like molecules. Top-scoring hits were further filtered out depending on their ability to show constant good binding affinities towards the molecular dynamic simulation (MDS)-derived enzyme's conformers. Final MDS experiments were performed on the ligand–protein complexes (compounds **1–12**, Table S1) to verify their binding modes and calculate their binding free energy. Consequently, a final selection of six compounds (**1–6**) was proposed to possess high potential as anti-SARS-CoV-2 drug candidates. Our study provides insight into the role of the M^{pro} structural flexibility during interactions with the possible inhibitors and sheds light on the structure-based design of anti-coronavirus disease 2019 (COVID-19) therapeutics targeting SARS-CoV-2.

Keywords: SARS-CoV-2; Covid-19; M^{pro}; microbial natural products; docking; molecular dynamic simulation

1. Introduction

In 2002, the first spread of coronavirus associated with severe acute respiratory syndrome coronavirus (SARS-CoV) emerged in southern China. This outbreak successfully subsided by the summer of 2003 [1], with no more than 8500 confirmed infections and just over 900 deaths worldwide [2]. Upon the spread of this outbreak, the global response was immediate to characterize its causative agent as a novel coronavirus (SARS-CoV) [3]. The recurrence of SARS in the Guangdong province of China in December 2003 [4] illustrated the need to continue efforts to study this virus and its key molecular targets to develop appropriate therapeutics for its treatment. In 2012, another coronavirus wave originating in Jeddah, Saudi Arabia emerged and spread within and beyond the Middle East. The reported strain (MERS-CoV) was associated with severe pneumonia and multi-organ failure [5]. The limited number of infected cases during the previous coronaviruses waves did not encourage a serious worldwide development of effective treatments.

Recently and until June 2020, the outbreak of a new coronavirus (SARS-CoV-2) that originated from Wuhan, China, in late 2019, has led to more than 9.3 million infections and more than 479,000 deaths throughout 216 countries without any proven antiviral agents or effective vaccine according to WHO official updates. However, repurposing of previous medications has shown some reported clinical improvements [6]. This time, worldwide efforts are being made to characterize molecular targets, pivotal for the development of anti-coronavirus therapies. The term coronavirus was coined according to its corona-like appearance in the electron microscope, due to its spikes that radiate outwards from the viral envelope. The spherical capsid envelops a positive-strand RNA genome of about 30 kb, which is considered the largest of its kind. The viral genome is predominated by two open reading frames that are connected by a ribosomal frameshift site and encode the two replicase proteins, pp1a and pp1ab [7]. These polyproteins are cleaved by the viral main protease (M^{pro}), also called chymotrypsin-like protease, $3CL^{Pro}$ [8,9]. M^{pro} is considered a highly conserved molecular target across coronaviruses, and hence, it was designated as a potential target for anti-coronavirus drug development [10]. Earlier reports on viral protease inhibitors (i.e., the HIV protease inhibitor lopinavir) revealed significant in vitro anti-SARS-CoV-2 activities [11].

Natural products along with natural product-inspired synthetic and semisynthetic compounds are still an excellent structural motif for the discovery of new therapeutics, including antiviral agents. Natural products derived from microbial sources are considered unique in their chemical diversity in comparison with plant-derived ones. Approximately 53% of the FDA-approved natural products-based drugs are of microbial origin, notably the antiviral ones [12]. For example, Ara-A (**9**) (also known as vidarabine) is considered one of the earliest antiviral nucleoside analogues that was reported from *Streptomyces antibioticus* [13]. Afterwards, several nucleoside-based antiviral agents of microbial origin were developed [14]. Additionally, several ansamycins-based antibiotics (e.g., rifamycin, **10**) have shown interesting antiviral properties against a wide range of infectious viruses [15]. Furthermore, the well-known immunomodulatory drug of fungal origin mycophenolic acid (**11**) has also shown a broad antiviral activity [15]

Recently, the microbial-derived FDA-approved anti-parasitic drug ivermectin (**12**), a semisynthetic pentacyclic sixteen-membered lactone derived from the soil bacterium *Streptomyces avermitilis*, proved to be an effective in vitro inhibitor of SARS-CoV-2 replication [16,17]. In this context, several in silico techniques that have recently gained a lot of attention in drug discovery campaigns (e.g., structure and ligand-based virtual screening, docking and molecular dynamics) [18,19]. Hence, we initiated a virtual screening of a big library of microbial natural products (more than 24,000 compounds) aimed at the discovery of potential drug candidates against the SARS-CoV-2 M^{pro}, taking into account the drug-likeness properties to select only druggable candidates. Simple docking protocols do not take into consideration the flexible nature of proteins; therefore, their success rates in most cases are between 60–70% [20]. To increase the docking performance, top hits retrieved from the primary pharmacophore-based screening were further docked on a series of receptor conformers (i.e., ensemble docking) [21] taken from the molecular dynamic simulation (MDS) to consider the factor of the binding pockets' flexibility. Finally, further MDS of the protein–ligand complexes were performed to verify our docking experiments and calculate their binding free energy (ΔG). Drug candidates proposed in this study could provide a promising starting point for the in vitro and in vivo testing and further development of potential drug leads against the newly emerged coronavirus disease 2019 (COVID-19). The procedure of the current investigation is depicted in Figure 1.

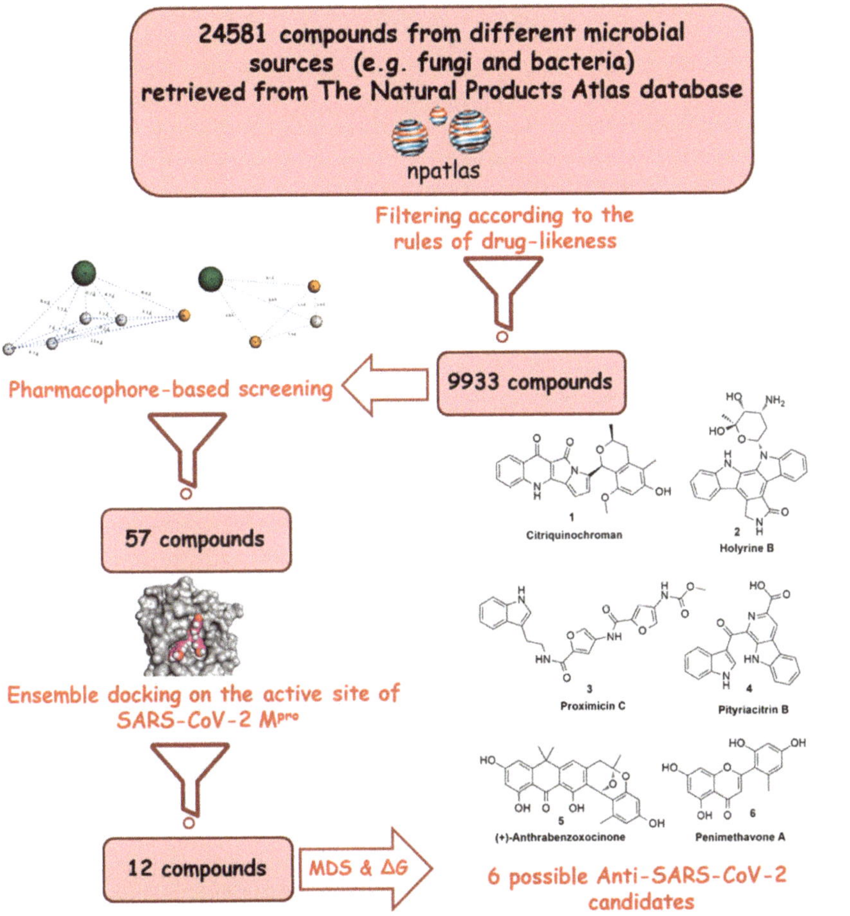

Figure 1. Applied strategy in the present study.

2. Materials and Methods

2.1. Preparation of SARS-CoV-2 M^pro and the Compounds Dataset

Crystal structures of M^pro (PDB code: 6LU7 and 6M2N) were obtained from the Protein Data Bank (http://www.pdb.org), and all heteroatoms and water molecules were removed for MDS and molecular docking studies. The chemical structures of the tested microbial specialized metabolites were retrieved from the online dataset; The Natural Products Atlas (https://www.npatlas.org/joomla/index.php) [22] with a final compiled dataset consisting of 24,581 compounds. Subsequently, this library of compounds was subjected to LigandScout software [23] to select 9933 compounds that showed drug-like properties (i.e., obey Lipinski's role of five) [24].

2.2. Molecular Dynamic Simulation

Molecular dynamic simulations (MDS) for the free M^pro enzyme and ligand–enzyme complexes were performed using the Nanoscale Molecular Dynamics (NAMD) 2.6 software [25], employing the CHARMM27 force field [26]. Hydrogen atoms were added to initial coordinates for M^{pro} using the psfgen plugin included in the Visual Molecular Dynamic (VMD) 1.9 software [27]. Subsequently, the protein system was solvated using TIP3P water particles and 0.15 M NaCl. The equilibration procedure comprised 1500 minimization steps followed by 30 ps of MDS at 10 k with fixed protein atoms. Then, the entire system was minimized over 1500 steps at 0 K, followed by gradual heating from 10 to 310 K using temperature reassignment during the initial 60 ps of the 100 ps equilibration MDS. The final step involved NTP simulation (30 ps) using the Nose–Hoover Langevin piston pressure control at 310 K and 1.01325 bars for density (volume) fitting [28]. Thereafter, the MDS was continued for 25 ns for the entire system (20 ns for the enzyme–ligand complexes). The trajectory was stored every 0.1 ns and further analyzed with the VMD 1.9 software. The MDS output over 25 ns provided several structural conformers that were sampled every 0.1 ns to evaluate the conformational changes of the entire protein structure to analyze the root mean square deviation (RMSD) and root mean square fluctuation (RMSF). All parameters and topologies of the compounds selected for MDS (1–12, Table S1) were prepared using the online software Ligand Reader & Modeler (http://www.charmm-gui.org/?doc=input/ligandrm) [29] and the VMD Force Field Toolkit (ffTK) [27]. Binding free energy calculations (ΔG) were performed using the free energy perturbation (FEB) method through the web-based software Absolute Ligand Binder [30] together with the MDS software NAMD 2.6 [25]. Additionally, they were calculated using another web-based software, namely K_{DEEP} (https://www.playmolecule.org/Kdeep/), which applies a neural-networking algorithm during its computations [31].

2.3. Pharmacophore-Based Virtual Screening and Molecular Docking

The pharmacophore-based screening was performed by the online sever Pharmit (http://pharmit.csb.pitt.edu/) [32]. The pharmacophore models were constructed from the SARS-M^{pro} enzymes co-crystallized with baicalein (7) and N3 (8). All the pre-installed Pharmit parameters remained unchanged. The resulting two models were used for the virtual screening of the compounds filtered from the prepared microbial natural products library (9933 compounds). Afterwards, compounds with RMSD > 2Å were excluded. Docking experiments were performed using AutoDock Vina software [33]. All compounds that fitted into the predetermined pharmacophore models (57 compounds) were then subjected to molecular docking against the M^{pro}'s active site conformers that were sampled from the MDS every 5 ns (i.e., ensemble docking). We set an average docking score of −10 kcal/mol (the average docking score of compound **7**) as a cut-off to select the top-scoring hits. Afterwards, the retrieved top hits (1–12, Table S1) were ranked according to their binding energies (Table 1 and Table S1). The generated docking poses were visualized and analyzed using Pymol software [34].

Table 1. M^{pro} top-scoring ligands alongside their binding energies using different calculation methods and their molecular interactions inside the active site.

Ligand	ΔG_{Vina} (kcal/mol)	ΔG * $_{FEP}$ (kcal/mol)	ΔG ** K_{DEEP} (kcal/mol)	$\Delta G_{average}$ (kcal/mol)	Hydrogen Bonding Interactions	Hydrophobic Interactions
Citriquinochroman (1)	−14.7	−11.9	−10.5	−12.4	THR-26, ASN-142, GLY-143, CYS-145, GLU-166, ASP-187, ARG-188, GLN-189, THR-190, GLN-192	HID-41, MET-49, PRO-168
Holyrine B (2)	−14.5	−11.5	−10.9	−12.3	LEU-141, ASN-142, GLY-143, SER-144, CYS-145, HID-163, HIE-164, GLU-166, PRO-168, ASP-187, ARG-188, GLN-189, THR-190, GLN-192	HID-41, MET-49, MET-165, PRO-168

Table 1. *Cont.*

Ligand	ΔG_{Vina} (kcal/mol)	ΔG * FEP (kcal/mol)	ΔG ** K_{DEEP} (kcal/mol)	$\Delta G_{average}$ (kcal/mol)	Hydrogen Bonding Interactions	Hydrophobic Interactions
Proximicin C (3)	−14.1	−12.1	−10.3	−12.2	GLY-143, SER-144, CYS-145, GLU-166, PRO-168, ASP-187, ARG-188, GLN-189, THR-190	Leu-27, HID-41, MET-49, MET-165, PRO-168
Pityriacitrin B (4)	−13.4	−12.1	−11.1	−12.2	PHE-140, LEU-141, GLY-143, SER-144, CYS-145, HID-163, HIE-164, MET-165, GLU-166, GLN-189	HID-41, MET-49, GLN-189
Anthrabenzoxocinone (5)	−13.2	−10.3	-9.5	−11	THR-26, HID-41, CYS-44, ASN-142, GLY-143, CYS-145, HIE-164, HIE-164, MET-165, GLU-166, VAL-186, ASP-187, ARG-188, GLN-189, THR-190, GLN-192	HID-41, MET-49, MET-165, GLN-189
Penimethavone A (6)	−12.1	−11.4	-8.9	−10.8	LEU-141, GLY-143, SER-144, CYS-145, HIE-164, HIE-164, MET-165, GLU-166, HID-172, VAL-186, ASP-187, ARG-188, GLN-189, GLN-192	HID-41, MET-49, MET-165, GLN-189
Co-crystalized ligand (7)	−10.1	-9.2	-8.9	-9.4	LEU-141, ASN-142, GLY-143, GLU-166, GLN-189.	HID-41, MET-49, GLN-189
Co-crystalized ligand (8)	−10.9	−11.4	-9.4	−10.6	PHE-140, GLY-143, CYS-145, HIE-164, GLU-166, GLN-189, THR-190.	HID-41, MET-49, GLN-189

* Binding free energy calculated by the free energy perturbation (FEB) method [30], ** Binding free energy calculated by a neural networking method (K_{DEEP}) [31].

3. Results and Discussion

3.1. Structure and Dynamics of the SARS-CoV-2 M^{pro}

The catalytic site of the SARS-CoV-2 M^{pro} was found to be the largest cavity on the whole protein (static volume = 385.56 Å^3) and is located in domains I and II (residues 11 to 99 and 100 to 182, respectively, Figure 2A). Additionally, it is smaller than the earlier SARS-CoV M^{pro} (static volume = 447.7 Å^3) [35]. It is worth noting that the enzyme's domain III, which consists of a globular cluster of five helices, is present only in coronaviruses and is responsible for the regulation of M^{pro} dimerization [36]. MDS of the reported SARS-CoV-2 M^{pro} (PDB: 6LU7) [10] was performed to study the conformational changes in the active site using clustering analysis (conformation was sampled every 0.1 ns). The carbon alpha (Cα) root main square deviation (RMSD) values with respect to the initial structure were calculated for 25 ns of the simulation and were found to oscillate from 0.78 to 2.78 Å with a median value of (2.51 Å), reaching equilibrium (i.e., plateau) at 4.8 ns (Figure 2D). Regarding root mean square fluctuation (RMSF) values, they demonstrated that M^{pro} had moderate flexibility (average RMSF = 2.43 Å, Figure 2E), where the active site showed the highest conformational changes (average RMSF = 3.1 Å), particularly at the THR-45 to ILE-59 loop (Figure 2B), which showed high fluctuations (i.e., RMSF values ranged from 5.2 to 8.9 Å, Figure 2E).

The volume of the active site was 399.56 Å^3 at the beginning of the simulation (Figure 2C), and during the MDS, it gradually increased to reach 479.65 Å^3 at 10 ns, and then began to decrease to reach 314.63 Å^3 at the end of the MDS. Docking and virtual screening studies on such flexible catalytic active sites using only their crystalized static form would lead to poor prediction results. Hence, considering multiple structural conformations (i.e., ensemble docking) derived from the MDS study for our virtual screening in the present investigation could significantly improve the predicted results.

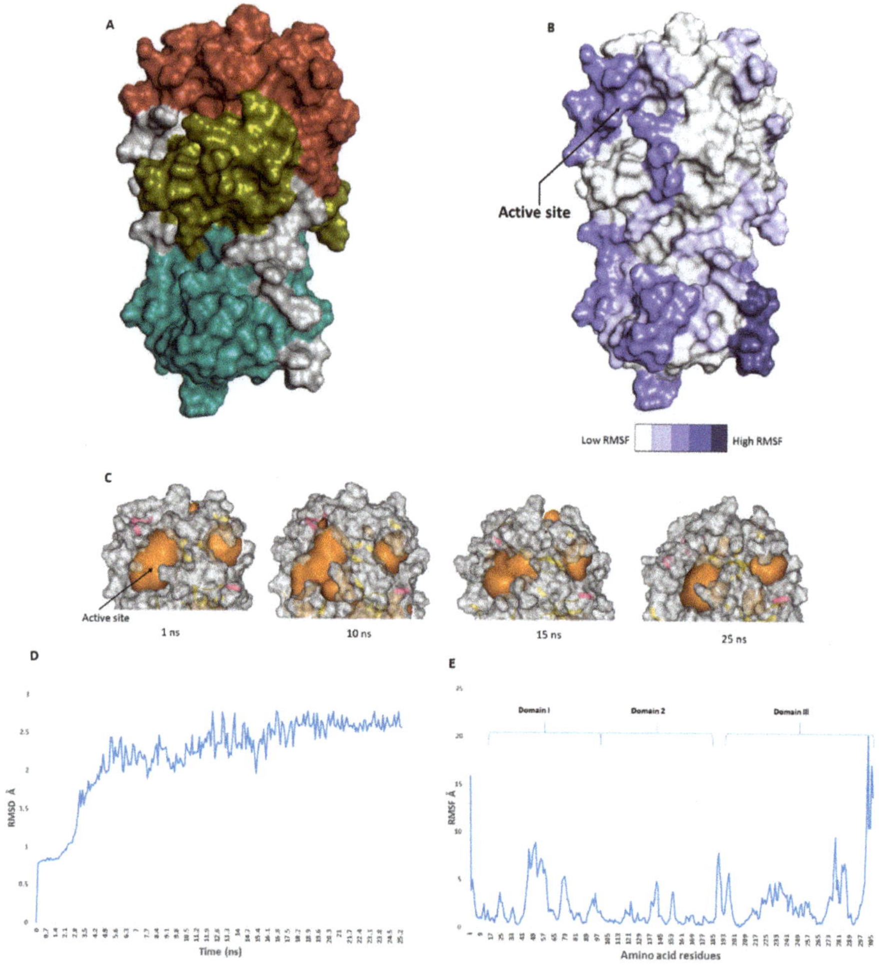

Figure 2. (**A**): The main domains (M^pro^) in severe acute respiratory syndrome coronavirus 2 (SARS-CoV-2) (Brick red: Domain I, Golden yellow: Domain II, Cyan: Domain III), (**B**): Heat-map illustrating the flexible regions on SARS-CoV-2 M^{pro}, (**C**): The active site volume changes during the course of molecular dynamic simulation (MDS) (**D** and **E**): root main square deviation (RMSD) and root main square fluctuation (RMSF) of SARS-CoV-2 M^{pro} after 25 ns of MDS.

3.2. Pharmacophore-Based Modeling and Screening

To discover potential naturally occurring ligands that could block the M^{pro} active site, an extensive specialized microbial natural product database (The Natural Product Atlas) containing more than 24,000 different compounds was utilized for. Firstly, the database was filtered according to drug-likeness (Lipinski's rules [24]) to get 9933 drug-like candidates.

The crystal structure of SARS-CoV-2 M^{pro} (PDB id: 6LU7) was reported earlier, along with its co-crystallized peptide inhibitor N3 [10]. N3 (**8**) is fitted inside the M^{pro} active site through multiple strong H-bonds (e.g., GLY-143, HIS-164, GLU-166, GLN-189, and THR-190) alongside a covalent bond with CYS-145. Additionally, its isopropyl group is imbedded inside a hydrophobic pocket that consists of HIS-41, MET-49, and GLN-189 (Figure 3A, Table 1).

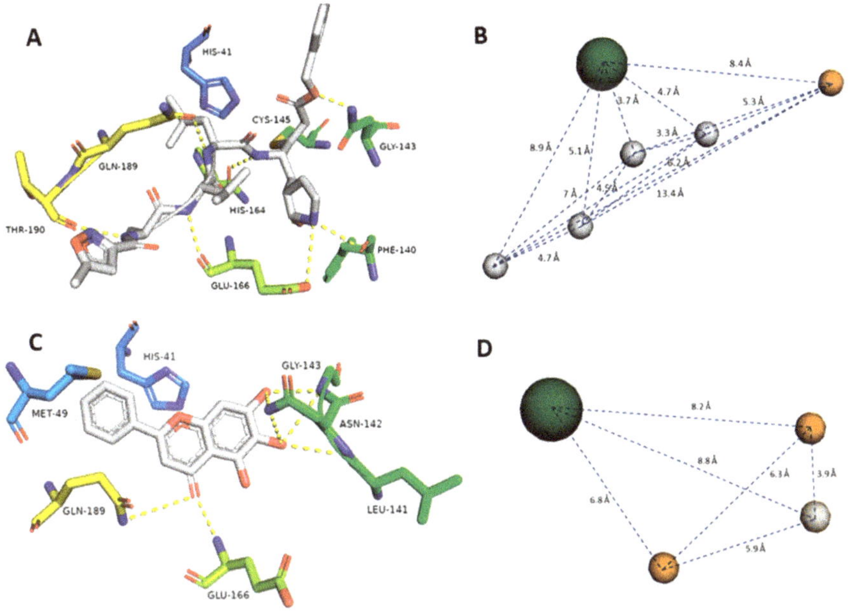

Figure 3. Generated pharmacophore models (**B** and **D**) according to the previously reported co-crystalized ligands (**7** and **8**, **A** and **C**). Gray spheres indicate hydrogen bond donors, orange spheres indicate hydrogen bond acceptors, and green spheres indicate hydrophobic centers. Green amino acid residues represent the S1 pocket, blue amino acid residues represent the S2 pocket, yellow amino acid residues represent the S3 and S4 pockets in the M^{pro} active site (**A** and **C**).

Recently, a novel flavonoid-based non-covalent inhibitor (baicalein, **7**) (PDB code: 6M2N) was found to accept seven H-bonds from LEU-141, ASN-142, GLY-143, Glu-166, and GLN-189. Moreover, its phenyl moiety was also fitted inside the hydrophobic pocket of HIS-41, MET-49, and GLN-189 (Figure 3C, Table 1). The predetermination of pharmacophoric characteristics prior to structure-based virtual screening would help in selecting the best hits with the best interaction inside the binding pocket [37].

To define the essential features of the interaction inside the M^{pro}'s active site [32], two structure-based pharmacophore models were constructed depending on the two previously described inhibitors. The N3 (**8**)-based-pharmacophore model had the following features: four H-bond donors derived from four amide groups, one carboxyl group-derived oxygen atom as an H-bond acceptor, and the isopropyl group to represent a hydrophobic center (Figure 3B). On the other hand, the baicalein (**7**)-based pharmacophore model showed the following features: two H-bond acceptors derived from one hydroxyl group and the ketonic oxygen, one H-bond donor derived from another hydroxyl group, and the phenyl group to represent a hydrophobic center (Figure 3D).

Subsequently, these binding site-derived pharmacophore models were used in our virtual screening against the MND-selected compounds (the 9933 compounds that obey Lipinski's rules) using the online server Pharmit [32]. This allowed us to select the compounds with predetermined pharmacophoric features capable of interacting with the reported key residues. This filtration step led to the recognition of 363 compounds that met the predetermined model features, of which we selected only 57 compounds (i.e., showed RMSD values lower than 2 Å with respect to the co-crystalized ligands **7** and **8**) to undergo a subsequent docking-based virtual screening.

3.3. Molecular Docking and Binding Mode Investigation

Docking of the selected compounds inside the M^{pro} active site was performed on AutoDock Vina, which was able to reproduce the binding mode of the co-crystalized ligands [37], N3 (**8**) and baicalein (**7**), with an RMSD values of 1.22 and 0.51 Å, respectively. Fifty seven compounds were filtered according to the predetermined pharmacophoric features and drug-likeness properties, and thereafter docked separately on the SARS-CoV-2 M^{pro} active site using several snapshots (every 5 ns) derived from the MDS (i.e., ensemble docking). This allowed us to further select the best binding compounds taking into consideration the flexibility of the active site. Top-scoring hits (those with an average binding energy score > −10 kcal/mol, Table S1 and Figure S1) with binding modes comparable with both N3 (**8**) and baicalein (7), were then subjected to MDS and binding free energy computation to further verify the suggested pharmacophore models, docking-derived binding poses, and binding affinities. Only compounds **1–6** (Figure 4 and Table 1) exhibited stable binding orientations inside the enzyme binding pocket throughout the MDS (Figures S2–S13) and constant binding energies higher than that of the co-crystallized inhibitors, i.e., compounds **7** and **8**.

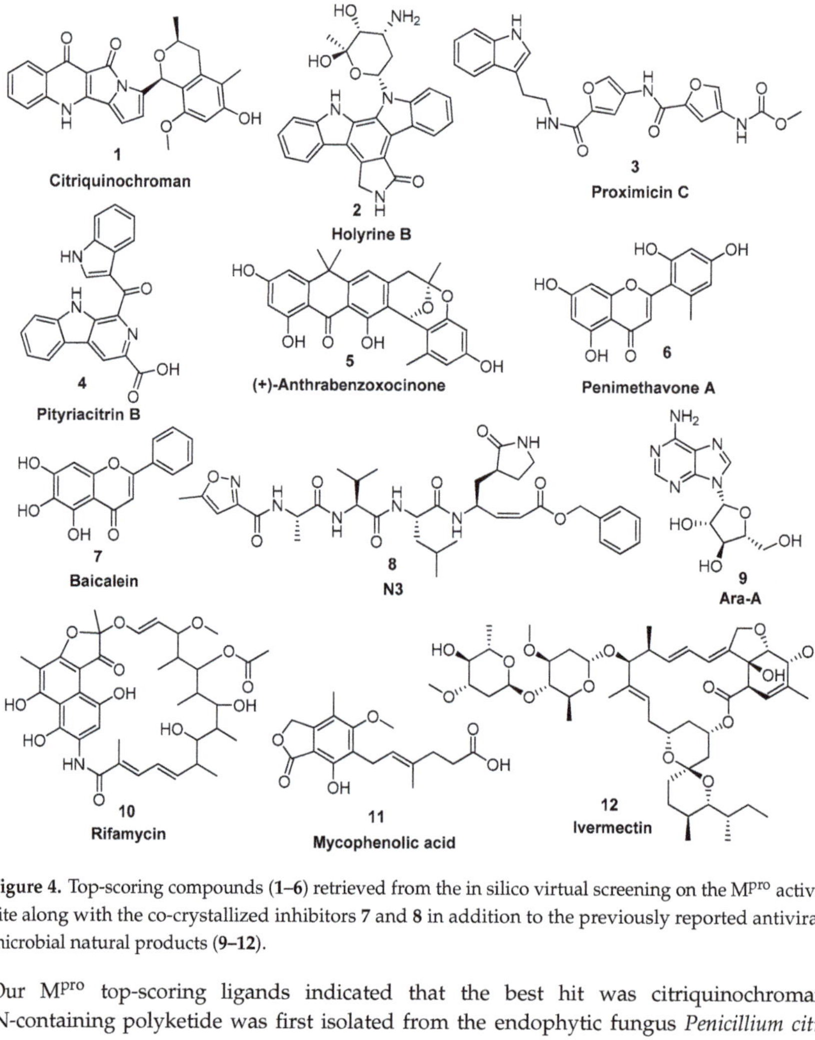

Figure 4. Top-scoring compounds (**1–6**) retrieved from the in silico virtual screening on the M^{pro} active site along with the co-crystallized inhibitors **7** and **8** in addition to the previously reported antiviral microbial natural products (**9–12**).

Our M^{pro} top-scoring ligands indicated that the best hit was citriquinochroman (**1**). This N-containing polyketide was first isolated from the endophytic fungus *Penicillium citrinum* in 2013 and showed moderate anticancer activity [38]. It showed the least binding energy ($\Delta G_{average} = -12.4$ kcal/ mol), with perfect fitting inside the enzyme active site in the crystallized form, where it anchored itself via a network of H-bond interactions with the reported key binding residues (7 H-bonds) [10]. Despite the flexibility of the enzyme active site, citriquinochroman (**1**) was able to keep its orientation during the course of MDS (Figure 5 and Figure S2) with a transient drop in its binding affinity at 3–5.5 ns ($\Delta G_{Vina} = -8.9$ kcal/mol). Afterwards, both THR-26 and GLN-192 stabilized compound **1** with additional H-bond interactions until the end of the MDS (Figure 5).

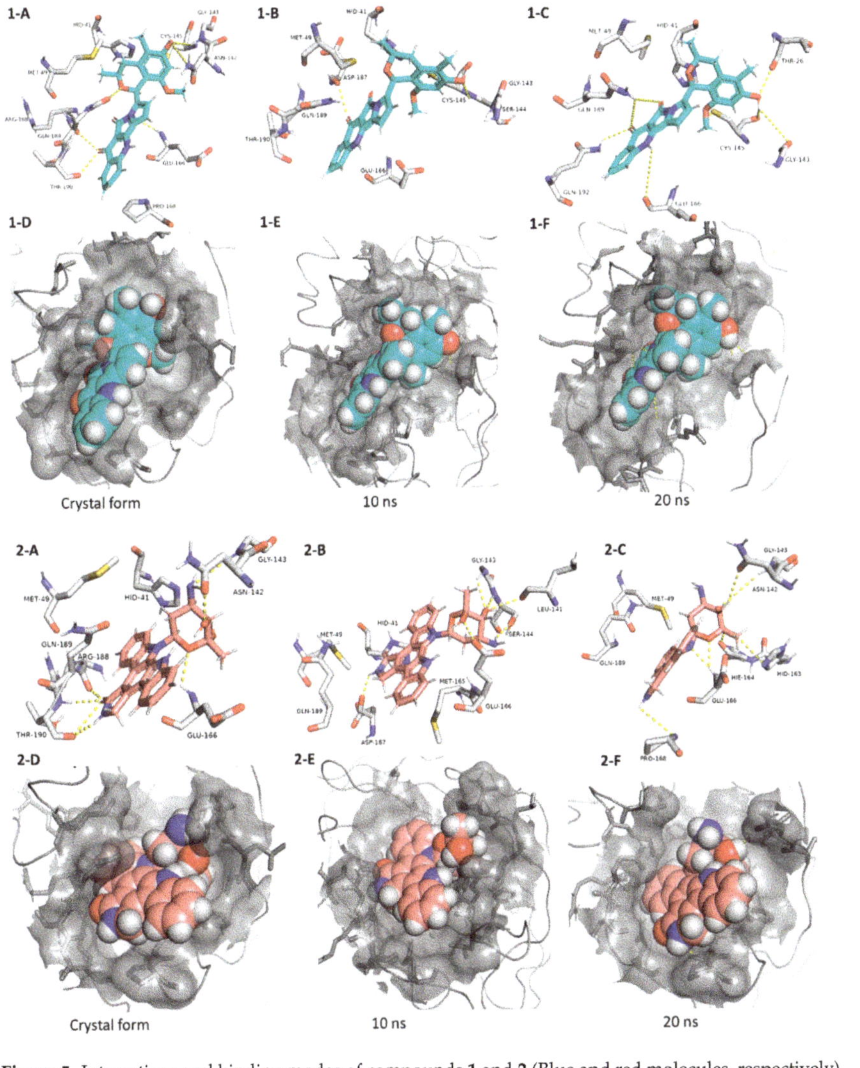

Figure 5. Interactions and binding modes of compounds **1** and **2** (Blue and red molecules, respectively) inside the M^{pro} active site in the crystal form and during MDS (**1-A–2-F**).

The second most promising hit was holyrine B (**2**), an indolocarbazole alkaloid that was previously isolated from a marine-derived actinomycete [39]. Holyrine B (**2**) exhibited binding modes similar to that of citriquinochroman (**1**), where it also interacted through H-bonding or hydrophobic interactions with the reported key amino acid residues [10]. Additionally, it was able to keep these strong interactions throughout the MDS and form additional H-bonds and hydrophobic interactions with extra amino acid residues like HID-163, HIE-164, PRO-168, and MET-165 (Figure 5 and Table 1).

The third best hit was the aminofuran antibiotic proximicin C (**3**), which was isolated from the marine actinomycete *Verrucosispora* MG-37 [40]. Proximicin C (**3**) showed an interesting binding pose inside the enzyme active site in the crystalized form where it interacted with several amino acid residues including the reported ones (9 H-bonds and 2 hydrophobic interactions, Figure 6). During the MDS and in contrast to the previous candidates (**1** and **2**), proximicin C's (**3**) binding affinity remained constant for 5 ns and started to increase afterwards until the end of the simulation. Such stable fitting inside the binding site could be attributed to the molecular flexibility of this compound that enabled it to accommodate itself well inside the changing active site of SARS-CoV-2 M^{pro} (Figure 6).

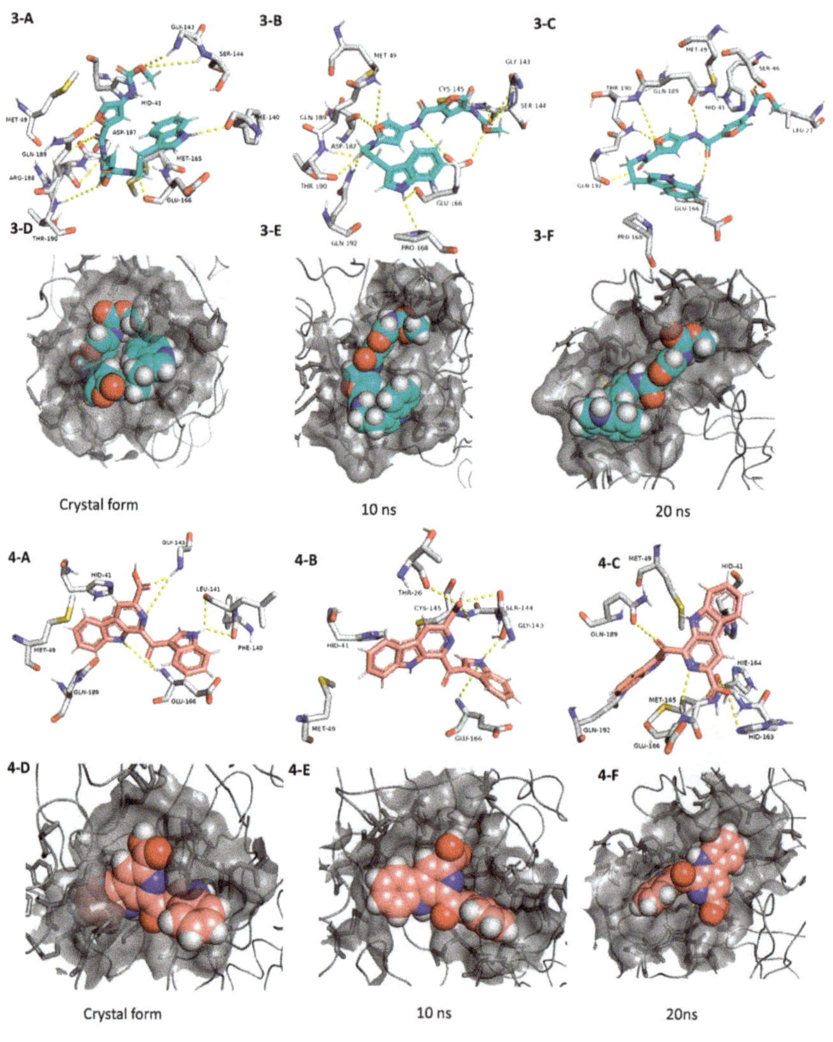

Figure 6. Interactions and binding modes of compounds **3** and **4** (Blue and red molecules, respectively) inside the M^{pro} active site in the crystal form and during MDS (**3-A–4-F**).

Pityriacitrin B (**4**) showed excellent fitting inside the M^pro^ active site, where the phenyl moiety of the β-carboline part was impeded inside the hydrophobic pocket (HID-41, MET-94, GLN-189), while the other indole arm together with the β-carboline's pyridine moiety interacted with the key binding amino acid residues (PHE-140, LEU-141, GLY-143, and GLU-166, Figure 6). Moreover, it was able to not only keep these interactions throughout the MDS, but also to form additional strong H-bonds with THR-26, SER-144, and CYS-145. Pityriacitrin B (**4**) was first isolated from the human pathogenic yeast *Malassezia furfur* [41]. Later on, it was identified as an efficient UV absorbing agent [42].

Coming to our next hit, (+)-anthrabenzoxocinone (**5**), it showed an interaction pattern similar to the previous candidates (**1–4**, and **7**, **8**) inside the crystallized form of the enzyme active site. However, from the beginning of the MDS, this compound gradually detached itself from the binding site, and starting from 7.2 ns, it took a different orientation with better interactions (Figure 7 and Table 1). At the end of the MDS, (+)-anthrabenzoxocinone (**5**) was able to form a wide network of H-bonds (10 H-bonds) and hydrophobic interactions (4 hydrophobic interactions) (Figure 7). Anthrabenzoxocinone (**5**) was isolated from a soil-derived *Streptomyces* sp. in 2014 [43].

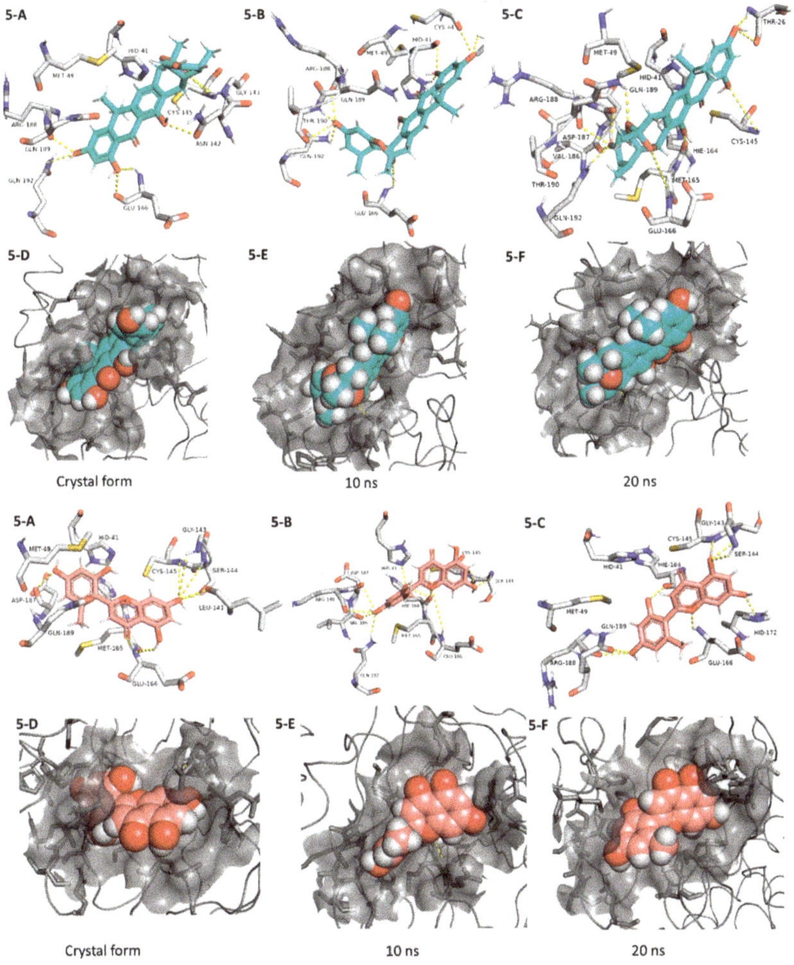

Figure 7. Interactions and binding modes of compounds **5** and **6** (Blue and red molecules, respectively) inside the M^{pro} active site in the crystal form and during MDS (**5-A–6-F**).

Finally, penimethavone A (**6**), which has a flavone structure similar to the co-crystalized ligand (baicalein, **7**), was able to form molecular interactions almost identical to those of the co-crystalized ligand (**7**) (Figures 3C and 7, Table 1). Additionally, it adopted a stable binding mode during the MDS, particularly near the end of simulation when it expanded its H-bonds network to involve extra binding residues like HIE-164 and HID-172. Penimethavone A (**6**) is an unusual flavone derivative with a methylated B-ring that was isolated from the gorgonian marine soft coral-derived *Penicillium chrysogenum* [44].

4. Conclusions

The SARS-CoV-2 pandemic crisis has inspired scientists with diverse backgrounds to help with a speedy discovery of potential treatments or vaccines. In the present virtual screening and molecular modelling study, we suggested that the active site of the newly emerged SARS-CoV-2 M^{pro} is quite flexible. Thus, its utilization in just simple docking experiments could lead to inaccurate results. Consequently, this catalytic active site was utilized in a combination of ligand-based followed by structural-based virtual screening against a big library of microbial-derived specialized metabolites, which was initially filtered according to the drug-likeness of its molecules. Top-scoring hits were further subjected to an ensemble docking protocol depending on the enzyme-generated conformers during the MDS. This step allowed us to select only ligands with stable binding affinity and modes for considering the flexibility of the active binding site. MDS, together with binding energy and affinity computations were performed for the selected hits as a final validation step to nominate six molecules with possible high potential to modulate/inhibit the SARS-CoV-2 M^{pro} active site. This study emphasized the power of computer-aided drug design and modelling in speeding up the process of drug discovery, which is currently an urgent need under the spread of COVID-19. It also highlighted the ability of natural products, particularly those of high structural diversity like microbial-derived metabolites, to provide potential drug-leads. Further in vitro testing of the drug candidates retrieved from our study is highly recommended as a promising starting point for the rapid development of drug leads against newly emerged COVID-19.

Supplementary Materials: The following are available online at http://www.mdpi.com/2076-2607/8/7/970/s1, Figure S1: Structures of the top-scoring compounds, Figures S2–S13: RMSDs of the M^{pro} enzyme-ligand complexes and the docking binding poses of compounds (**1–12**) inside the binding pocket of the M^{pro}'s enzyme, Table S1: Top-scoring hits retrieved from the structural-based virtual screening.

Author Contributions: Conceptualization, A.M.S., U.R.A., H.M.H., and M.E.R.; methodology, A.M.S., H.A.A., A.O.E.-G., and Y.I.S.; data curation, A.M.S., H.A.A., and L.B.; original draft preparation, A.M.S.; writing, review and editing, All authors. All authors have read and agreed to the published version of the manuscript.

Funding: This project was funded by Ministry of Health, Kingdom of Saudi Arabia, under grant number 16581700012—project number 581. The authors, therefore, gratefully acknowledge Ministry of Health support.

Acknowledgments: The authors thank Nahda University (Egypt) and Ministry of Health (Kingdom of Saudi Arabia) for supporting this work.

Conflicts of Interest: The authors declare there is no conflict of interest.

References

1. World Health Organization. Alert, Verification and Public Health Management of SARS in the Post-Outbreak Period. Available online: http://www.who.int.csr/sars/postoutbreak/en/ (accessed on 11 June 2020).
2. World Health Organization. Summary Table of SARS cases by Country, 1 November 2002 to 31 July 2003. Available online: http://www.who.int/csr/sars/country/2003_08_15/en/ (accessed on 11 June 2020).
3. Ksiazek, T.G.; Erdman, D.; Goldsmith, C.S.; Zaki, S.R.; Peret, T.; Emery, S.; Tong, S.; Urbani, C.; Comer, J.A.; Lim, W.; et al. A novel coronavirus associated with Severe Acute Respiratory Syndrome. *N. Engl. J. Med.* **2003**, *348*, 1953–1966. [CrossRef] [PubMed]
4. World Health Organization. New Case of Labo- Ratory-Confirmed SARS in Guangdong, China, Update 5. Available online: https://www.who.int/csr/don/2004_01_31/en/ (accessed on 11 June 2020).

5. Zumla, A.; Chan, J.F.; Azhar, E.I.; Hui, D.S.; Yuen, K.Y. Coronaviruses—Drug discovery and therapeutic options. *Nat. Rev. Drug Discov.* **2016**, *15*, 327. [CrossRef] [PubMed]
6. Ciliberto, G.; Cardone, L. Boosting the arsenal against COVID-19 through computational drug repurposing. *Drug Discov. Today* **2020**, in press. [CrossRef]
7. Thiel, V.; Ivanov, K.A.; Putics, A.; Hertzig, T.; Schelle, B.; Bayer, S.; Weißbrich, B.; Snijder, E.J.; Rabenau, H.; Doerr, H.W.; et al. Mechanisms and enzymes involved in SARS coronavirus genome expression. *J. Gen. Virol.* **2003**, *84*, 2305–2315. [CrossRef]
8. Sayed, A.M.; Khattab, A.R.; AboulMagd, A.M.; Hassan, H.M.; Rateb, M.E.; Zaid, H.; Abdelmohsen, U.R. Nature as a treasure trove of potential anti-SARS-CoV drug leads: A structural/mechanistic rationale. *RSC Adv.* **2020**, *10*, 19790–19802. [CrossRef]
9. Hoffmann, M.; Kleine-Weber, H.; Schroeder, S.; Krüger, N.; Herrler, T.; Erichsen, S.; Schiergens, T.S.; Herrler, G.; Wu, N.-H.; Nitsche, A.; et al. SARS-CoV-2 cell entry depends on ACE2 and TMPRSS2 and is blocked by a clinically proven protease inhibitor. *Cell* **2020**, *181*, 271–280.e8. [CrossRef]
10. Jin, Z.; Du, X.; Xu, Y.; Deng, Y.; Liu, M.; Zhao, Y.; Duan, Y. Structure of M^{pro} from COVID-19 virus and discovery of its inhibitors. *Nature* **2020**, *582*, 289–293. [CrossRef]
11. Choy, K.T.; Wong, A.Y.L.; Kaewpreedee, P.; Sia, S.F.; Chen, D.; Hui, K.P.Y.; Chu, D.K.W.; Chan, M.C.W.; Cheung, P.P.-H.; Huang, X.; et al. Remdesivir, lopinavir, emetine, and homoharringtonine inhibit SARS-CoV-2 replication in vitro. *Antivir. Res.* **2020**, 104786. [CrossRef]
12. Patridge, E.; Gareiss, P.; Kinch, M.S.; Hoyer, D. An analysis of FDA-approved drugs: Natural products and their derivatives. *Drug Discov. Today* **2016**, *21*, 204–207. [CrossRef]
13. Farmer, P.B.; Suhadolnik, R.J. Nucleoside antibiotics. Biosynthesis of arabonofuranosyladenine by Streptomyces antibioticus. *Biochemistry* **1972**, *11*, 911–916. [CrossRef]
14. Seley-Radtke, K.L.; Yates, M.K. The evolution of nucleoside analogue antivirals: A review for chemists and non-chemists. Part 1: Early structural modifications to the nucleoside scaffold. *Antivir. Res.* **2018**, *154*, 66–86. [CrossRef] [PubMed]
15. El Sayed, K.A. Natural products as antiviral agents. In *Studies in Natural Products Chemistry*; Elsevir: Amsterdam, The Netherlands, 2000; Volume 24, pp. 473–572.
16. Abdelmohsen, U.R.; Bayer, K.; Hentschel, U. Diversity, abundance and natural products of marine sponge-associated actinomycetes. *Nat. Prod. Rep.* **2014**, *31*, 381–399. [CrossRef] [PubMed]
17. Caly, L.; Druce, J.D.; Catton, M.G.; Jans, D.A.; Wagstaff, K.M. The FDA-approved Drug Ivermectin inhibits the replication of SARS-CoV-2 in vitro. *Antivir. Res.* **2020**, 104787. [CrossRef] [PubMed]
18. Rahman, N.; Basharat, Z.; Yousuf, M.; Castaldo, G.; Rastrelli, L.; Khan, H. Virtual Screening of Natural Products against Type II Transmembrane Serine Protease (TMPRSS2), the Priming Agent of Coronavirus 2 (SARS-CoV-2). *Molecules* **2020**, *25*, 2271. [CrossRef]
19. Ngwa, W.; Kumar, R.; Thompson, D.; Lyerly, W.; Moore, R.; Reid, T.E.; Lowe, H.; Toyang, N. Potential of Flavonoid-Inspired Phytomedicines against COVID-19. *Molecules* **2020**, *25*, 2707. [CrossRef]
20. Slynko, I.; Scharfe, M.; Rumpf, T.; Eib, J.; Metzger, E.; Schüle, R.; Jung, M.; Sippl, W. Virtual screening of PRK1 inhibitors: Ensemble docking, rescoring using binding free energy calculation and QSAR model development. *J. Chem. Inform. Model.* **2014**, *54*, 138–150. [CrossRef]
21. Amaro, R.E.; Baudry, J.; Chodera, J.; Demir, Ö.; McCammon, J.A.; Miao, Y.; Smith, J.C. Ensemble docking in drug discovery. *Biophys. J.* **2018**, *114*, 2271–2278. [CrossRef]
22. Van Santen, J.A.; Jacob, G.; Singh, A.L.; Aniebok, V.; Balunas, M.J.; Bunsko, D.; Neto, F.C.; Castaño-Espriu, L.; Chang, C.; Clark, T.N.; et al. The natural products atlas: An open access knowledge base for microbial natural products discovery. *ACS Cent. Sci.* **2019**, *5*, 1824–1833. [CrossRef]
23. Wolber, G.; Langer, T. LigandScout: 3-D pharmacophores derived from protein-bound ligands and their use as virtual screening filters. *J. Chem. Inform. Model.* **2005**, *45*, 160–169. [CrossRef]
24. Lipinski, C.A. Lead-and drug-like compounds: The rule-of-five revolution. *Drug Discov. Today Technol.* **2004**, *1*, 337–341. [CrossRef]
25. Phillips, J.C.; Braun, R.; Wang, W.; Gumbart, J.; Tajkhorshid, E.; Villa, E.; Schulten, K. Scalable molecular dynamics with NAMD. *J. Comput. Chem.* **2005**, *26*, 1781–1802. [CrossRef] [PubMed]
26. MacKerell, A.D., Jr.; Bashford, D.; Bellott, M.L.; Dunbrack, R.L., Jr.; Evanseck, J.D.; Field, M.J.; Fischer, S.; Gao, J.; Guo, H.; Ha, S.; et al. All-atom empirical potential for molecular modeling and dynamics studies of proteins. *J. Phys. Chem. B* **1998**, *102*, 3586–3616. [CrossRef] [PubMed]

27. Humphrey, W.; Dalke, A.; Schulten, K. VMD: Visual molecular dynamics. *J. Mol. Graph.* **1996**, *14*, 33–38. [CrossRef]
28. Martyna, G.J.; Tobias, D.J.; Klein, M.L. Constant pressure molecular dynamics algorithms. *J. Chem. Phys.* **1994**, *101*, 4177–4189. [CrossRef]
29. Jo, S.; Kim, T.; Iyer, V.G.; Im, W. CHARMM-GUI: A web-based graphical user interface for CHARMM. *J. Comput. Chem.* **2008**, *29*, 1859–1865. [CrossRef]
30. Jo, S.; Jiang, W.; Lee, H.S.; Roux, B.; Im, W. CHARMM-GUI Ligand Binder for absolute binding free energy calculations and its application. *J. Chem. Inf. Model.* **2013**, *53*, 267–277. [CrossRef]
31. Jiménez, J.; Skalic, M.; Martinez-Rosell, G.; De Fabritiis, G. K deep: Protein–ligand absolute binding affinity prediction via 3d-convolutional neural networks. *J. Chem. Informat. Mod.* **2018**, *58*, 287–296. [CrossRef]
32. Sunseri, J.; Koes, D.R. Pharmit: Interactive exploration of chemical space. *Nucleic. Acids Res.* **2016**, *44*, W442–W448. [CrossRef]
33. Seeliger, D.; de Groot, B.L. Ligand docking and binding site analysis with PyMOL and Autodock/Vina. *J. Comput. Mol. Des.* **2010**, *24*, 417–422. [CrossRef]
34. Lill, M.A.; Danielson, M.L. Computer-aided drug design platform using PyMOL. *J. Comp. Mol. Des.* **2011**, *25*, 13–19. [CrossRef]
35. Gentile, D.; Patamia, V.; Scala, A.; Sciortino, M.T.; Piperno, A.; Rescifina, A. Putative inhibitors of SARS-CoV-2 main protease from a library of marine natural products: A virtual screening and molecular modeling study. *Mar. Drugs* **2020**, *18*, 225. [CrossRef]
36. Zhang, L.; Lin, D.; Sun, X.; Curth, U.; Drosten, C.; Sauerhering, L.; Becker, S.; Rox, K.; Hilgenfeld, R. Crystal structure of SARS-CoV-2 main protease provides a basis for design of improved α-ketoamide inhibitors. *Science* **2020**, *368*, 409–412. [CrossRef] [PubMed]
37. Yang, S.Y. Pharmacophore modeling and applications in drug discovery: Challenges and recent advances. *Drug Discov. Today* **2010**, *15*, 444–450. [CrossRef]
38. El-Neketi, M.; Ebrahim, W.; Lin, W.; Gedara, S.; Badria, F.; Saad, H.E.A.; Lai, D.; Proksch, P. Alkaloids and polyketides from *Penicillium citrinum*, an endophyte isolated from the Moroccan plant Ceratonia siliqua. *J. Nat. Prod.* **2013**, *76*, 1099–1104. [CrossRef] [PubMed]
39. Williams, D.E.; Bernan, V.S.; Ritacco, F.V.; Maiese, W.M.; Greenstein, M.; Andersen, R.J. Holyrines A and B, possible intermediates in staurosporine biosynthesis produced in culture by a marine actinomycete obtained from the North Atlantic Ocean. *Tetrahedron Lett.* **1999**, *40*, 7171–7174. [CrossRef]
40. Fiedler, H.P.; Bruntner, C.; Riedlinger, J.; Bull, A.T.; Knutsen, G.; Goodfellow, M.; Jones, A.; Maldonado, L.; Pathom-Aree, W.; Beil, W.; et al. Proximicin A, B and C, novel aminofuran antibiotic and anticancer compounds isolated from marine strains of the actinomycete *Verrucosispora*. *J. Antibiot.* **2008**, *61*, 158–163. [CrossRef] [PubMed]
41. Irlinger, B.; Bartsch, A.; Krämer, H.J.; Mayser, P.; Steglich, W. New tryptophan metabolites from cultures of the lipophilic yeast *Malassezia furfur*. *Helv. Chim. Acta* **2005**, *88*, 1472–1485. [CrossRef]
42. Gambichler, T.; Krämer, H.J.; Boms, S.; Skrygan, M.; Tomi, N.S.; Altmeyer, P.; Mayser, P. Quantification of ultraviolet protective effects of pityriacitrin in humans. *Arch. Dermatol. Res.* **2007**, *299*, 517–520. [CrossRef]
43. Chen, H.; Liu, N.; Huang, Y.; Chen, Y. Isolation of an anthrabenzoxocinone 1.264-C from *Streptomyces* sp. FXJ1. 264 and absolute configuration determination of the anthrabenzoxocinones. *Tetrahedron Asymmetry* **2014**, *25*, 113–116. [CrossRef]
44. Hou, X.M.; Wang, C.Y.; Gu, Y.C.; Shao, C.L. Penimethavone A, a flavone from a gorgonian-derived fungus *Penicillium chrysogenum*. *Nat. Prod. Res.* **2016**, *30*, 2274–2277. [CrossRef]

microorganisms

MDPI

Article

Comparative Study of the Proteins Involved in the Fermentation-Derived Compounds in Two Strains of *Saccharomyces cerevisiae* during Sparkling Wine Second Fermentation

María del Carmen González-Jiménez [1], Teresa García-Martínez [1], Juan Carlos Mauricio [1,*], Irene Sánchez-León [1], Anna Puig-Pujol [2], Juan Moreno [1] and Jaime Moreno-García [1]

1 Department of Agricultural Chemistry, Edaphology and Microbiology, Microbiology Area, Agrifood Campus of International Excellence ceiA3, University of Cordoba, 14014 Cordoba, Spain; b02gojim@uco.es (M.d.C.G.-J.); mi2gamam@uco.es (T.G.-M.); b32salei@uco.es (I.S.-L.); qe1movij@uco.es (J.M.); b62mogaj@uco.es (J.M.-G.)

2 Department of Enological Research, Institute of Agrifood Research and Technology-Catalan Institute of Vine and wine (IRTA-INCAVI), 08720 Barcelona, Spain; anna.puig@irta.cat

* Correspondence: mi1gamaj@uco.es; Tel.: +34-957-218-640; Fax: +34-957-218-650

Received: 13 July 2020; Accepted: 6 August 2020; Published: 8 August 2020

Abstract: Sparkling wine is a distinctive wine. *Saccharomyces cerevisiae* flor yeasts is innovative and ideal for the sparkling wine industry due to the yeasts' resistance to high ethanol concentrations, surface adhesion properties that ease wine clarification, and the ability to provide a characteristic volatilome and odorant profile. The objective of this work is to study the proteins in a flor yeast and a conventional yeast that are responsible for the production of the volatile compounds released during sparkling wine elaboration. The proteins were identified using the OFFGEL fractionator and LTQ Orbitrap. We identified 50 and 43 proteins in the flor yeast and the conventional yeast, respectively. Proteomic profiles did not show remarkable differences between strains except for Adh1p, Fba1p, Tdh1p, Tdh2p, Tdh3p, and Pgk1p, which showed higher concentrations in the flor yeast versus the conventional yeast. The higher concentration of these proteins could explain the fuller body in less alcoholic wines obtained when using flor yeasts. The data presented here can be thought of as a proteomic map for either flor or conventional yeasts which can be useful to understand how these strains metabolize the sugars and release pleasant volatiles under sparkling wine elaboration conditions.

Keywords: sparkling wine; second fermentation; fermentation by-products; *Saccharomyces cerevisiae* flor yeast; proteins

1. Introduction

Sparkling wine is a very distinctive wine with a unique winemaking process. Its peculiarity is mainly due to a second fermentation performed in closed bottles, where wines acquire an effervescent characteristic. This is followed by a long aging process, in which the wine is in contact with the yeast lees and thereby affecting its organoleptic properties. Its production, despite being lower compared to that of still wines, has an extensive economic impact on the enology industry. This is due to the relatively high economic value of most sparkling wines [1].

Sparkling wine elaboration by the *champenoise* or traditional method (like champagne in France and cava in Spain) involves two main steps. First, a fermentation where the grape must be converted to wine and second, a process called "*prise de mousse*" [2]. The latter consists of a secondary fermentation process in sealed bottles after adding sugar and yeast, followed by at least nine months of aging on lees at low temperature (12–16 °C). During the *"prise de mousse"*, yeasts are subjected to several stress factors, such as high ethanol content, nitrogen deficiency, low pH values, low temperature and CO_2 overpressure [3]. These affect yeast metabolism and contribute to important modifications of sparkling wine organoleptic properties [4]. During fermentation, the yeast produces ethanol and carbon dioxide, among others, which, despite being toxic, the yeasts are able to cope with. It is during the aging of the wine in contact with the lees when mannoproteins are released as well as compounds derived from autolysis and enzymes involved in reactions that affect some aroma precursors [5,6].

A large number of studies have reported the metabolic/enzymatic potential of certain non-conventional yeasts and their role in improving some technological and sensory aspects of wine [7–10], such as the positive effect on aroma, glycerol, polysaccharides, mannoproteins, and volatile acid [11,12]. Therefore, the use of non-conventional yeasts in wine fermentations has become a current trend in the wine industry. These unconventional yeast species are used in winemaking with objectives such as (i) control the acidity [11], (ii) improve color extraction and mouthfeel [13], (iii) reduce the ethanol content [8,14]; and, more recently, and (iv) improve foam properties in sparkling wines [15].

Due to the capacity of this yeast to support high concentrations of ethanol, the use of a non-conventional yeast, such as *Saccharomyces cerevisiae* flor yeast strains for sparkling wine elaboration is suitable and is a possible advantage for the industry of wine. Further, flor yeast strains possess distinctive characteristics compared to other fermentative *S. cerevisiae* strains, such as their capacity to form a biofilm on the air-liquid interface of the wine for the elaboration of Sherry wines [16]. These cell adhesive properties allow the winemakers to easily remove yeasts and sediment in the "degüelle" phase during the production of cava by the traditional method. Moreover, previous studies have demonstrated that flor yeast is a good candidate for sparkling wine elaboration because they produce volatile compounds, such as higher alcohols, aldehydes, esters and ketones, and its influence on the wine final aroma [17,18]. The use of flor yeast could reduce production cost and time, marking a significant step forward in the sparkling wine industry. At the same time, it mitigates the current situation of low diversity of commercially available yeasts for winemaking, in this particular case for the production of sparkling wines by the traditional method.

Here, we aim to reveal the yeast proteins responsible for the production of fermentation compounds released in the second fermentation during the elaboration of cava by a flor *S. cerevisiae*, and compare it with a conventional strain used in the production of this type of wine. The data presented can be seen as a proteomic map of either the flor or conventional yeasts which can be useful to understand how these strains metabolize the sugars and release pleasant volatiles under *prise de mousse* conditions. This knowledge can shed light on the molecular mechanism behind the production of characteristic volatile compounds that will determine the odorant profiles of this type of wine.

2. Materials and Methods

2.1. Microorganism and Experimental Conditions

The microorganisms used were two strains of *S. cerevisiae*. The first strain, *S. cerevisiae* G1 (ATCC: MYA-2451), is an industrial flor wine yeast from the collection of the Department of Microbiology of the University of Cordoba (Spain). It was isolated from Fine Sherry wine of Montilla-Moriles designation of origin (DO) (Spain). This strain forms a thick biofilm (velum) about 30 days after inoculation with a cell viability higher than 90% [19]. The second strain, *S. cerevisiae* P29 (CECT 11770), was used as the control strain and isolated in the Penedès grape-growing area (Spain) by the Catalan Institute of Vines and Wines (INCAVI). INCAVI recommends P29 for the elaboration of "cava" Spanish sparkling wine. A standardized commercial base wine, obtained by fermenting musts from Macabeo and Chardonnay grapes in a proportion 6:4, was used for the second fermentation. After settling, the base wine was subjected to a second fermentation inside 750 mL bottles at 14 °C. Sucrose and yeast cells were added to the base wine to reach 22 g/L per bottle and 1.5×10^6 cells/mL.

The changes caused by yeast during the second fermentation were monitored at three sampling times: (i) the base wine (T0) (ii) at the middle of fermentation stage, when CO_2 pressure reached 3 bar (MF); and (iii) at the end of the second fermentation (EF) one month after when CO_2 pressure reached 6.5 bar. Data shown in Figure 1. All the samples were analyzed in triplicate.

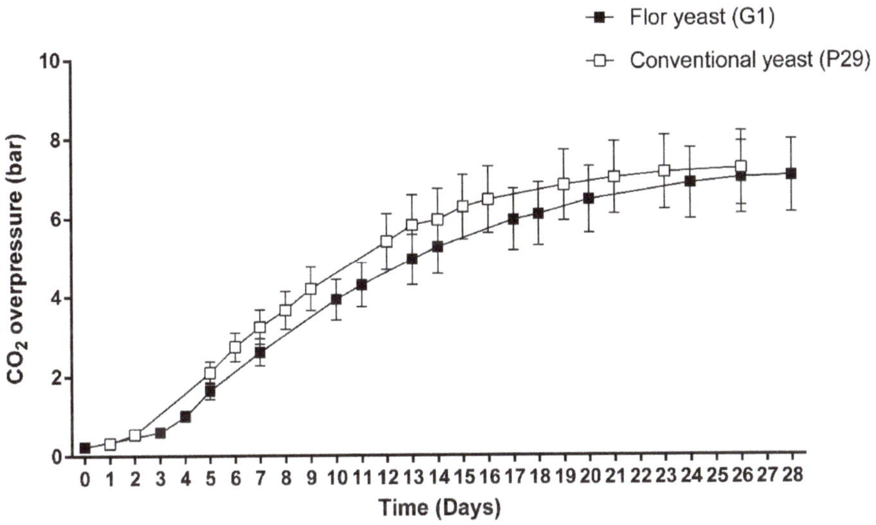

Figure 1. Evolution of endogenous CO_2 released by flor yeast and conventional yeast during the second fermentation in Spanish sparkling wine (cava) elaboration (Porras-Agüera et al., 2019) [20].

2.2. Proteome Analysis

The cells were collected from each bottle by centrifugation at 4500× *g* for 10 min by a centrifuge (Hettich® ROTINA 38/38R, Kirchlengern, Germany) and the sediment was washed twice with sterile distilled cold water. Afterwards, cells were broken by a mechanical technique in Vibrogen Cell Mill V6 (Edmund Bühler, Bodelshausen, Germany) using 500 µm diameter glass balls (Sigma-Aldrich, Darmstadt, Germany). Once the cells were broken, the protein pull was extracted. For this purpose, extraction buffer and protease inhibitors cocktails were used. A total of 500 µg of protein of each condition and replica was loaded. The OFFGEL high-resolution kit, pH 3–10 (Agilent Technologies, Palo Alto, CA, USA) was used for protein preparative isoelectric focusing (IEF) in solution. Protein samples were solubilized in protein OFFGEL fractionation buffer (Agilent Technologies, Part number 5188–6444, Santa Clara, CA, USA), and aliquots evenly distributed in 12-well 3100 OFFGEL fractionator trays according to the supplier's instructions. Proteins from each well were scanned and fragmented on an LTQ Orbitrap XL mass spectrometer (Thermo Fisher Scientific, San Jose, CA, USA) equipped with a nano LC Ultimate 3000 system (Dionex, Germering, Germany). To obtain the concentration of a protein in the sample, Exponentially Modified Protein Abundance Index (emPAI) was used [21]. These procedures and methods are described in more detail by Moreno-Garcia et al. (2015) [22] and Porras-Agüera et al. (2020) [23].

The quantified aroma compounds were related to proteins directly involved in their metabolism using the following databases: YMDB (yeast metabolome database; http://www.ymdb.ca/), SGD (*Saccharomyces* genome database; http://www.yeastgenome.org/), and Uniprot (http://www.uniprot.org/).

2.3. Statistical Analysis

The software package Statgraphics Centurion XVI.II, (STSC, Inc., Rockville, MD, USA) was used for statistical analysis of the proteins. A multiple-sample comparison procedure (MSC) was used to compare two or more independent samples via ANOVA and Fisher's test to establish homogenous groups at a level of significance of 95% (p-value < 0.05). Data were previously normalized according to root square and Pareto scaling, to avoid the differences introduced by the measurement units [24]. All treatments were evaluated in triplicate.

In addition, a correlation analysis to establish significant relationships between metabolites and proteins were carried out according to Metaboanalyst (https://www.metaboanalyst.ca/).

3. Results and Discussion

A total of 50 proteins and 43 proteins related to the metabolism of fermentation metabolites (ethanol, glycerol, acetic acid, acetaldehyde, acetoin, and 2,3-butanediol) have been identified in flor yeast and conventional yeast, respectively (Table 1, Supplementary Table S1). The fermentation related proteins have been sorted in subpathways. Each subpathway is commented and discussed separately.

Table 1. Composition of the base wine (T0) and the Spanish sparkling wine at the middle of the second fermentation (MF) and at the end of the second fermentation (EF) in flor yeast and conventional yeast strains. Data provided by Martínez-García et al. (2017) and (2020) [17,25].

	Flor Yeast			Conventional Yeast		
	T0	MF	EF	T0	MF	EF
Ethanol (% v/v)	10.23 ± 0.02	10.76 ± 0.04	11.4 ± 0.1	10.23 ± 0.02	10.85 ± 0.04	11.60 ± 0.03
Acetaldehyde (mg/L)	87 ± 1	132 ± 1	87 ± 16	87.2 ± 1.1	133.9 ± 7.3	85.2 ± 0.2
Acetoin (mg/L)	19 ± 1	61 ± 1	31 ± 2	19.3 ± 1.3	127.5 ± 11.8	24.5 ± 6.3
2,3-butanediol (mg/L)	171 ± 7	221 ± 4	200 ± 33	171 ± 6	166 ± 4	192 ± 12
Acetic acid (g/L)	0.23 ± 0.02	0.20 ± 0.00	0.28 ± 0.02	0.23 ± 0.02	0.20 ± 0.00	0.22 ± 0.00
Glycerol (mg/L)	4020 ± 656	4493 ± 164	4227 ± 297	4019 ± 655	4557 ± 212	3513 ± 163

3.1. *Glycolysis/Gluconeogenesis Proteome*

A total of 25 and 23 proteins involved with the glycolysis/gluconeogenesis pathway out of a total of 38 proteins currently documented in *S. cerevisiae*, were identified in the flor yeast and conventional strain, respectively. The contents of the different proteins were analyzed at different times in the second fermentation in the production of sparkling wine (T0, MF, and EF).

In general, the proteomic profiles obtained in both strains for the proteins involved in these pathways were not remarkably different. Content of proteins like glyceraldehyde-3-phosphate dehydrogenases (Tdh1p, Tdh2p and Tdh3p), Pgk1p and enolases (Eno1p and Eno2p), progressively increased during the second fermentation in both strains but the increase was more abrupt in the flor yeast (Supplementary Table S1). These proteins catalyze the reversible steps 1, 2, and 4 of the subpathway that synthesizes pyruvate from D-glyceraldehyde 3-phosphate, steps shared by glycolysis and gluconeogenesis pathways. An increased synthesis of glyceraldehyde-3-phosphate dehydrogenases during the second fermentation could be related to the recycling of NAD^+/NADH for the continuation of glycolysis; otherwise, the glycolytic flow would decrease, which could lead to exhaustion of the ATP energy charge, making it lethal for yeast [26]. Most of the NADH produced during glycolysis is used by yeast for the formation of ethanol from acetaldehyde. Figures 2 and 3 prove a significant inverse correlation between glycolysis/gluconeogenesis protein content and glucose concentration, indicating that enzymes are degrading glucose. However, at EF, when fermentable carbon sources are depleted (~0.3 g/L) and the major carbon sources are ethanol or glycerol, yeast Tdhps, Pgk1p and Enops can catalyze for gluconeogenesis [16,22]. Recently, Porras-Agüera et al. (2019) postulated that the increase in Tdhps content could be related to cell death or stress response, and thus proposing them as possible cell death biomarkers during the second fermentation [20]. A higher abundance of gluconeogenesis-related proteins in the flor yeast versus the conventional sparkling yeast may be related to an evolutionary adaptation of the first strain to media with high concentrations of non-carbon sources where flor yeasts are predominant [27].

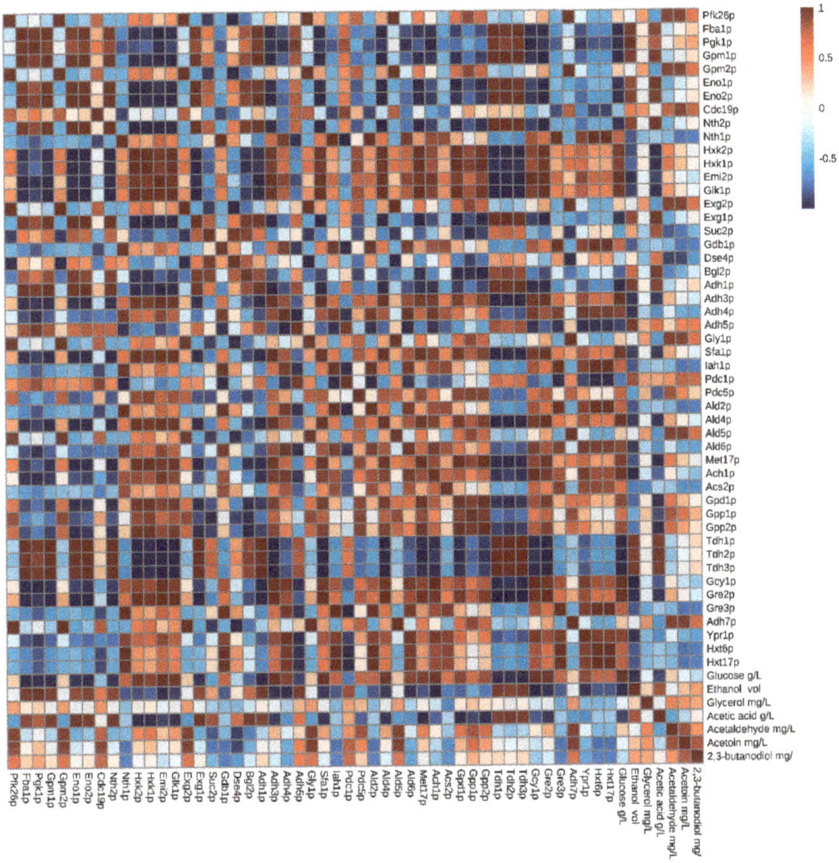

Figure 2. Matrix of correlations established between proteins and compounds released in the second fermentation in the flor yeast G1 strain. Metabolome data extracted from Martínez-García et al. (2020) [17].

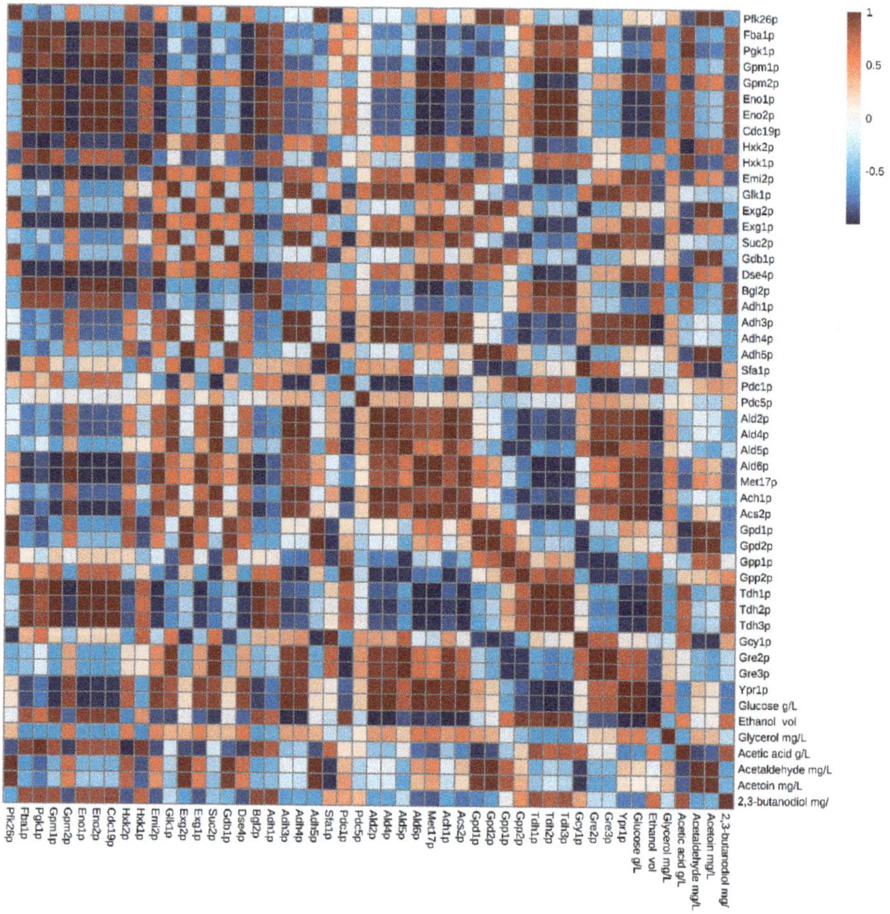

Figure 3. Matrix of correlations established between proteins and compounds released in the second fermentation in the conventional yeast P29 strain. Metabolome data extracted from Martínez-García et al. (2017) [25].

Cell wall proteins glucanases Exg1p, Exg2p, and Bgl2p that hydrolyze β-glucan chains in the cell wall leading to the release of glucose, were also reported in higher concentrations in the flor yeast. The presence of these proteins at early stages of the fermentation may be related to cell expansion during growth while at the end of fermentation can be involved in cell wall degradation [5]. One more protein that reported an increase in its concentration at EF in flor yeast is Suc2p, which was found 5-fold higher than in the conventional strain. This protein is capable of transforming sucrose into glucose and fructose. Its non-glycosylated form is expressed constitutively, while its glycosylated form is regulated by the repression of glucose [28]. The absence of glucose in the medium at the end of the second fermentation causes the yeasts to synthesize the non-glycosylated form. This enzyme is excreted into the periplasmic space, where the hydrolysis occurs, and letting the monosaccharide products of the reaction, glucose and fructose, be transported into the cell. Higher Suc2p contents at EF in the flor yeast may be associated to a higher cell wall degradation that correlates with higher autolysis proteins compared to the conventional strain [18,23].

3.2. Proteins Related to the Metabolism of Pyruvate to Ethanol and Acetic Acid

A total of 16 and 12 proteins involved in the formation of ethanol from pyruvate (out of a total of 22 proteins currently documented in *S. cerevisiae*) were identified in the flor yeast and conventional strain, respectively (Supplementary Table S1).

Adh1p stood out by ranging in content from 0.4 to 1.8 (mol%) with a positive trend towards EF. Adh1p and Adh5p contents were reported two-fold more in concentration for the flor yeast compared to the conventional strain in MF and EF; where EF > MF. These enzymes are responsible for reversible exchange of acetaldehyde and ethanol during glucose fermentation. This increase is related to the drastic decrease in the amount of acetaldehyde quantified at the end of the second fermentation and the increase in the amount of ethanol produced, indicating a reaction direction towards ethanol (Figures 2 and 3). At concentrations below 70 mg/L, acetaldehyde can provide a fruity flavor to wine, which often occurs in freshly fermented wines. However, it can be pungent and negative over 100 mg/L and is often associated with bruised apple, Sherry, walnut and oxidation [29]. In this study, both the flor yeast and the conventional yeast presented concentrations lower than 100 mg/L at the end of the second fermentation [17,25] supporting the use of flor yeast in sparkling wine second fermentation. Furthermore, pyruvate decarboxylases Pdc1p and Pdc5p highlighted in MF for high protein content in flor yeast compared to the conventional strain. Both pyruvate decarboxylases are key enzymes in alcoholic fermentation and are responsible for the decarboxylation of pyruvate to acetaldehyde. Pdc1p is the main active form during glucose catabolism, while Pdc5p is a secondary form that is only expressed under thiamine starvation [30]. This higher protein content was reflected in an increased acetaldehyde concentration in the middle of the second fermentation in flor yeast (Figure 2). High concentration of Adhps in flor yeast at EF, when no sugars remain, may be related to an adaptive proteomic response of these yeasts to environments with only non-fermentable carbon sources [16,31]. However, the anoxia inside the bottles does not allow ethanol consumption so the alcohol dehydrogenases, although abundant, may remain inactive.

The metabolism of acetic acid in *S. cerevisiae* is primarily synthesized as an intermediate by cytosolic pyruvate dehydrogenase bypass. This involves the conversion of pyruvate to acetaldehyde by pyruvate decarboxylase, which is subsequently oxidized to acetate by the action of aldehyde dehydrogenase (*ALD*) [32–34]. This acetic acid is key for the formation of fatty acids, acetyl-CoA, through the action of acetyl-CoA synthetase (*ACS*), so there must be a balance between *ALD* and *ACS* activity. The optimal concentration in wine is less than 0.20 g/L [35]. In excessive amounts, acetic acid gives wine a pungent taste and an unpleasant aroma of vinegar [36].

A total of seven proteins related to the acetic acid metabolism were identified in both strains, out of a total of 14 documented proteins in *S. cerevisiae*. However, in quantitative terms, the proteomic profile obtained for each strain was different (Supplementary Table S1). At T0, Ald4p stands out for higher protein content in conventional strain versus flor yeast, however in MF, this protein was two-fold more in protein content in flor yeast. As mentioned above, Ald4p together with Ald5p, participate in a pathway of mitochondrial pyruvate dehydrogenase, in which pyruvate is first decarboxylated to acetaldehyde in the cytosol by pyruvate decarboxylase and then converted to acetate by mitochondrial acetaldehyde dehydrogenases [37]. In general, the protein content of Acs2p reported was lower compared to Ald4p. According to Verduyn et al. (1990) [38] less acetic acid is produced if *ALD* activity is lower than *ACS* activity. This can explain the significant increase in acetic acid during the course of the second fermentation in flor yeast [17].

Even though no protein was identified at the end of the second fermentation in flor yeast and almost not quantifiable in the conventional yeast, the amount of acetic acid produced was higher at this sampling time. This fact could be correlated to the decrease in the amount of acetaldehyde, previously described [17,25]. This conversion could have taken place when the cells are still performing the alcoholic fermentation when the proteins were detectable, and the acetic acid remains until the sampling time.

3.3. Proteins Related to the Metabolism of Pyruvate-Acetoin-2,3-Butanediol

Acetoin and 2,3-butanediol are by-products generated by *S. cerevisiae* during alcoholic fermentation that confer buttery and cream aromas, when present over concentrations of 0.03 and 0.67 g/L, respectively. Acetoin can also be a precursor of some off-odor compounds, such as diacetyl. High fermentative wine yeasts generally produce low acetoin levels [39]. In this study, the conventional strain produced less acetoin than the flor yeast, however, the differences were not significant. The acetoin/2,3-butanediol pathway in yeasts contributes to detoxification of acetaldehyde because acetoin is a weaker inhibitor than acetaldehyde [40]. Out of the five proteins documented in *S. cerevisiae* (Bdh1p, Bdh2p, Pdc1p, Pdc5p, and Pdc6p) involved in the formation of 2,3-butanediol, only two were identified (Pdc1p and Pdc5p) in both strains (Supplementary Table S1). In MF, higher contents were observed for Pdc1p and Pdc5p in flor yeast. In both cases a direct correlation could be established between the quantity of acetoin quantified and Pdc1p and Pdc5p, and this correlation was stronger for flor yeast except for the conventional yeast Pdc5p (Figures 2 and 3).

On the other hand, Bdh1p and Bdh2p were not reported in any of the strains involved in the reversible oxidation of acetoin to 2,3-butanediol and the irreversible reduction of 2,3-butanediol to (S)-acetoin, respectively. 2,3-Butanediol represents an important source of aroma [41] although it has a very high odor threshold value (~150 mg/L). In wine, its concentration varies from approximately 0.2 to 3 g/L, with an average value of approximately 0.57 g/L. This high content can have some effect on the wine bouquet due to its slightly bitter taste and also on the body of the wine due to its viscosity [41]. The changes in 2,3-butanediol and the absence of Bdh1p and Bdh2p could be attributed to limitation in the detection method that could not quantify very low protein content or to a potential activity of another enzyme that catalyzes the same reaction or another reaction known to produce this compound.

3.4. Proteins Related to the Metabolism of Dihydroxyacetone Phosphate-Glycerol.

Glycerol is quantitatively the most important fermentation product after ethanol and carbon dioxide, its concentration depends on environmental factors, such as temperature, aeration, sulfite level, and yeast strain [42]. Glycerol contributes positively to the sensory quality of the wine, providing smoothness and viscosity [43]. In *S. cerevisiae*, this polyol plays two main roles in physiological processes: it fights osmotic stress and controls intracellular redox balance, and [44–46] converts excess NADH generated during biomass formation to NAD^+. Glycerol is synthesized by reducing dihydroxyacetone phosphate to glycerol 3-phosphate and is catalyzed by an NAD-dependent cytosolic G3P dehydrogenase (*GPD*), followed by dephosphorylation of glycerol 3-phosphate by a specific phosphatase (*GPP*).

In this work, 11 proteins have been identified (out of 19 documented proteins in *S. cerevisiae*) related to glycerol metabolism in both strains (Supplementary Table S1). In general, the proteomic profile obtained for each sampling time was similar in both strains, but a higher protein content was reported in the case of flor yeast. Gpd1p, Gpp1p, and Gpp2p almost doubled in quantity in T0 and MF in the conventional yeast versus flor yeast. Remize et al. (2003) detected that Gpd1p increases during the growth phase. The beginning of the second fermentation in sparkling wine involves anaerobiosis and osmotic stress that influence the expression of *GPD* genes [47]. Under these conditions, the respiratory chain does not function and the production of glycerol is the only possible mechanism of re-oxidation of NADH. Gpd1p, Gpd2p (not identified), Gpp1p and Gpp2p play a major role in glycerol formation. Depending on the way in which they are combined, they have been related to the production of glycerol during osmotic stress (Gpd1p-Gpp2p combination) or to the adjustment of the NADH-NAD^+ redox balance under anaerobic conditions (Gpd2p-Gpp1p combination) [47]. In this work, the first combination (Gpd1p-Gpp2p) has been reported for both yeast strains. Both yeasts increased the synthesis of both proteins at T0 and MF which would promote the accumulation of glycerol inside the cell to withstand the osmotic stress. However, in flor yeast these two proteins were not identified at EF. Further, no significant differences were obtained in the extracellular glycerol concentration in flor yeast while in conventional yeast there was a significant decrease in concentration from MF to EF [17,25]. It was not possible to establish any significant correlation between the concentration of this metabolite and the content of the proteins involved in any of the strains (Figures 2 and 3). Also, glyceraldehyde-3-phosphate dehydrogenases, previously commented, can influence the glycerol concentration (maybe producing at MF and consuming at EF). These results suggest that there is a prevalence of the metabolic pathway of ethanol production versus that of glycerol formation since considerable increase in the ethanol concentration was obtained while the glycerol concentration remained stable, possibly due to an accumulation of this compound inside the cell. Another possible explanation for this fact is that these proteins have not been activated by yeast, causing a change in the coenzyme requirement during the synthesis of glutamate from NADPH to NADH, and decreasing the availability of NADH for the synthesis of glycerol and an increase in the yield of ethanol [48]. The balance between the concentration of ethanol and glycerol results in pleasant and stable wines.

The metabolites and proteins that displayed the highest concentration in each strain are highlighted in a schematic figure (Figures 4 and 5) to provide a better understanding of the results obtained in this work.

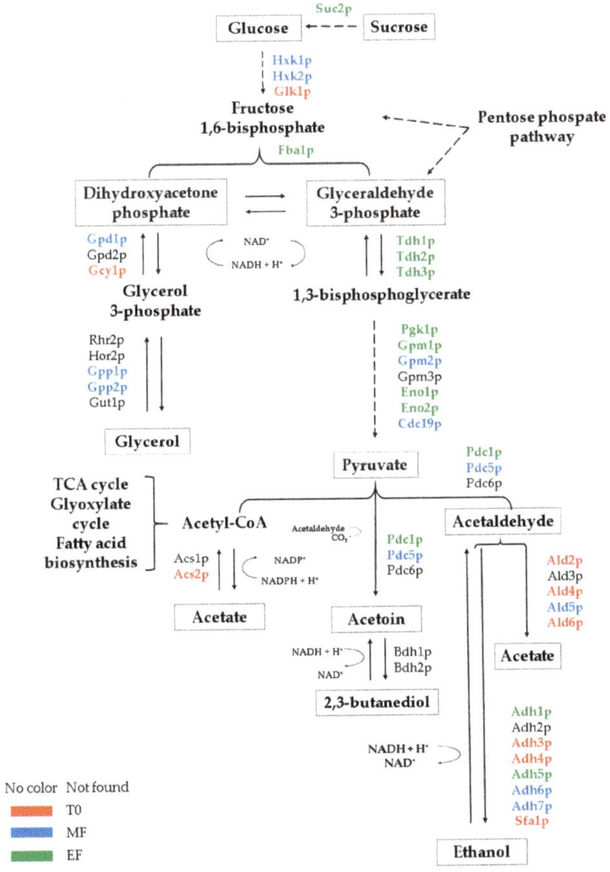

Figure 4. Summary of the scheme of proteins involved in the compounds derived from *Saccharomyces cerevisiae* flor yeast fermentation during the second fermentation in the production of sparkling wine. The color of the protein names represents the condition in which the highest protein content of the proteins was identified. Each condition is represented by a color: red for the base wine, T0; blue for the middle of the fermentation, MF; green for the end of the second fermentation, EF.

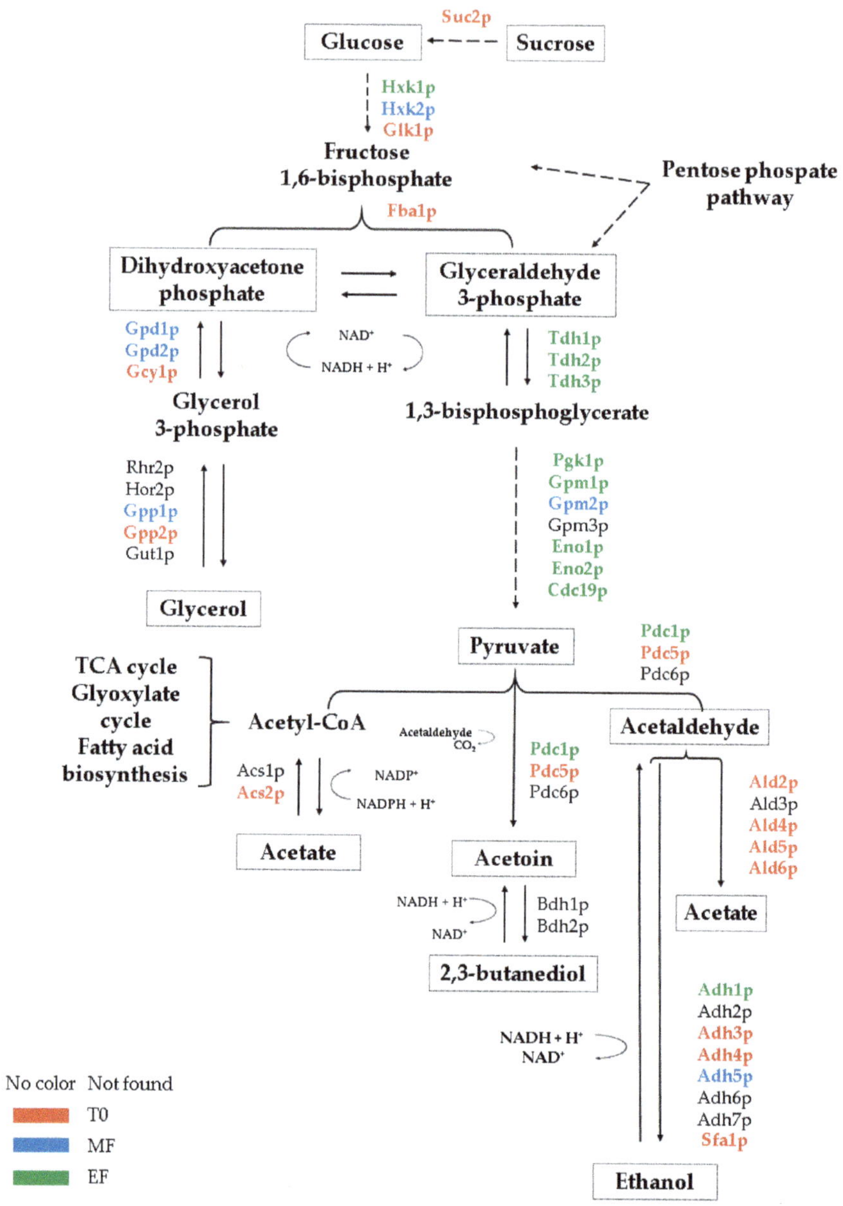

Figure 5. Summary of the scheme of proteins involved in the compounds derived from fermentation in *Saccharomyces cerevisiae* conventional yeast during the second fermentation in the production of sparkling wine. The color of the protein names represents the condition in which the highest protein content of the proteins was identified. Each condition is represented by a color: red for the base wine, T0; blue for the middle of the fermentation, MF; green for the end of the second fermentation, EF.

4. Conclusions

This work is focused on the use of an unconventional flor yeast to produce sparkling wine by comparing with a typical wine yeast strain. The relationship of the yeast proteome with the exo-metabolites excreted in the medium during the second fermentation in the production of sparkling wine has been established.

Fifty proteins and 43 proteins related to the metabolism and transport of fermentation metabolites (ethanol, glycerol, acetic acid, acetaldehyde, acetoin, and 2,3-butanediol) have been identified in flor yeast and conventional yeast, respectively. Not remarkable differences were found among the tested strains, but a lower concentration of most proteins was reported in the conventional yeast. Consequently, the concentration of the related metabolites was different in each strain and all above their odor threshold.

This study highlights that flor yeasts generally used to produce Sherry wine, can perform autolysis at high levels during the second fermentation and improve the quality and diversity of sparkling wine. In view of the results obtained, the use of this type of flor yeast is suggested for the production of sparkling wine. In addition, this type of yeast can resist high content of ethanol and has high adhesion properties. These characteristics make it an ideal candidate for the production of sparkling wines.

Supplementary Materials: The following are available online at http://www.mdpi.com/2076-2607/8/8/1209/s1, Table S1: List of the proteins involved in the fermentation-derived compounds identified in *Saccharomyces cerevisiae* G1 and P29 under typical second fermentation conditions in three sampling times (T0: base wine; MF: middle of the second fermentation; EF: end of the second fermentation).

Author Contributions: All authors designed the research; M.d.C.G.-J., I.S.-L., J.M.-G., J.M., J.C.M. and T.G.-M. read and approved the final manuscript. M.d.C.G.-J. conducted the study, analyzed the experimental data, and drafted the manuscript. I.S.-L. helped analyze the experimental data. J.M. helped analyze and review the experimental data. J.C.M., J.M.-G. and T.G.-M. participated in the coordination of the study and carried out the experimental part. A.P.-P. crafted the methodology. All authors reviewed the drafts and contributed to writing the manuscript. All authors have read and agreed to the published version of the manuscript.

Funding: This study was funded by the "XXIII Programa Propio de Fomento de la Investigación 2018" (MOD 4.2. SINERGIAS, Ref XXIII. PP Mod 4.2, JC Mauricio) from University of Cordoba (Spain), and by Spain's Ministry of Science, Innovation and Universities and the European Fund of Regional Development (FEDER): Grant No. RTA2014-00016-C03-03.

Acknowledgments: Kind help from the staff at the Central Research Support Service (SCAI) of the University of Cordoba (Spain) with the protein analyses is gratefully acknowledged and Minami Ogawa for the language revision of this paper.

Conflicts of Interest: The authors declare no conflict of interest.

References

1. Pozo-Bayón, M.Á.; Martínez-Rodríguez, A.; Pueyo, E.; Moreno-Arribas, M.V. Chemical and biochemical features involved in sparkling wine production: From a traditional to an improved winemaking technology. *Trends Food Sci. Technol.* **2009**, *20*, 289–299. [CrossRef]
2. Buxaderas, S.; López-Tamames, E. Sparkling wines: Features and trends from tradition. In *Advances in Food and Nutrition Research*; Academic Press: Cambridge, MA, USA, 2012; Volume 66, pp. 1–45.
3. Penacho, V.; Valero, E.; González, R. Transcription profiling of sparkling wine second fermentation. *Int. J. Food Microbiol.* **2012**, *153*, 176–182. [CrossRef] [PubMed]
4. Garofalo, C.; Arena, P.M.; Laddomada, B.; Cappello, S.M.; Bleve, G.; Grieco, F.; Beneduce, L.; Berbegal, C.; Spano, G.; Capozzi, V. Starter cultures for sparkling wine. *Fermentation* **2016**, *2*, 21. [CrossRef]
5. Alexandre, H.; Guilloux–Nematier, M. Yeast autolysis in sparkling wine—A review. *Aust. J. Grape Wine Res.* **2006**, *12*, 119–127. [CrossRef]
6. Sartor, S.; Toaldo, I.M.; Panceri, C.P.; Caliari, V.; Luna, A.S.; de Gois, J.S.; Bordignon-Luiz, M.T. Changes in organic acids, polyphenolic and elemental composition of rosé sparkling wines treated with mannoproteins during over-lees aging. *Food Res. Int.* **2019**, *124*, 34–42. [CrossRef] [PubMed]
7. Ruiz, J.; Ortega, N.; Martín-Santamaría, M.; Acedo, A.; Marquina, D.; Pascual, O.; Rozès, N.; Zamora, F.; Santos, A.; Belda, I. Occurrence and enological properties of two new non-conventional yeasts (*Nakazawaea ishiwadae* and *Lodderomyces elongisporus*) in wine fermentations. *Int. J. Food Microbiol.* **2019**, *305*, 108255. [CrossRef] [PubMed]

8. Moreno, J.; Moreno-García, J.; López-Muñoz, B.; Mauricio, J.C.; García-Martínez, T. Use of a flor velum yeast for modulating colour, ethanol and major aroma compound contents in red wine. *Food Chem.* **2016**, *213*, 90–97. [CrossRef]
9. Quincozes, L.; Santos, P.; Vieira, L.; Gabbardo, M.; Eckhardt, D.P.; Cunha, W.; Costa, V.; Zigiotto, L.; Schumacher, R. Influence of yeasts of the genus *Saccharomyces* and not *Saccharomyces* in elaboration of white wines. *BIO Web Conf.* **2019**, *12*, 02014. [CrossRef]
10. Alonso-del-Real, J.; Lairón-Peris, M.; Barrio, E.; Querol, A. Effect of temperature on the prevalence of *Saccharomyces* non *cerevisiae* species against a *S. cerevisiae* wine strain in wine fermentation: Competition, physiological fitness, and influence in final wine composition. *Front. Microbiol.* **2017**, *8*, 150. [CrossRef]
11. Ciani, M.; Comitini, F. Non-*Saccharomyces* wine yeasts have a promising role in biotechnological approaches to winemaking. *Ann. Microbiol.* **2011**, *61*, 25–32. [CrossRef]
12. Nuñez-Guerrero, M.A.; Paez-Lerma, J.B.; Rutiaga-Quiñones, O.M.; González-Herrera, S.M.; Soto-Cruz, N.O. Performance of mixtures of *Saccharomyces* and non-*Saccharomyces* native yeasts during alcoholic fermentation of *Agave duranguensis* juice. *Food Microbiol.* **2016**, *54*, 91–97. [CrossRef]
13. Suárez-Lepe, J.A.; Morata, A. New trends in yeast selection for winemaking. *Trends Food Sci. Technol.* **2012**, *23*, 39–50. [CrossRef]
14. Ciani, M.; Morales, P.; Comitini, F.; Tronchoni, J.; Canonico, L.; Curiel, J.A.; Oro, L.; Rodrigues, A.J.; Gonzalez, R. Non-conventional yeast species for lowering ethanol content of wines. *Front. Microbiol.* **2016**, *7*, 642. [CrossRef] [PubMed]
15. Medina-Trujillo, L.; González-Royo, E.; Sieczkowski, N.; Heras, J.; Canals, J.M.; Zamora, F. Effect of sequential inoculation (*Torulaspora delbrueckii/Saccharomyces cerevisiae*) in the first fermentation on the foaming properties of sparkling wine. *Eur. Food Res. Technol.* **2017**, *243*, 681–688. [CrossRef]
16. Alexandre, H. Flor yeasts of *Saccharomyces cerevisiae*-their ecology, genetics and metabolism. *Int. J. Food Microbiol.* **2013**, *167*, 269–275. [CrossRef]
17. Martínez-García, R.; Roldán-Romero, Y.; Moreno, J.; Puig-Pujol, A.; Mauricio, J.C.; García-Martínez, T. Use of a flor yeast strain for the second fermentation of sparkling wines: Effect of endogenous CO_2 over-pressure on the volatilome. *Food Chem.* **2020**, *308*, 125555. [CrossRef]
18. González-Jiménez, M.C.; Moreno-García, J.; García-Martínez, T.; Moreno, J.J.; Puig-Pujol, A.; Capdevilla, F.; Mauricio, J.C. Differential analysis of proteins involved in ester metabolism in two *Saccharomyces cerevisiae* strains during the second fermentation in sparkling wine elaboration. *Microorganisms* **2020**, *8*, 403. [CrossRef]
19. Mauricio, J.C.; Moreno, J.J.; Ortega, J.M. *In vitro* specific activities of alcohol and aldehyde dehydrogenases from two flor yeast during controlled wine ageing. *J. Agric. Food Chem.* **1997**, *45*, 1967–1971. [CrossRef]
20. Porras-Agüera, J.A.; Moreno-García, J.; Mauricio, J.C.; Moreno, J.; García-Martínez, T. First proteomic approach to identify cell death biomarkers in wine yeasts during sparkling wine production. *Microorganisms* **2019**, *7*, 542. [CrossRef]
21. Ishihama, Y.; Oda, Y.; Tabata, T.; Sato, T.; Nagasu, T.; Rappsilber, J.; Mann, M. Exponentially modified protein abundance index (emPAI) for estimation of absolute protein amount in proteomics by the number of sequenced peptides per protein. *Mol. Cell. Proteomics* **2005**, *4*, 1265–1272. [CrossRef]
22. Moreno-García, J.; García-Martínez, T.; Moreno, J.; Mauricio, J.C. Proteins involved in flor yeast carbon metabolism under biofilm formation conditions. *Food Microbiol.* **2015**, *46*, 25–33. [CrossRef] [PubMed]
23. Porras-Agüera, J.A.; Moreno-García, J.; González-Jiménez, M.C.; Mauricio, J.C.; Moreno, J.; García-Martínez, T. Autophagic proteome in two *Saccharomyces cerevisiae* strains during second fermentation for sparkling wine elaboration. *Microorganisms* **2020**, *8*, 523. [CrossRef] [PubMed]
24. Seisonen, S.; Vene, K.; Koppel, K. The current practice in the application of chemometrics for correlation of sensory and gas chromatography data. *Food Chem.* **2016**, *210*, 530–540. [CrossRef] [PubMed]
25. Martínez-García, R.; García-Martínez, T.; Puig-Pujol, A.; Mauricio, J.C.; Moreno, J. Changes in sparkling wine aroma during the second fermentation under CO_2 pressure in sealed bottle. *Food Chem.* **2017**, *237*, 1030–1040. [CrossRef] [PubMed]
26. Valadi, H.; Valadi, Å.; Ansell, R.; Gustafsson, L.; Adler, L.; Norbeck, J.; Blomberg, A. NADH-reductive stress in *Saccharomyces cerevisiae* induces the expression of the minor isoform of glyceraldehyde-3-phosphate dehydrogenase (*TDH1*). *Curr. Genet.* **2004**, *45*, 90–95. [PubMed]
27. Esteve-Zarzoso, B.; Peris-Torán, M.J.; Garcıa-Maiquez, E.; Uruburu, F.; Querol, A. Yeast population dynamics during the fermentation and biological aging of sherry wines. *Appl. Environ. Microb.* **2001**, *67*, 2056–2061. [CrossRef]

28. Carlson, M.; Botstein, D. Two differentially regulated mRNAs with different 5′ ends encode secreted and intracellular forms of yeast invertase. *Cell* **1982**, *28*, 145–154. [CrossRef]
29. Byrne, S.; Howell, G. *Acetaldehyde: How to Limit its Formation during Fermentation*; Aust, N.Z., Ed.; Grapegrower Winemaker: Sydney, NSW, Australia, 2017; pp. 68–69.
30. Mojzita, D.; Hohmann, S. Pdc2 coordinates expression of the *THI* regulon in the yeast *Saccharomyces cerevisiae*. *Mol. Genet. Genom.* **2006**, *276*, 147–161. [CrossRef]
31. Moreno-García, J.; García-Martínez, T.; Millán, M.C.; Mauricio, J.C.; Moreno, J. Proteins involved in wine aroma compounds metabolism by a *Saccharomyces cerevisiae* flor-velum yeast strain grown in two conditions. *Food Microbiol.* **2015**, *51*, 1–9. [CrossRef]
32. Ugliano, M.; Henschke, P.A. Yeasts and Wine Flavour. In *Wine Chemistry and Biochemistry*; Springer: Berlin/Heidelberg, Germany, 2009; pp. 313–392.
33. Remize, F.; Andrieu, E.; Dequin, S. Engineering of the pyruvate dehydrogenase bypass in *Saccharomyces cerevisiae: Role* of the cytosolic Mg^{2+} and mitochondrial K^{+} acetaldehyde dehydrogenases Ald6p and Ald4p in acetate formation during alcoholic fermentation. *Appl. Environ. Microbiol.* **2000**, *66*, 3151–3159. [CrossRef]
34. Pigeau, G.M.; Inglis, D.L. Response of wine yeast (*Saccharomyces cerevisiae*) aldehyde dehydrogenases to acetaldehyde stress during Icewine fermentation. *J. Appl. Microbiol.* **2007**, *103*, 1576–1586. [CrossRef] [PubMed]
35. González, V.F. Aroma, aromas y olores del vino. In *Análisis Sensorial y Cata de los Vinos de España*; Dialnet: Logroño, Spain, 2001; pp. 207–233.
36. Vilela-Moura, A.; Schuller, D.; Mendes-Faia, A.; Silva, R.D.; Chaves, S.R.; Sousa, M.J.; Corte-Real, M. The impact of acetate metabolism on yeast fermentative performance and wine quality: Reduction of volatile acidity of grape musts and wines. *Appl. Microbiol. Biotechnol.* **2011**, *89*, 271–280. [CrossRef] [PubMed]
37. Boubekeur, S.; Camougrand, N.; Bunoust, O.; Rigoulet, M.; Guérin, B. Participation of acetaldehyde dehydrogenases in ethanol and pyruvate metabolism of the yeast *Saccharomyces cerevisiae*. *Eur. J. Biochem.* **2001**, *268*, 5057–5065. [CrossRef] [PubMed]
38. Verduyn, C.; Postma, E.; Scheffers, W.A.; Van Dijken, J.P. Physiology of *Saccharomyces cerevisiae* in anaerobic glucose-limited chemostat cultures. *J. Gen. Microbiol.* **1990**, *136*, 395–403. [CrossRef] [PubMed]
39. Romano, P.; Suzzi, G. Origin and Production of Acetoin during Wine Yeast Fermentation. *Appl. Environ. Microbiol.* **1996**, *62*, 309–315. [CrossRef]
40. Michnick, S.; Roustan, J.L.; Remize, F.; Barre, P.; Dequin, S. Modulation of glycerol and ethanol yields during alcoholic fermentation in *Saccharomyces cerevisiae* strains overexpressed or disrupted for *GPD1* encoding glycerol 3-phosphate dehydrogenase. *Yeast* **1997**, *13*, 783–793. [CrossRef]
41. Romano, P.; Brandolini, V.; Ansaloni, C.; Menziani, E. The production of 2,3-butanediol as a differentiating character in wine yeasts. *World J. Microbiol. Biotechnol.* **1998**, *14*, 649–653. [CrossRef]
42. Borrull, A.; Poblet, M.; Rozès, N. New insights into the capacity of commercial wine yeasts to grow on sparkling wine media. Factor screening for improving wine yeast selection. *Food Microbiol.* **2015**, *48*, 41–48. [CrossRef]
43. Kemp, B.; Alexandre, H.; Robillard, B.; Marchal, R. Effect of production phase on bottle-fermented sparkling wine quality. *J. Agric. Food Chem.* **2015**, *63*, 19–38. [CrossRef]
44. Goold, H.D.; Kroukamp, H.; Williams, T.C.; Paulsen, I.T.; Varela, C.; Pretorius, I.S. Yeast's balancing act between ethanol and glycerol production in low-alcohol wines. *Microb. Biotechnol.* **2017**, *10*, 264–278. [CrossRef]
45. Hohmann, S. An integrated view on a eukaryotic osmoregulation system. *Curr. Genet.* **2015**, *61*, 373–382. [CrossRef] [PubMed]
46. García-Mauricio, J.C.; García-Martínez, T. Role of Yeasts in Sweet Wines. In *Book Sweet, Reinforced and Fortified Wines: Grape Biochemistry, Technology and Vinification*; John Wiley & Sons: Hoboken, NJ, USA, 2013.
47. Remize, F.; Cambon, B.; Barnavon, L.; Dequin, S. Glycerol formation during wine fermentation is mainly linked to Gpd1p and is only partially controlled by the HOG pathway. *Yeast* **2003**, *20*, 1243–1253. [CrossRef] [PubMed]
48. Nissen, T.L.; Kielland Brandt, M.C.; Nielsen, J.; Villadsen, J. Optimization of ethanol production in *Saccharomyces cerevisiae* by metabolic engineering of the ammonium assimilation. *Metab. Eng.* **2000**, *2*, 69–77. [CrossRef] [PubMed]

microorganisms

Article

Heterologous Production of β-Caryophyllene and Evaluation of Its Activity against Plant Pathogenic Fungi

Fabienne Hilgers [1,†], Samer S. Habash [2,†], Anita Loeschcke [1], Yannic Sebastian Ackermann [1], Stefan Neumann [2], Achim Heck [3], Oliver Klaus [1], Jennifer Hage-Hülsmann [1], Florian M. W. Grundler [2], Karl-Erich Jaeger [1,3], A. Sylvia S. Schleker [2,*] and Thomas Drepper [1,*]

1 Institute of Molecular Enzyme Technology, Heinrich-Heine-University Düsseldorf, Forschungszentrum Jülich, Wilhelm-Johnen-Straße, 52428 Jülich, Germany; f.hilgers@fz-juelich.de (F.H.); a.loeschcke@fz-juelich.de (A.L.); y.ackermann@fz-juelich.de (Y.S.A.); o.klaus@fz-juelich.de (O.K.); j.hage-huelsmann@fz-juelich.de (J.H.-H.); karl-erich.jaeger@fz-juelich.de (K.-E.J.)
2 INRES—Molecular Phytomedicine, University of Bonn, Karlrobert-Kreiten-Str. 13, 53115 Bonn, Germany; samer@uni-bonn.de (S.S.H.); sneuman1@uni-bonn.de (S.N.); grundler@uni-bonn.de (F.M.W.G.)
3 Institute of Bio- and Geosciences (IBG-1: Biotechnology) Forschungszentrum Jülich, Wilhelm-Johnen-Straße, 52428 Jülich, Germany; a.heck@fz-juelich.de
* Correspondence: sylvia.schleker@uni-bonn.de (A.S.S.S.); t.drepper@fz-juelich.de (T.D.)
† These authors contributed equally to this work.

Abstract: Terpenoids constitute one of the largest and most diverse groups within the class of secondary metabolites, comprising over 80,000 compounds. They not only exhibit important functions in plant physiology but also have commercial potential in the biotechnological, pharmaceutical, and agricultural sectors due to their promising properties, including various bioactivities against pathogens, inflammations, and cancer. In this work, we therefore aimed to implement the plant sesquiterpenoid pathway leading to β-caryophyllene in the heterologous host *Rhodobacter capsulatus* and achieved a maximum production of 139 ± 31 mg L^{-1} culture. As this sesquiterpene offers various beneficial anti-phytopathogenic activities, we evaluated the bioactivity of β-caryophyllene and its oxygenated derivative β-caryophyllene oxide against different phytopathogenic fungi. Here, both compounds significantly inhibited the growth of *Sclerotinia sclerotiorum* and *Fusarium oxysporum* by up to 40%, while growth of *Alternaria brassicicola* was only slightly affected, and *Phoma lingam* and *Rhizoctonia solani* were unaffected. At the same time, the compounds showed a promising low inhibitory profile for a variety of plant growth-promoting bacteria at suitable compound concentrations. Our observations thus give a first indication that β-caryophyllene and β-caryophyllene oxide are promising natural agents, which might be applicable for the management of certain plant pathogenic fungi in agricultural crop production.

Keywords: terpenoids; sesquiterpene production; *Rhodobacter capsulatus*; β-caryophyllene; bioactivity; phytopathogens; plant pathogenic fungi; plant growth-promoting bacteria

Citation: Hilgers, F.; Habash, S.S.; Loeschcke, A.; Ackermann, Y.S.; Neumann, S.; Heck, A.; Klaus, O.; Hage-Hülsmann, J.; Grundler, F.M.W.; Jaeger, K.-E.; et al. Heterologous Production of β-Caryophyllene and Evaluation of Its Activity against Plant Pathogenic Fungi. *Microorganisms* **2021**, *9*, 168. https://doi.org/10.3390/microorganisms9010168

Received: 10 December 2020
Accepted: 10 January 2021
Published: 14 January 2021

1. Introduction

Among secondary metabolites, terpenoids including the class of sesquiterpenoids represent one of the largest and most diverse groups with over 80,000 known compounds, mostly isolated from plants [1–4]. Based on their number of carbon atoms, they can be divided into the subclasses of hemi- (C_5), mono- (C_{10}), sesqui- (C_{15}), di- (C_{20}), tri- (C_{30}), tetra- (C_{40}) and polyterpenes (>C_{40}) [5,6]. In general, the terpenoid synthesis starts from the two isoprene intermediates isopentenyl pyrophosphate (IPP) and dimethylallyl pyrophosphate (DMAPP), which are provided either by the mevalonate (MVA) pathway or by the 1-deoxy-D-xylulose 5-phosphate (DXP) pathway, also known as the 2-C-methyl-D-erythritol 4-phosphate (MEP) pathway. While the MVA pathway uses acetyl-Coenzyme A (acetyl-CoA) as a substrate and is predominantly found in eukaryotes (e.g., mammals, plants, and fungi), archaea and a few bacteria [7], the DXP pathway starts from glyceraldehyde-3-phosphate (GAP) and pyruvate and primarily occurs in bacteria, cyanobacteria, and green algae [8]. Starting from IPP and DMAPP, the elongation of linear prenyl pyrophosphates is catalyzed by prenyltransferases via head-to-tail condensations and results in C_{10} geranyl pyrophosphate (GPP), C_{15} farnesyl pyrophosphate (FPP), and C_{20} geranylgeranyl pyrophosphate (GGPP). Finally, GPP is used as a precursor molecule for the synthesis of monoterpenoids, FPP for sesqui- and triterpenoid production, and GGPP for di- and tetraterpenoid biosynthesis. Terpenes exhibit manifold functions in plant physiology and development, including photoprotection (carotenoids), communication (e.g., pinene), or repellant activity against predators and parasites (e.g., verbenone, β-caryophyllene) [9–11]. Furthermore, terpenes are of commercial interest for the pharmaceutical sector due to their various bioactivities suitable for the treatment of pathogen infections, inflammation, or cancer [12,13]. For example, the sesquiterpene farnesol shows inhibitory effects against antibiotic-resistant *Staphylococci*, not only inhibiting the growth of planktonic cells in free suspension but also suppressing biofilm formation of *Staphylococcus aureus*, *Staphylococcus epidermidis*, and *Burkholderia pseudomallei* [14–17]. In the past, these compounds were exclusively obtained from essential oils of natural plant sources, requiring complex and time-consuming downstream processing. β-caryophyllene, for example, was extracted from *Cannabis sativa* [18], clove basil, *Ocimum gratissimum* [19], or representatives of the plant genus *Cordia*, such as *Cordia verbenaceae* [20]. However, the application of microorganisms as heterologous hosts allows the establishment of alternative, cost-effective, and sustainable biotechnological production processes [21–26]. As the efficiency of such processes strongly depends on the achieved production titers, metabolic engineering of the applied hosts together with the optimization of the respective secondary metabolite pathways has gained more attention in the recent past [1,27–31]. So far, terpenoids were mostly produced in the heterologous hosts *Escherichia coli* and *Saccharomyces cerevisiae* [32–36]. However, in recent studies, the terpene production in less common microbes such as phototrophs has also been established and optimized, as for example documented by the *Rhodobacter*-based production of β-farnesene, nootkatone, valencene, and amorphadiene [23,37–40], or the production of various terpenes in cyanobacteria [41].

The phototrophic non-sulfur α-proteobacteria of the genus *Rhodobacter* feature some unique physiological properties, making them interesting microbial hosts for heterologous terpene production: (i) the cell membrane is commonly considered to be a critical determinant in terpenoid production since it can function as a storage compartment for the involved enzymes and metabolites [42,43]. In this context, *Rhodobacter* seems to be particularly suited for terpene production since the bacterium can form an extended intracytoplasmic membrane system (ICM), thereby providing a naturally enlarged reservoir for membrane-bound enzymes and terpenes [44,45]. (ii) As these phototrophic bacteria produce the carotenoids spheroidene and spheroidenone using the DXP pathway [46,47], they further offer a robust and effective isoprenoid metabolism that can be engineered for efficient terpenoid production. (iii) *Rhodobacter* species enable photo(hetero)tropic growth in low-cost minimal media at relatively high growth rates, allowing the utilization of sunlight as an energy source for sustainable cultivation and production processes. Recent studies could demonstrate that engineering the isoprenoid precursor biosynthesis can lead to a strong increase of sesqui- and triterpenoid formation in *R. capsulatus* [39,48,49] and *R. sphaeroides* [38,40,50–52]. In particular, the co-expression of a terpene synthase with the FPP synthase IspA, and/or enzymes constituting the heterologous MVA pathway, resulted in enhanced production of the corresponding plant terpenoids.

A major problem in agricultural crop production is the large number of plant-damaging animals such as insects, mites, and nematodes or pathogens including viruses, bacteria, and fungi, some of which lead to high economic losses of around 60% globally [53]. One of the most widely distributed and destructive pathogens of plants causing white mold disease in more than 400 host plants all over the world is the fungus *Sclerotinia sclerotiorum* (Lib.) de Bary [54]. Another devastating example of fungal diseases is the plant vascular wilt caused by the *Fusarium* species [55]. Other fungal pathogens such as *Phoma lingam* [56,57], *Alternaria brassicicola* [58], and *Rhizoctonia solani* [59,60] also cause major yield reduction in important crops. To control these pathogens and due to the rapidly growing world population and the resulting increase of food consumption, farmers are using synthetic and biological substances as fertilizers, pesticides, or growth regulators side by side with the cultivation of resistant or tolerant plant varieties [61–64]. Each of these methods has its limitations, but so far, the use of pesticides is the most convenient and commonly used method. Nevertheless, these can have numerous severe side effects on the environment, including the soil [65]. The soil is inhabited by an enormous diversity of organisms that are important players in maintaining a functional ecosystem and that comprise microorganisms with beneficial properties for plant development and health. For that reason, effective and sustainable alternatives are needed. Firstly, plants, as a part of a complex ecosystem, can produce enormous amounts of secondary metabolites for their survival and maintenance. Phenolics and terpenes are examples of metabolites that are produced by plants and act as antimicrobial agents and feeding deterrents [66–72]. The presence of a wide range of terpenes encouraged their use as nature-inspired plant protection agents in agriculture or their use for drug development. One of the commonly stress-associated terpenes is the sesquiterpene β-caryophyllene [73–75]. As mentioned in the previous section, various studies showed that β-caryophyllene exhibits diverse biological activities against many organisms. From a plant protection perspective β-caryophyllene was reported to promote plant growth, to induce plant defense genes, to attract entomopathogenic nematodes, and to be active against certain plant pathogenic bacteria and fungi [76–80].

In this study, we therefore aimed to use the modular co-expression of DXP/MVA genes in combination with the strictly controlled P_{nif} promoter to reconstitute the pathway of the plant sesquiterpene β-caryophyllene in *R. capsulatus* and to optimize the production under phototrophic growth conditions. For heterologous sesquiterpene production, the β-caryophyllene synthase QHS1 from *Artemisia annua* was used. Since this terpene offers a variety of beneficial bioactivities, we further evaluated the potential use of β-caryophyllene and its oxygenated derivative β-caryophyllene oxide as nature-derived fungicides. To this end, the bioactivity of β-caryophyllene/oxide against both representative plant growth-promoting bacteria and phytopathogenic fungi was investigated.

2. Materials and Methods

2.1. Bacterial Strains and Cultivation Conditions

The *Escherichia coli* strain DH5α and strain S17-1 were used for cloning and conjugation of plasmid DNA [81,82]. *E. coli* cells were cultivated at 37 °C using LB agar plates or liquid medium (Luria/Miller, Carl Roth®, Karlsruhe, Germany), containing kanamycin (25 μg mL^{-1}) when appropriate. *R. capsulatus* SB1003 [83] and SB1003-MVA [39], encompassing the chromosomally located genes *mvaA, idi, hsc, mvk, pmk* and *mvd* (also designated as MVA gene cluster) from *Paracoccus zeaxanthinifaciens*, were used for plant terpene production. All *R. capsulatus* strains used in this study were either cultivated on PY agar plates [84] containing 2% (*w*/*v*) Select Agar (Thermo Fisher Scientific, Waltham, MA, USA) or in RCV liquid medium [85] at 30 °C. Both media were supplemented with rifampicin (25 μg mL^{-1}). For cultivation of the recombinant *Rhodobacter* strain SB1003-MVA, gentamicin (4 μg mL^{-1}) was further added to the medium. If not stated otherwise, photoheterotrophic cultivation was conducted under anaerobic conditions and permanent illumination with bulb light (2500 lx), as described previously [39]. All bacterial strains and plasmids used in this study are listed in Table S1 (Supplementary Materials). All strains for bioactivity and minimum inhibitory concentration (MIC) evaluation are listed in the respective results section.

2.2. Construction of Expression Vectors

The expression vectors used in this study are based on the pRhon5Hi-2 vector carrying the promoter of the *nifH* gene for heterologous gene expression [39]. The sequence of β-caryophyllene synthase QHS1 from *A. annua* (UniProt: Q8SA63) was used to generate an appropriate synthetic gene whose DNA sequence is suitable for the codon-usage of *R. capsulatus*. For DNA sequence adaptation, the Codon Optimization Tool by IDT Integrated DNA Technologies and the Graphical Codon Usage Analyzer tool were used [86]. The 1.7-kb *QHS1* gene was obtained from Eurofins Genomics. The synthetic DNA fragment was flanked by appropriate restriction endonuclease recognition sequences (*XbaI*/*Hind*III). The final sequence of the synthetic DNA fragment is shown in the Supplementary Materials. For the construction, the *XbaI*/*Hind*III hydrolyzed *QHS1* fragment was inserted into likewise hydrolyzed pRhon5Hi-2 as well as a variant, providing the additional isoprenoid biosynthetic gene *ispA*. Thereby, the expression vectors pRhon5Hi-2-QHS1 and pRhon5Hi-2-QHS1-ispA were constructed, carrying the terpene synthase gene immediately downstream of the P_{nif} promoter of the vector. Correct nucleotide sequences of all constructs were confirmed by Sanger sequencing (Eurofins Genomics, Ebersberg, Germany). The *QHS1* expression vectors are summarized in Table S1 (Supplementary Materials).

2.3. Cultivation of *R. capsulatus* for Heterologous Terpene Production

For the expression of the heterologous terpene biosynthetic genes, respective pRhon5Hi-2-based plasmids were transferred to cells of different *R. capsulatus* strains via conjugational transfer employing *E. coli* S17-1 as donor [84]. Thereafter, transconjugants were selected and further cultivated on PY agar containing kanamycin (25 µg mL^{-1}) and rifampicin (25 µg mL^{-1}). Subsequently, *Rhodobacter* cells were cultivated in airtight 4.5 mL screw neck vials (Macherey-Nagel, Düren, Germany) or airtight 15 mL hungate tubes [87] in liquid RCV medium containing kanamycin (25 µg mL^{-1}) and rifampicin (25 µg mL^{-1}). Precultures were cultivated in 15 mL RCV medium containing 0.1% $(NH_4)_2SO_4$ inoculated with cells from a freshly grown PY agar plate and incubated for 48 h at 30 °C and with bulb light illumination. Expression cultures were inoculated from precultures to an optical density at 660 nm of 0.05 in 4.5 mL or 15 mL RCV medium containing 0.1% serine as an exclusive nitrogen source. Subsequently, cells were incubated at 30 °C under permanent illumination with bulb light (3.6 mW cm^{-2} at 850 nm) or IR light (5.6 mW cm^{-2} at 850 nm) for 3–5 days. For microaerobic expression cultures, cells were cultivated in 20–60 mL RCV medium containing 0.1% serine in 100 mL flasks at 30 °C and 130 rpm in the dark. The absence of ammonium and the cultivation under oxygen-limited conditions led to the induction of the P_{nif}-dependent target gene expression. For the extraction of the produced sesquiterpenes, the cultures were overlaid with 150 µL or 500 µL *n*-dodecane, respectively, during inoculation [88].

2.4. Extraction, GC Analysis and Quantification of Sesquiterpenes

Basically, analysis of produced sesquiterpenes was conducted as described in Troost et al., 2019 [39]. In the following, the procedure is briefly described. To facilitate terpene extraction into the organic phase (*n*-dodecane) after cultivation, screw neck vials or hungate tubes were incubated in a horizontal position under permanent shaking (130 rpm, 30 °C, 24 h, in the dark) using a Multitron Standard incubation shaker (Infors HT). The *n*-dodecane samples were analyzed by gas chromatography (GC) using the Agilent *6890N* gas chromatograph equipped with a (5%-phenyl)-methylpolysiloxane *HP-5* column (length, 30 m; inside diameter, 0.32 mm; film thickness, 0.25 µm; Agilent Technologies) and a flame ionization detector (FID). The temperatures of the injector and FID were set to 240 and 300 °C, respectively. The GC was loaded with a 4-µL sample of each *n*-dodecane layer using a split ratio of 100:1 with helium as carrier gas. The following column temperatures were used during analysis: (i) 100 °C for 5 min, (ii) increased of temperature with a heating rate of 10 °C per min up to 180 °C, (iii) increased of temperature with a heating rate of 20 °C per min up to 300 °C. The signal of β-caryophyllene produced in *R. capsulatus* was verified by comparison of its retention times to a corresponding reference (β-caryophyllene from Sigma Aldrich, product number: 22075, retention time: 10.13 min). In order to determine the final product titers, the transfer efficiency from producing cells into the *n*-dodecane phase was determined as described in Supplementary Method section "Analysis of *n*-dodecane-mediated β-caryophyllene extraction from phototrophically grown *R. capsulatus*". In brief, accumulated terpenes were extracted from cell lysates using *n*-dodecane. Subsequently, products were quantified using calibration curves of the reference compound, taking into account the specific transfer efficiencies of β-caryophyllene.

2.5. Effect of β-Caryophyllene and β-Caryophyllene Oxide on Plant Pathogenic Fungi

Isolates of the plant pathogenic fungi *P. lingam*, *S. sclerotiorum* and *A. brassicicola* were obtained from the Leibniz-Institut DSMZ (Deutsche Sammlung von Mikroorganismen und Zellkulturen GmbH, Braunschweig, Germany), while isolates of *F. oxysporum*, *R. solani* were obtained from the INRES, Plant Diseases and Plant Protection, University of Bonn. All isolates were sub-cultured on potato dextrose agar (PDA) at 24 °C and were used in this study to evaluate the bioactivity of the compounds on hyphal growth.

To test the bioactivities of β-caryophyllene and β-caryophyllene oxide, compounds were dissolved in a mixture of DMSO and Tween 20 (ratio of 1:2) to prepare differently concentrated stock solutions. These were mixed with PDA to gain the final concentrations 62.5, 125, and 250 μg mL^{-1} and to prepare PDA agar plates with terpenoids. The final DMSO and Tween 20 concentrations were always 1% (v/v) and 0.5% (v/v), respectively. Fungal discs with a diameter of 0.5 cm were cut from the culture media of freshly grown agar plates without terpenoids and placed upside down in the middle of PDA plates containing the chemicals. PDA plates with 0.5% (v/v) DMSO and 1% (v/v) Tween 20 alone were used as control. All plates were incubated for 7 days at 24 °C. Subsequently, the diameter of the fungal colony was measured, and the percentage of growth inhibition compared to the solvent control was calculated. Differences between the treatments were statistically analyzed using SigmaPlot software by one-way analysis of variance (ANOVA) and multiple comparisons for significance were performed at ($p < 0.05$) using the Holm-Sidak method.

2.6. Determination of the Minimum Inhibitory Concentration (MIC) of β-Caryophyllene and β-Caryophyllene Oxide in Liquid Cultures of Bacteria

The minimum inhibitory concentration of β-caryophyllene and β-caryophyllene was determined according to reference [89]. For the precultures, 10 mL Müller Hinton (MH) medium (Merck, Germany) was first inoculated in 100 mL flasks with four single bacterial colonies. For *R. capsulatus*, RCV was used. The liquid cultures were incubated for 18 h at 130 rpm and 37 °C (*R. capsulatus* at 30 °C). For the main cultures, MH or RCV medium was supplemented with differently concentrated stock solutions of β-caryophyllene and β-caryophyllene oxide in a mixture of DMSO and Tween 20 (ratio 1:2) to gain final concentrations of 62.5, 125, and 250 μg mL^{-1}. All bacterial cultures were adjusted to a cell density corresponding to an optical density at 625 nm of 0.1 and then diluted 50-fold with medium for *B. subtilis*, *P. putida*, *P. fluorescens*, *R. rhizogenes* and *P. polymyxa* and 2-fold for *R. capsulatus*. For the inoculation of 96-well microtiter plates (Greiner Bio-One GmbH, Frickenhausen, Germany), 50 μL MH medium with the corresponding concentration of the substance to be tested and the solvent controls were mixed with 50 μL of previously diluted bacterial culture, resulting in an end optical density at 625 nm of 0.001 and 0.025, respectively. The solvent control contained 1% (v/v) Tween 20 and 0.5% (v/v) DMSO. After inoculation, the microtiter plates (MTPs) were first shaken in a SpectraMax i3x (Molecular Devices, San Jose, CA, USA) plate photometer for 20 s to mix the solution and then incubated for 20 h at 37 °C. *R. capsulatus* was incubated at 30 °C and 300 rpm. For subsequent determination of the MICs, the optical density of cell cultures was determined at 625 nm in a plate photometer. The MIC was defined based on the European Committee on Antimicrobial Susceptibility Testing (EUCAST) guidelines as the compound concentration at which an optical density at 625 nm minus the background absorbance equals 0 [90].

3. Results

In the past, terpenoids were exclusively obtained from natural plant sources, e.g., by extracting them from essential oils, requiring a complex and time-consuming downstream processing. The heterologous production of sesquiterpenes in a suitable microbial host, however, bears many benefits. For example, it offers the possibility to solely produce a desired compound so that it can be rather easily purified without the need of removing closely related constituents [21,34]. Thus, we here aimed to reconstitute the plant sesquiterpene pathway of β-caryophyllene in *R. capsulatus* and optimize the production under phototrophic growth conditions. Since many sesquiterpenoids exhibit promising antimicrobial activities, the antifungal efficacy of β-caryophyllene and its oxidized form against phytopathogenic fungi were evaluated.

3.1. Establishment of β-Caryophyllene Production in R. capsulatus via Overexpression of Isoprenoid Precursor Genes

Recently, we described the heterologous synthesis of the plant sesquiterpenoids valencene and patchoulol in the phototrophic bacterium *R. capsulatus* and its modular improvement by engineering the biosynthesis of the central precursor FPP [39]. To evaluate if *R. capsulatus* and the modular engineering principle can analogously be applied for the synthesis of the plant-derived sesquiterpene β-caryophyllene, we expressed the gene encoding β-caryophyllene synthase QHS1 from *A. annua* in the bacterial host. To this end, the expression vectors pRhon5Hi-2-QHS1 and pRhon5Hi-2-QHS1-ispA, carrying an additional copy of the intrinsic FPP synthase gene *ispA*, were transferred to the *R. capsulatus* wild type strain SB1003 and strain SB1003-MVA. The latter strain additionally contains the chromosomally integrated MVA pathway genes derived from *Paracoccus zeaxanthinifaciens* and thus offers a second isoprenoid biosynthesis pathway. To compare the β-caryophyllene production in all *Rhodobacter* strains grown under phototrophic conditions, cells were incubated in the absence of molecular oxygen and ammonium under constant bulb light illumination. Terpene accumulation was determined in the late stationary growth phase by analyzing *n*-dodecane samples via GC-FID measurements. The increase of β-caryophyllene production in tested *R. capsulatus* strains is shown in Figure 1 as relative values using *R. capsulatus* SB1003 solely carrying the plasmid-encoded *QHS1* gene as reference strain.

As shown in Figure 1, the expression of the β-caryophyllene synthase gene *QHS1* in *R. capsulatus* strain SB1003 led to a measurable production of β-caryophyllene. Remarkably, the co-expression of *QHS1* and *ispA* in the *R. capsulatus* strain SB1003 as well as *QHS1* expression in the engineered SB1003-MVA strain did not result in increased β-caryophyllene synthesis. However, concerted expression of *QHS1* and *ispA* in *R. capsulatus* SB1003-MVA led to a considerable increase of sesquiterpenoid production of about 300% in comparison to the reference strain.

3.2. Optimization of β-Caryophyllene Production in R. capsulatus via Modification of Cultivation Conditions

In the above-described experiments, we could demonstrate that modular engineering of the isoprenoid biosynthesis can also be applied to improve β-caryophyllene production in *R. capsulatus*. Next, we analyzed whether the modification of cultivation conditions including a prolonged cultivation time or the change of illumination parameters can further improve the product yield in the better-performing strain SB1003-MVA. First, β-caryophyllene accumulation was comparatively analyzed over five days in photoheterotrophically-grown cultures of *R. capsulatus* strain SB1003-MVA carrying pRhon5Hi-2-QHS1 or pRhon5Hi-2-QHS1-ispA. Product formation was determined by analyzing the overlaid *n*-dodecane samples via GC-FID measurements (Figure 2, blue bars).

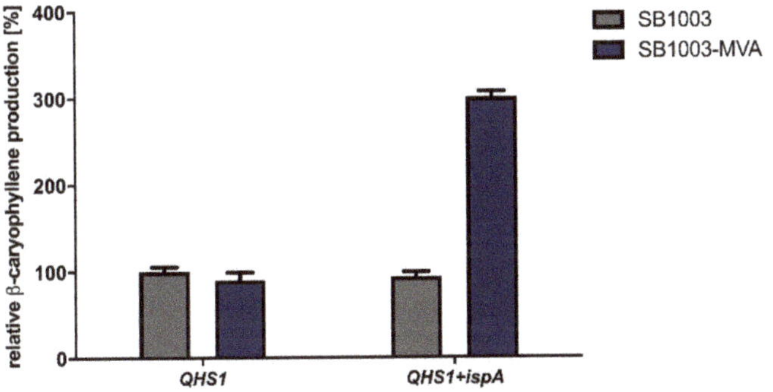

Figure 1. Heterologous β-caryophyllene production in the *R. capsulatus* strains SB1003 and SB1003-MVA. The β-caryophyllene synthase gene *QHS1* from *A. annua* was expressed in *R. capsulatus* SB1003 wild type (grey bars) and SB1003-MVA (blue bars), which additionally carries the MVA gene cluster from *P. zeaxanthinifaciens* to enable a second isoprenoid biosynthesis route. Moreover, the *ispA* gene encoding the *R. capsulatus* FPP synthase was co-expressed on the same plasmid to further enhance sesquiterpene production titers. Product formation was determined in cell cultures after three days of photoheterotrophic cultivation (gas-tight hungate tubes, 30 °C) under ammonium depletion and constant illumination with bulb light (3.6 mW cm^{-2} at 850 nm). The produced β-caryophyllene was sampled in overlaid *n*-dodecane phases for GC-FID analysis. The increase of β-caryophyllene production in engineered *R. capsulatus* strains is shown as relative values. To this end, the *R. capsulatus* SB1003 carrying the plasmid-encoded *QHS1* gene was used as a reference strain. Values are means of three independent biological replicates (n = 3) and error bars indicate the respective standard deviations.

The highest product levels could be detected after three days of cultivation, where cells have typically reached the beginning of the stationary growth phase. The elongation of the cultivation time did not show increased product accumulation so that all further production experiments were carried out for three days. Under standard phototrophic conditions, conventional light bulbs are used for the illumination of *R. capsulatus* cells [39,49]. This conventional light source offers a broad emission spectrum with a relatively high proportion in the infrared (IR) light range (>750 nm; 3.6 mW cm^{-2} at 850 nm) suitable for excitation of bacteriochlorophyll *a* (BChl *a*) exhibiting excitation maxima at 800 and 860 nm (Figure S6, Supplementary Materials, Reference [91]). To improve the illumination conditions for sesquiterpene production under phototrophic conditions, we subsequently analyzed if the use of (i) alternative cultivation vessels offering a better light penetration of cell cultures by a more favorable surface-area-to-volume ratio (Table S2, Supplementary Materials) or (ii) a customized IR-LED array (5.6 mW cm^{-2} at 850 nm) suitable for specific excitation of the photopigment BChl *a* with high light intensities can help to increase product formation.

To investigate the influence of illumination conditions on the heterologous production of β-caryophyllene, the strain SB1003-MVA carrying the expression vector pRhon5Hi-2-QHS1-ispA was cultivated over three days under photoheterotrophic conditions and constant illumination with bulb light or IR light in an ammonium-depleted medium in screw neck vials. As shown in Figure 2, the change of cultivation vessel geometry resulted only in a slight increase of β-caryophyllene production of *R. capsulatus* SB1003-MVA (pRhon5Hi-2-QHS1-ispA), whereas high irradiation with IR light led to a 1.9-fold increase of the final product accumulation. These results indicate that the applied illumination conditions should be taken into account to reach high product yields when *R. capsulatus* is used as an alternative terpene production host. This assumption is further supported by the observation that product levels were much lower in *R. capsulatus* SB1003-MVA (pRhon5Hi-2-QHS1-ispA) cultures that have been grown under non-phototrophic, i.e., microaerobic conditions (Figure 2, green bars). For non-phototrophic cultivation, *R. capsulatus* SB1003-MVA (pRhon5Hi-2-QHS1-ispA) was grown in 100 mL, unbaffled shake flasks containing different volumes of medium in the dark to implement different aeration conditions (green bars). Those conditions could lead to the formation of β-caryophyllene oxide, the oxygenated derivative of β-caryophyllene. As previously described, a filling volume of 60 mL is most suitable for the induction of intrinsic terpene formation and P_{nif}-mediated target gene expression in *R. capsulatus* [39], which is corroborated by the observed β-caryophyllene production levels. Nevertheless, only a quarter of the product yield could be achieved under microaerobic, non-phototrophic growth conditions when compared to the corresponding values of phototrophically grown cells (*R. capsulatus* SB1003-MVA, pRhon5Hi-2-QHS1-ispA, 3 days, bulb light), and only traces of the oxygenated derivative were detectable (data not shown). However, to fully understand the effects of varying cultivation conditions on β-caryophyllene production, further experiments have to be performed in future studies.

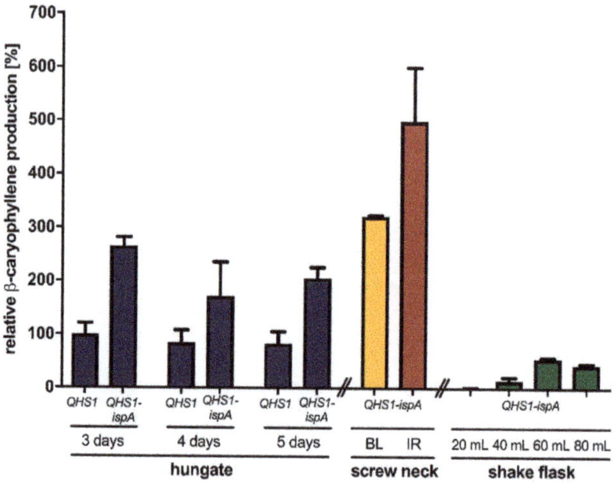

Figure 2. Heterologous β-caryophyllene production in the *R. capsulatus* strain SB1003-MVA with dependence on the cultivation time and illumination conditions. The β-caryophyllene accumulation was determined in *R. capsulatus QHS1* expression strains SB1003-MVA (pRhon5Hi-2-QHS1) and SB1003-MVA (pRhon5Hi-2-QHS1-ispA). First, product formation was determined in cell cultures after three, four, and five days of photoheterotrophic cultivation in 15 mL hungate tubes using standard illumination conditions (bulb lights, 3.6 mW cm^{-2} at 850 nm) and RCV medium supplemented with 0.1% serine. Blue bars represent the results of this experiment. Second, illumination conditions were changed by cultivating *R. capsulatus* strain SB1003-MVA (pRhon5Hi-2-QHS1-ispA) for three days under photoheterotrophic conditions using either constant illumination with bulb lights (BL; 3.6 mW cm^{-2} at 850 nm, yellow bar) or IR-emitting diodes (IR; 5.6 mW cm^{-2} at 850 nm, red bar). Here, 4.5-mL screw neck vials were used to improve light penetration due to a more favorable surface-area-to-volume ratio of this cultivation vessel. For non-phototrophic cultivation, the same strain was grown in 100-mL, unbaffled shake flasks containing different volumes of serine-supplemented RCV medium (shake flask, green bars). In all cultures, the produced β-caryophyllene was sampled in overlaid *n*-dodecane phases for GC-FID analysis. The increase of β-caryophyllene production is shown as relative values using *R. capsulatus* SB1003-MVA carrying the QHS1 expression plasmid pRhon5Hi-2-QHS1 as a reference strain. Values are the means of three independent biological replicates ($n = 3$) and the error bars indicate the respective standard deviations.

To accurately determine the final product titers, we analyzed (i) the individual transfer efficiency of β-caryophyllene from intact cells into the *n*-dodecane phase, (ii) the effect of the ICM, which is formed by *R. capsulatus* cells under phototrophic conditions, on sesquiterpenoid extraction, (iii) the differences in terpene transfer efficiencies when comparing single and repeated *n*-dodecane extraction, and finally (iv) the effect of the presence and absence of organic solvent on the final product titers (Supplementary Method section "Analysis of *n*-dodecane-mediated β-caryophyllene extraction from phototrophically grown *R. capsulatus*"). Finally, we were able to determine a product titer of 90 ± 19 mg L^{-1} β-caryophyllene for *R. capsulatus* SB1003-MVA with pRhon5Hi-2-QHS1-ispA after 3 days of cultivation in hungate tubes under bulb light. This titer could be further increased by using IR light and screw neck vials for cultivation, reaching a final product titer of 139 ± 31 mg L^{-1}. Based on these values and the reached cell densities, the respective productivities were further calculated (Table S3, Supplementary Materials).

In summary, we showed that *R. capsulatus* can efficiently synthesize the sesquiterpene β-caryophyllene. Furthermore, the modular adaptation of precursor gene expression under phototrophic growth conditions as well as the adjustment of cultivation conditions resulted in an increased sesquiterpenoid formation.

3.3. Evaluation of Bioactivities of β-Caryophyllene and β-Caryophyllene Oxide against Different Phytopathogenic Organisms

The agricultural industry is affected by a dwindling number of effective antimicrobial substances. On the other hand, farmers have to control plant pathogenic organisms without damaging non-target organisms. As β-caryophyllene and β-caryophyllene oxide offer a variety of beneficial bioactivities [70,79,92–95], we evaluated the potential use of those two sesquiterpenes as a nature-derived fungicide. To this end, we analyzed the activity against different phytopathogenic fungi, as well as various plant growth-promoting bacteria (PGPB).

3.3.1. Bioactivities of β-Caryophyllene and β-Caryophyllene Oxide against Phytopathogenic Fungi

We investigated the bioactivity of β-caryophyllene and β-caryophyllene oxide, which can be formed spontaneously by uncatalyzed processes [96,97], against the plant pathogenic fungi *S. sclerotiorum*, *F. oxysporum*, *A. brassicicola*, *P. lingam*, and *R. solani*. This analysis would additionally reveal whether the compound's oxygenation influences potential antifungal properties. Therefore, PDA agar plates were supplemented with increasing concentrations of both compounds, fungal discs were transferred onto these plates and fungal growth was determined. Evaluation revealed that the degree of growth inhibition due to direct terpene exposure varied depending on the compound and the fungus (Figure 3).

Both compounds inhibited the hyphal growth of *S. sclerotiorum* when compared to the solvent control. The inhibition reached up to 30% when the fungus was exposed to β-caryophyllene, while it was up to 40% when the fungus was cultivated on medium containing β-caryophyllene oxide. The effect against *F. oxysporum* was less pronounced. Around 20% inhibition was observed when the fungus was cultivated on the β-caryophyllene-supplemented medium, while it was around 30% in the case of β-caryophyllene oxide. Finally, the presence of β-caryophyllene in the growth medium slightly inhibited the growth of *A. brassicicola* while inhibition was higher and reached a maximum of 10% when β-caryophyllene oxide was used. No significant effect of both compounds was observed against *P. lingam* and *R. solani* (Figure S7, Supplementary Materials). Our results thus reveal that β-caryophyllene and its oxidized form possess antifungal activity against certain phytopathogenic fungi and that β-caryophyllene oxide tends to be more effective in inhibiting fungal growth.

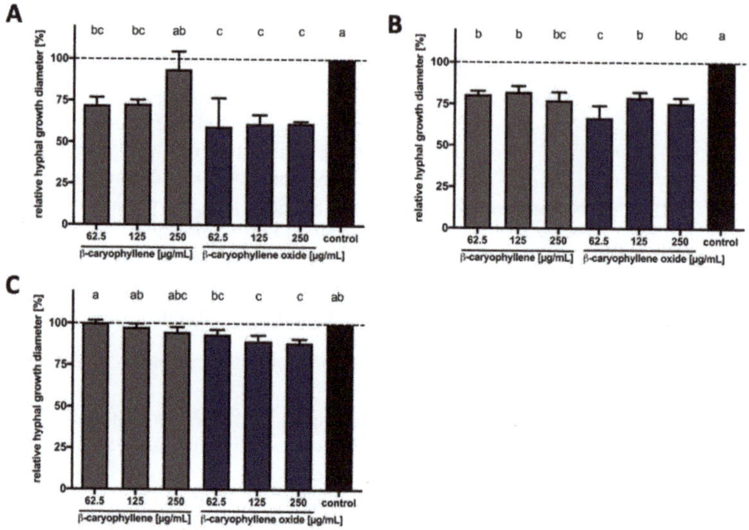

Figure 3. The effect of β-caryophyllene and β-caryophyllene oxide on the hyphal growth of plant pathogenic fungi. The effect of β-caryophyllene and β-caryophyllene oxide against *S. sclerotiorum* (**A**), *F. oxysporum* (**B**) and *A. brassicicola* (**C**). Final concentrations of 62.5 µg mL^{-1}, 125 µg mL^{-1}, and 250 µg mL^{-1} of β-caryophyllene (grey bars) and β-caryophyllene oxide (blue bars) in PDA growth medium were used. Medium mixed with the solvents DMSO and Tween 20 (final concentrations, 0.5% and 1% v/v, respectively) was used as the control (black bars). An equally sized disk with fungal mycelium was placed in the center of each plate and incubated for seven days at 24 °C. Subsequently, the diameter of each fungal colony was measured, and the relative growth compared to the solvent control was calculated. Each bar represents the mean ± standard deviation of three independent biological measurements with three technical replicates each (n = 9). Different letters on the top of the bars indicate significant differences between the treatments based on ANOVA and Holm-Sidak post-hoc method ($p < 0.05$), while the same letters represent no significant differences.

3.3.2. Antimicrobial Activities against Plant Growth-Promoting Bacteria

As the previous investigations showed antifungal properties against several phytopathogenic fungi, the use of the sesquiterpenoids as natural compound-based plant protection products could be considered. To investigate potential toxic off-target effects on bacteria that promote plant growth, we next examined whether the addition of β-caryophyllene/oxide affects the growth of bacteria at concentrations used in the hyphal growth assay. For this purpose, the growth of representatives of the plant growth-promoting bacteria (PGPB) group [98–100], including the two diazotrophic bacteria *Rhizobium rhizogenes* and *Rhodobacter capsulatus* [101,102], the two bacilli *Bacillus subtilis* [103,104] and *Paenibacillus polymyxa* [105], as well as the pseudomonads *Pseudomonas fluorescens* [104] and *Pseudomonas putida* [106] was analyzed in presence of β-caryophyllene and β-caryophyllene oxide. Both compounds were added to diluted bacterial cultures in increasing concentrations. After overnight incubation, the MICs were determined according to the respective optical density of the cell cultures (Figure 4).

The bacteria *R. rhizogenes*, *R. capsulatus*, *B. subtilis*, and *P. polymyxa* did not show reduced cell growth in comparison to the solvent control upon the addition of the two terpenes β-caryophyllene and β-caryophyllene oxide (Figure 4A–D). These bacteria showed an increase in growth, which could be explained by the metabolization of the terpenes. For *P. putida*, no effect of β-caryophyllene oxide was detected compared to the solvent control (Figure 4F). β-caryophyllene showed an influence on *P. putida*, which was concentration independent since all tested concentrations led to comparable cell growth. The cell densities were about 40% lower compared to the solvent control. This effect was also observed for *P. fluorescens*, where the two terpenes reduced growth by up to 40% (Figure 4E).

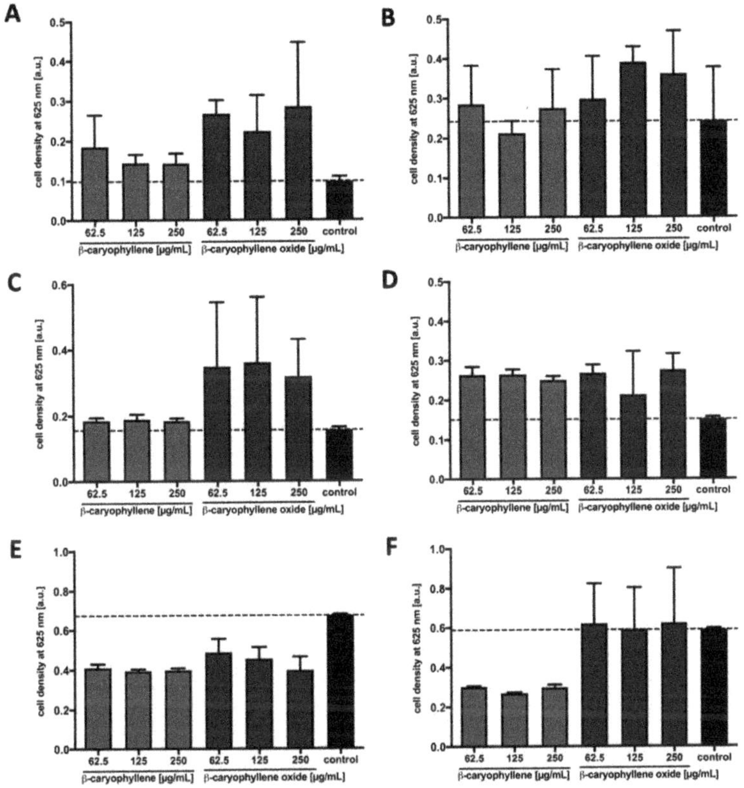

Figure 4. The influence of β-caryophyllene and β-caryophyllene oxide on the growth of plant growth-promoting bacteria. Final concentrations of 62.5 µg mL^{-1}, 125 µg mL^{-1} and 250 µg mL^{-1} of β-caryophyllene (grey bars) and β-caryophyllene oxide (blue bars) were added to cultures of *R. rhizogenes* (**A**), *R. capsulatus* (**B**), *B. subtilis* (**C**), *P. polymyxa* (**D**), *P. fluorescens* (**E**) and *P. putida* (**F**) in 100 µL MH medium (*R. capsulatus* in RCV medium) in MTPs. The final solvent concentration was 1% (*v*/*v*) Tween 20 and 0.5% (*v*/*v*) DMSO. To determine the influence of the terpenes on the growth of the bacteria, the cells were incubated stationary for 20 h at 37 °C (*R. capsulatus* at 30 °C) and the cell density was measured at 625 nm using a plate photometer. The solvent control (control, black bars) was MH or RCV medium containing 1% (*v*/*v*) Tween 20 and 0.5% (*v*/*v*) DMSO. Values are means of three independent biological replicates (n = 3) and error bars indicate the respective standard deviations.

In summary, for β-caryophyllene and β-caryophyllene oxide, no MIC could be determined for any of the tested PGPB, but a reduction of cell growth could be observed for both Pseudomonads. As a diverse group of different representative soil bacteria was tested, the results nevertheless indicate that β-caryophyllene and β-caryophyllene oxide do not exhibit strong broad-spectrum antibacterial activities at concentrations which considerably inhibit the hyphal growth of *S. sclerotiorum* and *F. oxysporum* (63 µg mL^{-1}).

4. Discussion

The management of plant pathogens in the process of crop production is crucial, no matter whether organic, integrated, or conventional farming practices are applied.

For many decades, synthetic pesticides were considered the fastest and most effective pest and pathogen control method. Recently, due to the rise of public health concerns about pesticide toxicity and harm to the environment, many of these effective chemicals were banned, thereby markedly limiting the options for plant protection. Therefore, it is important to find environmentally safe and sustainable natural products to control pathogens and thus ensure yield and food quality. In this context, plant metabolites are a rich source of bioactive compounds explorable for the use of preventing and controlling plant pathogenic microbes. In the last decade, several studies investigated terpenoids as potential antiphytopathogenic compounds [67,71,76,77,107,108]. β-caryophyllene is a natural bicyclic sesquiterpene that is a constituent of many essential oils. Many studies showed that these essential oils, which are containing β-caryophyllene as one of the main ingredients, are active against plant pathogens [109–111]. For example, methanol extracts from *Artemisia annua* leaves, one of the common β-caryophyllene producers, strongly inhibit the growth of the plant pathogenic fungi *F. oxysporum* and *Fusarium solani* [79]. In another study, the essential oil from *Murraya paniculata* leaves showed inhibitory activities on the mycelial growth of *S. sclerotiorum*, a fungus that poses a high risk to several crops. The gas chromatography analysis of the essential oil composition introduced β-caryophyllene as one of the main constituents (23.8%) [109]. Furthermore, essential oils from *Piper aduncum*, which also has β-caryophyllene as one of its main constituents (7.2%), inhibited the mycelial growth of the fungus *S. sclerotiorum* [110]. As a second alternative or complementary means for plant protection, there is also a multitude of important and useful microorganisms that support plant growth, which are called plant growth-promoting bacteria and plant growth-promoting fungi (PGPF). To offset the negative effects of chemical substances or make their use superfluous, more and more PGPB are now being used in agriculture [112]. Microorganisms can fulfill different functions in this process. *Bacillus subtilis*, for example, accumulates at the root system during the germination of various plants and prevents competing harmful fungi from spreading [103]. Diazotrophic organisms can supply plants with biologically available nitrogen by fixing atmospheric dinitrogen, thus making it available to the plants [113]. When fighting phytopathogens, it is important to consider and ideally avoid negative off-target effects on the above-mentioned beneficial microorganisms. Corresponding tests are therefore now frequently included in the first evaluation of antimicrobial activities.

So far, no studies were investigating the effect of pure β-caryophyllene and β-caryophyllene oxide against a selection of phytopathogenic fungi aiming to determine and compare the potential antifungal properties of the two compounds and species-specific differences in sensitivity. In our current study, we show the potential of sustainable production of β-caryophyllene in the heterologous host *Rhodobacter capsulatus* and the species-specific promoting or inhibitory effects for selected plant growth-promoting bacteria for both of the tested sesquiterpenoids at appropriate compound concentrations. Furthermore, we tested the bioactivities of both β-caryophyllene and β-caryophyllene oxide against several plant pathogenic fungi and showed that both substances were active against certain fungi. Interestingly, the oxidized form tended to be even more effective, and additionally has a more beneficial activity profile concerning the PGPB. These results are supported by previous reports which are introducing β-caryophyllene as a bioactive compound in its purified form [79] and as a component of several essential oils [109,110]. The purified β-caryophyllene showed a MIC of 130 $\mu g\ mL^{-1}$ for *F. oxysporum* [79], which is below the maximal concentration tested in this study. However, our plate-based approach is not completely comparable with the method used for MIC determination in liquid medium. According to our results, the inhibitory effect of the hyphal growth was different depending on the tested fungus. Such a difference is dependent on the fungal species and frequently described by previous studies showing that the novel fungicide 3-[5-(4-chlorophenyl)-2,3-dimethyl-3-isoxazolidinyl] pyridine (SYP-Z048) affected several pathogenic fungi in different ways [114]. Overall, our current results demonstrate that both β-caryophyllene and β-caryophyllene oxide exhibit bioactivity against plant pathogenic fungi and therefore could be suitable as potential fungicides in agriculture as, in contrast to many broad-spectrum pesticides, they do not harm many species of plant growth-promoting bacteria. However, despite the sesquiterpenes being natural compounds, which are often associated with non-harmful ecotoxicological profiles, effects against *Pseudomonas* species were corroborated and will need to be taken into account. So far, there is only limited information about the individual activities of terpenes against plant pathogens and the underlying molecular mechanisms. To be able to explain the differences we observed in the activity of the two terpenes against the different organisms and to get more data on the activity spectrum, our investigations need to be extended by including more target and non-target organisms. In addition, the respective modes of action on the molecular level have to be determined. Besides additional plant pathogens, this not only includes analyzing further plant growth-promoting bacteria, but it must be tested if plant growth-promoting fungi react sensitively to the terpenes, as indicated by a previous promising study [80]. In particular, fungi of the genus *Trichoderma*, which are said to have many advantageous properties for plants, could be investigated more closely [115,116].

To be able to provide appropriate quantities of an active antifungal substance, the heterologous production of promising sesquiterpenes in a suitable microbial host bears various benefits, such as the possibility to solely produce the desired compound without complex downstream processing and in high amounts. Therefore, we established the biosynthesis of the plant sesquiterpene β-caryophyllene in the heterologous production host *R. capsulatus* under phototrophic and non-phototrophic conditions. For this purpose, the intrinsic isoprenoid biosynthesis pathway was optimized in terms of its precursor supply. In particular, the P_{nif}-based co-expression of *ispA* and the genetically integrated MVA pathway resulted in a substantial increase in sesquiterpenoid production of around 300%. These results are in agreement with previous studies, where engineering of isoprenoid precursor supply was a valuable tool to increase the terpenoid production in *Rhodobacter* [38,40,48] and other bacterial hosts [23,117–125]. Also, we were able to increase the terpene production level further by changing the cultivation conditions from bulb light in a 14 mL hungate tube to IR light in a 4.5 mL screw neck vial, achieving a final β-caryophyllene titer of 139.29 ± 31.35 mg L^{-1} and a specific productivity of 1.30 ± 0.32 mg g^{-1} dry cells h^{-1}. In recent studies, production titers around 220 mg L^{-1} [126] and specific productivities of 1.15 mg g^{-1} dry cells h^{-1} [127] were achieved in *E. coli*. Thus, we attained yields comparable to the current literature and successfully established *R. capsulatus* as a heterologous host for the production of β-caryophyllene. Furthermore, the β-caryophyllene yields achieved in *R. capsulatus* could be sufficient to use this host as a microbial system for in situ agent delivery. In the future, sesquiterpenoid producing *R. capsulatus* might thus be applicable as cell extracts with biocontrol activities for plant protection or as engineered antiphytopathogenic PGPB that can be added as live cultures to soils contained in vertical farming.

Supplementary Materials: The following are available online at https://www.mdpi.com/2076-2607/9/1/168/s1. Figure S1: Transfer efficiency of the β-caryophyllene reference compound from cultivation medium into the *n*-dodecane phase in the presence of intact *R. capsulatus* cells in hungate and screw neck vials, Figure S2: Transfer efficiency of the β-caryophyllene reference compound from cultivation medium into the *n*-dodecane phase in the presence of intact and disrupted *R. capsulatus* cells, Figure S3: Extraction efficiency of the β-caryophyllene reference compound from cultivation medium in the presence of disrupted and intact *R. capsulatus* cells by repeatedly using *n*-dodecane as organic solvent over four days, Figure S4: Comparison of relative β-caryophyllene formation in *R. capsulatus* production strains cultivated with and without an *n*-dodecane layer, Figure S5: Quantification of extracted β-caryophyllene via a calibration curve of β-caryophyllene reference signals in GC-FID analyses, Figure S6: Emission range of different light sources and the absorption spectrum of phototrophically cultivated *R. capsulatus* cells [128], Figure S7: Effect of β-caryophyllene and β-caryophyllene oxide on the hyphal growth of plant pathogenic fungi, Table S1: Bacterial strains and plasmids used in this study, Table S2: Cultivation vessel specifications, Table S3: Productivities of β-caryophyllene in *R. capsulatus* SB1003 cultures.

Author Contributions: Conceptualization, T.D., A.L., K.-E.J., A.S.S.S., F.M.W.G.; methodology, F.H., Y.S.A., J.H.-H., A.H., S.S.H.; validation, F.H., S.S.H.; formal analysis, S.S.H., F.H.; investigation, F.H., S.S.H., Y.S.A., S.N., O.K.; writing—original draft preparation, F.H., S.S.H.; writing—review and editing, T.D., A.S.S.S., A.L., K.-E.J., F.M.W.G.; visualization, F.H.; supervision, T.D., A.L., A.S.S.S.; project administration, A.L., T.D., A.S.S.S., K.-E.J., F.M.W.G.; funding acquisition, A.L., T.D., A.S.S.S., K.-E.J., F.M.W.G. All authors have read and agreed to the published version of the manuscript.

Funding: The work was supported by grants from the Bioeconomy Science Center, and the European Regional Development Fund (ERDF: 34.EFRE-0300096) within the project CLIB-Kompetenzzentrum Biotechnologie CKB). The scientific activities of the Bioeconomy Science Center were financially supported by the Ministry of Innovation, Science and Research of the German federal state of North Rhine-Westphalia MIWF within the framework of the NRW Strategieprojekt BioSC (No. 313/323-400-00213).

Institutional Review Board Statement: Not applicable.

Informed Consent Statement: Not applicable.

Data Availability Statement: The datasets generated and/or analyzed during the current study are available from the corresponding authors on reasonable request.

Acknowledgments: The authors would like to thank Olaf Cladders, Edwin Thiemann, Volker Neu, and Helmut Guenther from Vossloh-Schwabe Lighting Solutions GmbH & Co. KG (Kamp-Lintfort, Germany) for their time, efforts and tremendous expertise in the LED array development thereby making the "illumination project" possible.

Conflicts of Interest: The authors declare no conflict of interest. The funders had no role in the design of the study, in the collection, analyzes, or interpretation of data, in the writing of the manuscript, or in the decision to publish the results.

References

1. Bian, G.; Deng, Z.; Liu, T. Strategies for terpenoid overproduction and new terpenoid discovery. *Curr. Opin. Biotechnol.* **2017**, *48*, 234–241. [CrossRef] [PubMed]
2. Christianson, D.W. Structural and Chemical Biology of Terpenoid Cyclases. *Chem. Rev.* **2017**, *117*, 11570–11648. [CrossRef] [PubMed]
3. Pemberton, T.A.; Chen, M.; Harris, G.G.; Chou, W.K.W.; Duan, L.; Köksal, M.; Genshaft, A.S.; Cane, D.E.; Christianson, D.W. Exploring the Influence of Domain Architecture on the Catalytic Function of Diterpene Synthases. *Biochemistry* **2017**, *56*, 2010–2023. [CrossRef] [PubMed]
4. Wink, M. Modes of Action of Herbal Medicines and Plant Secondary Metabolites. *Medicines* **2015**, *2*, 251–286. [CrossRef]
5. Ruzicka, L. The isoprene rule and the biogenesis of terpenic compounds. *Experientia* **1953**, *9*, 357–367. [CrossRef]
6. Croteau, R.; Kutchan, T.M.; Lewis, N.G. Secondary Metabolites. *Biochem. Mol. Biol. Plants* **2000**, *7*, 1250–1318.
7. Boucher, Y.; Doolittle, W.F. The role of lateral gene transfer in the evolution of isoprenoid biosynthesis pathways. *Mol. Microbiol.* **2000**, *37*, 703–716. [CrossRef]
8. Frank, A.; Groll, M. The Methylerythritol Phosphate Pathway to Isoprenoids. *Chem. Rev.* **2017**, *117*, 5675–5703. [CrossRef]
9. Langenheim, J.H. Higher plant terpenoids: A phytocentric overview of their ecological roles. *J. Chem. Ecol.* **1994**, *20*, 1223–1280. [CrossRef]
10. Gershenzon, J.; Dudareva, N. The function of terpene natural products in the natural world. *Nat. Chem. Biol.* **2007**, *3*, 408–414. [CrossRef]
11. Pichersky, E.; Raguso, R.A. Why do plants produce so many terpenoid compounds? *New Phytol.* **2018**, *220*, 692–702. [CrossRef] [PubMed]
12. Efferth, T. From ancient herb to modern drug: *Artemisia annua* and artemisinin for cancer therapy. *Semin. Cancer Biol.* **2017**, *46*, 65–83. [CrossRef] [PubMed]
13. Mahizan, N.A.; Yang, S.-K.; Moo, C.-L.; Song, A.A.-L.; Chong, C.-M.; Chong, C.-W.; Abushelaibi, A.; Lim, S.-H.E.; Lai, K.-S. Terpene Derivatives as a Potential Agent against Antimicrobial Resistance (AMR) Pathogens. *Molecules* **2019**, *24*, 2631. [CrossRef] [PubMed]
14. Walencka, E.; Rozalska, S.; Wysokinska, H.; Rozalski, M.; Kuzma, L.; Rozalska, B. Salvipisone and aethiopinone from *Salvia sclarea* hairy roots modulate staphylococcal antibiotic resistance and express anti-biofilm activity. *Planta Med.* **2007**, *73*, 545–551. [CrossRef] [PubMed]
15. Jabra-Rizk, M.A.; Meiller, T.F.; James, C.E.; Shirtliff, M.E. Effect of farnesol on *Staphylococcus aureus* biofilm formation and antimicrobial susceptibility. *Antimicrob. Agents Chemother.* **2006**, *50*, 1463–1469. [CrossRef] [PubMed]
16. Gomes, F.I.A.; Teixeira, P.; Azeredo, J.; Oliveira, R. Effect of farnesol on planktonic and biofilm cells of *Staphylococcus epidermidis*. *Curr. Microbiol.* **2009**, *59*, 118–122. [CrossRef] [PubMed]

17. Castelo-Branco, D.S.C.M.; Riello, G.B.; Vasconcelos, D.C.; Guedes, G.M.M.; Serpa, R.; Bandeira, T.J.P.G.; Monteiro, A.J.; Cordeiro, R.A.; Rocha, M.F.G.; Sidrim, J.J.C.; et al. Farnesol increases the susceptibility of *Burkholderia pseudomallei* biofilm to antimicrobials used to treat melioidosis. *J. Appl. Microbiol.* **2016**, *120*, 600–606. [CrossRef]
18. Malingre, T.; Hendriks, H.; Batterman, S.; Bos, R.; Visser, J. The Essential Oil of *Cannabis sativa*. *Planta Med.* **1975**, *28*, 56–61. [CrossRef]
19. De Vasconcelos Silva, M.; Craveiro, A.; Abreu Matos, F.; Machado, M.I.; Alencar, J. Chemical variation during daytime of constituents of the essential oil of *Ocimum gratissimum* leaves. *Fitoterapia* **1999**, *70*, 32–34. [CrossRef]
20. Matias, E.F.F.; Alves, E.F.; Silva, M.K.N.; Carvalho, V.R.A.; Figueredo, F.G.; Ferreira, J.V.A.; Coutinho, H.D.M.; Silva, J.M.F.L.; Ribeiro-Filho, J.; Costa, J.G.M. Seasonal variation, chemical composition and biological activity of the essential oil of *Cordia verbenacea* DC (*Boraginaceae*) and the sabinene. *Ind. Crops Prod.* **2016**, *87*, 45–53. [CrossRef]
21. Marienhagen, J.; Bott, M. Metabolic engineering of microorganisms for the synthesis of plant natural products. *J. Biotechnol.* **2013**, *163*, 166–178. [CrossRef]
22. Kallscheuer, N.; Classen, T.; Drepper, T.; Marienhagen, J. Production of plant metabolites with applications in the food industry using engineered microorganisms. *Curr. Opin. Biotechnol.* **2019**, *56*, 7–17. [CrossRef]
23. Schempp, F.M.; Drummond, L.; Buchhaupt, M.; Schrader, J. Microbial Cell Factories for the Production of Terpenoid Flavor and Fragrance Compounds. *J. Agric. Food Chem.* **2018**, *66*, 2247–2258. [CrossRef] [PubMed]
24. Pham, J.V.; Yilma, M.A.; Feliz, A.; Majid, M.T.; Maffetone, N.; Walker, J.R.; Kim, E.; Cho, H.J.; Reynolds, J.M.; Song, M.C.; et al. A Review of the Microbial Production of Bioactive Natural Products and Biologics. *Front. Microbiol.* **2019**, *10*, 1–27.
25. Cravens, A.; Payne, J.; Smolke, C.D. Synthetic biology strategies for microbial biosynthesis of plant natural products. *Nat. Commun.* **2019**, *10*, 2142.
26. Sgobba, E.; Wendisch, V.F. Synthetic microbial consortia for small molecule production. *Curr. Opin. Biotechnol.* **2020**, *62*, 72–79. [CrossRef] [PubMed]
27. Kirby, J.; Keasling, J.D. Metabolic engineering of microorganisms for isoprenoid production. *Nat. Prod. Rep.* **2008**, *25*, 656–661. [PubMed]
28. Mitchell, W. Natural products from synthetic biology. *Curr. Opin. Chem. Biol.* **2011**, *15*, 505–515. [CrossRef] [PubMed]
29. Chen, Y.; Zhou, Y.J.; Siewers, V.; Nielsen, J. Enabling technologies to advance microbial isoprenoid production. *Adv. Biochem. Eng. Biotechnol.* **2015**, *148*, 143–160.
30. Wong, J.; Rios-Solis, L.; Keasling, J.D. Microbial Production of Isoprenoids. In *Consequences of Microbial Interactions with Hydrocarbons, Oils, and Lipids: Production of Fuels and Chemicals*; Lee, S.Y., Ed.; Springer International Publishing: Cham, Switzerland, 2016; pp. 1–24. ISBN 978-3-319-31421-1.
31. Helfrich, E.J.N.; Lin, G.-M.; Voigt, C.A.; Clardy, J. Bacterial terpene biosynthesis: Challenges and opportunities for pathway engineering. *Beilstein J. Org. Chem.* **2019**, *15*, 2889–2906.
32. Chemler, J.A.; Koffas, M.A.G. Metabolic engineering for plant natural product biosynthesis in microbes. *Curr. Opin. Biotechnol.* **2008**, *19*, 597–605. [CrossRef] [PubMed]
33. Paddon, C.J.; Westfall, P.J.; Pitera, D.J.; Benjamin, K.; Fisher, K.; McPhee, D.; Leavell, M.D.; Tai, A.; Main, A.; Eng, D.; et al. High-level semi-synthetic production of the potent antimalarial artemisinin. *Nature* **2013**, *496*, 528–532. [CrossRef] [PubMed]
34. Atanasov, A.G.; Waltenberger, B.; Pferschy-Wenzig, E.-M.M.; Linder, T.; Wawrosch, C.; Uhrin, P.; Temml, V.; Wang, L.; Schwaiger, S.; Heiss, E.H.; et al. Discovery and resupply of pharmacologically active plant-derived natural products: A review. *Biotechnol. Adv.* **2015**, *33*, 1582–1614. [CrossRef]
35. Zhang, C.; Hong, K. Production of Terpenoids by Synthetic Biology Approaches. *Front. Bioeng. Biotechnol.* **2020**, *8*, 347. [CrossRef] [PubMed]
36. Yu, Y.; Rasool, A.; Liu, H.; Lv, B.; Chang, P.; Song, H.; Wang, Y.; Li, C. Engineering *Saccharomyces cerevisiae* for high yield production of α-amyrin via synergistic remodeling of α-amyrin synthase and expanding the storage pool. *Metab. Eng.* **2020**, *62*, 72–83. [CrossRef]
37. Bauer, K.; Garbe, D.; Surburg, H. *Common Fragrance and Flavor Materials*; Wiley-VCH: Weinheim, Germany, 2001.
38. Beekwilder, J.; van Houwelingen, A.; Cankar, K.; van Dijk, A.D.J.; de Jong, R.M.; Stoopen, G.; Bouwmeester, H.; Achkar, J.; Sonke, T.; Bosch, D. Valencene synthase from the heartwood of Nootka cypress (*Callitropsis nootkatensis*) for biotechnological production of valencene. *Plant Biotechnol. J.* **2014**, *12*, 174–182. [CrossRef]
39. Troost, K.; Loeschcke, A.; Hilgers, F.; Özgür, A.Y.; Weber, T.M.; Santiago-Schübel, B.; Svensson, V.; Hage-Hülsmann, J.; Habash, S.S.; Grundler, F.M.W.; et al. Engineered *Rhodobacter capsulatus* as a Phototrophic Platform Organism for the Synthesis of Plant Sesquiterpenoids. *Front. Microbiol.* **2019**, *10*, 1998. [CrossRef]
40. Orsi, E.; Folch, P.L.; Monje-López, V.T.; Fernhout, B.M.; Turcato, A.; Kengen, S.W.M.; Eggink, G.; Weusthuis, R.A. Characterization of heterotrophic growth and sesquiterpene production by *Rhodobacter sphaeroides* on a defined medium. *J. Ind. Microbiol. Biotechnol.* **2019**, *46*, 1179–1190. [CrossRef]
41. Lin, P.-C.; Pakrasi, H.B. Engineering cyanobacteria for production of terpenoids. *Planta* **2019**, *249*, 145–154. [CrossRef]
42. Das, A.; Yoon, S.-H.; Lee, S.-H.; Kim, J.-Y.; Oh, D.-K.; Kim, S.-W. An update on microbial carotenoid production: Application of recent metabolic engineering tools. *Appl. Microbiol. Biotechnol.* **2007**, *77*, 505–512. [CrossRef]
43. Arendt, P.; Miettinen, K.; Pollier, J.; De Rycke, R.; Callewaert, N.; Goossens, A. An endoplasmic reticulum-engineered yeast platform for overproduction of triterpenoids. *Metab. Eng.* **2017**, *40*, 165–175. [CrossRef] [PubMed]

44. Tucker, J.D.; Siebert, C.A.; Escalante, M.; Adams, P.G.; Olsen, J.D.; Otto, C.; Stokes, D.L.; Hunter, C.N. Membrane invagination in *Rhodobacter sphaeroides* is initiated at curved regions of the cytoplasmic membrane, then forms both budded and fully detached spherical vesicles. *Mol. Microbiol.* **2010**, *76*, 833–847. [CrossRef] [PubMed]
45. Drews, G. The intracytoplasmic membranes of purple bacteria—Assembly of energy-transducing complexes. *J. Mol. Microbiol. Biotechnol.* **2013**, *23*, 35–47. [CrossRef] [PubMed]
46. Armstrong, G.A.; Alberti, M.; Leach, F.; Hearst, J.E. Nucleotide sequence, organization, and nature of the protein products of the carotenoid biosynthesis gene cluster of *Rhodobacter capsulatus*. *Mol. Gen. Genet. MGG* **1989**, *216*, 254–268. [CrossRef]
47. Armstrong, G.A. Genetics of eubacterial carotenoid biosynthesis: A colorful tale. *Annu. Rev. Microbiol.* **1997**, *51*, 629–659. [CrossRef]
48. Khan, N.E.; Nybo, S.E.; Chappell, J.; Curtis, W.R. Triterpene hydrocarbon production engineered into a metabolically versatile host—*Rhodobacter capsulatus*. *Biotechnol. Bioeng.* **2015**, *112*, 1523–1532. [CrossRef]
49. Loeschcke, A.; Dienst, D.; Wewer, V.; Hage-Hülsmann, J.; Dietsch, M.; Kranz-Finger, S.; Hüren, V.; Metzger, S.; Urlacher, V.B.; Gigolashvili, T.; et al. The photosynthetic bacteria *Rhodobacter capsulatus* and *Synechocystis* sp. PCC 6803 as new hosts for cyclic plant triterpene biosynthesis. *PLoS ONE* **2017**, *12*, e0189816. [CrossRef]
50. Orsi, E.; Beekwilder, J.; Peek, S.; Eggink, G.; Kengen, S.W.M.; Weusthuis, R.A. Metabolic flux ratio analysis by parallel 13C labeling of isoprenoid biosynthesis in *Rhodobacter sphaeroides*. *Metab. Eng.* **2020**, *57*, 228–238. [CrossRef]
51. Orsi, E.; Mougiakos, I.; Post, W.; Beekwilder, J.; Dompè, M.; Eggink, G.; Van Der Oost, J.; Kengen, S.W.M.; Weusthuis, R.A. Growth-uncoupled isoprenoid synthesis in *Rhodobacter sphaeroides*. *Biotechnol. Biofuels* **2020**, *13*. [CrossRef]
52. Orsi, E.; Beekwilder, J.; van Gelder, D.; van Houwelingen, A.; Eggink, G.; Kengen, S.W.M.; Weusthuis, R.A. Functional replacement of isoprenoid pathways in *Rhodobacter sphaeroides*. *Microb. Biotechnol.* **2020**, *13*, 1082–1093. [CrossRef]
53. Oerke, E.-C. Crop losses to pests. *J. Agric. Sci.* **2006**, *144*, 31–43. [CrossRef]
54. Bolton, M.D.; Thomma, B.P.H.J.; Nelson, B.D. *Sclerotinia sclerotiorum* (Lib.) de Bary: Biology and molecular traits of a cosmopolitan pathogen. *Mol. Plant Pathol.* **2006**, *7*, 1–16. [CrossRef] [PubMed]
55. Okungbowa, F.I.; Shittu, H.O. *Fusarium* wilts: An overview. *Environ. Res. J.* **2012**, *6*, 83–102.
56. West, J.S.; Kharbanda, P.D.; Barbetti, M.J.; Fitt, B.D.L. Epidemiology and management of *Leptosphaeria maculans* (phoma stem canker) on oilseed rape in Australia, Canada and Europe. *Plant Pathol.* **2001**, *50*, 10–27. [CrossRef]
57. Fitt, B.D.L.; Brun, H.; Barbetti, M.J.; Rimmer, S.R. World-Wide Importance of Phoma Stem Canker (*Leptosphaeria maculans* and *L. biglobosa*) on Oilseed Rape (*Brassica napus*). *Eur. J. Plant Pathol.* **2006**, *114*, 3–15. [CrossRef]
58. Singh, H.K.; Singh, R.B.; Kumar, P.; Singh, M.; Yadav, J.K.; Singh, P.K.; Chauhan, M.P.; Shakywar, R.C.; Maurya, K.N.; Priyanka, B.S.; et al. *Alternaria* blight of rapeseed mustard–A Review. *J. Environ. Biol.* **2017**, *38*, 1405–1420. [CrossRef]
59. Verma, P.R. Biology and control of *Rhizoctonia solani* on rapeseed: A Review. *Phytoprotection* **2005**, *77*, 99–111. [CrossRef]
60. Paulitz, T.C.; Okubara, P.A.; Schillinger, W.F. First Report of Damping-Off of Canola Caused by *Rhizoctonia solani* AG 2-1 in Washington State. *Plant Dis.* **2006**, *90*, 829. [CrossRef]
61. Bridge, J. Nematode management in sustainable and subsistence agriculture. *Annu. Rev. Phytopathol.* **1996**, *34*, 201–225. [CrossRef]
62. Heydari, A.; Pessarakli, M. A Review on Biological Control of Fungal Plant Pathogens Using Microbial Antagonists. *J. Biol. Sci.* **2010**, *10*, 273–290. [CrossRef]
63. Habash, S.; Al-Banna, L. Phosphonate fertilizers suppressed root knot nematodes *Meloidogyne javanica* and *M. incognita*. *J. Nematol.* **2011**, *43*, 95–100. [PubMed]
64. Timper, P. Conserving and enhancing biological control of nematodes. *J. Nematol.* **2014**, *46*, 75–89. [PubMed]
65. Lu, C.; Tian, H. Global nitrogen and phosphorus fertilizer use for agriculture production in the past half century: Shifted hot spots and nutrient imbalance. *Earth Syst. Sci. Data* **2017**, *9*, 181–192. [CrossRef]
66. Cheng, A.-X.; Xiang, C.-Y.; Li, J.-X.; Yang, C.-Q.; Hu, W.-L.; Wang, L.-J.; Lou, Y.-G.; Chen, X.-Y. The rice (E)-β-caryophyllene synthase (OsTPS3) accounts for the major inducible volatile sesquiterpenes. *Phytochemistry* **2007**, *68*, 1632–1641. [CrossRef] [PubMed]
67. Echeverrigaray, S.; Zacaria, J.; Beltrão, R. Nematicidal Activity of Monoterpenoids against the Root-Knot Nematode *Meloidogyne incognita*. *Phytopathology* **2010**, *100*, 199–203. [CrossRef]
68. Zengin, H.; Baysal, A.H. Antibacterial and antioxidant activity of essential oil terpenes against pathogenic and spoilage-forming bacteria and cell structure-activity relationships evaluated by SEM microscopy. *Molecules* **2014**, *19*, 17773–17798. [CrossRef]
69. Dambolena, J.S.; Zunino, M.P.; Herrera, J.M.; Pizzolitto, R.P.; Areco, V.A.; Zygadlo, J.A. Terpenes: Natural Products for Controlling Insects of Importance to Human Health—A Structure-Activity Relationship Study. *Psyche A J. Entomol.* **2016**, *2016*, 1–17. [CrossRef]
70. Araniti, F.; Sánchez-Moreiras, A.M.; Graña, E.; Reigosa, M.J.; Abenavoli, M.R. Terpenoid trans -caryophyllene inhibits weed germination and induces plant water status alteration and oxidative damage in adult *Arabidopsis*. *Plant Biol.* **2017**, *19*, 79–89. [CrossRef]
71. Pungartnik, C. Antifungal Potential of Terpenes from *Spondias Purpurea* L. Leaf Extract against *Moniliophthora perniciosa* that causes Witches Broom Disease of *Theobroma cacao*. *Int. J. Complement. Altern. Med.* **2017**, *7*. [CrossRef]
72. Habash, S.S.; Könen, P.P.; Loeschcke, A.; Wüst, M.; Jaeger, K.-E.; Drepper, T.; Grundler, F.M.W.; Schleker, A.S.S. The Plant Sesquiterpene Nootkatone Efficiently Reduces *Heterodera schachtii* Parasitism by Activating Plant Defense. *Int. J. Mol. Sci.* **2020**, *21*, 9627. [CrossRef]

73. Kigathi, R.N.; Unsicker, S.B.; Reichelt, M.; Kesselmeier, J.; Gershenzon, J.; Weisser, W.W. Emission of Volatile Organic Compounds After Herbivory from *Trifolium pratense* (L.) Under Laboratory and Field Conditions. *J. Chem. Ecol.* **2009**, *35*, 1335–1348. [CrossRef] [PubMed]
74. Pazouki, L.; Kanagendran, A.; Li, S.; Kännaste, A.; Rajabi Memari, H.; Bichele, R.; Niinemets, Ü. Mono- and sesquiterpene release from tomato (*Solanum lycopersicum*) leaves upon mild and severe heat stress and through recovery: From gene expression to emission responses. *Environ. Exp. Bot.* **2016**, *132*, 1–15. [CrossRef] [PubMed]
75. Muchlinski, A.; Chen, X.; Lovell, J.T.; Köllner, T.G.; Pelot, K.A.; Zerbe, P.; Ruggiero, M.; Callaway, L.; Laliberte, S.; Chen, F.; et al. Biosynthesis and Emission of Stress-Induced Volatile Terpenes in Roots and Leaves of Switchgrass (*Panicum virgatum* L.). *Front. Plant Sci.* **2019**, *10*, 1144. [CrossRef] [PubMed]
76. Huang, M.; Sanchez-Moreiras, A.M.; Abel, C.; Sohrabi, R.; Lee, S.; Gershenzon, J.; Tholl, D. The major volatile organic compound emitted from *Arabidopsis thaliana* flowers, the sesquiterpene (E)-β-caryophyllene, is a defense against a bacterial pathogen. *New Phytol.* **2012**, *193*, 997–1008. [CrossRef] [PubMed]
77. Rasmann, S.; Köllner, T.G.; Degenhardt, J.; Hiltpold, I.; Toepfer, S.; Kuhlmann, U.; Gershenzon, J.; Turlings, T.C.J. Recruitment of entomopathogenic nematodes by insect-damaged maize roots. *Nature* **2005**, *434*, 732–737. [CrossRef]
78. Degenhardt, J.; Hiltpold, I.; Kollner, T.G.; Frey, M.; Gierl, A.; Gershenzon, J.; Hibbard, B.E.; Ellersieck, M.R.; Turlings, T.C.J. Restoring a maize root signal that attracts insect-killing nematodes to control a major pest. *Proc. Natl. Acad. Sci. USA* **2009**, *106*, 13213–13218. [CrossRef]
79. Ma, Y.-N.; Chen, C.-J.; Li, Q.-Q.; Xu, F.-R.; Cheng, Y.-X.; Dong, X. Monitoring Antifungal Agents of *Artemisia annua* against *Fusarium oxysporum* and *Fusarium solani*, Associated with *Panax notoginseng* Root-Rot Disease. *Molecules* **2019**, *24*, 213. [CrossRef]
80. Yamagiwa, Y.; Inagaki, Y.; Ichinose, Y.; Toyoda, K.; Hyakumachi, M.; Shiraishi, T. *Talaromyces wortmannii* FS2 emits β-caryophyllene, which promotes plant growth and induces resistance. *J. Gen. Plant. Pathol.* **2011**, *77*, 336–341. [CrossRef]
81. Hanahan, D. Studies on transformation of *Escherichia coli* with plasmids. *J. Mol. Biol.* **1983**, *166*, 557–580. [CrossRef]
82. Simon, R.; Priefer, U.; Pühler, A. A Broad Host Range Mobilization System for In Vivo Genetic Engineering: Transposon Mutagenesis in Gram Negative Bacteria. *Bio/Technology* **1983**, *1*, 784–791. [CrossRef]
83. Strnad, H.; Lapidus, A.; Paces, J.; Ulbrich, P.; Vlcek, C.; Paces, V.; Haselkorn, R. Complete genome sequence of the photosynthetic purple nonsulfur bacterium *Rhodobacter capsulatus* SB1003. *J. Bacteriol.* **2010**, *192*, 3545–3546. [CrossRef] [PubMed]
84. Klipp, W.; Masepohl, B.; Pühler, A. Identification and mapping of nitrogen fixation genes of *Rhodobacter capsulatus*: Duplication of a nifA-nifB region. *J. Bacteriol.* **1988**, *170*, 693–699. [CrossRef] [PubMed]
85. Weaver, P.F.; Wall, J.D.; Gest, H. Characterization of *Rhodopseudomonas capsulata*. *Arch. Microbiol.* **1975**, *105*, 207–216. [CrossRef] [PubMed]
86. Fuhrmann, M.; Hausherr, A.; Ferbitz, L.; Schödl, T.; Heitzer, M.; Hegemann, P. Monitoring dynamic expression of nuclear genes in *Chlamydomonas reinhardtii* by using a synthetic luciferase reporter gene. *Plant Mol. Biol.* **2004**, *55*, 869–881. [CrossRef] [PubMed]
87. Hungate, R.E. *Chapter IV A Roll Tube Method for Cultivation of Strict Anaerobes*; Norris, J.R., Ribbons, D.W.B.T.-M., Eds.; Academic Press: Cambridge, MA, USA, 1969; Volume 3, pp. 117–132.
88. Rodriguez, S.; Kirby, J.; Denby, C.M.; Keasling, J.D. Production and quantification of sesquiterpenes in *Saccharomyces cerevisiae*, including extraction, detection and quantification of terpene products and key related metabolites. *Nat. Protoc.* **2014**, *9*, 1980–1996. [CrossRef]
89. Wiegand, I.; Hilpert, K.; Hancock, R.E.W. Agar and broth dilution methods to determine the minimal inhibitory concentration (MIC) of antimicrobial substances. *Nat. Protoc.* **2008**, *3*, 163–175. [CrossRef]
90. European Committee for Antimicrobial Susceptibility Testing (EUCAST) of the European Society of Clinical Microbiology and Infectious Diseases (ESCMID). Determination of minimum inhibitory concentrations (MICs) of antibacterial agents by broth dilution. *Clin. Microbiol. Infect.* **2003**, *9*, ix–xv. [CrossRef]
91. Kim, S.; Jahandar, M.; Jeong, J.H.; Lim, D.C. Recent Progress in Solar Cell Technology for Low-Light Indoor Applications. *Curr. Altern. Energy* **2019**, *3*, 3–17. [CrossRef]
92. Ruberto, G.; Baratta, M.T. Antioxidant activity of selected essential oil components in two lipid model systems. *Food Chem.* **2000**, *69*, 167–174. [CrossRef]
93. Medeiros, R.; Passos, G.F.; Vitor, C.E.; Koepp, J.; Mazzuco, T.L.; Pianowski, L.F.; Campos, M.M.; Calixto, J.B. Effect of two active compounds obtained from the essential oil of *Cordia verbenacea* on the acute inflammatory responses elicited by LPS in the rat paw. *Br. J. Pharmacol.* **2007**, *151*, 618–627. [CrossRef]
94. Fidyt, K.; Fiedorowicz, A.; Strządała, L.; Szumny, A. β-caryophyllene and β-caryophyllene oxide-natural compounds of anticancer and analgesic properties. *Cancer Med.* **2016**, *5*, 3007–3017. [CrossRef] [PubMed]
95. Paula-Freire, L.I.G.; Andersen, M.L.; Gama, V.S.; Molska, G.R.; Carlini, E.L.A. The oral administration of trans-caryophyllene attenuates acute and chronic pain in mice. *Phytomedicine* **2014**, *21*, 356–362. [CrossRef] [PubMed]
96. Sköld, M.; Karlberg, A.-T.; Matura, M.; Börje, A. The fragrance chemical β-caryophyllene—Air oxidation and skin sensitization. *Food Chem. Toxicol.* **2006**, *44*, 538–545. [CrossRef] [PubMed]
97. Steenackers, B.; Neirinckx, A.; De Cooman, L.; Hermans, I.; De Vos, D. The strained sesquiterpene β-caryophyllene as a probe for the solvent-assisted epoxidation mechanism. *ChemPhysChem* **2014**, *15*, 966–973. [CrossRef]
98. De Souza, R.; Ambrosini, A.; Passaglia, L.M.P. Plant growth-promoting bacteria as inoculants in agricultural soils. *Genet. Mol. Biol.* **2015**, *38*, 1678–4685. [CrossRef]

99. Nath Yadav, A. Plant Growth Promoting Bacteria: Biodiversity and Multifunctional Attributes for Sustainable Agriculture. *Adv. Biotechnol. Microbiol.* **2017**, *5*. [CrossRef]
100. Singh, V.K.; Singh, A.K.; Singh, P.P.; Kumar, A. Interaction of plant growth promoting bacteria with tomato under abiotic stress: A review. *Agric. Ecosyst. Environ.* **2018**, *267*, 129–140. [CrossRef]
101. Çakmakçi, R.; Dönmez, F.; Aydin, A.; Şahin, F. Growth promotion of plants by plant growth-promoting rhizobacteria under greenhouse and two different field soil conditions. *Soil Biol. Biochem.* **2006**, *38*, 1482–1487. [CrossRef]
102. Çakmakçi, R.; Dönmez, M.F.; Erdoğan, Ü. The effect of plant growth promoting rhizobacteria on Barley seedling growth, nutrient uptake, some soil properties, and bacterial counts. *Turk. J. Agric. For.* **2007**, *31*, 189–199.
103. Lahlali, R.; Peng, G.; Gossen, B.D.; McGregor, L.; Yu, F.Q.; Hynes, R.K.; Hwang, S.F.; McDonald, M.R.; Boyetchko, S.M. Evidence that the Biofungicide Serenade (*Bacillus subtilis*) Suppresses Clubroot on Canola via Antibiosis and Induced Host Resistance. *Phytopathology* **2012**, *103*, 245–254. [CrossRef]
104. Berg, G. Plant-microbe interactions promoting plant growth and health: Perspectives for controlled use of microorganisms in agriculture. *Appl. Microbiol. Biotechnol.* **2009**, *84*, 11–18. [CrossRef] [PubMed]
105. El-Howeity, M.A.; Asfour, M.M. Response of some varieties of canola plant (*Brassica napus* L.) cultivated in a newly reclaimed desert to plant growth promoting rhizobacteria and mineral nitrogen fertilizer. *Ann. Agric. Sci.* **2012**, *57*, 129–136. [CrossRef]
106. Bertrand, H.; Nalin, R.; Bally, R.; Cleyet-Marel, J.-C. Isolation and identification of the most efficient plant growth-promoting bacteria associated with canola (*Brassica napus*). *Biol. Fertil. Soils* **2001**, *33*, 152–156. [CrossRef]
107. Ntalli, N.; Ferrari, F.; Giannakou, I.O.; Menkissoglu-Spiroudi, U. Synergistic and antagonistic interactions of terpenes against *Meloidogyne incognita* and the nematicidal activity of essential oils from seven plants indigenous to Greece. *Pest. Manag. Sci.* **2011**, *67*, 341–351. [CrossRef] [PubMed]
108. Jiménez-Reyes, M.F.; Carrasco, H.; Olea, A.F.; Silva-Moreno, E. Natural Compounds: A Sustainable Alternative to the Phytopathogens Control. *J. Chil. Chem. Soc.* **2019**, *64*, 4459–4465. [CrossRef]
109. Da Silva, F.; Alves, C.; Oliveira Filho, J.; Vieira, T.; Crotti, A.E.; Miranda, M. Chemical constituents of essential oil from *Murraya paniculata* leaves and its application to in vitro biological control of the fungus *Sclerotinia sclerotiorum*. *Food Sci. Technol.* **2019**, *39*. [CrossRef]
110. Valadares, A.C.F.; Alves, C.C.F.; Alves, J.M.; de Deus, I.P.B.; de Oliveira Fi, J.G.; Dos Santos, T.C.L.; Dias, H.J.; Crotti, A.E.M.; Miranda, M.L.D. Essential oils from *Piper aduncum* inflorescences and leaves: Chemical composition and antifungal activity against *Sclerotinia sclerotiorum*. *An. Acad. Bras. Cienc.* **2018**, *90*, 2691–2699. [CrossRef]
111. Yang, C.; Yang, C.; Gao, X.; Jiang, Y.; Sun, B.; Gao, F.; Yang, S. Synergy between methylerythritol phosphate pathway and mevalonate pathway for isoprene production in *Escherichia coli* Synergy between methylerythritol phosphate pathway and mevalonate pathway for isoprene production in *Escherichia coli*. *Metab. Eng.* **2016**, *37*, 79–91. [CrossRef]
112. Syed, S.; Prasad Tollamadugu, N.V.K.V. Chapter 16—Role of Plant Growth-Promoting Microorganisms as a Tool for Environmental Sustainability. In *Recent Developments in Applied Microbiology and Biochemistry*; Buddolla, V., Ed.; Academic Press: Cambridge, MA, USA, 2019; pp. 209–222. ISBN 978-0-12-816328-3.
113. Dobbelaere, S.; Vanderleyden, J.; Okon, Y. Plant Growth-Promoting Effects of Diazotrophs in the Rhizosphere. *CRC Crit. Rev. Plant Sci.* **2003**, *22*, 107–149. [CrossRef]
114. Chen, F.; Han, P.; Liu, P.; Si, N.; Liu, J.; Liu, X. Activity of the novel fungicide SYP-Z048 against plant pathogens. *Sci. Rep.* **2014**, *4*, 6473. [CrossRef]
115. Guzmán-Guzmán, P.; Porras-Troncoso, M.D.; Olmedo-Monfil, V.; Herrera-Estrella, A. *Trichoderma* Species: Versatile Plant Symbionts. *Phytopathology* **2018**, *109*, 6–16. [CrossRef] [PubMed]
116. Finkel, O.M.; Castrillo, G.; Herrera Paredes, S.; Salas González, I.; Dangl, J.L. Understanding and exploiting plant beneficial microbes. *Curr. Opin. Plant. Biol.* **2017**, *38*, 155–163. [CrossRef] [PubMed]
117. Anthony, J.R.; Anthony, L.C.; Nowroozi, F.; Kwon, G.; Newman, J.D.; Keasling, J.D. Optimization of the mevalonate-based isoprenoid biosynthetic pathway in *Escherichia coli* for production of the anti-malarial drug precursor amorpha-4,11-diene. *Metab. Eng.* **2009**, *11*, 13–19. [CrossRef] [PubMed]
118. Ajikumar, P.K.; Xiao, W.-H.; Tyo, K.E.J.; Wang, Y.; Simeon, F.; Leonard, E.; Mucha, O.; Phon, T.H.; Pfeifer, B.; Stephanopoulos, G. Isoprenoid Pathway Optimization for Taxol Precursor Overproduction in *Escherichia coli*. *Science (80-)* **2010**, *330*, 70–74. [CrossRef]
119. Henke, N.; Wichmann, J.; Baier, T.; Frohwitter, J.; Lauersen, K.; Risse, J.; Peters-Wendisch, P.; Kruse, O.; Wendisch, V. Patchoulol Production with Metabolically Engineered *Corynebacterium glutamicum*. *Genes* **2018**, *9*, 219. [CrossRef]
120. Frohwitter, J.; Heider, S.A.E.; Peters-Wendisch, P.; Beekwilder, J.; Wendisch, V.F. Production of the sesquiterpene (+)-valencene by metabolically engineered *Corynebacterium glutamicum*. *J. Biotechnol.* **2014**, *191*, 205–213. [CrossRef]
121. Chen, H.; Zhu, C.; Zhu, M.; Xiong, J.; Ma, H.; Zhuo, M.; Li, S. High production of valencene in *Saccharomyces cerevisiae* through metabolic engineering. *Microb. Cell Fact.* **2019**, *18*, 195.
122. Westfall, P.J.; Pitera, D.J.; Lenihan, J.R.; Eng, D.; Woolard, F.X.; Regentin, R.; Horning, T.; Tsuruta, H.; Melis, D.J.; Owens, A.; et al. Production of amorphadiene in yeast, and its conversion to dihydroartemisinic acid, precursor to the antimalarial agent artemisinin. *Proc. Natl. Acad. Sci. USA* **2012**, *109*, E111–E118.
123. Englund, E.; Shabestary, K.; Hudson, E.P.; Lindberg, P. Systematic overexpression study to find target enzymes enhancing production of terpenes in *Synechocystis* PCC 6803, using isoprene as a model compound. *Metab. Eng.* **2018**, *49*, 164–177. [CrossRef]

124. Bentley, F.K.; Zurbriggen, A.; Melis, A. Heterologous expression of the mevalonic acid pathway in cyanobacteria enhances endogenous carbon partitioning to isoprene. *Mol. Plant.* **2014**, *7*, 71–86. [CrossRef]
125. Krieg, T.; Sydow, A.; Faust, S.; Huth, I.; Holtmann, D. CO_2 to Terpenes: Autotrophic and Electroautotrophic α-Humulene Production with *Cupriavidus necator*. *Angew. Chem. Int. Ed.* **2018**, *57*, 1879–1882. [CrossRef] [PubMed]
126. Yang, J.; Li, Z.; Guo, L.; Du, J.; Bae, H.-J. Biosynthesis of β-caryophyllene, a novel terpene-based high-density biofuel precursor, using engineered *Escherichia coli*. *Renew. Energy* **2016**, *99*, 216–223. [CrossRef]
127. Yang, J.; Nie, Q. Engineering *Escherichia coli* to convert acetic acid to β-caryophyllene. *Microb. Cell Fact.* **2016**, *15*, 74. [CrossRef] [PubMed]
128. Hogenkamp, F.; Hilgers, F.; Knapp, A.; Klaus, O.; Bier, C.; Binder, D.; Jaeger, K.-E.; Drepper, T.; Pietruszka, J. Effect of Photocaged Isopropyl β-D-1-Thiogalactopyranoside Solubility on Light-Responsiveness of LacI-controlled Expression Systems in Different Bacteria. *ChemBioChem* **2020**. [CrossRef] [PubMed]

microorganisms

MDPI

Article

Fermentative *N*-Methylanthranilate Production by Engineered *Corynebacterium glutamicum*

Tatjana Walter [1], Nour Al Medani [1], Arthur Burgardt [1], Katarina Cankar [2], Lenny Ferrer [1], Anastasia Kerbs [1], Jin-Ho Lee [3], Melanie Mindt [1,2], Joe Max Risse [4] and Volker F. Wendisch [1,*]

1 Genetics of Prokaryotes, Faculty of Biology and CeBiTec, Bielefeld University, 33615 Bielefeld, Germany; t.walter@uni-bielefeld.de (T.W.); mohamad.al@uni-bielefeld.de (N.A.M.); arthur.burgardt@uni-bielefeld.de (A.B.); lferrer@cebitec.uni-bielefeld.de (L.F.); anastasia.kerbs@uni-bielefeld.de (A.K.); melanie.mindt@wur.nl (M.M.)

2 BU Bioscience, Wageningen University & Research, 6700AA Wageningen, The Netherlands; katarina.cankar@wur.nl

3 Major in Food Science & Biotechnology, School of Food Biotechnology & Nutrition, Kyungsung University, Busan 48434, Korea; jhlee83@ks.ac.kr

4 Fermentation Technology, Technical Faculty and CeBiTec, Bielefeld University, 33615 Bielefeld, Germany; jrisse@uni-bielefeld.de

* Correspondence: volker.wendisch@uni-bielefeld.de

Received: 22 May 2020; Accepted: 5 June 2020; Published: 8 June 2020

Abstract: The *N*-functionalized amino acid *N*-methylanthranilate is an important precursor for bioactive compounds such as anticancer acridone alkaloids, the antinociceptive alkaloid *O*-isopropyl *N*-methylanthranilate, the flavor compound *O*-methyl-*N*-methylanthranilate, and as a building block for peptide-based drugs. Current chemical and biocatalytic synthetic routes to *N*-alkylated amino acids are often unprofitable and restricted to low yields or high costs through cofactor regeneration systems. Amino acid fermentation processes using the Gram-positive bacterium *Corynebacterium glutamicum* are operated industrially at the million tons per annum scale. Fermentative processes using *C. glutamicum* for *N*-alkylated amino acids based on an imine reductase have been developed, while *N*-alkylation of the aromatic amino acid anthranilate with *S*-adenosyl methionine as methyl-donor has not been described for this bacterium. After metabolic engineering for enhanced supply of anthranilate by channeling carbon flux into the shikimate pathway, preventing by-product formation and enhancing sugar uptake, heterologous expression of the gene ANMT encoding anthranilate *N*-methyltransferase from *Ruta graveolens* resulted in production of *N*-methylanthranilate (NMA), which accumulated in the culture medium. Increased SAM regeneration by coexpression of the homologous adenosylhomocysteinase gene *sahH* improved *N*-methylanthranilate production. In a test bioreactor culture, the metabolically engineered *C. glutamicum* C1* strain produced NMA to a final titer of 0.5 $g{\cdot}L^{-1}$ with a volumetric productivity of 0.01 $g{\cdot}L^{-1}{\cdot}h^{-1}$ and a yield of 4.8 $mg{\cdot}g^{-1}$ glucose.

Keywords: *N*-functionalized amines; *N*-methylanthranilate; *Corynebacterium glutamicum*; metabolic engineering; sustainable production of quinoline precursors; acridone; quinazoline alkaloid drugs

1. Introduction

N-Functionalization of natural products as well as fine and bulk chemicals includes *N*-hydroxylation, *N*-acetylation, *N*-phosphorylation, or *N*-alkylation. These amine and amino acid modifications are found in all domains of life, and they fulfill various physiological roles such as resistance of bacteria to the antibiotic rifampicin by its *N*-hydroxylation [1], biosynthesis of the hormone melatonin via *N*-acetylated serotonin in plants and mammals [2], or assimilation of methylamine as carbon and energy source in methylotrophic bacteria [3].

The biotechnological and chemical interest in N-functionalized amines, especially in N-alkylated amino acids, has increased recently because of their beneficial impact as building blocks when incorporated into peptide-based drugs. Better membrane permeability, increased stability against proteases, stabilization of discrete confirmations, prevention of peptide aggregation by reduced formation of hydrogen bonds, or increased receptor subtype selectivity were shown for peptide-based drugs as consequence of amino acid N-alkylation [4]. For example, N-methylation of the N–Cα peptide bonds of transition state mimetics developed to inhibit malarial protease, which is required for infecting erythrocytes, improved their lipophilicity and stability against proteolysis, thus enhancing activity against *Plasmodium* parasites [5]. Free N-alkylated amines such as the N-ethylated glutamine derivative L-theanine, which prominently occurs in green tea, or O-methyl-N-methylanthranilate of grapes are flavoring compounds with applications in the food, cosmetics, flavor, and fragrances industries.

Chemical synthesis of free N-alkylated amino acids is well studied, and various routes are known, such as by nucleophilic substitution of α-bromo acids, N-methylation of sulfonamides, carbamates or amides, reduction of Schiff bases generated with an amino acid and formaldehyde or other aldehydes, by direct alkylation of protected amino acids or by ring-opening of 5-oxazolidinones [6–9]. However, these processes are often limited by low product yields, over-methylation, toxic reagents, or their incomplete enantiopurity [10,11]. Recently, enzyme catalysis routes with N-methyltransferases, dehydrogenases, ketimine reductases, or imine reductases that depend on cofactor regeneration systems have been described [12]. Fermentation processes using simple mineral salts media have been developed for three different routes for de novo production of N-alkylated amino acids. Two metabolic engineering strategies for reductive alkylamination of 2-oxo acids with monomethylamine that either make use of a C1-assimilation pathway present in methylotrophic bacteria [13] or of the imine reductase DpkA [14] have been established. S-Adenosyl-L-methionine (SAM)-dependent methylation of aromatic amino acids by N-methyltransferases has also been described [15].

N-methylanthranilate (NMA) is an intermediate of the acridone alkaloid biosynthesis in plants. The SAM-dependent transfer of a methyl group to anthranilate initiates the biosynthesis of NMA-dependent biosynthesis of N-methylated acridone alkaloids and avenacin in plants [16,17]. Until now only one N-methyltransferase enzyme ANMT was characterized from the common rue, *Ruta graveolens* L., which accumulates N-methylated acridones exclusively. This enzyme shows narrow specificity for anthranilate, not accepting methylated catechol, salicylate, caffeate, 3- and 4-hydroxybenzoate, and anthraniloyl-CoA as substrates [16]. The acridone alkaloids and avenacin pathways diverge after SAM-dependent N-methylation of acridone anthranilate with regard to activation for transfer to the respective alkaloid intermediate. An ATP-dependent transfer of CoA is postulated for the acridone alkaloid biosynthesis [18], while UDP glucose-dependent O-glycosylation was shown as second step of the avenacin biosynthesis [17]. Acridone alkaloids and avenacin are known as bioactive compounds with cytotoxic, anticancer, antimicrobial, or antiparasitic properties and are, therefore, used for pharmaceutical and therapeutic purposes. Several N-methylated acridones, namely citrusamine, evoxanthine, arborinine, or normelicopine, were identified in diverse plants [19]. Arborinine, as an example, was found in ethyl acetate extracts from *Glycosmis parva,* and it showed anticancer activity against human cervical cancer cells since activation of caspase-dependent apoptosis without inducing the DNA damage response was observed [20]. N-methylanthranilate also serves as precursor for the flavoring agent O-methyl-N-methylanthranilate, which has an orange blossom and grape-like odor, the antinociceptive alkaloid O-isopropyl-N-methylanthranilate, or the anti-inflammatory active compound O-propyl-N-methylanthranilate [21–23].

Safe production of amino acids for the food and feed industry has been established at the annual million-ton scale for decades with *Corynebacterium glutamicum* as the dominant production host [24]. *C. glutamicum* grows on simple mineral salts media and can utilize various sugars [25,26], acids such as citrate [27], and alcohols such as ethanol [28]. A well-established toolbox enabled metabolic engineered-based approaches for production of diverse value-added compounds. Besides the production of proteinogenic amino acids, also a broad range of non-proteinogenic amino acid products like γ-aminobutyrate [29], 5-aminovalerate [30,31], pipecolic acid [32,33], *N*-methylated amino acids like *N*-methylalanine (NMeAla) [34] and sarcosine [35], aromatic compounds like 4-hydroxybenzoate [36,37] or protocatechuic acid [38], and functionalized aromatics like 7-chloro- or 7-bromo-tryptophan [39,40] and *O*-methylanthranilate [41] have been demonstrated.

Here, we describe fermentative *N*-methylanthranilate production by metabolic engineering of genome-reduced chassis strain *C. glutamicum* C1*, a robust basic strain for synthetic biology and industrial biotechnology [42]. Fermentative NMA production from glucose involved SAM-dependent ANMT from *R. graveloens* combined with metabolic engineering for efficient supply of the precursor anthranilate (Figure 1).

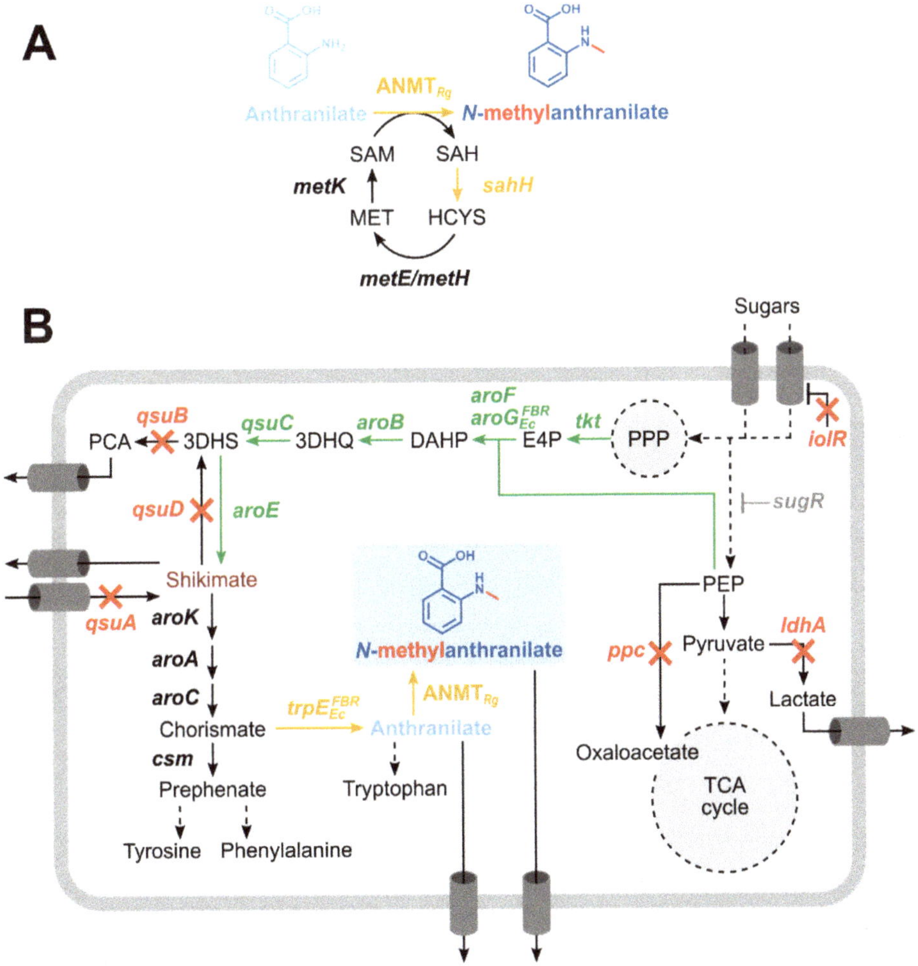

Figure 1. Schematic representation of *N*-methylanthranilate (NMA) biosynthesis (**A**) embedded into aromatic amino acid metabolism of engineered *C. glutamicum* (**B**). Continuous arrows indicate single reactions, dashed arrows indicate multiple reactions. Green arrows and gene names indicate genome-based overexpression, yellow arrows and gene names indicate vector-based expression, crossed arrows and red gene names indicate gene deletion. (**A**) *N*-methylation of anthranilate by *N*-methylanthranilate transferase (ANMT) from *R. graveolens* under consumption of *S*-adenosylmethionine (SAM). The SAM regeneration cycle is depicted with overexpression of *sahH*,

S-adenosylhomocysteine hydrolase. SAH, *S*-adenosylhomocysteine; HCYS, L-homocysteine; MET, L-methionine; *metE/metH*, methionine synthase; *metK*, methionine adenosyltransferase. (**B**) Strain engineering towards production of NMA. Grey *sugR* indicates reversion of deleted *sugR* back to wild type *sugR*. PEP, phosphoenolpyruvate; TCA, tricarboxylic acid; PPP, pentose phosphate pathway; E4P, erythrose-4-phosphate; DAHP, 3-deoxy-D-arabinoheptulosonate-7-phosphate; 3DHQ, 3-dehydroquinate; 3DHS, 3-dehydroshikimic acid; PCA, protocatechuic acid; *iolR*, transcriptional regulator; *sugR*, transcriptional regulator; *ppc*, phosphoenolpyruvate carboxylase; *ldhA*, lactate dehydrogenase; *tkt*, transketolase; *aroF*, DAHP synthase; *aroG^FBR^*, feedback-resistant DAHP synthase from *Escherichia coli*; *aroB*, 3-dehydroquinate synthase; *qsuC*, 3-dehydroquinate dehydratase; *qsuB*, 3-dehydroshikimate dehydratase; *qsuD*, shikimate dehydrogenase; *aroE*, shikimate dehydrogenase; *qsuA*, putative shikimate importer; *aroK*, shikimate kinase; *aroA*, 5-enolpyruvylshikimate-3-phosphate synthase; *aroC*, chorismate synthase; *csm*, chorismate mutase; *trpE^FBR^*, feedback-resistant anthranilate synthase from *E. coli*.

2. Materials and Methods

2.1. Bacterial Strains and Culture Conditions

All bacterial strains used are listed in Table 1. *Escherichia coli* DH5α [43] was used for plasmid construction. *C. glutamicum* C1* was used as host organism for shikimate, anthranilate, and NMA production. Pre-cultures of *E. coli* and *C. glutamicum* were performed in lysogeny broth (LB) and brain heart infusion (BHI) medium at 37 or 30 °C in baffled shake flasks on a rotary shaker (160 rpm or 120 rpm). Cultures were inoculated freshly from LB agar plates. When necessary, spectinomycin (100 $\mu g \cdot mL^{-1}$) and kanamycin (25 $\mu g \cdot mL^{-1}$) were added to the medium. For induction of gene expression from vectors pEKEx3 and pGold, isopropyl-β-D-1-thiogalactopyranoside (IPTG) was added to the medium. For the performance of growth or production experiments of *C. glutamicum*, pre-cultures were inoculated as described above. After cell harvesting (3200× *g*, 7 min), cells were washed with TN-buffer pH 6.3 (50 mM Tris-HCL, 50 mM NaCl) and inoculated to an optical density at 600 nm (OD_{600}) of 1 in CGXII minimal medium [44] and 40 g glucose as sole carbon source. *C. glutamicum* grown in 500 mL baffled shake flasks was followed by measuring OD_{600} using a V-1200 spectrophotometer (VWR, Radnor, PA, USA). An OD_{600} of 1 was determined to be equivalent to a biomass concentration of 0.25 g cell dry weight per liter.

Table 1. Bacterial strains used in this study.

Strains	Description	Source
Corynebacterium glutamicum		
WT	*C. glutamicum* wild-type strain ATCC13032	ATCC
C1*	Genome-reduced chassis strain derived from	[42]
ARO01	Δ*vdh*::P_{ilvC}-*aroG*^{D146N} mutant of C1*	This work
ARO02	Δ*ldhA* mutant of ARO01	This work
ARO03	Δ*sugR* mutant of ARO02	This work
ARO04	Δ*aroR*::P_{ilvC}-*aroF* mutant of ARO03	This work
ARO05	Δ*qsuABCD*::P_{tuf}-*qsuC* mutant of ARO04	This work
ARO06	Δ*ppc*::P_{sod}-*aroB* mutant of ARO05	This work
ARO07	ΔP_{tkt}::P_{tuf}-*tkt* mutant of ARO06	This work
ARO08	Δ*iolR*::P_{tuf}-*aroE* mutant of ARO07	This work
ARO09	Δ*sugR*::*sugR* mutant of ARO08	This work
NMA100	ARO09 carrying pEKEx3 and pGold	This work
NMA101	ARO09 carrying pEKEx3 and pGold-ANMT	This work
NMA102	ARO09 carrying pEKEx3 and pGold-ANMT-*sahH*	This work
NMA103	ARO09 carrying pEKEx3-*trpE*^{FBR} and pGold	This work
NMA104	ARO09 carrying pEKEx3-*trpE*^{FBR} and pGold-ANMT	This work
NMA105	ARO09 carrying pEKEx3-*trpE*^{FBR} and pGold-ANMT-*sahH*	This work
Escherichia coli		
S17-1	*recA pro hsdR* RP4-2-Tc::Mu-Km::Tn7	[45]
DH5α	*F-thi-1 endA1 hsdr17(r-, m-) supE44 1lacU169 (Φ80lacZ1M15) recA1 gyrA96*	[43]

Evaluation of the effects of anthranilate and NMA on *C. glutamicum* growth was performed in the microbioreactor system Biolector (m2p-labs; Aachen, Germany). Pre-cultures were grown in BHI-rich medium overnight and transferred to second pre-culture of CGXII minimal medium with 40 g·L^{-1} glucose until the early exponential phase before inoculating to the main medium of CGXII minimal medium and 40 g·L^{-1} glucose with addition of varying anthranilate (solved in water) and NMA (solved in methanol) concentrations. Each condition with NMA contained 1.65 M methanol. Growth experiments in the Biolector were carried out using 48-well flower plates (MTP-48-B; m2p-labs) with a filling volume of 1 mL, at 30 °C, and 1200 rpm shaking frequency. Humidity was kept constant at 85%, and online biomass measurements of scattered light were monitored with backscatter gain of 20.

2.2. Fed-Batch Cultivation

Fed-Batch fermentation of *C. glutamicum* NMA105 was performed in an initial volume of 2 L in a bioreactor (3.7 L KLF, Bioengineering AG, 8636 Wald, Switzerland) at 30 °C, 0.2 bar overpressure, and an aeration rate of 2 NL·min^{-1}. We did not perform off-gas analysis. To maintain relative dissolved oxygen saturation at 30%, stirrer speed was controlled during growth. The pH was maintained at pH 7.0 due to controlled addition of KOH (4 M) and phosphoric acid (10% (*w*/*w*)). To avoid foaming, the antifoam Sruktol® J647 was added manually when necessary. Feeding with 400 g·L^{-1} glucose and 150 g·L^{-1} $(NH_4)_2SO_4$ (total volume: 500 mL) was activated when the relative dissolved oxygen saturation (rDOS) signal rose above 60% and stopped when rDOS fell below 60%. Samples were taken automatically every 4 h during the whole cultivation and cooled down to 4 °C until further use. *C. glutamicum* NMA105 cells were transferred from a first pre-culture grown in LB in shake flasks to a second pre-culture in standard CGXII (pH 7.0) medium with 40 g·L^{-1} glucose (without IPTG) and the required antibiotics. For the bioreactor culture, standard CGXII medium without addition of 3-(*N*-morpholino)propanesulfonic acid (MOPS) and antibiotics was used. The fermenter was inoculated with the second pre-culture to an OD of 1.5 and immediately induced with 1 mM of IPTG.

2.3. Molecular Genetic Techniques and Strain Construction

Standard molecular genetic techniques were performed as described [46]. Competent *E. coli* DH5α [43] was performed with the RbCl method and transformed by heat shock [46]. Transformation of *C. glutamicum* was performed by electroporation [44]. The gene *trpE*^{FBR} was amplified using specific primers (Table 2) with ALLin™ HiFi DNA Polymerase (highQu GmbH, Kraichtal, Germany). The PCR products were assembled with *Bam*HI restricted pEKEx3 via Gibson Assembly [44].

For heterologous expression of the *N*-methylanthranilate transferase gene, firstly, the pEC-XK99E vector was modified to be suitable for Golden Gate based modular assembly of multiple genes simultaneously. To this end, the three *Bsa*I sites present in the vector located in the *rrnB* terminator, the vector backbone, and the *repA* ORF were removed. Next, a linker containing two *Bsa*I sites (CAGATGAGACCGCATGCCTGCAAGGTCTCAGTAT) was added to the MCS between *Eco*RI and *Sac*I restriction sites. The resulting vector was named pGold (GenBank: MT521917). The coding sequence (CDS) of the plant gene ANMT (GenBank: DQ884932.1) encoding the *N*-methylanthranilate transferase of *Ruta graveolens* was codon-harmonized to the natural codon frequency of *C. glutamicum* ATCC13032 with the codon usage table of kazusa database [47] and synthesized with Golden Gate assembly compatible flanking regions including recognition site for the restriction enzyme type 2 *BsaI* and pGold complementary sequences and an optimized RBS [48,49] (Supplementary Data Table S1). The gene ANMT was amplified using specific primers (Table 2) with ALLin™ HiFi DNA Polymerase according to the manufacturer (highQu GmbH, Kraichtal, Germany). The PCR products were assembled with digested pGold-ANMT with *Bam*HI via Gibson Assembly [44].

Table 2. Oligonucleotides used in this study.

Name	Oligonucleotide Sequence (5′ to 3′)
vdh-conf-fw	GACCTCTAGGGCAGCAGTG
vdh-conf-rv	CTGTTCAGCGGATTAGCG
ldhA-conf-fw	TGATGGCACCAGTTGCGATGT
ldhA-conf-rv	CCATGATGCAGGATGGAGTA
sugR-conf-fw	CGAGATGCTGTGGTTTTGAG
sugR-conf-rv	GCTTATCGGGTGTGGGAATG
US-aroR-fw	CCTGCAGGTCGACTCTAGAGCGATGCAGAATAATGCAGTTAG
US-aroR-rv	CGGAGCTTGCCTGGGAGTTTGGAACCTTAACACACTTTC
PilvC-aroR-fw	GAAAGTGTGTTAAGGTTCCAAACTCCCAGGCAAGCTCCGCGC
PilvC-aroR-rv	**GAAAAAACCTCCTTTAGTGTGTAGTTAAGTT**ATGGTGATGGGAGAAAATCTCGCCTTTCG
DS-aroR-fw	AT**CACCATAACTTAACTACACACTAAAGGAGGTTTTTTC**ATGAGTTCTCCAGTCTCACTCGAAAAC
DS-aroR-rv	GAATTCGAGCTCGGTACCCGGGCAATGCGCAAGCCCTCTGGG
aroR-conf-fw	GGAACTCCCGTTGAGGTG
aroR-conf-rv	GTGGTACGAGCGCCGATTG
US-qsuA-fw	CCTGCAGGTCGACTCTAGAGGTTGGCAGCGCAACCAGTC
US-qsuA-rv	CTACTGACACGCTAAAACGCTGTCGATCCTGTTCATCG
Ptuf-qsuC-fw	CGATGAACAGGATCGACAGCGTTTTAGCGTGTCAGTAG
Ptuf-qsuC-rv	**CTGAAGGGCCTCCTTTC**TCCTCCTGGACTTCGTGG
qsuC-fw	GGA**GAAAGGAGGCCCTTCAG**ATGCCTGGAAAAATTCTCCTCC
qsuC-rv	GTCGAGGTTTTACTGACTCTTCTACTTTTTGAGATTTGCCAGG
DS-qsuD-fw	CTCAAAAAGTAGAAGAGTCAGTAAAACCTCGACGC
DS-qsuD-rv	GAATTCGAGCTCGGTACCCGGGATTTCGCGGATGGGTCTAAGTATG
qsu-conf-fw	GTTCGTGGACAAGTGTGGTGG
qsu-conf-rv	GTTCGTGGACAAGTGTGGTGG
US-ppc-fw	GCCTGCAGGTCGACTCTAGAGCGCTCAGGAAGTGTGCAAGGC
US-ppc-rv	GTACTACCCAGCCGGCTGGGGATCCCTACTTTAAACACTCTTTCACATTGAGGGTG
Psod-aroB-fw	AATGTGAAAGAGTGTTTAAAGTAGGAAGCGCCTCATCAGCGGTAAC
Psod-aroB-rv	**CTCCTTTAAAAATAAGTCGCCTACC**AAAATCCTTTCGTAGGTTTCCGC
aroB-fw	GCGGAAACCTACGAAAGGATTTT**GGTAGGCGACTTATTTTTAAAGGAGGTTTTTT** ATGAGCGCAGTGCAGATTTTC
aroB-rv	CTTCTCTCATCCGCCAAAATTAGTGGCTGATTGCCTCATAAG
Term-aroB-fw	CTTATGAGGCAATCAGCCACTAATTTTGGCGGATGAGAGAAG
Term-aroB-rv	AGTACTACCCAGCCGGCTGGGGATCCAAAAGAGTTTGTAGAAACGC
DS-ppc-fw	TGAAAGAGTGTTTAAAGTAGGGATCCCCAGCCGGCTGGGTAGTAC
DS-ppc-rv	GAATTCGAGCTCGGTACCCGGGCAGTGGGGAGACAACAGGTCG
ppc-conf-fw	CCGTCGGGAAACAGTTCCCC
ppc-conf-rv	GCAGACCCGTAAGTCCCTTGC
US-tkt-fw	GCATGCCTGCAGGTCGACTCTAGAGTGACCCAGGTGGACGCCAAC
US-tkt-rv	GTGGACATTCGCAGGGTAACGGCCAAGGTGTGATCAATCTTAAGTC
Ptuf-tkt-fw	GACTTAAGATTGATCACACCTTGGCCGTTACCCTGCGAATGTCCAC
Ptuf-tkt-rv	CGTCAAGGTGGTCATCTGAAGGGCCTCCTTTCTGTATGTCCTCCTGGACTTC
DS-tkt-fw	CAGGAGGACATACA**GAAAGGAGGCCCTTCAG**ATGACCACCTTGACGCTGTC
DS-tkt-rv	GAATTCGAGCTCGGTACCCGGGTGGCGGTACTCAGGGTGTCC
tkt-conf-fw	GTTCCCGAATCAATCTTTTTAATG
tkt-conf-rv	GACCCTGGCCAAGAGGGCCAGTG
US-iolR-fw	GCCTGCAGGTCGACTCTAGAGCGACCCTCACGATCGCATG
US-iolR-rv	CTACTGACACGCTAAAACGCGATGTCTCCTTTCGTTGCCC
Ptuf-aroE-fw	GGGCAACGAAAGGAGACATCGCGTTTTAGCGTGTCAGTAG
Ptuf-aroE-rv	CCCA**TCTGAAGGGCCTCCTTTC**TCCTCCTGGACTTCGTGGTG
aroE-fw	GGA**GAAAGGAGGCCCTTCAG**ATGGGTTCTCACATCACTCACCG
aroE-rv	CAGAAGGGCTCTTTGGTTTATTTCTTAGTGTTCTTCTGAGATGCCTAAAGACTC
DS-iolR-fw	GAGTCTTTAGGCATCTCAGAAGAACACTAAGAAATAAACCAAAGAGCCCTTCTG
DS-iolR-rv	GAATTCGAGCTCGGTACCCGGGCGCTCTCCATCCGCTGGAC
iolR-conf-fw	CAGATAGAGGAACCCAAGGCG
iolR-conf-rv	GGACTTCGTGAGTGCTCGTC
sugR_reintegr-fw	CTGCAGGTCGACTCTAGAGCCTGCGCAGGGACCCTAATAAG
sugR_reintegr-rv	GAATTCGAGCTCGGTACCCGGGCCTGCAGTAAAAGATTCCCGC
x3-trpE-fw	CCTGCAGGTCGACTCTAGAG**GAAAGGAGGCCCTTCAG** ATGCAAACACAAAAACCGACTCTCGAACTG
x3-trpE-rv	AAAACGACGGCCAGTGAATTTCAGAAAGTCTCCTGTGCATGATGCGC
pGANMT-sahH-fw	ATGAGCTCGGTACC**CGGGCGGGACGAAGAGAACCGTTACAAGAATAAAGGAGGTTTTTT** ATGGCACAGGTTATGGACTTC
pGANMT-sahH-rv	CTGCAGGTCGACTCTAGAGTTAGTAGCGGTAGTGCTCCGG

Ribosomal binding sites are in bold, and binding regions of Gibson oligonucleotides are underlined.

Chromosomal gene deletions and replacements in C1*-derived strains were performed by two-step homologous recombination [44] using the suicide vector pK19*mobsacB* [50]. The genomic regions flanking the respective gene for homologous recombination were amplified from *C. glutamicum* WT as described elsewhere [51] using the respective Primer pairs containing artificial RBS ([48,49], Table 2). The purified PCR products were assembled and simultaneously cloned into restricted pK19*mobsacB* by Gibson Assembly resulting in the plasmids listed in Table 3. Transfer of the suicide vectors was carried out by trans-conjugation using *E. coli* S17 as donor strain [33]. For the first recombination event, integration of the vector in one of the targeted flanking regions was selected via kanamycin resistance. The resulting clones showed sucrose sensitivity due to the levansucrase gene *sacB*. Suicide vector excision was selected by sucrose resistance. Gene deletions or replacements were verified by PCR and sequencing with respective primers (Table 2).

Table 3. List of plasmids used in this study.

Plasmids	Description	Source
pK19*mobsacB*	Km^R; *E. coli/C. glutamicum* shuttle vector for construction of insertion and deletion mutants in *C. glutamicum* (pK19 $oriV_{Ec}$ *sacB lacZα*)	[50]
pK19-Δ*vdh*::P_{ilvC}-*aroG*^{D146N}	pK19*mobsacB* with a construct for replacement of *vdh* (cg2953) by *aroG*^{D146N} from *E. coli* under control of *C. glutamicum* promoter P_{ilvC}	[36]
pK19-Δ*ldhA*	pK19*mobsacB* with a construct for deletion of *ldhA* (cg3219)	[52]
pK19-Δ*sugR*	pK19*mobsacB* with a construct for deletion of *sugR* (cg2115)	[53]
pK19-Δ*aroR*::P_{ilvC}	pK19*mobsacB* with a construct for replacement of *aroR* and the native promoter of *aroF* by *C. glutamicum* promoter P_{ilvC} and an artificial RBS	This work
pK19-Δ*qsuABCD*::P_{tuf}-*qsuC*	pK19*mobsacB* with a construct for replacement of *qsuABCD* (cg0501-cg0504) by *qsuC* (cg0503) with an artificial RBS under control of *C. glutamicum* promoter P_{tuf}	This work
pK19-Δ*ppc*::P_{sod}-*aroB*	pK19*mobsacB* with a construct for replacement of *ppc* (cg1787) by *aroB* (cg1827) with an artificial RBS under control of *C. glutamicum* promoter P_{sod}	This work
pK19-ΔP_{tkt}::P_{tuf}	pK19*mobsacB* with a construct for replacement of the *tkt* (cg1774) promoter by *C. glutamicum* promoter P_{tuf} and artificial RBS	This work
pK19-Δ*iolR*::P_{tuf}-*aroE*	pK19*mobsacB* with a construct for replacement of *iolR* (cg0196) by *aroE* (cg1835) with an artificial RBS under control of *C. glutamicum* promoter P_{tuf}	This work
pK19-Δ*sugR*::*sugR*	pK19*mobsacB* with a construct for reintegration of *sugR* (cg2115) into its native locus	This work
pEKEx3	$Spec^R$, $P_{tac}lacI^q$, pBL1 $oriV_{Cg}$, *C. glutamicum/E. coli* expression shuttle vector	[54]
pEKEx3-*trpE*^{FBR}	$Spec^R$, pEKEx3 overexpressing *trpE*^{S40F} from *E. coli* K12 containing an artificial RBS	This work
pEC-XK99E	Km^R, $P_{trc}lacI^q$, pGA1 $oriV_{Ec}$, *C. glutamicum/E. coli* expression shuttle vector	[55]
pGold	Km^R, $P_{trc}lacI^q$, pGA1 $oriV_{Ec}$, *C. glutamicum/E. coli* expression shuttle vector with *BsaI* recognition site for Golden Gate assembly	This work
pGold-ANMT	Km^R, pGold overexpressing codon harmonized ANMT from *Ruta graveolens* with an artificial RBS	This work
pGold-ANMT-*sahH*	Km^R, pGold overexpressing a synthetic operon with codon harmonized ANMT from *R. graveolens* with an artificial RBS and *sahH* from *C. glutamicum* with an artificial RBS	This work

2.4. *Quantification of Amino Acids and Organic Acids*

Extracellular amino acids and carbohydrates were quantified by high-performance liquid chromatography (HPLC) (1200 series, Agilent Technologies Deutschland GmbH, Böblingen, Germany). The culture supernatants were collected at different time points and centrifuged (20,200× *g*) for HPLC analysis.

For the detection of α-ketoglutarate (α-KG), trehalose, and lactate, an amino exchange column (Aminex, 300 mm × 8 mm, 10 µm particle size, 25 Å pore diameter, CS Chromatographie Service GmbH, 52379 Langerwehe, Germany) was used. The measurements were performed under isocratic conditions for 17 min at 60 °C with 5 mM sulfuric acid and a flow rate of 0.8 mL·min $^{-1}$. The detection was carried out with a Diode Array Detector (DAD, 1200 series, Agilent Technologies, Santa Clara, CA 95051, USA) at 210 nm.

Separation of shikimate, anthranilate, and NMA was performed with a pre-column (LiChrospher 100 RP18 EC-5μ (40 × 4 mm), CS Chromatographie Service GmbH, Langerwehe, Germany) and a main column (LiChrospher 100 RP18 EC-5μ (125 × 4 mm), CS Chromatographie Service GmbH). A mobile phase of buffer A (0.1% trifluoroacetic acid dissolved in water) and buffer B (acetonitrile) was used with a flow rate of 1 $mL{\cdot}min^{-1}$. The following gradient was applied: 0–1 min 10% B; 1–10 min a linear gradient of B from 10% to 70%; 10–12 min 70% B; 12–14 min a linear gradient of B from 70% to 10%; 14–18 min 10% B [41]. The injection volume was 20 μL, and detection was performed with DAD at 210, 280, and 330 nm.

3. Results

3.1. *Corynebacterium glutamicum as Suitable Host for NMA Production*

C. glutamicum is widely used in amino acid fermentation, which operates at a million tons per annum scale [56]; however, it has not been engineered so far for NMA production. As expected, inspection of the genome revealed that there was no gene(s) encoding for a native enzyme that may *N*-methylate anthranilate to yield NMA. To study the growth responses of *C. glutamicum* to anthranilate and NMA, the wild-type strain ATCC13032 (WT) was cultivated with addition of varying anthranilate and NMA concentrations to CGXII minimal medium and 40 $g{\cdot}L^{-1}$ glucose. Neither anthranilate nor NMA were utilized or converted by *C. glutamicum* WT, since their concentrations in supernatants analyzed at the beginning and the end of cultivation were comparable. Maximal biomass concentrations (expressed as ΔOD_{600}) were hardly affected by addition of anthranilate or NMA. By extrapolation, the concentrations of anthranilate (about 36 mM) and NMA (about 34 mM), which reduced the specific growth rate in glucose minimal medium to half-maximal, were determined (Figure 2). Based on the observed tolerance, *C. glutamicum* is a suitable candidate for production of anthranilate and NMA.

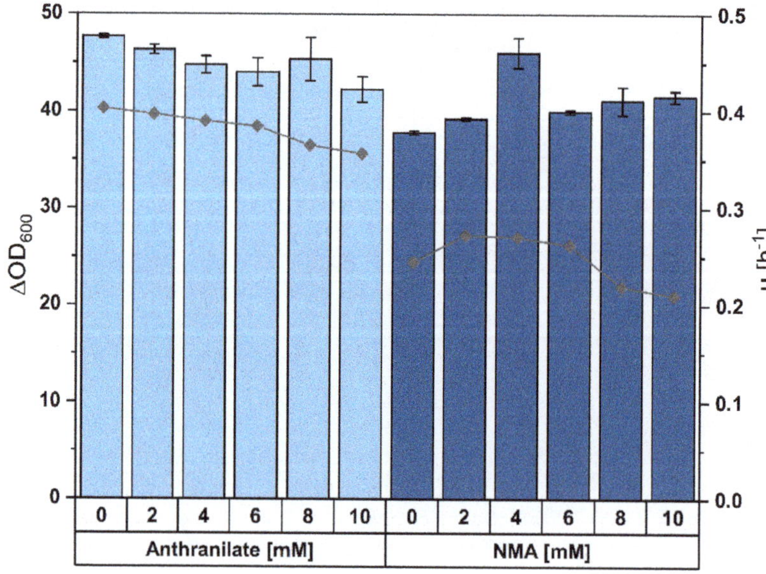

Figure 2. Effect of externally added NMA (bright blue) and anthranilate (dark blue) on biomass formation (columns) and specific growth rate (lines) of *C. glutamicum* strain ATCC13032. Each condition with NMA contained the same amount of methanol (1.65 M) in minimal media. Averages and standard deviation of triplicate cultivations are shown.

3.2. *Construction of a C. glutamicum Platform Strain for Production of Anthranilate*

Since anthranilate, an intermediate of the tryptophan branch in the shikimate pathway, is a direct precursor of NMA, *C. glutamicum* C1* was engineered for increased supply of shikimate pathway intermediates by eliminating bottlenecks and minimizing formation of by-products (Figure 1). Hence, in sequential steps, *aroG*^{D146} encoding feedback resistant 3-deoxy-D-arabino-heptulosonate-7-phosphate (DAHP) synthase from *E. coli* [57] was inserted into the locus of *vdh* coding for vanillin dehydrogenase, which oxidizes vanillin and other aromatic aldehydes such as protocatechic aldehyde [58]. Next, an in-frame deletion of *ldhA* to reduce L-lactate formation (ARO02) and an *sugR* deletion to increase glycolytic gene expression and sugar uptake [59] were introduced to yield strain ARO03.

Upon transformation with pEKEx3 as an empty vector control and pEKEx3-*trpE*^{FBR} for expression of feedback-resistant anthranilate synthase from *E. coli* [60], strains were evaluated regarding their growth behavior, anthranilate production, and formation of by-products. After 48 h of shake flask cultivation, ARO03(pEKEx3) exhibited decreased biomass formation and increased trehalose and α-ketoglutarate accumulation as compared to ARO01(pEKEx3). Expression of *trpE*^{FBR} further decreased biomass formation (i.e., 16.4% less than in empty vector). Comparing strains C1* to ARO03 carrying pEKEx3-*trpE*^{FBR} revealed a stepwise increase both in anthranilate and in shikimate production (Figure 3). For example, ARO03 strain harboring pEKEx3-*trpE*^{FBR} produced 17.6 ± 1.0 mM anthranilate and 6.8 ± 0.8 mM shikimate as compared to C1*(pEKEx3-*trpE*^{FBR}) that accumulated only 9.0 ± 0.2 mM anthranilate and 1.7 ± 0.1 mM shikimate.

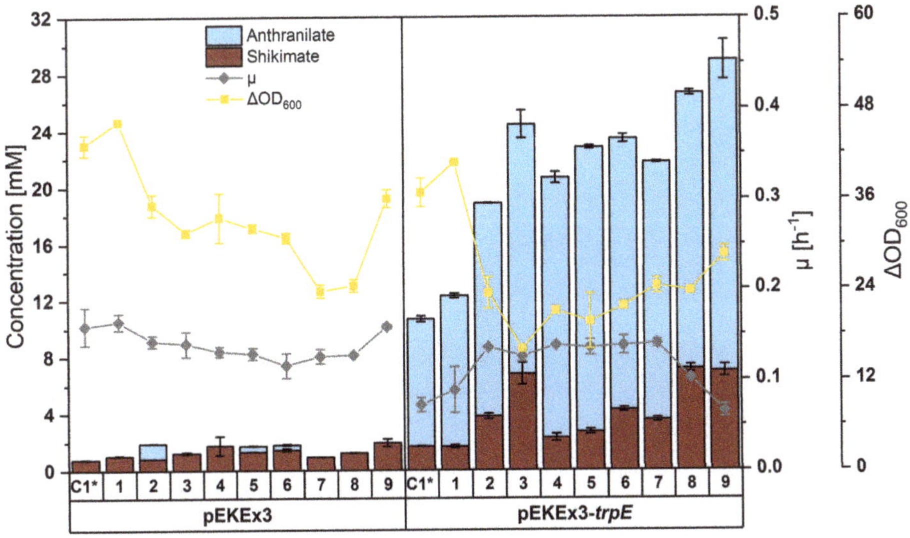

Figure 3. Production of shikimate (maroon bars) and anthranilate (light blue bars), maximal specific growth rate (gray diamonds) and biomass formation (yellow squares) by *C. glutamicum* strains C1* and ARO01 to ARO09 carrying either pEKEx3 (left panel) or pEKEx3-*trpE*^{FBR} (right panel) were grown in shake flasks in CGXII minimal medium with 40 g·L^{-1} glucose for 48 h. Means and arithmetic errors of duplicate cultures are shown.

To further increase the carbon flux towards shikimate, several further metabolic engineering steps were undertaken. In ARO04, the gene *aroR*, which codes for a translational regulatory leader peptide and is located upstream of DHAP synthase gene *aroF* [61], was replaced by an *ilvC* promoter followed by an optimized RBS in order to relieve negative translational control of *aroF* by phenylalanine and tyrosine. As described previously [36], the *qsuABCD* operon was replaced by *qsuC* transcribed from the constitutive strong *tuf* promoter in strain ARO05. This blocked conversion of

3-dehydroshikimate (3-DHS) to the unwanted by-product protocatechuate (PCA) on the one hand and increased the flux from 3-dehydroquinate (3-DHQ) to 3-DHS on the other hand. The replacement of *ppc* encoding phosphoenolpyruvate (PEP) carboxylase by a second copy of endogenous *aroB* encoding 3-DHQ synthase in ARO06 probably increased supply of PEP as precursor for the shikimate pathway, and overexpression of *aroB* increased conversion of DHAP to 3-DHQ. To increase supply of erythrose-4-phosphate (E4P) as second precursor of the shikimate pathway [62], the native promoter upstream of transketolase gene *tkt* was exchanged by the constitutive strong promoter P*tuf* with an artificial RBS. Since *tkt* is co-transcribed with other genes of the pentose phosphate pathway as operon *tkt-tal-zwf-opcA-pgl*, this promoter exchange is expected to increase flux into the pentose phosphate pathway towards E4P in strain ARO07.

Upon transformation with pEKEx3-*trpE*^{FBR}, ARO07 produced only slightly more anthranilate (18.2 ± 0.1 mM) than ARO03(pEKEx3-*trpE*^{FBR}), but less shikimate, trehalose, and α-ketoglutarate (Figure 4). Growth was comparably fast (μ of $0.14 \pm 0.01\ h^{-1}$ compared to $0.13 \pm 0.01\ h^{-1}$), but a higher biomass was reached (OD_{600} of 24.4 ± 1.0 compared with 16.1 ± 0.1) (Figure 3).

In ARO08, shikimate dehydrogenase gene *aroE* was overexpressed from the strong constitutive promoter P*tuf* and used to replace *iolR*. In the absence of IolR, the inositol catabolism operon (cg0197-cg0207), *cg1268*, and PEP carboxykinase gene *pck* are deregulated [63,64], and *iolT1*, which codes for a non-phosphoenolpyruvate dependent phosphotransferase transporter (non-PTS) inositol uptake system, is derepressed. Non-PTS uptake of glucose is known to improve availability of PEP. The final strain, ARO09, is a *sugR*-positive derivative of ARO08. ARO09(pEKEx3-*trpE*^{FBR}) grew faster than ARO7(pEKEx3-*trpE*^{FBR}) (Figure 3) and accumulated less trehalose as unwanted by-product. The maximum anthranilate titer of 22.0 ± 1.4 mM (equivalent to about $3.1\ g{\cdot}L^{-1}$ anthranilate) was achieved with ARO09(pEKEx3-*trpE*^{FBR}) after 48 h of shake flask cultivation. This titer was 2.5 times more than that obtained with C1*(pEKEx3-*trpE*^{FBR}). Taken together, an anthranilate producing *C. glutamicum* strain converting 12.7% of carbon from glucose (Figure 4) to about $3.1\ g{\cdot}L^{-1}$ of anthranilic acid, the direct precursor for NMA, was constructed.

3.3. Establishing Fermentative Production of NMA by C. glutamicum

NMA is synthesized from anthranilate in a single SAM-dependent methylation reaction at its amino group (Figure 1). Therefore, the anthranilate producing *C. glutamicum* strain ARO09(pEKEx3-*trpE*^{FBR}) was used for heterologous expression of the anthranilate *N*-methyltransferase gene ANMT from *R. graveolens*. Transformation of ARO09(pEKEx3-*trpE*^{FBR}) with pGold-ANMT yielded strain NMA104. To improve SAM regeneration, the endogenous *S*-adenosylhomocysteinase gene *sahH* was expressed as synthetic operon with ANMT from plasmid pGold-ANMT-*sahH* and used to transform ARO09(pEKEx3-*trpE*^{FBR}) yielding strain NMA105. As negative control, pGold was introduced into ARO09(pEKEx3-*trpE*^{FBR}) yielding strain NMA103 (Table 1). For comparison, the shikimate producing strain ARO9(pEKEx3) was transformed with pGold, pGold-ANMT, and pGold- ANMT-*sahH* yielding strains NMA100, NMA101, and NMA102, respectively (Table 1).

In order to test for NMA production, strains NMA100 to NMA105 were cultivated in CGXII minimal medium supplemented with $40\ g{\cdot}L^{-1}$ glucose as carbon source. HPLC analysis of supernatants after cultivation for 48 h revealed that NMA100 and NMA103 did not produce NMA, which was expected since they lacked ANMT from *R. graveolens* (Figure 5). Expression of ANMT alone or in combination with endogenous *sahH* resulted in production of about 0.5 mM NMA by strains NMA101 and NMA102, respectively. This indicated functional expression of ANMT from *R. graveolens* in *C. glutamicum*.

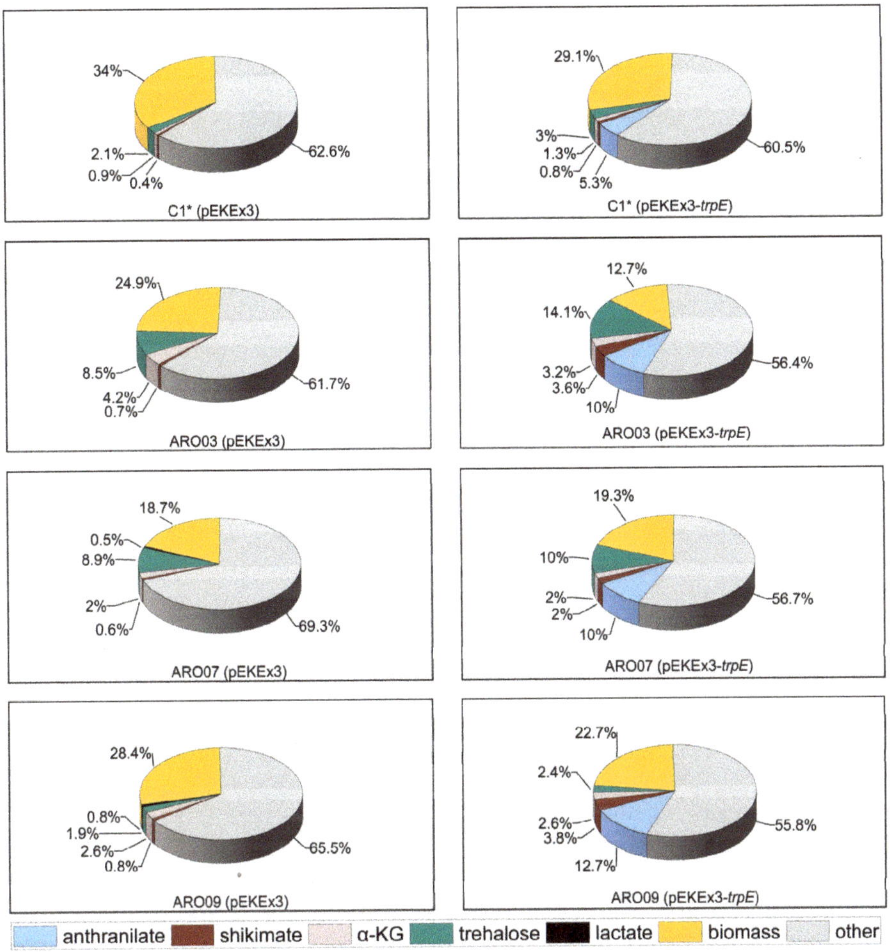

Figure 4. Fate of carbon from glucose in cultivations of *C. glutamicum* ARO strains carrying the empty vector (left) or pEKEx3-$trpE^{FBR}$ (right). Carbon (given in mol%) derived from glucose found after 48 h in secreted products anthranilate (blue), shikimate (maroon), α-ketoglutarate (light red), trehalose (green), lactate (black) as well as in the formed biomass (yellow) are shown for *C. glutamicum* strains C1*, ARO03, ARO07, and ARO09 harboring either pEKEx3 (left) or pEKEx3-$trpE^{FBRfbr}$ (right). Carbon that could not be accounted for is depicted in gray (other). Values were determined from duplicate cultures. Experimental error was less than 20%. Abbreviations used: α-KG, α-ketoglutarate. Carbon distribution of all ARO strains can be found in the Supplementary Data (Figure S1; Figure S2).

Coexpression of $trpE^{FBR}$ to boost anthranilate production with ANMT alone (strain NMA104) resulted in production of 1.7 ± 0.1 mM (0.25 ± 0.02 g·L^{-1}) NMA. The finding that the anthranilate concentration was reduced from 20.8 ± 0.0 mM as obtained with NMA103 to 17.3 ± 0.9 mM (NMA104) indicated that conversion of anthranilate to NMA was incomplete (at about 10 mol%). Upon coexpression of $trpE^{FBR}$ with both ANMT and *sahH* in strain NMA105, 15.8 ± 1.9 mM anthranilate remained as unconverted precursor (Figure 5), and a significantly increased NMA titer of 2.2 ± 0.2 mM was obtained. This maximal titer in shake flasks corresponds to 0.34 ± 0.02 g·L^{-1}. Thus, metabolic engineering of *C. glutamicum* for NMA production was achieved.

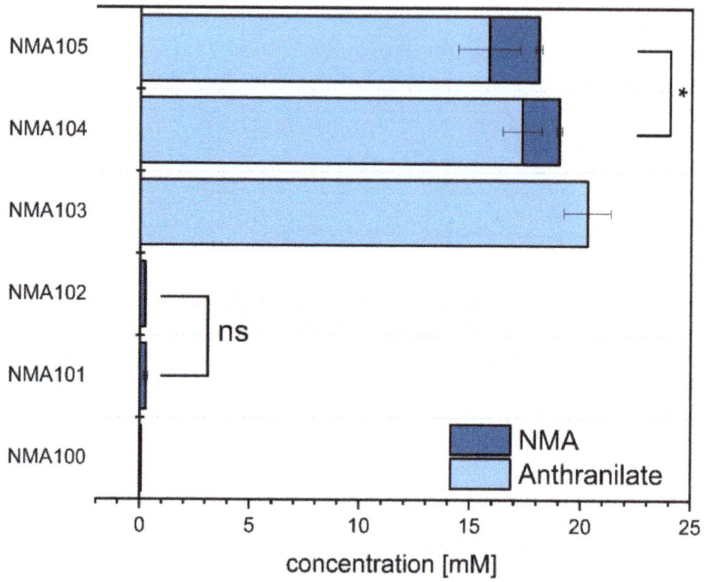

Figure 5. Production of anthranilate (light blue) and NMA (dark blue) by *C. glutamicum* strains NMA100 to NMA105. Cultivation was performed in minimal medium supplemented with 40 $g{\cdot}L^{-1}$ glucose as carbon source. 1 mM IPTG was added for induction of gene expression. Means and standard deviations of triplicate cultures determined after 48 h cultivation are depicted. Significance has been determined for NMA concentrations based on a two-sided unpaired Student's t-test (*: $p < 0.05$; ns: not significant).

3.4. *Fed-Batch Production of NMA in Bioreactors*

For industrial applications, a production in larger volumes is preferable, which runs under controlled conditions to obtain a constant production titer. The stability of the NMA production of the metabolically engineered strain NMA105 was investigated in a fed-batch cultivation. Starting with a working volume of 2 L CGXII minimal medium containing 40 $g{\cdot}L^{-1}$ glucose as carbon source, 160 mL feed (400 $g{\cdot}L^{-1}$ and 150 $g{\cdot}L^{-1}$ $(NH_4)_2SO_4$) was added in a controlled manner depending on the rDOS (see Section 2.2). In total, 104 g glucose was consumed during 48 h fed-batch cultivation with no residual substrate concentrations detectable in the cultivation broth. The strain showed slow growth to OD_{600} 5 in the first 24 h. In the following phase, growth was faster (growth rate of 0.12 h^{-1}, which was comparable to the growth rate observed in shaking flasks), and a maximal optical density of 53 was reached (Figure 6). High concentrations of by-products accumulated, i.e., 1.4 $g{\cdot}L^{-1}$ of the intermediate shikimate and 2.6 $g{\cdot}L^{-1}$ of the direct precursor anthranilate (Figure 6). Compared to production in shaking flasks (Figure 5), a reduced product yield on glucose (4.8 $mg{\cdot}g^{-1}$ as compared to 8.4 $mg{\cdot}g^{-1}$ in shaking flask) and a comparable volumetric productivity were observed, but NMA accumulated to an about 1.5-fold higher titer (0.5 $g{\cdot}L^{-1}$ as compared to 0.34 $g{\cdot}L^{-1}$). Taken together, the fed-batch fermentation with the newly constructed *C. glutamicum* strain NMA105 showed stable production of NMA in bioreactors at the 2 L scale (Figure 6). A final titer of 0.5 $g{\cdot}L^{-1}$ with a volumetric productivity of 0.01 $g{\cdot}L^{-1}{\cdot}h^{-1}$ and a yield of 4.8 $mg{\cdot}g^{-1}$ glucose was achieved.

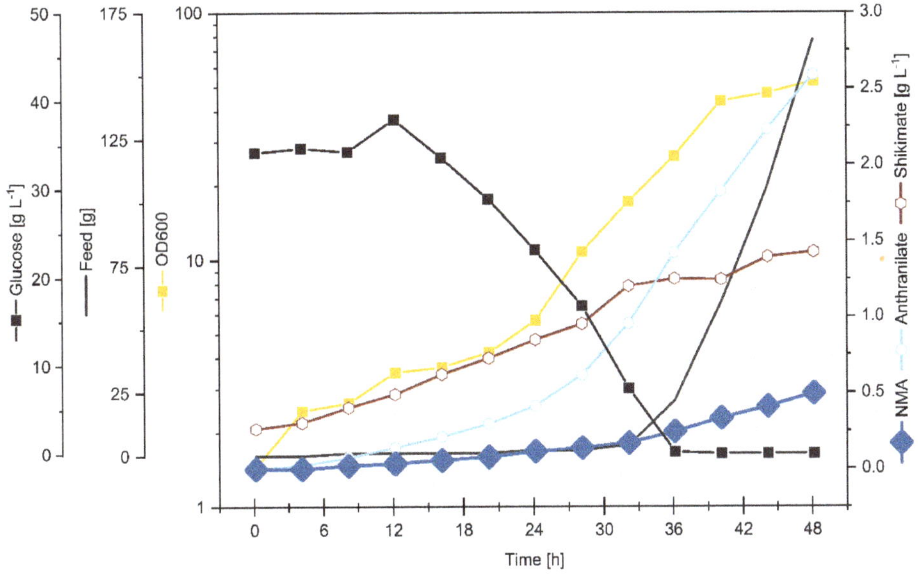

Figure 6. NMA production by *C. glutamicum* strains NMA105 in bioreactors operated in fed-batch mode. The cultivation (initial volume of 2 L) was performed in minimal medium supplemented with 40 g·L^{-1} glucose (dark grey line with squares). The feed (dark grey line) contained 400 g·L^{-1} glucose and 150 g·L^{-1} $(NH_4)_2SO_4$. 1 mM IPTG was added for induction of gene expression during inoculation. OD_{600} (yellow) and concentrations of anthranilate (light blue), shikimate (maroon), and NMA (dark blue) in the culture broth are depicted. One of two representative fermentations is shown.

4. Discussion

N-methylanthranilate production was achieved by applying the plant enzyme *N*-methylanthranilate transferase ANMT of *R. graveolens* in a newly metabolically engineered *C. glutamicum* anthranilate overproducer. *N*-methylanthranilate is known as precursor for several industrially and medically relevant compounds. ANMT of *R. graveolens* showed a narrow substrate specificity when various amino benzoic or benzoic acids or phenolic derivatives were tested as substrates [16]. However, feeding *O*-methylanthranilate (OMA) to *E. coli* expressing ANMT led to production of the flavoring compound *O*-methyl-*N*-methylanthranilate [15]. Hypothetically, ANMT could also be an interesting candidate to produce the pharmaceutically interesting compounds *O*-propyl- or *O*-isopropyl-*N*-methylanthranilate [22,23]. In the biosynthesis of acridone alkaloids, e.g., in *R. graveolens*, *N*-methylation of anthranilate catalyzed by ANMT is a key step preceding CoA activation and, thus, separating primary metabolism (here tryptophan synthesis) from secondary metabolism [16,19]. Recently, production of about 26 mg·L^{-1} 1,3-dihydroxy−10-methylacridone [65] and about 18 mg·L^{-1} 4-hydroxy-1-methyl-2(1H)-quinolone [66] were established in *E. coli* coexpressing ANMT from *R. graveloens*, anthranilate coenzyme A ligase from *P. aeruginosa*, and acridone synthase of *R. graveolens* or the anthraniloyl-CoA anthraniloyltransferase from *P. aeruginosa*. In these biosynthesis pathways, one molecule of NMA is required per one molecule 1,3-dihydroxy-10-methylacridone or 4-hydroxy-1-methyl-2(1H)-quinolone [65,66]. The NMA-producing *C. glutamicum* strain NMA105 developed here may in the future be used in combination with this engineered *E. coli* strain, possibly as synthetic consortium [67,68], or *C. glutamicum* NMA105 itself may be engineered for production of acridone alkaloids.

Biosynthesis of *N*-alkylated amino acids can be catalyzed by other enzymes besides *N*-methyltransferases. However, while reductive amination using free ammonia is known for many enzymes, only few enzyme classes accept alkyl amines for *N*-alkylation, e.g., opine dehydrogenases, *N*-methyl amino acid dehydrogenases, ketimine reductases, pyrroline-5-carboxylate reductases, or imine reductases [12]. These processes differ regarding the substrate spectra of the involved enzymes. For example, anthranilate *N*-methylation described here as well as *N*-methylglutamate production established in *Pseudomonas putida* using *N*-methylglutamate synthase and γ-glutamylmethylamide synthetase of the methylamine assimilation pathway of *Methylobacterium extorquens* [13] have narrow substrate spectra (e.g., GMAS from *Methylovorus mays* also forms γ-glutamylethylamide, also known as theanine [69]) compared with *N*-alkylation using the imine reductase DpkA of *Pseudomonas putida* [12]. Several methylated or ethylated amino acids could be produced by *C. glutamicum* using the wild-type or a mutant version of DpkA and either MMA or ethylamine as substrates [14,34,35]. With respect to aromatic amino acids, *N*-methyl-L-phenylalanine could be obtained from phenylpyruvate by enzyme catalysis using DpkA and MMA [12]; however, production of NMA via DpkA by *N*-alkylamination of a carbonyl precursor of NMA has not been described.

The NMA process described here showed lower titers (0.5 $g{\cdot}L^{-1}$) than the processes depending on reductive alkylamination using MMA (about 32 $g{\cdot}L^{-1}$ *N*-methylalanine [34] and about 9 $g{\cdot}L^{-1}$ sarcosine [14]). This may be due to (a) higher activity of DpkA compared with ANMT, (b) better provision of the precursors pyruvate and glyoxalate than of anthranilate, and/or (c) the requirement of SAM for ANMT as compared to MMA for DpkA. Indeed, purified DpkA has a much higher activity (about 40 $U{\cdot}mg^{-1}$) [70] than purified ANMT (about 0.04 $U{\cdot}mg^{-1}$) [16]. Moreover, while ARO09(pEKEx3-*trpE*^{FBR}) produced 3 $g{\cdot}L^{-1}$ anthranilate (Figure 3), the precursor strains used for production of *N*-methylalanine and sarcosine produced up to 45 $g{\cdot}L^{-1}$ pyruvate [71] and about 5 $g{\cdot}L^{-1}$ glycolate [72], respectively. Third, reductive methylamination using DpkA requires addition of MMA as methyl donor to the medium. This is beneficial since MMA has a low price, is readily available, is tolerated well by *C. glutamicum* [34], and because stoichiometric excess of MMA can be used to drive reductive *N*-methylation by mass action law.

Compared to NMA production by an engineered *E. coli* strain expressing the *N*-methyltransferase of *R. graveloens* [15], the NMA production by engineered *C. glutamicum* using the same enzyme described here resulted in about 12 times higher titers in shaking flask cultivation (370 $mg{\cdot}L^{-1}$ as compared to 29 $mg{\cdot}L^{-1}$). This may be due to the fact that, in this study, *C. glutamicum* was metabolically engineered for improved supply of the direct NMA precursor anthranilate as, e.g., strain ARO09(pEKEx3-*trpE*^{FBR}) produced about 3 $g{\cdot}L^{-1}$ anthranilate. Moreover, while the *E. coli* relied on native SAM regeneration [15], in *C. glutamicum* the endogenous gene for SAM regeneration *sahH* was overexpressed to increase SAM regeneration, and NMA production was improved 1.36-fold (compare 0.34 ± 0.02 $g{\cdot}L^{-1}$ for NMA105 with 0.25 ± 0.02 $g{\cdot}L^{-1}$ for NMA104 in Figure 5). Two bottlenecks observed with the *C. glutamicum* strain engineered here may be overcome by future metabolic engineering: incomplete conversion of shikimate to anthranilate and incomplete *N*-methylation of anthranilate by SAM-dependent ANMT. To improve conversion of shikimate to anthranilate from about half to full conversion (compare about 1.4 $g{\cdot}L^{-1}$ of shikimate and 2.6 $g{\cdot}L^{-1}$ anthranilate produced by NMA105 in bioreactor cultivation; Figure 6), expression of the operon *aroCKB* encoding chorismate synthase, shikimate kinase, and 3-dehydroquinate synthase may be boosted, e.g., by changing the endogenous promoter for the strong promoter P*tuf* and using shikimate kinase from *Methanocaldococcus jannaschii* as shown previously [36]. In addition, various studies have shown that deletion of the chorismate mutase will increase the carbon flux towards tryptophan biosynthesis [36,40,73].

SAM-dependent *N*-methylation of anthranilate by ANMT from *R. graveloens* represents the second bottleneck. ANMT from *R. graveolens* shows high affinity for its substrates (K_M of 7.1 μM for anthranilate and K_M of 3.3 μM for SAM), and inhibition by its product NMA has not been described [16]. On the other hand, the inherently low activity of ANMT as compared, e.g., to DpkA (see above) may limit conversion of anthranilate to NMA. Importantly, regeneration of the methyl donor SAM (Figure 1A) is critical in all SAM-dependent methylation reactions. This is even more important for ANMT from *R. graveolens* because it is inhibited by SAH with a K_I value of 37.2 μM [74]. As shown here and elsewhere [41], overexpression of one gene of the SAM regeneration system (Figure 1A), *S*-adenosylhomocysteine (SAH) hydrolase gene *sahH*, partly overcame SAM limitation since conversion of anthranilate to NMA was improved 1.36-fold (Figure 5). This may be due to reduced inhibition of ANMT from *R. graveolens* by SAH (see above) and/or better SAM regeneration. Irrespective of *sahH* overexpression, not more than about 14 mol% of anthranilate was *N*-methylated to NMA (Figure 5). As shown for OMA production [41], overexpression of SAM synthetase gene *metK* in addition to *sahH* improved SAM regeneration, whereas deletion of cystathionin-γ-synthase gene *metB* and of *mcbR* and *cg3031* that code for transcriptional regulators involved in regulation of methionine biosynthesis were not beneficial. Addition of methionine even reduced the production [41]. These changes and abolishing pathways competing for SAM and its precursor by deletion of homoserine kinase gene *thrB* along with overexpression of *metK* and *vgb*, coding for methionine adenosyltransferase and *Vitreoscilla* hemoglobin, led to a *C. glutamicum* strain secreting about 0.2 $g{\cdot}L^{-1}$ SAM within 48 h [75]. In addition to improving SAM regeneration (as shown here by *sahH* overexpression), it may be beneficial for NMA production to increase SAM biosynthesis and, therefore, the intracellular concentration of SAM. Thus, possibly, NMA production may be improved by overexpression of SAM biosynthesis genes such as *metK*, or by de-repression of SAM biosynthesis, e.g., via deletion of *mcbR*, or by deletion of genes for enzymes competing with use of SAM or of SAM biosynthetic precursors such as *thrB*.

NMA may inhibit anthranilate biosynthesis since NMA was not produced in addition to anthranilate, while the combined titer of NMA and anthranilate remained similar when comparing strains NMA103, NMA104, and NMA105 (Figure 5). Enzymes that are inhibited by NMA have not been described to date. However, product inhibition of anthranilate synthase by anthranilate is known, e.g., in *Streptomyces* [76], which belongs to the actinobacteria as *C. glutamicum*, and in *Salmonella typhimurium* with a K_I of 0.06 mM anthranilate [77]. Here, we used the *E. coli* enzyme TrpE, which is known to be inhibited by tryptophan, which binds at a site distant from the active center (allosteric regulation) [78]. In the mutant $TrpE^{S40F}$, Trp binding is lost as well as allosteric inhibition by Trp [78]. Product inhibition by anthranilate is expected to involve binding to the active center. Since NMA differs from anthranilate just by the *N*-methyl group, it is conceivable that NMA inhibits in a similar way as anthranilate. This may explain that upon NMA production the anthranilate titer decreased (Figure 5).

NMA also affected growth of *C. glutamicum* (34 mM or 5 $g{\cdot}L^{-1}$ reduced the growth rate to half-maximal; Figure 2), but to a lesser extent than OMA, for which a complete growth inhibition was observed at 2 $g{\cdot}L^{-1}$ OMA [41]. Inhibition of growth by OMA was overcome by application of a tributyrin-based extraction method [41]. This approach likely cannot be transferred directly to the NMA process since OMA contains a methylated carboxy group, whereas the amino group is methylated in NMA. Adaptive laboratory evolution (ALE) is an efficient method to select more tolerant strains and has been applied to *C. glutamicum* to select strains with improved tolerance to methanol [79–81] or lignocellulose-derived inhibitors [82].

Taken together, this study characterized NMA production by metabolically engineered *C. glutamicum*, and a first bioreactor process leading to a final titer of 0.5 $g{\cdot}L^{-1}$ NMA with a volumetric productivity of 0.01 $g{\cdot}L^{-1}{\cdot}h^{-1}$ and a yield of 4.8 $mg{\cdot}g^{-1}$ glucose was achieved. This strain provides the basis to develop an industrially competitive NMA process and shows potential to enable access to a fermentative route to pharmaceutically relevant secondary metabolites such as the acridone alkaloids.

Supplementary Materials: The following figures are available online at http://www.mdpi.com/2076-2607/8/6/866/s1, Figure S1: Carbon flux analysis of anthranilate producing *C. glutamicum* ARO strains, Figure S2: Carbon flux analysis of *C. glutamicum* ARO strains harboring the plasmid pEKEx3, Table S1: Codon harmonized nucleotide sequence (5′-3′) of the plant gene ANMT.

Author Contributions: T.W., N.A.M., A.B., K.C., L.F., A.K., and M.M. constructed strains. T.W., N.A.M., A.B., K.C., L.F., A.K., and M.M. performed the experiments. T.W., A.B., K.C., L.F., A.K., M.M., J.M.R., and V.F.W. analyzed the data. J.-H.L., K.C., and V.F.W. provided resources. T.W., A.B., K.C., L.F., A.K., M.M., and J.M.R. drafted the manuscript. T.W. and V.F.W. reviewed and edited the manuscript. V.F.W. finalized the manuscript. All authors agreed to the final version of the manuscript.

Funding: Funding by ERACoBiotech via grant INDIE (BMEL 22023517) is gratefully acknowledged. Support for the Article Processing Charge by the Deutsche Forschungsgemeinschaft and the Open Access Publication Fund of Bielefeld University is acknowledged.

Acknowledgments: We want to thank Anne-Laure Ricord for cloning expression plasmids and Thomas Schäffer for his support during fed-batch bioreactor cultivation.

Conflicts of Interest: The authors declare no conflicts of interest. The funders had no role in the design of the study; in the collection, analyses, or interpretation of data; in the writing of the manuscript, or in the decision to publish the results.

References

1. Liu, L.-K.; Abdelwahab, H.; del Campo, J.S.M.; Mehra-Chaudhary, R.; Sobrado, P.; Tanner, J.J. The Structure of the Antibiotic Deactivating, N-hydroxylating Rifampicin Monooxygenase. *J. Boil. Chem.* **2016**, *291*, 21553–21562. [CrossRef] [PubMed]
2. Zhao, D.; Yu, Y.; Shen, Y.; Liu, Q.; Zhao, Z.; Sharma, R.; Reiter, R.J. Melatonin Synthesis and Function: Evolutionary History in Animals and Plants. *Front. Endocrinol.* **2019**, *10*, 249. [CrossRef] [PubMed]
3. Chen, Y.; Scanlan, J.; Song, L.; Crombie, A.T.; Rahman, T.; Schäfer, H.; Murrell, J.C. γ-Glutamylmethylamide Is an Essential Intermediate in the Metabolism of Methylamine by *Methylocella silvestris*. *Appl. Environ. Microbiol.* **2010**, *76*, 4530–4537. [CrossRef] [PubMed]
4. Chatterjee, J.; Rechenmacher, F.; Kessler, H. N-Methylation of Peptides and Proteins: An Important Element for Modulating Biological Functions. *Angew. Chem. Int. Ed.* **2012**, *52*, 254–269. [CrossRef] [PubMed]
5. Gazdik, M.; O'Neill, M.; Lopaticki, S.; Lowes, K.N.; Smith, B.J.; Cowman, A.F.; Boddey, J.A.; Gasser, R. The effect of N-methylation on transition state mimetic inhibitors of the *Plasmodium protease*, plasmepsin V. *MedChemComm* **2015**, *6*, 437–443. [CrossRef]
6. Aurelio, L.; Brownlee, R.T.C.; Hughes, A.B. Synthetic Preparation of N-Methyl-α-amino Acids. *Chem. Rev.* **2004**, *104*, 5823–5846. [CrossRef]
7. de Marco, R.; Leggio, A.; Liguori, A.; Marino, T.; Perri, F.; Russo, N. Site-Selective Methylation ofNβ-Nosyl Hydrazides of N-Nosyl Protected α-Amino Acids. *J. Org. Chem.* **2010**, *75*, 3381–3386. [CrossRef]
8. Belsito, E.; di Gioia, M.L.; Greco, A.; Leggio, A.; Liguori, A.; Perri, F.; Siciliano, C.; Viscomi, M.C. N-Methyl-N-nosyl-β3-amino Acids. *J. Org. Chem.* **2007**, *72*, 4798–4802. [CrossRef]
9. Freidinger, R.M.; Hinkle, J.S.; Perlow, D.S. Synthesis of 9-fluorenylmethyloxycarbonyl-protected N-alkyl amino acids by reduction of oxazolidinones. *J. Org. Chem.* **1983**, *48*, 77–81. [CrossRef]
10. di Gioia, M.L.; Leggio, A.; Malagrinò, F.; Romio, E.; Siciliano, C.; Liguori, A. N-Methylated α-Amino Acids and Peptides: Synthesis and Biological Activity. *Mini-Rev. Med. Chem.* **2016**, *16*, 1. [CrossRef]
11. Sharma, A.; Kumar, A.; Monaim, S.A.H.A.; Jad, Y.E.; El-Faham, A.; de la Torre, B.G.; Albericio, F. N-methylation in amino acids and peptides: Scope and limitations. *Biopolymers* **2018**, *109*, e23110. [CrossRef] [PubMed]
12. Hyslop, J.F.; Lovelock, S.L.; Watson, A.J.B.; Sutton, P.W.; Roiban, G.-D. N-Alkyl-α-amino acids in Nature and their biocatalytic preparation. *J. Biotechnol.* **2019**, *293*, 56–65. [CrossRef] [PubMed]
13. Mindt, M.; Walter, T.; Risse, J.M.; Wendisch, V. Fermentative Production of N-Methylglutamate From Glycerol by Recombinant Pseudomonas putida. *Front. Bioeng. Biotechnol.* **2018**, *6*, 159. [CrossRef] [PubMed]
14. Mindt, M.; Hannibal, S.; Heuser, M.; Risse, J.M.; Sasikumar, K.; Nampoothiri, K.M.; Wendisch, V. Fermentative Production of N-Alkylated Glycine Derivatives by Recombinant *Corynebacterium glutamicum* Using a Mutant of Imine Reductase DpkA From *Pseudomonas putida*. *Front. Bioeng. Biotechnol.* **2019**, *7*, 232. [CrossRef]
15. Lee, H.L.; Kim, S.-Y.; Kim, E.J.; Han, D.Y.; Kim, B.-G.; Ahn, J.-H. Synthesis of Methylated Anthranilate Derivatives Using Engineered Strains of *Escherichia coli*. *J. Microbiol. Biotechnol.* **2019**, *29*, 839–844. [CrossRef]
16. Rohde, B.; Hans, J.; Martens, S.; Baumert, A.; Hunziker, P.; Matern, U. Anthranilate N-methyltransferase, a branch-point enzyme of acridone biosynthesis. *Plant J.* **2007**, *53*, 541–553. [CrossRef]
17. Mugford, S.T.; Louveau, T.; Melton, R.; Qi, X.; Bakht, S.; Hill, L.; Tsurushima, T.; Honkanen, S.; Rosser, S.J.; Lomonossoff, G.P.; et al. Modularity of plant metabolic gene clusters: A trio of linked genes that are collectively required for acylation of triterpenes in oat. *Plant Cell* **2013**, *25*, 1078–1092. [CrossRef]
18. Baumert, A.; Schmidt, J.; Gröger, D. Synthesis and mass spectral analysis of coenzyme a thioester of anthranilic acid and itsN-methyl derivative involved in acridone alkaloid biosynthesis. *Phytochem. Anal.* **1993**, *4*, 165–170. [CrossRef]

19. Michael, J.P. Acridone Alkaloids. In *The Alkaloids: Chemistry and Biology*; Elsevier BV: Amsterdam, The Netherlands, 2017; pp. 1–108.
20. Piboonprai, K.; Khumkhrong, P.; Khongkow, M.; Yata, T.; Ruangrungsi, N.; Chansriniyom, C.; Iempridee, T. Anticancer activity of arborinine from Glycosmis parva leaf extract in human cervical cancer cells. *Biochem. Biophys. Res. Commun.* **2018**, *500*, 866–872. [CrossRef]
21. *Flavours and Fragrances*; Springer Science and Business Media LLC: Berlin/Heidelberg, Germany, 2007.
22. Radulović, N.S.; Miltojevic, A.; McDermott, M.; Waldren, S.; Parnell, J.A.; Pinheiro, M.M.G.; Fernandes, P.D.; Menezes, F.D.S. Identification of a new antinociceptive alkaloid isopropyl N-methylanthranilate from the essential oil of *Choisya ternata* Kunth. *J. Ethnopharmacol.* **2011**, *135*, 610–619. [CrossRef]
23. Pinheiro, M.M.G.; Miltojevic, A.; Radulović, N.S.; Abdul-Wahab, I.R.; Boylan, F.; Fernandes, P.D. Anti-Inflammatory Activity of *Choisya ternata* Kunth Essential Oil, Ternanthranin, and Its Two Synthetic Analogs (Methyl and Propyl N-Methylanthranilates). *PLoS ONE* **2015**, *10*, e0121063. [CrossRef] [PubMed]
24. Wendisch, V.F. Metabolic engineering advances and prospects for amino acid production. *Metab. Eng.* **2020**, *58*, 17–34. [CrossRef] [PubMed]
25. Ikeda, M. Sugar transport systems in Corynebacterium glutamicum: Features and applications to strain development. *Appl. Microbiol. Biotechnol.* **2012**, *96*, 1191–1200. [CrossRef] [PubMed]
26. Blombach, B.; Seibold, G. Carbohydrate metabolism in *Corynebacterium glutamicum* and applications for the metabolic engineering of l-lysine production strains. *Appl. Microbiol. Biotechnol.* **2010**, *86*, 1313–1322. [CrossRef] [PubMed]
27. Polen, T.; Schluesener, D.; Poetsch, A.; Bott, M.; Wendisch, V. Characterization of citrate utilization in *Corynebacterium glutamicum* by transcriptome and proteome analysis. *FEMS Microbiol. Lett.* **2007**, *273*, 109–119. [CrossRef] [PubMed]
28. Arndt, A.; Auchter, M.; Ishige, T.; Wendisch, V.F.; Eikmanns, B.J. Ethanol catabolism in Corynebacterium glutamicum. *J. Mol. Microbiol. Biotechnol.* **2008**, *15*, 222–233. [CrossRef]
29. Jorge, J.M.P.; Leggewie, C.; Wendisch, V. A new metabolic route to produce gamma-aminobutyric acid by *Corynebacterium glutamicum* from glucose. *Amino Acids* **2016**, *48*, 2519–2531. [CrossRef]
30. Jorge, J.M.; Pérez-García, F.; Wendisch, V. A new metabolic route for the fermentative production of 5-aminovalerate from glucose and alternative carbon sources. *Bioresour. Technol.* **2017**, *245*, 1701–1709. [CrossRef]
31. Shin, J.; Park, S.H.; Oh, Y.H.; Choi, J.W.; Lee, M.H.; Cho, J.S.; Jeong, K.J.; Joo, J.C.; Yu, J.; Park, S.J.; et al. Metabolic engineering of *Corynebacterium glutamicum* for enhanced production of 5-aminovaleric acid. *Microb. Cell Fact.* **2016**, *15*, 174. [CrossRef]
32. Pérez-García, F.; Risse, J.M.; Friehs, K.; Wendisch, V. Fermentative production of L-pipecolic acid from glucose and alternative carbon sources. *Biotechnol. J.* **2017**, *12*, 1600646. [CrossRef]
33. Pérez-García, F.; Peters-Wendisch, P.; Wendisch, V. Engineering *Corynebacterium glutamicum* for fast production of l-lysine and l-pipecolic acid. *Appl. Microbiol. Biotechnol.* **2016**, *100*, 8075–8090. [CrossRef] [PubMed]
34. Mindt, M.; Risse, J.M.; Gruß, H.; Sewald, N.; Eikmanns, B.J.; Wendisch, V. One-step process for production of N-methylated amino acids from sugars and methylamine using recombinant *Corynebacterium glutamicum* as biocatalyst. *Sci. Rep.* **2018**, *8*, 12895. [CrossRef] [PubMed]
35. Mindt, M.; Heuser, M.; Wendisch, V. Xylose as preferred substrate for sarcosine production by recombinant *Corynebacterium glutamicum*. *Bioresour. Technol.* **2019**, *281*, 135–142. [CrossRef] [PubMed]
36. Purwanto, H.S.; Kang, M.-S.; Ferrer, L.; Han, S.-S.; Lee, J.-Y.; Kim, H.-S.; Lee, J.-H. Rational engineering of the shikimate and related pathways in *Corynebacterium glutamicum* for 4-hydroxybenzoate production. *J. Biotechnol.* **2018**, *282*, 92–100. [CrossRef]
37. Kallscheuer, N.; Marienhagen, J. *Corynebacterium glutamicum* as platform to produce hydroxybenzoic acids. *Microb. Cell Factories* **2018**, *17*, 70. [CrossRef]
38. Okai, N.; Miyoshi, T.; Takeshima, Y.; Kuwahara, H.; Ogino, C.; Kondo, A. Production of protocatechuic acid by *Corynebacterium glutamicum* expressing chorismate-pyruvate lyase from *Escherichia coli*. *Appl. Microbiol. Biotechnol.* **2015**, *100*, 135–145. [CrossRef]
39. Veldmann, K.H.; Dachwitz, S.; Risse, J.M.; Lee, J.-H.; Sewald, N.; Wendisch, V. Bromination of L-tryptophan in a Fermentative Process with *Corynebacterium glutamicum*. *Front. Bioeng. Biotechnol.* **2019**, *7*, 219. [CrossRef]

40. Veldmann, K.H.; Minges, H.; Sewald, N.; Lee, J.-H.; Wendisch, V. Metabolic engineering of *Corynebacterium glutamicum* for the fermentative production of halogenated tryptophan. *J. Biotechnol.* **2019**, *291*, 7–16. [CrossRef]
41. Luo, Z.W.; Cho, J.S.; Lee, S.Y. Microbial production of methyl anthranilate, a grape flavor compound. *Proc. Natl. Acad. Sci. USA* **2019**, *116*, 10749–10756. [CrossRef]
42. Baumgart, M.; Unthan, S.; Kloß, R.; Radek, A.; Polen, T.; Tenhaef, N.; Müller, M.F.; Küberl, A.; Siebert, D.; Brühl, N.; et al. *Corynebacterium glutamicum* Chassis C1*: Building and Testing a Novel Platform Host for Synthetic Biology and Industrial Biotechnology. *ACS Synth. Boil.* **2017**, *7*, 132–144. [CrossRef]
43. Hanahan, D. Studies on transformation of *Escherichia coli* with plasmids. *J. Mol. Boil.* **1983**, *166*, 557–580. [CrossRef]
44. Eggeling, L.; Bott, M. *Handbook of Corynebacterium glutamicum*; CRC Press: Boca Raton, FL, USA, 2005; ISBN 978-1-4200-3969-6.
45. Simon, R.; Priefer, U.; Pühler, A. A Broad Host Range Mobilization System for In Vivo Genetic Engineering: Transposon Mutagenesis in Gram Negative Bacteria. *Biotechnology* **1983**, *1*, 784–791. [CrossRef]
46. Green, M.R.; Sambrook, J.; Sambrook, J. *Molecular Cloning: A Laboratory Manual*, 4th ed.; Cold Spring Harbor Laboratory Press: Cold Spring Harbor, NY, USA, 2012; ISBN 978-1-936113-41-5.
47. Codon Usage Database. Available online: https://www.kazusa.or.jp/codon/ (accessed on 20 May 2020).
48. Salis, H.M.; Mirsky, E.A.; Voigt, C.A. Automated design of synthetic ribosome binding sites to control protein expression. *Nat. Biotechnol.* **2009**, *27*, 946–950. [CrossRef] [PubMed]
49. De Novo DNA: The Future of Genetic Systems Design and Engineering. Available online: https://www.denovodna.com/software/design_rbs_calculator (accessed on 22 May 2020).
50. Schäfer, A.; Tauch, A.; Jäger, W.; Kalinowski, J.; Thierbach, G.; Pühler, A. Small mobilizable multi-purpose cloning vectors derived from the *Escherichia coli* plasmids pK18 and pK19: Selection of defined deletions in the chromosome of *Corynebacterium glutamicum*. *Gene* **1994**, *145*, 69–73. [CrossRef]
51. Heider, S.A.E.; Peters-Wendisch, P.; Netzer, R.; Stafnes, M.; Brautaset, T.; Wendisch, V. Production and glucosylation of C50 and C40 carotenoids by metabolically engineered *Corynebacterium glutamicum*. *Appl. Microbiol. Biotechnol.* **2013**, *98*, 1223–1235. [CrossRef] [PubMed]
52. Blombach, B.; Riester, T.; Wieschalka, S.; Ziert, C.; Youn, J.-W.; Wendisch, V.; Eikmanns, B.J. *Corynebacterium glutamicum* Tailored for Efficient Isobutanol Production. *Appl. Environ. Microbiol.* **2011**, *77*, 3300–3310. [CrossRef]
53. Engels, V.; Wendisch, V. The DeoR-Type Regulator SugR Represses Expression of ptsG in *Corynebacterium glutamicum*. *J. Bacteriol.* **2007**, *189*, 2955–2966. [CrossRef]
54. Stansen, C.; Uy, D.; Delaunay, S.; Eggeling, L.; Goergen, J.-L.; Wendisch, V. Characterization of a *Corynebacterium glutamicum* Lactate Utilization Operon Induced during Temperature-Triggered Glutamate Production. *Appl. Environ. Microbiol.* **2005**, *71*, 5920–5928. [CrossRef]
55. Kirchner, O.; Tauch, A. Tools for genetic engineering in the amino acid-producing bacterium *Corynebacterium glutamicum*. *J. Biotechnol.* **2003**, *104*, 287–299. [CrossRef]
56. Wendisch, V.; Lee, J.-H. *Metabolic Engineering in Corynebacterium Glutamicum*; Springer Science and Business Media LLC: Berlin/Heidelberg, Germany, 2020; pp. 287–322.
57. Kikuchi, Y.; Tsujimoto, K.; Kurahashi, O. Mutational analysis of the feedback sites of phenylalanine-sensitive 3-deoxy-D-arabino-heptulosonate-7-phosphate synthase of *Escherichia coli*. *Appl. Environ. Microbiol.* **1997**, *63*, 761–762. [CrossRef]
58. Lee, J.-H.; Wendisch, V. Production of amino acids—Genetic and metabolic engineering approaches. *Bioresour. Technol.* **2017**, *245*, 1575–1587. [CrossRef] [PubMed]
59. Hasegawa, S.; Tanaka, Y.; Suda, M.; Jojima, T.; Inui, M. Enhanced Glucose Consumption and Organic Acid Production by Engineered *Corynebacterium glutamicum* Based on Analysis of a pfkB1 Deletion Mutant. *Appl. Environ. Microbiol.* **2016**, *83*. [CrossRef] [PubMed]
60. Sander, T.; Farke, N.; Diehl, C.; Kuntz, M.; Glatter, T.; Link, H. Allosteric Feedback Inhibition Enables Robust Amino Acid Biosynthesis in *E. coli* by Enforcing Enzyme Overabundance. *Cell Syst.* **2019**, *8*, 66–75. [CrossRef]
61. Neshat, A.; Mentz, A.; Rückert, C.; Kalinowski, J. Transcriptome sequencing revealed the transcriptional organization at ribosome-mediated attenuation sites in *Corynebacterium glutamicum* and identified a novel attenuator involved in aromatic amino acid biosynthesis. *J. Biotechnol.* **2014**, *190*, 55–63. [CrossRef]

62. Draths, K.M.; Pompliano, D.L.; Conley, D.L.; Frost, J.W.; Berry, A.; Disbrow, G.L.; Staversky, R.J.; Lievense, J.C. Biocatalytic synthesis of aromatics from D-glucose: The role of transketolase. *J. Am. Chem. Soc.* **1992**, *114*, 3956–3962. [CrossRef]
63. Klaffl, S.; Brocker, M.; Kalinowski, J.; Eikmanns, B.J.; Bott, M. Complex Regulation of the Phosphoenolpyruvate Carboxykinase Gene pck and Characterization of Its GntR-Type Regulator IolR as a Repressor of myo-Inositol Utilization Genes in *Corynebacterium glutamicum*. *J. Bacteriol.* **2013**, *195*, 4283–4296. [CrossRef] [PubMed]
64. Hyeon, J.E.; Kang, D.H.; Kim, Y.I.; You, S.K.; Han, S.O. GntR-Type Transcriptional Regulator PckR Negatively Regulates the Expression of Phosphoenolpyruvate Carboxykinase in *Corynebacterium glutamicum*. *J. Bacteriol.* **2012**, *194*, 2181–2188. [CrossRef]
65. Choi, G.-S.; Choo, H.J.; Kim, B.-G.; Ahn, J.-H. Synthesis of acridone derivatives via heterologous expression of a plant type III polyketide synthase in *Escherichia coli*. *Microb. Cell Fact.* **2020**, *19*, 1–11. [CrossRef]
66. Choo, H.J.; Ahn, J.-H. Synthesis of Three Bioactive Aromatic Compounds by Introducing Polyketide Synthase Genes into Engineered *Escherichia coli*. *J. Agric. Food Chem.* **2019**, *67*, 8581–8589. [CrossRef]
67. Sgobba, E.; Stumpf, A.K.; Vortmann, M.; Jagmann, N.; Krehenbrink, M.; Dirks-Hofmeister, M.E.; Moerschbacher, B.M.; Philipp, B.; Wendisch, V. Synthetic *Escherichia coli-Corynebacterium glutamicum* consortia for l-lysine production from starch and sucrose. *Bioresour. Technol.* **2018**, *260*, 302–310. [CrossRef]
68. Sgobba, E.; Wendisch, V. Synthetic microbial consortia for small molecule production. *Curr. Opin. Biotechnol.* **2020**, *62*, 72–79. [CrossRef] [PubMed]
69. Yamamoto, S.; Morihara, Y.; Wakayama, M.; Tachiki, T. Theanine Production by Coupled Fermentation with Energy Transfer Using γ-Glutamylmethylamide Synthetase of *Methylovorus mays* No. 9. *Biosci. Biotechnol. Biochem.* **2008**, *72*, 1206–1211. [CrossRef] [PubMed]
70. Watanabe, S.; Tanimoto, Y.; Yamauchi, S.; Tozawa, Y.; Sawayama, S.; Watanabe, Y. Identification and characterization of trans-3-hydroxy-l-proline dehydratase and Δ(1)-pyrroline-2-carboxylate reductase involved in trans-3-hydroxy-l-proline metabolism of bacteria. *FEBS Open Biol.* **2014**, *4*, 240–250. [CrossRef] [PubMed]
71. Wieschalka, S.; Blombach, B.; Eikmanns, B.J. Engineering *Corynebacterium glutamicum* to produce pyruvate. *Appl. Microbiol. Biotechnol.* **2012**, *94*, 449–459. [CrossRef]
72. Zahoor, A.; Otten, A.; Wendisch, V. Metabolic engineering of *Corynebacterium glutamicum* for glycolate production. *J. Biotechnol.* **2014**, *192*, 366–375. [CrossRef]
73. Li, P.-P.; Liu, Y.-J.; Liu, S.-J. Genetic and biochemical identification of the chorismate mutase from *Corynebacterium glutamicum*. *Microbiology* **2009**, *155*, 3382–3391. [CrossRef]
74. Maier, W.; Baumert, A.; Gröger, D. Partial Purification and Characterization of S-Adenosyl-L- Methionine: Anthranilic Acid N-Methyltransferase from Ruta Cell Suspension Cultures. *J. Plant Physiol.* **1995**, *145*, 1–6. [CrossRef]
75. Han, G.; Hu, X.; Qin, T.; Li, Y.; Wang, X. Metabolic engineering of *Corynebacterium glutamicum* ATCC13032 to produce S -adenosyl- l -methionine. *Enzym. Microb. Technol.* **2016**, *83*, 14–21. [CrossRef]
76. Francis, M.M.; Vining, L.C.; Westlake, D.W. Characterization, and regulation of anthranilate synthetase from a chloramphenicol-producing streptomycete. *J. Bacteriol.* **1978**, *134*, 10–16. [CrossRef]
77. Henderson, E.J.; Nagano, H.; Zalkin, H.; Hwang, L.H. The anthranilate synthetase-anthranilate 5-phosphoribosylpyrophosphate phosphoribosyltransferase aggregate. Purification of the aggregate and regulatory properties of anthranilate synthetase. *J. Boil. Chem.* **1970**, *245*, 1416–1423.
78. Caligiuri, M.G.; Bauerle, R. Identification of amino acid residues involved in feedback regulation of the anthranilate synthase complex from *Salmonella typhimurium*. Evidence for an amino-terminal regulatory site. *J. Boil. Chem.* **1991**, *266*, 8328–8335.
79. Leßmeier, L.; Wendisch, V. Identification of two mutations increasing the methanol tolerance of *Corynebacterium glutamicum*. *BMC Microbiol.* **2015**, *15*, 216. [CrossRef]
80. Wang, Y.; Fan, L.; Tuyishime, P.; Liu, J.; Zhang, K.; Gao, N.; Zhang, Z.; Ni, X.; Feng, J.; Yuan, Q.; et al. Adaptive laboratory evolution enhances methanol tolerance and conversion in engineered *Corynebacterium glutamicum*. *Commun. Boil.* **2020**, *3*, 217. [CrossRef] [PubMed]

81. Hennig, G.; Haupka, C.; Brito, L.F.; Rückert, C.; Cahoreau, E.; Heux, S.; Wendisch, V.F. Methanol-Essential Growth of *Corynebacterium glutamicum*: Adaptive Laboratory Evolution Overcomes Limitation due to Methanethiol Assimilation Pathway. *Int. J. Mol. Sci.* **2020**, *21*, 3617. [CrossRef] [PubMed]
82. Wang, X.; Khushk, I.; Xiao, Y.; Gao, Q.; Bao, J. Tolerance improvement of *Corynebacterium glutamicum* on lignocellulose derived inhibitors by adaptive evolution. *Appl. Microbiol. Biotechnol.* **2017**, *102*, 377–388. [CrossRef] [PubMed]

microorganisms

MDPI

Article

Alternative Extraction and Characterization of Nitrogen-Containing Azaphilone Red Pigments and Ergosterol Derivatives from the Marine-Derived Fungal *Talaromyces* sp. 30570 Strain with Industrial Relevance

Juliana Lebeau [1], Thomas Petit [1,2], Mireille Fouillaud [1], Laurent Dufossé [1] and Yanis Caro [1,2,*]

1 Laboratoire de Chimie et de Biotechnologie des Produits Naturels, CHEMBIOPRO, Université de La Réunion, 15 Avenue René Cassin, CS 92003, F-97744 Saint-Denis, France; juliana.lebeau@univ-reunion.fr (J.L.); thomas.petit@univ-reunion.fr (T.P.); mireille.fouillaud@univ-reunion.fr (M.F.); laurent.dufosse@univ-reunion.fr (L.D.)

2 Département Hygiène Sécurité Environnement (HSE), IUT La Réunion, Université de La Réunion, 40 Avenue de Soweto, BP 373, F-97455 Saint-Pierre, France

* Correspondence: yanis.caro@univ-reunion.fr

Received: 1 November 2020; Accepted: 1 December 2020; Published: 3 December 2020

Abstract: Many species of *Talaromyces* of marine origin could be considered as non-toxigenic fungal cell factory. Some strains could produce water-soluble active biopigments in submerged cultures. These fungal pigments are of interest due to their applications in the design of new pharmaceutical products. In this study, the azaphilone red pigments and ergosterol derivatives produced by a wild type of *Talaromyces* sp. 30570 (CBS 206.89 B) marine-derived fungal strain with industrial relevance were described. The strain was isolated from the coral reef of the Réunion island. An alternative extraction of the fungal pigments using high pressure with eco-friendly solvents was studied. Twelve different red pigments were detected, including two pigmented ergosterol derivatives. Nine metabolites were identified using HPLC-PDA-ESI/MS as *Monascus*-like azaphilone pigments. In particular, derivatives of nitrogen-containing azaphilone red pigment, like PP-R, 6-[(Z)-2-Carboxyvinyl]-*N*-GABA-PP-V, *N*-threonine-monascorubramin, *N*-glutaryl-rubropunctamin, monascorubramin, and presumed *N*-threonyl-rubropunctamin (or acid form of the pigment PP-R) were the major pigmented compounds produced. Interestingly, the bioproduction of these red pigments occurred only when complex organic nitrogen sources were present in the culture medium. These findings are important for the field of the selective production of *Monascus*-like azaphilone red pigments for the industries.

Keywords: *Talaromyces*; azaphilone; marine fungi; *N*-threonyl-rubropunctamin; PP-R; greener extraction; red pigments; fungal pigments

1. Introduction

With the progress of biotechnologies, the investigation and exploitation of rich natural sources to isolate natural products with commercial applications has gained increasing interest. Interestingly, the quest for novel drugs has driven research back to look closer at what nature has to offer: biodiversity and untapped natural resources [1,2]. Microorganisms represent a vast repertoire of natural products, many of them with industrial importance. Industrially important fungal bioactive compounds, such as enzymes, organic acids, biochemicals and pigments (with shades of orange, yellow, red, etc.), can be produced from specific fungi [3–7]. As some synthetic colorants have carcinogenic and teratogenic effects, fungal pigments represent an alternative source of natural colorants that are independent of agro-climatic conditions [7,8]. Red colorants of fungal origin have become more and more valued and sought after in the industries, like textiles, food, cosmetics, and pharmaceutics [9,10]. Indeed, to this day, very few stable red colorants of natural origin are available for the industries. Consequently, fungal red pigments are now well established in the industry among the natural colorants, competing with plant and microalgae pigments [3,11].

Fungi of marine origin represent a source of active metabolites exerting pharmacological properties for drug applications [2,6]. In accordance with their genetic potential, some marine-derived fungal strains of *Talaromyces* produced toxin-free polyketide-based pigments and could then be exploited in the industries as a non-toxigenic fungal cell factory in future. Polyketide-based pigments are characterized by a multitude of complex and diverse chemical structures, including quinones (naphthoquinones, hydroxy-anthraquinones) and azaphilones [5]. They involve biosynthetic pathways catalyzed by multiple polyketide synthase enzymes (PKS). The biological properties of fungal azaphilone pigments with pyrone–quinone structures may open new avenues for their use in the production of valuable drugs for medical use. Since ancient times, the fermentation of *Monascus* species has been used to color food products (like meat, wine, cheese, rice and koji) in Asian far-east countries. These fungi produced well-known, yellow-orange-red, azaphilone-based pigments [3,5], but their use as food colorants is not allowed in the USA and in European countries due to the occasional occurrence of the mycotoxin citrinin, along with the undesirable compound mevinolin [9,12–15]. Recent studies have shown that some *Talaromyces/Penicillium* sp. non-pathogenic to humans, such as *Talaromyces aculeatus*, *T. pinophilus*, *T. funiculosus*, *T. atroroseus*, *T. minioluteus*, *T. marneffei* and *T. albobiverticillius*, naturally secrete soluble *Monascus*-like azaphilone red pigments and their amino acid derivatives, without side-production of mycotoxins [15–18]. *Talaromyces/Penicillium* species are promising sources of fungal polyketide-based red pigments (monascorubramin, rubropunctamin, PP-R, etc.), which can be safely applied in the industries (such as animal feed supplementation, foods, nutraceuticals, pharmaceuticals and cosmetics) [19]. More recently, studies performed by Chen et al. [20,21] and Liu et al. [22] have explained the biosynthetic pathway of *Monascus*-like azaphilone pigments in *Monascus* and *Talaromyces/Penicillium* genera. They described a common gene cluster responsible for the pigment production in these genera, as well as differences regarding the gene organization, copy numbers and allelic diversity.

Traditionally, the *Monascus*-like azaphilone pigments are being extracted from microbial biomass by conventional solid–liquid extraction processes and require extended extraction times, high temperatures and important volumes of various organic solvents (*n*-hexane, acetone, chloroform, ethyl acetate, etc.) [23]. Extracting the fungal pigments via green processing technologies presents a promising approach to pursue a more sustainable production of natural colorants [24]. Therefore, alternatives are assessed for pigment extraction by different technologies (e.g., extraction assisted by ultrasound, microwave, enzymatic or high-pressure treatments) [5,25–27].

The aim of this study is focused on the characterization of the target bioactive compounds (e.g., derivatives of nitrogen-containing azaphilone red pigments and ergosterol derivatives) produced by a wild type of *Talaromyces* sp. 30570 (CBS 206.89 B) marine-derived fungal strain isolated from the coral reef of the Réunion island. The influence of the nutrients' profile on the fungal pigments production in two submerged cultures, either with simple or complex sources of nitrogen, were also studied. Furthermore, we investigated the use of an alternative Pressurized Liquid Extraction (PLE) method based on the extraction procedure published in our previous work [25], using eco-friendly solvents (e.g., water, methanol and/or ethanol, which are allowed in the US and in the EU for the extraction of natural products), for advanced mycelial pigment extraction from the marine-derived *Talaromyces* sp. 30570 fungal strain. This alternative extraction consists of a high-pressure extraction process from the mycelial cells carried out at a high temperature and elevated pressure (>10 MPa) in order to maintain the solvents at liquid state when applied to the sample, as well as to maximize the extraction efficiency [5,25].

The pigment composition was characterized by high-performance liquid chromatography-diode array detection-electrospray ionization mass spectrometry (HPLC-PDA-ESI/MS).

2. Materials and Methods

2.1. Submerged Fermentation of Fungal Strain

The fungal strain was sampled from Réunion island coral-reef according to our previous study and was identified as *Talaromyces* sp. CBS 206.89 B (collection strain No. 30570 in our collection reference system) using morphological observation and gene sequencing [28]. Two submerged culture media, the Defined Minimal Dextrose broth (DMD) and the Potato Dextrose Broth (PDB), containing simple source of nitrogen (i.e., ammonium sulfate) and complex sources of nitrogen (like amino acids and proteins), respectively, were used for the comparison in terms of pigment production in submerged fungal cultures, as reported earlier [25]. Culture pH medium was adjusted to 6.0 ± 0.2 prior to sterilization. Pre-culture was prepared by taking a loop of fungus from 7-day old culture grown on PDA Petri plates and transferred into 60 mL sterilized PDB culture medium. The flasks were incubated at 26 °C for 72 h. Cultures were carried out in 250 mL Erlenmeyer flasks containing 100 mL of sterilized culture medium. The flasks were inoculated with 1% (*w/v*) 72 h-old seed culture and incubated at 26 °C for 7 days at 150 rpm (Infors Multitron HT) (Figure 1).

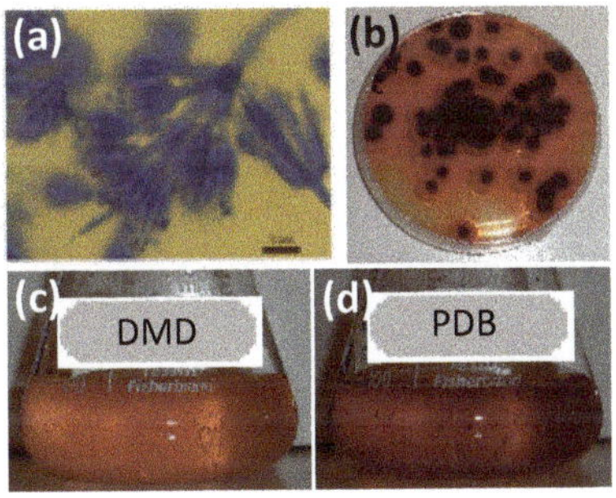

Figure 1. Morphological features of the marine-derived *Talaromyces* sp. 30570 strain: (**a**) Conidiophores produced on Potato Dextrose Agar (PDA) media, stained with lactophenol blue (scale bar 5 μm); (**b**) Reverse face of fungus grown on PDA; (**c**) Red pigment production in Defined Minimal Dextrose (DMD) medium incubated for 7 days at 24 °C; (**d**) Red pigment production in Potato Dextrose Broth (PDB) medium incubated for 7 days at 24 °C.

2.2. Biomass Separation, Extraction and Quantification of the Polyketide-Based Pigments

Fungal biomass was separated from fermentation broth by centrifugation at 10,000 rpm for 10 min (Centrifuge Sigma 3 K 3OH) and vacuum filtration. Biomass was lyophilized (FreeZone 2.5 Liter 50C Benchtop freeze dryer, LABCONCO, Kansas City, MO, USA) and then weighed. The fungal pigments were extracted and fractionated from the mycelial cells of *Talaromyces* sp. 30570 using an alternative pressurized liquid extraction (PLE) process with eco-friendly solvents (water, methanol and ethanol) according to the method recently published by Lebeau et al. [25]. The PLE process was performed on a Dionex ASE system (ASE^{TM} 350 apparatus, Dionex, Germering, Germany). The weighed sample (lyophilized biomass) was transferred to a 10-mL stainless steel extraction cell equipped with two cellulose filters on the bottom and containing glass balls (diameter 0.25–0.50 mm). Then, the sample was subjected to a six-stage extraction procedure under high pressure as an attempt to entirely extract the intracellular pigments from the mycelium. The sequence of solvents was set to display a decreasing polarity profile: purified water was used as the first extraction solvent, followed by 50% methanol, then 50% ethanol, >99.9% methanol, and MeOH:EtOH (1/1, *v/v*), and >99.9% ethanol as the last extraction solvent (Figure 2). The PLE extraction conditions were: temperature: 90 °C, pressure

>10 MPa, heating time: 5 min, static time: 18 min, flush: 100%, and purge: 5 min. Solvents (methanol and ethanol, 99.9%-HPLC quality) were obtained from Carlo Erba (Val de Reuil, France). Purified water was obtained from a Milli-Q system (EMD Millipore Co., Billerica, MA, USA).

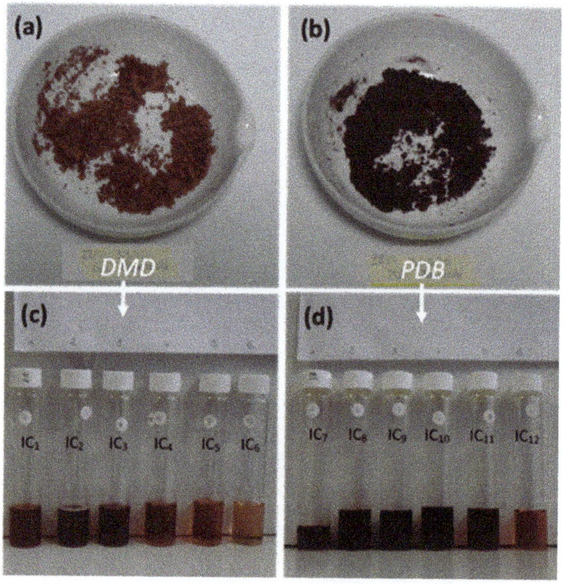

Figure 2. Red-colored dried residues obtained from submerged culture of the marine-derived *Talaromyces* sp. 30570 strain (**a**) in DMD, or (**b**) in PDB medium; (**c**,**d**) intracellular (IC) liquid extracts obtained using the pressurized liquid extraction method on mycelial cells of *Talaromyces* sp. 30570. The sequence of solvents was set to display a decreasing polarity profile (from IC_1 to IC_6; or from IC_7 to IC_{12}): purified water was used as the first extraction solvent, followed by 50% aqueous methanol, then 50% aqueous ethanol, >99.9% methanol, and MeOH:EtOH (1/1, *v/v*), and >99.9% ethanol as the last extraction solvent.

The pigment content extracted from the mycelial biomass was analyzed by spectral analysis using a UV-visible spectrophotometer (UV-1800, Shimadzu Corporation, Tokyo, Japan) at 276 nm (i.e., λ_{max} of the well-known *Monascus*-like pigments rubropunctamin and monascurubramin) according the method earlier reported [25], and expressed in terms of milli-equivalents of polyketide-based pigments per liter of culture medium (i.e., volumetric production in meqv·L^{-1} in the culture medium). All experiments were conducted in triplicate. The extracts were then stored at 4 °C in an amber vial prior to chromatographic analysis.

2.3. HPLC-DAD Analysis

Pigment composition was characterized by reverse phase HPLC-DAD using the Ultimate 3000 apparatus (Dionex, Germering, Germany) based on the analytical method reported by Lebeau et al. [25]. Analytical conditions: 25 µL injection; Hypersil Gold^{TM} column (2.1 mm i.d. × 150 mm, 5 µm; Thermo Scientific Inc., Waltham, MA, USA); temperature 30 °C; elution with a water-acetonitrile-formic acid gradient system [25]. Data were analyzed by the Chromeleon v.6.80 software (Dionex). Acetonitrile (99.9%-HPLC quality) and formic acid (purity 99%) were obtained from Carlo Erba (Val de Reuil, France).

2.4. UHPLC-HR-ESI-MS Analyses

The pigments were identified by UHPLC- High Resolution Electrospray Ionization (HR-ESI) MS analyses according to the method published by Klitgaard et al. [29], and using the Agilent 1290 Infinity

LC system with a DAD detector, coupled to an Agilent 6550 iFunnel Q-TOF with an electrospray ionization source (Agilent Technologies, Santa Clara, CA, USA), and a Poroshell 120 Phenyl–Hexyl column (2.1 mm i.d. × 250 mm, 2.7 μm; Agilent). The analytical conditions used in this study were those earlier reported by Klitgaard et al. [29]: the separation was performed at 60 °C with a water-acetonitrile gradient (with 20 mM formic acid) going from 10% (*v/v*) to 100% acetonitrile in 15 min, followed by 3 min with 100% acetonitrile. The flow rate was kept constant at 0.35 mL/min. Mass spectra were recorded as centroid data for *m/z* 85–1700 in positive and negative ESI-MS mode, with an acquisition rate of 10 spectra/s.

3. Results

3.1. Alternative Extraction and Characterization of Monascus-Like Azaphilone Pigments from the Marine-Derived Talaromyces sp. 30570 Strain

Results from the analysis of the pigmented extracts using the alternative PLE method revealed a great diversity of the chemical structures of the *Talaromyces* sp. 30570 pigments. A series of intracellular extracts (IC) (Figure 2) were collected based on the PLE extraction procedure investigated, and their compositions in terms of fungal pigments were characterized by HPLC-DAD chromatography (Figure 3). Twelve pigmented compounds (compounds **1** to **12**) and one other colorless secondary metabolite (compound **13**, identified as ergosterol, see below) were identified. In particular, our results highlighted that the multi-step PLE procedure gives encouraging results in terms of selectivity of the extraction of the polyketide-based red pigments. This can be shown by two elements.

First, the initial extraction using hot pressurized water (90 °C and 10 MPa) enables the extraction with high selectivity of a highly polar pigment from fungal mycelium, namely compound **1**, found only in the aqueous fraction (Figure 3A), without co-extraction of side metabolites. Indeed, only one single peak (Rt 1.71 min; Figure 3A) was observed on the LC chromatogram. This compound **1** exhibits two absorption maxima at ca. 423 and 514 nm, which is characteristic of the *Monascus*-like nitrogen-containing azaphilone pigments [5,30]. Unfortunately, further dereplication experiments using HPLC-PDA-ESI/MS were not conclusive enough to fully elucidate the chemical structure of this highly polar red pigmented compound. Presumably, this highly polar compound **1**, exhibiting UV-visible λ_{max} at 201, 216, 244, 276, 423, and 514 nm, was presumed to be to a highly polar diglycoside derivative of a *Monascus*-like azaphilone red pigment. However, it cannot be concluded here that it represents only one compound, and alternative polar stationary phases (i.e., amide) should be used in further works for more relevant analysis of that/those compound(s).

Then, the following extractions using hot pressurized (90 °C and 10 MPa) hydroalcoholic mixtures such as 50% aqueous methanol (Figure 3B) or 50% aqueous ethanol (Figure 3C) enabled the extraction of others major *Monascus*-like red pigments from mycelium (pigments **2–12**) with a certain selectivity. Indeed, the extraction of other non-pigmented compounds such as ergosterol **13** (present in the intracellular metabolites produced by the mold) occurred only when less polar solvents, like pure methanol, were used (Figure 3D).

The UV-visible absorption spectra of the major pigmented molecules (compounds **1**, **5**, **6**, **8**, **9** and **10** detected by HPLC-DAD, Figure 3) produced by the marine-derived *Talaromyces* sp. 30570 fungal strain are shown in Figure 4.

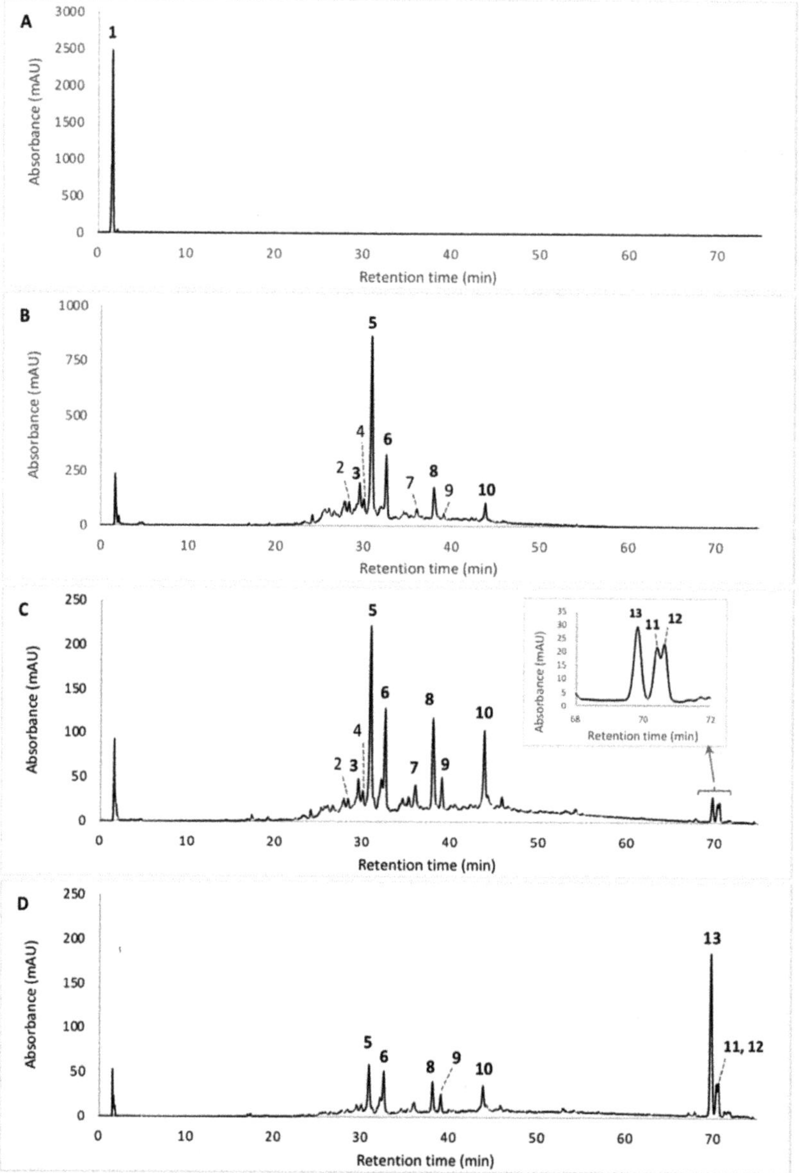

Figure 3. LC-DAD chromatograms of the fungal extracts from the mycelium of the marine-derived *Talaromyces* sp. 30570 strain cultivated in Potato Dextrose Broth (PDB). Pigments were extracted by pressurized liquid extraction: (**A**) water, (**B**) 50% aqueous methanol, (**C**) 50% aqueous ethanol, or (**D**) by pure methanol as extraction solvent. Assignment of the azaphilone red pigments **1–12** and ergosterol **13** were done by UV-visible and HRMS spectra. See Table 1 for the identification of the molecules.

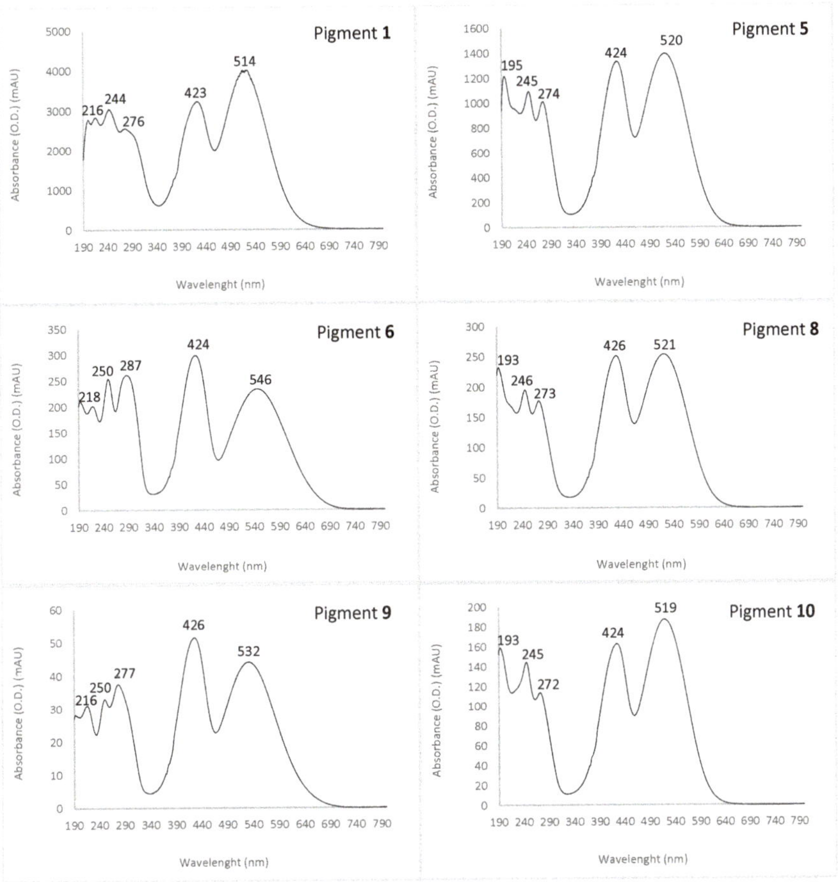

Figure 4. UV-visible absorption spectra of the major identified or assumed compounds **1**, **5**, **6**, **8**, **9** and **10** detected in intracellular extracts of the marine isolate *Talaromyces* sp. 30570 cultivated in PDB with reference to the chromatogram shown in Figure 3. Assignment of the nitrogen-containing azaphilone red pigments were done by UV-visible spectra and HRMS according to their mass to charge ratio. Pigments: 6-[(Z)-2-Carboxyvinyl]-N-GABA-PP-V **6**; *N*-threonine-monascorubramin **8**; *N*-glutaryl-rubropunctamin **9**; and monascorubramin **10**. Pigment **1**, not tentatively identified, was presumed to be to a highly polar compound, like a diglycoside derivative of a *Monascus*-like azaphilone red pigment. Then, it is presumed that the pigment **5** might reasonably be the molecule *N*-threonyl-rubropunctamin (or the acid form of the aforementioned PP-R), as recently reported by Rasmussen [30]. See Figure 5 for the chemical structure of the molecules.

The UV-Vis absorption spectra gave some indications on the chemical structure of the pigments produced. All the compounds (**1–12**) responsible for the pigmentation (with absorption in visible region) shown the same UV-visible spectral characteristics, i.e., a mountain-like spectrum with three or four UV λ_{max} in the range 193–201, 216–218, 244–250 and 272–287 nm. Furthermore, all these aromatic compounds displayed the characteristic nitrogen-containing *Monascus*-like azaphilone red pigments double visible peaks around 430 and 515 nm (range 423–430 and 514–546 nm) (Figure 4, Table 1) in accordance with the literature data [5,30]. This unique chemical fingerprint of these pigments would suggest the presence of the monascorubramin or rubropunctamin-type chromophore in the molecule, as reported earlier [30].

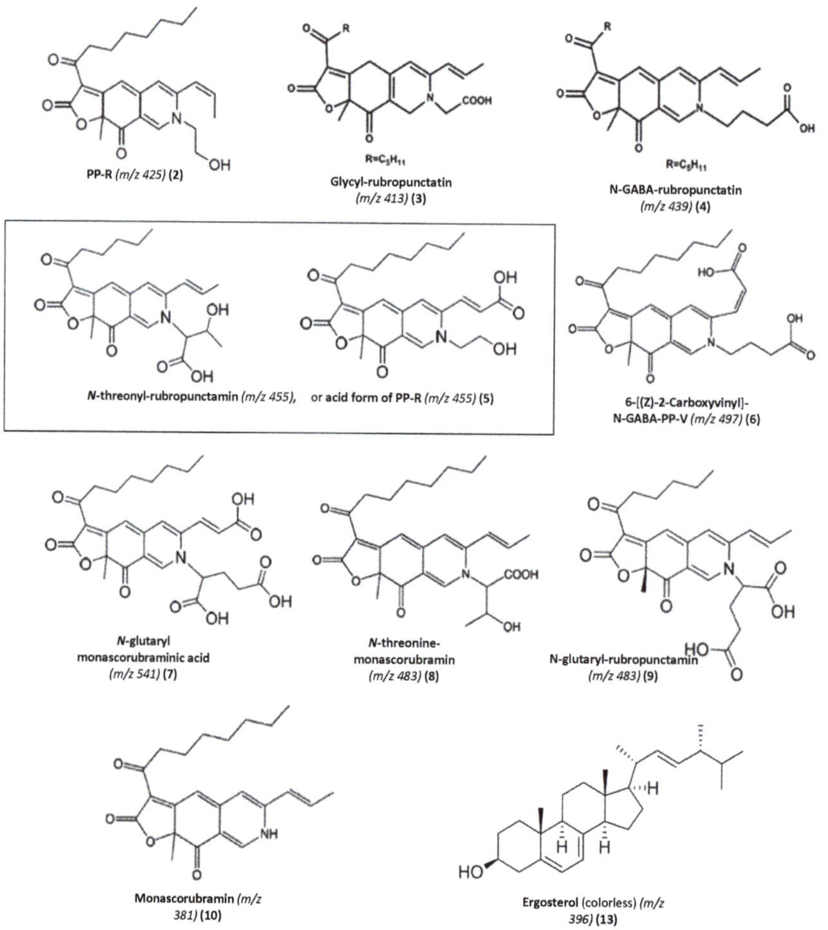

Figure 5. Main *Monascus*-like nitrogen-containing azaphilone pigments produced by the marine-derived *Talaromyces* sp. 30570 fungal strain. Assignment of the pigments (**2–10**) and ergosterol (**13**) were done by UV-visible spectra and HR-ESI-MS according to their mass to charge ratio. It is presumed that the major red pigment **5** produced by the fungal strain (see sidebar) might reasonably be the molecule *N*-threonyl-rubropunctamin ($C_{25}H_{29}NO_7$, *m/z* 455) or the acid form (R-COOH) of the pigment PP-R ($C_{25}H_{29}NO_7$, *m/z* 455) as recently reported by Rasmussen [30].

Thus, among the great number of compounds observed, twelve *Monascus*-type azaphilone red pigments (with absorption in the visible region due to the monascorubramin or rubropunctamin-type chromophore) were detected, and nine were tentatively identified as derivatives of nitrogen-containing azaphilone red pigment (see Figure 5 for the chemical structures of the major identified or assumed red pigments of *Talaromyces* sp. 30570). The retention time (Rt), UV-Vis λ_{max}, accurate masses of parent ion and of adduct ions, color, molecular formula and average mass for each compound identified in the extracts of the marine-derived *Talaromyces* sp. 30570 strain cultivated in PDB are gathered in Table 1.

Table 1. Overall compounds (12 derivatives of nitrogen-containing azaphilone red pigments and ergosterol **13**) identified from the fungal extracts of the marine-derived *Talaromyces* sp. 30570 fungal strain cultivated in Potato Dextrose Broth (PDB), with reference to the chromatograms shown in Figure 3.

No	Rt. (min)	UV-Vis λ_{max} (nm)	Observed Peak HR-ESI MS (*m/z*)	Tentative Identification (Identified or Assumed Compounds)	Proposed Molecular Formula	Monoisotopic Mass in Da (*Mass Error*) [(1)]	Ref.
1	1.71	201, 216, 244, 276, 423, 514	n.d.	Diglycoside derivative of a *Monascus*-like azaphilone red pigment (n.i.)	n.d.	n.d.	-
2	28.52	192, 245, 274, 421, 515	488.1820 [M + CAN + Na]$^+$	**PP-R** [7-(2-hydroxyethyl)-monascorubramin]	$C_{25}H_{31}NO_5$	425.22 *(0.0380)*	[14–16]
3	29.60	193, 245, 274, 421, 518	416.1960 [M + H]$^+$	**Glycyl-rubropunctatin**	$C_{23}H_{27}NO_6$	413.18 *(2.0160)*	[31–33]
4	30.15	193, 245, 274, 426, 515	440.1936 [M + H]$^+$	***N*-GABA-rubropunctatin** (GABA: γ-aminobutyric acid)	$C_{25}H_{29}NO_6$	439.51 *(0.3164)*	[20]
5	30.97	195, 245, 274, 424, 520	456.1543 [M + H]$^+$	***N*-threonyl-rubropunctamin** (or **acid form of PP-R**) (presumed)	$C_{25}H_{29}N0_7$	455.20 *(0.0457)*	[25,30]
6	32.66	193, 218, 250, 287, 424, 546	498.1665 [M + H]$^+$	**6-[(Z)-2-Carboxyvinyl]-*N*-GABA-PP-V**	$C_{27}H_{31}N0_8$	497.54 *(0.3735)*	[28,34]
7	36.11	196, 247, 288, 422, 522	542.1598 [M + H]$^+$	***N*-glutaryl-monascorubraminic acid**	$C_{28}H_{31}N0_{10}$	541.20 *(0.038)*	[30]
8	38.04	193, 246, 273, 426, 521	484.1910 [M+H]$^+$	***N*-threonine-monascorubramin**	$C_{27}H_{33}N0_7$	483.55 *(0.0402)*	[34]
9	39.10	193, 216, 250, 277, 426, 532	484.5110 [M + H]$^+$ 546.1556 [M + CAN + Na]$^+$	***N*-glutaryl-rubropunctamin**	$C_{26}H_{29}N0_8$	483.51 *(0.0010)*	[15,34–36]
10	43.95	193, 245, 272, 424, 519	381.1198 [M + H]$^+$	**Monascorubramin**	$C_{23}H_{27}NO_4$	381.19 *(1.0702)*	[15,16]
11	70.40	192, 248, 271, 282, 293, 434, 513	n.d.	Derivative of a *Monascus*-like azaphilone red pigment (n.i.)	n.d.	n.d	-
12	70.64	192, 248, 271, 282, 293, 434, 510	n.d.	Derivative of a *Monascus*-like azaphilone red pigment (n.i.)	n.d.	n.d.	-
13	69.78	192, 248, 271, 282, 293	393.2693	**Ergosterol** (colorless compound)	$C_{28}H_{44}O$	396.65 *(0.3807)*	[37,38]

n.d.: not determined.; n.i.: not identified; [(1)] the mass error (Da) between the observed MS peaks and proposed formula (for the molecular ion).

Among the derivatives of nitrogen-containing azaphilone red pigments identified, the compound **2** eluting at Rt. 28.52 min that presents UV-Vis λ_{max} at 192, 245, 274, 421 and 515 nm, was characteristic of the red pigment PP-R [7-(2-hydroxyethyl)-monascorubramin] which has previously been isolated from some other species of *Talaromyces* [14–16]. Indeed, the ACN-Na adduct ion $[M + CAN + Na]^+$ observed at *m/z* 488.1820 was in agreement with the calculated masses of the $C_{25}H_{31}NO_5$-CH_3CN-Na^+ adduct ion (*m/z* 488) and of molecular ion (*m/z* 425.22) to the red pigment PP-R suggesting a $C_{25}H_{31}NO_5$ molecular formula (Table 1, Figure 5; Figure S1) [14–16].

Additionally, our results suggested that the compound **3** (Rt. 29.60 min; λ_{max} at 193, 245, 274, 421 and 518 nm; *m/z* 416.1960 $[M + H]^+$) might reasonably correspond to the red pigment glycyl-rubropunctatin ($C_{23}H_{27}NO_6$; average mass *m/z* 413.18) (Figure S2) previously isolated from *Monascus* cultures [31–33].

The UV-visible and HR-ESI-MS spectra of the compound **4** (Rt. 30.15 min; λ_{max} at 193, 245, 274, 426 and 515 nm) were characteristics to the red pigment *N*-GABA-rubropunctatin (GABA: γ-aminobutyric acid) [20]: the protonated molecular ion $[M + H]^+$ observed at *m/z* 440.1936 was in agreement with the calculated mass of the *N*-GABA-rubropunctatin molecular ion (*m/z* 439.51) suggesting a $C_{25}H_{29}NO_6$ molecular formula (Table 1, Figure 5; Figure S3) [20].

Interestingly, our results demonstrated that the major pigment produced by the marine-derived *Talaromyces* sp. 30570 strain, i.e., the compound **5** eluting at Rt. 30.97 min with UV-Vis λ_{max} at 195, 245, 274, 424 and 520 nm (Figure 3, Table 1) was a derivative of nitrogen-containing azaphilone red pigment and it is presumed that this pigment might reasonably be the molecule *N*-threonyl-rubropunctamin [25], or the acid form of the aforementioned PP-R, as recently reported by Rasmussen [30] from another species of *Talaromyces* (i.e., *T. atroroseus*). Indeed, this compound **5**, displayed a protonated molecular ion $[M + H]^+$ at *m/z* 456.1543 (Figure S4), and the aforementioned derivatives *N*-threonyl-rubropunctamin and acid form of PP-R have the same nominal mass of 455.20, suggesting a $C_{25}H_{29}NO_7$ molecular formula [25,30] (Table 1, Figure 5), which should be in agreement with the protonated molecular ion observed in this study. Further works are needed to purify and fully characterize this red pigment produced by the fungus by NMR analysis.

The compound **6** (Rt. 32.66 min; λ_{max} at 193, 218, 250, 287, 424 and 546 nm) was identified as the derivative 6-[(Z)-2-Carboxyvinyl]-*N*-GABA-PP-V. Its $[M + H]^+$ ion, observed at *m/z* 498.1665, matched up well with the expected mass of the corresponding molecule (molecular ion *m/z* 497.54) suggesting a $C_{27}H_{31}NO_8$ molecular formula (Table 1, Figure 5; Figure S5). This derivative of azaphilone red pigment has recently been isolated from another marine-derived strain of *Talaromyces* sp. 30548 (e.g., strain CBS 206.89 A, identified as *T. albobiverticillius*) also collected from the coral reef of the Réunion island [28,34].

The compound **7** (Rt. 36.11 min; λ_{max} at 196, 247, 288, 422 and 522 nm) was characteristic of the red pigment *N*-glutaryl-monascorubraminic acid (acid form) according to the data reported earlier [30]. Its protonated molecular ion $[M + H]^+$ at *m/z* 542.1598 was consistent with the calculated mass of the molecular ion *m/z* 541.20 of the corresponding molecule, suggesting a $C_{28}H_{31}NO_{10}$ molecular formula (Table 1, Figure 5; Figure S6) [30].

The compound **8** (Rt. 38.04 min; λ_{max} at 193, 246, 273, 426 and 521 nm) was characteristic to the red pigment *N*-threonine-monascorubramin [34]. Its protonated molecular ion $[M + H]^+$ at *m/z* 484.1910 was in agreement with the calculated mass *m/z* 483.55 of the *N*-threonine-monascorubramin molecular ion, suggesting a $C_{27}H_{33}NO_7$ molecular formula (Table 1, Figure 5; Figure S7) [34].

The compound **9** (Rt. 39.10 min; λ_{max} at 193, 216, 250, 277, 426 and 532 nm) was characteristic to the red pigment *N*-glutaryl-rubropunctamin. Its protonated molecular ion $[M + H]^+$ at *m/z* 484.5110 supported by its ACN-Na adduct ion $[M + ACN + Na]^+$ at *m/z* 546.1556 (Figure S8) coincided nicely with the expected mass 483.51 of the *N*-glutaryl-rubropunctamin (with formula $C_{26}H_{29}NO_8$) isolated from other *Monascus* and *Talaromyces* species [15,34–36].

Then, the compound **10** (Rt. 43.95 min; λ_{max} at 193, 245, 272, 424 and 519 nm) seemed to correspond to the red pigment monascorubramin according to its protonated molecular ion *m/z* 381.1198 (Figure S9) relatively close to the calculated mass of the molecular ion *m/z* 381.19 of the corresponding molecule ($C_{23}H_{27}NO_4$) [15,16].

In addition to these *Monascus*-like azaphilone pigments, no known mycotoxins were reported in the extracts obtained from the PLE extraction investigated here. Finally, our results suggested that the apolar and colorless compound **13** (Figure 6, Table 1) eluting at Rt. 69.78 min is assumed to be the molecule ergosterol ($C_{28}H_{44}O$; 396 g/mol), according to its similar absorption spectrum and to the HR-ESI-MS characteristic ion $[M + H]^+$ at *m/z* 393.2693 (Figure S10). Indeed, ergosterol can undergo desaturation during LC-MS [37,38] (Table 1), consequently yielding a second molecular ion at *m/z* 393 in addition to the conventional molecular ion at *m/z* 397 $[M + H]^+$. The results described in the present study are consistent with several earlier investigations which have highlighted the presence of ergosterol and derivatives of ergosterol from fungi [38]. On top of everything, ergosterol and its derivates are proven, with interesting bioactivities with potential uses in pharmaceutics [38,39].

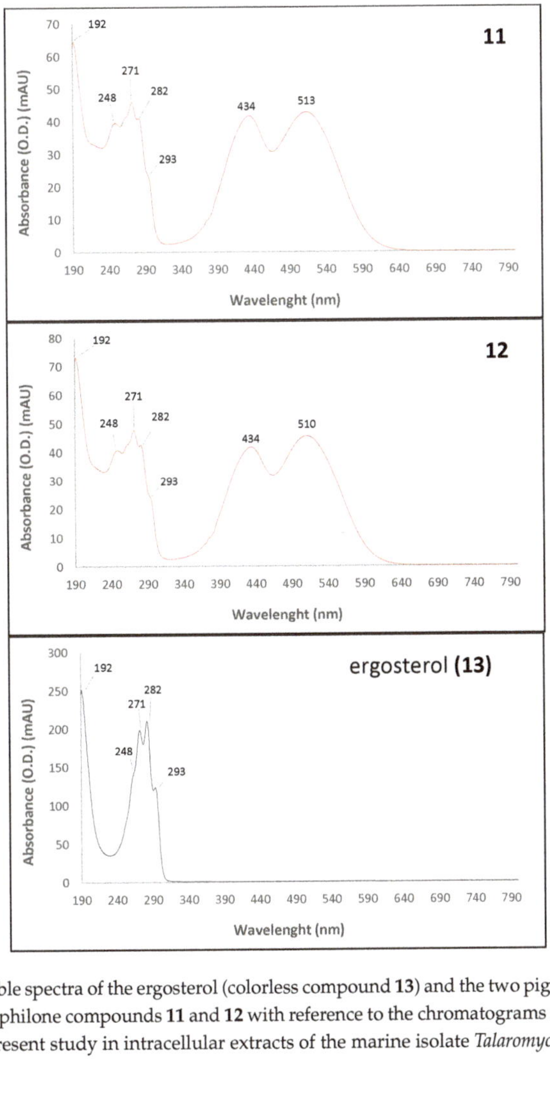

Figure 6. UV-visible spectra of the ergosterol (colorless compound **13**) and the two pigmented ergosterol derivatives of azaphilone compounds **11** and **12** with reference to the chromatograms shown in Figure 3, detected in the present study in intracellular extracts of the marine isolate *Talaromyces* sp. 30570.

Interestingly, based on the UV-visible spectra of the last molecules **11** and **12** (Figure 6), these two red pigments (with absorption at ca. 515 nm in the visible 'red' region) not tentatively identified by HPLC-PDA-ESI/MS (signal too weak) seemed to correspond to two pigmented ergosterol derivatives of azaphilone compounds. Indeed, they exhibited similar absorption spectra in the ultraviolet region to ergosterol molecule (i.e., λ_{max} at 248, 271, 282 and 293 nm). Surprisingly, they also displayed the characteristic nitrogen-containing *Monascus*-like azaphilone red pigments double visible peaks around 430 and 515 nm. To our knowledge, this is the first isolation of that kind of pigmented ergosterol derivatives of azaphilone red compounds from microorganisms. However, it was not possible to assign masses and chemical formulas to these two minor compounds.

3.2. *Influence of the Nutrients Profile on the Production of Monascus-Like Azaphilone Red Pigments by the Marine-Derived Talaromyces sp. 30570 Strain*

Surprisingly, among the twelve red pigments detected in this study from the marine-derived *Talaromyces* sp. 30570 when cultivating in PDB, our results shown that only three well-known red pigments, i.e., the glycyl-rubropunctatin **3**, N-GABA-rubropunctatin **4** and the *N*-threonyl-rubropunctamin **5**, are common to both submerged culture conditions: in PDB (Figure 7A) and defined minimal dextrose broth (DMD) (Figure 7B). Our results, reported in Table 2, indicated that the nutrients' profile of the fermentation broth has a clear impact on the pigment production by the marine-derived *Talaromyces* sp. 30570 fungal strain. These findings are corroborated by the results of earlier studies performed on *Talaromyces/Penicillium* species by Ogihara and Oishi [40] and Arai et al. [41], which have demonstrated that the fungal pigmentation will depend on the medium composition.

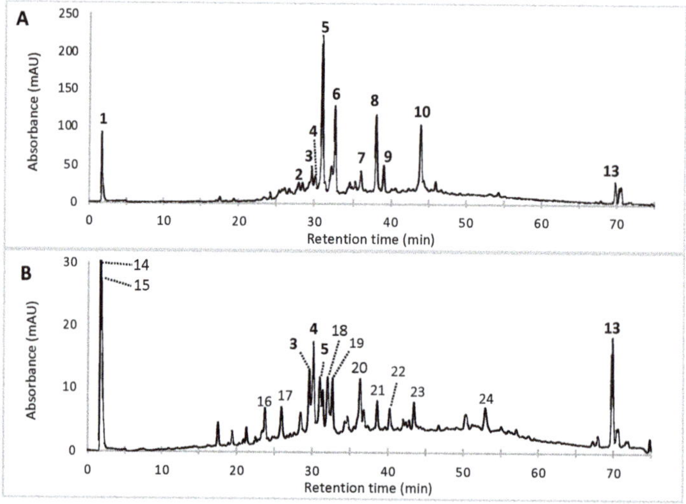

Figure 7. Chromatograms of the overall compounds detected by HPLC-DAD in the samples obtained by pressurized liquid extraction (PLE) using 50% aqueous ethanol as extraction solvent from the mycelium of the marine-derived *Talaromyces* sp. 30570 strain cultivated in submerged culture: (**A**) in Potato Dextrose Broth (PDB); and (**B**) in Defined Minimal Dextrose broth (DMD). Assignment of the polyketide-based compounds were done by UV-visible spectra and HRMS according to their mass to charge ratio. See Table 2 for the identification of the molecules.

Table 2. Overall compounds identified from the fungal extracts of the marine-derived *Talaromyces* sp. 30570 fungal strain cultivated in 2 different submerged cultures: in Potato Dextrose Broth (PDB) and in Defined Minimal Dextrose broth (DMD), with reference to the chromatograms shown in Figure 7.

No	Rt. (min)	UV-Vis λ_{max} (nm) (Bold: λ_{max} in Visible)	Tentative Identification (Identified or Assumed Compounds)	Polyketide-Based Compounds Content (meqv.L^{-1})	
				in PDB (1)	in DMD (1)
1	1.71	201, 216, 244, 276, **423, 514**	**Diglycoside derivative of a *Monascus*-like azaphilone red pigment**	124.8 ± 5.0	-
2	28.52	192, 245, 274, **421, 515**	**PP-R** [7-(2-hydroxyethyl)-monascorubramin]	6.7 ± 0.4	-
3	29.60	193, 245, 274, **421, 518**	**Glycyl-rubropunctatin**	22.1 ± 1.3	7.3 ± 0.3
4	30.15	193, 245, 274, **426, 515**	***N*-GABA-rubropunctatin** (GABA: γ-aminobutyric acid)	8.0 ± 0.3	4.9 ± 0.4
5	30.97	195, 245, 274, **424, 520**	***N*-threonyl-rubropunctamin** (or **acid form of PP-R**)	83.4 ± 4.1	9.0 ± 0.8
6	32.66	193, 218, 250, 287, **424, 546**	**6-[(Z)-2-Carboxyvinyl]-*N*-GABA-PP-V**	33.5 ± 1.3	-
7	36.11	196, 247, 288, **422, 522**	***N*-glutaryl-monascorubraminic acid**	7.3 ± 0.4	-
8	38.04	193, 246, 273, 426, 521	***N*-threonine-monascorubramin**	24.3 ± 0.6	-
9	39.10	193, 216, 250, 277, **426, 532**	***N*-glutaryl-rubropunctamin**	5.7 ± 0.2	-
10	43.95	193, 245, 272, **424, 519**	**Monascorubramin**	19.6 ± 0.9	-
11	70.40	192, 248, 271, 282, 293, **434, 513**	**Derivative of a *Monascus*-like azaphilone red pigment** (n.i.)	4.1 ± 0.2	-
12	70.64	192, 248, 271, 282, 293, **434, 510**	**Derivative of a *Monascus*-like azaphilone red pigment** (n.i.)	4.2 ± 0.4	-
13	69.78	192, 248, 271, 282, 293	**Ergosterol** (colorless compound)	24.0 ± 1.0	73.8 ± 1.9
14	1.63	198, 260	Colorless compound (n.i.)	-	31.1 ± 1.2
15	2.01	196, 258	Colorless compound (n.i.)	-	6.1 ± 0.3
16	23.79	203, 256, 298	Colorless compound (n.i.)	-	6.3 ± 0.3
17	25.93	196, 264, 278, **479**	Yellow pigment (n.i.)	-	5.2 ± 0.2
18	32.03	193, 252, 294, **428, 546**	Purple-red pigment (n.i.)	-	5.0 ± 0.3
19	32.73	192, 220, 246, 289, **415, 546**	Purple-red pigment (n.i.)	-	3.8 ± 0.3
20	36.30	193, 260, 274	Colorless compound (n.i.)	-	2.7 ± 0.2
21	38.57	192, 211, 243, 391	Colorless compound (n.i.)	-	1.5 ± 0.2
22	40.21	210, 292, 421	Colorless compound (n.i.)	-	0.4 ± 0.1
23	43.43	192, 280, **409, 431**	Yellow pigment (n.i.)	-	0.4 ± 0.1
24	52.93	192, 248, 271, 282, 293, **414**	Yellow pigment (n.i.)	-	17.4 ± 1.1

(1) mean (± standard deviation) expressed in meqv.L^{-1} of polyketide-based compounds produced in PDB and DMD broths.

Indeed, in minimal nutrient condition (e.g., DMD broth) containing glucose and inorganic nitrogen source (with salts and bio-elements), the extraction and recovery of derivatives of nitrogen-containing azaphilone red pigments from the marine-derived *Talaromyces* sp. 30570 mycelia was very low. In this minimal nutrient condition, the fungus yielded compounds, which were mostly unpigmented (e.g., ergosterol **13** and the not identified compounds **14–16** and **20–22**) (Table 2). Thus, the yield of nitrogen-containing azaphilone red pigments in this minimal culture medium was very poor compared to PDB medium containing complex organic nitrogen (such as amino acids and proteins) and carbon sources. When cultivated in the minimal medium, the fungal strain produced only the glycyl-rubropunctatin **3**, the *N*-GABA-rubropunctatin **4**, and the *N*-threonyl-rubropunctamin **5** as red pigments (Figure 7, Table 2), whereas in PDB the fungus yielded twelve derivatives of nitrogen-containing azaphilone red pigments (pigments **1–12**; Table 1).

More particularly, it is worth noticing that the fungal strain was unable to produce monascorubramin **10** and derivatives of monascorubramin like *N*-glutaryl-monascorubraminic acid **7**, *N*-threonine-monascorubramin **8** and 6-[(Z)-2-Carboxyvinyl]-*N*-GABA-PP-V **6** when cultivated in minimal nutrient condition without amino acids and proteins in the medium. These findings are consistent with the results described in previous reports, suggesting that organic sources of nitrogen favor high red pigment production by *Talaromyces/Penicillium* species [19]. Thus, the non-production of monascorubramin and its amino derivates, when the strain of *Talaromyces* sp. 30570 was cultivated in minimal medium could be explained by the unavailability of more suitable organic nitrogen sources (amino acid, peptides, etc.) for the biosynthesis of such nitrogen-based azaphilone compounds. This suggests that the presence of amino acids and peptides in PDB medium enables the functionality of this specific pathway.

4. Discussion

4.1. Efficiency and Selectivity of the Alternative Pressurized Liquid Extraction (PLE) of Azaphilone Red Pigments from the Mycelial Cells of the Marine-Derived Talaromyces sp. 30570

The *Monascus*-like azaphilone red pigments are water-soluble, thus they are readily extracted with polar solvents [42]. Our results revealed the alternative PLE technique to be highly efficient in removing *Monascus*-like azaphilone red pigments from mycelial biomass of the marine-derived *Talaromyces* sp. 30570 by using water, methanol and/or ethanol at 90 °C and 10 MPa as extraction solvents. The sole use of these eco-friendly solvents, which can be biosourced, adds to the novelty of our results. Indeed, solvents such as ethanol and methanol can be produced from carbon-neutral homoacetogenic gas fermentation [43] and biogas produced from wastes, respectively, strengthening the sustainability of such process. This alternative PLE technique should be considered as a promising eco-friendly extraction process for natural products from biological samples [5,24–26,39,44]. It also opens the way to further optimizations to the solvent mixture to use for isolating specific polyketide red pigments. Additionally, the use of pressurized nitrogen gas protects the target molecules (fungal pigments) from oxidation and ensures a higher quality of the recovered target molecules.

Although the different azaphilone red pigments identified isolated in this study have already been previously individually isolated in some species from *Monascus* (e.g., *M. ruber*) [20,21] and *Talaromyces* (e.g., *T. atroroseus* [30] and *T. albobiverticillius* [34]), this is the first report to our knowledge of the concomitant occurrence of these twelve azaphilone pigments in a fungal extract obtained from a culture of a wildtype marine strain of *Talaromyces* (e.g., *Talaromyces* sp. 30570). Azaphilonoids, and in particular, derivates of the pigmented monascorubramin and rubropunctamin produced by non-toxicogenic species from *Talaromyces* sp., are non-toxic compounds highly wanted in pharmaceutical industries due to their bioactivities (antibiotic, anti-inflammatory activities amongst others) [19,45,46]. Therefore, this ability to produce molecules with high industrial interest by the wildtype marine strain of *Talaromyces* sp. 30570 could be further expended and scaled up to commercial production. It is worthy of notice that other studies performed on strains of *T. atroroseus* [30] and *T. alboverticillius* [34] have reported the presence, in fungal extracts, of different pigmented compounds, such as monascusone

A, monascorubrin, PP-V, PP-Y, PP-O, as well as new pigmented azaphilone-like molecules, formerly known as atrorosins [47,48], and not detected in fungal extracts of *Talaromyces* sp. 30570 studied here. This observation clearly highlights the vast diversity of the polyketide-based pigments biosynthesized by the *Talaromyces* species, and in particular those from marine origin. This could also suggest the occurrence of an intraspecies diversity, as is already the case in other complexes of mold species, like in the *Talaromyces pinophilus* species complex [49] and in the *Fusarium oxysporum* species complex [39].

4.2. Putative Metabolic Pathway for the Production of Derivatives of Nitrogen-Containing Monascus-Like Azaphilone Red Pigments in the Marine-Derived Talaromyces sp. 30570

Monascus-like azaphilone pigments are colored metabolites with a pyrone–quinone structure [5]. They involve biosynthetic pathways catalyzed by multiple polyketide synthase enzymes (PKS). For over a decade now, a number of studies have attempted to assess the biosynthetic pathways of *Monascus*-like azaphilone pigments in the genera of *Monascus* and *Talaromyces/Penicillium* [17–21]. In the in vitro study of Chen et al. [20], the metabolic pathway of *Monascus*-like azaphilone pigments was elucidated in *Monascus ruber* M7. Then, Chen et al. [21] demonstrated that the biosynthetic gene clusters responsible for the pigment production in these fungi share orthologous genes for a conserved unitary trunk pathway [21]. They also described four physiological strategies responsible for the diversity of the *Monascus*-like azaphilone pigment structures in *Monascus* and *Talaromyces/Penicillium* genera, as well as differences regarding the gene organization, copy numbers and allelic diversity [20–22]. They mentioned that five and four gene clusters have been described to date from the *Monascus* and *Talaromyces* genera, respectively [20,21].

Thus, based on the previously reported models for *Monascus*-like azaphilone pigments biosynthesis in other *Monascus* and *Talaromyces/Penicillium* species [20–22], a putative pathway for the biosynthesis of derivatives of nitrogen-containing *Monascus*-like azaphilone red pigments and intermediates thereof in this marine-derived *Talaromyces* sp. 30570 strain was proposed in this study and described in Figure 8. Concerning the trunk pathway, the biosynthesis of yellow and orange azaphilone pigments is initiated via the polyketide pathway by a nonreducing PKS (known as *MrPigA* in *M. ruber* M7) [20,21] that features different domains like a starter unit acyl carrier protein transacylase (SAT), a ketoacyl synthase (KS), an acyltransferase (AT), a product template (PT), two acyl carrier proteins (ACP), a *C*-methyltransferase (MT) and a reductive release domain (R) [20–22] as shown in Figure 8. Then, the first stable azaphilone pigments intermediate (FK17-P2a) was synthetized by a ketoreductase (e.g., *MrPigC* in *M. ruber* M7) [20,21]. Next, an FAD-dependent monooxygenase (e.g., *MrpigN* in *M. ruber* M7) [20,21] is then critical to obtain the bicyclic pyran-containing azaphilone core. The polyketide-chromophore may come from further modifications to this azaphilone core by enzymatic or non-enzymatic reaction. Finally, the orange azaphilone pigments are formed by the esterification of a β-ketoacid (e.g., 3-oxo-octanoic acid and 3-oxo-decanoic acid resulted from the fatty acids biosynthetic pathway by a dedicated two-subunit fatty acid synthetase: e.g., MrPigJ/K in *M. ruber* M7) to the aforementioned polyketide-based chromophore, as shown in Figure 8.

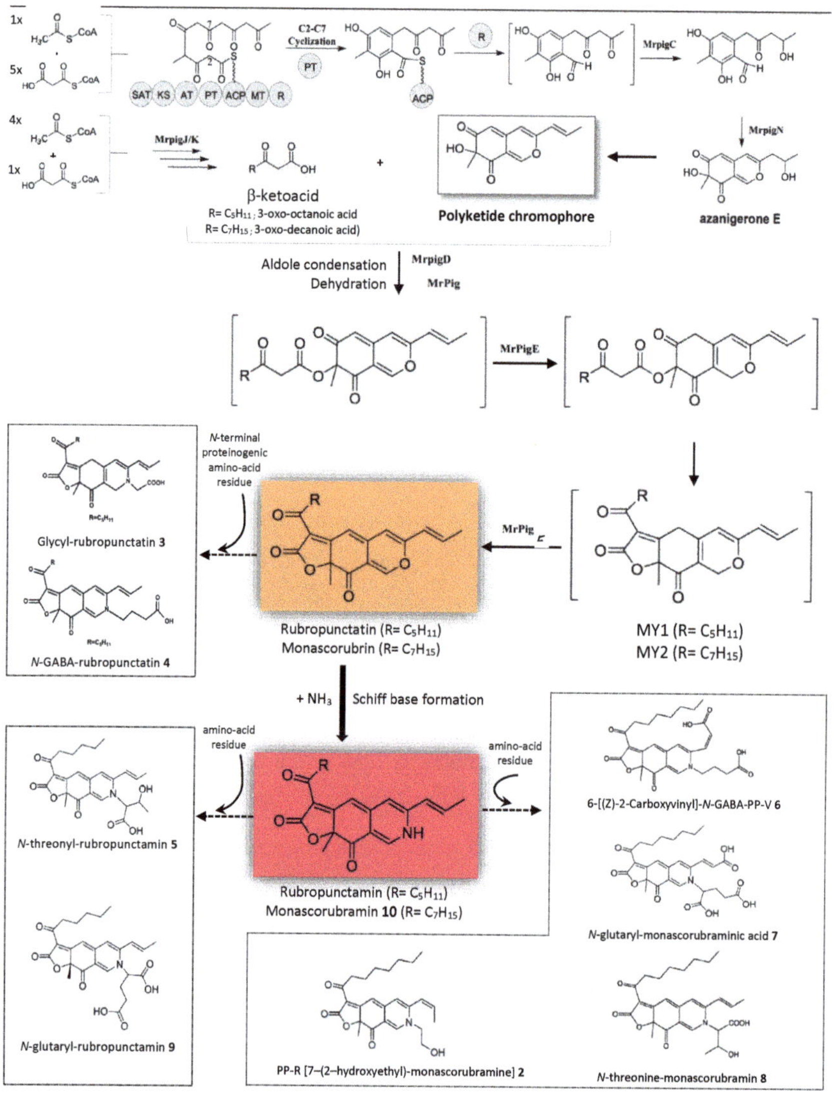

Figure 8. Putative metabolic pathway for the production of derivatives of nitrogen-containing azaphilone red pigments in the marine-derived *Talaromyces* sp. 30570 strain, based on the previously reported models for *Monascus*-like azaphilone pigments biosynthesis in other *Monascus* and *Talaromyces/Penicillium* species according to Chen et al. [20,21] and reviewed by Liu et al. [22]. Names of the enzymes are specified with reference to those identified in *M. ruber* M7 [20,21]. The non-reducing polyketide synthase *MrPigA* gene cluster encodes for a starter unit acyl carrier protein transacylase (SAT), a ketoacyl synthase (KS), an acyltransferase (AT), a product template (PT), two acyl carrier proteins (ACP), a C-methyltransferase (MT) and a reductive release domain (R). *MrPigC*: C-11-Ketoreductase; *MrPigD*: 4-*O*-Acyltransferase; *MrPigM*: *O*-Acyltransferase; *MrPigN*: FAD-dependent monooxygenase; *MrPigE*: NAD(P)H-dependent oxidoreductase; *MrPigF*: FAD-dependent oxidoreductase; *MrPigO*: Deacetylase. This figure was adapted from Chen et al. [20] and Liu et al. [22] with some modifications.

Woo et al. [17] also described similar findings in *P. marneffei* PM1: they demonstrated that the biosynthetic pathway for the production of azaphilone pigments is regulated by a gene cluster (*pks3*) that also encodes for KS, AT, ACP, MT and thiolester reductase (R) domains [17]. These authors suggested that the synthesis of the orange *Monascus*-like azaphilone pigments such as rubropunctatin and monascorubrin begins through the polyketide pathway, initially modulated by this *pks3* gene cluster in *P. marneffei* PM1 [17]. These orange pigments and their well-known derivatives, e.g., the pigments glycyl-rubropunctatin **3** and *N*-GABA-rubropunctatin **4** isolated in this study, may be formed by the esterification of 3-oxo-octanoic acid or 3-oxo-decanoic acid to the polyketide-based chromophore [17]. From there, red derivatives of *Monascus*-like azaphilone pigments can be synthetized by Schiff base formation reactions [17–19,22]. Indeed, the amination of the orange *Monascus*-like azaphilone pigments with proteins, amino acids or nucleic acids yields the azaphilone red pigments, including the derivatives of rubropunctamin (e.g., *N*-threonyl-rubropunctamin **5** and *N*-glutaryl-rubropunctamin **9**) and the derivatives of monascorubramin **10**, like PP-R **2**, 6-[(Z)-2-Carboxyvinyl]-*N*-GABA-PP-V **6**, *N*-glutaryl-monascorubraminic acid **7** and *N*-threonine-monascorubramin **8** identified in this study from the marine-derived *Talaromyces* sp. 30570 strain (Figure 8).

5. Conclusions

In this study, we demonstrated the potential of the marine-derived fungal strain *Talaromyces* sp. 30570 to produce a wide variety of water-soluble *Monascus*-like azaphilone red pigments with respect to the medium composition. Such environment-dependent responses confirmed that the manipulation of the culture conditions (in particular, the presence of organic nitrogen sources) may trigger the expression of certain biosynthetic pathways and the production of a high amount of nitrogen-containing red azaphilone pigments by the fungi. Among the twelve different pigments detected in the fungal extract, nine derivatives of nitrogen-containing azaphilone red pigments were identified. *N*-threonyl-rubropunctamin or the acid form of the pigment PP-R, 6-[(Z)-2-Carboxyvinyl]-*N*-GABA-PP-V, *N*-threonine-monascorubramin, *N*-glutaryl-rubropunctamin, and monascorubramin were the major pigmented compounds. Bioproduction of these molecules occurred only when complex organic nitrogen sources were present in the culture medium. These findings are important for the field of the selective production of these fungal red azaphilones. They may represent relevant metabolites for the industries. Indeed, among the natural colorants, the red ones are the most interesting, as they are increasingly used in human foods and in ingested drugs. These fungal red azaphilones are not only 'colored,' they often exhibit remarkable antibiotic and antitumoral activities, and are of interest due to their applications in the design of new pharmaceutical products.

Supplementary Materials: The following are available online at http://www.mdpi.com/2076-2607/8/12/1920/s1, Figure S1. Mass spectra of the assumed compound PP-R **2**; Figure S2. Mass spectra of the assumed compound Glycyl-rubropunctatin **3**; Figure S3. Mass spectra of the assumed compound *N*-GABA-rubropunctatin **4**; Figure S4. Mass spectra of the assumed compound *N*-threonyl-rubroptunctamin or acid form of PP-R **5**; Figure S5. Mass spectra of the assumed compound 6-[(Z)-2-Carboxylvinyl]-*N*-GABA-PP-V **6**; Figure S6. Mass spectra of the assumed compound *N*-glutaryl-monascorubraminic acid **7**; Figure S7. Mass spectra of the assumed compound *N*-threonine-monascorubramin **8**; Figure S8. Mass spectra of the assumed compound *N*-glutaryl-rubropunctamin **9**; Figure S9. Mass spectra of the assumed compound Monascorubramin **10**; Figure S10. Mass spectra of the assumed compound Ergosterol **13**.

Author Contributions: J.L. and Y.C. performed the experimental and laboratory work; J.L., T.P., M.F., L.D. and Y.C. worked on the analysis and interpretation of the data and contributed with valuable discussions; J.L. and Y.C. conceived the project, worked on the structure and wrote the paper. All authors have read and agreed to the published version of the manuscript.

Funding: This research was funded by Conseil Régional de La Réunion (Reunion Island, France), grant number No. DIRED 20140704 (COLORMAR Program). The funding body has no role in the design of the study and collection, analysis, and interpretation of data and in writing the manuscript.

Conflicts of Interest: The authors declare no conflict of interest.

References

1. Newman, D.J.; Cragg, G.M. Natural products as sources of new drugs from 1981 to 2014. *J. Nat. Prod.* **2016**, *79*, 629–661. [CrossRef] [PubMed]
2. Blunt, J.W.; Carroll, A.R.; Copp, B.R.; Davis, R.A.; Keyzers, R.A.; Prinsep, M.R. Marine natural products. *Nat. Prod. Rep.* **2018**, *35*, 8–53. [CrossRef] [PubMed]
3. Dufossé, L.; Fouillaud, M.; Caro, Y.; Mapari, S.A.S.; Sutthiwong, N. Filamentous fungi are large-scale producers of pigments and colorants for the food industry. *Curr. Opin. Biotech.* **2014**, *26*, 56–61. [CrossRef]
4. Akilandeswari, P.; Pradeep, B.V. Exploration of industrially important pigments from soil fungi. *Appl. Microbiol. Biotechnol.* **2016**, *100*, 1631–1643. [CrossRef]
5. Caro, Y.; Venkatachalam, M.; Lebeau, J.; Fouillaud, M.; Dufossé, L. Pigments and colorants from filamentous fungi. In *Fungal Metabolites. Reference Series in Phytochemistry*; Merillon, J.M., Ramawat, K.G., Eds.; Springer: Cham, Switzerland, 2017; pp. 499–568.
6. Fouillaud, M.; Venkatachalam, M.; Girard-Valenciennes, E.; Caro, Y.; Dufossé, L. Anthraquinones and derivatives from marine-derived fungi: Structural diversity and selected biological activities. *Mar. Drugs* **2016**, *14*, 64. [CrossRef]
7. Caro, Y.; Anamale, L.; Fouillaud, M.; Laurent, P.; Petit, T.; Dufossé, L. Natural hydroxyanthraquinoid pigments as potent food grade colorants: An overview. *Nat. Prod. Bioprospect.* **2012**, *2*, 174–193. [CrossRef]
8. Arikan, E.B.; Canli, O.; Caro, Y.; Dufossé, L.; Dizge, N. Production of bio-based pigments from food processing industry by-products (apple, pomegranate, black carrot, red beet pulps) using *Aspergillus carbonarius*. *J. Fungi* **2020**, *6*, 240. [CrossRef] [PubMed]
9. Mapari, S.A.S.; Thrane, U.; Meyer, A.S. Fungal polyketide azaphilone pigments as future natural food colorants? *Trends Biotechnol.* **2010**, *28*, 300–307. [CrossRef]
10. Lagashetti, A.C.; Dufossé, L.; Singh, S.K.; Singh, P.N. Fungal pigments and their prospects in different industries. *Microorganisms* **2019**, *7*, 604. [CrossRef]
11. Jannel, S.; Caro, Y.; Bermudes, M.; Petit, T. Novel insights into the biotechnological production of *Haematococcus pluvialis*-derived astaxanthin: Advances and key challenges to allow its industrial use as novel food ingredient. *J. Mar. Sci. Eng.* **2020**, *8*, 789. [CrossRef]
12. Jůzlová, P.; Martínková, L.; Křen, V. Secondary metabolites of the fungus *Monascus*: A review. *J. Ind. Microbiol.* **1996**, *16*, 163–170. [CrossRef]
13. Balakrishnan, B.; Karki, S.; Chiu, S.H.; Kim, H.J.; Suh, J.W.; Nam, B.; Yoon, Y.-M.; Chen, C.-C.; Kwon, H.-J. Genetic localization and in vivo characterization of a *Monascus* azaphilone pigment biosynthetic gene cluster. *Appl. Microbiol. Biotechnol.* **2013**, *97*, 6337–6345. [CrossRef] [PubMed]
14. Ogihara, J.; Kato, J.; Oishi, K.; Fujimoto, Y. PP-R, 7-(2-hydroxyethyl)-monascorubramine, a red pigment produced in the mycelia of *Penicillium* sp. AZ. *J. Biosci. Bioeng.* **2001**, *91*, 44–47. [CrossRef]
15. Mapari, S.A.; Meyer, A.S.; Thrane, U.; Frisvad, J.C. Identification of potentially safe promising fungal cell factories for the production of polyketide natural food colorants using chemotaxonomic rationale. *Microb. Cell Fact.* **2009**, *8*, 24. [CrossRef]
16. Frisvad, J.C.; Yilmaz, N.; Thrane, U.; Rasmussen, K.B.; Houbraken, J.; Samson, R.A. *Talaromyces atroroseus*, a new species efficiently producing industrially relevant red pigments. *PLoS ONE* **2013**, *8*, e84102. [CrossRef]
17. Woo, P.C.; Lam, C.-W.; Tam, E.W.T.; Lee, K.-C.; Yung, K.K.Y.; Leung, C.K.F.; Sze, K.-H.; Lau, S.K.P.; Yuen, K.-Y. The biosynthetic pathway for a thousand-year-old natural food colorant and citrinin in *Penicillium marneffei*. *Sci. Rep.* **2014**, *4*, 6728. [CrossRef]
18. Gomes, D.C.; Takahashi, J.A. Sequential fungal fermentation-biotransformation process to produce a red pigment from sclerotiorin. *Food Chem.* **2016**, *210*, 355–361. [CrossRef]
19. Morales-Oyervides, L.; Ruiz-Sánchez, J.P.; Oliveira, J.C.; Sousa-Gallagher, M.J.; Méndez-Zavala, A.; Giuffrida, D.; Dufossé, L.; Montañez, J. Biotechnological approaches for the production of natural colorants by *Talaromyces/Penicillium*: A review. *Biotechnol. Adv.* **2020**, *43*, 107601. [CrossRef]
20. Chen, W.; Chen, R.; Liu, Q.; He, Y.; He, K.; Ding, X.; Kang, L.; Guo, X.; Xie, N.; Zhou, Y.; et al. Orange, red, yellow: Biosynthesis of azaphilone pigments in *Monascus* fungi. *Chem. Sci.* **2017**, *8*, 4917–4925. [CrossRef]
21. Chen, W.; Feng, Y.; Molnár, I.; Chen, F. Nature and nurture: Confluence of pathway determinism with metabolic and chemical serendipity diversifies: *Monascus* azaphilone pigments. *Nat. Prod. Rep.* **2019**, *36*, 561–572. [CrossRef]

22. Liu, L.; Zhao, J.; Huang, Y.; Xin, Q.; Wang, Z. Diversifying of chemical structure of native *Monascus* pigments. *Front. Microbiol.* **2018**, *9*, 3143. [CrossRef] [PubMed]
23. Joshi, D.R.; Adhikari, N. An overview on common organic solvents and their toxicity. *J. Pharm. Res. Int.* **2019**, *28*, 1–18. [CrossRef]
24. Chemat, F.; Vian, M.A.; Cravotto, G. Green extraction of natural products: Concept and principles. *Int. J. Mol. Sci.* **2012**, *13*, 8615–8627. [CrossRef] [PubMed]
25. Lebeau, J.; Venkatachalam, M.; Fouillaud, M.; Petit, T.; Vinale, F.; Dufossé, L.; Caro, Y. Production and new extraction method of polyketide red pigments produced by ascomycetous fungi from terrestrial and marine habitats. *J. Fungi* **2017**, *3*, 34. [CrossRef]
26. Lebeau, J.; Petit, T.; Clerc, P.; Dufossé, L.; Caro, Y. Isolation of two novel purple naphthoquinone pigments concomitant with the bioactive red bikaverin and derivates thereof produced by *Fusarium oxysporum*. *Biotechnol. Prog.* **2019**, *35*, e2738. [CrossRef]
27. Martins, F.S.; Borges, L.L.; Paula, J.R.; Conceição, E.C. Impact of different extraction methods on the quality of *Dipteryx alata* extracts. *Rev. Bras. Farmacogn.* **2013**, *23*, 521–526. [CrossRef]
28. Fouillaud, M.; Venkatachalam, M.; Llorente, M.; Magalon, H.; Cuet, P.; Dufossé, L. Biodiversity of pigmented fungi isolated from marine environment in La Réunion island, Indian Ocean: New resources for colored metabolites. *J. Fungi* **2017**, *3*, 36. [CrossRef]
29. Klitgaard, A.; Iversen, A.; Andersen, M.R.; Larsen, T.O.; Frisvad, J.C.; Nielsen, K.F. Aggressive dereplication using UHPLC-DAD-QTOF: Screening extracts for up to 3000 fungal secondary metabolites. *Anal. Bioanal. Chem.* **2014**, *406*, 1933–1943. [CrossRef]
30. Rasmussen, K.B. *Talaromyces atroroseus*: Genome Sequencing, *Monascus* Pigments and Azaphilone Gene Cluster Evolution. Ph.D. Thesis, Technical University of Danemark, Kongens Lyngby, Danemark, 2015.
31. Izawa, S.; Harada, N.; Watanabe, T.; Kotokawa, N.; Yamamoto, A.; Hayatsu, H.; Arimoto-Kobayashi, S. Inhibitory effects of food-coloring agents derived from *Monascus* on the mutagenicity of heterocyclic amines. *J. Agric. Food Chem.* **1997**, *45*, 3980–3984. [CrossRef]
32. Yuliana, A.; Singgih, M.; Julianti, E.; Blanc, P.J. Derivates of azaphilone *Monascus* pigments. *Biocatal. Agric. Biotechnol.* **2017**, *9*, 183–194. [CrossRef]
33. Mukherjee, G.; Singh, S.K. Purification and characterization of a new red pigment from *Monascus purpureus* in submerged fermentation. *Process. Biochem.* **2011**, *46*, 188–192. [CrossRef]
34. Venkatachalam, M.; Zelena, M.; Cacciola, F.; Ceslova, L.; Girard-Valenciennes, E.; Clerc, P.; Dugo, P.; Mondello, L.; Fouillaud, M.; Rotondo, A.; et al. Partial characterization of the pigments produced by the marine-derived fungus *Talaromyces albobiverticillius* 30548. Towards a new fungal red colorant for the food industry. *J. Food Compos. Anal.* **2018**, *67*, 38–47.
35. Hajjaj, H.; Klaebe, A.; Loret, M.O.; Tzedakis, T.; Goma, G.; Blanc, P.J. Production and Identification of N-Glucosylrubropunctamine and N-Glucosylmonascorubramine from Monascus ruber and occurrence of electron donor-acceptor complexes in these red pigments. *Appl. Environ. Microbiol.* **1997**, *63*, 2671–2678. [CrossRef] [PubMed]
36. Jung, H.; Kim, C.; Kim, K.; Shin, C.S. Color characteristics of *Monascus* pigments derived by fermentation with various amino acids. *J. Agric. Food Chem.* **2003**, *51*, 1302–1306. [CrossRef]
37. Slominski, A.; Semak, I.; Zjawiony, J.; Wortsman, J.; Gandy, M.N.; Li, J.; Zbytek, B.; Li, W.; Tuckey, R.C. Enzymatic metabolism of ergosterol by cytochrome P450scc to biologically active 17α,24-dihydroxyergosterol. *Chem. Biol.* **2005**, *12*, 931–939. [CrossRef]
38. Dame, Z.T.; Silima, B.; Gryzenhout, M.; van Ree, T. Bioactive compounds from the endophytic fungus *Fusarium proliferatum*. *Nat. Prod. Res.* **2016**, *30*, 1301–1304. [CrossRef]
39. Lebeau, J.; Petit, T.; Dufossé, L.; Caro, Y. Putative metabolic pathway for the bioproduction of bikaverin and intermediates thereof in the wild *Fusarium oxysporum* LCP531 strain. *AMB Express* **2019**, *9*, 186. [CrossRef]
40. Ogihara, J.; Oishi, K. Effect of ammonium nitrate on the production of PP-V and monascorubrin homologues by *Penicillium* sp. AZ. *J. Biosci. Bioeng.* **2002**, *93*, 54–59. [CrossRef]
41. Arai, T.; Umemura, S.; Ota, T.; Ogihara, J.; Kato, J.; Kasumi, T. Effects of inorganic nitrogen sources on the production of PP-V [(10Z)-12-carboxyl-monascorubramine] and the expression of the nitrate assimilation gene cluster by *Penicillium* sp. *AZ*, *Biosci. Biotechnol. Biochem.* **2012**, *76*, 120–124. [CrossRef]
42. Padmavathi, T.; Prabhudessai, T. A solid liquid state culture method to stimulate *Monascus* pigments by intervention of different substrates. *Int. Res. J. Biol. Sci.* **2013**, *2*, 22–29.

43. Liew, F.; Martin, M.E.; Tappel, R.C.; Heijstra, B.D.; Mihalcea, C.; Köpke, M. Gas Fermentation—A flexible platform for commercial scale production of low-carbon-fuels and chemicals from waste and renewable feedstocks. *Front. Microbiol.* **2016**, *7*, 694. [CrossRef] [PubMed]
44. Vega, G.C.; Sohn, J.; Voogt, J.; Nilsson, A.E.; Birkved, M.; Olsen, S.I. Insights from combining techno-economic and life cycle assessment—A case study of polyphenol extraction from red wine pomace. *Resour. Conserv. Recycl.* **2020**, 100045. [CrossRef]
45. Gao, J.-M.; Yang, S.-X.; Qin, J.-C. Azaphilones: Chemistry and biology. *Chem. Rev.* **2013**, *113*, 4755–4811. [CrossRef] [PubMed]
46. Chen, C.; Tao, H.; Chen, W.; Yang, B.; Zhou, X.; Luo, X.; Liu, Y. Recent advances in the chemistry and biology of azaphilones. *RSC Adv.* **2020**, *10*, 10197–10220. [CrossRef]
47. Tolborg, G.; Ødum, A.S.R.; Isbrandt, T.; Larsen, T.O.; Workman, M. Unique processes yielding pure azaphilones in *Talaromyces atroroseus*. *Appl. Microbiol. Biotechnol.* **2020**, *104*, 603–613. [CrossRef]
48. Isbrandt, T.; Tolborg, G.; Ødum, A.; Workman, M.; Larsen, T.O. Atrorosins: A new subgroup of *Monascus* pigments from *Talaromyces atroroseus*. *Appl. Microbiol. Biotechnol.* **2020**, *104*, 615–622. [CrossRef]
49. Peterson, S.W.; Jurjevic, Z. The *Talaromyces pinophilus* species complex. *Fungal Biol.* **2019**, *123*, 745–762. [CrossRef]

microorganisms

MDPI

Article

Influence of Culture Conditions and Medium Compositions on the Production of Bacteriocin-Like Inhibitory Substances by *Lactococcus lactis* Gh1

Roslina Jawan [1,2], Sahar Abbasiliasi [3], Joo Shun Tan [4], Shuhaimi Mustafa [3,5], Murni Halim [1,6] and Arbakariya B. Ariff [1,6,*]

1 Bioprocessing and Biomanufacturing Research Centre, Faculty of Biotechnology and Biomolecular Sciences, Universiti Putra Malaysia, Serdang 43400, Malaysia; roslinaj@ums.edu.my (R.J.); murnihalim@upm.edu.my (M.H.)
2 Biotechnology Programme, Faculty of Science and Natural Resources, Universiti Malaysia Sabah, Kota Kinabalu 88400, Malaysia
3 Halal Products Research Institute, Universiti Putra Malaysia, Serdang 43400, Malaysia; sahar@upm.edu.my (S.A.); shuhaimimustafa@upm.edu.my (S.M.)
4 Bioprocess Technology, School of Industrial Technology, Universiti Sains Malaysia, Gelugor 11800, Malaysia; jooshun@usm.my
5 Department of Microbiology, Faculty of Biotechnology and Biomolecular Sciences, Universiti Putra Malaysia, Serdang 43400, Malaysia
6 Department of Bioprocess Technology, Faculty of Biotechnology and Biomolecular Sciences, Universiti Putra Malaysia, Serdang 43400, Malaysia
* Correspondence: arbarif@upm.edu.my; Tel.: +60-3-8946-7591

Received: 20 July 2020; Accepted: 24 August 2020; Published: 23 September 2020

Abstract: Antibacterial peptides or bacteriocins produced by many strains of lactic acid bacteria have been used as food preservatives for many years without any known adverse effects. Bacteriocin titres can be modified by altering the physiological and nutritional factors of the producing bacterium to improve the production in terms of yield and productivity. The effects of culture conditions (initial pH, inoculum age and inoculum size) and medium compositions (organic and inorganic nitrogen sources; carbon sources) were assessed for the production of bacteriocin-like inhibitory substances (BLIS) by *Lactococcus lactis* Gh1 in shake flask cultures. An inoculum of the mid-exponential phase culture at 1% (*v*/*v*) was the optimal age and size, while initial pH of culture media at alkaline and acidic state did not show a significant impact on BLIS secretion. Organic nitrogen sources were more favourable for BLIS production compared to inorganic sources. Production of BLIS by *L. lactis* Gh1 in soytone was 1.28-times higher as compared to that of organic nitrogen sources ($(NH_4)2SO_4$). The highest cell concentration ($X_{mX} = 0.69 \pm 0.026$ g·L^{-1}) and specific growth rate ($\mu_{max} = 0.14$ h^{-1}) were also observed in cultivation using soytone. By replacing carbon sources with fructose, BLIS production was increased up to 34.94% compared to BHI medium, which gave the biomass cell concentration and specific growth rate of 0.66 ± 0.002 g·L^{-1} and 0.11 h^{-1}, respectively. It can be concluded that the fermentation factors have pronounced influences on the growth of *L. lactis* Gh1 and BLIS production. Results from this study could be used for subsequent application in process design and optimisation for improving BLIS production by *L. lactis* Gh1 at larger scale.

Keywords: fermentation; *Lactococcus lactis*; bacteriocin; culture conditions; medium compositions

1. Introduction

The use of antimicrobials in shelf life enhancement of foods is a new branch of science. Bio-preservation is a technique used for extending the shelf life of food using natural or controlled microbiota or antimicrobials. The fermentation products as well as beneficial bacteria are generally selected in this process to control spoilage and render pathogen inactive. The special interest organism or central organism used for this purpose is lactic acid bacteria (LAB) and their metabolites [1]. LAB are capable of producing various antimicrobial compounds such as organic acids (lactic acid and acetic acid), diacetyl, ethanol, hydrogen peroxide, reuterin, acetaldehyde, acetoine, carbon dioxide and bacteriocins during fermentation processes [2].

LAB bacteriocins are considered good bio-preservative agents due to their non-toxic, non-immunogenic and thermo-resistance characteristics as well as broad bactericidal activity. These bacteriocins are most effective against Gram-positive bacteria and some damaged, Gram-negative bacteria including various pathogens such as *Listeria monocytogenes*, *Bacillus cereus*, *Staphylococcus aureus*, *Salmonella* in foods [1,3]. The use of bacteriocins as natural food preservatives fulfils consumer demands for high quality and safe foods without the use of chemical preservatives [4]. For the past years, bacteriocins have attracted considerable interest for their use as safe food preservatives since they are easily digested by the human gastrointestinal tract [5]. Several reports have showed that antimicrobial metabolites produced by *Lactococcus lactis* exhibit broad inhibitory property towards species that are closely related to LAB and other unrelated spoilage and pathogenic bacteria [6,7].

Recent research indicated that LAB and their natural products can offer promising opportunities for the development of efficient food bio preservation strategies [8–10]. The potential applications of bacteriocins from LAB as bio-preservatives for the inhibition of proliferation of *L. monocytogenes* in foods have been reported [11–14]. Generally, most LAB-bacteriocins act on pathogen cells by destabilisation and permeabilisation of the cytoplasmic membrane through the formation of transitory poration complexes or ionic channels that cause the reduction or dissipation of the proton motive force (PMF). Bacteriocins producing LAB strains protect themselves against the toxicity of their own bacteriocins by the expression of a specific immunity protein that is generally encoded in the bacteriocin operon namely induction factor (IF), histidine protein kinase (HPK) and a response regulator (RR) [15]. As food bio-preservative, bacteriocin could be used either as a food additive or through the application of bacteriocin-producing culture. However, the latter would necessitate the optimisation of specific fermentation conditions [16]. Formulation of the growth medium is an important factor that needs to be considered in producing any microbial products involving fermentation processes. On the other hand, formulation for industrial scale application should fulfil a number of criteria including cost-effectiveness, high product yield, short fermentation time and ease of downstream purification processes [17].

Specific requirements with reference to the production of bacteriocins through microbial fermentation have been reported by several studies [18–20]. Bacteriocin titres can be modified by altering the cultivation conditions of the producing bacterium and certain combinations of influencing factors. However, the influencing factors may be strain dependent and could vary with different types of bacteriocin. The optimal design of culture media is an important aspect to be considered when developing a fermentation process. The formulation of media containing complex nutrients is generally preferred for large-scale fermentations since it leads to the development of cost-effective processes that support maximum product yield. In the initial formulation of the medium in batch culture, an effort has been made to understand the best source of carbon and energy as well as the regulatory aspects of the enzyme [21].

The growth of bacteria and accumulation of their metabolites are strongly influenced by the environment and medium compositions such as culture pH, carbon and nitrogen sources, growth factors as well as minerals. It is difficult to detect these major factors and optimise them for biotechnological processes including multivariable [22]. The properties of the growth media including amino acid composition, carbon/nitrogen ratio, pH and lactose levels have a great influence on the change in biomass of the culture and the corresponding change in the level of bacteriocin production [23]. De Man, Rogosa and Sharpe (MRS) medium is usually the medium of choice for studying LAB fermentations, but it has its limitations. Using MRS broth or any commercially available laboratory-grade defined growth media for industrial production of bacteriocins would be prohibitively expensive besides involving unauthorised ingredients as food additives [24]. Therefore, the objective of the present study was to optimise the physiological (pH value, inoculum age and size) and nutritional (medium compositions) factors for improving the growth and ability of *Lactococcus lactis* Gh1 to secrete bacteriocin-like inhibitory substances (BLIS) in shake flask fermentation.

2. Materials and Methods

2.1. Materials

Culture media (Brain heart infusion (BHI), M17, MRS, tryptic soy broth (TSB) and LB media) and inorganic nitrogen sources ($(NH_4)_2SO_4$, NH_6PO_4, NH_4NO_3 and NH_4Cl) were purchased from Merck (Darmstadt, Germany). Carbon sources (fructose, glucose, galactose, sucrose, lactose, maltose, sorbitol and mannitol) were obtained from Fisher Chemical (Loughborough, United Kingdom), while organic nitrogen sources (yeast extract, meat extract, peptone and soytone) were from BD (Franklin Lakes, NJ, USA).

2.2. Microorganisms and Maintenance

A BLIS-producing *Lactococcus lactis* Gh1 isolated from a milk by-product of an Iranian traditional fermented milk was used throughout this study [25,26]. The indicator microorganism in antimicrobial activity was *Listeria monocytogenes* ATCC 15313. The stock cultures were maintained at −80 °C in MRS and BHI broth (Merck, Darmstadt, Germany), respectively, supplemented with 20% (*v*/*v*) glycerol (BDH Laboratory Supplies, Poole, UK), for use in the fermentation experiments.

2.3. Preparation of Inoculum and Culture Condition

The stock culture of *L. lactis* Gh1 was first revived on the BHI agar (Merck, Darmstadt, Germany) prior to the preparation of inoculum. A single colony of *L. lactis* Gh1 was cultured in 10 mL of BHI broth and incubated at 30 °C for 24 h. The 1% (*v*/*v*) of the culture was sub-cultured at 30 °C for 16–18 h before being used as an inoculum. The optical density (OD) of the culture at 650 nm was standardised at 1.89–2.00 ($\approx 2.68 \times 10^9$ CFU/mL) and used as an inoculum for all fermentations with the size of 1% (*v*/*v*). All experiments were conducted in 100 mL of BHI broth in 250 mL of Erlenmeyer flasks. The cultures were incubated at 30 °C in a horizontal shaker (B. Braun Biotech International, Melsungen, Germany) and agitated at 100 rpm for 16 h.

2.4. Factors Influencing the Production of Bacteriocin-Like Inhibitory Substances (BLIS)

In order to identify the factors that influence BLIS production, the one-factor-at-a-time (OFAT) approach was used. The selected factors were applied to the next experiment. The selection of the appropriate factors was based on the highest BLIS production.

2.5. Selection of Culture Media for BLIS Production

Five types of commercial culture media (MRS, M17, BHI, TSB and LB) were initially tested for production of BLIS.

2.6. Effects of Initial pH of Culture Media on BLIS Production

The initial pH of the selected media was adjusted to pH 2–9 by the addition of either 1.0 M of HCl or 1.0 M NaOH.

2.7. Effects of Inoculum Age, Size and Cultivation Conditions on BLIS Production

Based on the growth phase of *L. lactis* Gh1, three different inoculum ages namely early-exponential, late exponential and stationary phases were evaluated. The selected inoculum age with maximum

BLIS secretion was added into the media at different concentrations ranging from 0.5% to 10% (*v*/*v*). The optimum inoculum size and age was incubated in condition without and with agitation at 100 rpm.

2.8. Screening of Medium Composition for BLIS Production

To initiate the fermentation, a 250 mL Erlenmeyer flask containing 100 mL of BHI broth was inoculated with 1% (*v*/*v*) of inoculum at mid-exponential growth phase and incubated at 30 °C on a rotary shaker, which then agitated at 100 rpm for 16–18 h. BHI medium was selected in this study as this medium promoted significant BLIS production. The composition of BHI medium is shown in Table 1.

Table 1. Composition of commercial BHI medium.

Component	Amount (g/L)
Nutrient substrate (Extract of brain and heart, and peptones)	27.5
Sodium chloride	5
Di-Sodium hydrogen phosphate	2.5
D (+) glucose	2

To evaluate the effects of nitrogen sources on the production of BLIS, the nitrogen component in the BHI medium was replaced with different types of organic (yeast extract, meat extract, peptone and soytone) and inorganic ($(NH_4)_2SO_4$, NH_6PO_4, NH_4NO_3 and NH_4Cl) nitrogen sources. The amount of nitrogen was calculated based on the original quantity of nitrogen (N) in BHI medium. Nutrient substrate (27.5 g/L) of BHI medium is equivalent to 4.6 g of N. Therefore, nitrogen sources were prepared according to this nitrogen amount.

In the subsequent experiment to study the effects of different carbon sources namely monosaccharides (fructose, glucose, galactose), disaccharides (sucrose, lactose, maltose) and sugar alcohols (sorbitol, and mannitol) on BLIS production, soytone was used as a preferred nitrogen source.

2.9. Analytical Procedures

During the fermentation, samples were withdrawn at 2-h intervals and prepared for analysis. The changes in culture pH were measured using a pH meter (Mettler-Toledo, Switzerland). The culture samples were centrifuged (Eppendorf, Centrifuge 5810R, Hamburg, Germany) at 13,751× *g* for 10 min at 4 °C. The cell pellets were washed and resuspended twice with saline water (0.85%, *w*/*v* NaCl) for turbidity determination and read at 600 nm using a spectrophotometer (Biochrom, Libra S12, UK). The optical density (OD) was converted into dry cell weight (DCW) from a standard curve using an experimentally predetermined factor of 0.26 where one OD unit was equivalent to 0.26 of DCW per volume ($g \cdot L^{-1}$). Antibacterial activity (AU/mL) against *L. monocytogenes* ATCC 15313 was quantitatively performed by the agar well diffusion method as described by Abbasiliasi et al. [25]. Briefly, the culture of *L. lactis* Gh1 was centrifuged (Eppendorf, Centrifuge 5810R, Hamburg, Germany) at 13,751× *g* for 10 min at 4 °C. The supernatant (100 μL) was put in 6-mm agar plate wells that were previously seeded (1%, *v*/*v*) with an active-growing *L. monocytogenes* ATCC 15313. The plates were then placed at 4 °C for well diffusion of the sample into the agar media for 2 h prior to incubation at 37 °C. After 24 h of incubation, the inhibition zone of the supernatant against the indicator bacteria was measured using electronic calliper (in mm). The quantification of the antimicrobial activity (AU/mL) of BLIS was calculated using Equation (1):

$$\text{BLIS activity } \left(\frac{AU}{mL}\right) = \frac{A_Z - A_W}{V} \tag{1}$$

where, A_z = clear zone area (mm^2), A_w = well area (mm^2), V = volume of sample (mL).

2.10. Statistical Analysis

Analysis of variance (ANOVA) for mean data was performed by IBM SPSS Statistics 25 software. Duncan multiple range test was used to determine the significance among the treatments means with significant level at $p < 0.05$.

3. Results

3.1. Effects of Culture Media on the Production of BLIS

Among the culture media tested in this study, there were no significant differences ($p < 0.05$) in BLIS production between M17, BHI and TSB (Table 2). However, the highest BLIS production (672.86 ± 24.76) and μ_{max} (0.15 ± 0.0126 h^{-1}) were recorded in BHI. MRS medium promoted better cell growth (0.49 ± 0.002 $g \cdot L^{-1}$) compared to other media. The BLIS production was shown by the inhibition zones of *L. lactis* Gh1 supernatant against the *L. monocytogenes* ATCC 15313 in the antimicrobial assay. The increase in size of the inhibition zones was related to the inhibitory effect of the *L. lactis* Gh1.

Table 2. Growth of *L. lactis* Gh1 and bacteriocin-like inhibitory substances (BLIS) production in various types of culture media.

Media	Time for P_{mX} (h)	pH		Maximum BLIS Activity P_{mX} (AU. mL^{-1})	Maximum Cell Concentration X_{mX} ($g \cdot L^{-1}$)	Specific Growth Rate μ_{max} (h^{-1})
		Initial	Final			
MRS	16	5.43 ± 0.01	4.27 ± 0.04	375.48 ± 35.43 ^{b}	0.49 ± 0.002 ^{a}	0.09 ± 0.0057 ^{b}
M17	4	6.98 ± 0.01	6.49 ± 0.02	640.23 ± 21.06 ^{a}	0.43 ± 0.020 ^{b}	0.08 ± 0.0022 ^{b}
BHI	4	6.95 ± 0.02	5.95 ± 0.00	672.86 ± 24.76 ^{a}	0.46 ± 0.020 ^{ab}	0.15 ± 0.0126 ^{a}
TSB	4	6.92 ± 0.03	6.72 ± 0.02	637.84 ± 24.73 ^{a}	0.48 ± 0.040 ^{ab}	0.09 ± 0.0024 ^{b}
LB	-	6.01 ± 0.01	3.66 ± 0.02	0.00 ± 0.00 ^{c}	0.31 ± 0.002 ^{c}	0.09 ± 0.0042 ^{b}

Note: all values are expressed as means ± standard deviation in triplicate. Data followed by the same letters are not significantly different ($p < 0.05$) according to Duncan's multiple range test to evaluate the effect of investigated parameters. S.D: standard deviation.

The time course of BLIS production by *L. lactis* Gh1 is shown in Figure 1. The highest BLIS activity was reached at the late-exponential phase (4 h) and decreased afterwards when the cells entered the stationary phase. The BLIS activity was not detected at 18 h of incubation. The pH was decreased rapidly during the maximum production of BLIS and remained constant up to the end of incubation time. The specified pattern of cell growth and BLIS production indicated that the produced BLIS was a primary metabolite. BLIS production was reduced once cells entered the stationary phase. Since the use of BHI enhanced the growth of *L. lactis* Gh1 and BLIS production, this medium was selected for subsequent use in the statistical optimisation for further improving the fermentation performance.

3.2. Effects of Initial pH on BLIS Production

As shown in Table 3, the initial pH of the culture among the alkaline pH (7–9) did not give a significant difference ($p < 0.05$) on BLIS production by *L. lactis* Gh1. However, BLIS secretion started to decrease at pH lower than 6, whereas BLIS secretion and cell growth were suppressed at pH lower than 4. The BHI medium without pH alteration (pH: 7.04) recorded the highest BLIS production (599.45 ± 11.77 AU. mL^{-1}), cell concentration (0.36 ± 0.013 g L^{-1}) and specific growth rate (0.104 ± 0.0025 h^{-1}). The fermentation time to attain the maximum BLIS activity (P_{mX}) was the same (6 h) for all initial culture pH tested in this study.

3.3. Effects of Inoculum Age, Size and Sub-Culture Frequency on BLIS Production

Growth of *L. lactis* Gh1 and BLIS production at different inoculum ages is shown in Table 4. Inoculum at mid-exponential growth phase recorded the highest BLIS production (665.96 ± 1.83 AU. mL^{-1}) followed by the activity at stationary (627.98 ± 16.74 AU. mL^{-1}) and late-exponential (615.73 ± 5.34 AU. mL^{-1})

growth phases. In fermentation with inoculum from mid-exponential phase, cells tend to produce BLIS without growing.

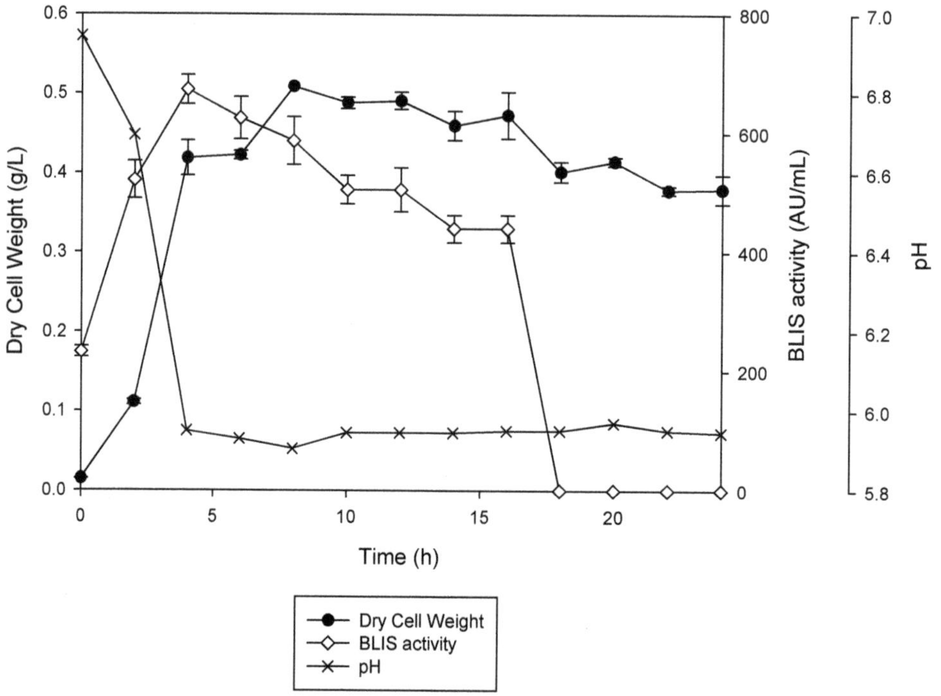

Figure 1. Growth profile of *L. lactis* Gh1 in Brain heart infusion (BHI) broth.

Table 3. Effects of initial culture pH on growth of *L. lactis* Gh1 and BLIS production using BHI medium.

Initial pH	Time for P_{mX} (h)	pH		Maximum BLIS Activity P_{mX} (AU. mL^{-1})	Maximum Cell Concentration X_{mX} ($g \cdot L^{-1}$)	Specific Growth Rate μ_{max} (h^{-1})
		Initial	Final			
2	-	2.10 ± 0.00	2.10 ± 0.00	0.00 ± 0.00 [d]	0.00 ± 0.000 [e]	0.000 ± 0.0000 [c]
3	-	3.02 ± 0.00	3.02 ± 0.00	0.00 ± 0.00 [d]	0.00 ± 0.000 [e]	0.000 ± 0.0000 [c]
4	8	4.13 ± 0.01	3.95 ± 0.01	464.42 ± 7.58 [c]	0.09 ± 0.000 [d]	0.016 ± 0.0000 [c]
5	6	5.00 ± 0.04	4.48 ± 0.02	475.17 ± 5.45 [c]	0.29 ± 0.002 [b]	0.050 ± 0.0005 [b]
6	6	6.37 ± 0.00	5.13 ± 0.02	548.67 ± 22.85 [b]	0.30 ± 0.000 [b]	0.051 ± 0.0003 [b]
7.04 *	6	7.10 ± 0.05	6.37 ± 0.00	599.45 ± 11.77 [a]	0.36 ± 0.013 [a]	0.104 ± 0.0025 [a]
8	6	7.66 ± 0.01	7.10 ± 0.02	581.99 ± 26.38 [ab]	0.27 ± 0.022 [b]	0.046 ± 0.0166 [b]
9	6	8.54 ± 0.02	7.89 ± 0.01	572.28 ± 26.64 [ab]	0.22 ± 0.020 [c]	0.043 ± 0.0142 [b]

Note: all values are expressed as means ± standard deviation in triplicate. Data followed by the same letters are not significantly different ($p < 0.05$) according to Duncan's multiple range test to evaluate the effect of investigated parameters. S.D: standard deviation. * Control (BHI broth without pH adjustment).

Table 4. Growth of *L. lactis* Gh1 and BLIS production using BHI medium, inoculated with different ages of inoculum. The inoculum age was set at 1% (*v/v*).

Growth Phase	Time for P_{mX} (h)	pH		Maximum BLIS Activity P_{mX} (AU. mL^{-1})	Maximum Cell Concentration X_{mX} ($g \cdot L^{-1}$)	Specific Growth Rate μ_{max} (h^{-1})
		Initial	Final			
Mid-exponential	6	6.91 ± 0.01	6.31 ± 0.00	665.96 ± 1.83 [a]	0.28 ± 0.033 [a]	0.113 ± 0.010 [a]
Late-exponential	6	6.90 ± 0.01	6.28 ± 0.01	615.73 ± 5.34 [b]	0.33 ± 0.017 [a]	0.110 ± 0.004 [a]
Stationary	8	6.90 ± 0.02	6.33 ± 0.01	627.98 ± 16.74 [b]	0.31 ± 0.033 [a]	0.100 ± 0.020 [a]

Note: all values are expressed as means ± standard deviation in triplicate. Data followed by the same letters are not significantly different ($p < 0.05$) according to Duncan's multiple range test to evaluate the effect of investigated parameters. S.D: standard deviation.

The effects of inoculum size on growth of *L. lactis* Gh1 and BLIS production is shown in Table 5. The results demonstrated that the size of inoculum has a strong relation with the production of BLIS. The fermentation with inoculum size of 1% (*v/v*) produced the highest BLIS compared to that with high inoculum size (>2% *v/v*). BLIS secretion was the lowest at the highest inoculum size (10%, *v/v*) so far studied. The inoculum size ranging from 4% (*v/v*) to 10% (*v/v*) produced more cells than BLIS production. The use of inoculum at a stationary growth phase and with the lowest or highest size influenced the duration of the lag phase, which gave a longer time to reach to P_{max} after 8–10 h of fermentation.

Table 5. Growth of *L. lactis* Gh1 and BLIS production using BHI medium, inoculated with different sizes of inoculum. The inoculum was set at mid-exponential phase.

Inoculum Size % (*v/v*)	Time for P_{mX} (h)	pH		Maximum BLIS Activity P_{mX} (AU. mL^{-1})	Maximum Cell Concentration X_{mX} (g L^{-1})	Specific Growth Rate μ_{max} (h^{-1})
		Initial	Final			
0.5	8	7.15 ± 0.01	6.33 ± 0.00	593.02 ± 57.36 ^{b}	0.33 ± 0.004 ^{d}	0.071 ± 0.001 ^{d}
1.0	6	7.13 ± 0.01	6.33 ± 0.01	700.24 ± 55.44 ^{a}	0.34 ± 0.002 ^{c}	0.071 ± 0.001 ^{d}
2.0	6	7.10 ± 0.00	6.29 ± 0.02	538.32 ± 72.05 ^{bc}	0.42 ± 0.002 ^{b}	0.095 ± 0.003 ^{c}
4.0	4	7.06 ± 0.01	6.25 ± 0.00	532.83 ± 7.69 ^{bc}	0.44 ± 0.005 ^{a}	0.151 ± 0.001 ^{a}
6.0	4	7.03 ± 0.04	6.22 ± 0.04	535.39 ± 17.98 ^{bc}	0.44 ± 0.004 ^{ab}	0.096 ± 0.002 ^{c}
8.0	8	7.00 ± 0.02	6.21 ± 0.03	566.08 ± 18.36 ^{b}	0.43 ± 0.002 ^{ab}	0.092 ± 0.010 ^{c}
10.0	10	6.95 ± 0.01	6.18 ± 0.02	479.60 ± 32.10 ^{c}	0.42 ± 0.016 ^{ab}	0.112 ± 0.003 ^{b}

Note: all values are expressed as means ± standard deviation in triplicate. Data followed by the same letters are not significantly different ($p < 0.05$) according to Duncan's multiple range test to evaluate the effect of investigated parameters. S.D: standard deviation.

BLIS production by *L. lactis* Gh1 was also influenced by the frequency in the sub-culturing of inoculum (Table 6). Good cell growth with enhanced BLIS productivity was observed in fermentation with inoculum sub-cultured for two times, which was two times higher (728.83 ± 37.80 AU. mL^{-1}) compared to that obtained in fermentation with a single time sub-cultured inoculum (553.45 ± 4.58 AU. mL^{-1}) with shorter lag phase length.

Table 6. Growth of *L. lactis* Gh1 and BLIS production using BHI medium, inoculated with different sub-culturing frequencies of inoculum. The inoculum was set at mid-exponential phase; at 1% (*v/v*) inoculum size.

Sub-Culturing Frequency	Time for P_{mX} (h)	pH		Maximum BLIS Activity P_{mX} (AU. mL^{-1})	Maximum Cell Concentration X_{mX} (g·L^{-1})	Specific Growth Rate μ_{max} (h^{-1})
		Initial	Final			
One time	10	7.15 ± 0.01	6.55 ± 0.01	553.45 ± 4.58	0.27 ± 0.004	0.05 ± 0.001
Two times	6	7.12 ± 0.01	6.49 ± 0.01	728.83 ± 37.80	0.21 ± 0.002	0.05 ± 0.001

3.4. Effects of Different Organic and Inorganic Nitrogen Sources on BLIS Production

The effects of organic and inorganic nitrogen sources on growth of *L. lactis* Gh1 and production of BLIS are shown in Table 7. Organic nitrogen sources were more favourable for BLIS production compared to inorganic sources. *L. lactis* Gh1 grown in soytone-supplemented medium produced the maximum BLIS production (764.71 ± 15.15 AU. mL^{-1}) and maximum cell (0.69 ± 0.026 g L^{-1}) significantly higher ($p < 0.5$) than those obtained by other organic (yeast extract, meat extract, peptone, tryptone, soytone) or inorganic ($(NH_4)_2SO_4$, NH_4NO_3, NH_6PO_4, NH_4Cl) nitrogen sources. The replacement of nitrogen source of BHI medium with soytone resulted in shorter lag phase length and time (4 h) in producing maximum BLIS compared to BHI and inorganic nitrogen sources (6 h).

3.5. Effects of Different Types of Carbon Source on BLIS Production

Effects of different types of carbon sources on BLIS production by *L. lactis* Gh1 are shown in Table 8. BLIS production was the highest (707.43 ± 16.83 AU. mL^{-1}) in fermentation using fructose compared to other carbon sources (galactose, glucose, sucrose, lactose, maltose, mannitol and sorbitol) tested in this study. In modified BHI medium, the replacement of carbon and nitrogen sources with fructose and soytone increased the BLIS production up to 34.94% compared to commercial BHI. The role of carbon sources was unenviable as long as the nitrogen was added into the media since the BLIS production (634.82 ± 26.40 AU. mL^{-1}) was still detected in the control treatment (without carbon sources). In addition, BLIS production in control was higher compared to that of BHI medium indicating the beneficial effect of soytone in supplying other minerals needed by *L. lactis* Gh1 than nitrogen components.

Table 7. Growth of *L. lactis* Gh1 and BLIS production by using modified BHI medium supplemented with different organic nitrogen (N) sources. All nitrogen sources were added at concentration of 4.6 g/L N.

Nitrogen Source	Time for P_{mX} (h)	pH		Maximum BLIS Activity P_{mX} (AU. mL^{-1})	Maximum Cell Concentration X_{mX} ($g{\cdot}L^{-1}$)	Specific Growth Rate μ_{max} (h^{-1})
		Initial	Final			
BHI	6	6.94 ± 0.01	6.28 ± 0.00	567.05 ± 47.01 [bc]	0.35 ± 0.004 [d]	0.074 ± 0.0012 [b]
No nitrogen sources	-	6.86 ± 0.00	6.29 ± 0.02	0.00 ± 0.00 [e]	0.00 ± 0.00 [i]	0.000 ± 0.0000 [d]
Organic Nitrogen						
Yeast extract	4	6.46 ± 0.01	4.98 ± 0.01	619.41 ± 26.02 [b]	0.45 ± 0.005 [b]	0.107 ± 0.0040 [c]
Meat extract	4	6.51 ± 0.02	5.38 ± 0.01	594.87 ± 8.69 [b]	0.42 ± 0.011 [c]	0.1062 ± 0.0050 [a]
Peptone	4	6.82 ± 0.03	6.58 ± 0.00	580.00 ± 14.14 [b]	0.36 ± 0.016 [d]	0.0547 ± 0.0025 [bc]
Soytone	4	6.80 ± 0.00	4.78 ± 0.01	764.71 ± 15.15 [a]	0.69 ± 0.026 [a]	0.0542 ± 0.0274 [c]
Tryptone	4	7.02 ± 0.01	6.71 ± 0.02	515.59 ± 33.57 [cd]	0.20 ± 0.001 [e]	0.0429 ± 0.0005 [c]
Inorganic Nitrogen						
$(NH_4)2SO_4$	6	6.86 ± 0.02	5.40 ± 0.00	599.08 ± 35.88 [b]	0.07 ± 0.0013 [f]	0.0120 ± 0.0004 [d]
NH_6PO_4	6	5.08 ± 0.02	5.10 ± 0.01	509.07 ± 26.23 [d]	0.01 ± 0.0007 h[i]	0.0007 ± 0.0005 [d]
NH_4NO_3	-	6.82 ± 0.01	6.25 ± 0.01	0.00 ± 0.00 [e]	0.03 ± 0.0004 [gh]	0.0060 ± 0.0010 [d]
NH_4Cl	-	6.44 ± 0.00	5.98 ± 0.02	0.00 ± 0.00 [e]	0.03 ± 0.0011 [g]	0.0076 ± 0.0001 [d]

Note: all values are expressed as means ± standard deviation in triplicate. Data followed by the same letters are not significantly different ($p < 0.05$) according to Duncan's multiple range test to evaluate the effect of investigated parameters. S.D: standard deviation.

Table 8. Growth of *L. lactis* Gh1 and BLIS production by using modified BHI medium supplemented with 4.6 g/L N of soytone at different carbon sources. All carbon sources were added at 2 g/L.

Carbon Source	Time for P_{mX} (h)	pH		Maximum BLIS Activity P_{mX} (AU. mL^{-1})	Maximum Cell Concentration X_{mX} ($g{\cdot}L^{-1}$)	Specific Growth Rate μ_{max} (h^{-1})
		Initial	Final			
BHI	8	7.11 ± 0.00	6.19 ± 0.01	530.12 ± 3.38 [d]	0.24 ± 0.0005 [g]	0.061 ± 0.0003 [f]
No carbon sources	10	6.89 ± 0.01	5.68 ± 0.00	634.82 ± 26.40 [bc]	0.56 ± 0.0018 [f]	0.046 ± 0.0003 [g]
Monosaccharides						
Fructose	10	6.86 ± 0.01	4.96 ± 0.00	715.36 ± 13.77 [a]	0.66 ± 0.0018 [c]	0.108 ± 0.0007 [e]
Glucose	10	6.79 ± 0.00	4.79 ± 0.02	674.03 ± 25.05 [abc]	0.67 ± 0.0110 [b]	0.139 ± 0.0066 [c]
Galactose	10	6.86 ± 0.02	5.15 ± 0.02	678.51 ± 1.23 [abc]	0.65 ± 0.0018 [d]	0.214 ± 0.0011 [a]
Disaccharides						
Sucrose	8	6.98 ± 0.00	4.85 ± 0.01	661.77 ± 17.54 [abc]	0.70 ± 0.0018 [a]	0.136 ± 0.0010 [c]
Lactose	8	6.98 ± 0.00	5.53 ± 0.02	622.99 ± 28.62 [c]	0.60 ± 0.0018 [e]	0.123 ± 0.0018 [d]
Maltose	8	6.99 ± 0.01	5.24 ± 0.02	687.45 ± 43.19 [ab]	0.67 ± 0.0018 [b]	0.198 ± 0.0010 [b]
Sugar Alcohol						
Sorbitol	8	7.01 ± 0.01	5.68 ± 0.00	636.48 ± 21.62 [bc]	0.56 ± 0.0018 [f]	0.206 ± 0.0027 [ab]
Mannitol	8	7.00 ± 0.02	5.16 ± 0.02	651.78 ± 9.69 [bc]	0.57 ± 0.0037 [f]	0.199 ± 0.0011 [b]

Note: all values are expressed as means ± standard deviation in triplicate. Data followed by the same letters are not significantly different ($p < 0.05$) according to Duncan's multiple range test to evaluate the effect of investigated parameters. S.D: standard deviation.

4. Discussion

The production of BLIS is reliant on the type of culture media and the composition. The inverse relationship between cell growth and bacteriocin production in this study was supported by Ünlü, Nielsen and Ionita [27] who stated that the good bacterial growth does not guarantee good bacteriocin production although bacteriocin production is associated with bacterial growth. BHI is one of the complex media for the cultivation of LAB available in today's market beside de Man Rogosa and Sharpe (MRS), NaLa (sodium lactate), M17 and trypticase soy broth yeast extract (TSBYE) [28]. However, contradicting the reported literature, BHI was rarely preferred as a good media for bacteriocin production due to low productivity [17]. To date, there were not many findings highlighting the suitability of BHI in the production of BLIS from LAB. In agreement with many reports [29–32] medium composition greatly influenced BLIS production of LAB. Sodium chloride (NaCl) was one of the major components in the medium used in this study (Table 1), which enhanced and stabilised BLIS secretion [33].

The specified pattern of cell growth and BLIS production in the current study indicated that the produced BLIS was a primary metabolite. This finding is in agreement with Taheri, Samadi, Ehsani, Khoshayand and Jamalifar [34] as well as Lv, Zhang and Cong [35]. BLIS production was reduced once cells entered the stationary phase, which might be due to the proteolytic degradation during lysis, aggregation and/or adsorption of bacteriocins on the cell wall of the producing microorganisms [36]. The inconsistencies of bacteriocin production kinetic, either growth associated or non-growth associated, are related to pH dependent phenomena such as the adsorption of bacteriocins onto cell surfaces and/or the post-translational processing of the pre-peptides to active forms [37,38]. Production of bacteriocin by LAB usually follows primary metabolite growth-associated kinetics in which the production occurs during exponential growth phase and ceases once stationary phase is reached [39]. This is however not always the case and the relationship between bacteriocin production and growth are strain dependent [40]. In some cases, a correlation exists between peptide- and biomass production [41,42] while in other cases, bacteriocin production only starts at the beginning of stationary phase [39,43–45]. Recently, the BLIS produced by *Lactococcus lactis* Gh1 has been characterised [26]. The activity of BLIS produced by *L. lactis* Gh1 did not change with changes in pH from pH 4.36 to pH 8, which confirmed the proteinaceous nature of BLIS as antimicrobial substance.

Specific requirements with reference to the production of bacteriocins by LAB have been reported. The type or composition of culture medium especially nitrogen and carbon source greatly influenced bacteriocin production [36]. Lowering the amount of the organic nitrogen sources in the medium while keeping the nisin yield constant is advantageous not only for bacteriocin purification, but also for lowering the production cost [28]. All of these media are good for neutralising lactic acid and improving cell growth, but do not consider the accumulation of bacteriocin and high content of nitrogen sources especially proteins and peptides that may bring about the difficulties of bacteriocin purification. Since the use of BHI enhanced the growth of *L. lactis* Gh1 and BLIS production, this medium was selected for subsequent use in the statistical optimisation for further improving fermentation performance.

BLIS secretion started to decrease at pH lower than 6, whereas BLIS secretion and cell growth were suppressed at pH lower than 4. In general, culture pH greatly influences cell growth and bacteriocin production by aggregation, adsorption of bacteriocin to the cells and/or proteolytic degradation of bacteriocin [17,46]. The dependency of bacteriocin production on culture pH indicates that pH could regulate the expression of biosynthetic genes similarly observed for several classes of genes [47]. Initial pH level of culture medium is one of the key factors influencing the growth of the bacteriocinogenic LAB strains and the adsorption of bacteriocins onto the cell wall of the producing microorganisms [48]. Reduced BLIS production at pH below 6 as observed in this study is in agreement with the report by Yang et al. [48] who claimed that BLIS was maximally adsorbed to LAB cells at pH ranging from 6.0 to 4.0. Whereas the stunted cell growth and BLIS production at pH lower than 4 have been reported by many researchers [32,48,49]. Nutrient transport, which is a pH-function, may be one of the growth rate limiting actions in LAB. Therefore, the failure to grow at acidic pH is likely to be caused by a limitation of cytoplasmic processes (acidification of the cytoplasm and collapse of the motive force) [50–52].

Inoculum at mid-exponential growth phase recorded the highest BLIS production. In batch fermentation, the highest growth rate is at mid-exponential phase. The rate of cell growth in the culture is proportionate to the number of cells present at any given time during exponential growth phase [53], an important role in bacteriocin production where the optimum inocula size is favourable for the highest productivity [54]. The results demonstrate a strong relation of the inoculum size with the production of BLIS. Addition of 1% (*v*/*v*) of inoculum was intensively applied in inoculum preparation of LAB as reported previously [39,55]. Optimal size of inocula may vary from strain producers due to their cell proliferation rate, ability to metabolise medium, mass transfers, medium size and nutrient composition. However, minimum inocula size is preferred due to easier inoculation in large scale fermentation [54]. BLIS secretion was the lowest at the highest inoculum size (10%, *v*/*v*) so far studied. The inoculum size ranging from 4% (*v*/*v*) to 10% (*v*/*v*) produced more cells than BLIS production. High inoculum size has been attributed to a substantial reduction in oxygen tension at high bacterial densities [56]. The use of inoculum at a stationary growth phase and with the lowest or highest size influenced the duration of the lag phase, which gave a longer time to reach to P_{max} after 8–10 h of fermentation.

Organic nitrogen sources were more favourable for BLIS production as compared to inorganic sources. *L. lactis* Gh1 grown in soytone-supplemented medium produced maximum BLIS production and cell significantly higher than those obtained by other organic or inorganic nitrogen sources. The replacement of nitrogen source of BHI medium with soytone also resulted in shorter lag phase length and time (4 h) for producing maximum BLIS compared to that with BHI and inorganic nitrogen sources (6 h). This finding is in line with that by Ramachandran et al. [57] who concluded that organic nitrogen gave higher bacteriocin yield compared to inorganic nitrogen sources. Lechiancole et al. [58] stated that the growth and bacteriocin production of *L. sakei* were markedly improved with the replacement of tryptone with bacteriological peptone or soytone. Soytone is rich in minerals such as magnesium, potassium, sodium, chloride, sulphate, phosphate as well as free amino acids such as alanine, arginine, asparagine, and aspartic acid. The growth of *L. lactis* Gh1 and BLIS production was supressed in the absence of nitrogen sources in the culture medium, indicating the crucial function of nitrogen supplementation in supporting the cell growth. Lactococcal strains are nutritionally fastidious microorganisms, in which cell growth and bacteriocin production are influenced by a rich medium with organic nitrogen source [36]. Nitrogen sources are required for all processes involving biological growth especially with reference to synthesis of cellular protein and nucleic acid production as well as for bacterial metabolism [17]. The ability of LAB to metabolise different carbon sources is based on the specific activities of the enzymes involved in carbohydrate degradation; for example, amylolytic LAB have the ability to secrete amylase, which hydrolyse starch to fermentable sugars [59]. Osmotic stress, which increases the energy demand, apparently reduces the maximum secretion of bacteriocin, indicating that the energy is required in excess for the synthesis [17]. In bacteriocin fermentation, glucose is the preferred carbon source to stimulate bacteriocin production. Many researchers demonstrated that high bacteriocin yield was associated with the presence of glucose in growth medium and no other monosaccharides [60]. The highest bacteriocin production was recorded with the addition of glucose, maltose, lactose, and sucrose [17]. In contrary to the results of this study, fructose was found to be the most appropriate carbon source for BLIS production by *L. lactis* Gh1. Fructose was not preferentially metabolised compared to other carbon sources for bacteriocin production [61,62]. Stolz et al. [63] reported the use of fructose as an energy source by *L. reuteri* LTH 3120 and *L. amylovorus* LTH 3122. Fructose was also considered as the preferred energy source for maintenance, especially at temperatures greater than 34 °C in which fructose was converted by *L. amylovorus* DCE 471 at a faster rate than maltose [64].

Generally, bacteriocin production is a growth associated process. Higher cell density accumulates more inducer peptides that function in the quorum sensing regulations to induce bacteriocin production [65]. However, these facts are inconsistent with the findings in this study. The selected parameter with the highest BLIS production exhibited contradictory results with the cell growth. This observation was found in most fermentation factors tested in this study (media, inoculum preparation and carbon sources). From the results of this study, it can be claimed that BLIS production by *L. lactis* Gh1 did not follow the general growth-dependent bacteriocin production. Some bacteriocins are favourably produced in unfavourable growth conditions [66,67].

5. Conclusions

Results from this study demonstrated that the growth of *L. lactis* Gh1 and BLIS production were influenced by the physiological (initial pH, inoculum age and inoculum size) and nutritional (medium compositions) factors. Optimal fermentation conditions for BLIS production were not necessarily appropriate for the good growth rate of *L. lactis* Gh1. The replacement of nitrogen and carbon sources to soytone and fructose as well as the mid-exponential age of inoculum at 1% (*v*/*v*) were the selected factors for high BLIS production by *L. lactis* Gh1. The modified BHI broth in the present work could represent an alternative medium for BLIS production since it permitted comparable BLIS level compared to conventional BHI broth. Results from this study could be used for subsequent application in process design and optimisation. In future, the integration of mathematical techniques such as response surface methodology (RSM) and artificial neural network (ANN) should be applied for systematic optimal medium formulation for BLIS production.

Author Contributions: Conceptualisation, S.A. and A.B.A.; data curation, R.J., S.A. and J.S.T.; formal analysis, R.J. and S.A.; funding acquisition, A.B.A.; Investigation, R.J. and S.A.; methodology, R.J., S.A. and J.S.T.; project administration, S.A., J.S.T. and A.B.A.; resources, S.M., M.H. and A.B.A.; supervision, S.A., J.S.T., S.M., M.H. and A.B.A.; validation, S.M., M.H. and A.B.A.; visualisation, S.M., M.H. and A.B.A.; writing—original draft, R.J.; writing—review and editing, S.A., J.S.T. and A.B.A. All authors have read and agreed to the published version of the manuscript.

Funding: The authors acknowledge the financial supports received by Ministry of Higher Education (MOHE), Malaysia under Prototype Research Grant Scheme (PRGS/2/2015/SG05/UPM/01/2). The APC was funded by Universiti Putra Malaysia.

Conflicts of Interest: The authors declare no conflict of interest. The funders had no role in the design of the study; in the collection, analyses, or interpretation of data; in the writing of the manuscript, or in the decision to publish the results.

References

1. Singh, V.P. Recent approaches in food bio-preservation—A review. *Open Vet. J.* **2018**, *8*, 104–111. [CrossRef]
2. Özogul, F.; Hamed, I. The importance of lactic acid bacteria for the prevention of bacterial growth and their biogenic amines formation: A review. *Crit. Rev. Food Sci. Nutr.* **2018**, *58*, 1660–1670. [CrossRef]
3. Abbasiliasi, S.; Ramanan, R.N.; Tengku Azmi, T.I.; Shuhaimi, M.; Mohammad, R.; Ariff, A.B. Partial characterization of antimicrobial compound produced by *Lactobacillus paracasei* LA 07, a strain isolated from Budu. *Minerva Biotecnol.* **2010**, *22*, 75–82.
4. Silva, C.C.G.; Silva, S.P.M.; Ribeiro, S.C. Application of bacteriocins and protective cultures in dairy food preservation. *Front. Microbiol.* **2018**, *9*, 594. [CrossRef]
5. Mills, S.; Serrano, L.; Griffin, C.; O'connor, P.M.; Schaad, G.; Bruining, C.; Hill, C.; Ross, R.P.; Meijer, W.C. Inhibitory activity of *Lactobacillus plantarum* LMG P-26358 against *Listeria innocua* when used as an adjunct starter in the manufacture of cheese. *Microb. Cell Fact.* **2011**, *10*, S7. [CrossRef]
6. Grazia, S.E.; Sumayyah, S.; Haiti, F.S.; Sahlan, M.; Heng, N.C.K.; Malik, A. Bacteriocin-like inhibitory substance (BLIS) activity of *Streptococcus macedonicus* MBF10-2 and its synergistic action in combination with antibiotics. *Asian Pac. J. Trop. Med.* **2017**, *10*, 1140–1145. [CrossRef]
7. Loh, J.Y.; Lim, Y.Y.; Ting, A.S.Y. Bacteriocin-like substances produced by *Lactococcus lactis* subsp. lactis CF4MRS isolated from fish intestine: Antimicrobial activities and inhibitory properties. *Int. Food Res. J.* **2017**, *24*, 394–400.

8. Chakchouk-Mtibaa, A.; Smaoui, S.; Ktari, N.; Sellem, I.; Najah, S.; Karray-Rebai, I.; Mellouli, L. Biopreservative efficacy of bacteriocin BacFL31 in raw ground Turkey meat in terms of microbiological, physicochemical, and sensory qualities. *Biocontrol Sci.* **2017**, *22*, 67–77. [CrossRef]
9. Leelavatcharamas, V.; Arbsuwan, N.; Apiraksakorn, J.; Laopaiboon, P.; Kishida, M. Thermotolerant bacteriocin-producing lactic acid bacteria isolated from Thai local fermented foods and their bacteriocin productivity. *Biocontrol Sci.* **2011**, *16*, 33–40. [CrossRef]
10. Settanni, L.; Corsetti, A. Application of bacteriocins in vegetable food biopreservation. *Int. J. Food Microbiol.* **2008**, *121*, 123–138. [CrossRef] [PubMed]
11. Barbosa, M.S.; Todorov, S.D.; Ivanova, I.; Chobert, J.-M.; Thomas Haertlé, T.; de Melo Franco, B.D.G. Improving safety of salami by application of bacteriocins produced by an autochthonous *Lactobacillus curvatus* isolate. *Food Microbiol.* **2015**, *46*, 254–262. [CrossRef] [PubMed]
12. Fontana, C.; Cocconcelli, P.S.; Vignolo, G.; Saavedra, L. Occurrence of antilisterial structural bacteriocins genes in meat borne lactic acid bacteria. *Food Control* **2015**, *47*, 53–59. [CrossRef]
13. Todorov, S.D.; Vaz-Velho, M.; de Melo Franco, B.D.G.; Holzapfel, W.H. Partial characterization of bacteriocins produced by three strains of *Lactobacillus sakei*, isolated from salpicao, a fermented meat product from North-West of Portugal. *Food Control* **2013**, *30*, 111–121. [CrossRef]
14. Szabóová, R.; Lauková, A.; Simonová, M.P.; Strompfová, V.; Chrastinova, L. Bacteriocin-producing enterococci' from rabbit meat. *Malays. J. Microbiol.* **2012**, *8*, 211–218.
15. Cintas, L.M.; Casaus, M.P.; Herranz, C.; Nes, I.F.; Hernández, P.E. Review: Bacteriocins of lactic acid bacteria. *Food Sci. Technol. Int.* **2001**, *7*, 281–305. [CrossRef]
16. Abbasiliasi, S.; Ramanan, R.N.; Ibrahim, T.A.T.; Mustafa, S.; Mohamad, R.; Daud, H.M.; Ariff, A.B. Effect of medium composition and culture condition on the production of bacteriocin-like inhibitory substances (BLIS) by *Lactobacillus Paracasei* LA07, a strain isolated from Budu. *Biotechnol. Biotechnol. Equip.* **2011**, *25*, 2652–2657. [CrossRef]
17. Abbasiliasi, S.; Tan, J.S.; Tengku Ibrahim, T.A.; Bashokouh, F.; Ramakrishnan, N.R.; Mustafa, S.; Ariff, A. Fermentation factors influencing the production of bacteriocins by lactic acid bacteria: A review. *Rsc Adv.* **2017**, *7*, 29395–29420. [CrossRef]
18. Abbasiliasi, S.; Tan, J.S.; Kadkhodaei, S.; Nelofer, R.; Tengku Ibrahim, T.A.; Shuhaimi, M.; Ariff, A.B. Enhancement of BLIS production by *Pediococcus acidilactici* kp10 in optimized fermentation conditions using an artificial neural network. *RSC Adv.* **2016**, *6*, 6342–6349. [CrossRef]
19. Castro, M.P.; Palavecino, N.Z.; Herman, C.; Garro, O.A.; Campos, C.A. Lactic acid bacteria isolated from artisanal dry sausages: Characterization of antibacterial compounds and study of the factors affecting bacteriocin production. *Meat Sci.* **2010**, *87*, 321–329. [CrossRef]
20. Motta, A.S.; Brandelli, A. Influence of growth conditions on bacteriocin production by *Brevibacterium linens*. *Appl. Microbiol. Biotechnol.* **2003**, *62*, 163–167. [CrossRef]
21. Dinarvand, M.; Rezaee, M.; Masomian, M.; Jazayeri, S.; Zareian, M.; Abbasi, S.; Ariff, A.B. Effect of C/N ratio and media optimization through response surface methodology on simultaneous productions of intra- and extracellular inulinase and invertase from *Aspergillus niger* ATCC 20611. *Biomed. Res. Int.* **2013**, *2013*, 1–13. [CrossRef] [PubMed]
22. Lee, Y.M.; Kim, J.S.; Kim, W.J. Optimization for the maximum bacteriocin production of *Lactobacillus brevis* DF01 using response surface methodology. *Food Sci. Biotechnol.* **2012**, *21*, 653–659. [CrossRef]
23. Guerra, N.; Pastrana, L. Enhanced Nisin and Pediocin production on whey supplemented with different nitrogen sources. *Biotechnol. Lett.* **2001**, *23*, 609–612. [CrossRef]
24. Arakawa, K.; Kawai, Y.; Fujitani, K.; Nishimura, J.; Kitizawa, H.; Komine, K.; Kai, K.; Saito, T. Bacteriocin production of probiotic *Lactobacillus gasseri* LA39 isolated from human feces in milk-based media. *Anim. Sci. J.* **2008**, *79*, 634–640. [CrossRef]
25. Abbasiliasi, S.; Tan, J.S.; Ibrahim, T.A.T.; Ramanan, R.N.; Vakhshiteh, F.; Mustafa, S.; Ariff, A. Isolation of *Pediococcus acidilactici* Kp10 with ability to secrete bacteriocin-like inhibitory substance from milk products for applications in food industry. *BMC Microbiol.* **2012**, *12*, 260. [CrossRef] [PubMed]
26. Jawan, R.; Abbasiliasi, S.; Mustafa, S.; Kapri, M.R.; Halim, M. In vitro evaluation of potential probiotic strain *Lactococcus lactis* Gh1 and its bacteriocin-like inhibitory substances for potential use in the food industry. *Probiotics Antimicrob. Proteins* **2020**, 1–19. [CrossRef]

27. Ünlü, G.; Nielsen, B.; Ionita, C. Production of antilisterial bacteriocins from lactic acid bacteria in dairy-based media: A comparative study. *Probiotics Antimicrob. Proteins* **2015**, *7*, 259–274. [CrossRef]
28. Li, C.; Bai, J.; Cai, Z.; Ouyang, F. Optimization of a cultural medium for bacteriocin production by *Lactococcus lactis* using response surface methodology. *J. Biotechnol.* **2002**, *93*, 27–34. [CrossRef]
29. Schirru, S.; Favaro, L.; Mangia, N.; Basaglia, M.; Casella, S.; Comunian, R.; Fancello, F.M.F.B.; de Souza Oliveira, R.; Todorov, S. Comparison of bacteriocins production from *Enterococcus faecium* strains in cheese whey and optimized commercial MRS medium. *Ann. Microbiol.* **2014**, *64*, 321–331. [CrossRef]
30. Mahrous, H.; Mohamed, A.; El-Mongy, M.; El-Batal, A.; Hamza, H. Study of bacteriocin production and optimization using new isolates of *Lactobacillus* ssp. isolated from some dairy products under different culture conditions. *Food Nutr. Sci.* **2013**, *4*, 342–356.
31. Liu, C.; Liu, Y.; Liao, W.; Wen, Z.; Chen, S. Application of statistically-based experimental designs for the optimization of Nisin production from whey. *Biotechnol. Lett.* **2003**, *25*, 877–882. [CrossRef] [PubMed]
32. Mataragas, M.; Metaxopoulos, J.; Galiotou, M.; Drosinos, E.H. Influence of pH and temperature by *Leuconostoc mesenteroides* L124 and *Lactobacillus curvatus* L442. *Meat Sci.* **2003**, *64*, 265–271. [CrossRef]
33. Leroy, F.; Vankrunkelsven, S.; De Greef, J.; De Vuyst, L. The stimulating effect of a harsh environment on the bacteriocin activity by *Enterococcus faecium* RZS C5 and dependency on the environmental stress factor used. *Int. J. Food Microbiol.* **2003**, *83*, 27–38. [CrossRef]
34. Taheri, P.; Samadi, N.; Ehsani, M.R.; Khoshayand, M.R.; Jamalifar, H. An evaluation and partial characterization of a bacteriocin produced by *Lactococcus lactis* subsp. lactis ST1 isolated from goat milk. *Braz. J. Microbiol.* **2012**, *43*, 1452–1462. [CrossRef]
35. Lv, W.; Zhang, X.; Cong, W. Modelling the production of nisin by *Lactococcus lactis* in fed-batch culture. *Appl. Microbiol. Biotechnol.* **2006**, *68*, 322–326. [CrossRef]
36. Cheigh, C.I.; Choi, H.J.; Park, H.; Kim, S.B.; Kook, M.C.; Kim, T.S.; Hwang, J.K.; Pyun, Y.R. Influence of growth conditions on the production of a nisin-like bacteriocin by *Lactococcus lactis* subsp. lactis A164 isolated from kimchi. *J. Biotechnol.* **2002**, *95*, 225–235. [CrossRef]
37. Biswas, S.R.; Ray, P.; Johnson, M.C.; Ray, B. Influence of growth conditions on the production of a bacteriocin, pediocin AcH, by *Pediococcus acidilactici* H. *Appl. Environ. Microbiol.* **1991**, *57*, 1265–1267. [CrossRef]
38. Yang, R.; Johnson, M.C.; Ray, B. Novel method to extract large amounts of bacteriocins from lactic acid bacteria. *Appl. Environ. Microbiol.* **1992**, *58*, 3355–3359. [CrossRef] [PubMed]
39. Abbasiliasi, S.; Tan, J.S.; Ibrahim, T.A.T.; Ramanan, R.N.; Kadkhodaei, S.; Mustafa, S.; Ariff, A.B. Kinetic modeling of bacteriocin-like inhibitory substance secretion by *Pediococcus acidilactici* Kp10 and its stability in food manufacturing conditions. *J. Food Sci. Technol.* **2018**, *55*, 1270–1284. [CrossRef]
40. Parente, E.; Brienza, C.; Ricciardi, A.; Addario, G. Growth and bacteriocin production by *Enterococcus faecium* DPC1146 in batch and continuous culture. *J. Ind. Microbiol. Biotechnol.* **1997**, *18*, 62–67. [CrossRef]
41. Abriouel, H.E.; Valdivia, E.; Martinez-Bueno, M.; Maqueda, M.; Gallvez, A. A simple method for semi-preparative-scale production and recovery of enterocin AS-48 derived from *Enterococcus faecalis* subsp. liquefaciens A-48-32. *J. Microbiol. Methods* **2003**, *55*, 599–605. [CrossRef]
42. Callewaert, R.; Vuyst, L.D. Bacteriocin production with *Lactobacillus amylovorus* DCE 471 is improved and stabilized by fed-batch fermentation. *Appl. Environ. Microbiol.* **2000**, *66*, 606–613. [CrossRef] [PubMed]
43. Coetzee, J.C.J. *Increased Production of bacST4SA by Enterococcus Mundtii in an Industrial-Based Medium with pH-Control*; University of Stellenbosch: Stellenbosch, South Africa, 2007.
44. Guerra, N.P.; Rua, M.L.; Pastrana, L. Nutritional factors affecting the production of two bacteriocins from lactic acid bacteria on whey. *Int. J. Food Microbiol.* **2001**, *70*, 267–281. [CrossRef]
45. Leroy, F.; De Vuyst, L. Bacteriocin production by *Enterococcus faecium* RZS C5 is cell density limited and occurs in the very early growth phase. *Int. J. Food Microbiol.* **2002**, *72*, 155–164. [CrossRef]
46. Kim, M.-H.; Kong, Y.-J.; Baek, H.; Hyun, H.-H. Optimization of culture conditions and medium composition for the production of micrococcin GO5 by *Micrococcus* sp. GO5. *J. Biotechnol.* **2006**, *121*, 54–61. [CrossRef]
47. Motta, A.S.; Brandelli, A. Evaluation of environmental conditions for production of bacteriocin-like substance by *Bacillus* sp. strain P34. *World J. Microbiol. Biotechnol.* **2008**, *24*, 641–646. [CrossRef]
48. Yang, E.; Fan, L.; Yan, J.; Jiang, Y.; Doucette, C.; Fillmore, S.; Walker, B. Influence of culture media, pH and temperature on growth and bacteriocin production of bacteriocinogenic lactic acid bacteria. *AMB Express* **2018**, *8*, 10. [CrossRef]

49. LeBlanc, J.G.; Garro, M.S.; Savoy de Giori, G. Effect of pH on *Lactobacillus fermentum* growth, raffinose removal, α-galactosidase activity and fermentation products. *Appl. Microbiol. Biotechnol.* **2004**, *65*, 119–123. [CrossRef]
50. Poolman, B.; Konings, W.N. Relation of growth of *Streptococcus lactis* and *Streptococcus cremoris* to amino acid transport. *J. Bacteriol.* **1988**, *170*, 700–707. [CrossRef]
51. Bibal, B.; Goma, G.; Vayssier, Y.; Pareilleux, A. Influence of pH, lactose and lactic acid on the growth of *Streptococcus cremoris*: A kinetic study. *Appl. Microbiol. Biotechnol.* **1988**, *23*, 340–344. [CrossRef]
52. Gonçalves, L.M.D.; Ramos, A.; Almeida, J.S.; Xavier, A.M.R.B.; Carrondo, M.J.T. Elucidation of the mechanism of lactic acid growth inhibition and production in batch cultures of *Lactobacillus rhamnosus*. *Appl. Microbiol. Biotechnol.* **1997**, *48*, 346–350. [CrossRef]
53. Charlebois, D.A.; Balázsi, G. Modeling cell population dynamics. *Silico Biol.* **2019**, *13*, 21–39. [CrossRef] [PubMed]
54. Lajis, A.F. Biomanufacturing process for the production of bacteriocins from Bacillaceae family. *Bioresour. Bioprocess.* **2020**, *7*, 2020. [CrossRef]
55. López-González, M.J.; Escobedo, S.; Rodríguez, A.; Neves, A.R.; Janzen, T.; Martínez, B. Adaptive evolution of industrial *Lactococcus lactis* under cell envelope stress provides phenotypic diversity. *Front. Microbiol.* **2018**, *9*, 1–17. [CrossRef]
56. Morrissey, I.; Smith, J.T. The importance of oxygen in the killing of bacteria by ofloxacin and ciprofloxacin. *Microbios* **1994**, *79*, 43–53. [PubMed]
57. Ramachandran, B.; Srivathsan, J.; Sivakami, V.; Harish, J.; Ravi Kumar, M.; Mukesh Kumar, D.J. Production and optimization of bacteriocin from *Lactococcus lactis*. *J. Acad. Ind. Res.* **2012**, *1*, 306–309.
58. Lechiancole, T.; Ricciardi, A.; Parente, E. Optimization of media and fermentation conditions for the growth of *Lactobacillus sakei*. *Ann. Microbiol.* **2002**, *52*, 257–274.
59. Dimov, S.; Peykov, S.; Raykova, D.; Ivanova, P. Influence of diverse sugars on BLIS production by three different *Enterococcus* strains. *Trakia J. Sci.* **2008**, *6*, 54–59.
60. Delgado, A.; Arroyo López, F.N.; Brito, D.; Peres, C.; Fevereiro, P.; Garrido-Fernández, A. Optimum bacteriocin production by *Lactobacillus plantarum* 17.2b requires absence of NaCl and apparently follows a mixed metabolite kinetics. *J. Biotechnol.* **2007**, *130*, 193–201. [CrossRef]
61. Todorov, S.D.; Dicks, L.M. Effect of growth medium on bacteriocin production by *Lactobacillus plantarum* ST194BZ, a strain isolated from boza. *Food Technol. Biotechnol.* **2005**, *43*, 165–173.
62. Todorov, S.D.; Dicks, L.M. Effect of medium components on bacteriocin production by *Lactobacillus plantarum* strains ST23LD and ST341LD, isolated from spoiled olive brine. *Microbiol. Res.* **2006**, *161*, 102–108. [CrossRef] [PubMed]
63. Stolz, P.; Vogel, R.F.; Hammes, W.P. Utilization of electron acceptors by lactobacilli isolated from sourdough. II. *Lactobacillus pontis, L. reuteri, L. amylovorus,* and *L. fermentum*. *Z. Lebensm. Unters. Forsch.* **1995**, *201*, 402–410. [CrossRef]
64. Messens, W.; Neysens, P.; Vansieleghem, W.; Vanderhoeven, J.; De Vuyst, L. Modeling growth and bacteriocin production by *Lactobacillus amylovorus* DCE 471 in response to temperature and pH values used for sourdough fermentations. *Appl. Environ. Microbiol.* **2002**, *68*, 1431–1435. [CrossRef] [PubMed]
65. De Vuyst, L.; Leroy, F. Bacteriocins from lactic acid bacteria: Production, purification, and food applications. *J. Mol. Microbiol. Biotechnol.* **2007**, *13*, 194–199. [CrossRef]
66. De Vuyst, L.; Callewaert, R.; Crabbe, K. Primary metabolite kinetics of bacteriocin biosynthesis by *Lactobacillus amylovorus* and evidence for stimulation ofbacteriocin production under un- favourable growth conditions. *Microbiology* **1996**, *142*, 817–827. [CrossRef]
67. Matsusaki, H.; Endo, N.; Sonomoto, K.; Ishizaki, A. Lantibiotic nisin Z fermentative production by *Lactococcus lactis* IO-1: Relationship between production of the lantibiotic and lactate and cell growth. *Appl. Microbiol. Biotechnol.* **1996**, *45*, 36–40. [CrossRef]

microorganisms

MDPI

Article

Integrated Stirred-Tank Bioreactor with Internal Adsorption for the Removal of Ammonium to Enhance the Cultivation Performance of *gdhA* Derivative *Pasteurella multocida* B:2

Siti Nur Hazwani Oslan [1,2], Joo Shun Tan [3], Sahar Abbasiliasi [4], Ahmad Ziad Sulaiman [2], Mohd Zamri Saad [5], Murni Halim [6] and Arbakariya B. Ariff [1,6,*]

1 Bioprocessing and Biomanufacturing Research Centre, Faculty of Biotechnology and Biomolecular Sciences, Universiti Putra Malaysia, Serdang, Selangor 43400, Malaysia; hazwanioslan@gmail.com
2 Faculty of Bioengineering and Technology, University Malaysia Kelantan, Jeli Campus, Jeli, Kelantan 17600, Malaysia; ziad@umk.edu.my
3 Bioprocess Technology, School of Industrial Technology, Universiti Sains Malaysia, Penang 11800, Malaysia; jooshun@usm.my
4 Halal Products Research Institute, Universiti Putra Malaysia, Serdang, Selangor 43400, Malaysia; s.abbasiliasi@gmail.com
5 Research Centre for Ruminant Diseases, Faculty of Veterinary Medicine, Universiti Putra Malaysia, Serdang, Selangor 43400, Malaysia; mzamri@upm.edu.my
6 Department of Bioprocess Technology, Faculty of Biotechnology and Biomolecular Sciences, Universiti Putra Malaysia, Serdang, Selangor 43400, Malaysia; murnihalim@upm.edu.my
* Correspondence: arbarif@upm.edu.my; Tel.: +603-89467591

Received: 10 September 2020; Accepted: 6 October 2020; Published: 24 October 2020

Abstract: Growth of mutant gdhA *Pasteurella multocida* B:2 was inhibited by the accumulation of a by-product, namely ammonium in the culture medium during fermentation. The removal of this by-product during the cultivation of mutant gdhA *P. multocida* B:2 in a 2 L stirred-tank bioreactor integrated with an internal column using cation-exchange adsorption resin for the improvement of cell viability was studied. Different types of bioreactor system (dispersed and internal) with resins were successfully used for ammonium removal at different agitation speeds. The cultivation in a bioreactor integrated with an internal column demonstrated a significant improvement in growth performance of mutant gdhA *P. multocida* B:2 (1.05×10^{11} cfu/mL), which was 1.6-fold and 8.4-fold as compared to cultivation with dispersed resin (7.2×10^{10} cfu/mL) and cultivation without resin (1.25×10^{10} cfu/mL), respectively. The accumulation of ammonium in culture medium without resin (801 mg/L) was 1.24-fold and 1.37-fold higher than culture with dispersed resin (642.50 mg/L) and culture in the bioreactor integrated with internal adsorption (586.50 mg/L), respectively. Results from this study demonstrated that cultivation in a bioreactor integrated with the internal adsorption column in order to remove ammonium could reduce the inhibitory effect of this by-product and improve the growth performance of mutant gdhA *P. multocida* B:2.

Keywords: cation-exchange resin; adsorption; removal; ammonium; cell viability; mutant gdhA *P. multocida* B:2

1. Introduction

Pasteurella multocida type B:2 known as Gram negative bacterium caused a hemorrhagic septicemia (HS) disease in cattle and buffaloes associated with morbidity leading to colossal losses to farmers and nation. An attenuated (mutant) *gdhA* derivative *P. multocida* B:2 has been developed by Sarah et al. [1] in order to control the HS disease. In addition, the mutant strain was stable and did not revert back to virulence (Hazwani et al., 2014) and established to maximize the survival rate, storage stability, and activity of the bacterial cells of live vaccine [2]. However, ammonium is accumulated, as a by-product, in the culture during the batch cultivation of mutant *P. multocida* B:2, which inhibits the growth and reduce the cell viability [2]. In pentose phosphate cycle, histidine metabolism is one of the amino acid biosynthesis and this metabolism might relate to the histidine utilization Hut pathways [3]. In this pathway, the bacterium is able to degrade histidine to ammonia. Furthermore, the Hut pathway is fundamentally a catabolic pathway that allows cells to use histidine as a source of carbon, energy, and nitrogen.

In environmental study, various methods have been developed to eliminate the ammonium ion in contaminated water. The methods include air stripping [4], ion exchange resins [5], adsorption [6], chemical precipitation [7], and nitrification reactions [8]. The use of cation-exchange resin for in situ removal of ammonium during batch cultivation of mutant *P. multocida* B:2 has been demonstrated to improve the viability of mutant *P. multocida* B:2 as discussed by Oslan et al. [2]. However, the use of dispersed resin in bioreactor for in situ adsorption of inhibitory metabolites may cause direct shear force from the impeller, which affects the stability and efficiency of the resins [9]. In addition, the bioreactor could carry out in situ adsorption of ammonium attached to the resin efficiency during fermentation and when the resin fluidized in the internal adsorbent provided better environmental conditions, such as pH level, temperature, aeration rate, and dissolved oxygen, as compared with external adsorbent could be achieved.

The stirred-tank bioreactor equipped with an internal adsorption column has been applied for improvement of the performance of fermentation subjected with by-product inhibition. In such a technique, the resin with the ability to adsorb the target by-product was entrapped in an internal adsorption column with no direct contact with the impeller in the bioreactor. The use of the internal column allows the fluidization of porous adsorbent resins, so in turn, the surface of adsorbent in the column can be increased. In addition, the use of an internal adsorption column allows the cells culture to flow freely through the column, while the product or by-product is captured at the same time [10].

In the previous study by Oslan et al. [11], they have shown the primary data on growth of mutant gdhA P. multocida B:2 were inhibited by the accumulation of by-product ammonium in the culture medium. The results from the study have demonstrated that ammonia accumulated in a culture of gdhA derivative *P. multocida* B:2 could be removed by the adsorption onto cation-exchange resins in shake flask. Hence, this study highlights the removal of by-product ammonium during the cultivation of mutant gdhA P. multocida B:2 in a 2 L stirred-tank bioreactor integrated with internal column using cation-exchange resin for improvement of cell viability mutant gdhA *P. multocida* B:2. The objective of this study was to develop an integrated stirred-tank bioreactor with an internal adsorption column for in situ removal of ammonium from the culture of mutant *P. multocida* B:2 for the enhancement of the growth performance in terms of cell count and viability. For comparison, the fermentations in a stirred-tank bioreactor without resin and with dispersed resin were also performed.

2. Materials and Methods

2.1. Microorganism and Inoculum Preparation

The *gdhA* derivative of *P. multocida* B:2 was cultured on blood agar base and incubated at 37 °C for 48 h. The single colony isolated from the petri dish was inoculated into flask containing 10 mL of yeast extract-dextrose with histidine (YDB-His) medium. This is the optimal medium for the production of mutant *P. multocida* B:2 [11]. The medium consisted of (g/L): yeast extract, 15.6; dextrose, 1.9; sodium chloride, 3; sodium dihydrogen phosphate, 2.5; and histidine, 3.1 supplemented with 60 mg/L streptomycin and 50 mg/L kanamycin. All chemicals were obtained from Sigma-Aldrich (St. Louis, MA, USA). A 5% of inoculum was inoculated into 100 mL YDB-His medium and incubated in an orbital shaker (Infors, GmbH, Fronreute, Germany) at 37 °C and agitated at 250 rpm for 16 h. The culture was standardized as an inoculum for the fermentation in bioreactor.

2.2. *Stirred-Tank Bioreactor Design and Internal Column Adsorption System*

A 2 L stirred-tank bioreactor with a working volume of 1 L was used in this study. The bioreactor consisted of a double jacketed borosilicate glass vessel, four baffles, and a stainless-steel top plate with several opening ports for sampling and electrodes (BIOSTAT, B. Braun Biotech International, GmbH, Munich, Germany). Relative dimensions of the stirred-tank bioreactor used in this study, with and without internal adsorption column, are shown in Figure 1. The total length between the sparger and the motor edge was 13.6 cm. The glass vessel was concave at the bottom to eliminate dead-spots. The bioreactor was equipped with a single six-bladed Rushton turbine (RT) impeller located at 3.0 cm from the oval bottom of the vessel. The impeller was spaced 2.0 cm between the air sparger.

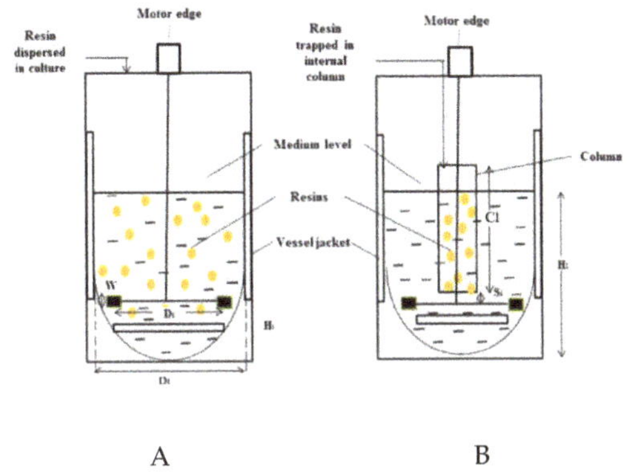

Figure 1. Schematic diagram of the stirred-tank bioreactor (**A**) without and (**B**) with internal column adsorption.

Table 1 shows the descriptions of the RT impeller and 2 L stirred-tank bioreactor. For experiments using internal column in bioreactor, the column was packed with the position of impeller below the column. The resins were then added into the column prior to autoclaving. The adsorption study was conducted simultaneously in batch fermentation from the culture throughout the whole fermentation process. The impeller was spaced 1.5 cm between the columns. The stirred-tank bioreactor was equipped with antifoam level, temperature, pH, and dissolved oxygen tension (DOT) control systems. The temperature of the culture was maintained at 37 °C, and the temperature was continuously measured by a platinum resistance thermometer with standard Pt100 probe (PT100, Mettler Toledo, Greifensee, Switzerland). The DOT in the culture was measured by a polarographic oxygen probe (InPro 6900, Mettler Toledo, Switzerland). After sterilization at 121 °C, 15 psi for 15 min, the probe was calibrated by sparging with nitrogen gas to set the DOT at 0% saturation, and then, the DOT was set to 100% saturation by sparging with filtered air for several hours until oxygen was saturated. The pH of the culture was monitored continuously by a pH ($Ag/AgCl_2$) electrode (InPro3253/225/PT100 Mettler Toledo, Switzerland), but the pH was not controlled during the fermentation. The dissolve oxygen tension (DOT) was automatically adjusted to the set level and the impeller speed was fixed according to the design of experiment. The different agitation was studied at 200 rpm, 300 rpm, 400 rpm, and 500 rpm for bioreactor without resin and dispersed resin, followed by 700 rpm for cation-exchange resin in column. All experiments were performed in triplicate, each being repeated at least three times.

Table 1. Descriptions of Rushton turbine (RT) impeller and 2 L stirred-tank bioreactor.

Descriptions	Value
D_i, impeller diameter(cm)	5.5
D_t, tank diameter (cm)	13.0
Hi, height impeller (cm)	5.6
Hl, height liquid (cm)	10.0
Li, impeller blade length (cm)	1.4
W_i, impeller width (cm)	1.0
N, number of blades	6
C_l, Column length (cm)	15

D_i impeller diameter; D_t, tank diameter; Hi, height impeller; Hl, height liquid; Li, impeller blade length; W_i, impeller width; C_l, Column length.

2.3. Cultivation Experiments

The Amberlite 86 cation-exchange resins (Sigma-Aldrich, St. Louis, MA, USA) were selected for this study based on the highest ammonium adsorption capacity from a previous study by Oslan et al. [11]. The Amberlite 86 cation-exchange resins at 10 g/L were prepared and washed with deionized water to remove the impurities. The washed resins were directly inserted into the column positioned in the bioreactor before being autoclaved at 121 °C for 15 min. The experiment was carried out after the cooling down of the bioreactor; the inoculum was transferred into the bioreactor with the supplementation of 60 mg/L streptomycin and 50 mg/L kanamycin. Kanamycin was added to prevent the release of kanamycin cassette by the cells, whereas streptomycin was added to avoid the growth of foreign cells in the culture.

2.4. Determination of Volumetric Oxygen Transfer Rate (k_La)

The dynamic gassing out method in non-fermentative system was used in this study for the determination of k_La [12]. Initially, the dissolved oxygen in test liquid medium, without microorganism, was purged with nitrogen gas. Once dissolved oxygen tension (DOT) readout stabilized at 0%, compressed air was sparged in, and the gradual rise of DOT was monitored and recorded at 5 s interval until equilibrium saturation achieved. The k_La was determined in a stirred-tank reactor with and without internal adsorption column with the presence and absence of resins. The following oxygen mass transfer model was used for the determination of k_La:

$$dC_L/dt = K_La(C_E - C_L) \quad (1)$$

where C_L is the dissolved oxygen concentration and C_E is the saturated dissolved oxygen concentration in the solution. k_La was estimated from the slope of the straight line obtained from a plot of ln $(C_E - C_L)$ against the time.

2.5. Analytical Methods

The bacterial cell viability was expressed via serial dilution plate count technique and determined the dry cell weight (DCW). To determine the DCW of the culture sample, 10 mL of the sample culture from the bioreactor was centrifuged at 10,000× *g* at 4 °C for 15 min (Eppendorf, Hamburg, Germany), and the supernatant was used for glucose determination. The pellet of the cell was dried in an oven at 70 °C until a constant weight was obtained. The glucose concentration was analyzed using biochemistry analyzer (YSI 2700; Yellow Spring Instruments, OH, USA). The concentration of ammonium in the fermentation culture was determined using a Nessler method [13].

The cation-exchange resins and samples of cell culture were collected after the fermentation and examined under scanning electron microscope (SEM) (LEO 1455 VPSEM, Kensington, UK). For electron microscopic study, the cell culture was centrifuged at 10,000× *g* at 4 °C for 15 min (Eppendorf, Hamburg, Germany), and the pellet cells were fixed in 4% (*v*/*v*) glutaraldehyde for 4 h at 4 °C. Then, the fixed cells were washed three times for 10 min using 0.1 M sodium cacodylate buffer.

After post-fixation in 1% (*w*/*v*) osmium tetroxide for 2 h at 4 °C, the cells were washed again and dehydrated with a series of increasing concentrations of acetone 35%, 45%, 55%, 75%, and 95% for 10 min; followed by 100% for 15 min for 3 times followed by sectioning using a Leica-Reichert Ultracut Ultramicrotome (Leica, GmbH, Wetzlar, Germany) and staining with uranyl acetate and lead citrate for 10 min. Then, the samples were mounting on stuck before viewing under SEM.

3. Results

3.1. Effect of Agitation Speed on Volumetric Oxygen Transfer Rate (k_La)

Impeller speed is an important factor that effects gas liquid mass transfer in an agitated bioreactor. Figure 2 represents experimentally obtained correlations for evaluation of k_La values for a 2 L stirred-tank bioreactor set up (i) without column and no resin, (ii) with dispersed resin (no column), (iii) with internal column without resin, and (iv) with resin in an internal column at different agitation speeds from 200 to 700 (rpm) with air flow rate fixed at 1 L/min in the 2 L stirred-tank bioreactor. The k_La values were found to increase with increasing impeller speed. The increasing values of k_La was due to the breakage of rising bubbles by the impeller blade and gave rise to more interfacial area for gas transfer, subsequently increasing oxygen transfer in the culture. For both bioreactor systems, with and without an internal adsorption column, k_La was significantly reduced with the presence of resins. At the same agitation speed, k_La for the bioreactor with an internal adsorption column was significantly lower as compared to the bioreactor without an internal column. In addition, the k_{La} values for the bioreactor with an internal adsorption column was not significantly different at all agitation speed tested. With the integration of the column and resins into the bioreactor, the mixing of gas bubbles in the broth was slowed down, and subsequently reduced the oxygen transfer to the cells.

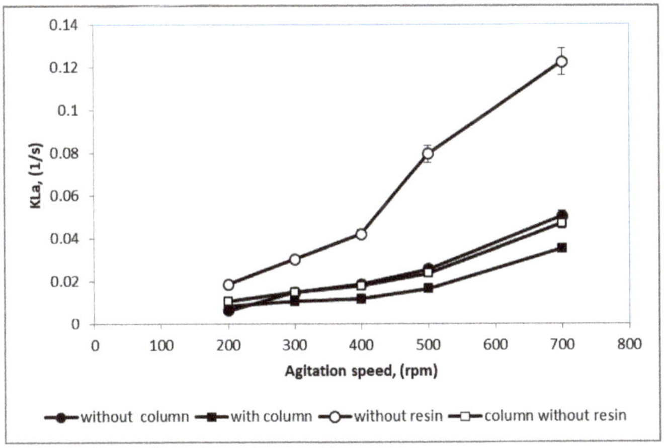

Figure 2. Comparable values of k_La for the 2 L stirred-tank bioreactor. (○) bioreactor without column no resin; (•) bioreactor with dispersed Amberlite IRC 86 resin (no column); (□) bioreactor with internal column without resin; (■) bioreactor with Amberlite IRC 86 resin in an internal column.

3.2. Effect of Amberlite IRC 86 Resins on the Cultivation Performance of Mutant P. multocida B:2

Figure 3 shows the time course of batch cultivation of mutant *P. multocida* B:2 in a 2 L stirred-tank bioreactor (i) without the addition of resin, (ii) with dispersed resin, and (iii) with resin in an internal adsorption column at different agitation speed (300, 400, 500, and 700 rpm). Based on the growth performance in Figure 3A, cell viability in the culture without Amberlite IRC 86 resins significantly decreased in cell viability as compared with dispersed resins and resins in internal adsorption column. The maximum cell viability (1.05×10^{11} cfu/mL) in a stirred-tank reactor with internal adsorption column, agitated at 500 rpm, was obtained at 16 h. To further study the effect of agitation speed on the cell viability in a stirred-tank reactor with internal adsorption column, the agitation speed was increased to 700 rpm. The cell viability was significantly decreased at higher agitation speed (700 rpm). The cell viability was significantly increased after 8 h in all fermentations tested.

(A)

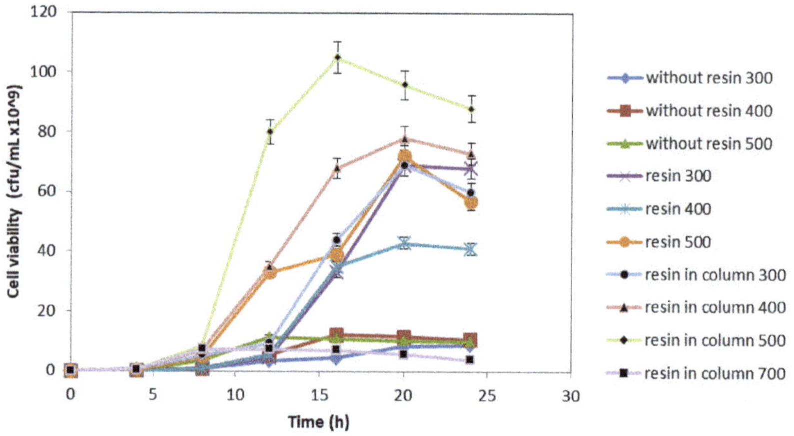

(B)

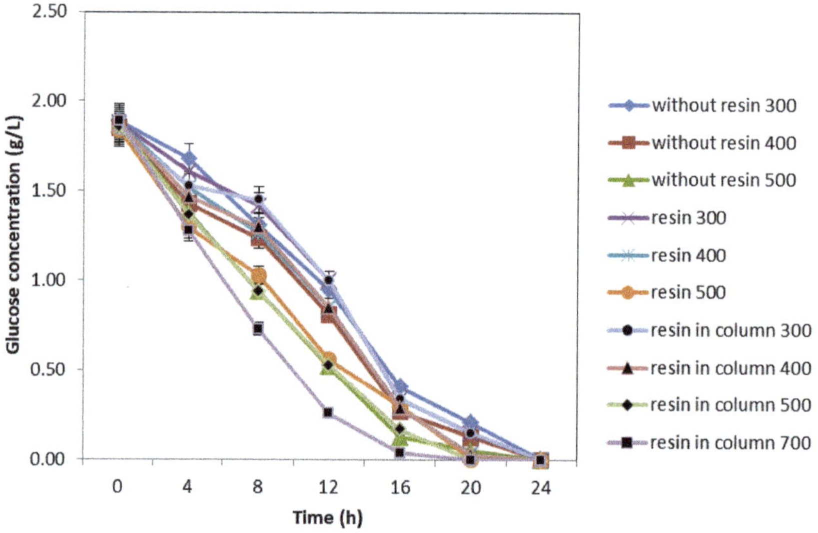

Figure 3. *Cont.*

(C)

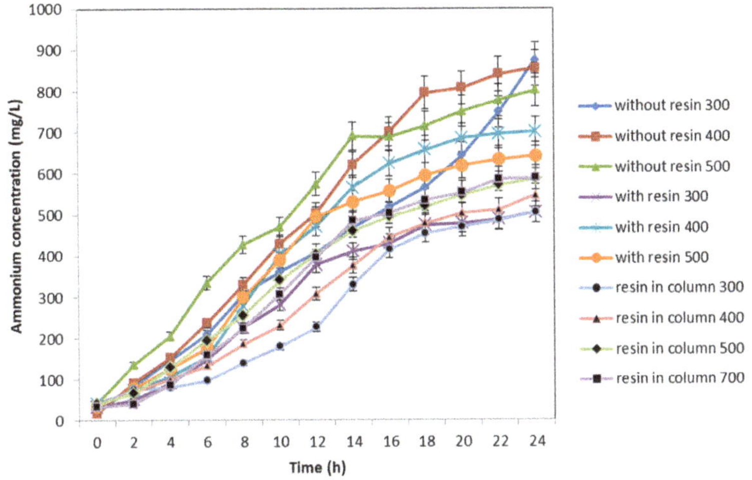

Figure 3. The time course of batch cultivation of *gdhA* derivative *Pasteurella multocida* B:2 with cell viability (**A**), glucose consumption (**B**), and ammonium concentration (**C**) in a 2 L stirred-tank reactor without the addition of resin, with dispersed Amberlite IRC 86 resin and Amberlite IRC 86 resin in internal column at different agitation speed.

Table 2 shows the comparison of the cultivation performance of mutant *P. multocida* B:2 in three different types of stirred-tank reactor with a different set up. The highest cell viability in the system without internal adsorption column was 7.2 × 10^{10} cfu/mL, while the culture with resins in column showed much higher viable cells (1.05 × 10^{11} cfu/mL) in the system at the agitation speed of 500 rpm. The result also indicated that the cell viability was significantly different ($p < 0.05$) when resins were used in this study. In terms of glucose consumption, Figure 3B shows the glucose was completely consumed at 20 h and the biomass production increased markedly with time at 16 h for all fermentations tested (Figure 3B). In addition, the fermentation in an integrated system of stirred-tank bioreactor and internal column shows that the growth cycle with maximum cell viability production was archived at 20 h and 16 h fermentation, respectively.

Table 2. Effect of culture *gdhA* derivative *P. multocida* B:2 with column and with resin in a 2 L stirred-tank bioreactor.

	Kinetic Parameters			
Media: YDB-His Agitation Speed: 500 rpm	**Max Viable Cell (cfu/mL)**	**X_{max} (mg/mL)**	**Specific Growth Rate, μ (h^{-1})**	**Productivity, *Pr* (mg/mL.h)**
without column and without resin	1.23 × $10^{10\,a}$	3.89 ± 0.05	0.505 ± 0.0882	0.243
without column and with resin	7.2 × $10^{10\,b}$	3.97 ± 0.08	0.554 ± 0.0925	0.1985
with column and with resin	1.05 × $10^{11\,c}$	4.04 ± 0.04	0.584 ± 0.095	0.2525

Values are mean ± SD (n = 3), $^{a–c}$ superscript with different letter is significantly different at $p < 0.05$.

Figure 3C shows the ammonium accumulated in the culture during the cultivation of mutant *P. multocida* B:2 in a 2 L stirred-tank reactor with different set up. The ammonium concentration in the bioreactor without resin was significantly increased with time, since a maximum concentration of 874.33 mg/L was achieved after 24 h at 300 rpm in which the culture pH was increased to pH 8.9. In the cultivation with resins in internal adsorption column, ammonium accumulation in culture medium was significantly reduced in which the highest concentration of ammonium was only 505.17 mg/L. In addition, the accumulation of ammonium at 16 h was 699.83 mg/L for reactor without resins, 557.67 mg/L in reactor with dispersed resin, and 494.5 mg/L for reactor with internal column. Hence, the cell viability increased with the decreased ammonium accumulation in the culture. This indicated that the addition of cation-exchange resins could reduce the ammonium accumulation in the culture medium by removal of the ammonium to a non-lethal level, as shown in Figure 3C. Hence, high concentration of cell viability of mutant *P. multocida* B:2 was achieved (1.05×10^{11} cfu/mL).

3.3. *Effect of Cation-Exchange Resin on Cell Morphology of Mutant P. multocida B:2*

At the end of the cultivation of mutant *P. multocida* B:2 with the in situ addition of cation-exchange resin, the cells were examined under the scanning electron microscopy (SEM). Figure 4 shows SEM photograph (magnification ×10,000) of cell morphology mutant *P. multocida* B:2 that were cultivated in a 2 L stirred-tank bioreactor without resin, (control) (A), (B) stirred-tank bioreactor with dispersed resin, and (C) stirred-tank bioreactor with resin in internal column. The morphology of mutant *P. multocida* B:2 cultivated with resin and control was similar, which was long rod. SEM of *P. multocida* showed that they exist as colonies of rod-shape bacteria in cells cultivated in bioreactor without resin with an average cell length of 1 μm and cell width of 0.4 μm. Overall, the morphology of the cells that cultivated in the bioreactor without resin was apparently normal (Figure 4A), but the cells deformed (Figure 4B) when cultivated in bioreactor with dispersed resin and elongated (Figure 4C) were observed for cells when cultivated in the bioreactor with resin packed in the internal column.

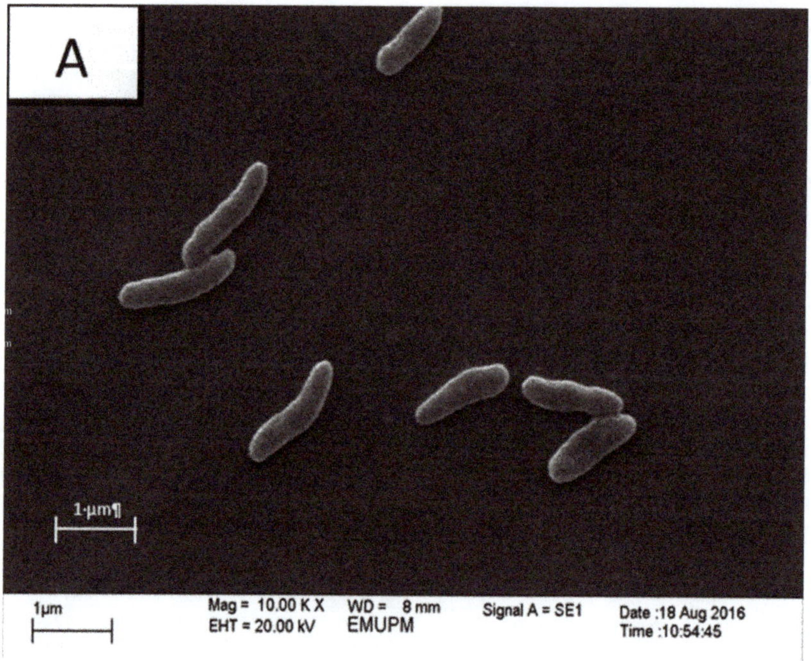

Figure 4. *Cont.*

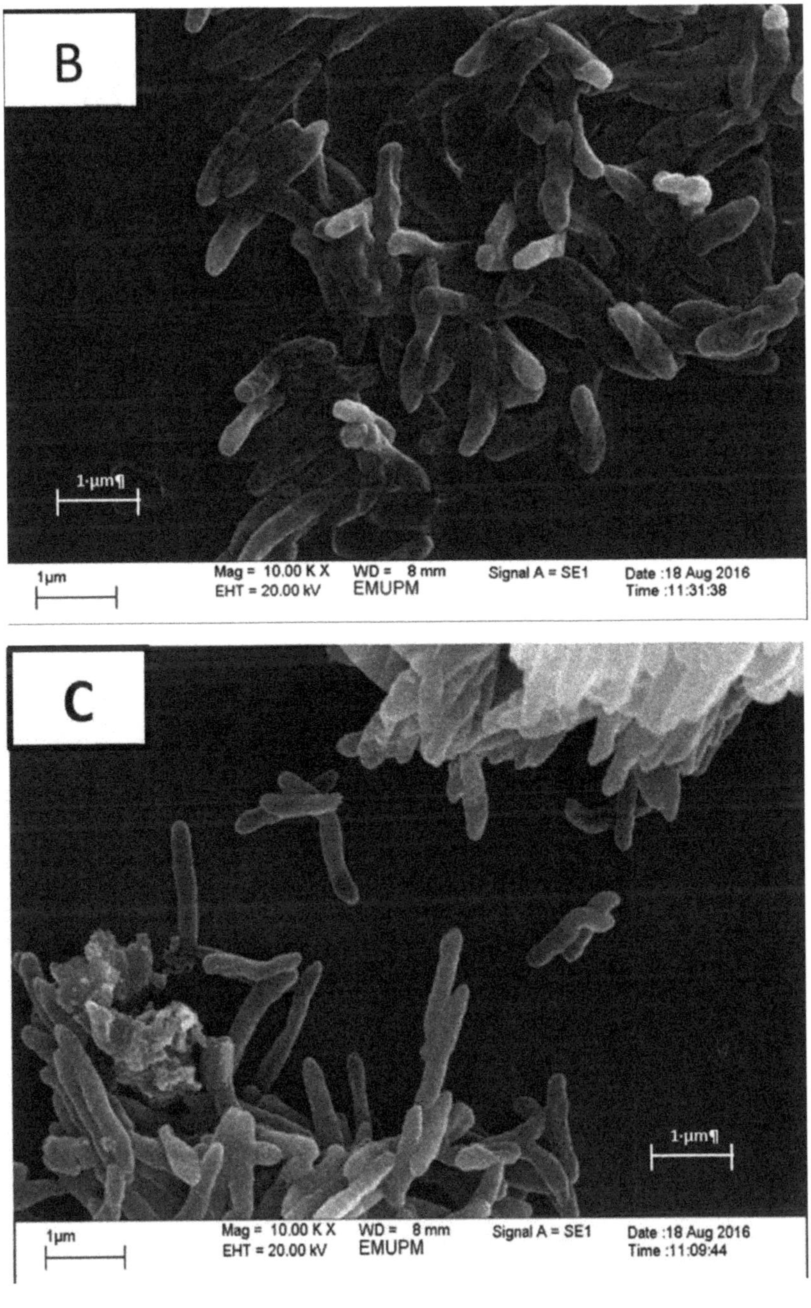

Figure 4. *Cont.*

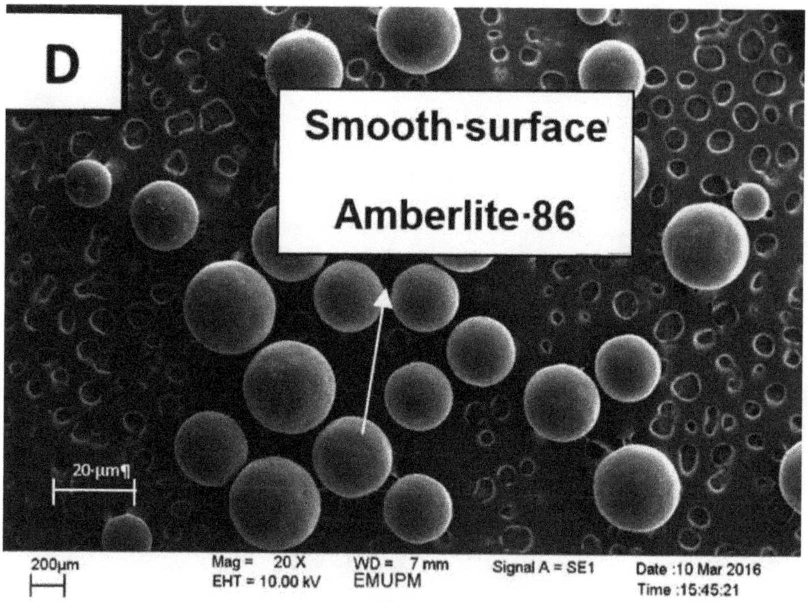

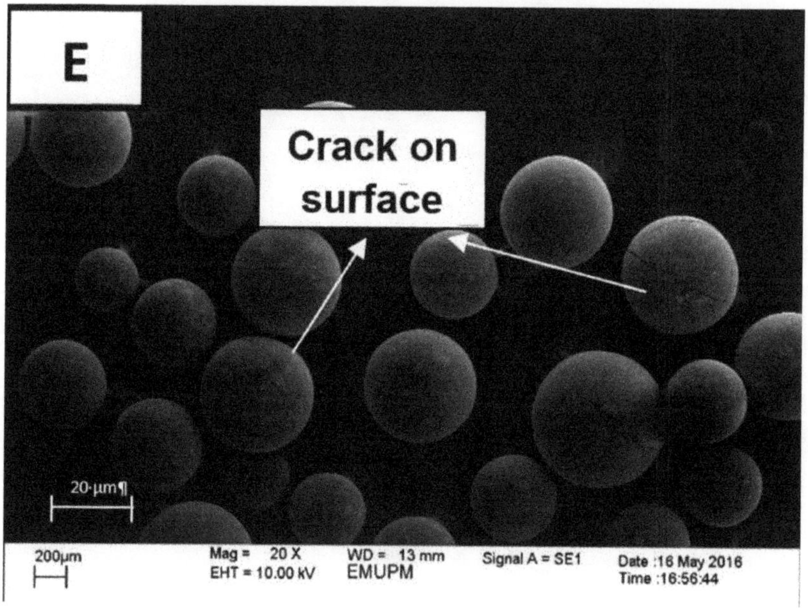

Figure 4. *Cont.*

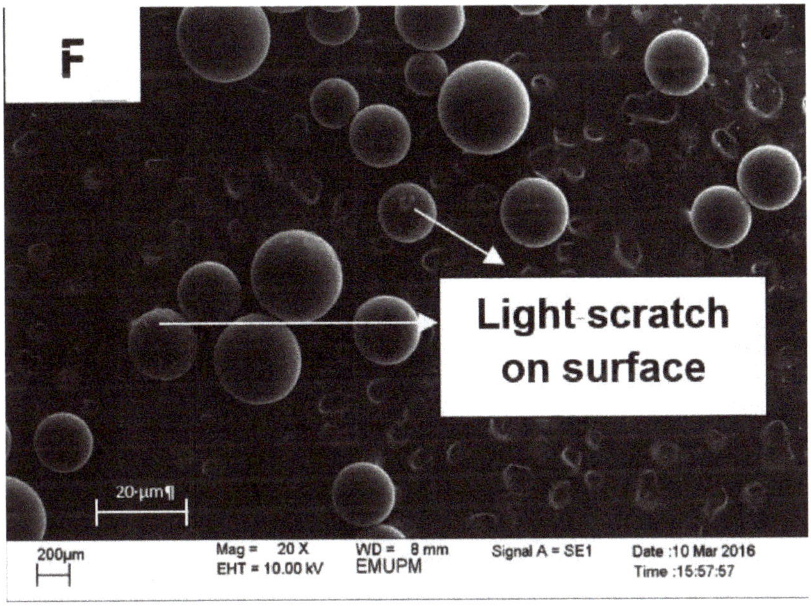

Figure 4. Scanning electron photograph of cell morphology mutant *gdhA P. multocida* B:2 (magnification ×10,000) in a 2 L stirred-tank bioreactor (**A**), without resin (**B**), with dispersed resin (**C**), with resin in internal column and surface structure resins Amberlite 86 (magnification ×200). (**D**) Control surface with smooth surface on resins. (**E**) Crack on the surface dispersed resins. (**F**) Light scratch on surface resins in internal column.

3.4. Effect of Agitation Speed on Surface Of Cation-Exchange Resin

Figure 4 shows scanning electron photograph (magnification ×200) on surface of cation exchange resin Amberlite IR86 (A), control resin (B), dispersed resin (C), in internal column at 500 rpm. Cracked surface was observed for resin used in cultivation with dispersed resin (Figure 4E). Meanwhile, only a very light scratch was detected on the surface of resins that were packed in an internal adsorption column (Figure 4F). Resins packed in an internal column were protected from experiencing direct shear force from the impeller that may damage the surface and consequently may enhance ion-exchange capacity and provide greater available surface area for adsorption site. This observation supports the assumption made in Section 2 on the lower cell viability (7.2×10^{10} cfu/mL) of culture in dispersed resin systems compared to culture with resins packed in a column (1.05×10^{11} cfu/mL) that may be due to the influence of direct shear force. The result is in agreement to the study of Tan et al. [9] in which the bioreactor with a column had better performance in terms of growth and productivity.

4. Discussion

Results from this study have demonstrated that the use of a stirred-tank bioreactor with internal adsorption column packed with resins has successfully improved the cultivation performance of mutant *gdhA P. multocida* B:2, by removing the ammonium accumulated in the culture. In this study, the experiment was shown the effect of agitation speed on volumetric oxygen transfer rate (k_La) with the integration of column and resins into the bioreactor, it slowed down the mixing of gas bubbles in the broth, subsequently reducing the oxygen transfer to the cells. Faccin et al. [14] reported that low oxygen supply may reduce the tricarboxylic acid (TCA) cycle activity, due to both thermodynamic reasons and metabolic control by reducing power NADPH, and allow that part of the acetyl-CoA available. The oxidation via TCA cycle predominates under balanced growth conditions, with NADH being generated and used in biosynthesis or energy generation. When biosynthesis decreases due to lack of a nutrient, the TCA cycle activity also decreases due to the high NADH concentration, resulting in a decrease in acetyl- CoA oxidation via TCA cycle. Under the limitation of oxygen, the reducing power NADPH, the by-product of ammonium ion could be reduced. In addition, the cultivation performance of aerobic microorganism will also be affected in oxygen-limited culture [15]. In order to provide sufficient oxygen transfer from the gas to the liquid phase, a suitable degree of agitation that exerted no or less damage to the microbial cell must be selected.

Despite the lower k_La value observed with the addition of resin and column compared to the bioreactor system without the presence of resin and column (Figure 2), in this study, the agitation speeds were sensitive in shear rate level, which may cause damage to the bacterial cell. Although oxygen is essential to growth and production, at elevated concentrations it could also be toxic for a variety of microbes [16]. Most aerobic microorganisms have developed protective responses to tolerate environmental oxygen concentrations. In *E. coli* cultures, when oxygen concentration surpasses the air saturation level, the oxygen species (ROS) such as accumulated byproducts of aerobic metabolism reactivated [17]. In addition, the use of oxygen-enriched air has become an accepted practice to support the aerobic growth of cells in bioreactors, which increases the cell density and improves process productivity [16]. However, exposing cells to high oxygen concentrations is known to enhance the accumulation of byproduct and reach to harmful levels that can overwhelm and cause stress to the cells [18]. In order to provide sufficient oxygen transfer from the gas to the liquid phase, the optimum agitation speed that did not cause damage to the microbial cells must be selected. The results shows constant k_La ensures equal oxygen transfer rates at the various scales of operation.

The in situ removal of ammonium accumulation in bioreactor system with selected resins provided certain conditions that influenced the production of viable cell of mutant *gdhA P. multocida* B:2. The integrated bioreactor system in this study enhance growth of mutant *gdhA P. multocida* B:2 with high cell viability through the removal of ammonium by resin adsorption. In this system, the liquid enters the internal column through lateral perforations (flow through perforated column). It was shown that the continuous adsorption between resin and medium occurs during fermentation. This concept has been applied by Jahangiri-Rad et al. [19], where the continuous fixed bed study was carried out by using PAN-oxime-nano Fe_2O_3 as a sorbent for the removal of nitrate from aqueous solution. The other available adsorption methods using a column are packed-bed chromatography [20] and fluidized beds in a bioreactor [18]. However, in packed-bed mode, a long sample application time and initial removal of particulate material are required to prevent clogging of the bed. The integrated packed bed column in the bioreactor was also successfully applied by Yanagi et al. [21] to enable high density culture of hepatocytes for use as a hybrid artificial liver support system or a bioreactor system. In their system, a packed-bed reactor was packed with using collagen-coated reticulated polyvinyl formal (PVF) resin, and this system was applied to a primary cultivation of hepatocytes. In the aqueous solution, ammonia exists in two different forms, which are ammonium ion and ammonia molecule. Some of the ammonia will be adsorbed by the resins, but some will be released by the system to environment following to the oxygen stream. Hence, quantification of the ammonia adsorbed by the resin is recommended for future study.

In this study, the effect of macroporous Amberlite 86 resin on the morphology of mutant *P. multocida* B:2 was investigated under SEM. The morphology of cell culture mutant was not changed throughout the cultivation. Moreover, the adsorption capacity of ammonium by the resins was dependent on the size of resins [22]. This indicates that without the direct shear force from the impeller, the resin has a higher ion-exchange capacity and greater available surface area. On the other hand, larger and higher density particles move to the bottom of the column in a bioreactor, forming a stable and uniform expansion. In addition, the use of small diameter resins and enlarged pores could be reduced the transport resistance, resulting in an increase in the dynamic adsorption capacity at high flow velocities in culture [9]. Higher density and larger size particles increase the operation flow rate, resulting in a reduction in the processing time. In this study, the Amberlite granules were not covered by *P. multocida* cells, as the cell surface charge of *P. multocida* cells was a negative charge [23], which would be repelled by Amberlite cation-exchange resin.

During the in situ adsorption in a bioreactor, the direct shear force from the impeller may affect the resin surface. Dispersed resin in a bioreactor will be damaged due to the direct contact with impeller in a bioreactor as well as higher collision between one another at increasing impeller speed. In contrast, direct shear effect can be avoided with the application of an internal column as the surface of resin can be fully covered by the column, and the adsorption may continuously take place in the bioreactor without direct contact between resins and the impeller. In addition, the influence of direct shear force on the cells resulted from the collision with resins, as in the bioreactor with the dispersed resins, could be avoided to yield higher viable cells in the integrated column system. The adsorption capacity in the culture could be increased due to the increment of the resin surface area in the trapped column.

5. Conclusions

Overall, this study has demonstrated that the concept of a stirred-tank bioreactor with an internal adsorption column packed with resins have successfully improved the cultivation performance of mutant gdhA *P. multocida* B:2 by continuously removing ammonium accumulated in the culture. The in situ removal of ammonium accumulation in a bioreactor system with selected resins influenced the production of viable cell of mutant gdhA *P. multocida* B:2. The accumulation of ammonium in the culture was significantly reduced with the addition of resin with the ability to adsorb ammonium during the cultivation process in the internal column with absorbent resin. The highest concentrations of ammonium accumulated in cultivation media without resin, with dispersed resin (without internal adsorption column), and in cultivation with internal adsorption column packed with resin were 801 mg/L, 642.50 mg/L, and 586.50 mg/L, respectively. The improvement in cell viability was significantly different at $p < 0.05$ as compared with cultivation with dispersed resin (without internal adsorption column) and without resin, where the highest viable cell numbers were 1.05×10^{11} cfu/mL, 7.2×10^{10} cfu/mL, and 1.25×10^{10} cfu/mL, respectively. To date, this kind of study has never been reported for any live bacterial vaccine aiming for the improvement of cell growth.

Author Contributions: Conceptualization, J.S.T., S.A. and A.B.A.; data curation, S.N.H.O., J.S.T. and S.A.; formal analysis, S.N.H.O., J.S.T. and S.A.; funding acquisition, A.B.A.; investigation, S.N.H.O.; methodology, S.N.H.O., J.S.T. and S.A.; project administration, J.S.T., S.A. and A.B.A.; resources, M.Z.S., M.H., A.Z.S. and A.B.A.; supervision, J.S.T., S.A., M.Z.S., M.H., A.Z.S. and A.B.A.; validation, M.Z.S., M.H., A.Z.S. and A.B.A.; visualization, M.Z.S., M.H., A.Z.S. and A.B.A.; writing—original draft, S.N.H.O.; writing—review and editing, J.S.T., S.A. and A.B.A. All authors have read and agreed to the published version of the manuscript.

Funding: This research was funded by Ministry of Higher Education (MOHE), Malaysia, under Prototype Research Grant Scheme (PRGS/2/2015/SG05/UPM/01/2). The APC was funded by Universiti Putra Malaysia.

Acknowledgments: The authors acknowledge the financial supports received by Universiti Putra Malaysia.

Conflicts of Interest: The authors declare no conflict of interest.

References

1. Sarah, S.; Zamri-Saad, M.; Zunita, Z.; Raha, A. Molecular cloning and sequence analysis of gdhA gene of Pasteurella multocida B: 2. *J. Anim. Vet. Adv.* **2006**, *5*, 1146–1149.
2. Oslan, S.N.H.; Halim, M.; Ramle, N.A.; Zamri-Saad, M.; Tan, J.S.; Kapri, M.R.; Ariff, A.B. Improved stability of live attenuated vaccine gdhA derivative *Pasteurella multocida* B:2 by freeze drying method for use as animal vaccine. *Cryobiology* **2017**, *79*, 1–8. [CrossRef] [PubMed]
3. Bender, R.A. Regulation of the Histidine Utilization (Hut) system in bacteria. *Microbiol. Mol. Biol. Rev.* **2012**, *76*, 565–584. [CrossRef] [PubMed]
4. Ozturk, I.; Altinbas, M.; Koyuncu, I.; Arikan, O.; Gomec-Yangin, C. Advanced physico-chemical treatment experiences on young municipal landfill leachates. *Waste Manag.* **2003**, *23*, 441–446. [CrossRef]

5. Sica, M.; Duta, A.; Teodosiu, C.; Draghici, C. Thermodynamic and kinetic study on ammonium removal from a synthetic water solution using ion exchange resin. *Clean Technol. Environ. Policy* **2013**, *16*, 351–359. [CrossRef]
6. Lin, L.; Lei, Z.; Wang, L.; Liu, X.; Zhang, Y.; Wan, C.; Lee, D.S.; Tay, J.H. Adsorption mechanisms of high-levels of ammonium onto natural and NaCl-modified zeolites. *Sep. Purif. Technol.* **2013**, *103*, 15–20. [CrossRef]
7. Yu, R.; Geng, J.; Ren, H.; Wang, Y.; Xu, K. Struvite pyrolysate recycling combined with dry pyrolysis for ammonium removal from wastewater. *Bioresour. Technol.* **2013**, *132*, 154–159. [CrossRef]
8. Zhang, Y.; Shi, Z.; Chen, M.; Dong, X.; Zhou, J. Evaluation of simultaneous nitrification and denitrification under controlled conditions by an aerobic denitrifier culture. *Bioresour. Technol.* **2015**, *175*, 602–605. [CrossRef]
9. Tan, J.S.; Ling, T.C.; Mustafa, S.; Tam, Y.J.; Ramanan, R.N.; Ariff, A.B. An integrated bioreactor-expanded bed adsorption system for the removal of acetate to enhance the production of alpha-interferon-2b by *Escherichia coli*. *Process Biochem.* **2013**, *48*, 551–558. [CrossRef]
10. Tolner, B.; Smith, L.; Begent, R.H.J.; A Chester, K. Expanded-bed adsorption immobilized-metal affinity chromatography. *Nat. Protoc.* **2006**, *1*, 1213–1222. [CrossRef] [PubMed]
11. Oslan, S.N.H.; Tan, J.S.; Saad, M.Z.; Halim, M.; Ariff, A.B. Improved cultivation of gdhA derivative *Pasteurella multocida* B:2 for high density of viable cells through in situ ammonium removal using cation-exchange resin for use as animal vaccine. *Process Biochem.* **2017**, *56*, 1–7. [CrossRef]
12. Bandyopadhyay, B.; Humphrey, A.E.; Taguchi, H. Dynamic measurement of the volumetric oxygen transfer coefficient in fermentation systems. *Biotechnol. Bioeng.* **1967**, *9*, 533–544. [CrossRef]
13. Jeong, H.; Park, J.; Kim, H. Determination of NH 4 + in environmental water with interfering substances using the modified nessler method. *J. Chem.* **2013**, *2013*, 359217. [CrossRef]
14. Faccin, D.J.L.; Rech, R.; Secchi, A.R.; Cardozo, N.S.M.; Ayub, M.A.Z. Influence of oxygen transfer rate on the accumulation of poly(3-hydroxybutyrate) by Bacillus megaterium. *Process Biochem.* **2013**, *48*, 420–425. [CrossRef]
15. Chang, G.; Wu, J.; Jiang, C.; Tian, G.; Wu, Q.; Chang, M.; Wang, X. The relationship of oxygen uptake rate and kLa with rheological properties in high cell density cultivation of docosahexaenoic acid by Schizochytrium sp. S31. *Bioresour. Technol.* **2014**, *152*, 234–240. [CrossRef]
16. Baez, A.; Shiloach, J. Effect of elevated oxygen concentration on bacteria, yeasts, and cells propagated for production of biological compounds. *Microb. Cell Factories* **2014**, *13*, 181. [CrossRef]
17. Baez, A.; Shiloach, J. Increasing dissolved-oxygen disrupts iron homeostasis in production cultures of *Escherichia coli*. *Antonie van Leeuwenhoek* **2016**, *110*, 115–124. [CrossRef]
18. Özkaya, B.; Kaksonen, A.H.; Sahinkaya, E.; Puhakka, J.A. Fluidized bed bioreactor for multiple environmental engineering solutions. *Water Res.* **2019**, *150*, 452–465. [CrossRef]
19. Jahangiri-Rad, M.; Jamshidi, A.; Rafee, M.; Nabizadeh, R. Adsorption performance of packed bed column for nitrate removal using PAN-oxime-nano Fe2O3. *J. Environ. Heal. Sci. Eng.* **2014**, *12*, 90. [CrossRef]
20. Lim, S.L.; Ng, H.W.; Akwiditya, M.A.; Ooi, C.W.; Chan, E.-S.; Ho, K.L.; Tan, W.S.; Chua, G.K.; Tey, B.T. Single-step purification of recombinant hepatitis B core antigen Y132A dimer from clarified Escherichia coli feedstock using a packed bed anion exchange chromatography. *Process Biochem.* **2018**, *69*, 208–215. [CrossRef]
21. Yanagi, K.; Miyoshi, H.; Fukuda, H.; Ohshima, N. A packed-bed reactor utilizing porous resin enables high density culture of hepatocytes. *Appl. Microbiol. Biotechnol.* **1992**, *37*, 316–320. [CrossRef]
22. Prajapati, J.C.; Syed, H.S.; Chauhan, J. Removal of ammonia from wastewater by ion exchange technology. *Int. J. Innov. Res. Technol.* **2014**, *1*, 9.
23. Morris, C.M.; George, A.; Wilson, W.W.; Champlin, F.R. Effect of Polymyxin B nonapeptide on daptomycin permeability and cell surface properties in pseudomonas aeruginosa, *Escherichia coli*, and *Pasteurella multocida*. *J. Antibiot.* **1995**, *48*, 67–72. [CrossRef]

microorganisms MDPI

Article

Statistical Optimization of the Physico-Chemical Parameters for Pigment Production in Submerged Fermentation of *Talaromyces albobiverticillius* 30548

Mekala Venkatachalam [1], Alain Shum-Chéong-Sing [1], Laurent Dufossé [1,2] and Mireille Fouillaud [1,2,*]

[1] Laboratoire de Chimie et Biotechnologie des Produits Naturels-EA 2212, Université de la Réunion, 15 Avenue René Cassin, CS 92003, CEDEX 9, F-97744 Saint-Denis, Ile de la Réunion, France; mekalavenkat@gmail.com (M.V.); alain.shum@univ-reunion.fr (A.S.-C.-S.); laurent.dufosse@univ-reunion.fr (L.D.)

[2] Ecole Supérieure d'Ingénieurs Réunion Océan Indien-ESIROI, 2 Rue Joseph Wetzell, F-97490 Sainte-Clotilde, Ile de la Réunion, France

* Correspondence: mireille.fouillaud@univ-reunion.fr; Tel.: +262-262-483-363

Received: 5 April 2020; Accepted: 8 May 2020; Published: 11 May 2020

Abstract: *Talaromyces albobiverticillius* 30548 is a marine-derived pigment producing filamentous fungus, isolated from the La Réunion island, in the Indian Ocean. The objective of this study was to examine and optimize the submerged fermentation (SmF) process parameters such as initial pH (4–9), temperature (21–27 °C), agitation speed (100–200 rpm), and fermentation time (0–336 h), for maximum production of pigments (orange and red) and biomass, using the Box–Behnken Experimental Design and Response Surface Modeling (BBED and RSM). This methodology allowed consideration of multifactorial interactions between a set of parameters. Experiments were carried out based on the BBED using 250 mL shake flasks, with a 100 mL working volume of potato dextrose broth (PDB). From the experimental data, mathematical models were developed to predict the pigments and biomass yields. The individual and interactive effects of the process variables on the responses were also investigated (RSM). The optimal conditions for maximum production of pigments and biomass were derived by the numerical optimization method, as follows—initial pH of 6.4, temperature of 24 °C, agitation speed of 164 rpm, and fermentation time of 149 h, respectively.

Keywords: *Talaromyces albobiverticillius* 30548; submerged fermentation; pigments; biomass; optimization; Box–Behnken experimental design; response surface modeling

1. Introduction

In recent years, development of alternative sources for the production of natural pigments has been focused on overcoming the unlimited usage of synthetic pigments, which are found to be hazardous to the human health and environment [1]. Natural colorants obtained from plants, animals, and microbial sources are alternatives to synthetic pigments [2,3]. Nowadays, the shift towards natural colorants is of great interest worldwide and is gaining its industrial importance in food, cosmetics, pharmaceuticals, paint, or dying applications. Microorganisms including bacteria, algae, yeasts, and filamentous fungi from marine or terrestrial origins are capable of producing natural dyes, which pave the way to develop the industrial production of natural compounds [4–6]. Indeed, microbial fermentations present the crucial advantages of independence from geographic or climatic constraints. Additionally, massive production of pigments using bioreactors can be totally controlled and optimized. The use of fungi for the production of commercially important products has been increasing rapidly over the past decades [4,7]. Taxonomically, it represents a large group, often prevailing in seawater, rocks, sediments, sand, soils, and diverse marine macro-organisms [5]. In general, the colorants produced by fungi are associated with increased yields, and huge array of compounds due to the abundance and vast diversity of fungal strains in the environment [8]. Majority of the pigments produced by fungi are quinones, flavonoids, melanins, and azaphilones, which belong to the aromatic polyketide chemical group [9–11] and have been widely described for their medicinal uses or potential utilization as dyes [12,13]. More specifically, filamentous fungi are considered to be a promising source of natural pigments. Moreover, some of the produced molecules also possess bioactive properties like anti-cancer, immunomodulatory, anti-proliferative, antibiotic properties, and so on [14].

Fungi isolated from the marine environments reveal a plethora of known or new compounds with still relatively unexplored bioactivities [15–17]. Marine fungi of several genera, such as *Aspergillus, Cladosporium, Eurotium, Monascus, Penicillium, Talaromyces,* and *Trichoderma* are potent pigment producers of different classes, with wide range of hues [18–22]. For fungal pigment production, submerged fermentation (SmF) appears to be a convenient and economical fermentation method. Indeed, SmF provides the space to produce pigments, owing to their deep commitment to sophisticated control devices and easy monitoring methods during production [23–25]. Besides this, SmF processes can also be affected by various abiotic factors, such as pH of the medium, temperature, agitation speed (gas diffusion), and fermentation time. The mentioned abiotic factors create a strong pressure on the pattern of fungal growth that brings out the interesting secondary metabolites like pigments [26–29].

Over the past few years, a number of studies identified that several species of *Penicillium* and *Talaromyces* produce *Monascus*-like red pigments without the co-production of mycotoxin citrinin, which is produced in *Monascus spp.* fermentations [30]. Some species of *Talaromyces* secrete large amounts of red pigment, notably species such as *Talaromyces purpurogenus, T. albobiverticillius, T. marneffei,* and *T. minioluteus* (often referenced in the literature under earlier *Penicillium* names). Isolates identified as *T. purpurogenus* have been considered to be industrially interesting for their red pigment production. Unfortunately, the pigment production comes along with the co-production of mycotoxins such as rubratoxin a and b and luteoskyrin. Whereas, a novel strain, *Talaromyces atroroseus* CBS 133442 produces diffusible red pigments, without any mycotoxin production [19]. Certain strains of *Talaromyces albobiverticillius* were studied and proposed to produce large amount of red pigments, thus, can be used for coloring food [31]. However, considering *T. albobiverticillius* 30548, collecting large amount of pigments is a real challenge. First, at a lab scale it allows the purification, characterization, and structural elucidation of all molecules produced, including undesirable compounds, if any. In a second step, optimizing the production of compounds of interest is crucial to determining the ultimate potential of the fungus, for a successful production of pigments at an industrial scale.

The present study investigated and optimized the SmF process crucial parameters such as initial pH (4–9), temperature (21–27 °C), agitation speed (100–200 rpm), and fermentation time (24–336 h) for the maximum production of pigments (orange and red) and biomass by *Talaromyces albobiverticillius* 30548. To achieve this goal, Box–Behnken Experimental Design (BBED) and response surface modeling (RSM) were used. BBED is a methodology based on a three interlocking 2^2 factorial experimental design. It is also described as a spherical revolving surface surrounding a cube constructed from the responses of the center and middle points of the experimental design [32–37]. This experimental design allows consideration of the multifactorial interactions between a set of crucial parameters. Derived from this mathematical approach, it is possible to optimize the culture conditions for maximum production of pigments or biomass and its interactive effects [38]. This is the first report to investigate and optimize the SmF process parameters by BBED and RSM, for maximal pigment and biomass production, using the marine-derived *T. albobiverticillius* 30548.

2. Materials and Methods

2.1. Microorganism and Maintenance

The filamentous fungus used in this study was isolated from the sediments collected from the marine environment of the La Réunion island, Indian Ocean (GPS coordinates: 21°06′22.11″ S, 55°14′ 15.78″E) [39]. From taxonomic identification of the fungal biodiversity center (Westerdijk Fungal Biodiversity Institute, Utrecht, The Netherlands), the fungus was identified as *Talaromyces albobiverticillius* and named as *T. albobiverticillius* 30548 (GenBank accession number MK937814). The isolated fungus was cultured in potato dextrose agar plates (PDA, BD-Difco, Sparks, NV, USA) and maintained at 4 °C, as well as sub-cultured at regular intervals for the experiments.

2.2. Preparation of Pre-Culture

Laboratory grade potato dextrose broth (PDB) (BD-Difco)) was used to prepare the seed medium. For inoculum preparation, 100 mg of mycelial spores was scrapped from actively grown culture on potato dextrose agar petri plate (PDA), (7 days old) (composition: PDB added with 2% Bacto agar (BD-Difco) and transferred into 250 mL Erlenmeyer flask containing 80 mL of sterilized PDB medium (pre-culture). The flasks were incubated at 25 °C for 48 h in an orbital shaker, at 200 rpm (Multitron Pro, Infors HT, Bottmingen, Switzerland) before being transferred to the main fermentation medium [19].

2.3. Fermentation Conditions

After 48 h of growth, the pre-culture broth was centrifuged at 8000 rpm for 6 min at room temperature (Sigma 4K15, Sigma Laborzentrifugen GmbH, Osterode am Harz, Germany) to separate the mycelia and the culture filtrate. The harvested mycelium (200 mg) was transferred as seed inoculum into a 500 mL Erlenmeyer flask containing 200 mL of sterile fermentation medium (PDB). Twenty-nine (29) series of submerged fermentations were carried out according to the experimental design (BBED) presented in Table 1. The BBD matrix (Table 2) allowed to fix a set of selected values for initial pH, temperature, agitation speed, and fermentation time, for each fermentation. The flasks were incubated in a controlled orbital shaker with the fixed conditions (Multitron Pro, Infors HT, Bottmingen, Switzerland) during the whole culture period. All experiments were performed in triplicates for each condition. At predefined intervals of 24, 168, 336 h (Table 1), fermented broth samples were taken out from each culture flask, in order to measure the concentration of orange and red pigments, as well as the increase in biomass weight.

2.4. Spectrophotometric Quantification of Extracellular Pigments

Five milliliters of the fermented culture broth were sampled at predefined regular intervals and filtered through pre-weighed nylon mesh of 48 μm pore size (Nitex 07401, Sefar AG, Heiden, Switzerland) to separate the mycelia (Table 1). The concentration of extracellular pigments was determined using UV-visible spectrophotometer (UV-VIS spectrophotometer UV-1800, Shimadzu, Tokyo, Japan) at 470 (orange pigments) and 500 nm (red pigments), as the color in the extracellular culture filtrate of *T. albobiverticillius* 30548 was intense dark red [39]. Quinizarin (Sigma-Aldrich, St. Louis, MO, USA) and Red Yeast rice pigments (RYrp) (Wuhan Jiacheng Biotechnology Co., Ltd., Wuhan, China) were used as the reference standards, since the color and absorbance profile of the pigments produced by *T. albobiverticillius* 30548 were similar to those of these two compounds, at a wavelength of 470 and 500 nm. Thus, the OD values at 470 and 500 nm were interpolated using the standard curve equations of the two chosen standards, respectively (Figures A1 and A2 in Appendix A), to obtain the pigment concentration in terms of gram per liter equivalents (*g/L* equivalent). The concentrations of red and orange pigments (*g/L* equivalent) were consequently calculated using Equations (1) and (2), as shown below:

$$\text{Orange pigment concentration, } C\left(\frac{g}{L}equiv.\right) = \frac{Abs_{470} - 0.1102}{0.0148} * 0.001, \quad (1)$$

$$\text{Red pigment concentration, } C\left(\frac{g}{L}equiv.\right) = \frac{Abs_{500} - 0.0968}{0.0069} * 0.001, \quad (2)$$

2.5. Dry Biomass Concentration

The amount of wet biomass was obtained through filtration using 48 m Nitex filter cloth (SEFAR AG, Switzerland) and was weighed precisely using an analytical balance. The filters with wet biomass was dried in a hot air oven (SNB 100, Memmert GMBH, Schwabach, Germany) at 105 °C for 17 h. The dried filters were weighed after letting them cool in a desiccator for 30 min to bring it down to room temperature. The amount of dry biomass in the culture broth was inferred from the following, Equation (3):

$$[C_{DB}]\frac{g}{L} = \frac{(W2 - W_1)}{V}, \quad (3)$$

where W_2 is the weight of the filter with biomass obtained after drying (g), W_1 is the weight of the corresponding empty filter (g), and V is the volume of sample (L).

2.6. Experimental Design and Statistical Analysis (BBED and RSM)

Box–Behnken experimental design (BBED) and response surface modeling (RSM) with four factors (pH, temperature, agitation speed, and fermentation time) at three levels were used in this study to investigate and optimize the effect of process parameters on the production of pigments (orange and red pigments) and fungal biomass. The process parameters and their ranges were chosen, based on preliminary "one-factor-at-a-time" studies (Table 1) (data not shown). Each independent variable was coded at three levels as −1 (low), 0 (central point), and +1 (high), which were—pH (4–9), temperature (21–27 °C), agitation speed (100–200 rpm), and fermentation time (24–336 h). The process variables were denoted as X_1, X_2, X_3, and X_4 and the actual value of process parameters were converted into the uncoded form, using the following Equation (4) [40].

$$x = \frac{X - ((X_{\max} + X_{\min})/2)}{(X_{\max} - X_{\min})/2}, \quad (4)$$

Table 1. Coded and actual values of the variables for the four factor Box–Behnken experimental design.

Variables	Symbol	Coded and Actual Values		
		−1	0	+1
pH	X_1	4	6.5	9
Temperature (°C)	X_2	21	24	27
Agitation speed (rpm)	X_3	100	150	200
Fermentation time (h)	X_4	24	168	336

In total, twenty-nine experiments were carried out including five center points. Design-Expert® Software (Version 9, Stat-Ease Inc., Minneapolis, MN, USA) was used in this study to construct the experimental design and statistically analyze the experimental data.

The obtained experimental results should fit well into the empirical second-order polynomial model (Equation (5)). The second-order polynomial equation correlated the relationship between independent variables and the responses. The mathematical form of the equation is given below:

$$Y = \beta_0 + \sum_{j=1}^{k} \beta_j x_j + \sum_{j=1}^{k} \beta_{jj} x_j^2 + \sum \sum_{i<j=2}^{k} \beta_{ij} x_i x_j + e_i, \quad (5)$$

where Y is the response; x_i and x_j are variables (i and j range from 1 to k); β_0 is the model intercept coefficient; β_j, β_{jj} and β_{ij} are interaction coefficients of the linear, quadratic, and second-order terms, respectively; k is the number of independent parameters ($k = 4$ in this study); and e_i is the error [41–43].

The sequential model sum of squares and model summary statistics were carried out on the obtained experimental data to evaluate the adequacy of various models (linear, interactive, quadratic, and cubic) fitted to the experimental data. Pareto analysis of variance (ANOVA) was utilized in this present work in order to generate the ANOVA table. The experimental data were evaluated with various descriptive statistical analyses such as p value, F value, degrees of freedom (DF), coefficient of variation (CV), coefficient of determination (R^2), adjusted coefficient of determination (Adj-R^2), and predicted coefficient of determination (Pre-R^2) to reveal the numerical consequence of the constructed quadratic mathematical model.

2.7. Optimization and Validation

The effect of fermentation process parameters on the responses, such as the pigment yields and the biomass were optimized by multi-response analysis, using Derringer's desired function methodology [44]. In this method, the responses were transformed into a dimensionless individual desirability function (d_i) that varied from 0 to 1 (lowest to highest desirability). From the geometric means of individual desires, the overall desirability function (D) was obtained using the following equation (Equation (6)).

$$D = (d_1^{n_1} \times d_2^{n_2} \times d_3^{n_3} \times \ldots\ldots\ldots\ldots \times d_k^{n_k})^{1/k} \quad (6)$$

where d_i is the individual desirability ranged from 0 to 1, k is the number of considered responses, and n_i is the weight of each response.

The dimensionless desirability (d_i) value of the response using the maximized function was calculated from the equation below (Equation (7)).

$$d_i = \frac{Y_i - Y_{min}}{Y_{max} - Y_{min}}, \quad (7)$$

where, Y_i is the obtained response value, Y_{min} is the obtained response minimum value, and Y_{max} is the obtained response maximum value.

To determine the validity of the developed mathematical model equation, triplicate experiments were performed under the optimal condition, as predicted by the model. The average values of the experimental data were compared with the predicted values of the developed model to find out the accuracy and suitability of this model.

Table 2. Box–Behnken design matrix with the experimental responses over process parameters in *Talaromyces albobiverticillius* 30548.

Exp. Run	pH	Temperature (°C)	Agitation Speed (rpm)	Fermentation Time (h)	Orange Pigment Yield (g/L^1)	Red Pigment Yield (g/L^2)	Dry Biomass Weight (g/L)
1	0	0	0	0	0.30	0.83	3.35
2	0	0	1	−1	1.15	1.44	5.38
3	0	1	0	1	0.30	0.82	1.97
4	−1	0	0	1	0.70	1.08	3.05
5	0	−1	0	−1	0.90	1.23	3.76
6	0	1	1	0	1.10	1.37	6.17
7	−1	0	0	−1	0.65	1.04	2.39
8	0	0	0	0	1.25	1.47	5.47
9	0	0	0	0	0.65	0.95	3.34
10	0	1	−1	0	1.20	1.48	6.05
11	1	1	0	0	0.40	0.85	4.83
12	0	0	0	0	1.40	1.68	5.10
13	0	0	0	0	1.00	1.31	3.33

Table 2. *Cont.*

Exp. Run	pH	Temperature (°C)	Agitation Speed (rpm)	Fermentation Time (h)	Orange Pigment Yield (g/L[1])	Red Pigment Yield (g/L[2])	Dry Biomass Weight (g/L)
14	0	−1	−1	0	−0.10	0.43	0.21
15	−1	0	1	0	0.65	1.03	3.75
16	−1	−1	0	0	1.20	1.39	5.34
17	0	1	0	−1	0.10	0.57	2.66
18	1	0	−1	0	1.20	1.48	4.28
19	0	−1	1	0	1.20	1.50	4.61
20	1	0	1	0	1.55	1.73	6.15
21	1	0	0	−1	0.85	1.18	5.21
22	0	−1	0	1	0.70	1.12	2.10
23	0	0	−1	−1	0.75	1.12	3.87
24	1	−1	0	0	0.60	0.98	2.93
25	−1	1	0	0	1.70	1.92	6.67
26	1	0	0	1	1.71	1.92	6.7
27	−1	0	−1	0	1.76	1.93	6.79
28	0	0	1	1	1.77	1.93	7.17
29	0	0	−1	1	1.75	1.93	6.65

1: g/L equiv. quinizarin 2: g/L equiv. RYrp.

3. Results

3.1. Box–Behnken Experimental Design Analysis

The selected statistical methods measured the effects of change in the operating variables and their mutual interactions on a system or a process through experimental design [45]. In this study, using the Design-Expert® Software (v9), the Box–Behnken experimental design with four factors at three levels was employed to investigate and optimize the influence of the process variables on the yields of pigments (orange and red) and fungal biomass for *T. albobiverticillius* 30548. The results are listed in Table 2.

3.2. Development of Second-Order Polynomial Models

Model adequacy checking was performed on the experimental data to determine whether the fitted model would give poor or misleading results. Four-degree polynomial models viz., linear, interactive (2 factors interaction, 2FI), quadratic and cubic models were fitted to the experimental data. Two different tests namely the sequential model sum of squares and model summary statistics were carried out in this study, to conclude the adequacy of models among various models to represent the responses. The adequacy of model summary output indicated that, the quadratic model was statistically highly significant due to its higher R^2, compared to the other models (Table 3). Hence, the quadratic model was selected in this study to investigate the effect of the process variables on the responses, such as pigments and biomass production in the *T. albobiverticillius* 30548 strain.

3.3. Determination of Second-Order Polynomial Equations

An empirical relationship expressed by a second-order polynomial equation with interaction terms was fitted between the experimental results obtained on the basis of the Box–Behnken experimental design and the input variables. The observed experimental results in each run were subjected to multiple regression analysis to calculate the regression coefficients of the model. Calculated regression coefficients were substituted in (Equation (4)) to obtain a model for the orange pigment yield OPY (Equation (8)), red pigment yield RPY (Equation (9)), and dry biomass weight DBW (Equation (10)). The final equations obtained in terms of the coded factors are given below:

$$OPY = +1.74 + 0.35X_1 - 0.11X_2 + 0.27X_3 - 0.029X_4 - 0.11X_1X_2 - 0.19X_1X_3 + 0.11X_1X_4 \\ +0.41X_2X_3 + 2.145^{-17} X_2X_4 + 0.10X_3X_4 - 0.42X_1^2 - 0.68X_2^2 - 0.35X_3^2 - 0.38X_4^2, \quad (8)$$

$$RPY = +1.92 + 0.28X_1 - 0.09X_2 + 0.20X_3 - 0.015X_4 - 0.087X_1X_2 - 0.17X_1X_3 + 0.075X_1X_4 \\ +0.31X_2X_3 - 0.019X_2X_4 + 0.071X_3X_4 - 0.33X_1^2 - 0.54X_2^2 - 0.31X_3^2 - 0.32X_4^2, \quad (9)$$

$$DBW = +6.80 + 0.77X_1 - 0.77X_2 + 1.24X_3 - 0.17X_4 - 0.24X_1X_2 - 0.021X_1X_3 - 0.61X_1X_4 \\ +1.18X_2X_3 + 055X_2X_4 + 0.17X_3X_4 - 1.02X_1^2 - 2.31X_2^2 - 1.35X_3^2 - 0.97X_4^2, \quad (10)$$

where X_1, X_2, X_3, and X_4 are the coded values of the independent process variables, respectively, pH, temperature, agitation speed, and fermentation time.

Table 3. Model summary statistics for responses.

Source	Std. Dev.	R^2	Adjusted R^2	Predicted R^2	Press	Remarks
Model summary statistics for Orange Pigment Yield						
Linear	0.46	0.3306	0.2191	0.1387	6.54	
2FI	0.48	0.4574	0.1560	0.0518	7.20	
Quadratic	0.089	0.9853	0.9706	0.9172	0.63	Suggested
Cubic	0.072	0.9959	0.9809	0.4736	4.00	Aliased
Model summary statistics for Red Pigment Yield						
Linear	0.37	0.3146	0.2004	0.1230	4.29	
2FI	0.39	0.4328	0.1178	0.0127	4.82	
Quadratic	0.096	0.9734	0.9468	0.8469	0.75	Suggested
Cubic	0.058	0.9958	0.9806	0.402	2.92	Aliased
Model summary statistics for Dry Biomass Weight						
Linear	1.49	0.3828	0.2799	0.1902	69.91	
2FI	1.58	0.4820	0.1943	−0.0036	86.64	
Quadratic	0.42	0.9714	0.9429	0.8445	13.42	Suggested
Cubic	0.18	0.9977	0.9894	0.9790	1.81	Aliased

3.4. Statistical Analysis

The adequacy and fitness of the models were tested by multiple regression analysis using the least square method. Significance of the developed models could be determined through ANOVA and the results are shown in Table A1 (Appendix B). The results of ANOVA indicated that, the developed models adequately represented the actual relationship between the independent variables and responses, in the chosen range. Analysis of variance followed by Fisher's statistical test (F-test) was applied to evaluate the significance of each variable. The high F- values, 67.05 for the orange pigment yield (OPY) measured at 470 nm, 36.61 for red pigment yield (RPY) measured at 500 nm, and 34.02 for the dry biomass weight (DBW) indicated that most of the variation in the response could be explained by the developed regression equations. The associated *p*-values were used to estimate whether F is large enough to indicate statistical significance, and *p*-values lower than 0.05 indicated that the developed model and the terms were statistically significant. In our study, the *p*-values were lower than 0.0001 for all responses (OPY, RPY, DBW). It exhibited the precision and the accuracy of the developed models.

Determination coefficient (R^2), adjusted R^2 ($AdjR^2$), predicted R^2 ($PreR^2$), and coefficient of variation (CV%) were calculated to check the adequacy (Adeq. Pre.) and accuracy of the developed models. The R^2 gave the proportion of total variation in the responses predicted by the models. The values of R^2 (0.9853 for OPY, 0.9734 for RPY, and 0.9714 for DBW) ensured a satisfactory fit of the quadratic model to the experimental data. For example, R^2 values indicated that the sample variation of 98.5% for OPY was attributed to the independent variables and only 1.5% of the total variations were not explained by the model. The adjusted determination coefficient ($AdjR^2$) corrected the R^2 value of the sample size and the number of terms in the model. In this study, the values of $AdjR^2$ (0.9706 for OPY at 470 nm, 0.9468 for RPY at 500 nm, and 0.942 for DBW) were also high and very close to the R^2

values, and indicated a better prediction of the model. However, when irrelevant variables are added to the model, the adjusted R^2 decreases. In general, AdjR^2 will always be less than or equal to R^2. PreR^2 is a measure of how good the model predicts a response value. In our case, the PreR^2 (0.9172 for OPY at 470 nm, 0.8469 for RPY at 500 nm, and 0.8445 for DBW) are in reasonable agreement with the AdjR^2.

The CV%, indicating the relative dispersion of the experimental points from the predictions of the second-order polynomial (SOP) models, were found to be 9.12 for OPY at 470 nm, 7.41 for RPY at 500 nm, and 9.41 for DBW, respectively. The low values of CV% clearly indicated a very high degree of precision and a good reliability of the experimental values. The high R^2 value and a small CV% value indicated that the developed model would be able to give a good estimate of response of the system over the ranges studied.

3.5. *Effect of Process Variables on Pigment Yield*

In our study, the growth of the fungus in PDB leads to the production of secondary metabolites, which are visualized through the formation of a colored broth during the fermentation process. Production of pigments started at 48 h, attained a maximum at 140 h, and showed a slight increase, thereafter, until the end of fermentation [39,46]. In pigment production, temperature and pH acts as a whole active mechanism that is probably related to the genetic and metabolic control of defense mechanisms, thus influencing the level of pigment concentration and high productivity of pigments [47].

3.5.1. Combined Effect of pH, Temperature, Agitation Speed, and Time on Orange Pigment Yield (OPY)

The effect of independent variables on orange pigment yield was studied by changing the levels of any two independent variables, while keeping the other two at their constant middle level. The response surface plots (RSM) were used to locate the optimal values for this fermentation process. Therefore, three response surface plots were obtained by considering all possible combinations.

Figure 1a shows the effect of interaction between pH and temperature on the orange pigment production (maximum absorbance at 470 nm). It revealed that both factors at their lower levels had a negative impact on the orange pigment production. Increase in pH and temperature led to a gradual increase in pigment production, of up to pH 7.5 and temperature 24 °C. Increasing the value of both independent variables (pH above 7.8 and temperature above 27 °C) also showed a negative effect on orange pigment production. As shown in Figure 1b, increasing the agitation speed (up to 155 rpm) and pH (up to 7.5) led to high pigment production (1.76 g/L of orange pigments quinizarin equivalent), while agitation speed above 155 rpm and pH above 7.5 showed a negative effect on orange pigment production.

The interaction between fermentation time and pH also play an important role in pigment production (Figure 1c); pH in the range of up to 8.1 (alkaline pH) at higher levels of fermentation time, was found to be significant for orange pigment production. It was noticed that fermentation time (234 h) and pH (<8.1) up to a certain level supported orange pigment production but both factors negatively applied at higher levels and affected the pigment production. Temperature and agitation speed showed significant influence on each other and also on OPY. Both parameters showed a linear and quadratic response for pigment yield (Figure 1d). Maximum yield was observed in the above mid-values for both parameters (temperature of 25 °C and agitation speed of 165 rpm) and above that level, yields of the pigment decreased. The interactive effects of temperature and fermentation time against OPY are depicted in Figure 1e. When temperature increased from 21 to 24 °C and fermentation time was between 24 and 259 h, it resulted in a gradual increase in pigment production, up to a maximum. Further increase in temperature (above 24 °C) and fermentation time (above 259 h) led to a decrease in pigment yield.

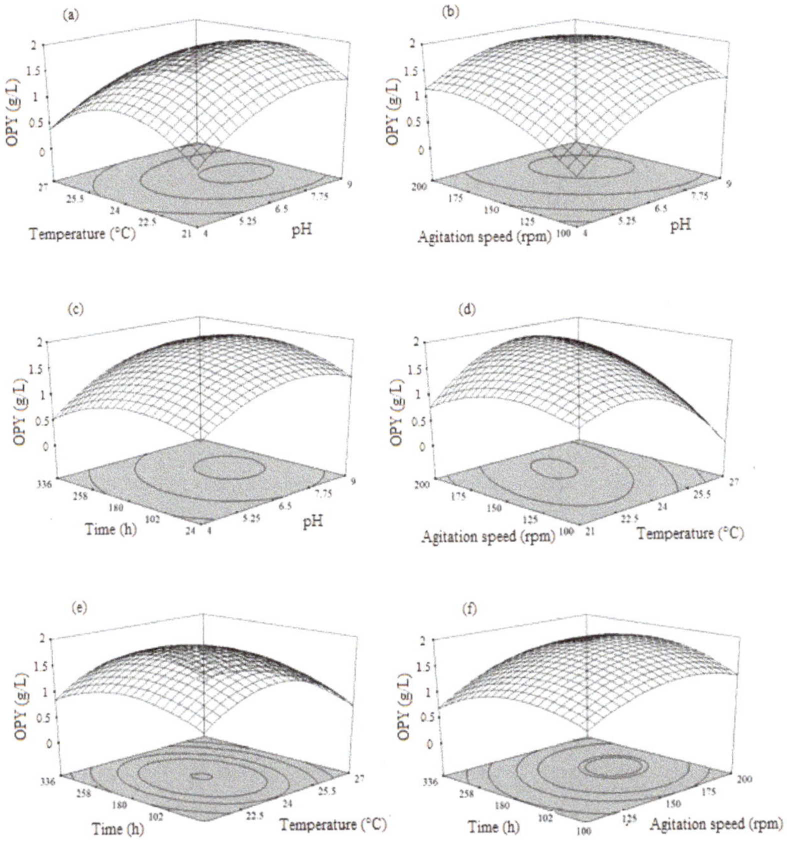

Figure 1. Response surface plots showing the influence of the process variables on the orange pigment yield (OPY): (**a**) Interaction between temperature and pH; (**b**) Interaction between agitation speed and pH; (**c**) Interaction between time and pH; (**d**) Interaction between agitation speed and temperature; (**e**) Interaction between time and temperature; (**f**) Interaction between time and agitation speed.

The three-dimensional (3D) response interactive plot of agitation speed and fermentation time illustrated that, OPY had increased with increase in agitation speed and fermentation time, up to 180 rpm and 224 h, respectively (Figure 1f). The yield of orange pigment was decreased with an increase in the agitation speed and fermentation time above 180 rpm and 247 h, respectively.

Combining the effects of all process variables investigated in this work, the initial pH and temperature of the medium played a key role in orange pigment synthesis. It was reported that a more acidic pH is favorable for most fungi for pigment synthesis in a submerged culture [4,48,49]. Similarly, acidic pH around 6.5 combined with temperature of up to 23.9 °C, agitation speed of 154.9 rpm, and fermentation time of 229 h was considered to be the optimal condition to produce maximal OPY of 1.76 g/L quinizarin equivalent.

3.5.2. Combined Effects of pH, Temperature, Agitation Speed, and Time on Red Pigment Yield (RPY)

Data obtained from the experiments were used to study the effect of process variables on red pigment production. From the results, it was confirmed that increasing pH and temperature from 4.0–6.5 and 22–25 °C enhanced the intensity of red pigment production in *T. albobiverticillius* 30548 and then decreased it (Figure 2a).

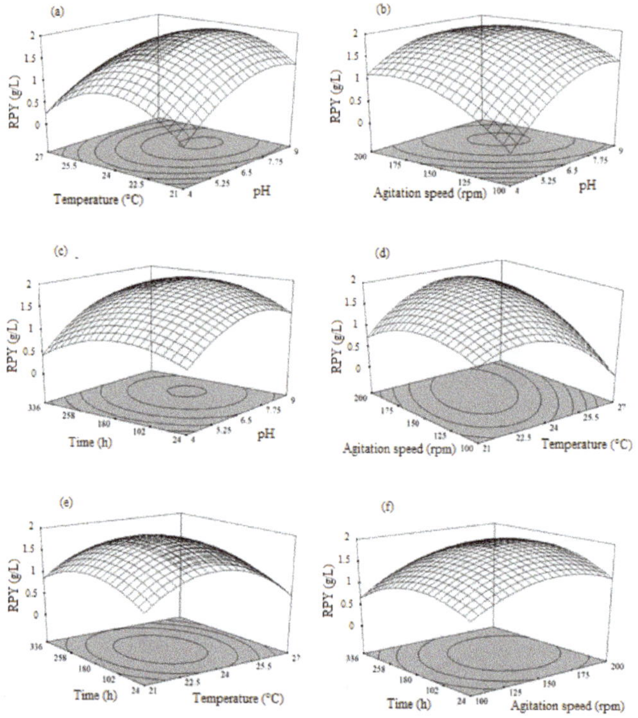

Figure 2. Response surface plots showing the influence of the process variables on the red pigment yield (RPY): (**a**) Interaction between temperature and pH; (**b**) Interaction between agitation speed and pH; (**c**) Interaction between time and pH; (**d**) Interaction between agitation speed and temperature; (**e**) Interaction between time and temperature; (**f**) Interaction between time and agitation speed.

Additionally, increasing the agitation speed from 100–158 rpm and the pH value from 4.0–6.5 enhanced the red pigment production (Figure 2b). The maximum production of red pigments was attained at 120–168 h of fermentation and further increase in incubation time led to a decrease in the production (Figure 2c,e,f). The mycelium of the species grew rapidly at the start of fermentation (12–48 h), matured in the period of 48–96 h, and more extracellular red metabolites were released for up to 168 h, in the fermentation medium. Further increasing incubation time led to the bleaching of the extracellular red color and then a decrease in the pigment yield. To examine the influence of agitation speed on red pigment production, experiments were carried out in various agitation speeds (100–200 rpm). The results showed that, red pigment production increased from 100–158 rpm and above that, it decreased rapidly (Figure 2d,f).

Maximum pigment production was exhibited in the middle level of the process variables and strong interaction was found between the variables, since the shape of the response surface and contour plot was elliptical in nature. By applying the numerical optimization method, the optimal condition was attained as follows—initial pH of 6.5, temperature of 24 °C, agitation speed of 158.4 rpm, and fermentation time of 198.6 h, with a maximal red pigment yield of 1.92 g/L, in terms of RYrp equivalent, respectively.

3.5.3. Combined Effects of pH, Temperature, Agitation Speed, and Time on Dry Biomass Weight (DBW)

The influence of the process variables over the dry biomass weight was similarly examined and the outcomes were depicted in Figure 3. The graph (Figure 3a) demonstrated the biomass yield was linearly increased with increasing levels of pH up to 7.8 and temperature up to 25 °C, and then the yields decreased gradually.

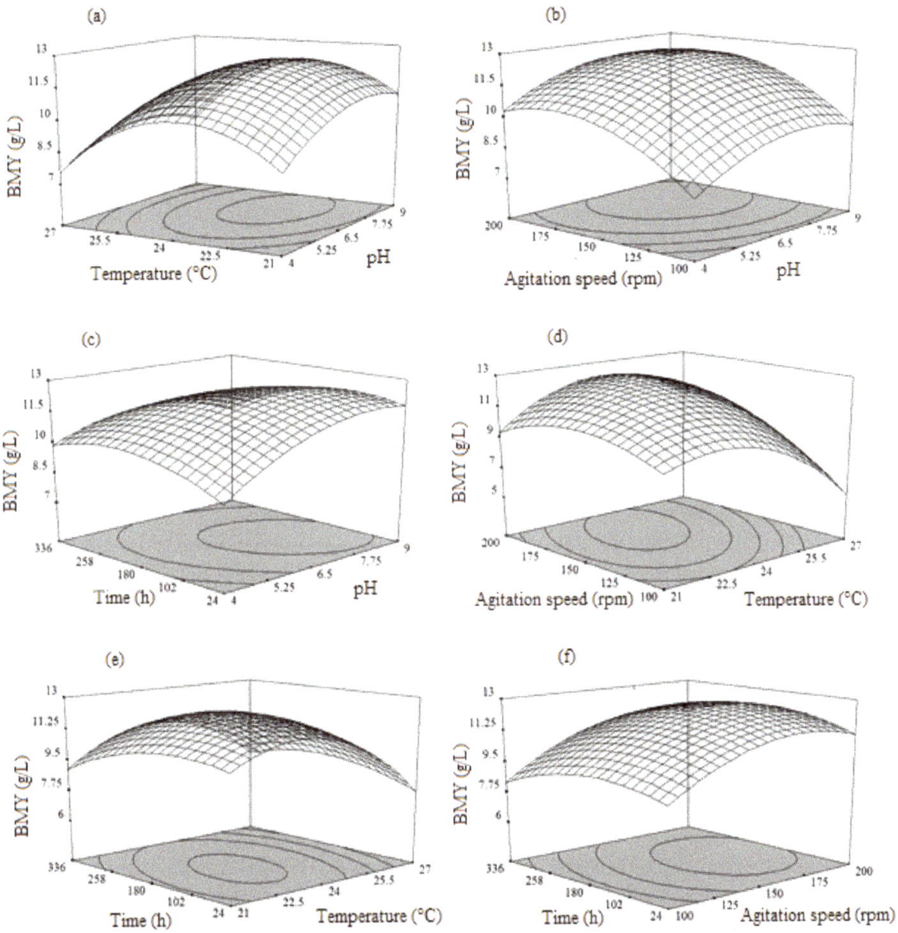

Figure 3. Response surface plots showing the influence of the process variables on dry biomass yield: (**a**) Interaction between temperature and pH; (**b**) Interaction between agitation speed and pH; (**c**) Interaction between time and pH; (**d**) Interaction between agitation speed and temperature; (**e**) Interaction between time and temperature; (**f**) Interaction between time and agitation speed.

Figure 3b shows the effect of interaction between pH and agitation speed on biomass production, which revealed that both factors at their lower level had no significant effect on the biomass production, but increase in pH (up to pH 7.7) and agitation speed (up to 180 rpm) led to a gradual increase in the biomass production, while increase in the value of both independent variables beyond pH 7.7 and agitation speed of 180 rpm showed a negative effect on biomass production. From Figure 3c, it can be seen that the biomass yield increased with increasing pH and fermentation time, up to its middle level. However, further increasing pH and fermentation time would decrease the biomass yield. The 3D response surface plots based on independent variables such as temperature and agitation speed were developed, while other variables were kept at middle levels (Figure 3d) and exhibited that, the biomass yield was enhanced with increasing temperature (24.5 °C) and agitation speed (185 rpm), and the yield decreased beyond these levels. Biomass production was also increased with increase in both variables (temperature (up to 25 °C) and fermentation time (up to 245 h)) (Figure 3e). However, there was a sharp convergence of the curve near the boundary, explaining that in the presence of temperature and fermentation time above certain limit (above temperature of 25 °C and fermentation time of 245 h), it would not contribute to increasing biomass production.

The interaction between agitation speed and fermentation time also played an important role in biomass production (Figure 3f). Agitation speed and fermentation time at above middle level was found to be significant for biomass production. It was noticed that agitation speed up to a certain level (190 rpm), as well as fermentation time (245 h) supported the biomass production but both affected the biomass production negatively, at higher levels. The analysis of response surface was performed in order to determine the optimal condition to produce maximal biomass yield.

The optimal condition was an initial pH of 6.6, temperature of 23.9 °C, agitation speed of 168.9 rpm, and fermentation time of 145.1 h, respectively, and the maximum predicted yield of biomass (7.16 g/L) was attained in the optimal condition.

3.6. Multi Response Optimization

The second-order polynomial models developed in this study were utilized for each response, in order to obtain specified optimum conditions. An optimal condition for the maximum pigment production (orange and red) and biomass yield in *T. albobiverticillius* 30548 under submerged fermentation was derived by Derringer's desired function methodology. This function searches for a combination of factor levels that simultaneously satisfies the requirements for each response in the design. Individual desirability (d_i) evaluates how the process parameters optimize a single response. This numerical optimization evaluates a point that maximizes the desirability function. One main objective of optimization could be to maximize the final pigment yield and biomass production and recalculating all responsible independent factors by using desirability functions, therefore, the goal for pH, temperature, agitation speed, and fermentation time was assigned, as in range.

The goal for the orange pigment yield (OPY), red pigment yield (RPY), and biomass production (DBW) was assigned to get the maximum. A weight factor of 1 was chosen for all individual desirability in this work. The "importance" of a goal could be changed in relation to the other goals. It could range from 1 (least importance) to 5 (most important). The default is for all goals to be equally important in a setting of 3. The optimization procedure was conducted under these settings and boundaries. Under the optimal conditions, the predicted orange pigment yield was 1.76 g/L quinizarin equivalent, red pigment yield was 1.92 g/L RYrp equivalent, and dry biomass yield was 7.16 g/L, with a desirability value of 0.983. The maximized overall desirability ($D = 0.983$) was calculated from the geometric means of the individual desirability functions (d_i) of each response.

3.7. Validation of the Optimized Condition

The validation was carried out in shake flasks in triplicates under the optimized conditions of the media, predicted by the polynomial model. The mean of experimental values of OPY, RPY, and DBW were compared with the predicted values. The existence of a good correlation between the experimental and predicted values indicated the reliability and validity of the proposed model (Figure 4a–c). Due to the applicability of optimal extraction condition in a practical manner, the attained optimal condition was altered as follows—initial pH of 6.4, temperature of 24 °C, agitation speed of 164 rpm, and fermentation time of 149 h, respectively. The experimental efficiency of the production of pigments and dry biomass under the optimum condition was found to be 1.76 ± 0.58 g/L quinizarin equivalent for OPY, 1.92 ± 0.76 g/L RYrp equivalent for RPY, and 7.18 ± 0.41 g/L for DBW, respectively.

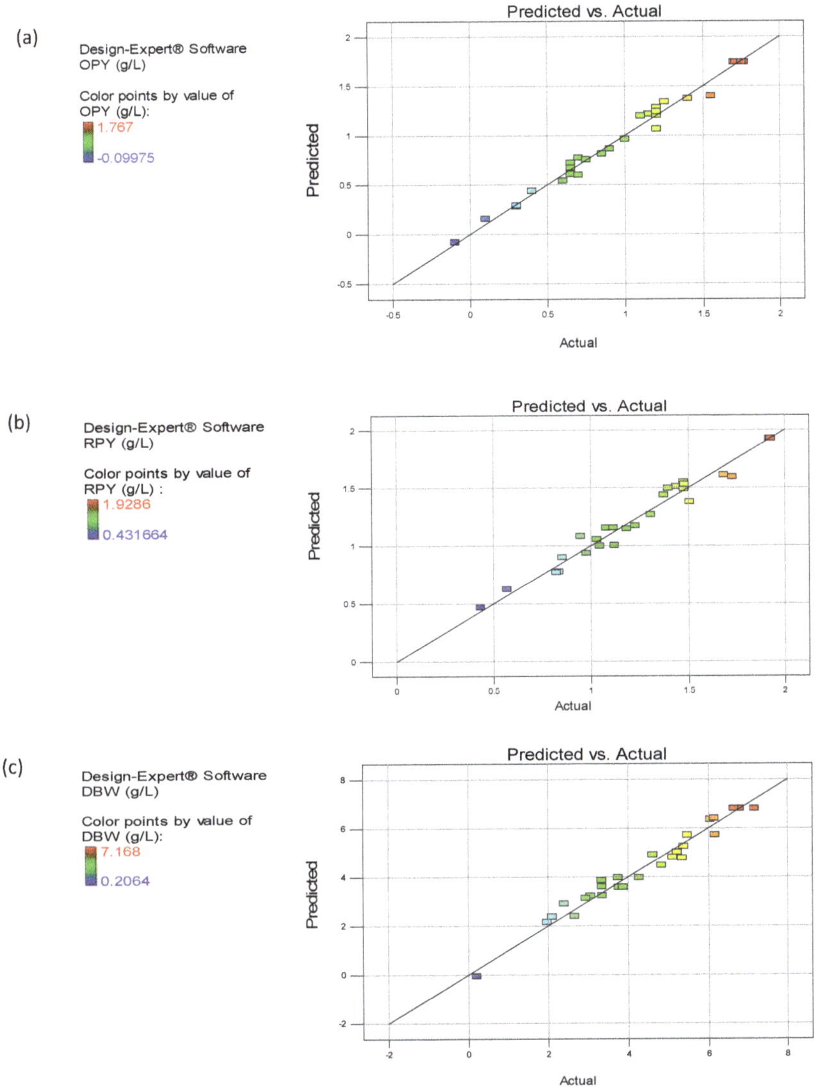

Figure 4. Validation of the polynomial model for (**a**) OPY, (**b**) red pigment yield (RPY), and (**c**) dry biomass weight (DBW) productions from *Talaromyces albobiverticillius* 30548.

4. Discussion

The results obtained in this work for the cultivation of *T. albobiverticillius* 30548 were consistent with Babitha et al. (2007), who reported that maximum production of pigments was noticed in *M. purpureus* at pH 4.5 to 7.5 [50]. Studies by many researchers have revealed that among the numerous environmental factors, medium pH and the source of nitrogen determine pigment production in a submerged culture. The optimal pH for *T. albobiverticillius* 30548 was found to be 6.4 in this study, which is close to the optimum pH for pigment production in *Talaromyces verruculosus* mentioned by Chadni et al. (2017) [51]. Indeed, acidic pH might enhance the hydrolysis of the substrates and subsequently favors the generation of the metabolite production [52]. From Yongsmith et al. (1993) it was demonstrated that the compounds excreted in the culture broth could react with ammonium ion or free amino groups to transform to red amine derivatives towards neutral pH and beyond [53]. This might happen in *T. albobiverticillius* 30548 as there is an evolution of the color hue in the culture broth (from yellow to red), along the entire fermentation period [39].

Temperature, another key parameter, controls the growth and metabolic activity of fungi. To find the optimal temperature for mycelial growth and pigment production, *T. albobiverticillius* 30548 was cultivated under various temperatures (21–27 °C). Fermentation at 24 °C is regarded as favorable for fungal growth in combination with an acidic pH and a medium agitation speed, which aid good production of pigments. Mendez et al. (2011) found a similar result in obtaining the highest level of pigment production at 24 °C in *P. purpurogenum GH2*, combined with an acidic pH [28]. However, in other studies with *Monascus spp., Penicillium spp.,* temperature at 30 °C is considered to be the best temperature for red pigment production. However, the microbial growth was hindered at 30 °C but a maximum growth was attained by increasing the temperature to 34 °C [26,54,55]. For the marine-derived *T. albobiverticillius* 30548, it is interesting to note that the optimal temperature for both mycelial growth and pigment production was found to be 24 °C.

Fermentation of any substrate is influenced by the time of incubation, so the pigment synthesis was monitored at predefined intervals. Pigment production, as measured by absorbance, was slightly delayed when compared to biomass growth, and was higher at 149 h (day 6) of fermentation time, until the stationary phase [39,46]. After 149 h, a decrease in pigment production was observed, which might be due to the decomposition of pigments (degradation of the chromophore pigment group) or changes in the pigment structure. A similar outcome was observed in the pigment production by *Monascus purpureus* [56].

Basically, in liquid fermentation, agitation speed increases the amount of dissolved oxygen and makes the oxygen more accessible to cells, leading to greater growth [57]. Similarly, in this present work, increasing agitation speed from 100–158 rpm enhanced the supply of oxygen in the growth phase and thus enhanced the biomass and pigment yield. However, above 158 rpm, agitation speed might damage the mycelial growth by high shear force and lead to a decrease in the pigment yields. Mohamed et al. (2012) reported similar results, as the highest pigment production was obtained at agitation speed of 150 rpm using *Monascus purpureus* [58]. Additionally, the highest level of yellow pigments (1.38 g/L) was produced by *Penicillium aculeatum* ATCC 10409 at an agitation speed of 100 and 150 rpm [47].

5. Conclusions

Box–Behnken experimental design (BBED) coupled with response surface modeling (RSM) was successfully applied in the present work. The methodology allowed to locate the optimal process parameters in the chosen range (initial pH (4–9), temperature (21–27 °C), agitation speed (100–200 rpm), and fermentation time (0–336 h)), to maximize pigment and biomass production in submerged fermentation of *T. albobiverticillius* 30548. The outcomes of this study indicated that, all process variables had a significant influence on the responses. The actual values obtained through the experimental studies were used to construct second-order quadratic model for the responses to predict the observed data. The optimal condition was attained through Derringer's desired function methodology (initial pH of 6.4, temperature of 24 °C, agitation speed of 164 rpm, and fermentation time of 149 h). It was materially validated by the experiment (1.76 ± 0.58 g/L quinizarin equivalent for OPY, 1.92 ± 0.76 g/L RYrp equivalent for RPY, and 7.18 ± 0.41 g/L for DBW).

The results of this study revealed that *T. albobiverticillius* 30548 could be an avenue for industrial application of pigment production. Considering the potential toxicity of compounds produced by *T. albobiverticillius* 30548 in SmF, complementary studies need to be conducted, based on the characterization of the 12 major compounds synthetized [59], and on the elucidation of the biosynthetic pathway, before using them for applications. In addition, the mathematical models constructed in this study and the methodology applied could be useful for the selection of appropriate parameters for the production of pigments at a large scale.

This study allowed for the consideration of other factors of pigment production, besides the physical process factors. Generally, carbon and nitrogen sources in the culture medium play a major role in the growth and production of metabolites. Further work on optimizing various nutrient types and their concentration in the culture media is in progress and will be developed in a further study.

Author Contributions: Conceptualization, methodology, writing-original draft preparation, M.V. and M.F.; investigation and visualization, M.V.; formal analysis, M.V. and A.S.-C.-S.; writing—review and editing, M.V., M.F., L.D., and A.S.-C.-S.; supervision, L.D. and M.F.; funding acquisition and project administration, M.F. All authors have read and agreed to the published version of the manuscript.

Funding: We would like to thank the Conseil Régional de La Réunion, Reunion Island, France, for the financial support of research activities dedicated to microbial pigments (DIRED 20140704 - COLORMAR Program and allocation régionale de recherche n° 2016020577/181733).

Acknowledgments: Authors are thankful to J. Prakash Maran for his valuable advices about the design of experiments and Cathie Milhau for technical support.

Conflicts of Interest: The authors declare no conflict of interest. The funders had no role in the design of the study; in the collection, analyses, or interpretation of data; in the writing of the manuscript, or in the decision to publish the results.

Appendix A

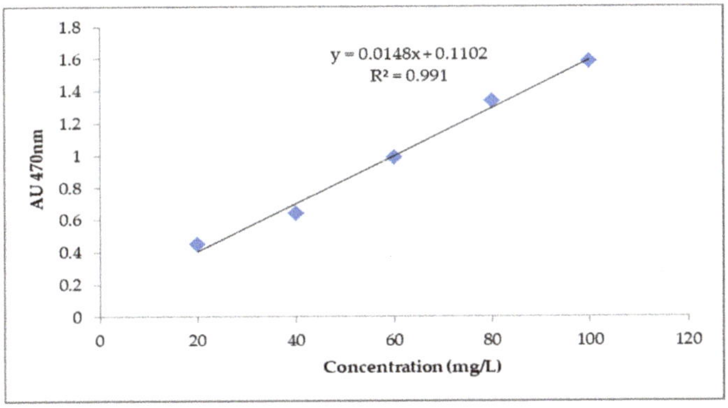

Figure A1. Standard curve for quinizarin at 470 nm.

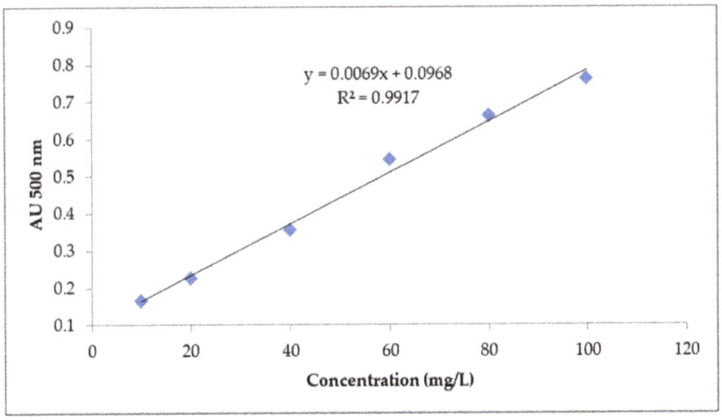

Figure A2. Standard curve for red yeast rice pigment (RYrp) at 500 nm.

Appendix B

Table A1. Regression coefficient (RC) of the models and their statistical parameters on the responses (Xi: coded values of independent variables (i = 1: pH; i = 2: temperature, i = 3: agitation speed; i = 4: fermentation time) and their interactions.

Source	DF	Orange Pigment Yield		Red Pigment Yield		Dry Biomass Weight	
		RC	*p* Value *	RC	*p* Value *	RC	*p* Value *
Model	14	1.74	<0.0001	1.92	<0.0001	6.80	<0.0001
X_1	1	0.35	<0.0001	0.28	<0.0001	0.77	<0.0001
X_2	1	−0.11	0.0009	−0.090	0.0058	−0.77	<0.0001
X_3	1	0.27	<0.0001	0.20	<0.0001	1.24	<0.0001
X_4	1	−0.029	0.2767	−0.015	0.5901	−0.17	0.1837
X_{12}	1	−0.11	0.0245	−0.087	0.0927	−0.24	0.2788
X_{13}	1	−0.19	0.0009	−0.17	0.0031	−0.02	0.9224
X_{14}	1	0.11	0.0245	0.075	0.1394	−0.61	0.0117
X_{23}	1	0.41	<0.0001	0.31	<0.0001	1.18	<0.0001
X_{24}	1	0.000	1.000	−0.019	0.7011	0.55	0.0210
X_{34}	1	0.10	0.0418	0.071	0.1615	0.17	0.4359
X_1^2	1	−0.42	<0.0001	−0.33	<0.0001	−1.02	<0.0001
X_2^2	1	−0.68	<0.0001	−0.54	<0.0001	−2.31	<0.0001
X_3^2	1	−0.35	<0.0001	−0.31	<0.0001	−1.35	<0.0001
X_4^2	1	−0.38	<0.0001	−0.32	<0.0001	−0.97	<0.0001
R^2		0.9853		0.9734		0.9714	
$AdjR^2$		0.9706		0.9468		0.9429	
$PreR^2$		0.9172		0.8469		0.8445	
CV%		9.12		7.41		9.41	
Adeq. Pre.		28.30		21.04		22.71	

R^2: Determination coefficient; $AdjR^2$: adjusted R^2; $PreR^2$: predicted R^2; CV%: coefficient of variation; and Adeq. Pre.: predicted adequacy. $p < 0.01$ highly significant; $0.01 < p < 0.05$ significant; and $p > 0.05$ not significant.

References

1. Weiss, B. Synthetic Food Colors and Neurobehavioral Hazards: The View from Environmental Health Research. *Environ. Health Perspect.* **2012**, *120*, 1–5. [CrossRef]
2. Dufossé, L. Microbial production of food grade pigments. *Food Technol. Biotechnol.* **2006**, *44*, 313–323.
3. Dufosse, L. Anthraquinones, the Dr Jekyll and Mr Hyde of the food pigment family. *Food Res. Int.* **2014**, *65*, 132–136. [CrossRef]
4. Kumar, A.; Vishwakarma, H.S.; Singh, J.; Dwivedi, S.; Kumar, M. Microbial pigments: Production and their applications in various industries. *IJPCBS* **2015**, *5*, 203–212.
5. Kamarudin, K.R. Microbial production of food grade pigments–screening and metabolic pathway analysis. *Food Technol. Biotechnol.* **2006**, *44*, 13–23.
6. Caro, Y.; Anamale, L.; Fouillaud, M.; Laurent, P.; Petit, T.; Dufosse, L. Natural hydroxyanthraquinoid pigments as potent food grade colorants: An overview. *Nat. Prod. Bioprospect.* **2012**, *2*, 174–193. [CrossRef]
7. Duran, N.; Teixeira, M.F.; De Conti, R.; Esposito, E. Ecological-friendly pigments from fungi. *Crit. Rev. Food Sci. Nutr.* **2002**, *42*, 53–66. [CrossRef]
8. Ventura, S.P.; Santos-Ebinuma, V.C.; Pereira, J.F.; Teixeira, M.F.; Pessoa, A.; Coutinho, J.A. Isolation of natural red colorants from fermented broth using ionic liquid-based aqueous two-phase systems. *J. Ind. Microbiol. Biotechnol.* **2013**, *40*, 507–516. [CrossRef]
9. Pastre, R.; Marinho, A.M.; Rodrigues-Filho, E.; Souza, A.Q.; Pereira, J.O. Diversity of polyketides produced by *Penicillium* species isolated from *Melia azedarach* and *Murraya paniculata*. *Química Nova* **2007**, *30*, 1867–1871. [CrossRef]
10. Bugni, T.S.; Ireland, C.M. Marine-derived fungi: A chemically and biologically diverse group of microorganisms. *Nat. Prod. Rep.* **2004**, *21*, 143–163. [CrossRef] [PubMed]

11. Osmanova, N.; Schultze, W.; Ayoub, N. Azaphilones: A class of fungal metabolites with diverse biological activities. *Phytochem. Rev.* **2010**, *9*, 315–342. [CrossRef]
12. Kongruang, S. Growth kinetics of biopigment production by Thai isolated *Monascus purpureus* in a stirred tank bioreactor. *J. Ind. Microbiol. Biotechnol.* **2011**, *38*, 93–99. [CrossRef] [PubMed]
13. Teixeira, M.F.; Martins, M.S.; Da Silva, J.; Kirsch, L.S.; Fernandes, O.C.; Carneiro, A.L.; De Conti, R.; Durán, N. Amazonian biodiversity: Pigments from *Aspergillus* and *Penicillium*–characterizations, antibacterial activities and their toxicities. *Curr. Trends Biotechnol. Pharm.* **2012**, *6*, 300–311.
14. Torres, F.A.E.; Zaccarim, B.R.; De Lencastre Novaes, L.C.; Jozala, A.F.; Santos, C.A.d.; Teixeira, M.F.S.; Santos-Ebinuma, V.C. Natural colorants from filamentous fungi. *Appl. Microbiol. Biotechnol.* **2016**, *100*, 2511–2521. [CrossRef]
15. Bhatnagar, I.; Kim, S.-K. Immense Essence of Excellence: Marine Microbial Bioactive Compounds. *Mar. Drugs* **2010**, *8*, 2673–2701. [CrossRef]
16. Kjer, J.; Debbab, A.; Aly, A.H.; Proksch, P. Methods for isolation of marine-derived endophytic fungi and their bioactive secondary products. *Nat. Protoc.* **2010**, *5*, 479–490. [CrossRef]
17. Imhoff, J.F. Natural products from marine fungi—Still an underrepresented resource. *Mar. drugs* **2016**, *14*, 19. [CrossRef]
18. Yilmaz, N.; Houbraken, J.; Hoekstra, E.; Frisvad, J.C.; Visagie, C.; Samson, R. Delimitation and characterisation of *Talaromyces purpurogenus* and related species. *Pers. Mol. Phylogeny Evol. Fungi* **2012**, *29*, 39–54. [CrossRef]
19. Frisvad, J.C.; Yilmaz, N.; Thrane, U.; Rasmussen, K.B.; Houbraken, J.; Samson, R.A. Talaromyces atroroseus, a new species efficiently producing industrially relevant red pigments. *PLoS ONE* **2013**, *8*, e84102. [CrossRef]
20. Mapari, S.A.; Meyer, A.S.; Thrane, U.; Frisvad, J.C. Identification of potentially safe promising fungal cell factories for the production of polyketide natural food colorants using chemotaxonomic rationale. *Microbial Cell Factories* **2009**, *8*, 1. [CrossRef]
21. Nicoletti, R.; Trincone, A. Bioactive compounds produced by strains of *Penicillium* and *Talaromyces* of marine origin. *Mar. Drugs* **2016**, *14*, 37. [CrossRef]
22. Caro, Y.; Venkatachalam, M.; Lebeau, J.; Fouillaud, M.; Dufossé, L. Pigments and colorants from filamentous fungi. In *Fungal Metabolites*, 1st ed.; Merillon, J.-M., Ramawat, K.G., Eds.; Springer: Berlin/Heidelberg, Germany, 2016; pp. 1–70.
23. Musoni, M.; Destain, J.; Thonart, P.; Bahama, J.-B.; Delvigne, F. Bioreactor design and implementation strategies for the cultivation of filamentous fungi and the production of fungal metabolites: From traditional methods to engineered systems/Conception de bioréacteurs et mise en oeuvre de stratégies pour la culture de champignons filamenteux et la production de métabolites d'origine fongique: Des méthodes traditionnelles aux technologies actuelles. *Biotechnol. Agron. Société Environ.* **2015**, *19*, 430.
24. General, T.; Kim, H.-J.; Prasad, B.; Ngo, H.T.A.; Vadakedath, N.; Cho, M.-G. Fungal utilization of a known and safe macroalga for pigment production using solid-state fermentation. *J. Appl. Phycol.* **2014**, *26*, 1547–1555. [CrossRef]
25. Hajjaj, H.; Blanc, P.; Goma, G.; Francois, J. Sampling techniques and comparative extraction procedures for quantitative determination of intra-and extracellular metabolites in filamentous fungi. *FEMS Microbiol. Lett.* **1998**, *164*, 195–200. [CrossRef]
26. Santos-Ebinuma, V.C.; Roberto, I.C.; Teixeira, M.F.S.; Pessoa, A., Jr. Improvement of submerged culture conditions to produce colorants by *Penicillium purpurogenum*. *Braz. J. Microbiol.* **2014**, *45*, 731–742. [CrossRef]
27. Lee, B.-K.; Park, N.-H.; Piao, H.Y.; Chung, W.-J. Production of red pigments by *Monascus purpureus* in submerged culture. *Biotechnol. Bioprocess Eng.* **2001**, *6*, 341–346. [CrossRef]
28. Méndez, A.; Pérez, C.; Montañéz, J.C.; Martínez, G.; Aguilar, C.N. Red pigment production by *Penicillium purpurogenum* GH2 is influenced by pH and temperature. *J. Zhejiang Univ. Sci. B* **2011**, *12*, 961–968. [CrossRef]
29. Orozco, S.F.B.; Kilikian, B.V. Effect of pH on citrinin and red pigments production by *Monascus purpureus* CCT3802. *World J. Microbiol. Biotechnol.* **2008**, *24*, 263–268. [CrossRef]
30. Liu, L.; Zhao, J.; Huang, Y.; Xin, Q.; Wang, Z. Diversifying of chemical structure of native *Monascus* pigments. *Front. Microbiol.* **2018**, *9*, 3143. [CrossRef]
31. Yilmaz, N.; Visagie, C.M.; Houbraken, J.; Frisvad, J.C.; Samson, R.A. Polyphasic taxonomy of the genus Talaromyces. *Studies Mycol.* **2014**, *78*, 175–341. [CrossRef]

32. Zhang, G.; He, L.; Hu, M. Optimized ultrasonic-assisted extraction of flavonoids from *Prunella vulgaris* L. and evaluation of antioxidant activities in vitro. *Innov. Food Sci. Emerg. Technol.* **2011**, *12*, 18–25. [CrossRef]
33. Yang, Z.; Zhai, W. Optimization of microwave-assisted extraction of anthocyanins from purple corn (*Zea mays* L.) cob and identification with HPLC–MS. *Innov. Food Sci. Emerg. Technol.* **2010**, *11*, 470–476. [CrossRef]
34. Parajó, J.; Santos, V.; Domínguez, H.; Vázquez, M. NH4OH-Based pretreatment for improving the nutritional quality of single-cell protein (SCP). *Appl. Biochem. Biotechnol.* **1995**, *55*, 133–149. [CrossRef]
35. Huang, W.; Xue, A.; Niu, H.; Jia, Z.; Wang, J. Optimised ultrasonic-assisted extraction of flavonoids from *Folium eucommiae* and evaluation of antioxidant activity in multi-test systems in vitro. *Food Chem.* **2009**, *114*, 1147–1154. [CrossRef]
36. Hayouni, E.A.; Abedrabba, M.; Bouix, M.; Hamdi, M. The effects of solvents and extraction method on the phenolic contents and biological activities in vitro of Tunisian *Quercus coccifera* L. and *Juniperus phoenicea* L. fruit extracts. *Food Chem.* **2007**, *105*, 1126–1134. [CrossRef]
37. Ge, Y.; Ni, Y.; Yan, H.; Chen, Y.; Cai, T. Optimization of the supercritical fluid extraction of natural vitamin E from wheat germ using response surface methodology. *J. Food Sci.* **2002**, *67*, 239–243. [CrossRef]
38. Sehrawat, R.; Panesar, P.S.; Swer, T.L.; Kumar, A. Response surface methodology (RSM) mediated interaction of media concentration and process parameters for the pigment production by *Monascus purpureus* MTCC 369 under solid state fermentation. *Pigment. Resin Technol.* **2017**, *46*, 14–20. [CrossRef]
39. Venkatachalam, M.; Magalon, H.; Dufossé, L.; Fouillaud, M. Production of pigments from the tropical marine-derived fungi *Talaromyces albobiverticillius*: New resources for natural red-colored metabolites. *J. Food Compos. Anal.* **2018**, *70*, 35–48. [CrossRef]
40. Bas, D.; Boyaci, I.H. Modeling and optimization I: Usability of response surface methodology. *J. Food Eng.* **2007**, *78*, 836–845. [CrossRef]
41. Bezerra, M.A.; Santelli, R.E.; Oliveira, E.P.; Villar, L.S.; Escaleira, L.A. Response surface methodology (RSM) as a tool for optimization in analytical chemistry. *Talanta* **2008**, *76*, 965–977. [CrossRef]
42. Ferreira, S.C.; Bruns, R.; Ferreira, H.; Matos, G.; David, J.; Brandao, G.; Da Silva, E.P.; Portugal, L.; Dos Reis, P.; Souza, A. Box-Behnken design: An alternative for the optimization of analytical methods. *Anal. Chim. Acta* **2007**, *597*, 179–186. [CrossRef]
43. Karichappan, T.; Venkatachalam, S.; Jeganathan, P.M. Analysis of efficiency of *Bacillus subtilis* to treat bagasse based paper and pulp industry wastewater-A novel approach. *J. Korean Chem. Soc.* **2014**, *58*, 198–204. [CrossRef]
44. Derringer, G.; Suich, R. Simultaneous Optimization of Several Response Variables. *J. Qual. Technol.* **1980**, *12*, 214–219. [CrossRef]
45. Ekren, O.; Ekren, B.Y. Size optimization of a PV/wind hybrid energy conversion system with battery storage using response surface methodology. *Appl. Energy* **2008**, *85*, 1086–1101. [CrossRef]
46. Venkatachalam, M.; Gérard, L.; Milhau, C.; Vinale, F.; Dufossé, L.; Fouillaud, M. Salinity and Temperature Influence Growth and Pigment Production in the Marine-Derived Fungal Strain *Talaromyces albobiverticillius* 30548. *Microorganisms* **2019**, *7*, 10. [CrossRef] [PubMed]
47. Afshari, M.; Shahidi, F.; Mortazavi, S.A.; Tabatabai, F.; Es' haghi, Z. Investigating the influence of pH, temperature and agitation speed on yellow pigment production by *Penicillium aculeatum* ATCC 10409. *Nat. Prod. Res.* **2015**, *29*, 1300–1306. [CrossRef] [PubMed]
48. Cho, Y.; Park, J.; Hwang, H.; Kim, S.; Choi, J.; Yun, J. Production of red pigment by submerged culture of *Paecilomyces sinclairii*. *Lett. Appl. Microbiol.* **2002**, *35*, 195–202. [CrossRef]
49. Bae, J.-T.; Sinha, J.; Park, J.-P.; Song, C.-H.; Yun, J.-W. Optimization of submerged culture conditions for exo-biopolymer production by *Paecilomyces japonica*. *J. Microbiol. Biotechnol.* **2000**, *10*, 482–487.
50. Babitha, S. Microbial Pigments. In *Biotechnology for Agro-Industrial Residues Utilisation: Utilisation of Agro-Residues*; Singh nee' Nigam, P., Pandey, A., Eds.; Springer: Berlin/Heidelberg, Germany, 2009; pp. 147–162.
51. Chadni, Z.; Rahaman, M.H.; Jerin, I.; Hoque, K.; Reza, M.A. Extraction and optimisation of red pigment production as secondary metabolites from *Talaromyces verruculosus* and its potential use in textile industries. *Mycology* **2017**, *8*, 48–57. [CrossRef]
52. Wong, H.; Lin, Y.; Koehler, P. Regulation of growth and pigmentation of *Monascus purpureus* by carbon and nitrogen concentrations. *Mycologia* **1981**, 649–654. [CrossRef]

53. Yongsmith, B.; Tabloka, W.; Yongmanitchai, W.; Bavavoda, R. Culture conditions for yellow pigment formation by *Monascus* sp. KB 10 grown on cassava medium. *World J. Microbiol. Biotechnol.* **1993**, *9*, 85–90. [CrossRef]
54. Babitha, S.; Soccol, C.R.; Pandey, A. Solid-state fermentation for the production of *Monascus* pigments from jackfruit seed. *Bioresour. Technol.* **2007**, *98*, 1554–1560. [CrossRef]
55. Gunasekaran, S.; Poorniammal, R. Optimization of fermentation conditions for red pigment production from *Penicillium* sp. under submerged cultivation. *Afr. J. Biotechnol.* **2008**, *7*, 1894–1898. [CrossRef]
56. Chen, M.-H.; Johns, M.R. Effect of pH and nitrogen source on pigment production by *Monascus purpureus*. *Appl. Microbiol. Biotechnol.* **1993**, *40*, 132–138. [CrossRef]
57. Jafari, A.; Sarrafzadeh, M.; Alemzadeh, I.; Vosoughi, M. Effect of stirrer speed and aeration rate on the production of glucose oxidase by *Aspergillus niger*. *J. Biol. Sci.* **2007**, *7*, 270–275.
58. Mohamed, M.S.; Mohamad, R.; Manan, M.A.; Ariff, A.B. Enhancement of red pigment production by *Monascus purpureus* FTC 5391 through retrofitting of helical ribbon impeller in stirred-tank fermenter. *Food Bioprocess. Technol.* **2012**, *5*, 80–91. [CrossRef]
59. Venkatachalam, M.; Zelena, M.; Cacciola, F.; Ceslova, L.; Girard-Valenciennes, E.; Clerc, P.; Dugo, P.; Mondello, L.; Fouillaud, M.; Rotondo, A. Partial characterization of the pigments produced by the marine-derived fungus *Talaromyces albobiverticillius* 30548. Towards a new fungal red colorant for the food industry. *J. Food Compos. Anal.* **2018**, *67*, 38–47. [CrossRef]

microorganisms

MDPI

Article

Evaluation of Filamentous Fungi and Yeasts for the Biodegradation of Sugarcane Distillery Wastewater

Graziella Chuppa-Tostain [1], Melissa Tan [2,3], Laetitia Adelard [4], Alain Shum-Cheong-Sing [2], Jean-Marie François [5], Yanis Caro [2,3] and Thomas Petit [2,3,*]

1 Competitiveness Cluster Qualitropic, 5 rue André Lardy, 97438 Sainte-Marie, Réunion Island, France; graziella.tostain@qualitropic.fr

2 Laboratoire de Chimie et Biotechnologies des Produits Naturels, Université de la Réunion, CHEMBIOPRO (EA 2212), 15 Avenue René Cassin, 97490 Sainte Clotilde, Réunion Island, France; melissa.tan@univ-reunion.fr (M.T.); alain.shum@univ-reunion.fr (A.S.-C.-S.); yanis.caro@univ-reunion.fr (Y.C.)

3 Département HSE, IUT de la Réunion, 40 Avenue de Soweto Terre-Sainte, BP 373, 97455 Saint-Pierre CEDEX, Réunion Island, France

4 Laboratoire de Physique et Ingénierie Mathématique pour l'Energie et l'EnvironnemeNT (PIMENT), Université de la Réunion, 117 rue Général Ailleret, 97430 Le Tampon, Réunion Island, France; laetitia.adelard@univ-reunion.fr

5 Laboratoire d'Ingénierie des Systèmes Biologiques et des Procédés, INSA de Toulouse, UMR INSA/CNRS 5504—UMR INSA/INRA 792, 135 Avenue de Rangueil, CEDEX 4, 31077 Toulouse, France; fran_jm@insa-toulouse.fr

* Correspondence: thomas.petit@univ-reunion.fr; Tel.: +33-262-692-65-1148

Received: 20 September 2020; Accepted: 4 October 2020; Published: 15 October 2020

Abstract: Sugarcane Distillery Spent Wash (DSW) is among the most pollutant industrial effluents, generally characterized by high Chemical Oxygen Demand (COD), high mineral matters and acidic pH, causing strong environmental impacts. Bioremediation is considered to be a good and cheap alternative to DSW treatment. In this study, 37 strains of yeasts and filamentous fungi were performed to assess their potential to significantly reduce four parameters characterizing the organic load of vinasses (COD, pH, minerals and OD_{475nm}). In all cases, a pH increase (until a final pH higher than 8.5, being an increase superior to 3.5 units, as compared to initial pH) and a COD and minerals removal could be observed, respectively (until 76.53% using *Aspergillus terreus* var. *africanus* and 77.57% using *Aspergillus niger*). Depending on the microorganism, the OD_{475nm} could decrease (generally when filamentous fungi were used) or increase (generally when yeasts were used). Among the strains tested, the species from *Aspergillus* and *Trametes* genus offered the best results in the depollution of DSW. Concomitant with the pollutant load removal, fungal biomass, with yields exceeding 20 $g \cdot L^{-1}$, was produced.

Keywords: sugarcane; distillery waste water; molasses spent wash; vinasse; fungi; yeasts; bioremediation; COD; discoloration

1. Introduction

In 2012, 83 billion liters of ethanol was produced worldwide, from which a third was from Brazil [1,2]. The European directive 2009/28/CE, relating to the promotion of the use of energy from renewable resources, also called RED (Renewable Energy Directive), sets, for each Member State, a binding target of a 10% share of renewable energy in the transport field in 2020. Biofuels production from waste, residues, non-food cellulosic material and lignocellulosic origin is particularly incited [3]. In Reunion Island, the sugarcane industry is one of the most important agricultural and economic activities of the French oversea department, and is located in the Indian Ocean [4].

In 2011, the three rum distilleries still active on the island produced 106,430 Hectoliters of Pure Alcohol (HPA) from sugarcane fermentation and molasses distillation [5]. The rum production is accompanied by the generation of a stillage called vinasse or distillery spent wash (DSW). Owing to the nature of the process, DSW are effluents with a high pollutant load, meaning high Chemical Oxygen Demand (COD, generally in the range of 100 to 150 g O_2 L^{-1}), a low pH (between 4.5 and 5.5) and a high optical density [6–8]. DSW also generally contains a high amount of potential nutrients such as nitrogen, phosphorous, potassium, sulfur and a large amount of micronutrients like calcium or magnesium [8]. Moreover, it is characterized by a dark brown color due to the presence of colored molecules such as phenolic acids, caramels from overheated sugars and furfurals from acid hydrolysis and melanoidins [9–11].

Melanoidins are dark brown polymers of low and high molecular weights that result from the Maillard reaction. This reaction, which happens at high temperatures and low pH, is a non-enzymatic browning reaction that results from condensation between reducing sugars and amino compounds [12], leading to molecules with complex structures [13]. It has been shown that less than 10% of the melanoidins present in the stillage can be degraded through conventional anaerobic–aerobic treatments [7,14]. Godshall (1999) showed that the amount of phenolic acids is higher in cane molasses stillage in comparison to beet molasses [15]. Depending on the sugarcane nature and the industrial processes used, intrinsic composition of DSW can vary significantly. Indeed, recent studies concerning COD show that vinasses from Indian and Mexican distilleries have COD around 104 and 121 g O_2 L^{-1}, respectively, whereas the COD of vinasses from Brazilian distilleries are two and three times less, i.e., 42 g O_2 L^{-1}. Moreover, proportions of potassium contained in DSW can range from 2.3 to 8.77 g·L^{-1} [16–19]. These characteristics make DSWs hazardous compounds when, for example, they are discharged in natural waterways. DSW can cause significant environmental problems by reducing the oxygenation of the water, causing eutrophication of contaminated waterways and creating toxic effects on aquatic organisms. Due to the presence of putrescible organics like skatole, indole and other sulfur compounds, DSW produces an obnoxious smell [20]. European and French regulations describe the strict specifications for industrial effluents. Their COD value must be less than 125 mg O_2 L^{-1} [21].

Among the different ways of treatment, bioremediation offers a good perspective as it constitutes a cheaper and easier technique compared to physico-chemical technologies. Different types of microorganisms can be used for bioremediation, namely bacteria, microalgae, yeasts or filamentous fungi [7,8,22]. Due to their rapid growth, yeasts are widely used in DSW treatment. Among the 203 yeast strains tested, Akaki and collaborators showed that strains from *Hansenula, Debaryomyces* and *Rhodotorula* genus could remove a range of 32–38% of COD contained in a diluted and supplemented DSW medium [23]. Unlike yeasts, filamentous fungi have slower growth but their broad extracellular hydrolytic enzymes allow for the assimilation of complex carbohydrates without prior hydrolysis by another technique. Moreover, they are less sensitive to nutrients, aeration, temperature and pH variations [24]. In their study, Sirianuntapiboon and collaborators described the potential of 228 fungal strains to discolor molasses. Among them, nine strains, including four species from *Aspergillus* genus, a strain of *Trametes versicolor* and four other unidentified strains, showed discoloration yields above 50% [25,26].

A literature review indicated that species of filamentous fungi such as Penicillium, Aspergillus (A. niger, A. oryzae and A. terreus), Galactomyces geotrichum, Trametes versicolor, Phanerochaete chrysosporium and Flavodon flavus and the yeasts Candida tropicalis, P. jadinii and Issatchenkia orientalis had already been highlighted because of their ability to remove refractory compounds from distillery wastewater [7,8,20,27–36]. However, most of these strains were tested in different and heterogeneous conditions and not all on vinasse from sugarcane distilleries. In this study, we report a comprehensive and standardized screening program, including 37 strains of yeasts and filamentous fungi selected from the abovementioned literature and from our own selection (provided from laboratory fungal strain collections) for their ability to degrade complex compounds of DSW from sugarcane. The capacity of each strain to grow and metabolize the substrates contained in vinasse was evaluated by following the evolution of a number of physicochemical parameters such as pH, COD, mineral matter, optical density and microbial biomass production.

2. Materials and Methods

2.1. Biological Material and Growth Conditions

Sugarcane Distillery Spent Wash (DSW) was collected directly at the column outlet (between 85 °C and 100 °C) from "Rivière du Mât" sugarcane distillery, Saint-Benoit, Reunion Island. After cooling, DSW was stored in small sterile bags at −20 °C until used. The yeasts and filamentous fungi strains used in this study were purchased from BCCM (Brussels, Belgium) strain collections (Table 1). The 37 selected strains used in this study were chosen according to their specific properties to degrade complex molecules and particularly DSW.

In order to be able to measure the depollution potential of the strains, independently of their capacity to grow on this medium, we decided to uncouple the depollution ability of the strain and the biomass production. Pre-cultures were therefore prepared by inoculating a full loop of 48 h growing cells on basal agar plate (PDA—Potatoes Dextrose Agar from Biotop) in sterile Malt Agar broth (MA—Merck, Germany) and incubated for 72 h. At the end of the incubation in MA broth, the total biomass formed was aseptically harvested by centrifugation, washed twice in sterile milliQ water and used for the inoculation of the main cultures. The main culture experiments contained 50 mL of autoclaved DSW (121 °C, 20 min). Yeast and filamentous fungi biomass harvested from pre-cultures were inoculated at a concentration of 10^5 cells (yeasts) or spores (filamentous fungi) per mL in the main culture flasks and incubated under aerobic conditions at 110 rpm and 25 °C for 10 days. Biomass and culture broth were then separated by filtration (filamentous fungi) or centrifugation (yeasts). Filtrations were performed on a Buchner system using Whatman filter No. 1. All centrifugations were carried out for 15 min at 8500 rpm.

2.2. Physico-Chemical Analysis

Cells harvested by centrifugation or filtration were incubated for 24 h at 105 °C for biomass determination. Supernatants and filtrates were stored at −20 °C before being used. The pH of broth medium during fermentation was measured using a pH-meter Denver Instrument (Germany). To evaluate the mineral content, ashes were measured by the incineration of 10 g of broth medium at 550 °C for 3 h in a muffle furnace Nabertherm Controller B170 (Lilienthal, Germany) [37]. COD measurements were carried out using Hanch Lange diagnostic kits (LCK 914) and measured spectrophotometrically with a DR 2800 spectrophotometer (Hach Lange, Dusseldorf, Germany). When necessary, the samples were adequately diluted with sterile deionized water and analyzed according to the manufacturer's instructions.

Absorbance of filtrates or supernatants was measured at 475 nm (corresponding to melanoidins) using a spectrophotometer Genesys 10 UV Scanning (Waltham, MA, USA) according to [19]. The discoloration yield was calculated according to the following equation:

$$\text{Discoloration } (\%) = \frac{I - F}{I}, \tag{1}$$

with I = Initial absorbance (Control) and F = Absorbance after aerobic fermentation. All assays were performed in triplicates.

2.3. Chemometrics

Multivariate statistics, including Principal Component Analysis (PCA) and Hierarchical Cluster Analysis (HCA), were employed to investigate the relationships among species with similar performances concerning biomass production and variations of pH, mineral content, COD and OD_{475nm}. PCA using Pearson correlation is a statistical method used in order to combine the original parameters (physicochemical variables) into several new uncorrelated components without losing significant information. The aim of this statistical method is to explain the variance–covariance structure of an experimental data set using a new set of coordinate systems. Every new principal component consisted of the linear combination of the original variables [38]. This method enabled us to define the characteristics of specific groups of strains. Hierarchical cluster analysis (HCA) was then used to identify the strains belonging to these groups. HCA is a statistical method to search for homogeneous clusters based on measured parameters. The hierarchical clustering process is represented by a dendrogram, in which each step of the clustering process is illustrated by a connection in the tree. Differences between these classes were tested with average Euclidean distances using the Ward method based on a variance approach. This Ward method provides a simple approach to approximate, for any given number of clusters, the partition minimizing the within-cluster inertia or "error sum of squares". In this study, the method was performed with the aim of minimizing the sum of the squares of any two clusters that could be formed at each step. The clusters were then fused in order to reduce the variability within a cluster. Further, the fusion of two clusters resulted in a minimum increase of the "error sum of squares" [39]. These analyses were performed thanks to XLSTAT programs (Addinsoft, Inc., Paris, France).

3. Results and Discussion

3.1. Effect of Aerobic Treatment on Chemical Oxygen Demand (COD)

The effect of the treatment of DSW on COD was found to be highly dependent of the strains used for aerobic treatment (Table 1). Aerobic fermentation of DSW by *Phanerochaete chrysosporium, Flavodon flavus, Fusarium proliferatum* and *Gibberella fujikuroi* appeared to be less efficient strains for COD reduction, with 23.5%, 28%, 34% and 38%, respectively, whereas *Aspergillus terreus var africanus, A. parasiticus, Trametes hirsuta, T. versicolor* and *A. terreus var. terreus* showed the highest decrease in COD (76.53%, 74.60%, 74.01%, 73.64% and 73.5%, respectively). Notably, the 9 *Aspergillus* and anamorph strains used in this study were among the most effective strains for COD reduction (COD reduction was higher than 65% for all 8 *Aspergillus* strains and 58.65% for *Fennellia flavipes*), indicating that these strains are particularly interesting for their reduction of the pollution load of DSW. COD reduction by *Pichia jadinii* and *Penicillium* sp. could reach 40.91% and circa 62%, respectively.

Some of our results were consistent with other published works. For instance, Gonzalez et al. (2000) reported a high COD reduction (62%) on diluted molasses spent wash treated by *Trametes* spp. [14]. Benito et al. (1997) also found that *T. versicolor* was able to reduce COD by more than 70% on supplemented sugar beet molasses [40]. Similarly, a reduction of 46% and 65% of COD was found for *P. jadinii* and *Penicillium* sp., respectively [30,41]. Aerobic treatment of cane molasses stillage with *A. niger* and *A. oryzae* led to a COD reduction of up to 78% and 88%, respectively [31–33]. On the contrary, Garcia et al. (1997) found that *A. terreus* lowered the COD of DSW by only 29% [34]. Surprisingly, our results using *P. chrysosporium* and *F. flavus* were found to be well below the observed values from the literature with, respectively, a 73% COD reduction on DSW supplemented with yeast extract and 80% on diluted DSW [42,43]. This difference may be explained by the fact that these studies were carried out on diluted and supplemented DSW, while we used crude DSW in our study. Moreover, COD reduction is generally concomitant with the discoloration of the vinasse. In their study, Fahy and collaborators (1997) showed that a sugar addition in the medium could significantly improve the depollution rate of vinasse by *P. chrysosporium* [44]. From these results, the efficiency of the strains to reduce COD is strongly dependent of the origin of the vinasse used (beet or cane for example) and their complementation with other sources of nutriments.

Table 1. Strains used in this study, effects of aerobic treatment of DSW on physicochemical parameters and biomass production.

Strain Number	Strains (Genera/Specie)	MUCL Reference Number	Reduction of COD [1] (%)	Effect on OD at 475 nm [2] (%)	Reduction of Minerals Content [3] (%)	Final pH [4]	Biomass Production [5] ($g \cdot L^{-1}$)
S1	*Arthroderma otae*	MUCL 41713	59.22	98.26	36.99	6.91	18.11
S2	*Aspergillus alutaceus*	MUCL 39539	69.23	58.12	73.49	7.91	21.7
S3	*Aspergillus flavus*	MUCL 19006	70.86	80.00	20.88	8.72	19.77
S4	*Aspergillus itaconicus*	MUCL 31306	73.23	64.64	40.94	7.64	21.41
S5	*Aspergillus niger*	MUCL 19001	70.11	73.04	77.57	8.31	21.25
S6	*Aspergillus oryzae*	MUCL 19009	65.98	77.97	66.62	8.86	24.35
S7	*Aspergillus parasiticus*	MUCL 14491	74.6	57.54	53.66	8.46	24.48
S8	*Aspergillus terreus var africanus*	MUCL 38960	76.53	110.14	66,00	9,00	29.19
S9	*Aspergillus terreus var terreus*	MUCL 38640	73.5	61.16	72.4	9.05	24.90
S10	*Candida albicans*	MUCL 30114	56.56	118.01	29.3	8.69	19.34
S11	*Candida dubliniensis*	MUCL 41201	45.98	152.53	32.18	8.45	10.29
S12	*Candida glabatra*	MUCL 29833	57.34	130.64	26.62	8.12	12.43
S13	*Candida tropicalis*	MUCL 29893	50.41	138.72	20.75	8.42	15.21
S14	*Clavispora lusitanea*	MUCL 29855	54.72	116.84	28.35	7.39	28.56
S15	*Colletotricum graminicola*	MUCL 44764	57.75	92.17	39.23	7.94	11.79
S16	*Cryptococcus albidus*	MUCL 30400	44.89	135.35	30.29	8.13	12.47
S17	*Fennellia flavipes*	MUCL 38811	58.65	75.65	61.46	8.74	17.98
S18	*Flavodon flavus*	MUCL 38427	28.99	77.10	37.48	6.17	14.98
S19	*Fusarium proliferatum*	MUCL 43482	34.44	91.59	53.83	6.37	6.38
S20	*Fusarium sporotrichioides*	MUCL 6133	55.13	140.00	43.81	8.25	8.6
S21	*Galactomyces geotrichum*	MUCL 43077	56.95	104.64	46.39	8.26	6.79
S22	*Gibberella fujikuroi*	MUCL 42883	37.89	106.96	36.58	6.76	4.12
S23	*Gibberella zeae*	MUCL 42841	55.8	109.28	33.52	8.04	20.9
S24	*Issatchenkia orientalis*	MUCL 29849	48.13	139.73	49.56	8.08	25.41
S25	*Komagatella pastoris*	MUCL 31260	69.7	107.41	56.84	7.99	9.1
S26	*Penicillium rugulosum*	MUCL 41583	62.48	86.38	56.28	8.72	2.36
S27	*Penicillium verrucosum*	MUCL 28674	62.09	168.99	49.05	9.03	8
S28	*Phanerochaete chrysosporium*	MUCL 38489	23.51	74.20	70.49	7.01	17
S29	*Pichia angusta*	MUCL 27761	49.52	114.14	20.78	6.53	4.04
S30	*Pichia guilliermondii*	MUCL 29837	54.78	138.38	28.21	7.54	12.23
S31	*Pichia jadinii*	MUCL 30058	40.91	141.08	46.33	8.21	14.67
S32	*Pseudozyma antarctica*	MUCL 47637	51.33	136.03	22.11	8.9	0.75
S33	*Rhizopus microsporus var oligosporus*	MUCL 31005	67.32	95.94	52.04	8.85	14.66
S34	*Saccharomyces cerevisiae*	MUCL 39449	55.84	190.91	41.17	8.88	11.64
S35	*Thanatephorus cucumeris*	MUCL 43254	65.28	105.22	44.63	6.66	16.32
S36	*Trametes hirsuta*	MUCL 40169	74.01	57.54	62.86	7.8	29.4
S37	*Trametes versicolor*	MUCL 44890	73.64	67.54	39.05	7.79	25.96

[1] Reduction of COD corresponded to ratio between COD removal during aerobic fermentation and initial COD of crude DSW based upon 100%; [2] Evolution of OD measured at 475 nm was the relation between OD after aerobic fermentation and OD of crude DSW based upon 100%; [3] Reduction of minerals content corresponded to ratio between minerals removal during aerobic fermentation and initial minerals content of crude DSW based upon 100%; [4] Final pH measured in DSW broth after 10 days of fermentation; [5] Biomass production was determined according to Materials and Methods (Sections 2.1 and 2.2) and corresponded to the difference between biomass obtained after DSW filtration at the end of 10 days fermentation and biomass inoculated on DSW at the beginning of the fermentation. Biomass inoculated was the sum between fungal cells obtained after growth on MA broth and total solids content naturally contained on DSW.

3.2. Effect of Aerobic Treatment on Colour

The effect of aerobic treatment of DSW on color was studied using optical density of DSW supernatant at 475 nm [19]. Consistent with the COD reduction, we found that the strains showing the highest reduction of color belong to *Aspergillus* and *Trametes* genus (Table 1). For instance, *Aspergillus parasiticus, A. alucateus, A. terreus var. terreus* and *A. itaconicus* led to a decrease of $OD_{475\ nm}$ up to 42.46%, 41.88%, 38.84% and 35.36%, respectively (Table 1). Similarly *Trametes hirsuta* and *T. versicolor* reached up to 42.46 and 32.46% of decolourisation of DSW. DSW treatment with *A. flavus, A. niger, A. oryzae, Fenellia flavipes, Flavodon flavus* and *Phanerochaete chrysosporium* also led to decolourization of DSW but to a lesser extent ($OD_{475\ nm}$ reduction was comprised of between 20 and 27%). Surprisingly, we observed that aerobic treatment of DSW by the yeasts (belonging to *Candida, Clavispora, Cryptococcus, Galactomyces, Issatchenkia, Komagatella, Pichia* and *Saccharomyces* genus) and by *Fusarium sporotrichoides* and *Penicillium verrucosum* resulted in small to high increase of $OD_{475\ nm}$. The most important increases of colourization were obtained for *Saccharomyces cerevisiae* (90.91%), *P. verrucosum* (68.99%), *C. dubliniensis* (52.53%), *P. jadinii* (41.08%), *F. sporotrichoides* (40%), *Issatchenkia orientalis* (39.73%), *C. tropicalis* (38.72%), *P. guilliermondii* (38.38%), *Pseudozyma antarctica* (36.03%), *Cryptococcus albidus* (35.35%) and *C. glabrata* (30.64%). The other yeasts species (*C. albicans, P. angusta, Komagatella pastoris, Galactomyces geotrichum, Rhizopus microsporus. var oligosporus* and *Thanatephorus cucumeris*) showed only limited colorization of the broth (less than 18%). A study has already noticed the increase of color after treatment. Kumar and collaborators (1998) reported that the optimum discoloration was closely related to the optimal growth and that the overall discoloration was obtained in the pH range of between five and eight, whereas at extreme pH levels, an increase in color was observed [42]. We can then hypothesize that the coloration observed in this study is probably due to the high final pH reached at the end of the process (Table 1). Finally, Kumar and collaborators (1998) reported that optimal discoloration was closely related to optimal growth and that overall discoloration was obtained in the pH range of five to eight, while at extreme pH levels, an increase in color was observed [42].

With respect to the color of DSW treated with *A. niger, F. flavus, T. versicolor* and *P. chrysosporium*, our results showed a lower impact, as compared to the literature. One of the most studied fungi for potential decolourization of distillery effluent was *Aspergillus sps. Aspergillus fumigatus* G-2-6, *Aspergillus niger, A. niveus, A. fumigatus* UB260 had an average of 55–79% decolourization [45–50]. Miranda et al. (1996) showed that, under optimal nutrient concentrations, aerobic treatment using *A. niger* allowed for a decolourization of beet molasses by 69%. Furthermore, they reported that 83% of the total color removed was eliminated biologically and 17% by adsorption on the mycelium [47]. Under optimal pH, Patil and collaborators (2003) showed that a melanoidin solution was decolourized from 60% to 72% by *A. niger* immobilized cells [51]. Raghukumar et al. (2001) reported that a diluted cane molasses stillage treated with *F. flavus* could reach up to 80% decolourization [43]. Further, aerobic treatment of a diluted molasses spent wash by *T. versicolor* had a decolourization yield of 53% [52]. When beet molasses were used, the decolourization yielded 58–81% $OD_{475\ nm}$ reduction. From 53.5 to 80% of decolourization of supplemented molasses spent wash treated by *P. chrysosporium* was reported [40,42]. Moreover, Fahy and coworkers (1997) demonstrated that the further addition of a carbon source like glucose in a 6.25% molasses spent wash medium strongly enhanced the decolourization yield from 49 to 80% by *P. chrysosporium* [44].

Some of our results were somewhat contradictory with other published works. For instance, a study showed that *C. tropicalis* could reach 75% decolourization level of a supplemented molasses spent wash when incubated at 45 °C [19]. Likewise, treatment of distillery spent wash with the ascomycetes of *Penicillium* genus resulted in about 50% reduction of the color [46]. With reference to *Thanatephorus cucumeris* (Rhizoctonia sp. D-90), Sirianuntapiboon and coworkers (1995) reported the decolourization of a melanoidin medium (molasses) by 87.5% thanks to an absorption mechanism. Indeed, the pigments were accumulated in cytoplasm and around the cell membrane before their degradation by intracellular enzymes [53]. To the best of our knowledge, no studies have focused on the decolourization of DSW by *Galactomyces geotrichum, Rhizopus microsporus, Giberella fujikuroi* and

Fusarium sp. Notwithstanding this, considering their use for molasses decolourization, *Galactomyces geotrichum* and *Rhizopus microsporus var. oligosporus* could achieve a color reduction of diluted molasses of up to 87% and 38%, respectively [36]. Similarly, Seyis and Subasioglu (2009) showed that molasses decolourization by *Gibberella fujikuroi and Fusarium* species were not successful [54]. The $OD_{475\ nm}$ increase could result from pigments repolymerization, from a higher rate of nutriment consumption and from production by the microorganism of molecules that also absorb at this wavelength [55–57].

3.3. Effect of Aerobic Treatment on pH

Compared to the initial pH of the DSW broth (in the range of 4.77–4.95), all microbial treatments of crude DSW led to a significant increase of final pH (Table 1). Alkalinisation of the medium may be the result of an ammonium release during the assimilation of nitrogen source like proteins for the microorganism growth or a consumption of organic acids or reducing sugar present in DSW [55]. Among the 37 strains tested in this study, 22 could achieve a pH final value above 8 units. Among the best alkalinising strains, maximum pH (>9 units) was reached for DSW incubated with *A. terreus var. africanus* (9.05), *P. verrucosum* (9.03) and *A. terreus var. terreus* (9.0). More generally, among the *Aspergillus* and anamorphs genera, seven strains were found to reach a pH of above 8.3 units.

Several studies have shown that the degradation of melanoidins, which is related to discoloration, tends to increase with alkaline pH. For instance, Hayase and collaborators (1984) reported that the discoloration of melanoidin occurred more rapidly at alkaline pH than at acidic or neutral pH and could reach up to 94% discoloration at pH 10 [58]. In addition, Mohana and coworkers (2007) reported that melanoidins are less soluble in acidic rather than in alkaline pHs and that pHs less than or greater than 7 units lead to a decrease of discoloration activity [59]. Similarly, Agarwal and collaborators (2010) claimed that melanoidins were more soluble at alkaline pH [60].

Contrary to these studies, we found no specific link between pH and (dis)colorisation of DSW was shown (see Table 1). Indeed, DSW aerobic fermentations using *A. terreus var. terreus* and *Penicillium verrucosum* led, in both cases, to an alkalinisation of the supernatant pH of DSW up to 9 units, but in the first case, an $OD_{475\ nm}$ decrease of 38.84% could be noticed, whereas an $OD_{475\ nm}$ increase of 68.99% was observed in the second case. Likewise, *A. oryzae* and *F. flavus* induced a decolourization of DSW by about 22%, but an alkalinisation of pH of 8.86 and 6.17, respectively.

As few sugar remain in residues like sugarcane molasses after sugar fabrication, the ethanol production from these residues conduced the use of harsher processing steps to depolymerize the structural polysaccharides. These processes result in side reaction products and in the acidification of the medium that are potentially inhibitory to microbial growth. Therefore, anaerobic digestion of the vinasse produced from sugarcane molasses may be fraught with problems [61]. As aerobic fermentation of DSW by yeasts and filamentous fungi bring about alkalinisation of DSW, the anaerobic digestion of the latter could be improved.

3.4. Biomass Production and Mineral Content of DSW after Aerobic Treatment

The biomass production of the 37 yeasts and filamentous fungi strains was measured during growth on crude DSW (Table 1). Microorganisms that presented the best production of biomass during aerobic treatment of DSW were *Trametes hirsuta* (29.40 $g{\cdot}L^{-1}$), *A. terreus var. africanus* (29.19 $g{\cdot}L^{-1}$), *Clavispora lusitaniae* (28.56 $g{\cdot}L^{-1}$), *T. versicolor* (25.96 $g{\cdot}L^{-1}$) and *Issatchenkia orientalis* (25.41 $g{\cdot}L^{-1}$). In the same way as COD, OD_{475nm} and pH, we again found that the *Aspergillus* genus was particularly efficient in biomass production on crude DSW. The 9 *Aspergillus* anamorphs strains showed that biomass productions, after 10 days incubation, were comprised of between 17.98 $g{\cdot}L^{-1}$ (*Fennellia flavipes*) and 29.19 $g{\cdot}L^{-1}$ (*Aspergillus terreus var africanus*). Smaller amounts of biomass were observed for aerobic fermentation of crude DSW by the yeasts such as *P. jadinii* and *S. cerevisiae* (14.67 and 11.67 $g{\cdot}L^{-1}$, respectively).

Several studies have concluded that the COD reduction and/or decolourisation of diluted and/or supplemented molasses spent wash from sugarcane or sugar beet feedstocks by strains of *Aspergillus*,

Penicillium, *Candida* and *Pichia* genus was accompanied by a fungal growth on the medium [62]. Biomass productions in DSW treated by *Aspergillus* and anamorphs strains were somewhat higher than those previously reported in literature. For instance, Rosalem and collaborators (1985) showed that biomass production of *Aspergillus niger* grown on DSW could vary from 8 to 13 g·L^{-1} [32]. Likewise, cellular concentration of *Aspergillus oryzae* grown on DSW were comprised between 12 and 17 g·L^{-1} dry weight [31,33]. In their study, Rolz and collaborators (1975) also demonstrated that biomass production by *Penicillium* sp. grown on DSW can reach up to 16 g·L^{-1} [30].

Data from the literature showed that *Issatchenkia orientalis* incubated in DSW supplemented with molasses, $MgSO_4$, urea and H_3PO_4 could only produce a biomass of up to 8 g·L^{-1} [63]. The growth of *S. cerevisiae* on molasses stillage reached a maximum biomass production of about 12.7 g·L^{-1} [64]. Similarly, growth of *P. jadinii* on DSW supplemented on molasses produced from 9 to 18 g·L^{-1} of dry biomass [65]. Our results therefore clearly indicate that aerobic treatment of crude DSW by these filamentous fungi and yeast strains could achieve a significant reduction of polluting loads of DSW concomitantly with a high production of dry biomass (Table 1) that could be further valuated into added value molecules. Unexpectedly, our study did not reveal a clear link between biomass production and COD reduction (Table 1). This was particularly true for the strains that grow poorly on DSW (biomass production of *P. antarctica*, *P. rugulosum*, *P. angusta* and *G. fujikuroi* were comprised between 0.75 and 4.12 g·L^{-1}), but showed a significant decrease in COD ranging from 38% to 62%. This result indicated that the enzymatic process of the reduction of polluting loads could work independently of the process of using nutriments from DSW for growth.

We also noticed that aerobic treatment by the 37 strains used in this study always resulted in a significant reduction of mineral content of DSW (Table 1). This decrease was considerable after treatment of DSW by *F. flavus* (61.5%), *A. terreus var. africanus* (66%), *A. oryzae* (66.6%), *P. chysosporium* (70.5%), *A. terreus var. terreus* (72.4%), *A. alutaceus* (73.5%) and *A. niger* (77.6%). In agreement with our results for COD, $OD_{475\ nm}$, pH and biomass production, we found that seven out of the nine *Aspergillus* and anamorphs strains showed a mineral reduction in the broth by at least 50%. This result confirmed the high potential of *Aspergillus* genus to efficiently reduce the polluting load of DSW concomitantly with a high valuable biomass production. Aerobic treatment conducted with *C. tropicalis* (20.8%), *P. angusta* (20.8%), *A. flavus* (20.9%), *P. antarctica* (22.1%), *C. glabrata* (26.6%), *P. guilliermondii* (28.2%), *C. lusitanea* (28.4%) and *C. albicans* (29.3%) led to a lesser, but significant decrease in mineral content. The growth of microorganisms is strongly dependent on micronutrients (such as iron, copper, manganese, zinc, and nickel) and macronutrients (like potassium, phosphorus, magnesium, nitrogen, sulphur, and calcium). These nutrients are involved in carbohydrate metabolism, amino-acids and vitamins production, Krebs cycle, nucleic acid production, pigments production and enzyme activities [66,67]. However, the absence of clear relationship between mineral content and biomass production may suggest that other phenomena are involved in the reduction of minerals in the media. For example, mineral content may decrease from precipitation as a consequence of DSW alkalinisation during aerobic treatment.

3.5. Statistical Relationships between Physico-Chemical Parameters

A Principal Component Analysis (PCA) was carried out to group the strains according to their performances on the physico-chemical parameters of DSW (biomass production and variations of pH, minerals content, COD and OD_{475nm}) and we investigated possible correlations between some of them. The Pearson correlation matrix showed that variables were moderately correlated between them (Table 2).

For instance, we detected some correlations for pH and COD reduction (with a Pearson correlation coefficient r of 0.508), reduction of minerals content and effect on $OD_{475\ nm}$ (r = 0.503), biomass production with COD reduction on the one hand (r = 0.466) and the effect on $OD_{475\ nm}$ on the other hand (r = 0.447). Applied to the five original variables, the Cattell's scree diagram [68] highlighted three significant Principal Components (PC) explaining 84.89% of the total variance, 45.62% for PC_1,

26.23% for PC_2 and 13.81% for PC_3 (Appendix A—Table A1). The active coordinates retained by PCA were used to create Figure 1A,B.

Table 2. Pearson correlation matrix.

Parameters	Reduction of COD (%)	Effect on OD at 475 nm (%)	Reduction of Minerals Content (%)	Final pH	Biomass Production (g·L^{-1})
Reduction of COD (%)	1	**0.344**	0.289	**0.508**	**0.466**
Effect on OD at 475 nm (%)	**0.344**	1	**0.503**	−0.197	**0.447**
Reduction of minerals content (%)	0.289	**0.503**	**1**	0.179	**0.355**
Final pH	**0.508**	−0.197	0.179	**1**	0.135
Biomass production (g·L^{-1})	**0.466**	**0.447**	**0.355**	0.135	**1**

Bold values are significantly different from 0 at a significance level α = 0.05.

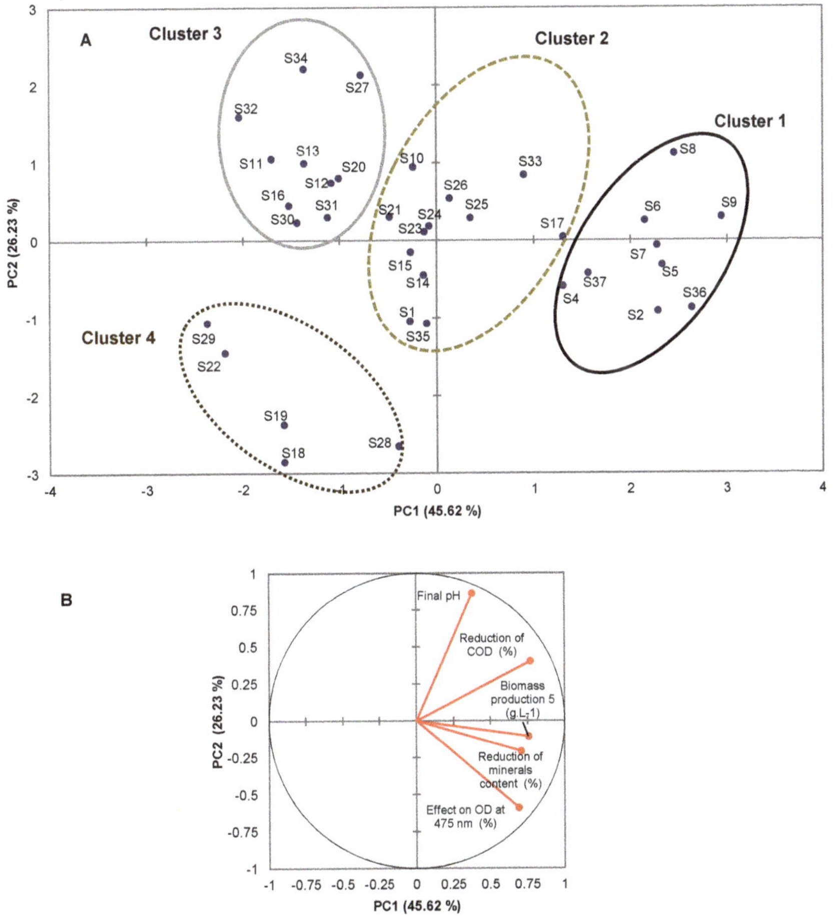

Figure 1. Score plot of PC_2 versus PC_1 (**A**,**B**) for fungal ability for DSW bioremediation.

Principal Component Analysis was performed using XLSTAT (Addinsoft). Predicted groups were correlated to CAH clusters. Cluster 1, consisted in strains S2, S3, S4, S5, S6, S7, S8, S9, S36 and S37 which had the most significant COD decreases and biomass production. Cluster 2 includes strains S1, S10, S14, S15, S17, S21, S23, S24, S25, S26, S33 and S35 that had a COD decrease and biomass production yields less higher than ones of the strains of cluster 1. Cluster 3, consisted in strains S11, S12, S13, S16, S20, S27, S30, S31, S32 and S34, which significantly increase $OD_{475\ nm}$. The remaining strains (S18, S19, S22, S28 and S29) constituted the last group (Cluster 4) and had a less important effect on the final pH.

The eigenvectors of the covariance calculated enabled the defining of three PCs (Table 3). Only the original variables, whose correlation values with the principal components were greater in absolute value than 0.5, were taken into account. The first axis PC_1 was representative of a global average level of the variables and strongly correlated with four of the five parameters (Appendix A—Table A2). These four parameters (COD reduction, biomass production, minerals content reduction and effect on $OD_{475\ nm}$) contributed for 93.77% to PC_1 construction. Additionally, the variables (final pH and the effect on the OD value at 475 nm) are in absolute value the original variables best correlated with the PC2 axis. PC_2 axis (Appendix A—Table A2). It can be noted that PC_2 was mainly built by the pH and the effect on $OD_{475\ nm}$ variables, i.e., 83.53% of contribution to PC_2 construction. Surprisingly, we found that some fungal species, such as *P. antarctica* (S32), had very little growth on DSW (0.75 $g{\cdot}L^{-1}$) despite a high COD consumption and a significant increase of pH, while species like *P. chrysosporium* (S28) showed significant biomass production (17 $g{\cdot}L^{-1}$), concomitant with small pH increase (7.01) and moderate COD consumption (23%). Then the third axis PC_3 was built mainly on biomass production and minerals content reduction (86.56% of the PC_3 construction). The variable reduction of mineral content also greatly contributed to the construction of the PC_3 axis (Appendix A—Table A2). By opposition to *A. flavus* (S3), which turned out to produce a high amount of biomass (19.77 $g{\cdot}L^{-1}$), but a weak minerals consumption (20.88%), *P. rugulosum* (S26) could consume a large amount of mineral content (56.26%) with very little growth on DSW (2.36 $g{\cdot}L^{-1}$) (Figure 1A). These results suggested that a part of the minerals was indeed used for fungal growth, while another part was precipitated due to the alkalinisation of the DSW.

Table 3. Correlations between the parameters and principal components.

Parameters	PC1	PC2	PC3
Reduction of COD (%)	0.769	0.402	−0.234
Effect on OD at 475 nm (%)	0.690	−0.591	−0.017
Reduction of minerals content (%)	0.707	−0.207	0.637
Final pH	0.377	0.864	0.195
Biomass production ($g{\cdot}L^{-1}$)	0.756	−0.107	−0.439

PCA indicated that the strains could be classified into three to four groups. According to hierarchical cluster analysis (HCA), four groups of strains with close characteristics had been defined, explaining 64.62% of the total inter-variance and 35.38% of the total intra-variance (Figure 2). The distribution of the clusters according PC_1 and PC_2 (Figure 1A,B) allowed us to define the common characteristics of strains belonging to the same cluster (Appendix A—Table A3). Cluster 1 including the 8 *Aspergillus* anamorphs strains and the 2 *Trametes* spp. was characterized by aerobic treatment resulting in both high biomass production, high COD and mineral content reductions and a strong impact on $OD_{475\ nm}$, resulting in significant decolourization. Cluster 3 included strains that led to a significant increase of $OD_{475\ nm}$ that could reach 190.9% in comparison to the $OD_{475\ nm}$ of crude DSW and conduced to the lesser mineral consumption. Cluster 4 consisted of strains whose effect on final pH was less important and that brought to a lesser biomass production and COD reduction. The final pH of DSW treated by strains defined in Clusters 1 and 3 were generally above pH = 8 whereas the pH of DSW treated by strains of Cluster 4 had a pH lower than seven. Cluster 2, which gathered all the other strains, was formed by strains that influenced COD and mineral contents and produced biomass on DSW, but less significantly than the strains of Cluster 1.

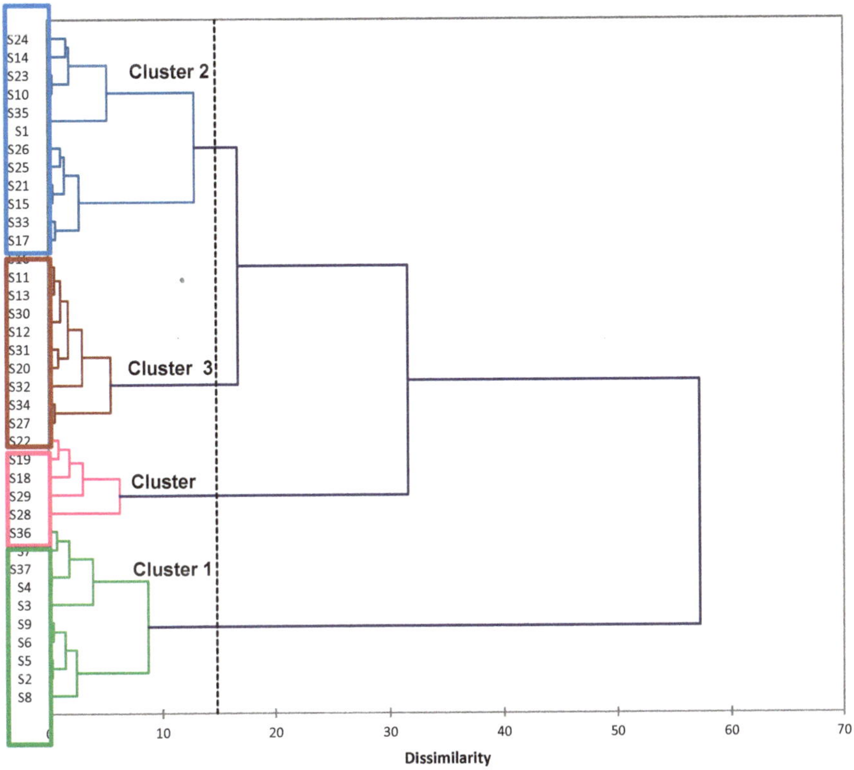

Figure 2. Hierarchical clustering of fungi species using Ward's method and XLSTAT (Addinsoft).

Automatic truncation based on entropy (dotted line) allowed identifying four consistent groups of fungi explaining 64.62% of the total inter-variance and 35.38% of the total intra-variance. Order of appearance of clusters (from top to down) was Cluster 2, Cluster 3, Cluster 4 and Cluster 1.

4. Conclusions

Among the 37 strains studied, we demonstrated that species from the *Aspergillus and Trametes* genus generally gave the best results for bioremediation purposes with COD reductions reaching until 77%, decolourization yields until 43% and a significant alkalizing ability (pH increase of 4 units). While the data from the literature concerned diluted and/or supplemented vinasse, our study first reported the depolluting potential of these same strains on raw vinasse. Mitigation of the pollution potential of aerobically treated effluent compared to crude vinasse was reflected in the significant increase of pH and high COD and mineral consumptions. Utilization of filamentous fungi and yeasts for sugarcane vinasse treatment turned out to be very promising. Mixed cultures should be performed in order to improve the depollution yields. Moreover, some strains were able to show striking growth on raw DSW. Seven of the nine *Aspergillus* sp. strains, the white-rot fungi *Trametes versicolor* and *T. hirsuta* and two yeasts *Clavispora lusitanea* and *Issatchenkia orientalis* could reach biomass yields higher than 20 $g \cdot L^{-1}$. The high yields of fungal biomass produced (until 29 $g \cdot L^{-1}$) could constitute an easily recoverable substrate for the production of renewable energy through anaerobic digestion.

Author Contributions: G.C.-T., L.A. and T.P. performed the experimental and laboratory work. G.C.-T., M.T., L.A., A.S.-C.-S., J.-M.F., Y.C. and T.P. worked on the analysis and interpretation of the data and contributed with valuable discussions. G.C.-T., M.T., L.A., A.S.-C.-S., J.-M.F., Y.C. and T.P. conceived the project, worked on the structure and wrote the paper. All authors have read and agreed to the published version of the manuscript.

Funding: This research was funded by Conseil Régional of Reunion Island. The funding body has no role in the design of the study and collection, analysis, and interpretation of data and in writing the manuscript.

Conflicts of Interest: The authors declare no conflict of interest.

Appendix A

Table A1. Eigenvalues obtained from PCA.

	PC_1	PC_2	PC_3	PC_4	PC_5
Eigen value	2.28	1.31	0.69	0.49	0.23
Variability (%)	45.62	26.23	13.81	9.72	4.61
Cumulative variability (%)	45.62	71.85	85.66	95.39	100.00

Table A2. Eigenvectors obtained from PCA.

Parameters	PC_1	PC_2	PC_3
Reduction of COD (%)	0.509	0.351	−0.281
Effect on OD at 475 nm (%)	0.457	−0.516	−0.021
Reduction of minerals content (%)	0.468	−0.181	0.766
Final pH	0.250	0.754	0.234
Biomass production ($g \cdot L^{-1}$)	0.501	−0.094	−0.528

Table A3. PCA and HCA data related to the five original parameters.

		Biomass Production ($g \cdot L^{-1}$)	Reduction of COD (%)	Reduction of Minerals Content (%)	Final pH	Effect on OD at 475 nm
All data	Min.	0.75	23.51	20.75	6.17	190.91
	Max.	29.4	76.53	77.57	9.05	57.94
	Aver.	15.85	57.5	45.07	8.02	105.24
	S.D.	7.69	13.24	16.25	0.81	33.39
Cluster 1	Min.	19.77	65.98	39.05	7.79	110.14
	Max.	29.4	76.53	77.57	9.05	42.46
	Aver.	24.24	72.17	57.35	8.35	70.77
	S.D.	3.3	3.1	18.36	0.54	16.13
Cluster 2	Min.	2.36	48.13	28.35	6.66	139.73
	Max.	28.56	69.7	61.46	8.74	75.65
	Aver.	15.94	59.38	44.55	8.02	104.13
	S.D.	7.55	5.97	11.12	0.72	16.6
Cluster 3	Min.	0.75	40.91	20.75	7.54	190.91
	Max.	15.21	62.09	49.05	9.03	130.64
	Aver.	10.63	51.87	34.05	8.39	147.26
	S.D.	4.17	6.46	10.26	0.45	18.83
Cluster 4	Min.	4.04	23.51	20.78	6.17	114.14
	Max.	17	49.52	70.49	7.01	74.2
	Aver.	9.30	34.87	43.83	6.57	92.8
	S.D.	6.22	9.84	18.94	0.33	17.68

References

1. GRFA Industry Issues. Available online: http://globalrfa.org/industry-issues/ (accessed on 11 February 2015).
2. UNICA and ApexBrasil Ethanol. Sugarcane Products. Available online: http://sugarcane.org/sugarcane-products/ethanol (accessed on 11 February 2015).
3. European Parliament and Council Directive. *The Promotion of the USE of Energy from Renewable Sources and Amending and Subsequently Repealing Directives*; 2009/28/EC, 2001/77/EC and 2003/30/EC; European Parliament and Council Directive: Strasbourg, France, 23 April 2009.
4. DAAF de la Réunion Présentation: Production, Superficies, Marché, Cadre Structurel .... Available online: http://daaf.reunion.agriculture.gouv.fr/Presentation-production#entete (accessed on 10 September 2020).
5. Patient, G.; Doligé, E. *Le Renouvellement du Régime Fiscal Applicable au Rhum Traditionnel des Départements D'outre-mer. [The Renewal of the tax Regime Applicable to the Traditional Rum of the Overseas Departments.] (No. 754)*; French Senate: Pair, France, 2012.
6. Rajagopal, V.; Paramjit, S.M.; Suresh, K.P.; Yogeswar, S.; Nageshwar, R.D.V.K.; Avinash, N. Significance of vinasses waste management in agriculture and environmental quality—Review. *Afr. J. Agric. Res.* **2014**, *9*, 2862–2873. [CrossRef]
7. Hoarau, J.; Caro, Y.; Grondin, I.; Petit, T. Sugarcane vinasse processing: Toward a status shift from waste to valuable resource. A review. *J. Water Process Eng.* **2018**, *24*, 11–25. [CrossRef]
8. Rodrigues Reis, C.E.; Hu, B. Vinasse from sugarcane ethanol production: Better treatment or better utilization? *Front. Energy Res.* **2017**, *5*, 7. [CrossRef]
9. Arimi, M.; Zhang, Y.; Götz, G.; Kiriamiti, K.; Geißen, S.-U. Antimicrobial colorants in molasses distillery wastewater and their removal technologies. *Int. Biodeterior. Biodegrad.* **2014**, *87*, 34–43. [CrossRef]
10. Francisca Kalavathi, L.; Uma, L.; Subrumanian, G. Degradation and metabolization of the pigment—Melanoidin in distillery effluent by the marine cyanobacterium *Oscillatoria boryana* BDU 92181. *Enzym. Microb. Technol.* **2001**, *29*, 246–251. [CrossRef]
11. Hoarau, J.; Grondin, I.; Caro, Y.; Petit, T. Sugarcane distillery spent wash, a new resource for third-generation biodiesel production. *Water* **2018**, *10*, 1623. [CrossRef]
12. Martins, S.I.F.S.; Van Boekel, M.A.J.S. A kinetic model for the glucose/glycine Maillard reaction pathways. *Food Chem.* **2005**, *90*, 257–269. [CrossRef]
13. Rivero-Pérez, M.D.; Pérez-Magariño, S.; González-San José, M.L. Role of melanoidins in sweet wines. *Anal. Chim. Acta* **2002**, *458*, 169–175. [CrossRef]
14. González, T.; Terron, M.C.; Yagüe, S.; Zapico, E.; Galetti, G.C.; González, A.E. Pyrolysis/gas chromatography/mass spectrometry monitoring of fungal-biotreated distillery wastewater using *Trametes* sp. I-62 (CECT 20197). *Rapid Commun. Mass Spectrom.* **2000**, *14*, 1417–1424.
15. Godshall, M.A. Removal of colorants and polysaccharides and the quality of white sugar. In Proceedings of the Association Andrew Van Hook, Reims, France, 29 December 1992; pp. 28–35.
16. España-Gamboa, E.I.; Mijangos-Cortés, J.O.; Hernández-Zárate, G.; Maldonado, J.A.D.; Alzate-Gaviria, L.M. Methane production by treating vinasses from hydrous ethanol using a modified UASB reactor. *Biotechnol. Biofuels* **2012**, *5*, 82. [CrossRef] [PubMed]
17. Ferreira, L.F.R.; Aguiar, M.M.; Messias, T.G.; Pompeu, G.B.; Lopez, A.M.Q.; Silva, D.P.; Monteiro, R.T. Evaluation of sugar-cane vinasse treated with *Pleurotus sajor-caju* utilizing aquatic organisms as toxicological indicators. *Ecotoxicol. Environ. Saf.* **2011**, *74*, 132–137. [CrossRef] [PubMed]
18. Hoarau, J.; Petit, T.; Grondin, I.; Marty, A.; Caro, Y. Phosphate as a limiting factor for the improvement of single cell oil production from *Yarrowia lipolytica* MUCL 30108 grown on pre-treated distillery spent wash. *J. Water Process Eng.* **2020**, *37*, 101392. [CrossRef]
19. Tiwari, S.; Gaur, R.; Singh, R. Decolorization of a recalcitrant organic compound (Melanoidin) by a novel thermotolerant yeast, *Candida tropicalis* RG-9. *BMC Biotechnol.* **2012**, *12*, 30. [CrossRef] [PubMed]
20. Pant, D.; Adholeya, A. Biological approaches for treatment of distillery wastewater: A review. *Bioresour. Technol.* **2007**, *98*, 2321–2334. [CrossRef] [PubMed]

21. Le Ministère Français de l'Ecologie, du Développement Durable et de l'Energie. *Arrêté du 21 Juillet 2015 Relatif aux Systèmes D'assainissement Collectif et aux Installations D'assainissement non Collectif, à L'exception des Installations D'assainissement non Collectif Recevant une Charge Brute de Pollution Organique Inférieure ou Egale à 1,2 kg/j de DBO5*; DEVL1519953N; The French Ministry of Ecology, Sustainable Development and Energy: Paris, France, 2015.
22. De Mattos, L.F.A.; Bastos, R.G. COD and nitrogen removal from sugarcane vinasse by heterotrophic green algae Desmodesmus sp. *Desalin. Water Treat.* **2016**, *57*, 9465–9473. [CrossRef]
23. Alaki, M.; Takahashi, T.; Ishiguro, K. Studies on microbiolofical treatment and utilization of cane molasses distillery wastes (Part 1): Screening of useful yeast strains. *Bull. Fac. Agric. Mie Univ.* **1981**, *62*, 155–161.
24. Kavanagh, K. *Fungi—Biology and Applications*; John Wiley & Sons Ltd.: West Sussex, UK, 2011.
25. Sirianuntapiboon, S.; Somchai, P.; Ohmomo, S.; Atthasampunna, P. Screening of Filamentous Fungi Having the Ability to Decolorize Molasses Pigments. *Agric. Biol. Chem.* **1988**, *52*, 387–392.
26. Santal, A.R.; Singh, N. Biodegradation of Melanoidin from Distillery Effluent: Role of Microbes and Their Potential Enzymes. In *Biodegradation of Hazardous and Special Products*; Chamy, R., Ed.; IntechOpen: London, UK, 2013; pp. 71–104.
27. Rulli, M.M.; Villegas, L.B.; Colin, V.L. Treatment of sugarcane vinasse using an autochthonous fungus from the northwest of Argentina and its potential application in fertigation practices. *J. Environ. Chem. Eng.* **2020**, *8*, 104371. [CrossRef]
28. Del Gobbo, L.M.; Villegas, L.B.; Colin, V.L. The potential application of an autochthonous fungus from the northwest of Argentina for treatment of sugarcane vinasse. *J. Hazard. Mater.* **2019**, *365*, 820–826. [CrossRef]
29. Del Gobbo, L.M.; Colin, V.L. Fungal technology applied to distillery effluent treatment. In *Approaches in Bioremediation. Nanotechnology in the Life Sciences*; Prasad, R., Aranda, E., Eds.; Springer: Cham, Switzerland, 2018. [CrossRef]
30. Rolz, C.; Cabrera, S.; de Espinosa, R.; Maldonado, O.; Menchu, J.F. The growth of filamentous fungi on rum distilling slops. *Ann. Technol. Agric.* **1975**, *24*, 445–451.
31. De Lamo, P.; De Menezes, T.J. Bioconversao da vinhaca para a producao de biomassa fungica [Bioconversion of vinasse for the production of fungal biomass]. *Coletânea Inst. Technol. Aliment.* **1978**, *9*, 281–312.
32. Rosalem, P.; Tauk, S.; Santos, M.C. Efeito da temperatura, ph, tempo de cultivo e nutrientes no crescimento de fungos imperfeitos em vinhaca [Effects of temperature, pH, cultivation time and nutrients in the growth of fungi imperfecti in vinasse]. *Rev. Microbiol.* **1985**, *16*, 299–304.
33. Araujo, N.; Visconti, A.; De Castro, H.; Barroso da Silva, H.; Ferraz, M.H.; Salles Filho, M. Producao de biomassa fungica de vinhoto [Fungal biomass production in vinasse]. *Bras. Acucar.* **1976**, *88*, 35–45.
34. Garcia, I.G.; Venceslada, J.L.B.; Peña, P.R.J.; Gómez, E.R. Biodegradation of phenol compounds in vinasse using *Aspergillus terreus* and *Geotrichum candidum*. *Water Res.* **1997**, *31*, 2005–2011. [CrossRef]
35. Chuppa-Tostain, G.; Hoarau, J.; Watson, M.; Adelard, L.; Shum Cheong Sing, A.; Caro, Y.; Grondin, I.; Bourven, I.; Francois, J.-M.; Girbal-Neuhauser, E.; et al. Production of *Aspergillus niger* biomass on sugarcane distillery wastewater: Physiological aspects and potential for biodiesel production. *Fungal. Biol. Biotechnol.* **2018**, *5*, 1. [CrossRef] [PubMed]
36. Kim, S.; Shoda, M. Decolorization of molasses and a dye by a newly isolated strain of the fungus *Geotrichum candidum* Dec 1. *Biotechnol. Bioeng.* **1999**, *62*, 114–119. [CrossRef]
37. Sluiter, A.; Hames, B.; Scarlata, C.; Sluiter, J.; Templeton, D. *Determination of Ash in Biomass*; No. NREL/TP-510-42622; National Renewable Energy Laboratory, U.S. Department of Energy: Golden, CO, USA, 2008.
38. Jollife, I.T. *Principal Component Analysis, Springer Series in Statistics*; Springer: New York, NY, USA, 2002.
39. Brown, C.E. *Applied Multivariate Statistics in Geohydrology and Related Sciences*; Springer: Berlin/Heidelberg, Germany, 1998.
40. Benito, G.G.; Miranda, M.P.; de los Santos, D.R. Decolorization of wastewater from an alcoholic fermentation process with *Trametes Versicolor*. *Bioresour. Technol.* **1997**, *61*, 33–37. [CrossRef]
41. Nudel, B.C.; Wahener, R.S.; Fraile, E.R.; Giuletti, A.M. The use of single and mixed cultures for aerobic treatment of cane sugar stillage and SCP production. *Biogolical Wastes* **1987**, *22*, 67–73. [CrossRef]
42. Kumar, V.; Wati, L.; Nigam, P.; Banat, I.M.; Yadav, B.S.; Singh, D.; Marchant, R. Decolorization and biodegradation of anaerobically digested sugarcane molasses spent wash effluent from biomethanation plants by white-rot fungi. *Process Biochem.* **1998**, *33*, 83–88. [CrossRef]

43. Raghukumar, C.; Rivonkar, G. Decolorization of molasses spent wash by the white-rot fungus *Flavodon flavus*, isolated from a marine habitat. *Appl. Microbiol. Biotechnol.* **2001**, *55*, 510–514. [CrossRef] [PubMed]
44. Fahy, V.; Fitzgibbon, F.J.; McMullan, G.; Singh, D.; Marchant, R. Decolourisation of molasses spent wash by *Phanerochaete chrysosporium*. *Biotechnol. Lett.* **1997**, *19*, 97–99. [CrossRef]
45. Angayarkanni, J.; Palaniswamy, M.; Swaminathan, K. Biotreatment of Distillery Effluent Using *Aspergillus niveus*. *Bull. Environ. Contam. Toxicol.* **2003**, *70*, 268–277. [CrossRef]
46. Jiménez, A.M.; Borja, R.; Martin, A. Aerobic–anaerobic biodegradation of beet molasses alcoholic fermentation wastewater. *Process Biochem.* **2003**, *38*, 1275–1284. [CrossRef]
47. Miranda, M.P.; Benito, G.G.; Cristobal, N.S.; Nieto, C.H. Color elimination from molasses wastewater by *Aspergillus niger*. *Bioresour. Technol.* **1996**, *57*, 229–235. [CrossRef]
48. Mohammad, P.; Azamidokht, H.; Fatollah, M.; Mahboubeh, B. Application of response surface methodology for optimization of important parameters in decolorizing treated distillery wastewater using *Aspergillus fumigatus* UB2 60. *Int. Biodeterior. Biodegrad.* **2006**, *57*, 195–199. [CrossRef]
49. Ohmomo, S.; Kaneko, Y.; Sirianuntapiboon, S.; Somchai, P.; Atthasampunna, P.; Nakamura, I. Decolorization of Molasses Waste Water by a Thermophilic Strain, *Aspergillus fumigatus* G-2-6. *Agric. Biol. Chem.* **1987**, *51*, 3339–3346.
50. Shayegan, J.; Pazouki, M.; Afshari, A. Continuous decolorization of anaerobically digested distillery wastewater. *Process Biochem.* **2005**, *40*, 1323–1329. [CrossRef]
51. Patil, P.U.; Kapadnis, D.P.; Dhamankar, V.S. Decolorisation of synthetic melanoidin and biogas effluent by immobilised fungal isolate of *Aspergillus niger* UM2. *Int. Sugar J.* **2003**, *105*, 10–13.
52. FitzGibbon, F.; Singh, D.; McMullan, G.; Marchant, R. The effect of phenolic acids and molasses spent wash concentration on distillery wastewater remediation by fungi. *Process Biochem.* **1998**, *33*, 799–803. [CrossRef]
53. Sirianuntapiboon, S.; Sihanonth, P.; Somchai, P.; Atthasampunna, P.; Hayashida, S. An absorption mechanism for the decolorization of melanoidin by *Rhizoctonia* sp. D-90. *Biosci. Biotechnol. Biochem.* **1995**, *59*, 1185–1189. [CrossRef]
54. Seyis, I.; Subasioglu, T. Screening of different fungi for decolorization of molasses. *Braz. J. Microbiol.* **2009**, *40*, 61–65. [CrossRef] [PubMed]
55. Adikane, H.V.; Dange, M.N.; Selvakumari, K. Optimization of anaerobically digested distillery molasses spent wash decolorization using soil as inoculum in the absence of additional carbon and nitrogen source. *Bioresour. Technol.* **2006**, *97*, 2131–2135. [CrossRef]
56. Jiranuntipon, S.; Chareonpornwattana, S.; Damronglerd, S.; Albasi, C.; Delia, M.-L. Decolorization of synthetic melanoidins-containing wastewater by a bacterial consortium. *J. Ind. Microbiol. Biotechnol.* **2008**, *35*, 1313–1321. [CrossRef]
57. Ramezani, A.; Darzi, G.N.; Mohammadi, M. Removal of melanoidin from molasses spent wash using fly ash-clay adsorbents. *Korean J. Chem. Eng.* **2011**, *24*, 1035–1041. [CrossRef]
58. Hayase, F.; Kim, S.B.; Kato, H. Decolorization and degradation products of the melanoidins by hydrogen peroxide. *Agric. Biol. Chem.* **1984**, *48*, 2711–2717.
59. Mohana, S.; Desai, C.; Madamwar, D. Biodegradation and decolourization of anaerobically treated distillery spent wash by a novel bacterial consortium. *Bioresour. Technol.* **2007**, *98*, 333–339. [CrossRef]
60. Agarwal, R.; Lata, S.; Gupta, M.; Singh, P. Removal of melanoidin present in distillery effluent as a major colorant: A Review. *J. Environ. Biol.* **2010**, *31*, 521–528.
61. Tian, Z.; Mohan, G.R.; Ingram, L.; Pullammanappalil, P. Anaerobic digestion for treatment of stillage from cellulosic bioethanol production. *Bioresour. Technol.* **2013**, *144*, 387–395. [CrossRef]
62. Singh, H. Fungal treatment of distillery and brewery wastes. In *Mycoremediation: Fungal Bioremediation*; Wiley-Interscience: Hoboken, NJ, USA, 2006.
63. Tauk, S.M. Culture of Candida in vinasse and molasses: Effect of acid and salt addition on biomass and raw protein production. *Eur. J. Appl. Microbiol. Biotechnol.* **1982**, *16*, 223–227. [CrossRef]
64. Selim, M.H.; Elshafei, A.M.; El-Diwany, A.I. Production of single cell protein from yeast strains grown in Egyptian vinasse. *Bioresour. Technol.* **1991**, *36*, 157–160. [CrossRef]
65. Cajo, L.; Nizama, L.; Carreño, C. Effect of inoculum and molasses concentration as supplement to vinasse of distillery for the production of biomass of native *Candida utilis*. *Sci. Agropecu.* **2011**, 65–72. [CrossRef]
66. Bushell, M.E. Chemical requirements for growth. In *Fungal Physiology*, 2nd ed.; Wiley & Sons, Inc.: New York, NY, USA, 1994; pp. 130–157.

67. Kavanagh, K. Introduction to fungal physiology. In *Fungi—Biology and Applications*; West Sussex: London, UK, 2011; pp. 1–34.
68. Cattell, R.B. The scree test for the number of factors. In *A History of Psychology in Autobiography*; Lindzey, G., Ed.; Prentice-Hall, Inc.: Englewood Cliffs, NJ, USA, 1974; Volume VI, pp. 61–100.

microorganisms

MDPI

Article

Optimized Production of a Redox Metabolite (pyocyanin) by *Pseudomonas aeruginosa* NEJ01R Using a Maize By-Product

Francisco Javier Bacame-Valenzuela [1,2], Jesús Alberto Pérez-Garcia [1], Mayra Leticia Figueroa-Magallón [1], Fabricio Espejel-Ayala [1], Luis Antonio Ortiz-Frade [1] and Yolanda Reyes-Vidal [1,2,*]

1 Centro de Investigación y Desarrollo Tecnológico en Electroquímica (CIDETEQ), Parque Tecnológico Querétaro s/n, Sanfandila, Pedro Escobedo, Querétaro C.P. 76703, Mexico; fbacame@cideteq.mx (F.J.B.-V.); jgarcia@cideteq.mx (J.A.P.-G.); mfigueroa@cideteq.mx (M.L.F.-M.); fespejel@cideteq.mx (F.E.-A.); lortiz@cideteq.mx (L.A.O.-F.)

2 Consejo Nacional de Ciencia y Tecnología (CONACYT)—Centro de Investigación y Desarrollo Tecnológico en Electroquímica (CIDETEQ), Parque Tecnológico Querétaro s/n, Sanfandila, Pedro Escobedo, Querétaro C.P. 76703, Mexico

* Correspondence: mreyes@cideteq.mx

Received: 13 August 2020; Accepted: 24 September 2020; Published: 10 October 2020

Abstract: *Pseudomonas aeruginosa* metabolizes pyocyanin, a redox molecule related to diverse biological activities. Culture conditions for the production of pyocyanin in a defined medium were optimized using a statistical design and response surface methodology. The obtained conditions were replicated using as substrate an alkaline residual liquid of cooked maize and its by-products. The untreated effluent (raw nejayote, RN) was processed to obtain a fraction without insoluble solids (clarified fraction, CL), then separated by a 30 kDa membrane where two fractions, namely, retentate (RE) and filtered (FI), were obtained. Optimal conditions in the defined medium were 29.6 °C, 223.7 rpm and pH = 6.92, which produced 2.21 $\mu g\ mL^{-1}$ of pyocyanin, and by using the wastewater, it was possible to obtain 3.25 $\mu g\ mL^{-1}$ of pyocyanin in the retentate fraction at 40 h. The retentate fraction presented the highest concentration of total solids related to the maximum concentration of pyocyanin (PYO) obtained. The pyocyanin redox behavior was analyzed using electrochemical techniques. In this way, valorization of lime-cooked maize wastewater (nejayote) used as a substrate was demonstrated in the production of a value-added compound, such as pyocyanin, a redox metabolite of *Pseudomonas aeruginosa* NEJ01R.

Keywords: pyocyanin; maize industry wastewater; *Pseudomonas aeruginosa* NEJ01R; redox metabolite; optimization; response surface analysis; bioproduction; valorization; electrochemical analysis

1. Introduction

Microorganisms can produce a large variety of extracellular compounds with different biological activities. Currently, there are more than 50,000 natural products used as drugs that are derived from microorganisms, mostly isolated from soil samples [1]. The genus *Pseudomonas*, Gram-negative aerobic Gammaproteobacteria belonging to the family Pseudomonadaceae, is known to produce a series of extracellular redox metabolites of the phenazine group. Species of the genus *Pseudomonas* are characterized by their metabolic versatility, giving them great relevance in environmental recycling processes, degrading a wide range of simple and complex organic compounds [2–4]. It has been demonstrated that *Pseudomonas aeruginosa* produces 5-methylphenazine-1-one or pyocyanin (PYO), a water-soluble blue-green pigment with a redox potential similar to menaquinone [5].

PYO is a biomolecule that participates in the quorum-sensing process of *P. aeruginosa*. This bacterial mechanism is used to coordinate population density, as well as for the regulation of gene expression and biofilm formation [6,7]. *Pseudomonas* integrates a bacterial strategy through PYO molecules for the inhibition of fungal phytopathogens of agricultural crops and cytostatic activity against cancer cell lines [8]. The antagonistic effect of the compounds derived from phenazines is associated with their redox behavior. Therefore, it is believed that many capabilities of PYO and other phenazine derivatives, in a variety of eukaryotic hosts and bacteria, are the result of oxidative activity or inactivation of important proteins in response to oxidative stress [9].

The redox process for PYO was studied through cyclic and square wave voltammetries. PYO and other redox metabolites included in the secretome of *P. aeruginosa* could be used as electrochemical biomarkers to detect its presence in a bacterial culture supernatant [10]. Furthermore, *P. aeruginosa* is considered an electrogenic microorganism and is used in bioelectrochemical systems for electrical energy production, carried out by direct electron transfer (through cytochromes in outer membrane or bacteria pili) and indirect electron transfer (through secondary metabolites excreted by bacteria). In this case, PYO of *P. aeruginosa* is the extracellular metabolite that facilitates electronic transfer between microorganism and electrode due to its reversible redox properties [11]. However, one disadvantage of PYO is its market cost since 5 mg (purity > 98%, HPLC grade) has a cost of 60–97 USD according to different suppliers [12–15]. Therefore, optimized processes must be developed that allow the metabolite to be obtained at high concentrations through operations less harmful to the environment, such as green chemistry, to extend the use of PYO.

An attractive approach is to use residues as substrates in bioprocesses to obtain value-added molecules. In this way, PYO production by submerged fermentation of *P. aeruginosa* has been carried out in glycerol (residues of biodiesel production) and a defined medium supplemented with a broad variety of raw materials [5,16,17]. One of the options to achieve this goal is the use of wastewaters as a carbon source, which could be considered toxic for the environment, but the use in culture media is an alternative that reduces its polluting and harmful potential, as well as its valorization.

Currently, among the agro-food-processing industrial effluents considered as environmental pollutants are by-products generated in the maize (*Zea mays*) cooking process. This traditional process is known as "nixtamalization" and includes boiling corn grains in a saturated solution of calcium hydroxide (0.5–2%) at 90 °C for 40 min; after this process, the corn grains are steeped for 12 h, then drained, and the resulting liquid is commonly known as nejayote [18,19]. Cooked maize grains (nixtamal) are used for the production of masa, tortillas and derived products in Mexico, Southern United States, Central and South America, Asia and parts of Europe. In Mexico, it is estimated that 14.4 million m^3 of nejayote is produced annually [20]. This wastewater is considered as a pollutant due to its contents extracted during the process: 0.5–14.5% of the corn's weight goes to the effluent, which has a pH of 12 [18]. Various compounds and some phytochemicals associated with the cell wall of corn grains are released by alkaline hydrolysis and are related to parameters such as total organic carbon (TOC; 2700–59,000 mg L^{-1}), biochemical oxygen demand (BOD; 2.69 mg O_2 L^{-1}) and chemical oxygen demand (COD; 7500–40,000 mg L^{-1}) [21–24]. Some of these parameters also indicate that the effluent has a high content of organic material (reducing and total sugars), which together with other compounds (protein, fiber, fat, calcium, arabinoxylans and polyphenols) define its content of nutrients (carbon source) and inorganic compounds, potentially applicable for biotechnological applications [25–29].

Likewise, this wastewater can be used as a raw substrate by microorganisms to obtain bioproducts. *Aspergillus oryzae* and different species of *Lactobacillus* produce a protease and bacteriocins, respectively, using nejayote as a substrate [30,31]. In other studies, two native isolated bacteria (*Bacillus flexus*) were used to biosynthesize amylases, xylanases, proteases and phenolic acid esterase [32]. Recently, *Bacillus megaterium* (a native nejayote strain) was able to transform ferulic acid present in the effluent into 4-vinylguaiacol [33]. Moreover, nejayote in combination with other residual effluents such as vinasse (a waste from tequila production) and swine wastewater allow for obtaining bioenergy or can used for the growth of microalgae [24,34]. Thus, it is clear that components of nejayote can be exploited for the production of different biotechnological products. Therefore, the objective of this work was to valorize lime-cooked maize wastewater as a culture medium to produce a redox metabolite using *P. aeruginosa* NEJ01R and to develop optimized culture conditions (temperature, pH and agitatio(n) in a defined culture medium by response surface methodology.

2. Materials and Methods

2.1. Microorganism

Bacteria were isolated from lime-cooked maize wastewater collected from a local mill situated in Pedro Escobedo, Querétaro (Mexico). Samples (1 mL) of nejayote at pH 11 (without other treatments) were plated on LB agar and incubated at 30 °C for 48 h. The composition of the LB agar was g L^{-1}) casein peptone (10), yeast extract (5), sodium chloride (0.5) and bacteriology agar (15). Single colonies with different morphologies were plated again in LB agar. Three colonies (NEJ01R, NEJR5 and NEJ03R) were selected because a green coloration was observed in plates. NEJ01R was used in this study because it showed pigment production in 24 h and growth in different substrates [16]. The strain NEJ01R was characterized using Gram stain (-) and the biochemical test API 20E (bioMérieux) as *Pseudomonas aeruginosa*. Genomic DNA was isolated from pure bacterial colonies using methodology described by [35]. A nucleotide sequence analysis (16 rRNA gene) was performed at Laboratorio Nacional de Biotecnología Agrícola, Médica y Ambiental (LANBAMA) (San Luis Potosi, México). The resulting rRNA sequences were submitted to the non-redundant nucleotide database at GenBank using the Basic Local Alignment Search Tool (BLAST) program to determine its identity. *P. aeruginosa* NEJ01R sequence showed 92.30% similarity to *P. aeruginosa* ACR20 with the NCBI Accession Number CP058333.1. The strain was stored on Luria-Bertani agar (LB) at 4 °C.

2.2. Statistical Design and Optimization by Response Surface

To optimize conditions of culture for biomass production and PYO generation in the defined culture medium, a central compound design was chosen with three experimental factors: temperature, pH and agitation speed. Design analysis and subsequent analysis of the experimental data were recorded in Statgraphics Technologies, Inc. (The Plains, Virginia, USA, version 15.0). Table S1 shows the design matrix. Experimental factors evaluated were obtained from preliminary experimental results. Nineteen experiments were achieved because five central points were selected to complete the lack-of-fit test. Biomass and PYO were the response variables and were measured randomly at 48 h, by duplication. The following equation shows the number of experiments achieved:

$$N = 2^k + 2k + cp \quad (1)$$

where N is the number of experiments, k is the number of experimental factors and cp denotes central points. To assure the rotatability and orthogonality in the experimental design, α was chosen with the next equation:

$$\alpha = \left(\frac{(FxN)^{1/2} - F}{2} \right)^{1/2} \quad (2)$$

where $F = 2k$. An analysis of variance (ANOVA) was performed to inspect the response surface model. The fitted polynomial equation was then expressed in the form of three-dimensional response surface plots to show the relationship among the responses and the experimental levels of each independent variable.

2.3. Culture Conditions

The inoculum was obtained from a liquid culture using LB broth. The composition of the LB broth was (g L^{-1}) casein peptone (10), yeast extract (5) and sodium chloride (0.5). The liquid culture

(50 mL in 250 mL Erlenmeyer flasks) was inoculated with a single fresh colony of *P. aeruginosa* NEJ01R (LB agar, 30 °C, for 24 h). The inoculated LB broth was incubated at 150 rpm, 30 °C, for 24 h.

All experimental units (EUs) generated by statistical design were made in 250 mL Erlenmeyer flasks with 50 mL of LB broth (defined medium, LB) and sterilized in an autoclave at 121 °C for 15 min. After this, each EU was inoculated with 1 mL of an inoculum culture. The EUs were incubated for 48 h using different combinations of conditions (pH, temperature and agitation) established by the statistical design.

2.4. Biomass Determination

The biomass was determined by the dry weight method. Aluminum dishes were pre-dried to constant weight in a drying oven (100 °C, for 1 h). Later, they were allowed to cool in a desiccator, and then their initial weight was recorded. After 48 h of incubation, cultures were centrifuged at 6600× *g* for 10 min. The supernatant was decanted, and the formed pellets were transferred to an aluminum dish and were placed in a drying oven at 100 °C for 1 h. Finally, the dishes were placed in a desiccator, and their final weight was measured. The biomass concentration was determined by the weight difference and reported in g L^{-1}.

2.5. PYO Determination

The decanted supernatants were used to quantify PYO by a liquid–liquid extraction method. Equal volumes of supernatant and chloroform (3 mL) were added and vigorously vortexed. The organic phase (blue) was separated, and a similar volume of 0.2 N HCl was added. The mix was vigorously stirred in a vortex, and the aqueous phase (red) was separated to measure it in a UV-Vis spectrophotometer (Genesys™ 10S, Thermo Scientific, WI, USA) at 520 nm. The absorbance obtained was multiplied by the factor 17.1 to obtain μg mL^{-1} of PYO [5].

2.6. Microbial Growth Kinetics and PYO Production

The kinetics of microbial growth were determined in cultures of *P. aeruginosa* NEJ01R under conditions optimized for growth and PYO production according to surface response. Samples were taken every two hours for 48 h of culture incubation. Serial dilutions (up to 10^{-8}) were made in 0.1 mM phosphate buffer (pH 7.0) for each sample. One milliliter of the dilutions was added on Petri dishes with agar standard methods and incubated (30 °C). After 24 h, a bacterial count (CFU mL^{-1}) was performed. In addition to bacterial count (CFU mL^{-1}), the determination of PYO was made following the methodology described in Section 2.5 [5].

2.7. Valorization of Maize Wastewater

The wastewater was used as a culture medium (raw nejayote, RN) without treatment under the conditions (composition and pH) used in a traditional process. Afterward, RN was processed using the methodology reported by [25] employing a flocculating agent and pH adjustment (7.2). The fraction obtained was labeled as clarified fraction (CL). The CL fraction was filtered using a Pellicon® system (30 kDa Biomax Mini polyethersulfone cassette, Merck Millipore, MA, USA). After this ultrafiltration process, the CL fraction was divided into retentate fraction (RE) and filtered fraction (FI). Fifty milliliters of RN and each fraction (CL, RE and FI) in 250 mL Erlenmeyer flasks was inoculated with 1 mL of an inoculum culture of *P. aeruginosa* NEJ01R. Submerged fermentations were incubated for 48 h using operational conditions (pH, temperature and agitation) established as optimal by the statistical design. All experiments were performed in triplicate.

2.8. Characterization of Maize Wastewater and Its By-Products

Before and after submerged fermentations with *P. aeruginosa* NEJ01R in optimized conditions (statistical design), RN and fractions were characterized. The concentration of total solids in the RN and three fractions (CL, RE and FI) was determined. Porcelain capsules (pre-dried to constant weight) were used in a drying oven at 100 °C for 1 h, and their initial weight was measured. One milliliter of the sample was placed and dried at 100 °C for 1 h and allowed to cool to room temperature, and the final weight was recorded. The concentration (g L^{-1}) of total solids was determined by the weight difference method. The concentration of insoluble solids was determined using porcelain Gooch crucibles and Whatman No. 1 filter; both materials were pre-dried to constant weight for 1 h in a drying oven. Subsequently, 2 mL of each sample was filtered in porcelain Gooch crucibles, and the crucibles with the filter paper were dried in an oven (100 °C, 1 h). Finally, their final weight was measured. The concentration of insoluble solids was determined by the weight difference method. The soluble solids were obtained by the difference between the weight of total solids and the weight of insoluble solids. The ashes in the fractions were also determined, and porcelain crucibles were pre-dried to constant weight in a drying oven (100 °C, 1 h). Then, 1 mL of the sample was placed in crucibles until dry on a heating plate (100 °C). The dry crucibles with samples were placed in a muffle (550 °C, 2 h). Finally, the final weight was measured. The ash concentration was determined by the weight difference method and reported in g L^{-1}.

2.9. Ferulic Acid Determination

The analysis of ferulic acid was carried out by an H-Class Acquity ultra-performance liquid chromatography (UPLC) system (Waters®, Mildford, MA, USA) using a C18 Waters UPLC BEH C18 column (50 mm × 2.1 mm i.d., 1.7 µm). Water/acetonitrile as a mobile phase was at 90:10 ratio, using a flow of 0.3 mL min^{-1}, maintaining the temperature of the column at 30 °C, with an injection volume of 5 µL and elution times of 5 min. The UPLC system was coupled to the quaternary pump, refrigerated autosampler and an extended wavelength photodiode array detector (PDA detector, Waters®, Mildford, MA, USA). In PDA detection, the system was employed by recording a wavelength of 320 nm. The chromatographic data were obtained and processed by the software Empower3 (Waters®, Mildford, MA, USA).

2.10. PYO Identification

An analysis was performed on a UPLC system with a cooling autosampler, a quaternary solvent manager, an oven for an analytical column, a PDA Detector and an Acquity QDa mass detector (Waters®, Mildford, MA, USA). The QDa mass detector is a compact, single, quadrupole mass detector equipped with an electrospray ionization (ESI) interface. A Waters UPLC BEH C18 column (50 mm × 2.1 mm i.d., 1.7 µm) was used at 30 °C. Water (HPLC grade) with 0.1% formic acid was used as mobile phase A, and mobile phase B was acetonitrile. The workflow was 0.3 mL min^{-1}, with a mobile phase A:B ratio of 95:5 *v*/*v*. The injection volume was 10 µL, with the autosampler kept at 15 °C. For mass detection, the QDa detector was operated in an electrospray positive-ion mode and the cone voltage was set at 10 V. The desolvation temperature was set at 600 °C. The Mass Spectrometry (MS) scan mode was used for a full mass spectrum between *m*/*z* 100 and 300, acquired with a sample rate of 5 points/s. In PDA detection, the system was employed by recording a multiwavelength set in the wavelength range of 210–800 nm. PYO (Sigma-Aldrich, St Louis, MO, USA) was used as the standard. All samples and standards were filtered by a 0.2 µm nylon membrane. Areas of peaks were determined using the Empower3 chromatography software (Waters®, Mildford, MA, USA).

2.11. Electrochemical Study

Cyclic voltammetry (CV) was performed at different scan rates on a conventional three-electrode cell employing a glassy carbon electrode as the working electrode, a Pt electrode as the counter electrode and Ag/AgCl as the reference electrode (3 M NaCl). Measurements were carried on a potentiostatic/galvanostatic device (Biologic VSP, Grenoble, France). Ten milliliters of sodium phosphate buffer (0.2 M, pH = 7) was used as the supporting electrolyte in the presence of PYO (0.14 mM) extracted from *P. aeruginosa* NEJ01R. Prior to each measurement, the solutions were purged with highly purified nitrogen for 10 min, and the compensation of the ohmic drop was carried out using electrochemical impedance spectroscopy. The resistance to the solution was measured at a frequency of 100 kHz with a sinus amplitude of 10 mV. The potential was established in its open-circuit value. The ohmic drop was set at 85%.

3. Results and Discussion

3.1. Culture Condition Optimization in Defined Medium

In order to determine the interaction of different variables (factors) with microbial growth and metabolite production by *P. aeruginosa* NEJ01R, a design experiment coupled to surface response methodology was employed. A central composite design also allowed us to evaluate the optimal conditions in the biomass and PYO produced in a defined culture medium (LB broth). At present, there are no studies that report optimization using statistical tools to determine the best combination of factors to obtain high concentrations of PYO in biological processes using LB broth. Table 1 shows the total experiments and results of biomass and PYO obtained.

Table 1. Experimental central composite design matrix; biomass and pyocyanin (PYO) experimental results.

Unit	Experimental Code Factors			Results			
	Temperature (X_1)	pH (X_2)	Agitation (X_3)	Biomass g L^{-1}		PYO μg mL^{-1}	
1	0	0	0	0.368	0.380	1.5903	1.4260
2	−1	−1	1	0.512	0.484	0.0001	0.0001
3	1	−1	−1	0.480	0.330	0.6600	0.5800
4	1	1	−1	0.318	0.278	0.4788	0.3420
5	0	0	−1.47119	0.600	0.710	2.8200	2.7800
6	0	0	0	0.388	0.380	1.5732	1.4312
7	0	0	0	0.368	0.366	1.6416	1.6758
8	−1.47119	0	0	0.440	0.470	0.5814	0.4275
9	1	−1	1	0.440	0.370	1.1900	1.2800
10	0	1.47119	0	0.036	0.020	0.0500	0.0400
11	1.47119	0	0	0.252	0.188	0.0342	0.0223
12	0	0	0	0.340	0.380	1.5390	1.4483
13	0	0	0	0.358	0.346	1.5323	1.4141
14	−1	1	1	0.320	0.398	1.4706	1.5561
15	−1	−1	−1	0.705	0.386	0.9498	1.0602
16	0	0	1.47119	0.974	0.484	0.6669	0.7011
17	1	1	1	0.370	0.240	0.1300	0.2200
18	−1	1	−1	0.364	0.356	1.4193	1.2996
19	0	−1.47119	0	0.380	0.342	2.3127	2.3598

Analysis of results was done considering the variability from experimental factors and total error (Table 2). Experimental conditions evaluated in this work allowed us to have several values of biomass and PYO. In some experiments, the biomass was preferred instead of PYO. Maximum biomass and PYO generated were 0.974 g L^{-1} and 2.82 μg mL^{-1}, respectively; and minimum values were: biomass = 0.02 g L^{-1} and PYO = 0.0001 μg mL^{-1}.

Table 2. ANOVA for biomass and PYO generated from the defined culture medium.

Source	Biomass		PYO	
	Coefficient	*p*-Value	Coefficient	*p*-Value
Constant (X_0)	0.3674		1.5832	
Temperature (X_1)	−0.0390	0.0732	−0.1539	0.01757
pH (X_2)	−0.0655	0.0042	−0.2053	0.0744
Agitation (X_3)	0.0227	0.2869	−0.2712	0.0210
Temperature:temperature ($X_1{}^2$)	−0.0138	0.5700	−0.6030	0.0001
Temperature:pH (X_1X_2)	0.0413	0.1236	−0.3620	0.0138
Temperature:agitation (X_1X_3)	0.0336	0.2074	0.1838	0.1919
pH:pH ($X_2{}^2$)	−0.0799	0.0026	−0.1760	0.1769
pH:agitation (X_2X_3)	0.0333	0.2107	0.0686	0.6215
Agitation:agitation ($X_3{}^2$)	0.1499	0.000	0.0787	0.5405

Moreover, the factors' levels were adequately separated to estimate the effects on the responses. Coefficients of regression equation are displayed in Table 3 considering the following equation:

$$R = \beta_0 + \beta_1 X_1 + \beta_2 X_2 + \beta_3 X_3 + \beta_4 X_1^2 + \beta_5 X_1 X_2 + \beta_6 X_1 X_3 + \beta_7 X_2^2 + \beta_8 X_2 X_3 + \beta_9 X_3^2 \quad (3)$$

where *R* is the estimated value of biomass or PYO and β is the polynomial coefficient.

Table 3. Calculated conditions for the maximum biomass and PYO production in the defined culture medium.

Response	Biomass Calculated, 0.71 g L^{-1} PYO Calculated, 2.21 µg mL^{-1} (Desirability = 0.9084)	
Factor	**Code**	**Real Value**
Temperature (°C)	−0.3220	29.6
pH	−0.5007	6.92
Agitation (rpm)	1.4711	223.7

The reduced model was achieved considering the factors with significant effect for biomass and PYO. Pareto charts show only the factors and interactions with effects in the responses (Figure 1). For biomass generation, the pH and the pH/pH interaction had a negative effect, while the interaction agitation/agitation was positive. Moreover, the last interaction had a higher value than pH and pH/pH interaction obtaining a high biomass value, indicating that moderate agitation and low pH values promoted an increase in the biomass. For the case of PYO, agitation and temperature/temperature and temperature/pH interaction had a negative effect on PYO generation. In the same case for biomass, to have a higher production of PYO, temperature, pH and agitation values should be at low levels (Table 1). Response surface plots (Figure 2) were constructed from the regression equations for biomass and PYO generation.

Response surface plots (Figure 2) confirmed the principal effect of temperature, pH and agitation rate. In the case of biomass, maximum values were obtained using low temperature, low pH and low agitation values. PYO response surfaces had a different behavior because low–medium temperatures also increased PYO generation. pH values maintained the PYO in a medium value, and the effect of agitation were similar to that of pH. Simultaneous analysis was done to obtain the maximum generation of biomass and PYO considering the temperature, pH and velocity of agitation. In this case, the desirability as well as factor conditions are shown in Table 3. The maximum biomass and PYO generation calculated from the lineal reduced regression equation are also shown. Simultaneous analysis showed that the factor values obtained the maximum biomass and PYO considering the desirability. The best conditions for PYO production obtained by the surface response methodology were used in cultures, reaching PYO = 2.05 μg mL^{-1}. These conditions were evaluated to continue the kinetic production of biomass and PYO.

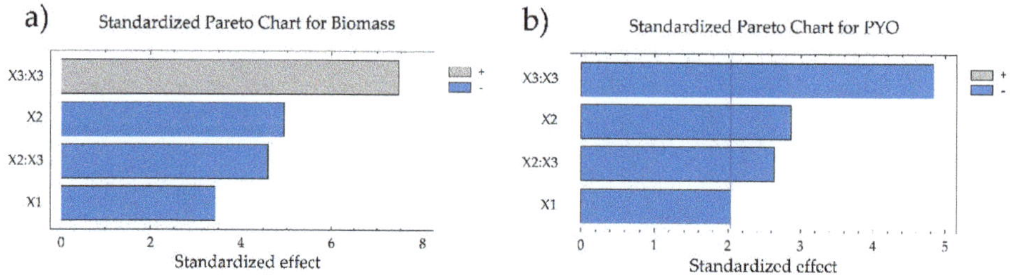

Figure 1. Pareto chart for (**a**) biomass and (**b**) pyocyanin (PYO) generation in the defined culture medium. The graphics show only significant effects. X1, temperature; X2, pH; X3, agitation.

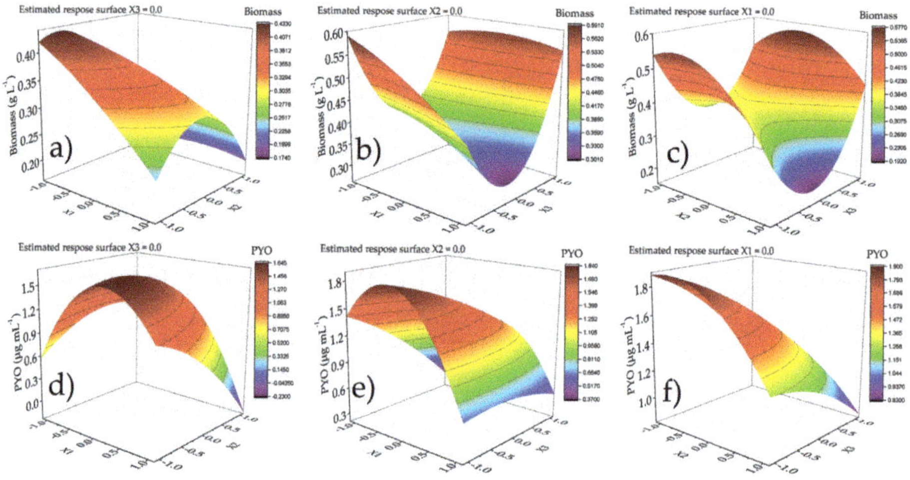

Figure 2. Response surface plots for biomass and PYO production estimated from regression equation in statistical analysis. (**a**) Interaction between temperature and pH; (**b**) interaction between temperature and agitation rate; (**c**) interaction between pH and agitation rate; (**d**) interaction between temperature and pH; (**e**) interaction between temperature and agitation rate; (**f**) interaction between pH and agitation rate.

3.2. Microbial Growth Kinetics and PYO Production

Figure 3a shows the kinetics of microbial growth and PYO production of *P. aeruginosa* NEJ01R, where the lag phase lasted for 5 h and subsequently started the exponential phase until 13 h. The stationary phase stayed until 43 h, after which it finally entered the decaying phase. PYO production was not associated with growth because production started at 16 h once the stationary phase began; then, it did not exhibit the same behavior as bacterial growth. The maximum production of PYO was at 40 h, three hours before starting the death phase. This microbial growth kinetics and PYO production were carried out in a defined LB medium, which is considered as a standard for the incubation period in the culture medium and is able to establish the maximum production time of PYO. Other studies report that the maximum production is after 48 and 72 h of growth, as is the case for several strains of *P. aeruginosa* isolated from surgical samples, minced meat and infected wounds [36]. The processing time is an important variable to study for the design of processes. In this work, it was possible to obtain a maximum concentration of PYO (3.3 μg mL^{-1}) in 40 h using a defined synthetic medium

(LB) under optimized conditions in the statistical design. While in other strains of *P. aeruginosa*, such as KU-BI02, concentrations of 2.560 μg mL^{-1} have been produced in 72 h using King's A medium supplemented with soybean seeds [17]; in strains *P. aeruginosa* R1 and *P. aeruginosa* U3, both in King's A medium, 4.5 μg mL^{-1} and 2.4 μg mL^{-1} of PYO were obtained, respectively, at 96 h of incubation [5].

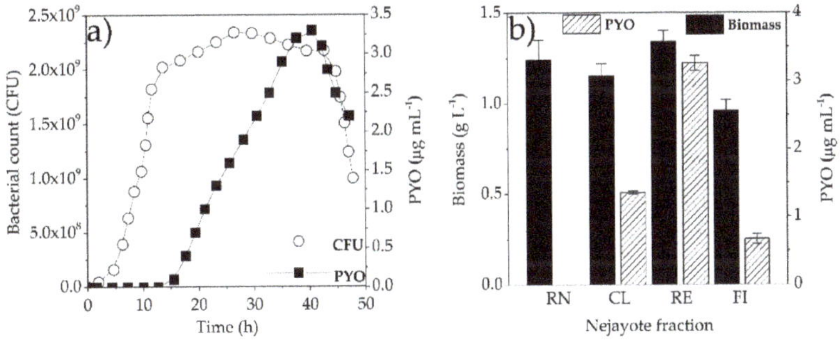

Figure 3. (**a**) Microbial growth kinetics and PYO production of *Pseudomonas aeruginosa* NEJ01R under optimized growth conditions; (**b**) biomass and PYO production of *P. aeruginosa* NEJ01R using raw nejayote (RN) and fractions (clarified (CL), retentate (RE) and filtered (FI) fractions) under optimized conditions.

3.3. Maize Wastewater (Nejayote) as a Substrate to PYO Production

Three fractions of lime-cooked maize wastewater (nejayote) were obtained (CL, RE and FI) in addition to raw nejayote (RN) to evaluate PYO production by *P. aeruginosa* NEJ01R. Figure 3b shows results concerning the samples used separately as the sole carbon source for fermentation of *P. aeruginosa* NEJ01R and PYO production, using optimal culture conditions established in the defined medium LB. RN substrate was only able to be used as a carbon source by *P. aeruginosa* NEJ01 because only biomass production was obtained (1.2 g L^{-1}). The PYO production was probably hindered due to the alkaline pH (11) of raw nejayote. These results can be compared with treatments in the experimental design, since in those that presented an alkaline pH of 11, a minimum PYO production was achieved, such as treatment 10 (Table 1). In the case of the CL fraction, values of biomass production were similar to those obtained in the RN substrate, but with PYO production. Contrary to the RN substrate, pH was adjusted to 7.2, which had a positive effect on metabolite production considering that the optimum pH to produce PYO was 6.92, according to the statistical design (Table 3).

The RE fraction had biomass production values higher than that of other fractions. RE and CL fractions were adjusted to pH 7.2. This condition was favorable for both biomass and PYO production. In the RE fraction, there were molecules larger than 30 kDa, which probably favored PYO production. Ultrafiltration (UF) membranes from 3 to 30 kDa concentrate high-molecular-weight components such as proteins, hydrolysates and phenolic fractions according to [37]. Nejayote contains polysaccharides such as feruloylated arabinoxylans, which are non-starchy compounds derived from endosperm cell walls of cereals and formed with a linear -(1→4)-xylopyranose backbone and -L-arabinofuranose residues as side chains on O3 and O2 and O3 [29]. Arabinoxylans can present some arabinose residues ester-linked on (O)-5 to ferulic acid (3-methoxy-4-hydroxycinnamic acid). These compounds display ferulic acid concentrations of 0.6 μg L^{-1}, as well as an arabinose-to-xylose ratio (A/X) of 0.57–0.65 with a molecular weight of 60 kDa [26]. In this work, the membrane used for the separation of fractions was 30 kDa, so it was possible that feruloylated arabinoxylans were in the RE fraction according to [25,27,28]. In that case, *P. aeruginosa* NEJ01R should have an appropriate enzymatic machinery to use this type of polysaccharide as a carbon source, such as polysaccharide-degrading enzymes. Therefore, it is feasible that *P. aeruginosa* NEJ01R produces enzymes to degrade the high volume of polysaccharides present in nejayote RE fraction with a positive effect on PYO production. Different strains of *P. aeruginosa* have demonstrated biosynthesis capacity for effluents of biotechnological processes (with complex composition) such as biodiesel production [16]; raw substrates, such as cottonseed meal, grape seeds, pea pods, taro leaves and olive wastes, hydrolyzed by acids [5]; and others, such as ground corn kernels, ground soybean seeds, potato cooking water, ground watermelon seeds and groundnut [17]. Different concentrations of PYO can be obtained in a defined medium such as nutrient broth and King's A supplemented with 1% of different raw substrates in submerged fermentations with *P. aeruginosa* KU-BI02. Ground corn kernel produces 0.3414 μg mL^{-1} of PYO when added to nutrient broth, while adding the same amount to King's A medium yields 1.877 μg mL^{-1} of PYO. Using the same base medium (nutrient broth), the addition of 1% ground soybean also produces 0.1702 μg mL^{-1}, which increases to 1.702 μg mL^{-1} when sweet potato cooking water is added. Similarly, minimum concentrations of PYO (0.5106 μg mL^{-1}) are obtained using King's A medium supplemented with ground watermelon seeds until reaching 2.560 μg mL^{-1} of PYO when the same King's A medium is supplemented with ground soybean [17].

On the other hand, the FI fraction displayed a lower production of biomass and PYO, which indicated that its components had a negative effect on its production. This fraction contains low-molecular-weight polysaccharides (molecules smaller than 30 kDa) and free hydroxycinnamic acids according to [25,27,28]. In nejayote, ferulic acid, *p*-coumaric acid and other oligomers have been reported [21,26], as well as antimicrobial capacity. *p*-Coumaric acid inhibited the growth of *Escherichia coli* (99.9%) at 1000 μg mL^{-1}, while *Staphylococcus aureus* and *Bacillus cereus* were inhibited at 500 μg mL^{-1} [38]. Other reports indicate that 0.1 g L^{-1} of ferulic acid is the minimum inhibitory concentration for *E. coli* and *P. aeruginosa*, and 1.1 g L^{-1} and 1.25 g L^{-1} for *S. aureus* and *Listeria monocytogenes*, respectively [39]. The highest proportion of hydroxycinnamic acids present in maize wastewater are linked to other molecules such as carbohydrates. However, a significant amount of hydroxycinnamic acid has also been found in free form. The concentration of ferulic acid in nejayote depends on the nixtamalization condition process and corn variety. In nixtamalization processes using different types of maize, ferulic acid concentration has been reported with values of 99.1 for white corn, 96.74 for yellow corn, 88.63 for red corn and 84.95 mg/100 g of dry matter for blue corn [23]. In this work, for the FI fraction, ferulic acid concentration was 2.1 g L^{-1}, a higher concentration than that reported previously. The lowest concentration of PYO obtained in the nejayote fraction (FI) was 0.67 μg L^{-1}. These results indicated that *P. aeruginosa* NEJ01R strain displayed resistance to the antimicrobial effect of ferulic acid and inhibition of PYO production.

In nixtamalization wastewater fractions, a similar ferulic acid concentration was found, which prevented an antimicrobial effect as shown in Figure 3b. The effect of hydroxycinnamic acids on PYO production by *P. aeruginosa* was reported previously, demonstrating that 4 mM of ferulic acid can inhibit about 20% of PYO production [40]. In another study, it was shown that phenolic compounds, such as methyl gallate, have a negative effect on PYO production related to quorum sensing system and other molecules that inhibited the formation of biofilm, motility, proteolytic, elastase or rhamnolipid production in *P. aeruginosa* PAO1 [41]. In our results, after treatment with *P. aeruginosa* NEJ01R, the initial concentration of ferulic acid decreased by 10% for CL and FI fractions. In the RE fraction and RN substrate, values decreased to 15% and 20%, respectively, suggesting that *P. aeruginosa* NEJ01R can use molecules of ferulic acid as a carbon source.

Furthermore, during the cooking process of corn, to obtain nixtamal, 0.5–2% of calcium hydroxide is added, so this compound may be present in the wastewater obtained. It has been reported that the concentration of this compound in nejayote can be up to 1526.21 mg L^{-1} [22]. The calcium hydroxide residues present in nixtamalization wastewater can have a negative impact on low PYO production using the FI fraction since it is likely that a high concentration of calcium was present. It was also demonstrated that a fraction with a concentration of 3155.3 ± 5.24 mg L^{-1} of calcium was obtained using clarified nejayote filtered by a 1 kDa membrane [42]. Therefore, according to our results, the FI fraction may have had a high calcium concentration that interfered with PYO production by *P. aeruginosa* NEJ01R. In a study about the effect of calcium chloride concentration on PYO production by *P. aeruginosa*

NRRL B-771 and a PaJC mutant, it was demonstrated that 50 mM of calcium chloride in the presence of 1% sucrose produced a PYO concentration of 3.53 μg mL^{-1} and 3.06 μg mL^{-1} for *P. aeruginosa* NRRL B-771 and the PaJC mutant, respectively. When the calcium chloride concentration was increased to 250 mM, the PYO production decreased to 1.66 μg mL^{-1} and 2.51 μg mL^{-1} for *P. aeruginosa* NRRL B-771 and the PaJC mutant, respectively [43]. This suggests that the concentration of calcium in the culture medium has a negative effect on the PYO production of *P. aeruginosa* NEJ01R. If we consider the membrane size used to obtain the fractions RE and FI, it is likely that a higher concentration of calcium accumulates in the FI fraction, and this would have a negative effect on PYO production related to lower concentration in the FI fraction compared with the RE fraction. Then, the ultrafiltration process utilized showed an efficient system to fractionate nixtamalization wastewater that is feasible in PYO production by *P. aeruginosa* NEJ01R in order to valorize maize by-products.

3.4. Characterization of Maize Wastewater Fractions

Figure 4 shows values of parameters evaluated for characterization of nejayote fractions before and after the treatment. The results indicated that total solid (TS) content in the RE fraction increased due to ultrafiltration (UF) treatment; it was possible to obtain a fraction concentrate with 34% more solids than that contained in the RN substrate. The composition of nejayote fractions separated by UF (100 kDa) is mainly ferulated arabinoxylans, which are phenolic compounds linked to long-chain carbohydrates [25,27]. The concentration of solids decreased by 20% after treatment with *P. aeruginosa* NEJ01R, with the highest percentage of total solids being removed, as well as the CL fraction, in which 20% of TS was also removed. Similarly, for soluble solids (SS) in the treatment of RE fraction, a higher concentration of SS (20%) was removed, and for the CL fraction, 18% was removed. The highest percentage of removal of insoluble solids (IS) was obtained in RN (50%). The crude nejayote did not receive any previous treatment as the fractions that displayed a higher initial content of IS. For CL and RE fractions, the removal percentages of IS were 22% and 29%, respectively.

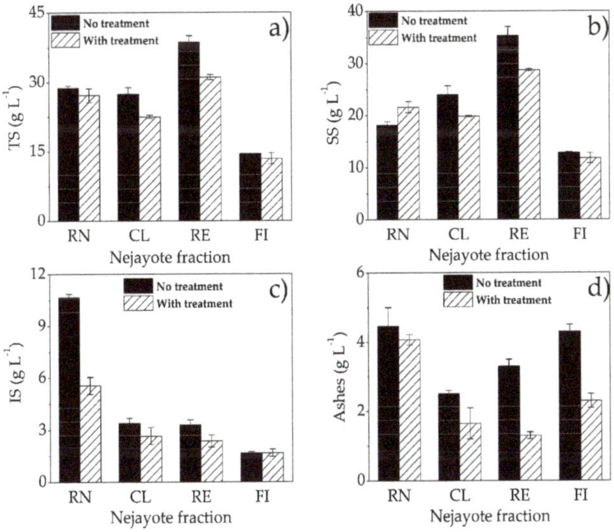

Figure 4. Determination of (**a**) total solids (ST), (**b**) soluble solids (SS), (**c**) insoluble solids (IS) and (**d**) ashes in nixtamalization wastewater crude (RN) and fractions (CL, RE and FI).

The FI fraction did not present IS removal, besides being the fraction that initially displayed the lowest IS concentration and was not used as a carbon source by *P. aeruginosa* NEJ01R. However, the FI fraction presented an ash removal of 45%, and the largest ash removal was in the RE fraction with 60%, followed by the CL fraction with 34% ash removal. These results showed that treatment with *P. aeruginosa* NEJ01R achieved a greater removal of solids in the RE fraction, which indicated that the concentration of solids favored the consumption of this as a carbon source by the strain used (*P. aeruginosa* NEJ01R). On the other hand, the fraction with the lowest solid removal was the FI fraction, which was also related to lower production of biomass and PYO, demonstrating that the type of molecules present in this fraction could have a negative effect, mainly in PYO production.

3.5. PYO Identification

The ultra-performance liquid chromatography system (UPLC system) was employed to analyze the sample extracted and compare it with a commercial standard (Sigma-Aldrich, St Louis, MO, USA). The sample extracted from the supernatant of the culture using *P. aeruginosa* NEJ01R showed two maximum peaks of 278 and 387 nm, similar to the standard commercial product (Figure S1, Supplementary Material). The UPLC chromatogram showed a single peak with a retention time of 1.45 min for PYO standard, which was comparable to that of the extracted sample, which suggested that it was the same molecule. On the other hand, the maximum at 278 nm of PYO was also reported for a culture of *P. aeruginosa* TBH2, an extracted sample obtained with chloroform and acidified with 1M HCl, and in the same way for PYO extracted from cultures of *P. aeruginosa* N11, *P. aeruginosa* D23 and a *P. aeruginosa* clinical isolate [3,44]. Figure S2 (Supplementary Material) displays the mass spectra of extract sample and standard PYO, giving a result of *m/z* 211 (M + H), which coincides with PYO mass of 210 g mol^{-1}. This result corresponds with that reported for a culture sample extract of *P. aeruginosa* BRp3 strain [45]. Our results demonstrated that the molecule extracted from the culture of *P. aeruginosa* NEJ01R was PYO.

3.6. Electrochemical Analysis of PYO

The redox behavior of PYO was analyzed with cyclic voltammetry. Figure 5a shows the electrochemical response of a PYO solution (0.14 mM) in the presence of 0.2 M phosphate buffer solution at pH 7. The analysis indicated a reduction process I_c with a peak potential E_{pc} of −0.269 V vs. Ag|AgCl 3 M NaCl value at 100 mV s^{-1}. In the reversal scan, a reduction process I_a was observed with a peak potential E_{pa} of −0.234 V vs. Ag|AgCl 3 M NaCl at the same scan rate.

The redox behavior of PYO was studied at different scanning rates. A normalization current analysis was carried out, coming from the following equation:

$$\frac{i}{nFAC_O^* D_O^{1/2}\left(\frac{nF}{RT}\right)^{1/2} v^{1/2}} = \pi^{1/2}x(\sigma t) \tag{4}$$

where n is the total number of electrons in the redox reaction, $D_o{}^{1/2}$ is the PYO diffusion coefficient, A is the electroactive area, C_o (PYO concentration) values are constants and $\pi^{1/2}\chi(\sigma t)$ is a dimensionless number.

This analysis suggested that the cathodic signal I_c did not have diffusional complications or coupled chemical reactions (Figure 5a). On the other hand, for signal I_a, the normalization of current values increased with $v^{1/2}$. The anodic peak potential E_{pa} varied significantly as the scan rate increased with shifts toward positive values. On the other hand, minimal changes in E_{pc} values were observed when $v^{1/2}$ was increased. This behavior suggested a reversible redox behavior with a possible adsorption for I_a. To confirm this idea, an analysis of the peak current i_p as a function of $v^{1/2}$ was also performed. According to the following equation (Randles Sevcik) [46], the value of i_p must present a linear relationship with $v^{1/2}$.

$$i_p = \left(2.69 \times 10^5\right) n^{3/2} A D_O{}^{1/2} C_O^* v^{1/2} \tag{5}$$

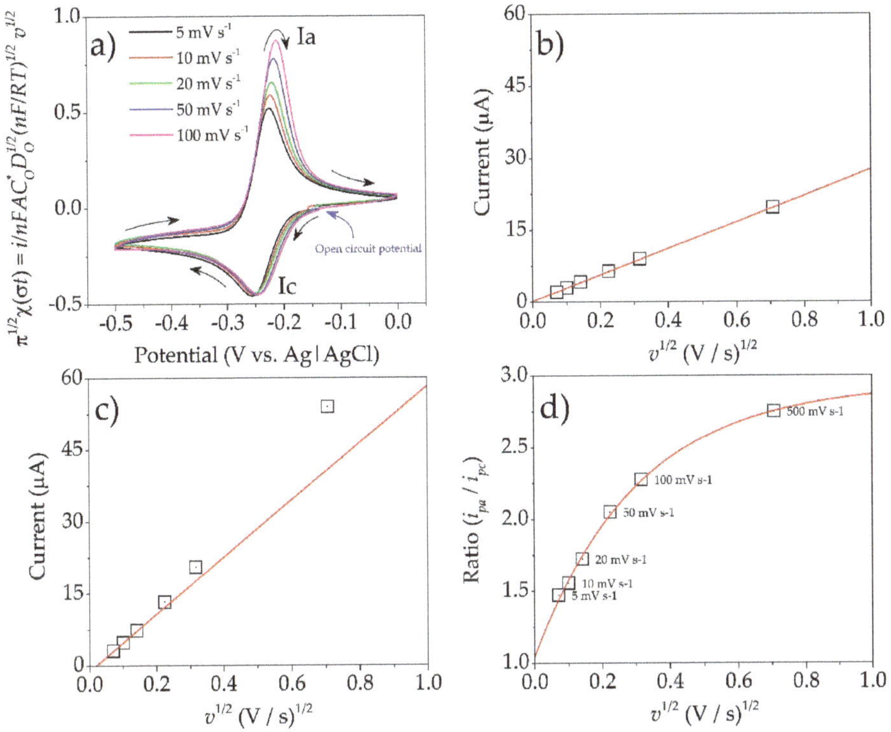

Figure 5. (**a**) Cyclic voltammetry of PYO (0.14 M) in current normalized at different scan rates (v). (**b**) i_{pc} values for PYO (0.14 M) as a function of $v^{1/2}$. (**c**) i_{pa} values for PYO (0.14 M) as a function of $v^{1/2}$. (**d**) i_{pa}/i_{pc} values of PYO (0.2 mM) as function of $v^{1/2}$.

Figure 5b presents the linear behavior of i_{pc} with respect to $v^{1/2}$, indicating that the electrochemical process *Ic* of PYO was controlled by diffusion. A change in linearity of i_{pa} vs $v^{1/2}$ was observed at high scan rates (Figure 5c). Hence, the ratio i_{pa}/i_{pc} at different scan rates was evaluated (Figure 5d). The behavior confirmed that the reversal oxidation process I_a was controlled by diffusion, with adsorption of PYO significantly reduced over the electrode surface [10,47]. Finally, considering low scan rates, where the adsorption process was not presented, a value of $|\Delta E_p| = 0.035$ V was calculated, which indicated a two-electron transfer, according to Equation (6). A value of $E_{1/2} = -0.251$ V vs. Ag|AgCl 3 M NaCl, related to the redox pair of PYO, was calculated according to Equation (7) [48,49]. Different applications of PYO have been described, in agriculture, medicine and the textile industry. However, the observation in two areas of application for this molecule is limited, namely in the extracellular transport of electrons in bioelectrochemical systems, specifically as biosensors, and in microbial electrochemical technologies. The phenomenon observed at high scanning speeds due to complications in the electrode related to absorption processes could be analyzed in future research to advance the application in the mentioned areas.

$$\left|E_{pa} - E_{pc}\right| = \frac{59 \text{ mV}}{n} \quad (6)$$

$$\frac{E_{pa} + E_{pc}}{2} = E_{1/2} \quad (7)$$

4. Conclusions

In this study, a coupled methodology of design of experiments and response surfaces was used to determine the optimal process conditions for obtaining PYO in submerged fermentation by *P. aeruginosa* NEJ01R. These conditions (29.6 °C, 223.7 rpm, pH 6.92) were evaluated experimentally, in LB medium, reaching values (2.05 $\mu g\ mL^{-1}$) close to those predicted statistically (2.21 $\mu g\ mL^{-1}$). We demonstrated that the by-product of the maize industry can be a culture medium, without supplementation, that supports the production of PYO, only in its pretreated fractions and separated by an ultrafiltration membrane. The highest PYO concentration (3.25 $\mu g\ mL^{-1}$) was obtained in the RE fraction at 48 h. Furthermore, the high concentration of ferulic acid (2.1 $g\ L^{-1}$) in the FI fraction did not have a negative effect on the growth of *P. aeruginosa*, which is related to resistance to its antimicrobial effect, although it did show a negative effect on the production of PYO. This is the first work that reports the use of wastewater as the only carbon source to synthesize PYO by a native strain of *P. aeruginosa* (NEJ01R). The molecule obtained showed redox behavior without diffusion complications at low scan rates.

Response surface plots indicated the relevance of factors with a significant effect for biomass and PYO production, indicating a positive effect using values in the low level (temperature, 21.5 °C; pH, 5; agitation, 76 rpm) for higher production in both responses. Optimized variables for PYO production in alkaline residual liquid of the maize industry and a detailed characterization establishing the composition of nejayote fractions and compounds that are involved in the metabolic induction of PYO are in progress.

Supplementary Materials: The following are available online at http://www.mdpi.com/2076-2607/8/10/1559/s1, Table S1: Experimental design matrix for biomass and PYO production in the defined culture medium, Figure S1: (a) Chromatograms and absorption spectra of PYO standard and (b) PYO extracted from *Pseudomonas aeruginosa* NEJ01R culture supernatant, Figure S2: (a) Mass spectra of PYO standard and (b) PYO extracted from *Pseudomonas aeruginosa* NEJ01R culture supernatant.

Author Contributions: Conceptualization, F.J.B.-V. and Y.R.-V.; investigation, M.L.F.-M. and J.A.P.-G.; formal analysis, F.E.-A.; validation, F.E.-A. and L.A.O.-F.; visualization, J.A.P.-G. and Y.R.-V.; writing—original draft preparation, F.J.B.-V. and Y.R.-V.; writing—review and editing, L.A.O.-F. and Y.R.-V.; funding acquisition, Y.R.-V. All authors have read and agreed to the published version of the manuscript.

Funding: This research was funded by Consejo Nacional de Ciencia y Tecnología (CONACYT), México (project 255468 Conacyt-Infraestructura, project 246052 Conacyt-SENER-Sustentabilidad Energética and project 258159 Conacy-SEP-Ciencia Básica).

Acknowledgments: F.J.B.-V., J.A.P.-G. and M.L.F.-M. are grateful to CONACYT for postdoctoral and doctoral scholarship support, and the authors are grateful to H.G. Cortes Cano for technical assistance and J. Peña-Castro for molecular biology assistance.

Conflicts of Interest: The authors declare no conflict of interest.

References

1. Samrot, A.V.; Rio, J.A.; Kumar, S.S.; Samanvitha, S.K. Bioprospecting studies of pigmenting *Pseudomonas aeruginosa* SU-1, *Microvirga aerilata* SU14 and *Bacillus megaterium* SU15 isolated from garden soil. *Biocatal. Agric. Biotechnol.* **2017**, *11*, 330–337. [CrossRef]
2. Yong, X.Y.; Yan, Z.Y.; Shen, H.B.; Zhou, J.; Wu, X.Y.; Zhang, L.J.; Zheng, T.; Jiang, M.; Wei, P.; Jia, H.H.; et al. An integrated aerobic-anaerobic strategy for performance enhancement of *Pseudomonas aeruginosa*-inoculated microbial fuel cell. *Bioresour. Technol.* **2017**, *241*, 1191–1196. [CrossRef] [PubMed]
3. Narenkumar, J.; Sathishkumar, K.; Sarankumar, R.K.; Murugan, K.; Rajasekar, A. An anticorrosive study on potential bioactive compound produced by *Pseudomonas aeruginosa* TBH2 against the biocorrosive bacterial biofilm on copper metal. *J. Mol. Liq.* **2017**, *243*, 706–713. [CrossRef]
4. Stancu, M.M. Production of some extracellular metabolites by a solvent-tolerant *Pseudomonas aeruginosa* strain. *Waste Biomass Valor.* **2018**, *9*, 1747–1755. [CrossRef]
5. El-Fouly, M.Z.; Sharaf, A.M.; Shahin, A.A.M.; El-Bialy, H.A.; Omara, A.M.A. Biosynthesis of pyocyanin pigment by *Pseudomonas aeruginosa*. *J. Radiat. Res. Appl. Sci.* **2015**, *8*, 36–48. [CrossRef]

6. Chatterjee, M.; D'Morris, S.; Vinod, P.; Warrier, S.; Vasudevan, A.K.; Vanuopadath, M.; Nair, S.S.; Paul-Prasanth, B.; Mohan, C.G.; Biswas, R. Mechanistic understanding of Phenyllactic acid mediated inhibition of quorum sensing and biofilm development in *Pseudomonas aeruginosa*. *Appl. Microbiol. Biotechnol.* **2017**, *101*, 8223–8236. [CrossRef]
7. Kang, H.; Gan, J.; Zhao, J.; Kong, W.; Zhang, J.; Zhu, M.; Li, F.; Song, Y.; Qin, J.; Liang, H. Crystal structure of *Pseudomonas aeruginosa* RsaL bound to promoter DNA reaffirms its role as a global regulator involved in quorum-sensing. *Nucleic Acids Res.* **2017**, *45*, 699–710. [CrossRef]
8. Patil, S.; Nikam, M.; Patil, H.; Anokhina, T.; Kochetkov, V.; Chaudhari, A. Bioactive pigment production by *Pseudomonas* spp. MCC 3145: Statistical media optimization, biochemical characterization, fungicidal and DNA intercalation-based cytostatic activity. *Process. Biochem.* **2017**, *58*, 298–305. [CrossRef]
9. Price-Whelan, A.; Dietrich, L.; Newman, D. Rethinking 'secondary' metabolism: Physiological roles for phenazine antibiotics. *Nat. Chem. Biol.* **2006**, *2*, 71–78. [CrossRef]
10. Oziat, J.; Gougis, M.; Malliaras, G.G.; Mailley, P. Electrochemical characterizations of four main redox–metabolites of *Pseudomonas aeruginosa*. *Electroanalysis* **2017**, *29*, 1332–1340. [CrossRef]
11. Qiao, Y.; Qiao, Y.J.; Zou, L.; Ma, C.X.; Liu, J.H. Real-time monitoring of phenazines excretion in *Pseudomonas aeruginosa* microbial fuel cell anode using cavity microelectrodes. *Bioresour. Technol.* **2015**, *198*, 1–6. [CrossRef] [PubMed]
12. Cayman Chemical Co. Home Page. Available online: https://www.caymanchem.com/product/10009594/pyocyanin (accessed on 4 September 2020).
13. Focus Biomolecules Home Page. Available online: https://focusbiomolecules.com/pyocyanin-ros-generator/ (accessed on 4 September 2020).
14. Santa Cruz Biotechnology Home Page. Available online: https://www.scbt.com/p/pyocyanin-85-66-5 (accessed on 4 September 2020).
15. Sigma-Aldrich Home Page. Available online: https://www.sigmaaldrich.com/catalog/product/sigma/p0046?lang=es&region=MX (accessed on 4 September 2020).
16. Bacame-Valenzuela, F.J.; Pérez-Garcia, J.A.; Castañeda-Zaldívar, F.; Reyes-Vidal, Y. Pyocyanin biosynthesis by *Pseudomonas aeruginosa* using a biodiesel byproduct. *Mex. J. Biotechnol.* **2020**, *5*, 1–16. [CrossRef]
17. DeBritto, S.; Gajbar, T.D.; Satapute, P.; Sundaram, L.; Lakshmikantha, R.Y.; Jogaiah, S.; Ito, S. Isolation and characterization of nutrient dependent pyocyanin from *Pseudomonas aeruginosa* and its dye and agrochemical properties. *Sci. Rep.* **2020**, *10*, 1542. [CrossRef] [PubMed]
18. Valderrama-Bravo, C.; Domínguez-Pacheco, F.; Hernández-Aguilar, C.; Flores-Saldaña, N.; Villagran-Ortiz, P.; Pérez-Reyes, C.; Sánchez-Hernández, G.; Oaxaca-Luna, A. Effect of nixtamalized maize with lime water (nejayote) on rheological and microbiological properties of masa. *J. Food Process. Preserv.* **2017**, *41*, e12748. [CrossRef]
19. Villada, J.A.; Sánchez-Sinencio, F.; Zelaya-Ángel, O.; Gutiérrez-Cortes, E.; Rodríguez-García, M.E. Study of the morphological, structural, thermal, and pasting corn transformation during the traditional nixtamalization process: From corn to tortilla. *J. Food Eng.* **2017**, *212*, 242–251. [CrossRef]
20. Rojas-García, C.; García-Lara, S.; Serna-Saldivar, S.O.; Gutiérrez-Uribe, J.A. Chemopreventive effects of free and bound phenolics associated to steep waters (Nejayote) obtained after nixtamalization of different maize types. *Plant. Foods Hum. Nutr.* **2012**, *67*, 94–99. [CrossRef]
21. Argun, M.S.; Argun, M.E. Treatment and alternative usage possibilities of a special wastewater: Nejayote. *J. Food Process. Eng.* **2017**, *48*, e12609. [CrossRef]
22. Castro-Muñoz, R.; Fila, V.; Durán-Páramo, E. A review of the primary by-product (Nejayote) of the nixtamalization during maize processing: Potential reuses. *Waste Biomass Valor.* **2019**, *10*, 13–22. [CrossRef]
23. Gutiérrez-Uribe, J.A.; Rojas-García, C.; García-Lara, S.; Serna-Saldivar, S.O. Phytochemical analysis of wastewater (nejayote) obtained after lime-cooking of different types of maize kernels processed into masa for tortillas. *J. Cereal Sci.* **2010**, *52*, 410–416. [CrossRef]
24. López-Pacheco, I.Y.; Carrillo-Nieves, D.; Salinas-Salazar, C.; Silva-Núñez, A.; Arévalo-Gallegos, A.; Barceló, D.; Afewerki, S.; Izbal, H.M.N.; Parra-Saldívar, R. Combination of nejayote and swine wastewater as a medium for *Arthrospira maxima* and *Chlorella vulgaris* production and wastewater treatment. *Sci. Total Environ.* **2019**, *676*, 356–367. [CrossRef]

25. Asaff-Torres, A.J.; Reyes-Vidal, M.Y. Un Método y un Sistema para el Tratamiento Integral de Aguas Residuales de una Industria del Maíz. WO/2014/119990. 2014. Available online: https://patentscope.wipo.int/search/es/detail.jsf?docId=WO2014119990 (accessed on 28 September 2020).
26. Ayala-Soto, F.; Serna-Saldívar, S.O.; García-Lara, S.; Pérez-Carrillo, E. Hydroxycinnamic acids, sugar composition and antioxidant capacity of arabinoxylans extracted from different maize fiber sources. *Food Hydrocoll.* **2014**, *35*, 471–475. [CrossRef]
27. Castro-Muñoz, R.; Cerón-Montes, G.I.; Barragán-Huerta, B.E.; Yáñez-Fernández, J. Recovery of carbohydrates from nixtamalization wastewaters (Nejayote) by ultrafiltration. *Rev. Mex. Ing. Quim.* **2015**, *14*, 735–744.
28. Castro-Muñoz, R.; Yáñez-Fernández, J. Valorization of nixtamalization wastewaters by integrated membrane process. *Food Bioprod. Process.* **2015**, *95*, 7–18. [CrossRef]
29. Niño-Medina, G.; Carvajal-Millán, E.; Lizardi, J.; Rascon-Chu, A.; Marquez-Escalante, J.A.; Gardea, A.; Martinez-López, A.L.; Guerrero, V. Maize processing waste water arabinoxylans: Gelling capability and cross-linking content. *Food Chem.* **2009**, *115*, 1286–1290. [CrossRef]
30. Rocha-Pizaña, M.R.; Chen, W.N.; Lee, J.J.L.; Bultimea-Cantúa, N.E.; González-Nimi, E.; Gutierrez-Uribe, J.A. Production of a potential collagenolytic protease by nejayote fermentation with *Aspergillus oryzae*. *Int. J. Food Sci. Technol.* **2020**, *55*, 3289–3296. [CrossRef]
31. Ramírez-Romero, G.; Reyes-Velazquez, M.; Cruz-Guerrero, A. Study of nejayote as culture medium for probiotics and production of bacteriocins. *Rev. Mex. Ing. Quim.* **2013**, *12*, 463–471.
32. Sanchez-Gonzalez, M.; Blanco-Gamez, A.; Escalante, A.; Valladares, A.G.; Olvera, C.; Parra, R. Isolation and characterization of new facultative alkaliphilic *Bacillus flexus* strains from maize processing waste water (nejayote). *Lett. Appl. Microbiol.* **2011**, *52*, 413–419. [CrossRef]
33. Baqueiro-Peña, I.; Contreras-Jácquez, V.; Kirchmayr, M.R.; Mateos-Díaz, J.C.; Valenzuela-Soto, E.M.; Asaff-Torres, A. Isolation and characterization of a new ferulic-acid-biotransforming *Bacillus megaterium* from maize alkaline wastewater (nejayote). *Curr. Microbiol.* **2019**, *76*, 1215–1224. [CrossRef]
34. García-Depraect, O.; Gómez-Romero, J.; León-Becerril, E.; López-López, A. A novel biohydrogen production process: Co-digestion of vinasse and nejayote as complex raw substrates using a robust inoculum. *Int. J. Hydrog. Energy.* **2017**, *42*, 5820–5831. [CrossRef]
35. Wilson, K. Preparation of genomic DNA from bacteria. *Curr. Protoc. Mol. Biol.* **2001**, *56*, 2–4. [CrossRef]
36. El-Shouny, W.A.; Al-Baidani, A.R.H.; Hamza, W.T. Antimicrobial activity of pyocyanin produced by *Pseudomonas aeruginosa* isolated from surgical wound-infections. *Intl. J. Pharm. Med. Sci.* **2011**, *1*, 1–7.
37. Castro-Muñoz, R.; Barragán-Huerta, B.E.; Fila, V.; Denis, P.C.; Ruby-Figueroa, R. Current role of membrane technology: From the treatment of agro-industrial by-products up to the valorization of valuable compounds. *Waste Biomass Valor.* **2018**, *9*, 513–529. [CrossRef]
38. Herald, P.J.; Davidson, P.M. Antibacterial activity of selected hydroxycinnamic acids. *J. Food Sci.* **1983**, *48*, 1378–1379. [CrossRef]
39. Borges, A.; Ferreira, C.; Saavedra, M.J.; Simões, M. Antibacterial activity and mode of action of ferulic and gallic acids against pathogenic bacteria. *Microb. Drug Resist.* **2013**, *19*, 256–265. [CrossRef] [PubMed]
40. Ugurlu, A.; Yagci, A.K.; Ulusoy, S.; Aksu, B.; Bosgelmez-Tinaz, G. Phenolic compounds affect production of pyocyanin, swarming motility and biofilm formation of *Pseudomonas aeruginosa*. *Asian Pac. J. Trop. Biomed.* **2016**, *6*, 698–701. [CrossRef]
41. Hossain, M.A.; Lee, S.J.; Park, N.H.; Mechesso, A.F.; Birhanu, B.T.; Kang, J.; Reza, A.; Suh, J.W.; Park, S.C. Impact of phenolic compounds in the acyl homoserine lactone-mediate quorum sensing regulatory pathways. *Sci. Rep.* **2017**, *7*, 10618. [CrossRef]
42. Castro-Muñoz, R.; Barragán-Huerta, B.E.; Yáñez-Fernández, J. The use of nixtamalization waste waters clarified by ultrafiltration for production of a fraction rich in phenolic compounds. *Waste Biomass Valor.* **2016**, *7*, 1167–1176. [CrossRef]
43. Özcan, D.; Kahraman, H. Pyocyanin production in the presence of calcium ion in *Pseudomonas aeruginosa* and recombinant bacteria. *Turkish J. Sci. Technol.* **2015**, *10*, 13–19.
44. Kerr, J.R.; Taylor, G.W.; Rutman, A.; HØiby, N.; Cole, P.J.; Wilson, R. *Pseudomonas aeruginosa* pyocyanin and 1-hydroxyphenazine inhibit fungal growth. *J. Clin. Pathol.* **1999**, *52*, 385–387. [CrossRef]
45. Yasmin, S.; Hafeez, F.Y.; Mirza, M.S.; Rasul, M.; Arshad, H.M.I.; Zubair, M.; Iqbal, M. Biocontrol of bacterial leaf blight of rice and profiling of secondary metabolites produced by rhizospheric *Pseudomonas aeruginosa* BRp3. *Front. Microbiol.* **2017**, *8*, 1895. [CrossRef]

46. Bard, A.J.; Faulkner, L.R. *Electrochemical Methods: Fundamentals and Applications*, 2nd ed.; Harris, D., Swain, E., Eds.; John Wiley & Sons, Inc.: New York, NY, USA, 2000.
47. Wopschall, R.H.; Shain, I. Effects of adsorption of electroactive species in stationary electrode polarography. *Anal. Chem.* **1967**, *39*, 1514–1527. [CrossRef]
48. Friedheim, E.; Michaelis, L. Potentiometric study of piocyanine. *J. Biol. Chem.* **1931**, *91*, 355–368.
49. Wang, Y.; Newman, D.K. Redox reactions of phenazine antibiotics with ferric (hydr)oxides and molecular oxygen. *Environ. Sci. Technol.* **2008**, *42*, 2380–2386. [CrossRef] [PubMed]

microorganisms

MDPI

Article

High-Titer Lactic Acid Production by *Pediococcus acidilactici* PA204 from Corn Stover through Fed-Batch Simultaneous Saccharification and Fermentation

Zhenting Zhang [1], Yanan Li [1], Jianguo Zhang [1,2], Nan Peng [1], Yunxiang Liang [1] and Shumiao Zhao [1,*]

[1] State Key Laboratory of Agricultural Microbiology, College of Life Science and Technology, Huazhong Agricultural University, Wuhan 430070, China; zhangzhenting@webmail.hzau.edu.cn (Z.Z.); lana463936202@163.com (Y.L.); zhangjianguo@hfut.edu.cn (J.Z.); nanp@mail.hzau.edu.cn (N.P.); fa-lyx@163.com (Y.L.)

[2] School of Food and Biological Engineering, Hefei University of Technology, Hefei 230009, China

* Correspondence: shumiaozhao@mail.hzau.edu.cn; Tel.: +86-27-8728-1267; Fax: +86-27-8728-0670

Received: 12 August 2020; Accepted: 22 September 2020; Published: 28 September 2020

Abstract: Lignocellulose comprised of cellulose and hemicellulose is one of the most abundant renewable feedstocks. Lactic acid bacteria have the ability to ferment sugar derived from lignocellulose. In this study, *Pediococcus acidilactici* PA204 is a lactic acid bacterium with a high tolerance of temperature and high-efficiency utilization of xylose. We developed a fed-batch simultaneous saccharification and fermentation (SSF) process at 37 °C (pH 6.0) using the 30 FPU (filter paper units)/g cellulase and 20 g/L corn steep powder in a 5 L bioreactor to produce lactic acid (LA). The titer, yield, and productivity of LA produced from 12% (*w/w*) NaOH-pretreated and washed stover were 92.01 g/L, 0.77 g/g stover, and 1.28 g/L/h, respectively, and those from 15% NaOH-pretreated and washed stover were 104.11 g/L, 0.69 g/g stover, and 1.24 g/L/h, respectively. This study develops a feasible fed-batch SSF process for LA production from corn stover and provides a promising candidate strain for high-titer and -yield lignocellulose-derived LA production.

Keywords: lactic acid fermentation; *Pediococcus acidilactici* PA204; simultaneous saccharification and fermentation; corn stover

1. Introduction

Lignocellulose, the most abundant global source of renewable biomass, is one of the most important raw materials for biofuel and biochemical production [1]. Corn stover, a lignocellulosic feedstock, is one of the most important agricultural residues available in high quantities with about 900 million tons produced in 2018 in China according to Ministry of Agriculture and Rural Affairs. Many studies have examined corn stover applied in different fields, such as generating electricity [2], biofuel [3] and biochemical production [4], and biological feed [5]. Especially, lactic acid (LA) is an important commodity chemical and also a monomer compound to produce biodegradable and biocompatible polylactic acid (PLA), which provides a sustainable alternative to petroleum-derived products [6]. The corn stover usually contains 37.5% cellulose, 22.4% hemicellulose, and 17.6% lignin, which can be hydrolyzed into hexose and pentose for fermentation [7].

However, lignocellulose hydrolysis and utilization remain considerable challenges in lignocellulose-derived biofuel and biochemical production [8]. At present, the main obstacles include lignocellulose pretreatment and hydrolysis, as well as the efficient fermentation of pentose derived from lignocellulosic hydrolysates into LA. Many studies have evaluated lignocellulose degradation and biofuel and biochemical production by using physical [9,10], chemical, and biological [8] pre-treatments. Chemical methods are the most common pretreatment methods, including acid treatment, alkaline treatment, alkaline/oxidative treatment, wet oxidation, and ozonolysis [11–13]. Xylose limits the conversion from the hydrolysate of lignocellulose into LA [14]. Therefore, it is necessary to isolate or construct the strains that can efficiently utilize pentose. Several lactic acid bacteria (LAB) such as *Lactobacillus pentosus* [4], *Lactobacillus brevis* [15], *Enterococcus mundtii* QU 25 [16], and *Enterococcus faceium* QU 50 [17] have been reported to ferment xylose via the phosphoketolase pathway (hetero-fermentation pathway). Recently, including *Lactobacillus* strains [4,15,18], *Bacillus coagulans* [12,19–21], and *Pediococcus acidilactici* [22,23], they have also been reported to produce high-titer LA from lignocellulosic materials. These strains with robust inhibition tolerance were found to be suitable for lignocellulose-derived LA production and were engineered for chemical production because of their thermophilic growth characteristics (except *Lactobacillus* strains) and strong pentose homofermentative activity.

Acid-pretreatment methods were used to efficiently produce LA. LA yield and titer produced from oil palm empty fruit bunch (OPEFB) acid hydrolysate reached 0.97 g/g and 59.2 g/L through *B. coagulans* fermentation, respectively [24]. LA yield and titer obtained from acid-pretreated wheat stover reached 0.46 g/g (wheat stover) and 38.73 g/L, respectively, via *B. coagulans* IPE22 fermentation [20]. Excellent LA production was obtained from sulfuric acid-pretreated and biodetoxified corn stover through *Pediococcus acidilactici* DQ2 fermentation [25] with a good LA titer of 101.9 g/L and poor yield of a mere 0.38 g/g stover as *P. acidilactici* DQ2 cannot utilize xylose. A high titer (104.4 g/L) of L-LA was obtained from dilute acid-pretreated and biodetoxified corn stover through an engineered *P. acidilactici* TY112 (CGMCC 8664) strain fermentation, and the yield of L-LA reached 0.72 g/g glucose from total corn stover regardless of xylose unavailability [26]. However, the acid-pretreated method can lead to hemicellulose loss and cannot efficiently remove lignin [27,28]. Based on this, alkali-pretreated methods were exploited and applied. In our previous studies, *B. coagulans* LA204 and *L. pentosus* FL0421 could efficiently utilize corn stover and corncob to produce LA [4,12,13]. By using NaOH-pretreated and washed corn stover, the LA yield and titer increased to 0.68 g/g substrate and 97.6 g/L, respectively, through *B. coagulans* LA204 fed-batch fermentation [12]. The LA yield and titer obtained from alkali-pretreated corn stover reached 0.66 g/g corn stover and 92.3 g/L, respectively, via *L. pentosus* FL0421 simultaneous saccharification and fermentation (SSF) [4]. By using *B. coagulans* LA204 fed-batch fermentation, the lactic acid titer and yield produced from 16% (*w/w*) NaOH-pretreated corncob were 122.99 g/L and 0.77 g/g, respectively, and those produced from 16% NH_3-H_2O_2-pretreated and washed corncob were 118.60 g/L and 0.74 g/g corncob, respectively [13].

Pediococcus acidilactici PA204 with high temperature tolerance (32–47 °C) and efficient xylose conversion into lactic acid was used for high titer lactic acid production at the high solids loading of corn stover through the simultaneous saccharification and fermentation (SSF). As probiotics, *P. acidilactici* PA204 produces bacteriocin, which has a good inhibitory effect on some pathogenic microorganisms in the intestinal tract [29]. Additionally, this strain has long been announced to be safe as a probiotic strain that can be used in food and drugs by the food and drug administration (FDA) [30].

In this study, we developed a fed-batch SSF process for LA production using corn stover as a carbon source through *P. acidilactici* PA204 fermentation, and found that this strain produced LA from 12% (*w/w* loading) NaOH-pretreated and washed corn stover with a titer of 92.01 g/L and yield of 0.77 g/g stover. It also produced acetic acid from the same corn stover with a titer of 10.03 g/L and yield of 0.08 g/g stover under nonsterile conditions. Our results provide a practical process for lactic acid production from lignocellulose residues. This study reveals that *P. acidilactici* PA204 is a promising candidate strain for high-yield and -titer lignocellulose-derived LA production.

2. Materials and Methods

2.1. Raw Materials and Enzyme

Corn stover was grown in the Northeast of China and harvested in the autumn of 2016. After harvesting, corn stover was cleaned, dried, and sieved with the 80-mesh and pretreated at 75 °C for 3 h with 5% sodium hydroxide (NaOH) at a 20% (*w*/*w*) loading. The resultant slurry was washed with water until the pH decreased to 8.0, and then filtered to a moisture content of 20% (*w*/*w*). The raw corn stover consisted of 37.12 ± 0.32 cellulose, 29.60 ± 0.26 hemicellulose, and 20.80 ± 0.56 lignin. The pretreated corn stover contained 51.34 ± 0.57 cellulose, 27.20 ± 0.34 hemicellulose, and 8.31 ± 0.89 lignin. The pretreated and washed corn stover contained 60.01 ± 0.51 cellulose, 27.33 ± 0.21 hemicellulose, and 6.87 ± 0.66 lignin. The cellulase used in this study was Cellic CTec2 (Novozymes, Denmark) containing cellulase, β-glucosidase, and xylanase, with its cellulose activity of 250 filter paper units (FPU)/mL.

2.2. Medium and Strain

The De Man-Rogosa-Sharpe (MRS) medium was used for seed culturing. MRS medium contained 10 g of peptone, 10 g of beef extract, 5 g of yeast extract, 20 g of dextrose, 5 g of sodium acetate trihydrate, 1 g of polysorbate 80, 2 g of dipotassium phosphate, 2 g of triammonium citrate, 0.25 g of magnesium sulfate heptahydrate, and 0.05 g of manganese sulfate tetrahydrate in 1 L deionized water [31]. The medium and water were autoclaved at 115 °C for 20 min. *Pediococcus acidilactici* PA204 was isolated and stored in our laboratory. Activation cultures were carried out in MRS medium at 37 °C for 24 h.

2.3. Simultaneous Saccharification and Fermentation (SSF), and NaOH-Pretreated Corn Stover

The mixture of 4% NaOH-pretreated and washed corn stover, 20 g/L corn steep powder, and cellulase at a concentration of 30 FPU/g stover was inoculated with 200 mL seed culture to establish the SSF process in a 2 L volume. Fermentation was carried out at 37 °C for 60 h with agitation at 150 rpm. The pH was maintained at 6.0 by automatic feeding of 10 M NaOH solution.

Fed-batch fermentations were performed with 8% pretreated and washed corn stover, a cellulase concentration of 30 FPU/g stover, 20 g/L corn steep powder, and 200 mL seed culture (*v*/*v*) in a total volume of 2 L. During fermentation, the washed corn stover was continuously fed at 24 h, and the final concentration of pretreated and washed corn stover reached 12% (*w*/*w*) or 15% (*w*/*w*) and total fermentation volume reached approximately 3 L. Enzyme feeding for all fermentations is illustrated in Figure 3 with the final enzyme concentration of 30 FPU/g stover. Samples were collected every 6 or 12 h. The concentrations of LA, acetic acid, glucose, and xylose were determined by HPLC. The fed corn stover was not sterilized.

2.4. Analysis of Sugars, Acids, and Biomass

Glucose, xylose, LA, acetic acid, and formic acid were analyzed using an Agilent 1200 HPLC system equipped with an RID-10A detector or an SPD-20A detector, and a Bio-Rad Aminex HPX-87H column with a column temperature of 40 °C. The mobile phase was 5 mM H_2SO_4, and the flow rate was set as 0.6 mL/min. The LA yield was defined as the produced LA (g) per total sugar or total corn stover (g). All samples were centrifuged and filtered through a 0.22 μm membrane prior to loading. The strain biomass was counted by the dilution plate method.

3. Results and Discussion

3.1. Effects of Temperature, pH, Carbon Source, and Nitrogen Source on LA Production

P. acidilactici PA204, which can efficiently utilize glucose to produce lactic acid, was isolated and stored in the Fermentation Engineering Laboratory at Huazhong Agriculture University. In this study, the fermentation abilities of *P. acidilactici* PA204 to glucose, xylose, and xylobiose were tested in flasks.

The sugar consumptions of glucose, xylose, and xylobiose media reached 94%, 86%, and 55% in 10 g/L sugar concentration, respectively. We found that the cell count in the glucose media was higher than those of other sugars (Figure 1A). These results suggested that *P. acidilactici* PA204 might have a strong ability to ferment lignocellulosic hydrolysates (glucose and xylose) into LA. We further examined the effect of temperature on *P. acidilactici* PA204 LA fermentation in glucose media. As shown in Figure 1B, the maximum LA yields (0.97 and 0.98 g/g sugar) were achieved at 37 °C and 42 °C, respectively, while lower yields were found at other temperatures. The value of OD600 reached the highest (5.82 and 5.52) at 37 °C and 42 °C, respectively. However, at 57 °C, the value of OD600 was only 1.34; thus, lactic acid can barely be produced. The effect of initial pH values on LA fermentation was also examined. The results indicated that an initial pH of 6.0 was optimal for lactic acid fermentation by this strain. As can be seen from Figure 1C, a higher LA titer was obtained at pH 6.0 (9.74 g/L vs. 10.27 g/L, pH 5.0 vs. 6.0). In addition, the effect of nitrogen source concentration on LA fermentation was examined with 10 g/L glucose as the carbon source and different concentrations of corn steep powder as the nitrogen source (Figure 1D). The results indicated that the highest LA concentration (9.92 g/L) was produced from corn steep powder (at the concentration of 20 g/L) and that the cell count in the media containing 20 and 30 g/L concentrations of corn steep powder was higher than that in other media, suggesting that the effects of nitrogen sources concentrations on lactic acid production efficiency coincided with those on cell growth efficiency. Moreover, the increase in nitrogen source concentration did not significantly affect the production of LA. These results indicated that corn steep powder can be used by *P. acidilactici* PA204 as an inexpensive nitrogen source for LA production and that the concentration of 20 g/L can be determined as the optimal nitrogen source concentrations.

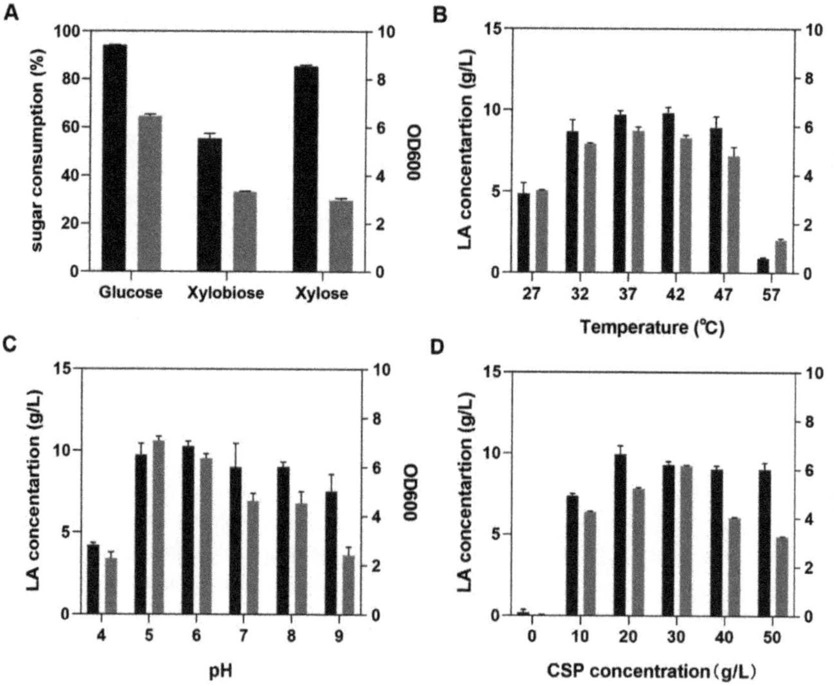

Figure 1. General features of *P. acidilactici* PA204 fermentation. **(A)** Lactic Acid (LA) production efficiency from glucose, xylose, and xylobiose media. Culture contains 10 g/L of each sugar and 5 g/L of

yeast extract, pH 6.0; fermentation was carried out at 37 °C. (**B**) Effect of temperature on LA fermentation. Culture contains 10 g/L glucose and 5 g/L yeast extract (pH 6.0). (**C**) Effect of pH on LA fermentation. Culture contains 10 g/L glucose and 5 g/L yeast extract; fermentation was carried out at 37 °C. (**D**) Effect of CPS at different concentrations on LA production. CSP: Corn steep powder. Culture contains 10 g/L glucose (pH 6.0). Fermentation was carried out at 37 °C. Black bars: Sugar consumption efficiency and LA concentration; gray bars: OD600.

3.2. High-Titer and High-Yield LA Production from NaOH-Pretreated Corn Stover through SSF

NaOH pretreatment was reported to be an efficient method to remove lignin and destroy the structure of lignocellulose in our previous studies [12]. In this study, we selected NaOH pretreatment to remove lignin from corn stover and determined the compositional changes of corn stover before and after NaOH pretreatments. The higher the solubilization and reduction of lignin, the greater the cellulose composition increase of pre-treated corn stover [28]. As shown in Table 1, the solid fraction from NaOH pre-treated and washed corn stover showed a significant 61.66% increase in cellulose composition due to the sharp 66.97% decline in the percentage of lignin, and a lower marked 7.67% decrease in hemicellulose composition. The results indicated that NaOH pretreatment is an efficient method to remove lignin and retain cellulose and hemicellulose.

Table 1. The composition of different corn stover in this study. A: Raw corn stover. B: NaOH pre-treated corn stover. C: NaOH pre-treated corn stover and washed corn stover.

Corn Stover	Cellulose (%)	Hemicellulose (%)	Lignin (%)
A	37.12 ± 0.32	29.60 ± 0.26	20.80 ± 0.56
B	51.34 ± 0.57	27.20 ± 0.34	8.31 ± 0.89
C	60.01 ± 0.51	27.33 ± 0.21	6.87 ± 0.66

In the initial SSF experiment, 4% (*w/w*) NaOH-pretreated and washed corn stover and 20 g/L corn steep powder were the carbon source and nitrogen source, respectively. Cellulase was added at the beginning of fermentation, and the final enzyme concentration used was 30 FPU/g stover. The 10 M NaOH solution was used as the neutralizer. Initially, LA was produced rapidly. At hour 6 during fermentation, the LA titer reached 15.35 g/L, and the productivity reached 2.59 g/L/h (Figure 2), suggesting that for the first 6 h, corn stover rapidly degraded by cellulase and xylanase into glucose and xylose. Then, *P. acidilactici* PA204 quickly utilized sugar to produce LA. However, at hour 48 during fermentation, the LA titer reached 25.92 g/L, the yield reached 0.65 g/g stover, and the average productivity was 0.54 g/L/h (Table 2). The average productivity at hour 48 was significantly lower than that at the early fermentation stage. At hour 33 during fermentation, glucose and xylose were hardly detected and no significant increase in LA titer was observed from hour 33 (25.48 g/L) to hour 48 (25.92 g/L). During fermentation, the amounts of bacteria was gradually increased, and reached 9.1×10^9 cfu/mL at hour 27. After that, the increase in bacterium count tended to be stable, indicating the end of fermentation.

Table 2. Summary of lactic acid fermentation by *P. acidilactici* PA204 using NaOH-pretreated and washed corn stover with washing as the carbon source. 4% stover: Simultaneous saccharification and fermentation (SSF) batch experiment. 12% and 15% stover: Fed-batch experiments.

Experiments	4% Stover	12% Stover	15% Stover
Lactic acid titer (g/L)	25.92	92.01	104.11
Lactic acid yield (g/g stover)	0.65	0.77	0.69
Lactic acid productivity (g/L/h)	0.54	1.28	1.24
Acetic acid titer (g/L)	6.57	10.03	10.28
Acetic acid yield (g/g stover)	0.16	0.08	0.07
Acetic acid productivity (g/L/h)	0.14	0.14	0.12

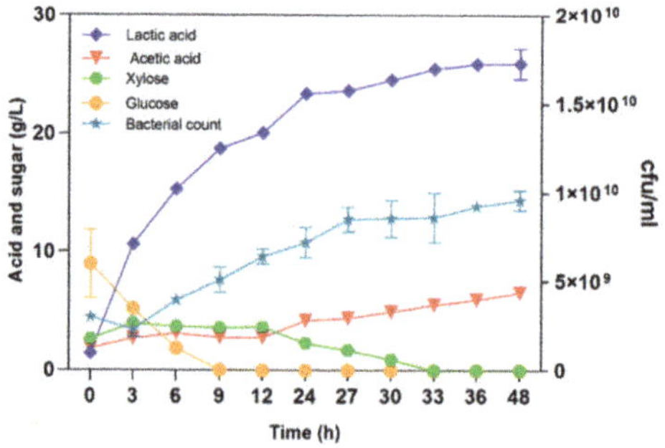

Figure 2. Lactic acid production from NaOH-pretreated and washed corn stover (4% *w/w*). The final cellulase concentration was 30 FPU/g stover. Diamonds indicate lactic acid. Inverted triangles represent acetic acid. Dots indicate glucose. Polygons represent xylose. Pentacle represent bacterial count. The three fermentation stages are presented in the figures.

3.3. High-Titer and High-Yield LA Production through Fed-Batch Fermentation

A good LA yield was obtained from 4% (*w/w*) NaOH-pretreated and washed corn stover, but the LA titer was not sufficient. Thus, a high solid loading of lignocellulosic materials was necessary to increase the LA titer. Considering the substance inhibition effect, the fed-batch experiment was carried out to increase the LA titer and yield. In two fed-batch experiments, 8% (*w/w*) NaOH-pretreated and washed corn stover (total of 2 L) was inoculated with 10% seed culture and supplemented with 30 FPU/g of the cellulase solution. At hour 24 during fermentation, the pretreated and washed corn stover was continuously fed to reach the final corn stover concentration of 12% (*w/w*) or 15% (*w/w*), and the enzyme was fed to reach the final enzyme concentration of 30 FPU/g stover (Figure 3). Finally, the titers of LA produced from high solid loadings of NaOH-pretreated and washed corn stover were tested. As Figure 3A shows, the LA yield from 12% solid loading reached (*w/w*) 0.77 g/g stover, LA titer reached 92.01 g/L, and the average productivity was 1.28 g/L/h. The acetic acid titer reached 10.03 g/L and its yield was 0.08 g/g stover (Table 2). However, the LA yield from 15% solid loading (*w/w*) reached 0.69 g/g stover, the titer reached 104.11 g/L, and the average productivity was 1.24 g/L/h (Figure 3B). The acetic acid titer reached 10.28 g/L and its yield was 0.07 g/g stover (Table 2).

Three basic stages were identified in the LA fed-batch fermentation process using corn stover as the substrate (Figure 3). During the first stage (0–12 h), LA was quickly produced in all two experiments, and during this period, the productivity was 3.05 g/L/h (12% *w/w* loading, Figure 3A), and 3.90 g/L/h (15% *w/w* loading, Figure 3B), respectively. These results indicated that LA productivity was dependent on the concentration of corn stover (8% *w/w* loading) at this stage. Therefore, both glucose and xylose were rapidly released, and then consumed (Figure 3). During the second stage (12–36 h) at 12% *w/w* corn stover loading (Figure 3A), LA production slowed down before being fed with corn stover (1.01 g/L/h, 12–24 h). However, after being fed with stover and enzyme, LA significantly increased (1.69 g/L/h, 24–36 h). Similarly, during the second stage (12–36 h) at 15% *w/w* corn stover loading (Figure 3B), LA production slowed down before corn stover feeding (1.04 g/L/h, 12–24 h) and LA increased (1.15 g/L/h, 24–36 h) after stover and enzymes feeding. Interestingly, LA productivity at 12% *w/w* stover feeding was slightly higher than that at 15% *w/w* stover feeding (1.69 g/L/h vs. 1.15 g/L/h, Figure 3), indicating that LA productivity was dependent on the concentration of corn stover, and that the inhibitors in the corn stover might inhibit LA production [4]. The sugars consumption was observed. Glucose was found to be quickly released and consumed completely. However, xylose was accumulated and was not utilized fully at the end of fermentation (Figure 3). The result may be produced due to substrate inhibition. When a large amount of xylose was continuously released, it could not be consumed and utilized by *P. acidilactici* PA204 in time. Phenolic compounds in corn stover might inhibit utilization of xylose by *P. acidilactici* PA204. The underlying mechanism needed to be further investigated.

During the third stage (36 h to the end of fermentation), LA titer was increased continuously and slowly in fed-batch experiments, and significant differences in LA titer and productivity were observed between the 12% *w/w* feeding group (92.01 g/L and 0.634 g/L/h) and 15% *w/w* feeding group (104.11 g/L and 0.646 g/L/h). The glucose was consumed completely at hour 36 and xylose remained at a certain concentration until the end of fermentation (5.29 g/L) in the 12% *w/w* feeding group (Figure 3A). However, in the 15% *w/w* feeding group, the glucose was consumed completely at hour 60 and xylose remained at the concentration of 9.66 g/L (Figure 3B). Although the concentration of lactic acid was increased, lactic acid yield was slightly decreased. These results suggested that a high solid loading of corn stover could release more sugars and increase the yield of lactic acid to some extent. However, a high solid loading caused the delayed consumption of glucose and retention of more xylose. The possible reason might lie in that more pretreated corn stover might result in more inhibitors, as well as a substrate effect.

As shown in Table 2, the LA titer, yield, and productivity at 12% (*w/w*) NaOH-pretreated and washed corn stover feeding were 92.01 g/L, 0.77 g/g stover, and 1.28 g/L/h, respectively, and at 15% (*w/w*) NaOH-pretreated and washed stover feeding, they were 104.11 g/L, 0.69 g/g stover, and 1.24 g/L/h, respectively. Generally, the chemical pretreatment method could induce inhibitors generation, whereas the washing method could obviously reduce the inhibitor concentration and improve LA production. However, the major disadvantage of the washing method is the substantial volume of waste washing water generated by inhibitor removal. In the industrial production process, wastewater should be strictly limited because of the high cost of wastewater treatment. Therefore, other methods including biological detoxification, pretreatment with few inhibitors, and screening of inhibitor-tolerant strains need to be developed in order to increase the LA titer and yield by using unwashed pretreated corn stover.

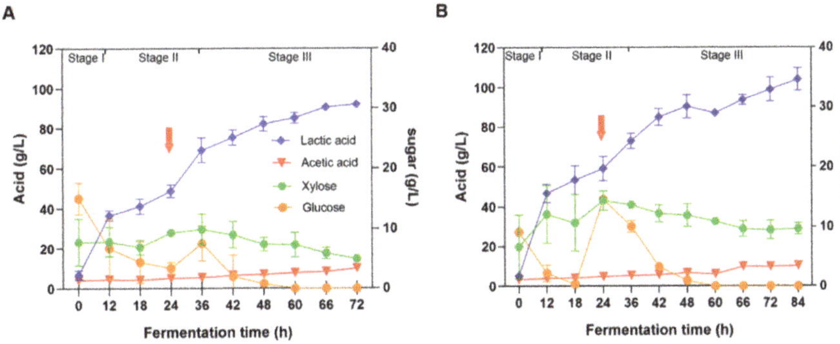

Figure 3. Lactic acid production from NaOH-pretreated and washed corn stover. (**A**) Initial 8% (*w/w*) NaOH-pretreated and washed corn stover was fed with the stover treated in the same manner at hour 24 to reach the final concentration of 12.0% (*w/w*). (**B**) Initial 8% (*w/w*) NaOH-pretreated and washed corn stover was fed with the stover treated in the same manner at hour 24 to reach the final concentration of 15.0% (*w/w*). Substrate and enzyme feedings are indicated by red arrows. The final cellulase concentration in both experiments was 30 FPU/g stover. Diamonds indicate lactic acid. Inverted triangles represent acetic acid. Dots indicate glucose. Polygons represent xylose. The three fermentation stages are presented in the figures.

4. Conclusions

In this study, we produced high-titer and high-yield LA from NaOH-pretreated and washed corn stover through *P. acidilactici* PA204 fermentation. Importantly, *P. acidilactici* PA204 was found to specifically ferment xylose to produce lactic acid. An efficient SSF fed-batch process for LA production from NaOH-pretreated and washed corn stover by *P. acidilactici* PA204 has been established at 37 °C and pH 6.0 with a cellulase activity of 30 FPU/g stover and 20 g/L corn steep powder in a 5 L bioreactor. The lactic acid titer, yield, and productivity were 92.01 g/L, 0.77 g/g stover, and 1.28 g/L/h at 12% (*w*/*w*) NaOH-pretreated and washed stover feeding, respectively, and they were 104.11 g/L, 0.69 g/g corncob, and 1.24 g/L/h at 15% NaOH-pretreated and washed stover feeding, respectively. What is more, fermentation without sterilization has advantages in reducing production cost and energy consumption. This study develops a feasible fed-batch SSF process for LA production from corn stover and provides a promising candidate strain for high-titer and -yield lignocellulose-derived LA production.

Author Contributions: Conceptualization, S.Z., J.Z., N.P. and Y.L. (Yunxiang Liang); methodology, Z.Z. and Y.L. (Yanan Li); formal analysis, Z.Z.; data curation, Z.Z. and Y.L. (Yanan Li); writing—original draft preparation, Z.Z.; writing—review and editing, S.Z.; visualization, Z.Z.; supervision, S.Z.; project administration, S.Z. and J.Z.; funding acquisition, S.Z. and J.Z. All authors have read and agreed to the published version of the manuscript.

Funding: This work was supported by the Fundamental Research Funds for the Central Universities (No. 2662018JC016) and State Key Laboratory of Agricultural Microbiology (grant AMLKF201808).

Conflicts of Interest: The authors declare no conflict of interest.

References

1. Lin, Y.; Tanaka, S. Ethanol fermentation from biomass resources: Current state and prospects. *Appl. Microbiol. Biotechnol.* **2006**, *69*, 627–642. [CrossRef] [PubMed]
2. Wang, X.; Feng, Y.; Wang, H.; Qu, Y.; Yu, Y.; Ren, N.; Li, N.; Wang, E.; Lee, H.; Logan, B.E. Bioaugmentation for Electricity Generation from Corn Stover Biomass Using Microbial Fuel Cells. *Environ. Sci. Technol.* **2009**, *43*, 6088–6093. [CrossRef] [PubMed]
3. Chen, S.; Xu, Z.; Li, X.; Yu, J.; Cai, M. Integrated bioethanol production from mixtures of corn and corn stover. *Bioresour. Technol.* **2018**, *258*, 18–25. [CrossRef] [PubMed]
4. Hu, J.; Lin, Y.; Zhang, Z.; Xiang, T.; Mei, Y.; Zhao, S.; Liang, Y.; Peng, N. High-titer lactic acid production by *Lactobacillus pentosus* FL0421 from corn stover using fed-batch simultaneous saccharification and fermentation. *Bioresour. Technol.* **2016**, *214*, 74–80. [CrossRef] [PubMed]
5. Li, Y.; Xu, F.; Li, Y.; Lu, J.; Li, S.; Shah, A.; Zhang, X.; Zhang, H.; Gong, X.; Li, G. Reactor performance and energy analysis of solid state anaerobic co-digestion of dairy manure with corn stover and tomato residues. *Waste Manag.* **2018**, *73*, 130–139. [CrossRef]
6. Abdel-Rahman, M.A.; Tashiro, Y.; Sonomoto, K. Recent advances in lactic acid production by microbial fermentation processes. *Biotechnol. Adv.* **2013**, *31*, 877–902. [CrossRef]
7. Zhu, Y.; Lee, Y.Y.; Elander, R.T. Conversion of aqueous ammonia-treated corn stover to lactic acid by simultaneous saccharification and cofermentation. *Appl. Microbiol. Biotechnol.* **2007**, *136*, 721–738.
8. Li, Y.; Lei, L.; Zheng, L.; Xiao, X.; Tang, H.; Luo, C. Genome sequencing of gut symbiotic *Bacillus velezensis* LC1 for bioethanol production from bamboo shoots. *Biotechnol. Biofuels* **2020**, *13*, 34. [CrossRef]
9. Zhang, J.; Shao, S.; Bao, J. Long term storage of dilute acid pretreated corn stover feedstock and ethanol fermentability evaluation. *Bioresour. Technol.* **2016**, *201*, 355–359. [CrossRef]
10. Talebnia, F.; Karakashev, D.; Angelidaki, I. Production of bioethanol from wheat straw: An overview on pretreatment, hydrolysis and fermentation. *Bioresour. Technol.* **2010**, *101*, 4744–4753. [CrossRef]
11. Lloyd, T.A.; Wyman, C.E. Combined sugar yields for dilute sulfuric acid pretreatment of corn stover followed by enzymatic hydrolysis of the remaining solids. *Bioresour. Technol.* **2005**, *96*, 1967–1977. [CrossRef] [PubMed]
12. Hu, J.; Zhang, Z.; Lin, Y.; Zhao, S.; Mei, Y.; Liang, Y.; Peng, N. High-titer lactic acid production from NaOH-pretreated corn stover by *Bacillus coagulans* LA204 using fed-batch simultaneous saccharification and fermentation under non-sterile condition. *Bioresour. Technol.* **2015**, *182*, 251–257. [CrossRef] [PubMed]
13. Zhang, Z.; Xie, Y.; He, X.; Li, X.; Hu, J.; Ruan, Z.; Zhao, S.; Peng, N.; Liang, Y. Comparison of high-titer lactic acid fermentation from NaOH- and NH_3-H_2O_2-pretreated corncob by *Bacillus coagulans* using simultaneous saccharification and fermentation. *Sci. Rep.* **2016**, *6*, 37245. [CrossRef] [PubMed]

14. Daiana, W.; Arias, J.M.; Modesto, L.F.; de França Passos, D.; Pereira, N., Jr. Lactic acid production from sugarcane bagasse hydrolysates by *Lactobacillus pentosus*: Integrating xylose and glucose fermentation. *Biotechnol. Prog.* **2018**, *35*, e2718.
15. Zhang, Y.; Vadlani, P.V. Lactic acid production from biomass-derived sugars via co-fermentation of *Lactobacillus brevis* and *Lactobacillus plantarum*. *J. Biosci. Bioeng.* **2015**, *119*, 694–699. [CrossRef] [PubMed]
16. Abdel-Rahman, M.A.; Tashiro, Y.; Zendo, T.; Hanada, K.; Shibata, K.; Sonomoto, K. Efficient homofermentative L-(+)-lactic acid production from xylose by a novel lactic acid bacterium, *Enterococcus mundtii* QU 25. *Appl. Environ. Microbiol.* **2011**, *77*, 1892–1895. [CrossRef]
17. Abdel-Rahman, M.A.; Tashiro, Y.; Zendo, T.; Sakai, K.; Sonomoto, K. *Enterococcus faecium* QU 50: A novel thermophilic lactic acid bacterium for high-yield l-lactic acid production from xylose. *FEMS Microbiol. Lett.* **2015**, *362*, 1–7. [CrossRef]
18. Cui, F.; Li, Y.; Wan, C. Lactic acid production from corn stover using mixed cultures of *Lactobacillus rhamnosus* and *Lactobacillus brevis*. *Bioresour. Technol.* **2011**, *102*, 1831–1836. [CrossRef]
19. Bischoff, K.M.; Liu, S.; Hughes, S.R.; Rich, J.O. Fermentation of corn fiber hydrolysate to lactic acid by the moderate thermophile *Bacillus coagulans*. *Biotechnol. Lett.* **2010**, *32*, 823–828. [CrossRef]
20. Zhang, Y.; Chen, X.; Luo, J.; Qi, B.; Wan, Y. An efficient process for lactic acid production from wheat straw by a newly isolated *Bacillus coagulans* strain IPE22. *Bioresour. Technol.* **2014**, *158*, 396–399. [CrossRef]
21. Ye, L.; Hudari, M.S.; Zhou, X.; Zhang, D.; Li, Z.; Wu, J.C. Conversion of acid hydrolysate of oil palm empty fruit bunch to L-lactic acid by newly isolated *Bacillus coagulans* JI12. *Appl. Microbiol. Biotechnol.* **2013**, *97*, 4831–4838. [CrossRef] [PubMed]
22. Qiu, Z.; Gao, Q.; Bao, J. Engineering *Pediococcus acidilactici* with xylose assimilation pathway for high titer cellulosic l-lactic acid fermentation. *Bioresour. Technol.* **2018**, *249*, 9–15. [CrossRef] [PubMed]
23. Qiu, Z.; Gao, Q.; Bao, J. Constructing xylose-assimilating pathways in *Pediococcus acidilactici* for high titer d-lactic acid fermentation from corn stover feedstock. *Bioresour. Technol.* **2017**, *245*, 1369–1376. [CrossRef] [PubMed]
24. Eom, I.Y.; Oh, Y.H.; Park, S.J.; Lee, S.H.; Yu, J.H. Fermentative l-lactic acid production from pretreated whole slurry of oil palm trunk treated by hydrothermolysis and subsequent enzymatic hydrolysis. *Bioresour. Technol.* **2015**, *185*, 143–149. [CrossRef] [PubMed]
25. Zhao, K.; Qiao, Q.; Chu, D.; Gu, H.; Dao, T.H.; Zhang, J.; Bao, J. Simultaneous saccharification and high titer lactic acid fermentation of corn stover using a newly isolated lactic acid bacterium *Pediococcus acidilactici* DQ2. *Bioresour. Technol.* **2013**, *135*, 481–489. [CrossRef] [PubMed]
26. Yi, X.; Zhang, P.; Sun, J.; Tu, Y.; Gao, Q.; Zhang, J.; Bao, J. Engineering wild-type robust *Pediococcus acidilactici* strain for high titer L- and D-lactic acid production from corn stover feedstock. *J. Biotechnol.* **2015**, *217*, 112–121. [CrossRef] [PubMed]
27. Guo, P.; Mochidzuki, K.; Cheng, W.; Zhou, M.; Gao, H.; Zheng, D.; Wang, X.; Cui, Z. Effects of different pretreatment strategies on corn stalk acidogenic fermentation using a microbial consortium. *Bioresour. Technol.* **2011**, *102*, 7526–7531. [CrossRef]
28. Toquero, C.; Bolado, S. Effect of four pretreatments on enzymatic hydrolysis and ethanol fermentation of wheat straw. Influence of inhibitors and washing. *Bioresour. Technol.* **2014**, *157*, 68–76. [CrossRef]
29. Millette, M.; Dupont, C.; Shareck, F.; Ruiz, M.T.; Lacroix, M. Purification and identification of the pediocin produced by *Pediococcus acidilactici* MM33, a new human intestinal Strain. *J. Appl. Microbiol.* **2008**, *104*, 269–275.
30. Porto, M.C.; Kuniyoshi, T.M.; Azevedo, P.O.; Vitolo, M.; Oliveira, R.P. *Pediococcus* spp.: An important genus of lactic acid bacteria and pediocin producers. *Biotechnol. Adv.* **2017**, *35*, 361–374. [CrossRef]
31. De Man, J.C. *Lactobacillus bulgaricus* (Luerssen et Kuehn) Holland. *Antonie Van Leeuwenhoek* **1960**, *26*, 77–80. [CrossRef] [PubMed]

microorganisms

MDPI

Perspective

Use of Permanent Wall-Deficient Cells as a System for the Discovery of New-to-Nature Metabolites

Shraddha Shitut [1,2,3,*], Güniz Özer Bergman [2], Alexander Kros [3], Daniel E. Rozen [2] and Dennis Claessen [2,*]

1 Origins Centre, Nijenborgh 7, 9747 AG Groningen, The Netherlands
2 Institute of Biology, Leiden University, 2333 BE Leiden, The Netherlands; g.ozer.bergman@umail.leidenuniv.nl (G.Ö.B.); d.e.rozen@biology.leidenuniv.nl (D.E.R.)
3 Leiden Institute of Chemistry, Leiden University, 2333 CC Leiden, The Netherlands; a.kros@chem.leidenuniv.nl
* Correspondence: shraddha.shitut@gmail.com (S.S.); d.claessen@biology.leidenuniv.nl (D.C.)

Received: 6 November 2020; Accepted: 28 November 2020; Published: 30 November 2020

Abstract: Filamentous actinobacteria are widely used as microbial cell factories to produce valuable secondary metabolites, including the vast majority of clinically relevant antimicrobial compounds. Secondary metabolites are typically encoded by large biosynthetic gene clusters, which allow for a modular approach to generating diverse compounds through recombination. Protoplast fusion is a popular method for whole genome recombination that uses fusion of cells that are transiently wall-deficient. This process has been applied for both inter- and intraspecies recombination. An important limiting step in obtaining diverse recombinants from fused protoplasts is regeneration of the cell wall, because this forces the chromosomes from different parental lines to segregate, thereby preventing further recombination. Recently, several labs have gained insight into wall-deficient bacteria that have the ability to proliferate without their cell wall, known as L-forms. Unlike protoplasts, L-forms can stably maintain multiple chromosomes over many division cycles. Fusion of such L-forms would potentially allow cells to express genes from both parental genomes while also extending the time for recombination, both of which can contribute to an increased chemical diversity. Here, we present a perspective on how L-form fusion has the potential to become a platform for novel compound discovery and may thus help to overcome the antibiotic discovery void.

Keywords: secondary metabolites; actinomycetes; protoplast fusion; novel compound discovery; cell wall-deficiency; heteroploidy

1. Introduction

Microorganisms have been used as a chassis to produce beneficial compounds like antibiotics, growth promoters, enzymes, inflammatory drugs, and protease inhibitors. Of these, antibiotics have made a tremendous impact on our lives due to their application in clinical and veterinary settings. Considering the global challenge of increased antimicrobial resistance, there is an urgent need for the discovery of new compounds. Antibiotics are produced via complex pathways encoded by large biosynthetic gene clusters (BGCs). These clusters not only include the necessary genes for biosynthesis of the antibiotic, but also those required for regulation of gene expression, export, and resistance to the antimicrobial compound. Streptomycetes are well-known for their ability to produce a large variety of antibiotics and may carry up to 70 BGCs per genome [1]. Notably, many of these gene clusters are silent or only poorly expressed, which so far has limited the successful exploitation of some of these metabolites as new antimicrobial agents. Genetic engineering of these BGCs for optimal production can be challenging, for instance, due to the presence of cryptic functions and the non-tractability of natural strains that harbor these BGCs. This has been one of the reasons behind the void in novel compound discovery with very few structurally new classes of antibiotics being introduced in the market in the past decades [2]. In contrast, genomic data analysis has unearthed 33,351 putative BGCs in 1154 microbial genomes [3]. The biosynthetic capabilities are thus plentiful, but our capacity to harness them is limited.

Protoplast fusion is a commonly used method to introduce a BGC or entire chromosome into a recipient cell for further genetic manipulation or directed evolution approaches. Its application for genome shuffling has resulted in strains with higher productivity, prominent examples being that of clavulanic acid [4], cephalosporin [5], diverse enzymatic activity [6], and novel compounds (indolizomycin [7]). This method has also led to the activation of silent BGCs [8]. However, protoplast fusion has potential disadvantages, such as the requirement for multiple fusion and regeneration phases, a short time frame during which recombination can occur due to the necessity for protoplasts to regenerate their cell wall, and lastly the instability of recombinants. From this perspective, we take a deeper look into the benefits and hurdles of protoplast fusion and then propose the use of permanent cell-wall-deficient forms as an alternative system for novel compound discovery.

2. History of Bacterial Protoplast Fusion

The need for protoplast fusion mainly arose from the difficulty of introducing DNA into cells, especially Gram-positive bacteria. The Gram-positive cell wall typically consists of many interconnected layers of peptidoglycan and teichoic acids [9]. Methods like electroporation and conjugation help alleviate the hurdles of passing through the multilayered cell wall but are restricted in terms of size of DNA that can be transformed. Hence, an alternative was found by stripping away the cell wall altogether. The basic set-up of protoplast fusion involves four consecutive steps: (i) protoplast formation, (ii) protoplast fusion, (iii) recombination between chromosomes within the fused cells, and (iv) reversion of protoplasts to walled cells (Figure 1) [10–12]. Protoplasts are obtained by treating walled cells with lysozyme, which degrades the major component of the cell wall, peptidoglycan (PG). The subsequent exposure of protoplasts to crowding agents, such as polyethylene glycol (PEG), causes protoplasts to aggregate. This in turn forces their membranes in close proximity with one another, causing them to fuse. Electrofusion and laser-induced fusion are also commonly used methods to obtain fused cells [12]. Fusion is followed by a brief period of recombination, during which the fusants are maintained in an osmotically protected environment, allowing them to regenerate their cell wall without lysis. In order to select for the desired recombinants, the cells are finally grown in a selective medium.

Because protoplast fusion is non-specific, it has been used for recombination within and between different species. Industrial strains like *Lactobacillus* species have been a common target for genome shuffling via protoplast fusion, resulting in increased yield of lactic acid and improved acid tolerance [13]. Within species fusion may also be combined with random mutagenesis to increase genetic variation, followed by screening for the desired phenotype [14]. For instance, improved degradation of the toxic pesticide pentachlorophenol was achieved in *Sphingobium chrolophenolicum* via protoplast fusion generating cells that had become more resistant [15]. Fusion between phylogenetically distant bacteria has also been performed with *Streptomyces griseus* and *Micromonospora* sp. where recombinants displayed characteristics of both parents. More specifically, recombinants had a colony morphology like *Micromonospora* but with the carbohydrate and amino acid utilization abilities of *S. griseus* [16]. Notably, protoplast fusion is not limited to microbial cells, but has also been successfully used with plants to improve particular traits [10,17].

Taken together, protoplast fusion as a means for genome shuffling has provided several new biological activities and strains with improved traits or growth dynamics [18].

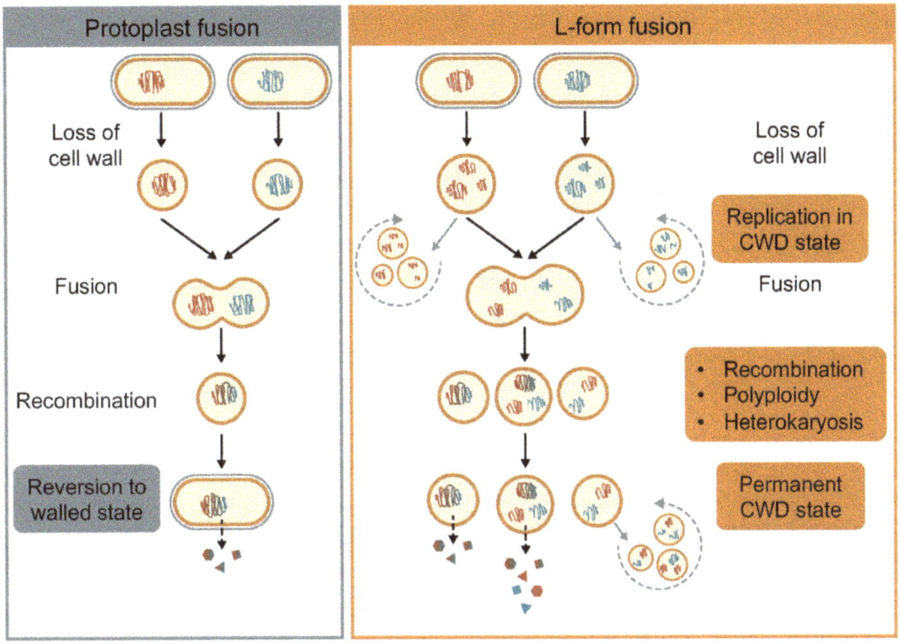

Figure 1. The general process of protoplast fusion compared with L-form fusion. Both methods share some steps like the loss of cell wall and cell–cell fusion but differ in other key steps. Fusion in L-forms can lead to multiple possibilities due to polyploidy in daughter cells, whereas reversion to a walled state is essential in protoplast fusion. Differences in chromosome size are for depiction purposes only and not to scale.

3. Disadvantages of Protoplast Fusion

The process of protoplast fusion introduces entire genomes into a single cellular compartment, allowing them to recombine. Protoplasts carrying multiples chromosomes are typically unable to be sustained for long periods of time and are thus unstable. This instability was observed when the mycaminose-producer *Streptomyces fradiae* was fused with a picronolide-producing *Streptomyces* strain [19]. The fusant transiently produced a novel macrolide antibiotic that was not formed by either parent that was hypothesized to have been due to a diploid state of the fusant. Unfortunately, the ability to produce this new compound was rapidly lost after subsequent culturing. Another disadvantage of fusing protoplasts is the limited period during which the chromosomes can fuse with one another to generate recombinants. This is due to the regeneration of a cell wall, which starts immediately after cells have fused. In some fusions, the efficiency of obtaining a recombinant with activity different compared to the parental lines is not quantitatively the same. Protoplasts of the raffinose utilizer *Micromonaspora carbonaceae* and *Streptomyces griseus* were fused by treatment with PEG [16]. Interestingly, only 2 out of 24 fusants revealed different antimicrobial activity compared to their parents. Moreover, protoplast regeneration can be unsuccessful due to inadequate media supplements, rendering the fusants non-viable. Fusants often die without the addition of sucrose, proline, and microelements in specific proportions to the selection media [20]. Lastly, rebuilding the cell wall in fusants producing cell wall-targeting agents may be complicated. This may lead to lower recombinant efficiencies following protoplast fusion [21]. An alternative method would thus be valuable, especially for generating increased phenotypic diversity and stabilizing fusant strains.

4. Cell-Wall-Deficient L-Forms as an Attractive Alternative for Protoplasts

While protoplasts are transiently wall-deficient, many bacteria can be forced into a permanent wall-deficient state as so-called L-forms [22,23]. They are variants of bacteria that have an altered cell wall organization, which when made irreversible on account of mutations in the genome result in stable cell wall deficiency. They may be referred to by other names in literature such as L-phase, L-variants, and L-organisms, with the "L" signifying the place of discovery, i.e., the Lister Institute [24]. Since their discovery, L-form research has contributed to the fields of bacterial cell biology [25,26], biotechnology [27], host-associated symbiotic and pathogenic interactions [28,29], and the origin of life [23,30]. Unlike protoplasts, cell-wall-deficient L-forms are able to propagate without their cell wall [31]. Proliferation of L-forms is based on biophysical principles, whereby an imbalance in the cell surface-to-volume ratio of cells causes membrane blebbing, tubulation, and vesicle formation. The imbalance required for this mode of proliferation can be obtained by increasing synthesis of fatty acids [32], which act as building blocks for the membrane, and by increasing the fluidity of the membrane [33]. Many (if not all) bacteria can be forced into an L-form state, including antibiotic-producing actinobacteria [34–40]. In most cases, loss of the cell wall is triggered by exposing walled cells to a combination of lytic enzymes and antibiotics in an osmotically protective environment. However, formation of wall-less cells has also been observed as a mechanism to escape stress [41], such as osmotic stress in filamentous actinomycetes [38], heat stress in *E. coli* [42], and nutrient stress in *Mycobacterium bovis* [43]. One of the important consequences of wall deficiency is that the resulting cells are polyploid. Because DNA segregation is likely unregulated during this phase, division of L-forms leads to the uneven distribution of chromosomes among progeny (Figure 1). While some cells inherit a single chromosome, others will inherit multiple chromosomes, with different ratios of the parental chromosomal types.

The structural similarity of L-forms to protoplasts makes then amenable to fusion using PEG or electrofusion. Increased membrane fluidity of L-form cells may also facilitate fusion in the absence of such inducing conditions. Indeed, we have observed spontaneous fusion between L-forms (our unpublished results). In contrast to protoplasts, L-form fusion is not followed by regeneration of the cell wall and the subsequent generation of haploid cells (Figure 1). Three potential outcomes are possible after fusion: (i) recombination between chromosomes, (ii) polyploidy where multiple copies of the recombinant genome are present, and (iii) heterokaryosis where both parental genomes are maintained. L-forms could thus overcome the problem of the short recombination time and fusants that only transiently produce a new metabolite due a diploid state prior to cell wall regeneration [19]. The use of L-forms instead allows for maintenance of heterokaryotic genomes in the same cytoplasmic compartment thereby prolonging opportunities for genetic variation to arise through recombination, resulting in novel production phenotypes. The presence of multiple chromosomes can have beneficial effects in terms of productivity (antibiotics, enzymes), fermentation, and biocontrol. This can happen via various routes such as (i) masking of deleterious alleles by a dominant or complimentary copy, (ii) interactions between genes that were previously not in the same environment, and (iii) gene dosage effects (multiple copies of a biosynthetic gene). It has been shown for *Streptomyces* specifically that interactions between different species can induce antibiotic production and activity [44,45]. By introducing genomes of different species in the same compartment using L-forms, one could potentially achieve a similar effect. Metabolite-inducing interactions that are governed by transcriptional regulators or epigenetic modifiers are good candidates to test [46,47]. The introduction of multiple copies of a gene cluster or induction of gene duplication is already in practice for increased antibiotic production [48,49]. For instance, actinorhodin production by *S. coelicolor* was increased 20-fold after tandem amplification of the *act* gene cluster, resulting in nine copies in the genome [48]. In the context of antibiotic production, L-forms are ideal for production of compounds targeting the cell wall since they are naturally resistant. This reduces loss of biomass during production. L-forms also do not show the formation of inclusion bodies, which can be a useful trait for overproducing proteins that are toxic to the cell [27]. Lastly, L-forms can be easily stored at −80 degrees for extended periods of time. It is yet unclear though if subculturing after revival has an effect on the genomes of the hybrid L-forms.

Collectively, the use of CWD bacteria alleviates most of the disadvantages of protoplast fusion. However, L-form fusion itself does have certain lose ends namely, the time and mutations required to achieve wall-deficient propagation, heterogeneity of fusants, and metabolite production levels. Obtaining a stable L-form strain from a wild-type requires multiple rounds of exposure to cell wall-targeting agents. During this exposure, mutations can arise that help alleviate the stress of surviving and growing without a cell wall such [41,50]. Recent studies are shedding light on the metabolic requirements for maintaining a wall-deficient state in a range of bacteria, thus accelerating the process of generating L-forms. The second issue of fusant heterogeneity when using L-forms can make phenotype screening and quantification difficult. Interestingly, the availability of single cell metabolomics, microfluidics, and encapsulation techniques can help solve this by isolating individual fusants immediately after fusion [51,52]. The fusants can then be tracked over time for certain biochemical activity to identify stable phenotypes. Lastly, studies attempting to use L-forms for overproduction of metabolites have resulted in mixed outcomes depending on the species of L-form used. The production of tetracycline by *S. viridifaciens* L-forms was lower than the walled parent [40], while on the other hand *P. mirabilis* L-forms engineered to express a variety of enzymes, antibodies, and polypeptides were able to produce a higher yield than their walled parent [27,53]. A better understanding of L-form growth and its physiology, division, and chromosome dynamics will help to develop this model as an attractive system for identifying novel compounds.

Taken together by providing the extended opportunities for recombination and maintenance of heterokaryons, the wall-deficient lifestyle provides an opportunity for novel compound discovery. The more we understand about growth, division, and DNA segregation in these systems, the better we can apply them for large scale bioprocessing as well.

Author Contributions: Conceptualization, S.S., D.E.R., A.K. and D.C.; literature review, S.S. and G.Ö.B.; writing-original draft, S.S.; writing-review and editing S.S., G.Ö.B., D.E.R., A.K. and D.C. All authors have read and agreed to the published version of the manuscript.

Funding: S.S. is supported by an NWA Startimpuls from the Origins Centre, the Netherlands. D.C. is supported by an NWO Vici grant.

Acknowledgments: We thank the members of the Claessen lab for fruitful discussions.

Conflicts of Interest: The authors declare no conflict of interest.

References

1. Belknap, K.C.; Park, C.J.; Barth, B.M.; Andam, C.P. Genome mining of biosynthetic and chemotherapeutic gene clusters in *Streptomyces* bacteria. *Sci. Rep.* **2020**, *10*, 2003. [CrossRef] [PubMed]
2. Silver, L.L. Challenges of antibacterial discovery. *Clin. Microbiol. Rev.* **2011**, *24*, 71–109. [CrossRef] [PubMed]
3. Cimermancic, P.; Medema, M.H.; Claesen, J.; Kurita, K.; Wieland Brown, L.C.; Mavrommatis, K.; Pati, A.; Godfrey, P.A.; Koehrsen, M.; Clardy, J.; et al. Insights into secondary metabolism from a global analysis of prokaryotic biosynthetic gene clusters. *Cell* **2014**, *158*, 412–421. [CrossRef] [PubMed]
4. Choi, K.P.; Kim, K.H.; Kim, J.W. Strain improvement of clavulanic acid producing *Streptomyces clavuligerus*. In Proceedings of the 10th International Symposium Biology Actinomycetes (ISBA), Beijing, China, 27–30 May 1997; Volume 12.
5. Hamlyn, P.F.; Ball, C. *Genetics of Industrial Microorganisms;* American Society for Microbiology: Washington, DC, USA, 1979.
6. Okanishi, M.; Suzuki, N.; Furuta, T. Variety of hybrid characters among recombinants obtained by interspecific protoplast fusion in Streptomycetes. *Biosci. Biotechnol. Biochem.* **1996**, *60*, 1233–1238. [CrossRef]
7. Gomi, S.; Ikeda, D.; Nakamura, H.; Naganawa, H.; Yamashita, F.; Hotta, K.; Kondo, S.; Okami, Y.; Umezawa, H.; Iitaka, Y. Isolation and structure of a new antibiotic, indolizomycin, produced by a strain SK2-52 obtained by interspecies fusion treatment. *J. Antibiot. (Tokyo)* **1984**, *37*, 1491–1494. [CrossRef]

8. Nguyen, C.T.; Dhakal, D.; Pham, V.T.; Nguyen, H.T.; Sohng, J.-K. Recent advances in strategies for activation and discovery/characterization of cryptic biosynthetic gene clusters in Streptomyces. *Microorganisms* **2020**, *8*, 616. [CrossRef]
9. Rajagopal, M.; Walker, S. Envelope structures of Gram-positive bacteria. In *Protein and Sugar Export and Assembly in Gram-positive Bacteria*; Springer: Heidelberg, Germany, 2015; pp. 1–44.
10. Peberdy, J.F. Protoplast fusion—A tool for genetic manipulation and breeding in industrial microorganisms. *Enzyme Microb. Technol.* **1980**, *2*, 23–29. [CrossRef]
11. Gokhale, D.V.; Puntambekar, U.S.; Deobagkar, D.N. Protoplast fusion: A tool for intergeneric gene transfer in bacteria. *Biotechnol. Adv.* **1993**, *11*, 199–217. [CrossRef]
12. Gong, J.; Zheng, H.; Wu, Z.; Chen, T.; Zhao, X. Genome shuffling: Progress and applications for phenotype improvement. *Biotechnol. Sustain. Hum. Soc.* **2009**, *27*, 996–1005. [CrossRef]
13. Patnaik, R.; Louie, S.; Gavrilovic, V.; Perry, K.; Stemmer, W.P.C.; Ryan, C.M.; del Cardayré, S. Genome shuffling of *Lactobacillus* for improved acid tolerance. *Nat. Biotechnol.* **2002**, *20*, 707–712. [CrossRef]
14. Hopwood, D.A.; Wright, H.M. Factors affecting recombinant frequency in protoplast fusions of *Streptomyces coelicolor*. *Microbiology* **1979**, *111*, 137–143. [CrossRef] [PubMed]
15. Dai, M.; Copley, S.D. Genome shuffling improves degradation of the anthropogenic pesticide pentachlorophenol by *Sphingobium chlorophenolicum* ATCC 39723. *Appl. Environ. Microbiol.* **2004**, *70*, 2391–2397. [CrossRef] [PubMed]
16. Imada, C.; Okanishi, M.; Okami, Y. Intergenus protoplast fusion between *Streptomyces* and *Micromonospora* with reference to the distribution of parental characteristics in the fusants. *J. Biosci. Bioeng.* **1999**, *88*, 143–147. [CrossRef]
17. Lazar, G.B. Recent developments in plant protoplast fusion and selection technology. In *Protoplasts 1983*; Springer: Berlin, Germany, 1983; pp. 61–67.
18. Adrio, J.-L.; Demain, A.L. Recombinant organisms for production of industrial products. *Bioeng. Bugs* **2010**, *1*, 116–131. [CrossRef] [PubMed]
19. Ikeda, H.; Inoue, M.; Tanaka, H.; Omura, S. Interspecific protoplast fusion among macrolide-producing streptomycetes. *J. Antibiot. (Tokyo)* **1984**, *37*, 1224–1230. [CrossRef]
20. Marcone, G.L.; Carrano, L.; Marinelli, F.; Beltrametti, F. Protoplast preparation and reversion to the normal filamentous growth in antibiotic-producing uncommon actinomycetes. *J. Antibiot. (Tokyo)* **2010**, *63*, 83–88. [CrossRef]
21. Baltz, R.H.; Matsushima, P. Protoplast fusion in Streptomyces: Conditions for efficient genetic recombination and cell regeneration. *Microbiology* **1981**, *127*, 137–146. [CrossRef]
22. Allan, E.J.; Hoischen, C.; Gumpert, J. Chapter 1 Bacterial L-Forms. In *Advances in Applied Microbiology*; Academic Press: Cambridge, MA, USA, 2009; Volume 68, pp. 1–39.
23. Errington, J.; Mickiewicz, K.; Kawai, Y.; Wu, L.J. L-form bacteria, chronic diseases and the origins of life. *Philos. Trans. R. Soc. B Biol. Sci.* **2016**, *371*, 20150494. [CrossRef]
24. Klieneberger, E. The natural occurrence of pleuropneumonia-like organism in apparent symbiosis with *Streptobacillus moniliformis* and other bacteria. *J. Pathol. Bacteriol.* **1935**, *40*, 93–105. [CrossRef]
25. Hoischen, C.; Gura, K.; Luge, C.; Gumpert, J. Lipid and fatty acid composition of cytoplasmic membranes from *Streptomyces hygroscopicus* and its stable protoplast-type L form. *J. Bacteriol.* **1997**, *179*, 3430–3436. [CrossRef]
26. Siddiqui, R.A.; Hoischen, C.; Holst, O.; Heinze, I.; Schlott, B.; Gumpert, J.; Diekmann, S.; Grosse, F.; Platzer, M. The analysis of cell division and cell wall synthesis genes reveals mutationally inactivated ftsQ and mraY in a protoplast-type L-form of *Escherichia coli*. *FEMS Microbiol. Lett.* **2006**, *258*, 305–311. [CrossRef] [PubMed]
27. Gumpert, J.; Hoischen, C. Use of cell wall-less bacteria (L-forms) for efficient expression and secretion of heterologous gene products. *Curr. Opin. Biotechnol.* **1998**, *9*, 506–509. [CrossRef]
28. Somerson, N.L.; Ehrman, L.; Kocka, J.P.; Gottlieb, F.J. Streptococcal L-forms isolated from *Drosophila paulistorum* semispecies cause sterility in male progeny. *Proc. Natl. Acad. Sci. USA* **1984**, *81*, 282–285. [CrossRef]
29. Markova, N. L-form bacteria cohabitants in human blood: Significance for health and diseases. *Discov. Med.* **2017**, *23*, 305–313. [PubMed]
30. Sladek, T.L. A hypothesis for the mechanism of mycoplasma evolution. *J. Theor. Biol.* **1986**, *120*, 457–465. [CrossRef]

31. Mercier, R.; Kawai, Y.; Errington, J. General principles for the formation and proliferation of a wall-free (L-form) state in bacteria. *eLife* **2014**, *3*, e04629. [CrossRef]
32. Mercier, R.; Kawai, Y.; Errington, J. Excess membrane synthesis drives a primitive mode of cell proliferation. *Cell* **2013**, *152*, 997–1007. [CrossRef]
33. Mercier, R.; Domínguez-Cuevas, P.; Errington, J. Crucial role for membrane fluidity in proliferation of primitive cells. *Cell Rep.* **2012**, *1*, 417–423. [CrossRef]
34. Dienes, L. L Organisms of Klieneberger and *Streptobacillus moniliformis*. *J. Infect. Dis.* **1939**, *65*, 24–42. [CrossRef]
35. Jass, J.; Phillips, L.E.; Allan, E.J.; Costerton, J.W.; Lappin-Scott, H.M. Growth and adhesion of *Enterococcus faecium* L-forms. *FEMS Microbiol. Lett.* **1994**, *115*, 157–162. [CrossRef]
36. Leaver, M.; Domínguez-Cuevas, P.; Coxhead, J.M.; Daniel, R.A.; Errington, J. Life without a wall or division machine in *Bacillus subtilis*. *Nature* **2009**, *457*, 849–853. [CrossRef] [PubMed]
37. Glover, W.A.; Yang, Y.; Zhang, Y. Insights into the molecular basis of L-Form formation and survival in *Escherichia coli*. *PLoS ONE* **2009**, *4*, e7316. [CrossRef] [PubMed]
38. Ramijan, K.; Ultee, E.; Willemse, J.; Zhang, Z.; Wondergem, J.A.J.; van der Meij, A.; Heinrich, D.; Briegel, A.; van Wezel, G.P.; Claessen, D. Stress-induced formation of cell-wall-deficient cells in filamentous actinomycetes. *Nat. Commun.* **2018**, *9*, 5164. [CrossRef] [PubMed]
39. Studer, P.; Staubli, T.; Wieser, N.; Wolf, P.; Schuppler, M.; Loessner, M.J. Proliferation of *Listeria monocytogenes* L-form cells by formation of internal and external vesicles. *Nat. Commun.* **2016**, *7*, 13631. [CrossRef]
40. Innes, C.M.J.; Allan, E.J. Induction, growth and antibiotic production of *Streptomyces viridifaciens* L-form bacteria. *J. Appl. Microbiol.* **2001**, *90*, 301–308. [CrossRef]
41. Claessen, D.; Errington, J. Cell wall deficiency as a coping strategy for stress. *Trends Microbiol.* **2019**, *27*, 1025–1033. [CrossRef]
42. Markova, N.; Slavchev, G.; Michailova, L.; Jourdanova, M. Survival of *Escherichia coli* under lethal heat stress by L-form conversion. *Int. J. Biol. Sci.* **2010**, *6*, 303–315. [CrossRef]
43. Slavchev, G.; Michailova, L.; Markova, N. Stress-induced L-forms of *Mycobacterium bovis*: A challenge to survivability. *New Microbiol.* **2013**, *36*, 157–166.
44. Abrudan, M.I.; Smakman, F.; Grimbergen, A.J.; Westhoff, S.; Miller, E.L.; van Wezel, G.P.; Rozen, D.E. Socially mediated induction and suppression of antibiosis during bacterial coexistence. *Proc. Natl. Acad. Sci. USA* **2015**, *112*, 11054–11059. [CrossRef]
45. Bertrand, S.; Bohni, N.; Schnee, S.; Schumpp, O.; Gindro, K.; Wolfender, J.-L. Metabolite induction via microorganism co-culture: A potential way to enhance chemical diversity for drug discovery. *Plants Pharm. Shelf Recent Trends Leads Find. Bioprod.* **2014**, *32*, 1180–1204. [CrossRef]
46. Williams, R.B.; Henrikson, J.C.; Hoover, A.R.; Lee, A.E.; Cichewicz, R.H. Epigenetic remodeling of the fungal secondary metabolome. *Org. Biomol. Chem.* **2008**, *6*, 1895–1897. [CrossRef] [PubMed]
47. Yang, X.-L.; Awakawa, T.; Wakimoto, T.; Abe, I. Induced biosyntheses of a novel butyrophenone and two aromatic polyketides in the plant pathogen *Stagonospora nodorum*. *Nat. Prod. Bioprospect.* **2013**, *3*, 141–144. [CrossRef]
48. Murakami, T.; Burian, J.; Yanai, K.; Bibb, M.J.; Thompson, C.J. A system for the targeted amplification of bacterial gene clusters multiplies antibiotic yield in *Streptomyces coelicolor*. *Proc. Natl. Acad. Sci. USA* **2011**, *108*, 16020–16025. [CrossRef] [PubMed]
49. Haginaka, K.; Asamizu, S.; Ozaki, T.; Igarashi, Y.; Furumai, T.; Onaka, H. Genetic approaches to generate hyper-producing strains of goadsporin: The relationships between productivity and gene duplication in secondary metabolite biosynthesis. *Biosci. Biotechnol. Biochem.* **2014**, *78*, 394–399. [CrossRef]
50. Kawai, Y.; Mercier, R.; Mickiewicz, K.; Serafini, A.; Sório de Carvalho, L.P.; Errington, J. Crucial role for central carbon metabolism in the bacterial L-form switch and killing by β-lactam antibiotics. *Nat. Microbiol.* **2019**, *4*, 1716–1726. [CrossRef]
51. Blainey, P.C. The future is now: Single-cell genomics of bacteria and archaea. *FEMS Microbiol. Rev.* **2013**, *37*, 407–427. [CrossRef]
52. Zenobi, R. Single-Cell metabolomics: Analytical and biological perspectives. *Science* **2013**, *342*, 1243259. [CrossRef]

53. Klessen, C.; Schmidt, K.H.; Gumpert, J.; Grosse, H.H.; Malke, H. Complete secretion of activable bovine prochymosin by genetically engineered L forms of *Proteus mirabilis*. *Appl. Environ. Microbiol.* **1989**, *55*, 1009–1015. [CrossRef]

MDPI
St. Alban-Anlage 66
4052 Basel
Switzerland
Tel. +41 61 683 77 34
Fax +41 61 302 89 18
www.mdpi.com

Microorganisms Editorial Office
E-mail: microorganisms@mdpi.com
www.mdpi.com/journal/microorganisms

www.ingramcontent.com/pod-product-compliance
Lightning Source LLC
LaVergne TN
LVHW070126100526
838202LV00016B/2237